DISTANCE AND MIDPOINT FORMULAS

The **distance** between $P(x_1, y_1)$ and $Q(x_2, y_2)$ is

$$PQ = \sqrt{(x_2 - x_1)^2 + (y_2 - y_1)^2}$$

and the coordinates of the **midpoint** of line segment $\overline{PQ}$ are

$$\left(\frac{x_1 + x_2}{2}, \frac{y_1 + y_2}{2}\right)$$

EQUATION OF A CIRCLE

The equation of a circle with center (h, k) and radius r is

$$(x - h)^2 + (y - k)^2 = r^2$$

LINES

1. The **slope** of a line through $P(x_1, y_1)$ and $Q(x_2, y_2)$ is

$$m = \frac{y_2 - y_1}{x_2 - x_1}$$

2. The **slope-intercept form** of a line with slope m and y-intercept b is

$$y = mx + b$$

3. The **point-slope form** of a line through $P(x_1, y_1)$ with slope m is

$$y - y_1 = m(x - x_1)$$

VERTEX FORMULA

The graph of $f(x) = ax^2 + bx + c$ is a *parabola*. The coordinates of its **vertex** are

$$\left(-\frac{b}{2a}, f\left(-\frac{b}{2a}\right)\right)$$

LOGARITHMS

1. $y = \log_b x$ is equivalent to $b^y = x$

2. $\log_b 1 = 0$ 3. $\log_b b = 1$

4. $\log_b b^x = x$ 5. $b^{\log_b x} = x$

6. $\log_b xy = \log_b x + \log_b y$ 7. $\log_b \frac{x}{y} = \log_b x - \log_b y$

8. $\log_b x^n = n \log_b x$ 9. $\log_b x = \dfrac{\log_a x}{\log_a b}$

GRAPHS OF TEN BASIC FUNCTIONS

1. **Constant Function**

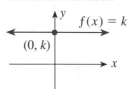

2. **Identity Function**

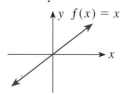

3. **Absolute Value Function**   4. **Squaring Function**

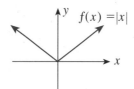
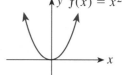

5. **Cubing Function**   6. **Reciprocal Function**

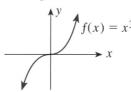
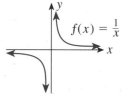

7. **Square Root Function**   8. **Cube Root Function**

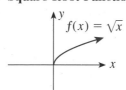
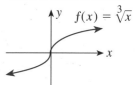

9. **Exponential Function**   10. **Logarithmic Function**

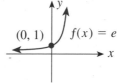
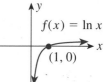

BINOMIAL THEOREM

$$(x + y)^n = \sum_{i=0}^{n} \binom{n}{i} x^{n-i} y^i$$

College Algebra

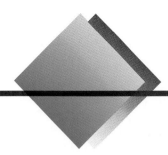

College Algebra

David E. Stevens
Wentworth Institute of Technology

WEST PUBLISHING COMPANY
Minneapolis/St. Paul ◆ New York
Los Angeles ◆ San Francisco

◆ PRODUCTION CREDITS

Copyediting: Katherine Townes/Tech*arts*
Design: Tech*arts*
Interior artwork: Scientific Illustrators
Cover photograph: Graeme Outerbridge

◆ WEST'S COMMITMENT TO THE ENVIRONMENT

In 1906, West Publishing Company began recycling materials left over from the production of books. This began a tradition of efficient and responsible use of resources. Today, up to 95 percent of our legal books and 70 percent of our college and school texts are printed on recycled, acid-free stock. West also recycles nearly 22 million pounds of scrap paper annually—the equivalent of 181,717 trees. Since the 1960s, West has devised ways to capture and recycle waste inks, solvents, oils, and vapors created in the printing process. We also recycle plastics of all kinds, wood, glass, corrugated cardboard, and batteries, and have eliminated the use of styrofoam book packaging. We at West are proud of the longevity and the scope of our commitment to the environment.

◆ PRODUCTION, PRINTING AND BINDING BY WEST PUBLISHING COMPANY.

◆ PHOTO CREDITS FOLLOW THE INDEX.

◆ COPYRIGHT © 1994 BY WEST PUBLISHING COMAPNY
610 Opperman Drive
P.O. Box 64526
St. Paul, MN 55164-0526

◆ PRINTED IN THE UNITED STATES OF AMERICA
01 00 99 98 97 96 95 94   8 7 6 5 4 3 2 1 0

◆ LIBRARY OF CONGRESS CATALOGING-IN-PUBLICATION DATA

◆ STEVENS, DAVID E.
    College algebra / David E. Stevens.
        p. cm.
    Includes index.
    ISBN 0-314-01221-4
    1. Algebra. I. Title.
QA154.2.S746 1994
512.9—dc20

∞   92-33548
CIP

CONTENTS

PREFACE

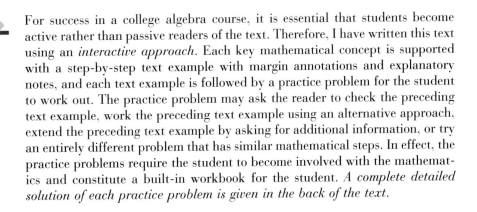

Intent

A course in college algebra must meet the needs of students with diverse mathematical backgrounds and goals. Science and engineering students enroll in this course before beginning a traditional three-semester calculus sequence. Business and management students enroll in this course before taking courses in finite mathematics, statistics, or a short course in calculus. Also, for other students, this is a terminal course that satisfies the core requirement for mathematics. In writing this text, I have tried to be sensitive to the needs of *all* these students by preparing them for higher-level mathematics courses and illustrating through real-life applied problems that the study of college algebra is a worthwhile educational experience.

Approach

For success in a college algebra course, it is essential that students become active rather than passive readers of the text. Therefore, I have written this text using an *interactive approach*. Each key mathematical concept is supported with a step-by-step text example with margin annotations and explanatory notes, and each text example is followed by a practice problem for the student to work out. The practice problem may ask the reader to check the preceding text example, work the preceding text example using an alternative approach, extend the preceding text example by asking for additional information, or try an entirely different problem that has similar mathematical steps. In effect, the practice problems require the student to become involved with the mathematics and constitute a built-in workbook for the student. *A complete detailed solution of each practice problem is given in the back of the text.*

Features

Written in a warm and user-friendly style, this is a traditional college algebra text with a contemporary flair in the sense that it starts to address the concerns of writing across the curriculum, group learning, critical thinking, and the use of modern technology in the math classroom. The following features distinguish this text from the existing texts in the market.

Applied Problems: To arouse student interest, each chapter opens with an applied problem and a related photograph. The solution to the applied problem occurs within the chapter after the necessary mathematics has been developed. Other applied problems from the fields of science, engineering, and business are introduced at every reasonable opportunity. They are clearly visible throughout the text and occur in separate subsections at the end of most sections. They are a highlight of this text and tend to show how college algebra relates to real-life situations.

Exercise Sets: The heart of any mathematics textbook is its end-of-section exercise sets. It is here that students are given an opportunity to practice the mathematics that has been developed. The exercise sets in this text are broken into three parts: *Basic Skills*, *Critical Thinking*, and *Calculator Activities*.

Basic Skills: These exercises are routine in nature and tend to mimic the text examples that are worked out in each section.

Critical Thinking: These exercises require the student to "think critically" and transcend the routine application of the basic skills to the next level of difficulty. Exercises in this group may require the student to draw upon skills developed in earlier chapters.

Calculator Activities: These exercises require the use of a calculator to solve basic-skill- and critical-thinking-type problems. Some of the exercises in this group ask the student to use a *graphing calculator* to solve problems in which standard algebraic methods do not apply. Problems that require the use of a graphing calculator are identified by the logo ⌨.

Some of the exercise sets also contain problems that are *calculus-related*. Designed for students who are taking a college algebra course as a prerequisite to calculus, these exercises illustrate the algebraic support that is needed for simplifying derivatives, solving optimization problems, finding the area bounded by two or more curves, and so on. In calculus, the ratio of the change in the variable y to the change in the variable x is designated by $\Delta y/\Delta x$. In this text, we use the logo $\boxed{\frac{\Delta y}{\Delta x}}$ to identify the problems in the exercise sets that are calculus-related.

Chapter Reviews: To help students prepare for chapter exams, each chapter in this text concludes with an extensive chapter review. The chapter reviews are broken in two parts: *Questions for Group Discussion* and *Review Exercises*.

Questions for Group Discussion: In keeping with the interactive approach, these questions allow students to state in their own words what they have learned in the chapter. Since many of these questions have open-ended answers, they are ideally suited for class or group discussions and are extremely valuable to those who believe in cooperative or collaborative learning.

Review Exercises: These exercises reinforce the ideas that are discussed in the chapter and allow the instructor to indicate to the student the types of problems that may appear on a chapter test.

Cumulative Reviews: To help students pull together ideas from several chapters, cumulative review exercises are strategically placed after Chapters 2, 6 and 8. The problems in these exercises are ungraded as far as difficulty and are presented in a random order. Some problems are basic and similar to those already studied, while others are more challenging and require some creative thinking.

Pedagogy: Every effort has been made to make this a text from which students can learn and succeed. The following pedagogical features attest to this fact:

Caution notes, flagged by the symbol ⬡(Caution), help eliminate misconceptions and bad mathematical habits by pointing out the most common errors that students make.

Introductory comments at the start of many sections introduce vocabulary and inform the reader of the purpose of the section.

Boxed definitions, formulas, laws, and properties state key mathematical ideas and provide the reader with quick and easy access to this information.

Step-by-step procedural boxes indicate the sequence of steps that a student can follow in order to simplify algebraic expressions, solve certain types of equations and inequalities, sketch the graphs of various functions, or find the inverse of a function or a matrix.

Development

Chapters 1 and 2 are ideally suited for students who have not taken a mathematics course in several years or who have not mastered the fundamental concepts in intermediate algebra. *Chapter 1* provides an extensive review of the basic algebraic topics such as exponents, radicals, factoring, and algebraic fractions. This chapter also expands on these topics and discusses some of the algebraic manipulations that are needed in calculus. With an early introduction to the complex numbers (Section 1.5), it is possible in Section 1.6 to discuss the complete factorization of certain polynomials over the integers, over the reals, and over the complex numbers. Hence, the groundwork has been set for the study of the factor theorem and the fundamental theorem of algebra (Chapter 5).

Chapter 2 discusses the methods of solving linear, fractional, quadratic, and radical equations. This chapter also expands on these methods and discusses some of the equation-solving techniques that are needed in calculus. Section 2.2 introduces a strategy for solving application problems, and this strategy is applied to all such problems throughout the chapter. Polynomial and rational inequalities are solved by finding their critical values and determining the algebraic sign to the left and to the right of each critical value. This work is organized in a table and interval notation is employed.

Chapters 3, 4, 5, and 6 represent the core of a college algebra course. *Chapter 3* introduces the coordinate plane, graphing techniques, and the functional concept. Section 3.4 lists eight basic functions and their graphs (constant, identity, absolute value, squaring, cubing, reciprocal, square root, and cube root functions) and then applies the vertical and horizontal shift rules, the x-and y-axis reflection rules, and the vertical stretch and shrink rules to sketch the graphs of several other related functions, These eight basic functions and their graphs are then used to discuss composition of functions, inverse functions, applied functions, and variation. The graphing calculator is introduced in this chapter and some of the exercises suggest using this tool to verify results or to explore new ideas.

Chapter 4 uses the ideas of shifting, reflecting, stretching and shrinking to develop the linear and quadratic functions from the identity and squaring functions. The vertex formula for a parabola is developed and used to solve some max-min applied problems. Section 4.4 introduces the conic sections in relation to the general quadratic equation in two unknowns, and Section 4.5 gives an introduction to solving a system of two equations in two unknowns by looking at the intersection points of their graphs. Also in Section 4.5, the graphing calculator is used to help solve some nonlinear systems that are not solvable by ordinary algebraic procedures.

Chapter 5 illustrates the relationship between factors, roots, and zeros and discusses methods for solving a polynomial equation of degree 3 or greater. The rational zero theorem is used to find any rational zeros of a polynomial function and the method of successive approximation or a graphing calculator is used to estimate any irrational zeros. The graph of a polynomial function is sketched by looking at its end behavior, examining the possible shapes for the middle portion of the graph, finding the x-and y-intercepts, and plotting a few points. The graph of a rational function is made by examining its vertical, horizontal, or oblique asymptotes, finding the x-and y-intercepts, and plotting a few points. The graphs of polynomial and rational functions are also used to solve some max-min applied problems with the aid of a graphing calculator.

Chapter 6 discusses the properties of real exponents, defines the exponential function, and develops the logarithmic function as the inverse of the exponential function. By letting the number of compounding periods in the compound interest formula increase without bound, the reader is shown how the number e develops in a real-life situation. The properties of logarithms are used to help graph functions containing logarithmic expressions (Section 6.4) and also to help solve exponential and logarithmic equations (Section 6.5). In Section 6.5, the graphing calculator is used to help solve some exponential equations that are not solvable by ordinary algebraic methods.

Chapters 7 and 8 include several other topics of interest in a college algebra course. *Chapter 7* extends the discussion of 2×2 systems (Section 4.5) to a systematic procedure for solving $n \times n$ linear systems by using Gaussian elimination. Matrices are first introduced as an aid for solving a system of linear equations and then applied to problems that require using the matrix operations of addition and multiplication. Determinants are evaluated by using minors and cofactors as well as by using row and column operations. Linear systems of inequalities are introduced and applied to linear programming problems in two unknowns.

Chapter 8 offers an introduction to sequences and series for the precalculus student. Section 8.2 introduces the idea of proof by mathematical induction. In this section, the reader is encouraged to use pattern recognition, guess a formula for the sum of a series, and then prove their guess by using mathematical induction. Similarly, pattern recognition and mathematical induction are used to develop a formula for the general element of an arithmetic sequence, the general element of a geometric sequence, and the expansion $(A + B)^n$. Sections 8.7 and 8.8 introduce the reader to the language of probability. These sections discuss sample spaces and events of random experiments, counting techniques, and some basic probability formulas.

Supplements

The following supplements are available for users of this text.

1. *Instructor's Solution Manual* by Eleanor Canter—includes complete worked-out solutions to all the even-numbered exercises.

2. *Student's Solution Manual* by Eleanor Canter—provides worked-out solutions for the odd exercises from the text.

3. *Instructor's Manual with Test Bank* by Cheryl Roberts—includes sample syllabi, suggested course schedules, chapter outlines with references to vid-

eos, homework assignments, chapter tests, and a test bank of multiple choice questions and open-ended problems.

4. *Graphing Calculator Manual*

5. *WESTEST* 3.0—computer-generated testing programs include algorithmically-generated questions and are available to qualified adopters. Macintosh and IBM-compatible versions are available.

6. *Video series*—"In Simplest Terms" produced by Annenberg/CPB Collection. The videos are referenced in the Instructor's Manual accompanying *College Algebra*.

7. *Mathens Tutorial Software*—generates problems of varying degrees of difficulty and guides students through step-by-step solutions.

Acknowledgments ◆

This text was class-tested with over 1000 students at Wentworth Institute of Technology. I thank these students and my colleagues Donald Filan, Michael John, Marcia Kemen, Anita Penta, and Charlene Solomon for their assistance throughout the development of this project. Special thanks go to my friend and colleague Eleanor Canter for her work in checking the answers and writing complete worked-out solutions to the more than 4500 exercises in this text. I also express my sincere thanks to the following reviewers. Their ideas were extremely helpful in shaping this text into its present form.

Haya Adner,
Queensborough Community College

Daniel D. Anderson,
University of Iowa

Thomas A. Atchison,
Stephen F. Austin State University

Jerald T. Ball,
Las Positas College

Kathleen Bavelas,
Manchester Community College

Louise M. Boyd,
Livingston University

Susan N. Boyer,
University of Maryland-Baltimore County

James R. Brasel,
Phillips County Community College

John E. Bruha,
University of Northern Iowa

Deborah A. Crocker,
Miami University

Judith Covington,
University of Southwest Louisiana

Terry Czerwinski,
University of Illinois-Chicago

Bettyann Daley,
University of Delaware

Ryness A. Doherty,
Community College of Denver

Michael W. Ecker,
Pennsylvania State University

Stuart Goff,
Keene State College

Sarita Gupta,
Northern Illinois University

Nancy Hyde,
Broward Community College

Louis Hoelzle,
Bucks County Community College

Ed Huffman,
Southwest Missouri State University

Sylvia M. Kennedy,
Broome Community College

Margret Kothmann,
University of Wisconsin-Stout

Keith Kuchar,
Northern Illinois University

Jeuel LaTorre,
Clemson University

Michael E. Mays,
West Virginia University

Myrna L. Mitchell,
Pima Community College

Maurice L. Monahan,
South Dakota State University

Jeri A. Nichols,
Bowling Green State University

Jean M. Prendergast,
Bridgewater State College

William Radulovich,
Florida Community College

James A. Reed, Sr.,
University of Hartford

Betty Rehfuss,
North Dakota State University

Cheryl V. Roberts,
Northern Virginia Community College

Stephen B. Rodi,
Austin Community College

William B. Rundberg,
College of San Mateo

Ned W. Schillow,
Lehigh County Community College

George W. Schultz,
St. Petersburg Junior College

Richard D. Semmler,
Northern Virginia Community College

Lynne B. Small,
University of San Diego

Charles Stone,
DeKalb College

Michael D. Taylor,
University of Southern Florida

Beverly Weatherwax,
Southwest Missouri State University

William H. White,
University of South Carolina-Spartanburg

Bruce Williamson,
University of Wisconsin-River Falls

Jim Wooland,
Florida State University

The production of a textbook is a team effort between the editorial staff and the author. My editor, Ron Pullins, always offered the support and guidance that I needed to complete this project. Denise Bayko organized our reviewers' comments into a format that revealed where extra work was needed. Kathi Townes copyedited the manuscript and prodded me to provide additional information that would benefit the reader, and Tamborah Moore kept the project on schedule despite some unusual circumstances. I thank each of you for the encouragement and enthusiasm you provided in the preparation of this book. I can't imagine working with a better team.

D.E. Stevens
Boston, Massachusetts, 1993

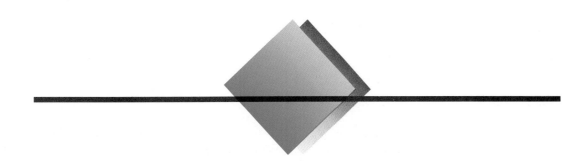

College Algebra

CHAPTER

1

Fundamental Algebraic Ideas

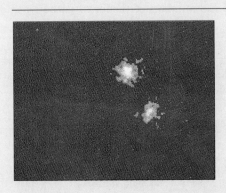

The distance from Earth to Alpha Centauri (the star closest to our Sun) is approximately 25,500,000,000,000 miles (mi). How many years does it take light from this star to reach Earth? Assume that light travels at 186,000 miles per second (mi/s).

(For the solution, see Example 3 in Section 1.2.)

1.1 The Real Numbers

◆ Introductory Comments

A chef requires 3 eggs and $\frac{2}{3}$ cup of sugar for a cake recipe. A meteorologist reports that the temperature is $-4\,°C$ and the barometric pressure is 29.35 inches. A student determines that the side of a right triangle is $\sqrt{5}$ units and the area of a circle is π square units. Numbers such as

$$3 \qquad \frac{2}{3} \qquad -4 \qquad 29.35 \qquad \sqrt{5} \quad \text{and} \quad \pi,$$

are elements of the **set of real numbers**.[1] These are the type of numbers that we see and work with every day. The set of real numbers has five important subsets:

1. Natural numbers
   or
   Positive integers $\qquad \{1, 2, 3, 4, \ldots\}$
2. Whole numbers $\qquad \{0, 1, 2, 3, \ldots\}$
3. Integers $\qquad \{\ldots, -3, -2, -1, 0, 1, 2, 3, \ldots\}$
4. Rational numbers $\qquad$ {All real numbers of the form $\frac{p}{q}$, where p and q are integers and $q \neq 0$.}

   or

   {All decimal numbers that either terminate or repeat the same block of digits.}

   Examples: $\frac{3}{5}, \frac{-8}{3}, \frac{16}{1}, \frac{0}{4}, 0.75, -5.343434\ldots$

5. Irrational numbers $\qquad$ {All real numbers that are not rational.}

   or

   {All decimal numbers that neither terminate nor repeat the same block of digits.}

   Examples: $\sqrt{2}, -\sqrt{3}, \pi, \sqrt[3]{6}, 3.050050005\ldots$

In algebra, real numbers such as 3, $\frac{2}{3}$, -4, 29.35, $\sqrt{5}$, and π are called **constants**, since each is a fixed number. Letters from the beginning of the alphabet, like a, b, c, and k, are often used as symbols for yet unspecified constants. For example, we may designate the sum of two fixed, yet unspecified, real numbers as

---

[1]The concept of a set is often used in mathematics. A *set* is a collection of objects, and these objects are called the *elements* of the set. A *subset* of a given set is formed by selecting particular elements of the set. Braces { } are used to enclose the elements of sets and subsets.

$$a + b$$

and the product of two fixed, yet unspecified, real numbers as

$$a \times b \qquad a \cdot b \qquad a(b) \qquad (a)b \qquad (a)(b) \quad \text{or} \quad ab$$

When we add two or more numbers, the numbers are called **terms**, and the result of the addition is the *sum*. When we multiply two or more numbers, the numbers are called **factors**, and the result is the *product*:

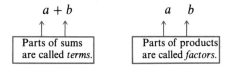

Throughout Chapter 1, we will introduce definitions and rules concerning terms and other definitions and rules concerning factors. It is important to learn to distinguish the difference between terms and factors.

◆   **Some Properties of the Real Numbers**

All the facts that we have learned in arithmetic and elementary algebra concerning the operations of addition and multiplication follow from just a few basic accepted statements called *axioms*. We refer to these axioms as the **fundamental properties of the real numbers**.

 Fundamental Properties of the Real Numbers

If a, b, and c are real numbers,

1. Closure Properties

   $a + b$ is a unique real number.

   ab is a unique real number.

2. Commutative Properties

   $a + b = b + a$

   $ab = ba$

3. Associative Properties

   $(a + b) + c = a + (b + c)$

   $(ab)c = a(bc)$

4. Identity Properties

   $a + 0 = 0 + a = a$

   $a(1) = (1)a = a$

5. Inverse Properties

   $a + (-a) = (-a) + a = 0$

   $a \cdot \dfrac{1}{a} = \dfrac{1}{a} \cdot a = 1 \quad (a \neq 0)$

6. Distributive Properties (over addition)

   $a(b + c) = ab + ac \quad \text{(left)}$

   $(a + b)c = ac + bc \quad \text{(right)}$

In property 5, the expression $-a$ is read "the additive inverse of a," "the opposite of a," or "the negative of a." It is important to note that if a is a negative number, then $-a$ is a *positive* number. For example, if $a = -4$ then $-a = -(-4) = 4$. We can summarize these facts as follows:

1. If a is positive, then $-a$ is negative.
2. If a is negative, then $-a$ is positive.
3. If a is 0, then $-a$ is 0.
4. The negative of $-a$ is written $-(-a)$, and $-(-a) = a$.

The *difference* of two real numbers, denoted $a - b$, can be defined in terms of addition.

 **Subtraction**

For all real numbers a and b,

$$a - b = a + (-b).$$

In other words, to subtract b from a, we add a with the negative of b. For example,

$$8 - 10 = 8 + (-10) = -2 \quad \text{and} \quad \sqrt{6} - (-5) = \sqrt{6} + [-(-5)] = \sqrt{6} + 5.$$

The *quotient* of two real numbers, denoted $a \div b$ or $\dfrac{a}{b}$, can be defined in terms of multiplication.

 Division

For all real numbers a and b ($b \neq 0$),

$$a \div b = \frac{a}{b} = a \cdot \frac{1}{b} = \frac{1}{b} \cdot a.$$

Thus, to divide a by b, we multiply a by the multiplicative inverse (or *reciprocal*) of b, which is $\dfrac{1}{b}$. Since 0 does not have a multiplicative inverse, division by zero is not allowed. Thus, some examples of division are

$$\frac{\sqrt{2}}{3} = \frac{1}{3}\sqrt{2} \quad \text{and} \quad \frac{\pi}{2} = \frac{1}{2}\pi,$$

but

$$\frac{3}{0} \text{ is } undefined.$$

The following *additional properties of the real numbers* can be derived from the preceding fundamental properties and definitions.

 Additional Properties of the Real Numbers

If a, b, and c are real numbers,

1. Multiplicative Property of 0 $a(0) = (0)a = 0$

2. Multiplicative Property of -1 $a(-1) = (-1)a = -a$

3. Negative of a Sum $-(a + b) = -a - b$

4. Negative of a Difference $-(a - b) = b - a$

5. Distributive Property (over subtraction) $a(b - c) = ab - ac$ (left)

    $(a - b)c = ac - bc$ (right)

6. Negatives in Products $(-a)b = -(ab) = a(-b)$ and

    $(-a)(-b) = ab$

7. Negatives in Quotients $\dfrac{-a}{b} = -\dfrac{a}{b} = \dfrac{a}{-b}$ and $\dfrac{-a}{-b} = \dfrac{a}{b}$

Quotients of the form $\dfrac{a}{b}$, where a and b are real numbers and $b \neq 0$, are called **fractions**. If two fractions $\dfrac{a}{b}$ and $\dfrac{c}{d}$ have the same numerical value, they are **equivalent fractions**, and we write $\dfrac{a}{b} = \dfrac{c}{d}$. *If two fractions are equivalent fractions, their cross products are equal.*

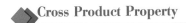 **Cross Product Property**

For real numbers a, b, c, and d, $(b \neq 0, d \neq 0)$,

$$\frac{a}{b} = \frac{c}{d} \quad \text{if and only if} \quad ad = bc.$$

Note: The phrase "if and only if" occurs frequently in mathematics. As used in the cross product property, it implies the following two statements:

1. If $a/b = c/d$, then $ad = bc$ and, conversely,

2. If $ad = bc$, then $a/b = c/d$.

To generate equivalent fractions or to reduce a fraction to lowest terms, we apply the **fundamental property of fractions**.

◆ **Fundamental Property of Fractions**

> For all fractions $\dfrac{a}{b}$ ($b \neq 0$) and all real numbers k ($k \neq 0$),
>
> $$\frac{a}{b} = \frac{ak}{bk}.$$

Some illustrations:

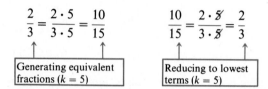

$$\frac{2}{3} = \frac{2 \cdot 5}{3 \cdot 5} = \frac{10}{15} \qquad\qquad \frac{10}{15} = \frac{2 \cdot \cancel{5}}{3 \cdot \cancel{5}} = \frac{2}{3}$$

Generating equivalent fractions ($k = 5$)   Reducing to lowest terms ($k = 5$)

The following *additional properties of fractions* are useful for adding, subtracting, multiplying, and dividing fractional expressions.

◆ **Additional Properties of Fractions**

For all fractions $\dfrac{a}{b}, \dfrac{c}{b}$, and $\dfrac{c}{d}$ ($b \neq 0, d \neq 0$),

1. **Addition Property** $\dfrac{a}{b} + \dfrac{c}{b} = \dfrac{a + c}{b}$

2. **Subtraction Property** $\dfrac{a}{b} - \dfrac{c}{b} = \dfrac{a - c}{b}$

3. **Multiplication Property** $\dfrac{a}{b} \cdot \dfrac{c}{d} = \dfrac{ac}{bd}$

4. **Division Property** $\dfrac{a}{b} \div \dfrac{c}{d} = \dfrac{a}{b} \cdot \dfrac{d}{c}$ ($c \neq 0$)

A **variable** is a symbol that represents *any* member from a given set of numbers. Letters from the end of the alphabet, like x, y, z, and t, are often used as variables. By assuming that a variable represents any member from the set of real numbers, we can manipulate and simplify any expression that contains constants and variables by using the previously listed definitions and properties. Since division by zero is undefined, we shall assume that any variable in a fractional expression is restricted to those values that give the expression a nonzero denominator.

E X A M P L E   1   Use the preceding definitions and properties to complete each statement. Assume that the variables in each fractional expression are restricted to those values that give a nonzero denominator.

(a) $2t + 4 = 4 + \boxed{}$ **(b)** $5(x + 2) = 5x + \boxed{}$

(c) $\dfrac{x + y}{3} = \boxed{}(x + y)$ **(d)** $x + y = x - \boxed{}$

(e) $\dfrac{5}{-(x - 6)} = \dfrac{\boxed{}}{x - 6}$ **(f)** $\dfrac{7x}{3} \div \dfrac{4y}{5} = \dfrac{7x}{3} \cdot \boxed{} = \boxed{}$

SOLUTION

(a) By the commutative property, $2t + 4 = 4 + \boxed{2t}$.

(b) By the distributive property, $5(x + 2) = 5x + \boxed{10}$.

(c) By the definition of division, $\dfrac{x + y}{3} = \boxed{\dfrac{1}{3}}(x + y)$.

(d) By the definition of subtraction, $x + y = x - \boxed{(-y)}$.

(e) By the property for negatives in quotients, $\dfrac{5}{-(x - 6)} = \dfrac{\boxed{-5}}{x - 6}$.

(f) By the division and multiplication properties of fractions,

$$\frac{7x}{3} \div \frac{4y}{5} = \frac{7x}{3} \cdot \boxed{\frac{5}{4y}} = \boxed{\frac{35x}{12y}}$$

♦

PROBLEM 1   Repeat Example 1 for each statement.

(a) $-(x + 2) + (x + 2) = \boxed{}$ **(b)** $\dfrac{x - 4}{3} = \dfrac{x}{3} - \boxed{}$

(c) $\dfrac{-(x + 4)}{-2x(x + 4)} = \dfrac{\boxed{}}{2x(x + 4)} = \boxed{}$

♦

♦ The Real Number Line and Inequality Symbols

A geometric interpretation of the real numbers can be shown on the *real number line*, as illustrated in Figure 1.1. From the figure we can see that this line has no gaps. For each point on this line there corresponds exactly one real number, and for each real number there corresponds exactly one point on this line. This type of relationship is called a **one-to-one correspondence**.

The real number associated with a point on the real number line is called the **coordinate** of the point. Referring to Figure 1.1, $-\frac{13}{4}$ is the coordinate of point A, $-\sqrt{2}$ is the coordinate of point B, $\frac{5}{3}$ is the coordinate of point C, and π is the coordinate of point D.

FIGURE 1.1

On the real number line there exists a *one-to-one correspondence* between the set of real numbers and the set of points on the line.

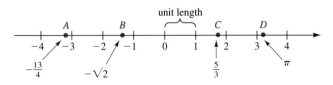

The real number line gives us a convenient way to compare two distinct real numbers a and b. If a is to the *right* of b on the real number line, then **a is greater than b**, and we write

$$a > b$$

Referring to Figure 1.1, since π is to the right of $\frac{5}{3}$, we have $\pi > \frac{5}{3}$.

If a is to the left of b on the real number line, then **a is less than b**, and we write

$$a < b$$

Referring to Figure 1.1, since $\frac{5}{3}$ is to the left of π, we have $\frac{5}{3} < \pi$. In general, for all real numbers a and b,

$$a < b \qquad \text{if and only if} \qquad b > a$$

The symbols $>$ and $<$ are called **inequality symbols**, and expressions like $a > b$ and $a < b$ are called **inequalities**. Two other inequality symbols are used frequently:

1. $\leq$ read "less than or equal to."
2. $\geq$ read "greater than or equal to."

Inequalities can be used to indicate if a number is *positive*, *negative*, *nonnegative*, or *nonpositive*, as shown in Table 1.1.

Table 1.1
Some inequalities and their meanings

Inequality	Meaning
$a > 0$ or $0 < a$	a is positive
$a < 0$ or $0 > a$	a is negative
$a \geq 0$ or $0 \leq a$	a is nonnegative
$a \leq 0$ or $0 \geq a$	a is nonpositive

FIGURE 1.2
Three distinct numbers on the real number line with $a < b < c$.

Figure 1.2 shows three distinct real numbers a, b, and c on a real number line. To indicate that b is between a and c on this line, we can write either

$$a < b < c \qquad \text{or} \qquad c > b > a$$

These expressions are called **double inequalities**. When using expressions like $a < b < c$ and $c > b > a$, be sure all the inequality symbols point in the same direction. For example, the expression $a < c > b$ is completely meaningless.

EXAMPLE 2   Rewrite each statement using inequality symbols:

(a) a is at most 6.   (b) b is at least -2.

(c) c is nonnegative and less than 10.

SOLUTION

(a) a is at most 6   is written as: (b) b is at least -2 is written as:

$$a \leq 6$$ $$b \geq -2$$

(c) c is nonnegative   and   less than 10   is written as:

$$0 \leq c \qquad and \qquad c < 10 \qquad \text{or, more compactly,} \quad 0 \leq c < 10.$$
◆

PROBLEM 2   Repeat Example 2 for each statement.

(a) a is not more than 8.   (b) b is negative and at least -4.   ◆

◆   **Absolute Value**

The distance between zero and a number a on the real number line, without regard to direction, is called the *absolute value* of a and is denoted $|a|$. Because distance is independent of direction and is always nonnegative, the absolute value of any real number is also nonnegative. That is, $|a| \geq 0$. A more formal definition of **absolute value** follows.

 Absolute Value

For any real number a,

$$|a| = \begin{cases} a & \text{if } a \geq 0 \\ -a & \text{if } a < 0. \end{cases}$$

To find $|4|$ from this definition, we use $|a| = a$, since $4 \geq 0$. Thus,

$$|4| = 4.$$

To find $|-4|$ from this definition, we use $|a| = -a$, since $-4 < 0$. Thus,

$$|-4| = -(-4) = 4.$$

Figure 1.3 illustrates $|4|$ and $|-4|$.

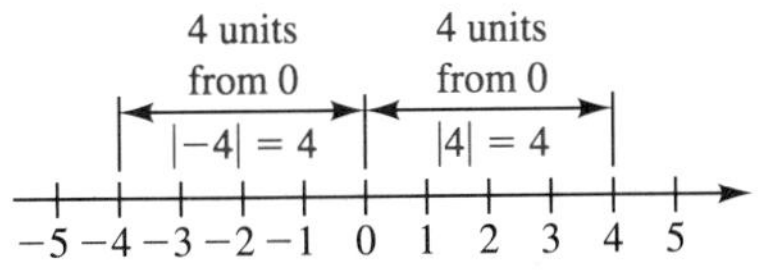

4 units   4 units
from 0   from 0
$|-4| = 4$ $|4| = 4$

$-5\ -4\ -3\ -2\ -1\ \ 0\ \ 1\ \ 2\ \ 3\ \ 4\ \ 5$

FIGURE 1.3

The absolute value of 4, denoted $|4|$, and the absolute value of -4, denoted $|-4|$, are both equal to 4.

If on the real number line the points A and B have coordinates a and b, respectively, then the **distance** between points A and B, denoted AB, is defined as follows:

Distance between Two Points on the Real Number Line

> For any points A and B with coordinates a and b, respectively,
>
> $$AB = |a - b|.$$

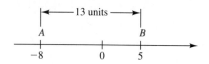

FIGURE 1.4

The distance between -8 and 5 is 13 units.

For example, if the coordinate of point A is -8 and the coordinate of point B is 5, then

$$AB = |-8 - 5| = |-13| = 13 \text{ units},$$

as shown in Figure 1.4.

Note also that $BA = |5 - (-8)| = |13| = 13$ units. In general, for any points A and B, the distance is the same in both directions, that is, $AB = BA$.

E X A M P L E 3 Find AB. Write each answer without absolute value bars.

(a) The coordinate of point A is $-\dfrac{13}{4}$, and the coordinate of point B is $\dfrac{5}{3}$.

(b) The coordinate of point A is 2, and the coordinate of point B is $\sqrt{3}$.

S O L U T I O N

(a) $AB = \left| \dfrac{-13}{4} - \dfrac{5}{3} \right| = \left| \dfrac{-39 - 20}{12} \right| = \left| \dfrac{-59}{12} \right| = \dfrac{59}{12}.$

(b) $AB = |2 - \sqrt{3}| = 2 - \sqrt{3}$ since $2 - \sqrt{3} > 0$ ◆

P R O B L E M 3 For Examples 3(a) and 3(b), verify that $BA = AB$. ◆

Exercises 1.1

Basic Skills

In Exercises 1–28, use the definitions and properties discussed in Section 1.1 to complete each statement. Assume that in each fractional expression variables are restricted to those values that give a nonzero denominator.

1. $3x + 6 = 6 + \square$

2. $(pq)3 = \square(pq)$

3. $\square + \lceil -(x + y) \rceil = 0$

4. $\square(x + 2) = 1$

5. $(2x + 3)\square = 0$

6. $3(x + 2) = 3x + \square$

7. $\dfrac{-(t + 5)}{2} = -\dfrac{\square}{2}$

8. $-2[-(x + y)] = \square(x + y)$

9. $9m + 6n = \square(3m + 2n)$

10. $-(5p - 4) = 4 - \square$

11. $5 - 3x = -(\square - 5)$

12. $\dfrac{4x+7}{4} = \square(4x+7)$

13. $(\pi - \sqrt{2})x = \square - \sqrt{2}\,x$

14. $(3x+4)+y = \square + (3x+y)$

15. $x[6(2+y)] = \square(y+2)$

16. $-x-(4-x) = -x-4+\square = \square$

17. $(x-3)(y+z) = (x-3)\square + (x-3)\square$

18. $\dfrac{0}{x+3} = \dfrac{1}{x+3}\cdot\square = \square$

19. $\dfrac{x-2}{\pi} = \dfrac{x}{\pi} - \square$

20. $\dfrac{x+2}{2} = \square(x+2) = \square + 1$

21. $\dfrac{x}{y} + \dfrac{m}{n} = \dfrac{\square}{yn} + \dfrac{\square}{yn} = \dfrac{\square}{yn}$

22. $\dfrac{5x}{-3x} = -\dfrac{\square}{3x} = \square$

23. $\dfrac{2x-2y}{3x-3y} = \dfrac{\square(x-y)}{\square(x-y)} = \square$

24. $-7+(x+7) = (-7+\square)+x = \square + x = \square$

25. $\dfrac{1}{y+2}[x(y+2)] = \left[\square(y+2)\right]x = \square x = \square$

26. $\dfrac{x}{3} - \dfrac{\pi}{y} = \dfrac{xy}{\square} - \dfrac{\square}{3y} = \square$

27. $\dfrac{2\sqrt{3}}{xt+t} = \dfrac{2}{x+1}\cdot\square$

28. $\dfrac{3x-3y}{2xy} = \dfrac{\square(x-y)}{2xy} = \dfrac{x-y}{2x} \div \square$

In Exercises 29–38, rewrite each statement using inequality symbols.

29. x is negative.   30. y is positive.

31. a is at most 7.   32. b is at least -9.

33. p is greater than 2 and less than or equal to 10.

34. q is less than or equal to 4 and greater than -1.

35. c is positive and less than 8.

36. d is negative and more than -5.

37. t is nonpositive and at least -2.

38. k is nonnegative and at most 4.

In Exercises 39–44, find the distance from A to B. Write each answer without absolute value bars.

39. The coordinate of point A is -9 and the coordinate of point B is 6.

40. The coordinate of point A is 13 and the coordinate of point B is -2.

41. The coordinate of point A is $\frac{7}{8}$ and the coordinate of point B is $-\frac{1}{3}$.

42. The coordinate of point A is $-\frac{3}{4}$ and the coordinate of point B is $-\frac{9}{10}$.

43. The coordinate of point A is $\sqrt{2}$ and the coordinate of point B is π.

44. The coordinate of point A is $-\sqrt{2}$ and the coordinate of point B is $\sqrt{3}$.

Critical Thinking

In Exercises 45–52, rewrite each expression so that it does not contain absolute value bars.

45. $-|\pi - 3|$ 46. $-|3 - \pi|$

47. $|\pi - x|$ if $x \geq \pi$ 48. $|\pi - x|$ if $x < \pi$

49. $|x-3| + |x-4|$ if $3 < x < 4$

50. $|x-3| - |x-4|$ if $x < 3$

51. $|x| < 5$ 52. $|y| > 3$

In Exercises 53–56, use absolute value notation to represent each statement.

53. The distance between a and 7 is at least 3 units.

54. c is less than 6 units from 0.

55. d is closer to 1 than to 0.

56. b is farther from -2 than from 5.

Calculator Activities

57. Use your calculator to help list the following real numbers in order from smallest to largest:

$$3.145 \qquad \pi \qquad \frac{22}{7} \qquad 3.2 \qquad \sqrt{10} \qquad \frac{157}{50}$$

58. Use your calculator to help list the following real numbers in order from largest to smallest:

$$\frac{7}{5} \qquad \sqrt{2} \qquad 1.414 \qquad \frac{71}{50} \qquad \frac{8\pi}{17} \qquad 1.5$$

1.2   Integer Exponents and Scientific Notation

◆ Introductory Comments

Repeated addition of a real number a may be described as multiplication:

$$\underbrace{a + a + a + \cdots + a}_{\boxed{n \text{ terms}}} = na.$$

To describe the repeated multiplication of a real number a, we use a *positive* integer exponent.

 Definition of a^n

> For all real numbers a and any positive integer n,
>
> $$a^n = \underbrace{(a)(a)(a)\cdots(a)}_{\boxed{n \text{ factors}}}$$

The expression a^n is read "a to the nth power" and is referred to as the **exponential form** of the repeated multiplication. In the expression a^n, a is called the **base** and n is the **exponent** or **power**.

Note: Parentheses can change the meaning of an exponential expression. For example,

$$-2^4 = -(2)(2)(2)(2) = -16, \qquad \text{whereas} \qquad (-2)^4 = (-2)(-2)(-2)(-2) = 16.$$

Also,

$$2y^4 = 2(y)(y)(y)(y), \qquad \text{whereas} \qquad (2y)^4 = (2y)(2y)(2y)(2y).$$

As you read through this section, notice how parentheses affect the meaning of each expression.

◆ Properties of Positive Integer Exponents

Using our definition of a^n, observe the following:

1. $a^2 \cdot a^3 = (a \cdot a)(a \cdot a \cdot a) = a^5$

$\boxed{2 + 3 \text{ or } 5 \text{ factors of } a}$

2. $(a^3)^2 = (a^3)(a^3) = (a \cdot a \cdot a)(a \cdot a \cdot a) = a^6$

$\boxed{3 \cdot 2 \text{ or } 6 \text{ factors of } a}$

3. **(a)** $\dfrac{a^5}{a^3} = \dfrac{(a \cdot a \cdot a) \cdot a \cdot a}{(a \cdot a \cdot a)} = a^2$ **(b)** $\dfrac{a^3}{a^5} = \dfrac{(a \cdot a \cdot a)}{(a \cdot a \cdot a) \cdot a \cdot a} = \dfrac{1}{a^2}$

$\boxed{5 - 3 \text{ or } 2 \text{ factors of } a}$ $\boxed{\begin{array}{l}5 - 3 \text{ or } 2 \text{ factors of } a \\ \text{in the denominator}\end{array}}$

4. $(ab)^3 = (ab)(ab)(ab) = (a \cdot a \cdot a)(b \cdot b \cdot b) = a^3 b^3$

$\boxed{3 \text{ factors of } a \text{ and } 3 \text{ factors of } b}$

5. $\left(\dfrac{a}{b}\right)^3 = \dfrac{a}{b} \cdot \dfrac{a}{b} \cdot \dfrac{a}{b} = \dfrac{a^3}{b^3}$

$\boxed{3 \text{ factors of } a \text{ and } 3 \text{ factors of } b}$

These examples suggest the following **properties of positive integer exponents**.

◆ Properties of Positive Integer Exponents

For all positive integers m and n and all real numbers a and b that yield nonzero denominators,

1. Product Property

$a^m a^n = a^{m+n}$

2. Power Property

$(a^m)^n = a^{mn}$

3. Quotient Properties

$$\frac{a^m}{a^n} = \begin{cases} a^{m-n} & \text{if } m > n \\ \dfrac{1}{a^{n-m}} & \text{if } m < n \\ 1 & \text{if } m = n \end{cases}$$

4. Power of a Product Property

$(ab)^n = a^n b^n$

5. Power of a Quotient Property

$\left(\dfrac{a}{b}\right)^n = \dfrac{a^n}{b^n}$

We can use these properties to write expressions that contain positive integer exponents in more compact, or *simpler*, form, as illustrated in Example 1.

EXAMPLE 1   Simplify each expression. Assume the variables are restricted to those real numbers that give nonzero denominators.

(a) $(-3xy^5)(7x^4y^3)$ (b) $[(x + y)^4]^2$

(c) $\dfrac{15x^4y^2z}{25x^3y^2z^3}$ (d) $\left(\dfrac{-3}{2t^4}\right)^3$

SOLUTION

(a) $(-3xy^5)(7x^4y^3) = (-3 \cdot 7)(x \cdot x^4)(y^5 \cdot y^3)$ **Rearrange factors**

$= (-3 \cdot 7)x^{1+4}y^{5+3}$ **Add exponents**

$= -21x^5y^8$ **Simplify**

(b) $[(x + y)^4]^2 = (x + y)^{4 \cdot 2} = (x + y)^8$ **Multiply exponents**

(c) $\dfrac{15x^4y^2z}{25x^3y^2z^3} = \dfrac{15}{25} \cdot x^{4-3} \cdot 1 \cdot \dfrac{1}{z^{3-1}}$ **Apply** *quotient properties*

$= \dfrac{3x}{5z^2}$ **Simplify**

(d) $\left(\dfrac{-3}{2t^4}\right)^3 = \dfrac{(-3)^3}{(2t^4)^3} = \dfrac{(-3)^3}{(2)^3(t^4)^3}$ **Apply** *power of a quotient* **and** *power of a product* **properties**

$= \dfrac{(-3)^3}{(2)^3t^{12}}$ **Multiply exponents**

$= \dfrac{-27}{8t^{12}}$ **Simplify**   ◆

Referring to Example 1(b), to rewrite $(x + y)^8$ as $x^8 + y^8$ is WRONG! It is not possible to bring a power into a sum or difference in the same manner as illustrated in properties 4 and 5 for products and quotients; that is,

$$(a + b)^n \neq a^n + b^n \qquad \text{and} \qquad (a - b)^n \neq a^n - b^n.$$

Note that

$$(3 \cdot 2)^2 = 3^2 \cdot 2^2, \qquad \text{whereas} \qquad (3 + 2)^2 \neq 3^2 + 2^2.$$

PROBLEM 1   Repeat Example 1 for each expression.

(a) $(-2x^5)^4$ (b) $\dfrac{3x + 4}{(3x + 4)^5}$ ◆

◆ **Zero and Negative Integer Exponents**

Our definition of a^n applies to positive integer exponents n. How should we define a^0 or a^{-n}? Certainly we would like the properties of positive integer exponents to hold for zero and negative integer exponents as well. If the product property, $a^m a^n = a^{m+n}$, is to hold for the zero exponent, then

$$a^0 \cdot a^n = a^{0+n} = a^n.$$

The only way $a^0 \cdot a^n$ can equal a^n is for a^0 to be the *identity element for multiplication*; that is, $a^0 = 1$. If the product property is to hold for negative integer exponents, then

$$a^{-n} \cdot a^n = a^{-n+n} = a^0 = 1.$$

The only way $a^{-n} \cdot a^n$ can equal 1 is for a^{-n} to be the *multiplicative inverse* or *reciprocal* of a^n; that is, $a^{-n} = \dfrac{1}{a^n}$.

 Definitions of a^0 and a^{-n}

> For all real numbers a ($a \neq 0$) and any integer n,
>
> $$a^0 = 1 \qquad \text{and} \qquad a^{-n} = \frac{1}{a^n}.$$

Some illustrations of zero and negative integer exponents:

$$5^0 = 1 \qquad\qquad 3^{-2} = \frac{1}{3^2} = \frac{1}{9}$$

$$(-2)^{-4} = \frac{1}{(-2)^4} = \frac{1}{16} \qquad\qquad -2^{-4} = -\frac{1}{2^4} = -\frac{1}{16}$$

Now, consider the quotient a^{-m}/b^{-n}, where m and n are positive integers. To write a^{-m}/b^{-n} *without* negative exponents, we can proceed as follows:

$$\frac{a^{-m}}{b^{-n}} = \frac{1/a^m}{1/b^n} = \frac{1}{a^m} \cdot \frac{b^n}{1} = \frac{b^n}{a^m}.$$

Notice that the factor a^{-m} in the numerator becomes a^m in the denominator, and the factor b^{-n} in the denominator becomes b^n in the numerator. In general, a factor can be moved from the numerator to the denominator, or from the denominator to the numerator, by simply changing the sign of the exponent.

Caution Only *factors* of the numerator or denominator can be moved in this fashion, never *terms* of the numerator or denominator. For example, to write

 as $\dfrac{3y^4}{x^2 z^3}$ is CORRECT!

However, to write

as $\dfrac{3 + y^4}{x^2 - z^3}$ is WRONG!

The expression $\dfrac{3 + x^{-2}}{y^{-4} - z^3}$ becomes the *complex fraction*

$$\frac{3 + \dfrac{1}{x^2}}{\dfrac{1}{y^4} - z^3}.$$

We will discuss and simplify such expressions in Section 1.7.

It can be shown that each of the five properties of positive integer exponents is valid for *all* integer exponents. In fact, since we now know the meaning of zero and negative integer exponents, the *quotient properties* can be combined into a single property and written as

$$\frac{a^m}{a^n} = a^{m-n} \quad (a \neq 0).$$

The important definitions and properties concerning integer exponents are summarized next. We refer to them as the **laws of exponents**.

 Laws of Exponents

1. $a^0 = 1$ 2. $a^{-n} = \dfrac{1}{a^n}$ 3. $a^m a^n = a^{m+n}$ 4. $(a^m)^n = a^{mn}$

5. $\dfrac{a^m}{a^n} = a^{m-n}$ 6. $(ab)^n = a^n b^n$ 7. $\left(\dfrac{a}{b}\right)^n = \dfrac{a^n}{b^n}$ 8. $\dfrac{a^{-m}}{b^{-n}} = \dfrac{b^n}{a^m}$

When using the laws of exponents to simplify exponential expressions, we always assume that the variables represent nonzero real numbers whenever they appear in a denominator or are raised to a zero or negative power.

EXAMPLE 2   Simplify each expression. Express the answer with positive exponents.

(a) $(x + y)^{-4}(x + y)^3$ (b) $\left(\dfrac{3}{y}\right)^{-2}$

(c) $(-3x^{-3}y^4)^{-2}$ (d) $\dfrac{4^{-2}x^{-3}y^4}{4x^{-4}y^{-3}}$

SOLUTION

(a) $(x + y)^{-4}(x + y)^3 = (x + y)^{-1}$ **Add exponents**

$$= \frac{1}{x + y}$$ **Change to a positive exponent**

(b) $\left(\dfrac{3}{y}\right)^{-2} = \dfrac{3^{-2}}{y^{-2}}$ Apply $(a/b)^n = a^n/b^n$ (law 7)

$\qquad\qquad = \dfrac{y^2}{3^2}$ or $\dfrac{y^2}{9}$ Change to positive exponents

(c) $(-3x^{-3}y^4)^{-2} = (-3)^{-2}(x^{-3})^{-2}(y^4)^{-2}$ Apply $(ab)^n = a^n b^n$

$\qquad\qquad = (-3)^{-2}x^6 y^{-8}$ Multiply exponents

$\qquad\qquad = \dfrac{x^6}{(-3)^2 y^8}$ or $\dfrac{x^6}{9y^8}$ Change to positive exponents

(d) $\dfrac{4^{-2}x^{-3}y^4}{4x^{-4}y^{-3}} = 4^{-2-1}x^{-3-(-4)}y^{4-(-3)}$ Subtract exponents

$\qquad\qquad = 4^{-3}x^1 y^7$ Simplify

$\qquad\qquad = \dfrac{xy^7}{4^3}$ or $\dfrac{xy^7}{64}$ Change to a positive exponent ◆

Note: Several alternate procedures may be used to simplify the expressions in Example 2. For example, you may wish to change to positive exponents immediately and then apply the properties of positive integer exponents.

PROBLEM 2 Referring to Example 2(d), first apply $a^{-m}/b^{-n} = b^n/a^m$ to eliminate the negative exponents. Then simplify by using the properties of positive integer exponents. You should obtain the same result. ◆

◆ **Application: Scientific Notation and Conversion Factors**

The laws of exponents may be used in conjunction with *scientific notation* to perform arithmetic calculations containing large and small numbers.

◆ **Scientific Notation**

> A number is written in **scientific notation** if it is of the form
>
> $$k \times 10^n \qquad \text{where} \quad 1 \le |k| < 10 \quad \text{and } n \text{ is an integer.}$$

Two examples of numbers written in scientific notation are

$$3.2 \times 10^3 \qquad \text{and} \qquad 9.84 \times 10^{-3}$$

A number written in scientific notation can easily be changed to *ordinary decimal form*. For 3.2×10^3, we know that

$$10^3 = (10)(10)(10) = 1000,$$

and multiplying by 1000 moves the decimal point *three* places to the *right*. Thus,

$$3.2 \times 10^3 = 3200.$$

right 3 places

The *positive* exponent tells us how many places to the *right* the decimal point is moved.

For 9.84×10^{-3}, we know that

$$10^{-3} = \frac{1}{10^3} = \frac{1}{1000},$$

and multiplying by $\frac{1}{1000}$ (or dividing by 1000) moves the decimal point *three* places to the *left*. Thus,

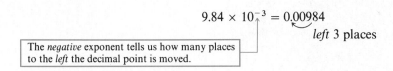

$$9.84 \times 10^{-3} = 0.00984$$

left 3 places

The *negative* exponent tells us how many places to the *left* the decimal point is moved.

To find the product

$$(0.00065)(800,000)$$

we first write the numbers in scientific notation and then apply the laws of exponents. To write 0.00065 in scientific notation, we move the decimal point four places to the *right* (between the 6 and 5) in order to obtain a number between 1 and 10. To accomplish this, we multiply 0.00065 by $(10^4 \times 10^{-4})$. Since $10^4 \times 10^{-4} = 10^0 = 1$, we obtain

$$0.00065 = 0.00065 \times 10^4 \times 10^{-4} = 6.5 \times 10^{-4}$$

Similarly, to write 800,000 in scientific notation, we move the decimal point 5 places to the left in order to obtain a number between 1 and 10. To accomplish this, we multiply by $(10^{-5} \times 10^5)$ as follows:

$$800,000 = 800,000. \times 10^{-5} \times 10^5 = 8 \times 10^5$$

Hence, we have

$$
\begin{aligned}
(0.00065)(800,000) &= (6.5 \times 10^{-4})(8 \times 10^5) \\
&= (6.5 \cdot 8) \times (10^{-4} \cdot 10^5) &&\textbf{Rearrange the factors} \\
&= \ 52.0 \ \times \ \ \ 10^1 \ \ \ = 520 &&\textbf{Add exponents}
\end{aligned}
$$

Scientists and engineers often use scientific notation along with *conversion factors* in their work. A **conversion factor** is formed by two equal measures: we place one of the equal measures in the numerator and the other in the denominator of the fraction. Hence, a conversion factor is equal to 1. Also, a conversion factor is considered an exact number and, therefore, does not affect the number of significant digits that we maintain in a calculation with measured data. (See Appendix A for a discussion of significant digits.)

EXAMPLE 3

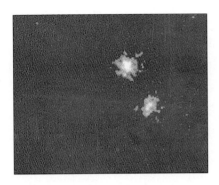

The distance from Earth to Alpha Centauri (the star closest to our Sun) is approximately 25,500,000,000,000 miles (mi). How many years does it take light from this star to reach Earth? Assume that light travels at 186,000 miles per second (mi/s).

SOLUTION First, we find the time in seconds (s) by dividing as follows:

$$\frac{25,500,000,000,000 \text{ mi}}{186,000 \text{ mi/s}} = \frac{2.55 \times 10^{13} \text{ mi}}{1.86 \times 10^5 \text{ mi/s}} = \frac{2.55}{1.86} \times \frac{10^{13}}{10^5} \text{ s}$$
$$\approx 1.371 \times 10^8 \text{ seconds}$$

Now, we apply three conversion factors and change from seconds to years (yr) as follows:

$$1.371 \times 10^8 \, \cancel{s} \cdot \frac{1 \, \cancel{hr}}{3600 \, \cancel{s}} \cdot \frac{1 \, \cancel{day}}{24 \, \cancel{hr}} \cdot \frac{1 \text{ yr}}{365 \, \cancel{days}}$$
$$= \frac{1.371 \times 10^8}{(3.6 \times 10^3)(2.4 \times 10^1)(3.65 \times 10^2)} \text{ yr}$$
$$= \frac{1.371}{(3.6)(2.4)(3.65)} \times \frac{10^8}{10^3 \cdot 10^1 \cdot 10^2} \text{ yr}$$
$$\approx 0.0435 \times 10^2 \text{ yr}$$

Thus, it takes approximately 4.35 years for light to reach Earth from Alpha Centauri. ◆

PROBLEM 3

The distance from Earth to our Sun is approximately 93,000,000 mi. How many minutes does it take light from the Sun to reach Earth? Assume that light travels at 186,000 mi/s. ◆

In engineering, prefixes such as *mega-*, *kilo-*, *centi-*, *milli-*, and *micro-* are attached to the basic units of measurement—meters, grams, volts, watts, and so on. Listed in Table 1.2 are the symbols and meanings of these commonly used prefixes.

Table 1.2
Symbols and meanings of the most commonly used prefixes.

Prefix	Symbol	Meaning
mega-	M	10^6
kilo-	k	10^3
centi-	c	10^{-2}
milli-	m	10^{-3}
micro-	μ	10^{-6}

Numerical quantities are usually substituted in their basic unit size into a formula. As illustrated in the next example, if a prefix is attached to a basic unit, then the prefix must be converted to a power of ten before substituting the unit into a formula. If necessary, the answer in basic units may then be converted to a specified prefix.

E X A M P L E   4   The formula $P = V^2/R$ gives the power P in watts (W) dissipated by a resistance of R ohms (Ω) when the voltage drop across the resistance is V volts (V). Find the power P in milliwatts (mW) if $V = 1.2$ kV and $R = 20$ MΩ.

S O L U T I O N   Converting to powers of ten, we have

$$V = 1.2 \text{ kV} = 1.2 \times 10^3 \text{ V} \qquad \text{and} \qquad R = 20 \text{ M}\Omega = 20 \times 10^6 \text{ } \Omega.$$

Thus,

$$P = \frac{V^2}{R} = \frac{(1.2 \times 10^3)^2}{20 \times 10^6} = \frac{(1.2)^2 \times (10^3)^2}{20 \times 10^6} \text{ W}$$

$$= \frac{1.44 \times 10^6}{20 \times 10^6} \text{ W}$$

$$= 0.072 \times 10^0 \text{ W}$$

$$= 0.072 \text{ W} \quad \text{or} \quad 72 \text{ mW} \qquad \blacklozenge$$

P R O B L E M   4   Using the power formula in Example 4, calculate P in megawatts (MW) if $R = 30 \text{ } \Omega$ and $V = 600{,}000$ V. $\qquad \blacklozenge$

Exercises 1.2

Basic Skills

In Exercises 1–20, use the definitions of a^n, a^{-n}, and a^0 to evaluate each expression.

1. $(-6)^2$
2. 5^3
3. -4^2
4. -2^5
5. 4^{-3}
6. 3^{-4}
7. -8^{-2}
8. -2^{-3}
9. $\left(\frac{2}{3}\right)^{-4}$
10. $\left(\frac{2}{5}\right)^{-1}$
11. $(-9)^{-2}$
12. $\left(\frac{-3}{4}\right)^{-3}$
13. $\frac{(-\frac{1}{3})^2}{2^{-1}}$
14. $\frac{4^{-1}}{-(\frac{2}{3})^4}$
15. $\frac{2^{-1} - 3^0}{2^0 + 3^0}$
16. $\frac{3^0}{2^0 + 3^{-2}}$
17. $3^{-1} + 4^{-1}$
18. $\frac{8^{-2} - 4^{-1}}{2^{-1}}$
19. $\frac{3 \cdot 2^{-3}}{5 \cdot 4^{-2}}$
20. $\frac{3 + 2^{-3}}{5 + 4^{-2}}$

In Exercises 21–40, use the laws of exponents to simplify each expression. Express the answer without zero or negative exponents.

21. $-8x^{-1}y^{-3}$
22. $2^{-1}xy^{-4}$
23. $-5(1 - 2x)^{-6}(-2)$
24. $\frac{\Delta y}{\Delta x}$ $-3(x^2 + 1)^{-4}(2x)$
25. $(-2x^2y^3)(5xy^5)^2$
26. $(2x)^4(-x^2y)(-3xy^{-1})$
27. $(-2y^{-1})^4$
28. $(-3x^2y^{-2})^3$
29. $\frac{8x^4yz^3}{4xy^{-3}z^3}$
30. $\frac{9x^{-1}y^2z}{12x^3y^{-2}}$
31. $(x + 3)^2(x + 3)^{-3}$
32. $(x + 1)(x + 1)^4$
33. $[(x - 2)^2]^4$
34. $[(t - 6)^{-1}]^{-2}$
35. $\frac{(2y + 3)^{-5}(2y + 3)}{(2y + 3)^{-6}}$
36. $\frac{(2a - 5)(2a + 5)^{-3}}{(2a + 5)^3(2a - 5)^2}$

37. $\dfrac{(3m^2n)^{-2}}{2m^{-3}}$

38. $\dfrac{(3x^{-1}y^2)^{-4}}{(-x^4y^3)^{-1}}$

39. $\dfrac{(2p)^{-3}(q-r)^3}{(q-r)^{-4}(4p^2)^{-3}}$

40. $\dfrac{[a^{-2}(b-c)^2]^{-1}}{a(b-c)^{-3}}$

In Exercises 41–46, express each number in ordinary decimal form.

41. 6.9×10^{-9}

42. 2.33×10^6

43. -1750×10^8

44. 2200×10^{-5}

45. 0.00392×10^{-1}

46. -0.0698×10^0

In Exercises 47–52, express each number in scientific notation.

47. 54,300

48. 0.00000294

49. 0.13×10^{-6}

50. 1730×10^5

51. 240×10^3

52. 0.005×10^{-8}

In Exercises 53–60, perform the indicated operations. Write each answer in scientific notation.

53. $(8,000,000,000)(0.00000025)$

54. $(170,000,000)(-20,000,000,000,000)$

55. $\dfrac{0.0000496}{16,000,000,000,000}$

56. $\dfrac{486,000,000}{0.000000006}$

57. $(0.0000000002)^{-5}$

58. $(25,000,000,000)^{-2}$

59. $(300,000,000)(0.00002)^3$

60. $\dfrac{(8,000,000)^2(2000)}{(0.0004)^4}$

In Exercises 61–66, use scientific notation to help solve each problem.

61. The distance from Earth to the planet Pluto is approximately 3,700,000,000 mi. If a space ship leaves from Earth and averages 35,000 mi/hr, how many years will it take to reach Pluto?

62. **Halley's comet** travels approximately 43,300,000,000 km in one orbit around the sun and averages 65,100 km/hr. How many years does it take Halley's comet to orbit the sun once?

63. The formula $I = V/R$ gives the current I in amps (A) through a resistance of R ohms (Ω) when the voltage drop across the resistance is V volts (V). Calculate the current in microamps (μA) if the resistance is 330 MΩ and the voltage drop is 2.2 kV.

64. When two resistances R_1 and R_2 are connected in parallel, their total resistance (R_t) is given by

$$R_t = \frac{R_1 R_2}{R_1 + R_2}.$$

If $R_1 = 4.2$ MΩ and $R_2 = 680$ kΩ, find their total resistance R_t in kΩ.

65. The land area of the earth is approximately 1.484×10^8 square kilometers (sq km). In 1980, the world population was approximately 4,483,000,000. At that time, what was the amount of land area in square meters (sq m) per person?

66. If our national debt is approximately one trillion dollars, and the population of the United States is approximately 250,000,000, what is the amount of debt per person?

Critical Thinking

67. Given that x, y, and z are integers and $a \neq 0$, find the value of $a^{x-y} \cdot a^{y-z} \cdot a^{z-x}$.

68. Given that n is an integer and $b \neq 0$, express each of the following as a power of b:

(a) $(b^{2-n} \cdot b^{n-4})^{-3}$ (b) $\dfrac{(b^2)^n}{b^n \cdot b^{n-2}}$

 In Exercises 69–72, rewrite each expression in the form cx^n, where c is a constant and n is an integer. This procedure is often used when working with derivatives and integrals in calculus.

69. $\dfrac{2}{x^2}$ 70. $\dfrac{3}{4x}$ 71. $\dfrac{1}{3x^5}$ 72. $\dfrac{-1}{(2x)^3}$

73. We have defined a^0 to be equal to 1, provided $a \neq 0$. If we erroneously define 0^0 to be equal to 1, then its reciprocal $1/0^0$ must also be equal to 1, and by the laws of exponents, we could write

$$1 = 0^0 = \frac{1}{0^0} = \frac{a^0}{0^0} = \left(\frac{a}{0}\right)^0.$$

Do you see the fallacy in this argument? Explain.

74. What is the value of $(-1)^n$ when n is an even integer? an odd integer? 0?

Calculator Activities

75. Use the *power key* ($\boxed{Y^X}$ or $\boxed{X^Y}$) on your calculator to evaluate each expression. Record each answer as a decimal number rounded to four significant digits.

(a) $(1.0025)^{45}$ (b) $(-0.936)^{12}$
(c) $(0.0287)^{-3}$ (d) $(1.806)^{-10}$

76. Use the *scientific notation key* ($\boxed{EXP}$ or $\boxed{EE}$) on your calculator to evaluate each expression. Record each answer in the scientific notation form $k \times 10^n$, rounding k to three significant digits.

(a) $(2.731 \times 10^{-11}) - (3.924 \times 10^{-13})$

(b) $\dfrac{(8.75 \times 10^{12})(167 \times 10^{14})}{0.000000000796}$

77. Suppose you have a home mortgage loan of $80,000 at an interest rate of 9% per year over 30 years. The monthly payment may be found by evaluating

$$\frac{(7.5 \times 10^{-3})(8 \times 10^4)(1.0075)^{360}}{(1.0075)^{360} - 1}.$$

(a) Determine, to the nearest cent, the amount of the monthly payment.
(b) Determine the total amount you will repay the mortgage company.

78. You may exceed the computational range of your calculator when working with the power key or scientific notation key. Most calculators will not display positive values less than 1×10^{-99} or greater than $9.999999999 \times 10^{99}$. Determine the smallest and largest integer n for which you are able to obtain a value for 25^n on the display of your calculator.

1.3 Rational Exponents and Radicals

◆ Introductory Comments

In Section 1.2, we defined the expression a^n for integer exponents. We now wish to extend our definition to include exponents that are rational numbers. That is, we would like to assign meaning to expressions such as these:

$$49^{1/2} \qquad 32^{2/5} \qquad 16^{-3/4}$$

In order to make the transition from integer exponents to rational exponents, we must first discuss the meaning of the *nth root* of a real number.

nth root

> If a and b are real numbers, n is a positive integer, and $b^n = a$, then b is an **nth root** of a.

Some observations about nth roots:

1. If a is *positive* and n is *even*, then there are two real nth roots of a—one is positive and one is negative, and they differ only in sign. For example, -7 and 7 are the two real square (2nd) roots of 49, since $(-7)^2 = 49$ and $(7)^2 = 49$. Also, -2 and 2 are the two real fourth (4th) roots of 16, since $(-2)^4 = 16$ and $(2)^4 = 16$.

2. If a is *negative* and n is *even*, then there is no real root of a, because a real number raised to an even power is always nonnegative. For example, -49 has no real square root, since no real number squared is -49. *Imaginary numbers* are needed to define an even-order root of a negative number (see Section 1.5).

3. If a is *positive* and n is *odd*, then there is one real nth root of a and it is positive. For example, 3 is the only real cube (3rd) root of 27, since only $(3)^3 = 27$.

4. If a is *negative* and n is *odd*, then there is one real nth root of a and it is negative. For example, -3 is the only real cube (3rd) root of -27, since only $(-3)^3 = -27$.

5. The nth root of zero is 0. For example, the square root, cube root, fourth root, and so on, of zero is 0, since $0^2 = 0^3 = 0^4 = 0$.

These observations are summarized in Table 1.3.

Table 1.3
The nature of the nth root(s) of a

n	a	n**th root(s)** *of* a
even	positive	two real roots: one positive and one negative
even	negative	no real root
odd	positive	one real root: a positive root
odd	negative	one real root: a negative root
odd or even	zero	one real root: 0

According to Table 1.3, if n is *even* and a is *positive*, then there is one positive real root and one negative real root of a. To avoid ambiguity under these conditions, we assign $\sqrt[n]{a}$ the *positive* real root of a and call it the **principal nth root of a**.

 **Definition of** $\sqrt[n]{a}$

1. If a is *positive* and n is *even*, then
    a. $\sqrt[n]{a}$ is the *positive* or *principal* nth root of a.
    b. $-\sqrt[n]{a}$ is the *negative* of the principal nth root of a.
    c. $\sqrt[n]{-a}$ is not a real number.
2. If a is either *positive* or *negative* and n is *odd*, then $\sqrt[n]{a}$ represents the nth root of a.
3. $\sqrt[n]{0} = 0$ for all positive integers n.

Note: In the expression $\sqrt[n]{a}$, the symbol $\sqrt{}$ is called the **radical**, n is the **index**, and a is the **radicand**. We usually write the principal square root as $\sqrt{a}$ instead of $\sqrt[2]{a}$, but for $n \geq 3$ the index must be shown.

Some illustrations of nth roots:

$$\sqrt{49} = 7 \qquad -\sqrt{49} = -7 \qquad \sqrt{-49} \text{ is not a real number}$$
$$\sqrt[3]{27} = 3 \qquad \sqrt[3]{-27} = -3 \qquad \sqrt[6]{0} = 0$$

◆ **Rational Exponents**

We now define $a^{1/n}$, where n is a positive integer. Certainly, we would like a definition of $a^{1/n}$ to be consistent with the laws of exponents listed in Section 1.2. If the power property for exponents is to hold true, then we must have

$$(a^{1/n})^n = a^{(1/n)(n)} = a^1 = a.$$

The only way $(a^{1/n})^n$ can equal a is for $a^{1/n}$ to be an nth root of a. We define the expression $a^{1/n}$ as the **exponential form** of $\sqrt[n]{a}$.

◆ **Definition of $a^{1/n}$**
_____

> For any positive integer $n \geq 2$, where $a \geq 0$ when n is even,
>
> $$a^{1/n} = \sqrt[n]{a}.$$

Note the difference between these expressions:

$$49^{1/2} = \sqrt{49} = 7 \quad \text{and} \quad -49^{1/2} = -\sqrt{49} = -7$$
$$27^{1/3} = \sqrt[3]{27} = 3 \quad \text{and} \quad (-27)^{1/3} = \sqrt[3]{-27} = -3$$

How might we define $a^{m/n}$, where m and n are positive integers with m/n reduced to lowest terms and $a \geq 0$ when n is even? Again, if the power property for exponents is to hold true, we may define $a^{m/n}$ as follows.

◆ **Definition of $a^{m/n}$**
_____

> If $a^{1/n}$ is a real number and m and n are positive integers such that m/n is reduced to lowest terms, then
>
> $$a^{m/n} = (a^{1/n})^m = (\sqrt[n]{a})^m$$
>
> or
>
> $$a^{m/n} = (a^m)^{1/n} = \sqrt[n]{a^m}.$$

Thus, for $9^{3/2}$ we have two methods of reasoning:

1. $9^{3/2} = (9^{1/2})^3 = (\sqrt{9})^3 = (3)^3 = 27$ or
2. $9^{3/2} = (9^3)^{1/2} = \sqrt{9^3} = \sqrt{729} = 27.$

As you can see, using $a^{m/n} = (\sqrt[n]{a})^m$ is the easier method to apply. That is, *first take the root, then raise to the power.*

EXAMPLE 1   Evaluate each exponential expression.

(a) $32^{2/5}$ **(b)** $16^{-3/4}$ **(c)** $(-8)^{2/6}$

SOLUTION

(a) $32^{2/5} = (32^{1/5})^2 = (\sqrt[5]{32})^2 = (2)^2 = 4$

(b) $16^{-3/4} = \dfrac{1}{16^{3/4}} = \dfrac{1}{(16^{1/4})^3} = \dfrac{1}{(\sqrt[4]{16})^3} = \dfrac{1}{2^3} = \dfrac{1}{8}$

(c) $(-8)^{2/6} = (-8)^{1/3} = \sqrt[3]{-8} = -2$

    Reduce first   ◆

PROBLEM 1 Repeat Example 1 for $(-27)^{-2/3}$. 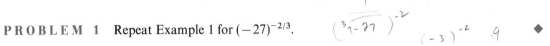   ◆

In calculus, when changing an expression from radical form to exponential form, or vice versa, we use

$$\sqrt[n]{a^m} = a^{m/n}$$

The power of the radicand is the numerator of the exponent, and the index of the radical is the denominator of the exponent.

EXAMPLE 2 (a) Change to exponential form: $\dfrac{1}{\sqrt{x^2 + 9}}$

       (b) Change to a radical expression: $(3x)^{2/3}$

       SOLUTION

       (a) $\dfrac{1}{\sqrt{x^2 + 9}} = \dfrac{1}{(x^2 + 9)^{1/2}}$ or $(x^2 + 9)^{-1/2}$

       (b) $(3x)^{2/3} = \sqrt[3]{(3x)^2} = \sqrt[3]{9x^2}$ ◆

PROBLEM 2 (a) Change to exponential form: $\sqrt[5]{(x^2 + 2)^2}$

       (b) Change to a radical expression: $(2x^2)^{3/7}$ ◆

To simplify expressions containing rational exponents, we use the laws of exponents (Section 1.2) with the understanding that *all variable bases represent positive real numbers.*

EXAMPLE 3 Simplify each expression. Assume the variables represent *positive* real numbers.

       (a) $(9x^2)^{1/2}$ (b) $(5x^{1/4})(-4x^{5/6})$ (c) $\left(\dfrac{x^{-3}}{-8y}\right)^{-2/3}$

SOLUTION

(a) $(9x^2)^{1/2} = 9^{1/2}(x^2)^{1/2}$ **Apply $(ab)^n = a^n b^n$**

$\qquad\qquad\quad = 9^{1/2}x^{(2)(1/2)}$ **Apply $(a^m)^n = a^{mn}$**

$\qquad\qquad\quad = 3x$ **Simplify**

(b) $(5x^{1/4})(-4x^{5/6}) = (5)(-4)x^{1/4}x^{5/6}$ **Regroup**

$\qquad\qquad\qquad\quad = -20x^{1/4 + 5/6}$ **Apply $a^m a^n = a^{m+n}$**

$\qquad\qquad\qquad\quad = -20x^{3/12 + 10/12}$ **Add the exponents and simplify**

$\qquad\qquad\qquad\quad = -20x^{13/12}$

(c) $\left(\dfrac{x^{-3}}{-8y}\right)^{-2/3} = \dfrac{(x^{-3})^{-2/3}}{(-8)^{-2/3}y^{-2/3}}$ **Apply $(a/b)^n = a^n/b^n$ and $(ab)^n = a^n b^n$**

$\qquad\qquad\qquad = \dfrac{x^2}{(-8)^{-2/3}y^{-2/3}}$ **Apply $(a^m)^n = a^{mn}$**

$\qquad\qquad\qquad = (-8)^{2/3}x^2 y^{2/3}$ **Change to positive exponents**

$\qquad\qquad\qquad = 4x^2 y^{2/3}$ **Simplify**   ◆

Note: Referring to Example 3(a), if we do *not* restrict x to *positive* real numbers, then we *cannot* state that $(9x^2)^{1/2}$ and $3x$ are equal. For example, suppose $x = -2$. Then

$$(9x^2)^{1/2} = [9(-2)^2]^{1/2} \qquad \text{whereas} \qquad 3x = 3(-2)$$
$$= [9(4)]^{1/2} \qquad\qquad\qquad\qquad\quad = -6$$
$$= 36^{1/2} = 6$$

If we allow x to be *any* real number, then we must write

$$(9x^2)^{1/2} = 3|x|.$$

In general, we have the following rule.

◆▶ **Rule for $(a^n)^{1/n}$**

> If n is a positive integer and a is *any* real number, then
>
> $$(a^n)^{1/n} = \sqrt[n]{a^n} = \begin{cases} a & \text{if } n \text{ is } \textit{odd} \\ |a| & \text{if } n \text{ is even.} \end{cases}$$

PROBLEM 3   Simplify $[(-2)^4 x^4]^{1/4}$ by assuming that

(a) x is a *positive* real number.

(b) x is *any* real number.   ◆

◆ Simplifying Radical Expressions

We can use the laws of exponents (Section 1.2) in conjunction with the definition of a rational exponent to derive the following **properties of radicals**.

Properties of Radicals

For any real numbers a and b and positive integers $m \geq 2$ and $n \geq 2$ such that $\sqrt[n]{a}$ and $\sqrt[n]{b}$ are *real numbers*,

1. $(\sqrt[n]{a})^n = a$

2. $\sqrt[m]{\sqrt[n]{a}} = \sqrt[mn]{a}$

3. $\sqrt[n]{ab} = \sqrt[n]{a} \cdot \sqrt[n]{b}$

4. $\sqrt[n]{\dfrac{a}{b}} = \dfrac{\sqrt[n]{a}}{\sqrt[n]{b}} \quad (b \neq 0)$

We can illustrate each of these properties:

1. $(\sqrt[3]{10})^3 = 10$

2. $\sqrt[3]{\sqrt{7}} = \sqrt[3 \cdot 2]{7} = \sqrt[6]{7}$

3. $\sqrt{5} \cdot \sqrt{2} = \sqrt{5 \cdot 2} = \sqrt{10}$

4. $\sqrt[3]{\dfrac{-27}{64}} = \dfrac{\sqrt[3]{-27}}{\sqrt[3]{64}} = \dfrac{\sqrt[3]{(-3)^3}}{\sqrt[3]{4^3}} = \dfrac{-3}{4}$

 It is not permissible to take the nth root of a sum or difference in the same manner as illustrated in properties 3 and 4 for products and quotients; that is,

$$\sqrt[n]{a+b} \neq \sqrt[n]{a} + \sqrt[n]{b} \quad \text{and} \quad \sqrt[n]{a-b} \neq \sqrt[n]{a} - \sqrt[n]{b}.$$

Notice that

$$\sqrt{9 \cdot 16} = \sqrt{9} \cdot \sqrt{16} \quad \text{but} \quad \sqrt{9 + 16} \neq \sqrt{9} + \sqrt{16}$$
$$\sqrt{144} = 3 \cdot 4 \qquad\qquad \sqrt{25} \neq 3 + 4$$
$$12 = 12 \qquad\qquad\qquad 5 \neq 7$$

The properties of radicals can be used to write radical expressions in *simplified form.*

 Simplified Form for Radical Expressions

A radical expression is in **simplified form** if

1. the radicand contains no factor to a power greater than or equal to the index of the radical; that is, $\sqrt[n]{a^m}$ has $m < n$.

2. the power of the radicand and the index of the radical have no common factor other than 1; that is, for $\sqrt[n]{a^m}$, the power m and the index n are relatively prime.

E X A M P L E   4   Write each radical expression in simplified form. Assume the variables represent *positive* real numbers.

(a) $\sqrt{16x^5 y^{13}}$

(b) $\sqrt{\sqrt[3]{16x^2}}$

(c) $\sqrt{\dfrac{8x^5}{y^6}}$

(d) $\sqrt[3]{3x^2 y} \cdot \sqrt[3]{9xy^2}$

SOLUTION

(a) $\sqrt{16x^5y^{13}} = \sqrt{4^2 \cdot (x^2)^2 x \cdot (y^6)^2 y}$

Express each factor in
terms of perfect squares

$= \sqrt{(4x^2y^6)^2 \cdot xy}$

Group the perfect square factors

$= \sqrt{(4x^2y^6)^2} \cdot \sqrt{xy}$

Apply $\sqrt[n]{ab} = \sqrt[n]{a} \cdot \sqrt[n]{b}$

$= 4x^2y^6\sqrt{xy}$

Simplify using $\sqrt[n]{a^n} = a$

(b) $\sqrt{\sqrt[3]{16x^2}} = \sqrt[6]{16x^2}$

Apply $\sqrt[m]{\sqrt[n]{a}} = \sqrt[mn]{a}$

$= \sqrt[6]{(4x)^2}$

Write the radicand as a perfect square

$= (4x)^{2/6}$

Change to exponential form

$= (4x)^{1/3} = \sqrt[3]{4x}$

Reduce the order of the radical

(c) $\sqrt{\dfrac{8x^5}{y^6}} = \dfrac{\sqrt{8x^5}}{\sqrt{y^6}}$

Apply $\sqrt[n]{a/b} = \sqrt[n]{a}\big/\sqrt[n]{b}$

$= \dfrac{\sqrt{(2)(2)^2 \cdot (x)(x^2)^2}}{\sqrt{(y^3)^2}}$

Rewrite in terms of perfect squares

$= \dfrac{\sqrt{(2x^2)^2} \cdot \sqrt{2x}}{\sqrt{(y^3)^2}}$

Regroup and apply $\sqrt[n]{ab} = \sqrt[n]{a} \cdot \sqrt[n]{b}$

$= \dfrac{2x^2\sqrt{2x}}{y^3}$

Simplify using $\sqrt[n]{a^n} = a$ when $a \geq 0$.

(d) $\sqrt[3]{3x^2y} \cdot \sqrt[3]{9xy^2} = \sqrt[3]{(3x^2y)(9xy^2)}$

Apply $\sqrt[n]{a} \cdot \sqrt[n]{b} = \sqrt[n]{ab}$

$= \sqrt[3]{27x^3y^3}$

Multiply

$= \sqrt[3]{(3xy)^3}$

Rewrite as a perfect cube

$= 3xy$

Simplify using $\sqrt[n]{a^n} = a$ ◆

PROBLEM 4   Repeat Example 4 for each radical expression.

(a) $\sqrt[3]{32(x + y)^3}$ (b) $\sqrt[4]{\sqrt[3]{x^2y^6}}$ ◆

◆ **Rationalizing Numerators and Denominators**

In algebra we must in some instances remove the radical symbol from the numerator or denominator of a fraction in order to perform further simplifications. The process is called **rationalizing the numerator or denominator**. To rationalize the numerator, we apply the fundamental property of fractions (Section 1.1) and multiply numerator and denominator by a *rationalizing factor* that produces a perfect *n*th power in the numerator. Similarly, to rationalize the denominator, we multiply both numerator and denominator by a *rationalizing factor* that produces a perfect *n*th power in the denominator.

EXAMPLE 5   (a) Rationalize the denominator: $\sqrt{\dfrac{2}{3xy^2}}$

(b)   Rationalize the numerator: $\dfrac{\sqrt[3]{5x^2}}{\sqrt[3]{2y}}$

Assume the variables represent *positive* real numbers.

SOLUTION

(a) $\sqrt{\dfrac{2}{3xy^2}} = \dfrac{\sqrt{2}}{\sqrt{3xy^2}}$

$\qquad\qquad = \dfrac{\sqrt{2}}{\sqrt{3xy^2}} \cdot \dfrac{\sqrt{3x}}{\sqrt{3x}} = \dfrac{\sqrt{6x}}{\sqrt{(3xy)^2}} = \dfrac{\sqrt{6x}}{3xy}$

> This rationalizing factor produces a perfect square in the denominator.

(b) $\dfrac{\sqrt[3]{5x^2}}{\sqrt[3]{2y}} = \dfrac{\sqrt[3]{5x^2}}{\sqrt[3]{2y}} \cdot \dfrac{\sqrt[3]{5^2 x}}{\sqrt[3]{5^2 x}} = \dfrac{\sqrt[3]{(5x)^3}}{\sqrt[3]{50xy}} = \dfrac{5x}{\sqrt[3]{50xy}}$

> This rationalizing factor produces a perfect cube in the numerator.

PROBLEM 5   Rationalize the denominator in Example 5(b), and simplify.   ◆

Exercises 1.3

 Basic Skills

In Exercises 1–20, evaluate each exponential expression.

1. $4^{1/2}$

2. $8^{1/3}$

3. $-64^{1/6}$

4. $-81^{1/2}$

5. $49^{3/2}$

6. $4^{5/2}$

7. $(-125)^{4/6}$

8. $(-64)^{6/9}$

9. $100^{-3/2}$

10. $27^{-2/3}$

11. $-81^{-3/4}$

12. $-36^{-3/2}$

13. $\left(-\dfrac{1}{8}\right)^{2/3}$

14. $\left(\dfrac{1}{16}\right)^{1/4}$

15. $(2.25)^{-1/2}$

16. $(-125)^{1/3} \cdot 25^{-3/2}$

17. $\dfrac{4^{-3/2} \cdot 32^{3/5}}{-27^{0/3}}$

18. $\dfrac{6^{-2}(0.04)^{1/2}}{(-3)^{-1}}$

19. $(2^{-2} - 8^{-2/3} + 4^{-3/2})^{-1}$

20. $4^{-3/2} - 4^{-1/2} + 4^0 - 4^{1/2} + 4^{3/2}$

In Exercises 21–28, rewrite each radical expression in exponential form. For Exercises 25 and 26 express the answer with negative exponents.

21. $\sqrt{2a}$

22. $\sqrt[4]{(x^2 + 1)^3}$

23. $\sqrt[7]{(y^2 - 3)^4}$

24. $\sqrt[10]{(2y)^5}$

25. $\dfrac{x}{\sqrt{x^2 + y^2}}$

26. $\dfrac{3}{\sqrt[3]{x^2}}$

27. $x\sqrt{x}$

28. $\sqrt[3]{y} \cdot \sqrt{y}$

In Exercises 29–36, rewrite each exponential expression in radical form.

29. $x^{1/2}$

30. $y^{3/4}$

31. $(3m^2)^{3/4}$

32. $(-2x)^{2/3}$

33. $(x + y)^{4/5}$

34. $[2(x - 3)^2]^{2/5}$

35. $2x^{-2/3}$

36. $(3p + q)^{-1/6}$

In Exercises 37–50, simplify each exponential expression. Assume the variables represent positive real numbers. Express the answer without zero or negative exponents.

37. $(4x^2)^{1/2} + (32x^5)^{1/5}$

38. $[16(x+1)^4]^{-1/4}$

39. $\left(\dfrac{x^3}{8}\right)^{-2/3}$

40. $\left(\dfrac{x^5 y^{10}}{32}\right)^{2/5}$

41. $(x+y)^{4/3}(x+y)^{3/2}$

42. $(x^2+1)^{-1/2}(x^2+1)^{3/2}$

43. $(-64x^3 y^{-3})^{-2/3}$

44. $(25x^4 y^{-6})^{3/2}$

45. $\dfrac{x^2(x^2+4)^{1/2}}{x^{2/3}(x^2+4)^{3/4}}$

46. $\dfrac{[36(m+8)^{-2}]^{-1/2}}{(m+8)^{2/3}}$

47. $\dfrac{(2^{-4/3}x^{2/3})^{3/4}}{x^{1/2}y^{-2}}$

48. $\dfrac{(3^{2/3}x^{-2}y^{1/2})^{-3/2}}{25x^{1/3}y^2}$

 49. $\frac{1}{2}(x^2-1)^{-1/2}(2x)$

 50. $\frac{1}{3}(3x-6x^3)^{-2/3}(3-18x^2)$

51. Rework Exercise 37 assuming x is *any* real number.

52. Rework Exercise 38 assuming x is *any* real number.

In Exercises 53–68, write each radical expression in simplified form. Assume the variables represent positive real numbers.

53. $\sqrt{(x+1)^2}$

54. $\sqrt[4]{16x^4 y^8}$

55. $\sqrt[3]{16x^4 y^3}$

56. $\sqrt{125x^7 y^2 z^3}$

57. $\sqrt{49(a+b)^2(a^2+b^2)}$

58. $\sqrt[3]{24(x^3-8)(x-8)^3}$

59. $\sqrt[3]{\dfrac{54x^3 y^4}{125z^3}}$

60. $\sqrt{\dfrac{b^2-4ac}{4a^2}}$

61. $\sqrt{\sqrt[3]{25x^2 y^4}}$

62. $\sqrt[4]{\sqrt[3]{36x^2}}$

63. $\dfrac{\sqrt{40x^4 yz}}{2\sqrt{5xy^3 z}}$

64. $\sqrt[3]{6x^2 y^2} \cdot \sqrt[3]{9xy^2}$

65. $\sqrt[3]{4}(\sqrt[3]{2} - \sqrt[3]{4} + \sqrt[3]{10})$

66. $\sqrt{abc}(a\sqrt{ab} + b\sqrt{bc} + c\sqrt{ac})$

67. $\sqrt{16^{-1}x^{-2}\sqrt[3]{(x+y)^2}}$

68. $\left\{\sqrt[3]{[(20xy)^2]^{-3/2}}\right\}^{-1}$

69. Rework Exercise 53 assuming x can be *any* real number.

70. Rework Exercise 54 assuming x and y can be *any* real numbers.

In Exercises 71–76, rationalize the numerator. Assume the variables represent positive real numbers.

71. $\dfrac{7\sqrt{6a}}{5}$

72. $\sqrt{\dfrac{x}{3}}$

73. $\sqrt[3]{\dfrac{(x+2)^2}{3x}}$

74. $\dfrac{4m\sqrt[3]{3m}}{\sqrt[3]{4n^2}}$

75. $\dfrac{\sqrt[4]{2p^2 q^3}}{\sqrt[4]{3}}$

76. $\dfrac{\sqrt[5]{16(x+y)^3}}{8x}$

In Exercises 77–82, rationalize the denominator. Assume the variables represent positive real numbers.

77. $\dfrac{1}{\sqrt{x+3}}$

78. $\dfrac{\sqrt{7}}{6x\sqrt{3x}}$

79. $\sqrt[3]{\dfrac{4}{9xy^2}}$

80. $\dfrac{1}{\sqrt[3]{(x+1)^2}}$

81. $\dfrac{\sqrt[4]{3}}{3\sqrt[4]{2(t+9)^3}}$

82. $\sqrt[7]{\dfrac{5z}{32x^3 y^4}}$

Critical Thinking

In Exercises 83–86, rewrite each product using only one radical sign.   Hint: Use rational exponents to change to radicals that have the same index.

83. $\sqrt{3} \cdot \sqrt[3]{5}$

84. $\sqrt[3]{2} \cdot \sqrt[4]{3}$

85. $\sqrt{x} \cdot \sqrt[3]{y^2}$

86. $\sqrt{2x} \cdot \sqrt[5]{3x^2}$

87. Given that n is an integer and $b \neq 0$, express each of the following as a power of b.

   (a) $(b^{3/n+1/2})^2 \cdot (b^{1/3-2/n})^3$

   (b) $\left(\dfrac{b^{2/n}}{b^{2/n+4}}\right)^{-1/2}$

88. Is $(-1)^{1/n}$ defined when n is an even integer? an odd integer? 0? If it is defined, state its value.

89. The time it takes a pendulum to swing back and forth in completing one cycle is called its *period*. If the length L of a pendulum is known, then its period may be determined by evaluating

$$2\pi\sqrt{\dfrac{L}{32}}.$$

Rationalize the denominator of this expression and then simplify.

90. Which of the following statements is incorrect? Explain.

   (a) $[(-2)^6]^{1/2} = (64)^{1/2} = 8$

   (b) $[(-2)^6]^{1/2} = (-2)^{6(1/2)} = (-2)^3 = -8$

Calculator Activities

91. Use the *power key* ($\boxed{Y^X}$ or $\boxed{x^Y}$) on your calculator to evaluate each expression. Record each answer as a decimal number rounded to four significant digits.

    (a) $(27.5)^{1/4}$ (b) $(-0.654)^{1/3}$
    (c) $(287)^{-6/5}$ (d) $(2.345 \times 10^{-3})^{3/2}$

92. Use the *power key* ($\boxed{Y^X}$ or $\boxed{x^Y}$) on your calculator to evaluate each expression. Record each answer as a decimal number rounded to four significant digits.

    (a) $\sqrt{275}$ (b) $\sqrt[3]{-8.32}$
    (c) $\sqrt[4]{0.1844}$ (d) $\sqrt{12.6 \times 10^8}$

93. When the depth of water in a circular above-ground swimming pool is 4 ft, the pool holds 6032 gallons of water, and the radius (in feet) of the pool may be found by computing

$$0.1825\left(\frac{6032}{\pi}\right)^{1/2}$$

Compute the radius of the pool.

94. The rational numbers 3, 3.1, 3.14, 3.141, 3.1415, 3.14159,... represent better and better approximations of the irrational number π. Fill in the table and observe whether 2^x appears to be approaching a certain value as x approaches π. How might we define 2^π?

x	3	3.1	3.14	3.141	3.1415	3.14159
2^x	—	—	—	—	—	—

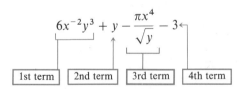

1.4 Operations with Algebraic Expressions

◆ Introductory Comments

A collection of variables and constants formed by the operations of addition, subtraction, multiplication, division, raising to a power, or taking a root is called an **algebraic expression**. The building blocks of algebraic expressions are called *terms*. A **term** consists of either a constant, a variable, or a product or quotient of constants and variables. The algebraic expression

$$6x^{-2}y^3 + y - \frac{\pi x^4}{\sqrt{y}} - 3$$

| 1st term | 2nd term | 3rd term | 4th term |

consists of four terms. Note that the operations of addition and subtraction separate the terms. If a term consists of the product of a real number and one or more variables, then the real number is called the **numerical coefficient** (or simply *coefficient*) of the term. In the preceding algebraic expression the numerical coefficient of the first term is 6; of the second term, 1; of the third term, $-\pi$. The fourth term is called a *constant term*, since it contains no variable.

◆ Combining Like Terms

Terms that differ only in their numerical coefficients are called **like terms**. If two or more like terms appear in an algebraic expression, we usually combine them into a single term by applying the commutative, associative, and distributive properties and adding their numerical cooefficients. The procedure is called *combining* or *collecting* like terms.

EXAMPLE 1   Collect like terms in each algebraic expression.

(a) $3x^2y + 2xy - 5x^2y$ (b) $2x^{1/2} + 3y^{-1} + 4x^{1/2} - y^{-1} + 9$

(c) $\sqrt{45x} + \sqrt{5x}$

SOLUTION

(a) Combining like terms, we have

$$3x^2y + 2xy - 5x^2y = (3 - 5)x^2y + 2xy$$
$$= -2x^2y + 2xy$$

(b) Combining like terms, we have

$$2x^{1/2} + 3y^{-1} + 4x^{1/2} - y^{-1} + 9 = (2 + 4)x^{1/2} + (3 - 1)y^{-1} + 9$$
$$= 6x^{1/2} + 2y^{-1} + 9$$

(c) The terms $\sqrt{45x}$ and $\sqrt{5x}$ are *unlike* and cannot be combined in their present form. However,

$$\sqrt{45x} = \sqrt{3^2 \cdot 5x} = 3\sqrt{5x}.$$

Hence,

$$\sqrt{45x} + \sqrt{5x} = 3\sqrt{5x} + \sqrt{5x} = (3 + 1)\sqrt{5x} = 4\sqrt{5x}$$

Only radicals with like radicands may be combined.

PROBLEM 1   Repeat Example 1 for each algebraic expression.

(a) $\frac{1}{2}xy + \frac{3}{4}xy + xy$ (b) $\sqrt[3]{8x} - \sqrt[3]{27x}$

◆ Identifying Polynomials

A *polynomial* is a special type of algebraic expression that occurs frequently in algebra.

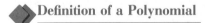
Definition of a Polynomial

> A **polynomial** is an algebraic expression in which no variable appears in any denominator or in any radicand, and any variable that does appear is raised to a nonnegative integer power.

If, after like terms are combined, a polynomial consists of only *one* term, it is called a **monomial**; if it consists of *two* terms, it is a **binomial**; and if it consists of *three* terms, it is a **trinomial**. Each of the following algebraic expressions is a polynomial:

$$5x^2y^4 + xy^3 - 7x^2 + 3 \qquad\qquad 3xy^3 \quad \text{(a monomial)}$$

$$2x^2 - 7 \quad \text{(a binomial)} \qquad x^2 - 3xy + 2y^2 \quad \text{(a trinomial)}$$

The **degree of a term** of a polynomial is the sum of all the exponents of the variables in the term. For the polynomial

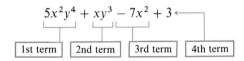

the degree of the first term is $2 + 4 = 6$; of the second term, $1 + 3 = 4$; and of the third term, 2. The degree of the fourth term, which is a constant term, is regarded as 0. The **degree of a polynomial** with unlike terms is the degree of the term with the *highest* degree in the polynomial. Since the polynomial $5x^2y^4 + xy^3 - 7x^2 + 3$ has terms of degree 6, 4, 2, and 0, respectively, the degree of this polynomial is 6.

EXAMPLE 2   Determine whether the given expression is a polynomial. If it is, give its degree.

(a) $5xy^{1/2}$ **(b)** $5x^2y - \dfrac{2x}{y}$

(c) $5x^2 - 2x + 7$ **(d)** $\sqrt{2}y^6 - x^3y^4 + \pi x + 4$

SOLUTION

(a) The term $5xy^{1/2}$ is *not* a polynomial because of the rational exponent on y.

(b) The expression $5x^2y - \dfrac{2x}{y}$ is *not* a polynomial because of the variable in the denominator.

(c) Note that each variable of this trinomial is raised to a positive integer power. The degree of the terms are 2, 1, and 0, respectively. Hence, $5x^2 - 2x + 7$ is a polynomial of degree 2.

(d) Note that each variable of this algebraic expression is raised to a positive integer power. The degree of the terms are 6, 7, 1, and 0, respectively. Hence, $\sqrt{2}y^6 - x^3y^4 + \pi x + 4$ is a polynomial of degree 7.   ◆

PROBLEM 2   Repeat Example 2 for each expression.

(a) $3x + 5y$ **(b)** $-7x^3y^2 + 5\sqrt{x} - 9$ ◆

◆   **Addition and Subtraction**

To find the *sum* of two algebraic expressions, we use the commutative and associative properties to group like terms, and then apply the distributive property to combine like terms. For example, to find the sum of the polynomials $3x + 5y$ and $5x^2 - 2x + 7y$, we write

$$(3x + 5y) + (5x^2 - 2x + 7y)$$
$$= 5x^2 + (3x - 2x) + (5y + 7y) \qquad \text{Group like terms}$$
$$= 5x^2 + x + 12y \qquad \text{Combine like terms}$$

Now recall from Section 1.1, to subtract b from a, we add a to the *additive inverse* of b. Thus, to find the *difference* $(3x + 5y) - (5x^2 - 2x + 7y)$, we add $(3x + 5y)$ to the additive inverse of $(5x^2 - 2x + 7y)$. Now the additive inverse of

$$(5x^2 - 2x + 7y) \qquad \text{is} \qquad (-5x^2 + 2x - 7y)$$

Note the change of sign in each term.

since

$$(5x^2 - 2x + 7y) + (-5x^2 + 2x - 7y)$$
$$= (5x^2 - 5x^2) + (-2x + 2x) + (7y - 7y) = 0.$$

We use this additive inverse to find the difference of the polynomials:

$$(3x + 5y) - (5x^2 - 2x + 7y)$$
$$= (3x + 5y) + (-5x^2 + 2x - 7y) \qquad \text{Definition of subtraction}$$
$$= -5x^2 + (3x + 2x) + (5y - 7y) \qquad \text{Group like terms}$$
$$= -5x^2 + 5x - 2y \qquad \text{Combine like terms}$$

EXAMPLE 3   Perform the indicated operations.

(a) $(x^3 + 4x^2y) - (xy^2 - 2yx^2) + (xy^2 + 2y^3)$

(b) $\left(\dfrac{3}{4x^{2/3}} + \dfrac{1}{y}\right) - \left(\dfrac{1}{4}x^{-2/3} - 3y^{-1} + 2\right)$

SOLUTION

(a) $(x^3 + 4x^2y) - (xy^2 - 2yx^2) + (xy^2 + 2y^3)$

$$= (x^3 + 4x^2y) + (-xy^2 + 2yx^2) + (xy^2 + 2y^3)$$

$$= x^3 + 6x^2y + 2y^3$$

(b) $\left(\dfrac{3}{4x^{2/3}} + \dfrac{1}{y}\right) - \left(\dfrac{1}{4}x^{-2/3} - 3y^{-1} + 2\right)$

$$= \left(\dfrac{3}{4}x^{-2/3} + y^{-1}\right) + \left(-\dfrac{1}{4}x^{-2/3} + 3y^{-1} - 2\right)$$

$$= \dfrac{1}{2}x^{-2/3} + 4y^{-1} - 2 \quad or \quad \dfrac{1}{2x^{2/3}} + \dfrac{4}{y} - 2$$

PROBLEM 3   Repeat Example 3 for $\left(5x^2 - 3x\sqrt{2x}\right) - \left(x^2 - x\sqrt{50x}\right)$.

◆ Multiplication

To find the *product* of two algebraic expressions, both of which do not contain just one term, we use the distributive property. First we multiply each term of the first algebraic expression by each term of the second algebraic expression, and then we collect like terms. The procedure is illustrated in the next example.

EXAMPLE 4   Find the product.

(a) $5\sqrt{x}(8\sqrt{x} - 2x - 1)$ **(b)** $(3x - 4y)(2x + 5y)$

(c) $(x + 2)(3x^3 - x^2 + 2x - 4)$ **(d)** $(3x^{-1}y^{1/2} + y^{5/2})(4xy^{3/2} - x^2y^{-1/2})$

SOLUTION

(a) $5\sqrt{x}(8\sqrt{x} - 2x - 1) = 40(\sqrt{x})^2 - 10x\sqrt{x} - 5\sqrt{x}$

$$= 40x - 10x\sqrt{x} - 5\sqrt{x}$$

(b) $(3x - 4y)(2x + 5y) = 6x^2 + 15xy - 8xy - 20y^2$

$$= 6x^2 + 7xy - 20y^2$$

(c) $(x + 2)(3x^3 - x^2 + 2x - 4) = 3x^4 - x^3 + 2x^2 - 4x + 6x^3 - 2x^2 + 4x - 8$

$$= 3x^4 + 5x^3 - 8$$

(d) $(3x^{-1}y^{1/2} + y^{5/2})(4xy^{3/2} - x^2y^{-1/2}) = 12x^0y^2 - 3x^1y^0 + 4xy^4 - x^2y^2$

$$= 12y^2 - 3x + 4xy^4 - x^2y^2$$

The product in Example 4(c) may also be found by arranging the polynomials *vertically* and multiplying:

$$3x^3 - x^2 + 2x - 4$$
$$x + 2$$

$$3x^4 - x^3 + 2x^2 - 4x \longleftarrow \boxed{\begin{array}{l}\text{Multiply each term of}\\(3x^3 - x^2 + 2x - 4)\text{ by } x.\end{array}}$$

$$6x^3 - 2x^2 + 4x - 8 \longleftarrow \boxed{\begin{array}{l}\text{Multiply each term of}\\(3x^3 - x^2 + 2x - 4)\text{ by } 2.\\\text{Align like terms with the row above.}\end{array}}$$

$$3x^4 + 5x^3 \qquad\qquad -8 \longleftarrow \boxed{\text{Add like terms from each row.}}$$

The advantage of the vertical method is that like terms are actually grouped at the same time the multiplication is performed.

PROBLEM 4   Use the vertical method to find the product $(m^3 - m^2 + 3m + 4)(m^2 + 2m - 3)$.

◆

◆ **Special Products**

There are three special products that occur quite frequently

Special Products

> If A and B are algebraic expressions, then
>
> 1. $(A + B)(A - B) = A^2 - B^2$
> 2. $(A + B)(A^2 - AB + B^2) = A^3 + B^3$
> 3. $(A - B)(A^2 + AB + B^2) = A^3 - B^3$

EXAMPLE 5   Use one of the special products to find each product.

   (a) $(3x + 2)(3x - 2)$ **(b)** $(4\sqrt{xy} - 3y^{-3})(4\sqrt{xy} + 3y^{-3})$

   (c) $(x + 2)(x^2 - 2x + 4)$ **(d)** $(2x^2 - 3)(4x^4 + 6x^2 + 9)$

SOLUTION

(a) Let $A = 3x$ and $B = 2$. Using

we have
$$(A + B)(A - B) = A^2 - B^2,$$
$$(3x + 2)(3x - 2) = (3x)^2 - (2)^2$$
$$= 9x^2 - 4.$$

(b) Let $A = 4\sqrt{xy}$ and $B = 3y^{-3}$. Using

we have
$$(A - B)(A + B) = A^2 - B^2,$$
$$(4\sqrt{xy} - 3y^{-3})(4\sqrt{xy} + 3y^{-3}) = (4\sqrt{xy})^2 - (3y^{-3})^2$$
$$= 16xy - 9y^{-6}.$$

(c) Let $A = x$ and $B = 2$. Then $A^2 = x^2$, $AB = 2x$, and $B^2 = 4$. Using

$$(A + B)(A^2 - AB + B^2) = A^3 + B^3,$$

we have

$$(x + 2)(x^2 - 2x + 4) = (x)^3 + (2)^3$$
$$= x^3 + 8.$$

(d) Let $A = 2x^2$ and $B = 3$. Then $A^2 = 4x^4$, $AB = 6x^2$, and $B^2 = 9$. Using

$$(A - B)(A^2 + AB + B^2) = A^3 - B^3,$$

we have

$$(2x^2 - 3)(4x^4 + 6x^2 + 9) = (2x^2)^3 - (3)^3$$
$$= 8x^6 - 27.$$ ◆

We can check each product by multiplying the expressions in the conventional manner. Checking Example 5(c), we find

$$(x + 2)(x^2 - 2x + 4) = x^3 - 2x^2 + 4x + 2x^2 - 4x + 8 = x^3 + 8.$$

PROBLEM 5   Check Example 5(d) by using the conventional method for multiplying algebraic expressions. ◆

◆ **The Square and Cube of a Binomial**

How do we *square* the binomial $A + B$? From the definition of an exponent, we have

$$(A + B)^2 = (A + B)(A + B) = A^2 + 2AB + B^2.$$

To *cube* a binomial, we can proceed as follows:

$$(A + B)^3 = (A + B)(A + B)^2 = (A + B)(A^2 + 2AB + B^2)$$
$$= A^3 + 2A^2B + AB^2 + A^2B + 2AB^2 + B^3$$
$$= A^3 + 3A^2B + 3AB^2 + B^3.$$

The Square and Cube of a Binomial

If A and B are algebraic expressions, then

$$(A + B)^2 = A^2 + 2AB + B^2$$

and

$$(A + B)^3 = A^3 + 3A^2B + 3AB^2 + B^3.$$

The process of squaring and cubing a binomial is called *expanding* the expression.

E X A M P L E 6   Expand each expression and simplify.

(a) $\left(4\sqrt{x} + 5\right)^2$ (b) $(3x - 2y)^3$

S O L U T I O N

(a) Let $A = 4\sqrt{x}$ and $B = 5$. Using

we have
$$(\ A\ + B)^2 = \ A^2\ + \ 2AB\ + B^2,$$
$$\left(4\sqrt{x} + 5\right)^2 = \left(4\sqrt{x}\right)^2 + 2\left(4\sqrt{x}\right)(5) + (5)^2$$
$$= 16x + 40\sqrt{x} + 25.$$

(b) By the definition of subtraction, $(3x - 2y)^3 = [3x + (-2y)]^3$. Letting $A = 3x$ and $B = -2y$, we have

$$(A + B)^3 = \ A^3\ + \ 3A^2B\ + \ 3AB^2\ + \ B^3$$
$$(3x - 2y)^3 = (3x)^3 + 3(3x)^2(-2y) + 3(3x)(-2y)^2 + (-2y)^3$$
$$= 27x^3 - 54x^2y + 36xy^2 - 8y^3$$

Note the alternating signs when B is negative.

◆

P R O B L E M 6   Repeat Example 6 for each expression.

(a) $(3x - 5y)^2$ (b) $(x^{-4} + 4)^3$ ◆

◆ **Grouping Symbols and the Order of Operations**

Grouping symbols, such as *parentheses* ( ), *brackets* [ ], and *braces* { }, are used to show that the terms contained within them represent a single quantity. Usually parentheses are innermost grouping symbols, then brackets, and then braces. To simplify an algebraic expression that contains more than one set of grouping symbols, start by removing the innermost ones first, following the order of operations.

◆ Order of Operations

1st:	Raise expressions to powers.
2nd:	Perform the multiplications and divisions, and
3rd:	Perform the additions and subtractions.

E X A M P L E 7   Simplify each expression.

(a) $2x[(8x + 2) - 2(2x + 1)^2]$ (b) $6x^2 + \{3[y - (2x - 1)(x + 3)]\}$

SOLUTION

Start with this power.

(a) $2x[(8x + 2) - 2(2x + 1)^2] = 2x[(8x + 2) - 2(4x^2 + 4x + 1)]$
$$= 2x[8x + 2 - 8x^2 - 8x - 2]$$
$$= 2x[-8x^2] = -16x^3$$

Start with this product.

(b) $6x^2 + \{3[y - (2x - 1)(x + 3)]\} = 6x^2 + \{3[y - (2x^2 + 5x - 3)]\}$

Retain this set of parentheses and subtract *all* terms of this product.

$$= 6x^2 + \{3[y - 2x^2 - 5x + 3]\}$$
$$= 6x^2 + \{3y - 6x^2 - 15x + 9\}$$
$$= 3y - 15x + 9$$ ◆

PROBLEM 7   Repeat Example 7 for $x^2 - [y(x + 2y) - (x + y)(2y - x)]$.   ◆

Exercises 1.4

Basic Skills

In Exercises 1–10, collect like terms.

1. $5x^5 - 3x^2 + 4x^2 - 6x^5$

2. $2xy^2 + 4y - 5x^2y - 4y$

3. $x^3 + 3x^2 + xy^2 - 5yx^2 - x^2y + y^3$

4. $3m^3 - 2m^2n + mn^4 - 3m^4n + m^2n$

5. $\frac{3}{4}x + \frac{1}{2}xy - \frac{7}{2}xy + \frac{1}{4}x$

6. $\frac{2}{3}x^2 - \frac{2}{3}xy^2 + \frac{3}{4}x^2 + \frac{3}{4}xy^2$

7. $5x^{2/3} - x^{-2/3} + \dfrac{1}{3x^{2/3}} - 9x^{2/3}$

8. $\dfrac{2}{t^2} + t^{-1/2} + 6t^{-2} - \dfrac{1}{4t^{1/2}}$

9. $\sqrt{10x} + 3\sqrt{40x} - 2\sqrt{90x}$

10. $3\sqrt{64p} - \sqrt{100p} - 5\sqrt{49p}$

In Exercises 11–20, determine whether the expression is a polynomial. If it is a polynomial, give its degree. If it is not a polynomial, explain why not.

11. $\pi x^2 y^4$

12. $\dfrac{18x^3 y^2}{z}$

13. $3x - y^{-1}$

14. $7x^2 y^3 - 4z^3$

15. $\sqrt{3}\, x^2 - 9x + \pi$

16. $5x^{-2}y + z + \sqrt{7}$

17. $9x^3 y^4 - 3xy^7 + 9xy - 3$

18. $3x^{1/2}y^3 - 7x + 4y - 8\pi$

19. $3x^2 + y^3 + z^4 + \sqrt{x^2 + y^2}$

20. $\dfrac{7x}{3} + \dfrac{9x^2}{7} + 2x^3 + 8x^4$

In Exercises 21–50, perform the indicated operations.

21. $(3m^2 + 2m - 9) + (4m^2 - 7m)$

22. $(-xy^2 + 4x^2y) - (2yx^2 - 4y^2x)$

23. $(3x^{-2} + y) - \left(\dfrac{1}{x^2} - y\right) + (x^2 + 4y)$

24. $(t^{1/2} - t^{-1/2}) - \left(t + \dfrac{1}{2t^{1/2}}\right) + (t - \sqrt{t})$

25. $-9x\sqrt[4]{9x^2} - (xy - x\sqrt{27x}) + (\sqrt{48x^3} + xy)$

26. $(\sqrt{20} - x\sqrt{2}) - (3\sqrt[4]{4x^4} - \sqrt{5}) - \sqrt{45}$

27. $-4m^2(8m^3n - 2m^2 + m - 1)$

28. $(4x^{-2}y^{-3} + x^{-1}y + y^{-2})5x^2y^3$

29. $\sqrt{1 + x}(\sqrt{x} - \sqrt{1 + x})$

30. $[(p + 8)^{2/3} - p^{1/3}](p + 8)^{1/3}$

31. $(x^2 - 9y)(3x + 7)$

32. $(2mn + m)(n - m)$

33. $(n - 2)(2n^2 + 5n - 2)$

34. $(p^2 + 4p - 2)(p^2 - p - 3)$

35. $(8x^{-1}y^{3/2} - y^{-1/2})(2xy^{1/2} + y^{3/2})$

36. $(x^{-2}y^{1/3} + 2y^{-2/3})(x^2y^{2/3} - 3x^3y^{-1/3} + x^4y^{5/3})$

37. $(x^2 + x\sqrt{3} + 3)(x\sqrt{3} - 3)$

38. $(\sqrt{a} - \sqrt{b} - \sqrt{c})(\sqrt{a} + \sqrt{b} + \sqrt{c})$

39. $(3x - y)(4x + y)$

40. $(9 - 2x^2)(5 - 3x^2)$

41. $(\sqrt{x} + 3)(\sqrt{x} - 5)$

42. $(x^{1/4} - y^{1/2})(x^{1/4} + 3y^{1/2})$

43. $n - 2(2n^2 + 5n - 2)$

44. $p^2 + 4p - 2(p^2 - p - 3)$

45. $7x^{1/3} - (x^{1/3} + 4)(x^{1/3} + 2)$

46. $15x + 2\sqrt{x} - (3\sqrt{x} - 2)(5\sqrt{x} + 4)$

47. $-m^3 - m[8m - 2m(m + 4)]$

48. $a^2 - [a(2a + b) - (2a - b)(a + b)]$

49. $a\{1 - [2b^2 + (a^{-2} - 1 - b^2)(a^{-2} + 1 + b^2)]\}$

50. $m^{1/2}n^{1/2}\{-[(m + n) - (m^{1/2} + m^{1/4}n^{1/4} + n^{1/2}) \\ \times (m^{1/2} - m^{1/4}n^{1/4} + n^{1/2})]\}$

In Exercises 51–60, use the special products to find each product. Check by multiplying in the conventional manner.

51. $(3x + y)(3x - y)$ 52. $(2x + 5)(2x - 5)$

53. $(x + 3)(x^2 - 3x + 9)$ 54. $(9x^2 - 6x + 4)(3x + 2)$

55. $(4x^2 + 2x + 1)(2x - 1)$

56. $(2x - 5)(4x^2 + 10x + 25)$

57. $(\sqrt{x} + \sqrt{3y})(\sqrt{x} - \sqrt{3y})$

58. $(2\sqrt{a} - \sqrt{4 - a})(2\sqrt{a} + \sqrt{4 - a})$

59. $(m^{2/3} + m^{1/3}n^{1/3} + n^{2/3})(m^{1/3} - n^{1/3})$

60. $(2p^{-1} + q^{-1})(4p^{-2} - 2p^{-1}q^{-1} + q^{-2})$

In Exercises 61–78, expand and simplify each expression.

61. $(m + 6)^2$ 62. $(y - 8)^2$

63. $(2a - 3b^2)^2$ 64. $(3x^3 + 4y)^2$

65. $(3x^{-3} + 5)^2$ 66. $(6a^{-1} - 5b^{-2})^2$

67. $(2\sqrt{x} - 7y)^2$ 68. $(2x^{-1/4} - x)^2$

69. $(\sqrt{x + 1} + \sqrt{x - 1})^2$ 70. $(3\sqrt{1 - x^2} - 3x)^2$

71. $(x + 3)^3$ 72. $(2 - y)^3$

73. $(4x - 3y^2)^3$ 74. $(2m^3 + 5)^3$

75. $\left(x^{-3} + \dfrac{1}{2y}\right)^3$ 76. $(8x^{-1/3} - x)^3$

77. $(\sqrt[3]{1 - x^3} + x)^3$ 78. $(\sqrt[3]{x + 1} + \sqrt[3]{x - 1})^3$

 Critical Thinking

79. Suppose one polynomial P has degree m and another polynomial Q has degree n, where $m > n$. What is the degree of the sum $P + Q$? the difference $P - Q$? the product PQ? the power P^2?

80. Develop a formula for the expansion of $(A + B)^4$. Use this formula to expand each binomial:

    (a) $(2x + 1)^4$ (b) $(x^2 - 2y)^4$

81. The area of a rectangle is the product of its length and width. Find a polynomial that describes the shaded area of the rectangle in the figure.

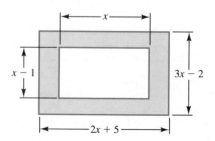

E X E R C I S E   8 1

82. The volume of a rectangular solid is the product of its length, width, and height. Find a polynomial that describes the volume of the rectangular solid shown in the figure.

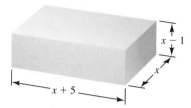

83. The area of a circle is the product of π and its radius squared. Find a polynomial that describes the shaded area of the circle shown in the figure.

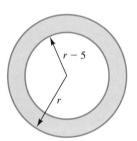

84. The volume of a sphere is the product of $4\pi/3$ and its radius cubed. Find a polynomial that describes the volume of the sphere in the figure.

85. Each teacher in a school system receives a 6% raise. Find a polynomial that describes a teacher's new salary if x is the previous salary.

86. Suppose a sum of $1000 earns simple interest at the rate of 9% per year and a sum of $500 earns simple interest at the rate of 6% per year. Find a polynomial that describes the total amount to which these sums will accumulate after t years.

 Calculator Activities

87. Most calculators are programmed to the order of operations discussed in this section. Evaluate each expression by applying the order of operations, and then compare the answer to the one you obtain by pressing the keys on your calculator in the order written in the expression. Do your answers agree?

(a) $28 + 12 \div 2^2 - 4 \cdot 3$ (b) $(28 + 12) \div (2^2 - 4 \cdot 3)$

88. Use the power key ($\boxed{Y^x}$ or $\boxed{X^Y}$) on your calculator to evaluate $(1.01)^3$. Then evaluate $(1 + .01)^3$ by using the expansion formula for $(A + B)^3$. Do your answers agree?

89. A steel beam 36.6 feet long and fixed at both ends carries a uniformly distributed load of 25,000 pounds (lb) as shown in the figure. The deflection of the beam (in inches) at any distance x from one of the fixed ends is given by

$$(4.72 \times 10^{-6}) \, x^2 (36.6 - x)^2.$$

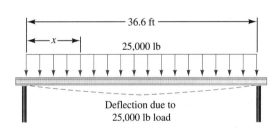

EXERCISE 89

(a) Write this expression as a trinomial, recording the numerical coefficients in scientific notation rounded to three significant digits.
(b) The maximum deflection for this beam occurs at midspan (18.3 ft). Determine the maximum deflection to the nearest hundredth of an inch.

90. A compressor powers the pneumatic tube system used at drive-through bank lanes to carry transactions between a customer and the teller. The horsepower of a compressor designed to carry a tube with a 4-inch diameter a distance of 60 feet is given by

$$0.032p^{3/2}(0.1815 + 0.00348p)$$

where p is the air pressure in the system, in pounds per square inch (psi). Find the product, recording the numerical coefficients in scientific notation rounded to three significant digits.

1.5 The Complex Numbers

◆ **Introductory Comments**

One of the basic properties of a real number is that its square is always *non-negative*. That is, if x is a real number, then

$$x^2 \geq 0.$$

Hence, the principal square root of a *negative* number *cannot* be a real number. For example, $\sqrt{-4}$ and $\sqrt{-9}$ *cannot* be real numbers, since no real number squared is -4 and no real number squared is -9. In this section, we extend the real number system to a larger system that includes the real numbers as well as those numbers whose squares are negative. We refer to this larger system as the **complex number system**.

◆ **Pure Imaginary Numbers**

In the seventeenth century, the French mathematician René Descartes used the word "imaginary" to describe roots like $\sqrt{-4}$ and $\sqrt{-9}$, and in the eighteenth century a Swiss mathematician, Leonhard Euler, introduced the **imaginary unit i**.

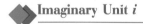
Imaginary Unit i

$$i = \sqrt{-1} \qquad \text{where} \qquad i^2 = -1$$

Thus, if a is a real number and $a > 0$, then we can find the **principal square root of $-a$** as follows:

$$\sqrt{-a} = \sqrt{-1 \cdot a} = \sqrt{-1} \cdot \sqrt{a} = i\sqrt{a}.$$

Principal Square Root of $-a$

If a is a real number and $a > 0$, then

$$\sqrt{-a} = i\sqrt{a}.$$

Thus, for $\sqrt{-4}$ and $\sqrt{-9}$, we have

$$\sqrt{-4} = i\sqrt{4} = 2i \qquad \text{and} \qquad \sqrt{-9} = i\sqrt{9} = 3i.$$

Note that the squares of $2i$ and $3i$ are -4 and -9, respectively:

$$(2i)^2 = 2^2 i^2 = 4(-1) = -4 \qquad \text{and} \qquad (3i)^2 = 3^2 i^2 = 9(-1) = -9.$$

Numbers such as $2i$ and $3i$, whose squares are negative, are called *pure imaginary numbers*.

 Pure Imaginary Number

> Any number of the form bi, where i is the imaginary unit and b is a real number such that $b \neq 0$, is a **pure imaginary number**.

We can add, subtract, multiply, and divide pure imaginary numbers by applying the commutative, associative, and distributive properties as we did with the real numbers. Some illustrations:

1. $\sqrt{-4} + \sqrt{-9} = 2i + 3i = (2 + 3)i = 5i$
2. $\sqrt{-4} - \sqrt{-9} = 2i - 3i = (2 - 3)i = -i$
3. $\sqrt{-4} \cdot \sqrt{-9} = (2i)(3i) = (2 \cdot 3)(i \cdot i) = 6i^2 = 6(-1) = -6$
4. $\dfrac{\sqrt{-4}}{\sqrt{-9}} = \dfrac{2i}{3i} = \dfrac{2}{3}$

From the preceding illustrations, note that the sum and difference of $2i$ and $3i$ are pure imaginary numbers, while the product and quotient of $2i$ and $3i$ are real numbers.

Caution The property of radicals that states $\sqrt{a}\,\sqrt{b} = \sqrt{ab}$ does not apply when both a and b are negative numbers. For example, to write

$$\sqrt{-4} \cdot \sqrt{-9} = \sqrt{(-4)(-9)} = \sqrt{36} = 6 \text{ is WRONG!}$$

When working with negative radicands, be sure to apply $\sqrt{-a} = i\sqrt{a}$ before using any of the properties of radicals discussed in Section 1.3.

Note that for the product $(\sqrt{-1})(-\sqrt{-1})$, we have

$$(\sqrt{-1})(-\sqrt{-1}) = (i)(-i) = -i^2 = -(-1) = 1.$$

Since the product $(i)(-i)$ equals 1, we know that i and $-i$ are *reciprocals* (or multiplicative inverses) of each other. Thus,

$$\frac{1}{i} = -i \qquad \text{and} \qquad \frac{1}{-i} = i$$

EXAMPLE 1   Simplify, and express each answer in the form b or bi, where b is a real number.

(a) $2\sqrt{-3} + \sqrt{-27}$ (b) $3\sqrt{5} \cdot \sqrt{\dfrac{-16}{5}}$

(c) $\sqrt{-2}(\sqrt{-18} + \sqrt{-8})$ (d) $\dfrac{8}{-\sqrt{-25}}$

SOLUTION

(a) $2\sqrt{-3} + \sqrt{-27} = 2i\sqrt{3} + 3i\sqrt{3} = 5i\sqrt{3}$

(b) $3\sqrt{5} \cdot \sqrt{\dfrac{-16}{5}} = 3\sqrt{5} \cdot \sqrt{\dfrac{16}{5}}\,i = 3\sqrt{5} \cdot \dfrac{4}{\sqrt{5}}\,i = 3 \cdot 4i = 12i$

(c) $\sqrt{-2}(\sqrt{-18} + \sqrt{-8}) = i\sqrt{2}(3i\sqrt{2} + 2i\sqrt{2})$
$$= 6i^2 + 4i^2 = 10i^2 = -10$$

(d) $\dfrac{8}{-\sqrt{-25}} = \dfrac{8}{-5i} = \dfrac{8}{5} \cdot \dfrac{1}{-i} = \dfrac{8}{5}\,i$

The reciprocal of $-i$ is i.

PROBLEM 1   Repeat Example 1 for each expression.

(a) $3\sqrt{-80} - \tfrac{1}{3}\sqrt{-45}$

(b) $\dfrac{9}{\sqrt{-36}}$

◆ **Powers of i**

When we raise i to successive positive integer powers, an interesting pattern develops:

$$i^1 = i \qquad\qquad i^2 = -1$$
$$i^3 = i^2 i = (-1)i = -i \qquad i^4 = i^2 i^2 = (-1)(-1) = 1$$
$$i^5 = i^4 i = 1(i) = i \qquad i^6 = i^4 i^2 = 1(-1) = -1$$
$$i^7 = i^4 i^3 = 1(-i) = -i \qquad i^8 = i^4 i^4 = 1(1) = 1$$
$$i^9 = i^8 i = 1(i) = i \qquad i^{10} = i^8 i^2 = 1(-1) = -1$$
$$i^{11} = i^8 i^3 = 1(-i) = -i \qquad i^{12} = i^8 i^4 = 1(1) = 1$$

Notice the values repeat in cycles of four according to the pattern i, -1, $-i$, 1. For higher powers of i, we use the fact that $i^4 = 1$ and apply the laws of exponents from Section 1.2.

EXAMPLE 2   Simplify, and express the answer in the form b or bi, where b is a real number.

(a) $3i^{100}$ (b) $\dfrac{1}{2i^{13}}$

SOLUTION

(a) $3i^{100} = 3(i^4)^{25} = 3(1)^{25} = 3$

(b) $\dfrac{1}{2i^{13}} = \dfrac{1}{2(i^4)^3 i} = \dfrac{1}{2(1)^3 i} = \dfrac{1}{2i} = \dfrac{1}{2}(-i) = -\dfrac{1}{2}i$

> The reciprocal of i is $-i$.

PROBLEM 2   Repeat Example 2 for each expression.

(a) i^{22} (b) $\dfrac{1}{i^{51}}$

◆ Operations with Complex Numbers

The *product* of a real number a ($a \neq 0$) and a pure imaginary number bi is the pure imaginary number $(ab)i$. However, the *sum* of a real number and a pure imaginary number is neither a real number nor a pure imaginary number. In the nineteenth century, the German mathematician Carl Friedrich Gauss denoted the sum of a real number a and a pure imaginary number bi by writing the complex number $a + bi$.

◆ Complex Number

> Any number of the form
>
> $$a + bi,$$
>
> where a and b are real numbers and i is the imaginary unit, is a **complex number**.

In the complex number $a + bi$, the number a is the *real part* and bi is the *imaginary part* of the number. If $b = 0$, then $a + bi$ becomes $a + 0i = a$. A complex number of the form $a + 0i$ is a real number. If $a = 0$, then $a + bi$ becomes $0 + bi = bi$. A complex number of the form $0 + bi$ is a *pure imaginary number*. If $b \neq 0$, then $a + bi$ is referred to simply as an **imaginary number**. Some examples of imaginary numbers are:

$$2 + 3i \qquad \pi - 6i \qquad 5i \qquad i\sqrt{2}$$

The set of imaginary numbers and the set of real numbers are the two major subsets of the set of complex numbers. The relationship between the various

FIGURE 1.5

Some subsets of the complex numbers.

Complex Numbers			
Imaginary Numbers		**Real Numbers**	
		Rational Numbers	**Irrational Numbers**

$2 + 3i$ $\pi - 6i$

$1 + i\sqrt{2}$ $\frac{3}{8}$ $-\frac{1}{2}$ π $\sqrt{2}$

Pure Imaginary Numbers

Integers $\sqrt[3]{7}$

$5i$ $-12i$ 0 32

$\frac{3}{7}i$ i -7 $0.020020002...$

$-i\sqrt{2}$ 1 -18

$\sqrt[4]{11}$

$1 + i$ $2.333...$

$-3 + \frac{3}{4}i$ 0.75 $\sqrt{3}$

sets of numbers is shown in Figure 1.5. All the numbers shown in Figure 1.5 are part of the *complex number system*.

The commutative, associative, and distributive properties are valid for complex numbers and are used when adding, subtracting, multiplying, and dividing complex numbers. For the *sum* of the complex numbers $a + bi$ and $c + di$, we have

$$(a + bi) + (c + di) = (a + c) + (bi + di)$$
$$= (a + c) + (b + d)i$$

| Sum of the real parts | Sum of the imaginary parts |

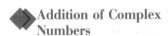

Addition of Complex Numbers

If $a + bi$ and $c + di$ are complex numbers, then

$$(a + bi) + (c + di) = (a + c) + (b + d)i.$$

The *additive identity* for any complex number $a + bi$ is $0 + 0i$, since

$$(a + bi) + (0 + 0i) = (a + 0) + (b + 0)i = a + bi.$$

The *additive inverse* for the complex number $c + di$ is $-c - di$, since

$$(c + di) + (-c - di) = (c - c) + (d - d)i = 0 + 0i.$$

Thus, for the *difference* $(a + bi) - (c + di)$, we add $a + bi$ with the additive inverse of $c + di$ as follows:

$$(a + bi) - (c + di) = (a + bi) + (-c - di)$$
$$= (a - c) + (b - d)i$$

| Difference of the real parts | Difference of the imaginary parts |

◆ **Subtraction of Complex Numbers**

> If $a + bi$ and $c + di$ are complex numbers, then
>
> $$(a + bi) - (c + di) = (a - c) + (b - d)i.$$

E X A M P L E   3   Perform the indicated operations, and express each answer in the form $a + bi$.

(a) $(6 - 3i) + (-4 + 7i)$ (b) $\left(3 + \sqrt{-25}\right) - \left(3 - \sqrt{-36}\right)$

S O L U T I O N

(a) $(6 - 3i) + (-4 + 7i) = (6 - 4) + (-3 + 7)i = 2 + 4i$

(b) $\left(3 + \sqrt{-25}\right) - \left(3 - \sqrt{-36}\right) = (3 + 5i) - (3 - 6i)$
$$= (3 - 3) + [5 - (-6)]i = 0 + 11i \quad \text{or} \quad 11i$$
◆

P R O B L E M   3   Repeat Example 3 for $(2 + 3i) - (5 - 4i) + (-3 + i)$.   ◆

For the product $(a + bi)(c + di)$, we apply the distributive property and multiply each part of the first complex number by each part of the second complex number. The procedure is similar to the way we multiplied two binomials in Section 1.4. Hence,

$$(a + bi)(c + di) = ac + ad\, i + bc\, i + bd\, i^2$$
$$= ac + (ad + bc)i + bd(-1)$$
$$= (ac - bd) + (ad + bc)i$$

◆ **Multiplication of Complex Numbers**

> If $a + bi$ and $c + di$ are complex numbers, then
>
> $$(a + bi)(c + di) = (ac - bd) + (ad + bc)i.$$

E X A M P L E   4   Perform the indicated operations, and express each answer in the form $a + bi$.

(a) $(3 + 4i)(5 - i)$ (b) $\left(1 - \sqrt{-4}\right)^2$

SOLUTION

(a) $(3 + 4i)(5 - i) = 15 - 3i + 20i - 4i^2$

$= 15 + 17i - 4(-1)$

$= 19 + 17i$

(b) $\left(1 - \sqrt{-4}\right)^2 = (1 - 2i)^2 = (1 - 2i)(1 - 2i)$

$= 1 - 2i - 2i + 4i^2$

$= 1 - 4i + 4(-1)$

$= -3 - 4i$ ◆

P R O B L E M 4   Repeat Example 4 for $(1 + 3i)(2 + 5i)$.   ◆

To find the quotient $\dfrac{a + bi}{c + di}$, we multiply numerator and denominator by the **complex conjugate** of the denominator, namely $c - di$. The *product* of complex conjugates is

$$(c + di)(c - di) = c^2 - cdi + cdi - d^2 i^2 = c^2 + d^2.$$

A *nonnegative* real number

The fact that the product of complex conjugates is a nonnegative real number enables us to express the quotient $\dfrac{a + bi}{c + di}$ as a complex number:

$$\frac{a + bi}{c + di} = \frac{(a + bi)(c - di)}{(c + di)(c - di)} = \frac{(ac + bd) + (bc - ad)i}{c^2 + d^2}$$

Complex conjugate of $c + di$

$$= \frac{ac + bd}{c^2 + d^2} + \frac{bc - ad}{c^2 + d^2}\, i$$

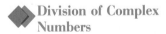
Division of Complex Numbers

If $a + bi$ and $c + di$ are complex numbers, then

$$\frac{a + bi}{c + di} = \frac{ac + bd}{c^2 + d^2} + \frac{bc - ad}{c^2 + d^2}\, i$$

EXAMPLE 5   Perform the indicated operations and express each answer in the form $a + bi$.

(a) $\dfrac{1}{2 + i}$ (b) $\dfrac{3 + 5i}{1 - 5i}$

SOLUTION

(a) $\dfrac{1}{2 + i} = \dfrac{1(2 - i)}{(2 + i)(2 - i)} = \dfrac{2 - i}{4 - i^2} = \dfrac{2 - i}{5} = \dfrac{2}{5} - \dfrac{1}{5}i$

Complex conjugate of $2 + i$

(b) $\dfrac{3 + 5i}{1 - 5i} = \dfrac{(3 + 5i)(1 + 5i)}{(1 - 5i)(1 + 5i)} = \dfrac{3 + 15i + 5i + 25i^2}{1 - 25i^2}$

Complex conjugate of $1 - 5i$

$= \dfrac{-22 + 20i}{26}$

$= \dfrac{-22}{26} + \dfrac{20}{26}i = -\dfrac{11}{13} + \dfrac{10}{13}i$ ◆

PROBLEM 5   Repeat Example 5 for $\dfrac{1 + \sqrt{-9}}{1 - \sqrt{-9}}$.   ◆

Exercises 1.5

Basic Skills

In Exercises 1–20, simplify each expression and write the answer in the form b or bi, where b is a real number.

1. $\sqrt{-16} + \sqrt{-25}$

2. $\sqrt{-36} - \sqrt{-49}$

3. $\sqrt{3} \cdot \sqrt{-12} \cdot \sqrt{-9}$

4. $\dfrac{\sqrt{-125} \cdot \sqrt{36}}{\sqrt{-5}}$

5. $5\sqrt{-5} + \sqrt{-20} - \frac{4}{3}\sqrt{-45}$

6. $2\sqrt{-3} - 2\sqrt{-27} + 6\sqrt{-48}$

7. $\sqrt{-3}(\sqrt{-12} - \sqrt{-27})$ 8. $\sqrt{2}(\sqrt{-18} - \sqrt{-50})$

9. $\dfrac{5\sqrt{18}}{10\sqrt{-2}}$ 10. $\dfrac{24}{2(\sqrt{-6})^2}$

11. $8\sqrt{7} \cdot \sqrt{\dfrac{36}{-7}}$ 12. $6 \cdot \sqrt{\dfrac{-4}{5}} \cdot \sqrt{5}$

13. $5i^{26}$ 14. $13i^{36}$

15. $2i^{-23}$ 16. $\dfrac{1}{6i^{37}}$

17. $i + i^2 + i^3 + i^4$ 18. $i^{-1} - i^{-2} + i^{-3} - i^{-4}$

19. $(3i)^3(2i)^2$ 20. $(2i)^4(-3i)^3$

In Exercises 21–42, perform the indicated operations and express each answer in the form a + bi.

21. $(9 - 3i) + (6 + 2i) + 4i$

22. $(12 + 6i) - (4 - 3i) + (1 + i)$

23. $(3 + 2\sqrt{-36}) - (5 - 3\sqrt{-49})$

24. $(8 - 4\sqrt{-9}) + (6 + 5\sqrt{-16})$

25. $8i(2 - 3i)$

26. $\sqrt{-25}(3\sqrt{-4} - 8)$

27. $(3 - \sqrt{-4})(4 + 3\sqrt{-9})$

28. $(8 - 2i)(1 - i)$

29. $\dfrac{1 - 2i}{1 + 2i}$

30. $\dfrac{4 - i}{2 - 3i}$

31. $\dfrac{\sqrt{-4}}{2 - \sqrt{-6}}$

32. $\dfrac{9}{1 + 2\sqrt{-3}}$

33. $\dfrac{4 + 2i}{i}$

34. $\dfrac{1 - \sqrt{-16}}{3i}$

35. $(-1 + 2\sqrt{-25})^2$

36. $(-4 - 5i)^2$

37. $(3 - 2i)^3$

38. $\dfrac{1}{(2 + \sqrt{-9})^3}$

39. $\dfrac{\sqrt{-9}(3 + \sqrt{-4})^2}{1 + \sqrt{-1}}$

40. $\dfrac{(2 - i)^2(1 + i)}{3 - 2i}$

41. $\dfrac{i^5 - i^2}{(2 + i)^2}$

42. $\dfrac{(3i^2 - i)^2}{2i^3 + i^{10}}$

Critical Thinking

43. Show that $1 + 0i$ is the *multiplicative identity* for the complex number $a + bi$.

44. Find the *multiplicative inverse*, or *reciprocal*, of the complex number $a + bi$.

45. Evaluate the polynomial $x^2 - 2x + 2$ when (a) $x = 1 + i$ and (b) $x = 1 - i$.

46. Evaluate the polynomial $x^2 - 6x + 13$ when (a) $x = 3 - 2i$ and (b) $x = 3 + 2i$.

47. Evaluate the polynomial $x^3 - 3x^2 + x + 5$ when (a) $x = 2 - i$ and (b) $x = 2 + i$

48. Evaluate the polynomial $x^3 - 4x^2 + 14x - 20$ when (a) $x = 1 + 3i$ and (b) $x = 1 - 3i$.

49. Show that $\dfrac{\sqrt{2} + i\sqrt{2}}{2}$ is a square root of i.

50. Show that $\dfrac{\sqrt{3} + i}{2}$ is a cube root of i.

Calculator Activities

51. When working with the *square root key* ($\boxed{\sqrt{}}$) on your calculator, you must enter nonnegative inputs. (A negative input causes an error message to appear in the display window.) Using your calculator in conjunction with the definition of the principal square root of $-a$, write each expression in the form bi, with b rounded to three significant digits.

    (a) $\sqrt{-18}$ (b) $\sqrt{-506}$
    (c) $\sqrt{-153.6}$ (d) $\sqrt{-0.8765}$

52. Evaluate

$$\frac{-b \pm \sqrt{b^2 - 4ac}}{2a}$$

for the given values of a, b, and c. Express each answer in the form $a + bi$ with a and b rounded to three significant digits.

    (a) $a = 2, b = 6, c = 7$ (b) $a = 13, b = -5, c = 10$

53. The magnitude of the current (in amperes, A) flowing through the series circuit shown in the sketch may be found by expressing

$$\frac{200}{150 + (188.5 - 76.4)i}$$

in the complex number form $a + bi$, and then computing $\sqrt{a^2 + b^2}$, where a is the real part and b the coefficient of the imaginary part of this complex number. Determine the magnitude of the current in this circuit.

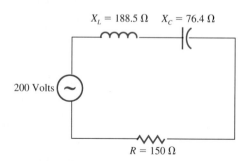

54. The magnitude of the total impedance (in ohms, Ω) for the parallel circuit shown in the sketch may be found by expressing

$$\frac{1}{\dfrac{1}{150} + \dfrac{1}{188.5i} - \dfrac{1}{76.4i}}$$

in the complex number form $a + bi$, and then computing $\sqrt{a^2 + b^2}$, where a is the real part and b the coefficient of the imaginary part of this complex number.

Determine the magnitude of the total impedance for this circuit.

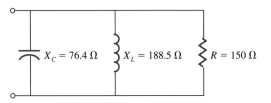

1.6   Factoring Techniques

◆ Introductory Comments

To **factor** an algebraic expression means to write it as the product of two or more algebraic expressions. Factoring forms the foundation for simplifying algebraic fractions (see Section 1.7) and for solving several types of equations and inequalities (see Chapter 2). In this section we discuss several *factoring techniques* that may be used to factor an algebraic expression.

◆ Factoring by Using the Distributive Property

We can use the distributive property from right to left to *factor out* common factors from a sum (or difference) as follows:

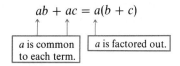

EXAMPLE 1   Use the distributive property to factor out *all* common factors from each expression.

(a) $8x^2y^5z + 6x^3y^4 - 10x^3y^3z^2$ **(b)** $6(2x + 5)^2 - 3x(2x + 5)$

SOLUTION

(a) The common factors of the three terms are 2, x^2, and y^3. Factoring out $2x^2y^3$, we have

$$8x^2y^5z + 6x^3y^4 - 10x^3y^3z^2 = 2x^2y^3(4y^2z + 3xy - 5xz^2).$$

(b) Algebraic expressions may contain common binomial factors. The common factors of $6(2x + 5)^2 - 3x(2x + 5)$ are 3 and $(2x + 5)$. Factoring out $3(2x + 5)$, we have

$$6(2x + 5)^2 - 3x(2x + 5) = 3(2x + 5)[2(2x + 5) - x]$$

Simplify

$$= 3(2x + 5)(3x + 10) \qquad \blacklozenge$$

P R O B L E M 1   Repeat Example 1 for $2y(x - 1)^3 + y(x - 1)^4$. $\blacklozenge$

◆ **Factoring by Grouping Terms**

From our knowledge of multiplying polynomials, we know that

$$(a + b)(c + d) = (c + d)(a + b) = ac + ad + bc + bd.$$

Now, how might we factor $ac + ad + bc + bd$? Notice that no factor is common to all four terms. However, by grouping the first two terms and the last two terms, we are able to factor by using the distributive property as follows:

$$ac + ad + bc + bd = (ac + ad) + (bc + bd) \qquad \text{**Group terms**}$$

$$= a(c + d) + b(c + d) \qquad \text{**Factor out common factors from each group**}$$

$$= (c + d)(a + b) \qquad \text{**Factor out the common binomial $(c + d)$**}$$

The key to factoring by grouping terms is to develop a common binomial that can be factored out of each group. As illustrated in the next example, this may require factoring a negative from one of the groups.

E X A M P L E 2   Factor each expression by grouping terms.

(a) $x^3 - 3x^2 + 5x - 15$ (b) $9 + 2xy - 6x - 3y$

S O L U T I O N

(a) $x^3 - 3x^2 + 5x - 15 = (x^3 - 3x^2) + (5x - 15)$
$$= x^2(x - 3) + 5(x - 3)$$
$$= (x - 3)(x^2 + 5)$$

(b) $9 + 2xy - 6x - 3y = (9 - 6x) + (2xy - 3y)$
$$= 3(3 - 2x) - y(3 - 2x)$$

> Note the factoring of $-y$ in order to obtain the common binomial $(3 - 2x)$.

$$= (3 - 2x)(3 - y) \qquad \blacklozenge$$

PROBLEM 2   There is usually more than one way to group the terms. Try grouping the first and last terms and the second and third terms of $9 + 2xy - 6x - 3y$, then factor. You should obtain the same result as in Example 2(b).   ◆

◆   **Factoring Trinomials**

If k_1 and k_2 are constants, then the product of the binomials $(x + k_1)$ and $(x + k_2)$ is a trinomial of the form $x^2 + bx + c$, where $b = k_1 + k_2$ and $c = k_1 k_2$. Thus, to factor trinomials of the form $x^2 + bx + c$, we find two numbers k_1 and k_2 whose product is c and whose sum is b. We then write

$$x^2 + bx + c = (x + k_1)(x + k_2).$$

EXAMPLE 3   Factor each trinomial.

(a) $x^2 - 9x + 14$ (b) $x^2 + 8xy - 20y^2$ (c) $x^6 + 5x^3 + 6$

SOLUTION

(a) For $x^2 - 9x + 14$, we know that $c = 14$ and $b = -9$. The two numbers whose product is 14 and whose sum is -9 are -2 and -7. Thus, we have $k_1 = -2$ and $k_2 = -7$:

$$x^2 - 9x + 14 = (x - 2)(x - 7).$$

(b) In a similar manner, we can factor trinomials of the form $x^2 + bxy + cy^2$. For $x^2 + 8xy - 20y^2$, we have $c = -20$ and $b = 8$. The two numbers whose product is -20 and whose sum is 8 are 10 and -2. Thus,

$$x^2 + 8xy - 20y^2 = (x + 10y)(x - 2y).$$

(c) We may treat $x^6 + 5x^3 + 6$ as $u^2 + 5u + 6$, where $u = x^3$. Since the two numbers whose product is 6 and whose sum is 5 are 3 and 2, we have

$$x^6 + 5x^3 + 6 = (x^3 + 3)(x^3 + 2).$$ ◆

PROBLEM 3   Repeat Example 3 for each trinomial.

(a) $x^2 - 2xy - 48y^2$ (b) $y^4 - 7y^2 + 10$ ◆

To factor trinomials of the form $ax^2 + bx + c$ $(a \neq 1)$, we find two numbers whose product is ac and whose sum is b. If these two numbers are k_1 and k_2, then we write

$$ax^2 + bx + c \qquad \text{as} \qquad ax^2 + k_1 x + k_2 x + c,$$

Replace bx with $k_1 x + k_2 x$

and factor this expression by grouping terms.

E X A M P L E   4   Factor each trinomial.

(a) $3x^2 - 10x + 8$ (b) $10x^2 - 17xy - 20y^2$ (c) $9x^4 + 24x^2 + 16$

S O L U T I O N

(a)   For $3x^2 - 10x + 8$, we have $ac = 3 \cdot 8 = 24$ and $b = -10$. The two numbers whose product is 24 and whose sum is -10 are -6 and -4. Thus,

$$3x^2 - 10x + 8 = 3x^2 - 6x - 4x + 8 \qquad \text{\textbf{Replace} } -10x \text{ \textbf{with}} \\ -6x \; -4x$$

$$= 3x(x - 2) - 4(x - 2) \qquad \text{\textbf{Factor by grouping terms}}$$

$$= (x - 2)(3x - 4)$$

(b)   Trinomials of the form $ax^2 + bxy + cy^2$ may also be factored in this manner. For $10x^2 - 17xy - 20y^2$, $ac = (10)(-20) = -200$ and $b = -17$. The two numbers whose product is -200 and whose sum is -17 are 8 and -25. Thus,

$$10x^2 - 17xy - 20y^2 = 10x^2 + 8xy - 25xy - 20y^2 \qquad \text{\textbf{Replace} } -17xy \\ \text{\textbf{with} } 8xy - 25xy$$

$$= 2x(5x + 4y) - 5y(5x + 4y) \qquad \text{\textbf{Factor by grouping terms}}$$

$$= (5x + 4y)(2x - 5y)$$

(c)   We may treat $9x^4 + 24x^2 + 16$ as $9u^2 + 24u + 16$, where $u = x^2$. Hence, we have

$$\boxed{ac = 9 \cdot 16 = 144}$$

$$9x^4 + 24x^2 + 16 = 9x^4 + 12x^2 + 12x^2 + 16 \qquad \text{\textbf{Now factor by grouping terms}}$$

$$\boxed{b}$$

$$= 3x^2(3x^2 + 4) + 4(3x^2 + 4)$$

$$= (3x^2 + 4)(3x^2 + 4) = (3x^2 + 4)^2 \qquad \blacklozenge$$

Note: We expressed the trinomial in Example 4(c) as the *perfect square* $(3x^2 + 4)^2$. Thus we refer to the trinomial, $9x^4 + 24x^2 + 16$, as a **perfect square trinomial**.

We may also use a *trial-and-error procedure* to factor trinomials of the form $ax^2 + bx + c$ or $ax^2 + bxy + cy^2$. Again, consider factoring

$$3x^2 - 10x + 8.$$

Since the first term's coefficient is 3, the middle term's coefficient is *negative*, and the last term's coefficient is *positive*, we begin by writing the two binomials

$$\left(3x - \square\right)\left(x - \square\right).$$

Now the two boxes must be filled in with two numbers whose product is 8 such that the sum of the inner and outer terms of this product is $-10x$. We have four possibilities:

$(3x - 8)(x - 1)$ sum of inner and outer product is $-11x$

$(3x - 1)(x - 8)$ sum of inner and outer product is $-25x$

$(3x - 2)(x - 4)$ sum of inner and outer product is $-14x$

$(3x - 4)(x - 2)$ sum of inner and outer product is $-10x$

Thus, by the trial-and-error procedure, we have

$$3x^2 - 10x + 8 = (3x - 4)(x - 2).$$

This procedure works well for trinomials of the forms $ax^2 + bx + c$ or $ax^2 + bxy + cy^2$ when either a or c is a prime number or when both a and c are prime numbers. Under these conditions there are usually very few trials to check, and the factors can be found quite easily.

PROBLEM 4   Factor $5x^2 + 13xy + 6y^2$.   ◆

◆   **Factoring Formulas and Further Techniques**

The special products listed in Section 1.4, when read from right to left, provide three useful factoring formulas.

Factoring Formulas

1.	Difference of squares	$A^2 - B^2 = (A + B)(A - B)$
2.	Sum of cubes	$A^3 + B^3 = (A + B)(A^2 - AB + B^2)$
3.	Difference of cubes	$A^3 - B^3 = (A - B)(A^2 + AB + B^2)$

EXAMPLE 5   Factor each expression.

(a) $9x^2 - 4y^2$ **(b)** $27x^3 - 125y^3$ **(c)** $x^6 + 8$

SOLUTION

(a) Applying the *difference of squares formula*, we obtain

$$9x^2 - 4y^2 = (3x)^2 - (2y)^2$$
$$= (3x + 2y)(3x - 2y).$$

(b) Applying the *difference of cubes formula*, we obtain

$$27x^3 - 125y^3 = (3x)^3 - (5y)^3$$
$$= (3x - 5y)[(3x)^2 + (3x)(5y) + (5y)^2]$$
$$= (3x - 5y)(9x^2 + 15xy + 25y^2).$$

(c) Applying the *sum of cubes formula*, we obtain

$$x^6 + 8 = (x^2)^3 + (2)^3$$
$$= (x^2 + 2)[(x^2)^2 - (x^2)(2) + (2)^2]$$
$$= (x^2 + 2)(x^4 - 2x^2 + 4). \qquad \blacklozenge$$

PROBLEM 5   Repeat Example 5 for each expression.

    (a) $(x + y)^2 - 25y^4$ **(b)** $64 - x^3$ $\blacklozenge$

For some polynomials with degree higher than 4, we may add and subtract a monomial in order to develop a difference of squares that is factorable. The procedure is illustrated in the next example.

EXAMPLE 6   Factor each polynomial.

    (a) $x^4 + 64$ **(b)** $x^4 + x^2 + 1$

SOLUTION

(a) $x^4 + 64 = (x^4 + 16x^2 + 64) - 16x^2$	**Add and subtract $16x^2$ and group terms as shown**
$= (x^2 + 8)^2 - (4x)^2$	**Factor the perfect square trinomial**
$= [(x^2 + 8) + 4x][(x^2 + 8) - 4x]$	**Factor the difference of squares**
$= (x^2 + 4x + 8)(x^2 - 4x + 8)$	**Rewrite**
(b) $x^4 + x^2 + 1 = (x^4 + 2x^2 + 1) - x^2$	**Add and subtract x^2 and group terms as shown**
$= (x^2 + 1)^2 - x^2$	**Factor the perfect square trinomial**
$= [(x^2 + 1) + x][(x^2 + 1) - x]$	**Factor the difference of squares**
$= (x^2 + x + 1)(x^2 - x + 1)$	**Rewrite** $\blacklozenge$

As with all factoring problems, it is a good idea to *check* the factoring process by finding the product of the factors.

PROBLEM 6   Check the factors found in parts (a) and (b) of Example 6. $\blacklozenge$

$\blacklozenge$ **Factoring Completely**

A polynomial with integer coefficients is **prime relative to the set of integers** if it cannot be written as the product of two polynomials of positive degree that

SECTION 1.6   Factoring Techniques

have *integer coefficients*. Five examples of polynomials that are prime relative to the set of integers are as follows:

1. $x^2 - 3$ Using $A^2 - B^2 = (A + B)(A - B)$, we could factor $x^2 - 3$ as $(x + \sqrt{3})(x - \sqrt{3})$, but $\sqrt{3}$ is *not* an integer.

2. $x^3 + 2$ Using $A^3 + B^3 = (A + B)(A^2 - AB + B^2)$, we could factor $x^3 + 2$ as

$$(x + \sqrt[3]{2})(x^2 - x\sqrt[3]{2} + \sqrt[3]{4}),$$

but $\sqrt[3]{2}$ and $\sqrt[3]{4}$ are *not* integers.

3. $x^4 + 25$ We could try adding and subtracting $10x^2$ and factoring as follows:

$$\begin{aligned} x^4 + 25 &= (x^4 + 10x^2 + 25) - 10x^2 \\ &= (x^2 + 5)^2 - (x\sqrt{10})^2 \\ &= (x^2 + 5 + x\sqrt{10})(x^2 + 5 - x\sqrt{10}), \end{aligned}$$

but $\sqrt{10}$ is *not* an integer.

4. $x^2 + 4x + 2$ We could try adding and subtracting 2 and factoring as follows:

$$\begin{aligned} x^2 + 4x + 2 &= (x^2 + 4x + 4) - 2 \\ &= (x + 2)^2 - (\sqrt{2})^2 \\ &= (x + 2 + \sqrt{2})(x + 2 - \sqrt{2}), \end{aligned}$$

but $\sqrt{2}$ is *not* an integer.

5. $x^2 + 1$ Writing $x^2 + 1$ as $x^2 - (-1) = x^2 - i^2$ and using $A^2 - B^2 = (A + B)(A - B)$, we could factor $x^2 + 1$ as $(x + i)(x - i)$, but i is *not* an integer.

Although $x^2 - 3$, $x^3 + 2$, $x^4 + 25$, and $x^2 + 4x + 2$ are prime relative to the set of integers, they are all factorable, or *nonprime relative to the set of real numbers* (or *reals*), because each can be expressed as the product of two polynomials of positive degree with *real coefficients*.

The polynomial $x^2 + 1$ is *prime relative to the set of real numbers*, since it cannot be expressed as the product of two or more polynomials with real coefficients. However, $x^2 + 1$ is factorable, or *nonprime, relative to the set of complex numbers*, because it can be written as the product of two polynomials of positive degree with *imaginary coefficients*.

A more detailed explanation of factoring polynomials is given in Chapter 5. For now, our concern is mainly with *factoring over the set of integers*.

When a polynomial is written as the product of prime factors, it is said to be **factored completely**. To factor a polynomial completely, we first remove any common factors, and then use the previously discussed methods of factoring to write the remaining factor as the product of two or more *prime polynomials*.

EXAMPLE 7   Factor completely over the set of integers.

(a) $6x^3y - 3x^2y - 18xy$ (b) $x^6 + x^3 - 2$

(c) $12x^3y + 12x^2y - 3xy - 3y$ (d) $2x^8 - 2$

SOLUTION

(a) $6x^3y - 3x^2y - 18xy = 3xy(2x^2 - x - 6)$ Factor out the common factor

$\qquad\qquad\qquad\qquad = 3xy(2x + 3)(x - 2)$ Factor the trinomial

(b) $x^6 + x^3 - 2 = (x^3 + 2)(x^3 - 1)$ Factor the trinomial

$\qquad\qquad\quad = (x^3 + 2)(x - 1)(x^2 + x + 1)$ Factor the difference of cubes

(c) $12x^3y + 12x^2y - 3xy - 3y$

$\qquad = 3y(4x^3 + 4x^2 - x - 1)$ Factor out the common factor

$\qquad = 3y[4x^2(x + 1) - 1(x + 1)]$ Factor by grouping terms

$\qquad = 3y(x + 1)(4x^2 - 1)$

$\qquad = 3y(x + 1)(2x + 1)(2x - 1)$ Factor the difference of squares

(d) $2x^8 - 2 = 2(x^8 - 1)$ Factor out the common factor

$\qquad = 2(x^4 + 1)(x^4 - 1)$ Factor the difference of squares

$\qquad = 2(x^4 + 1)(x^2 + 1)(x^2 - 1)$ Factor the difference of squares

$\qquad = 2(x^4 + 1)(x^2 + 1)(x + 1)(x - 1)$ Factor the difference of squares again   ◆

Suppose we wanted to factor $2x^8 - 2$ *over the reals*, instead of over the integers, as we did in Example 7(d). We could continue to "break down" the factor $x^4 + 1$ as follows:

$x^4 + 1 = (x^4 + 2x^2 + 1) - 2x^2$ Add and subtract $2x^2$

$\qquad = (x^2 + 1)^2 - (\sqrt{2}\,x)^2$ Factor the perfect square trinomial

$\qquad = (x^2 + 1 + \sqrt{2}\,x)(x^2 + 1 - \sqrt{2}\,x)$ Factor the difference of squares

Therefore, $2x^8 - 2$ can be factored over the reals:

$$2x^8 - 2 = 2(x^2 + \sqrt{2}\,x + 1)(x^2 - \sqrt{2}\,x + 1)(x^2 + 1)(x + 1)(x - 1)$$

PROBLEM 7   Factor $x^6 + x^3 - 2$ [Example 7(b)] completely over the reals.   ◆

Exercises 1.6

Basic Skills

In Exercises 1–88, factor completely over the integers.

1. $3a^2b - ab$

2. $48x^2y - 36xy^2$

3. $6x^3y - 9x^2y + 3xy^2$

4. $6x^2y^3 - 18x^3y^4 + 12x^3y^3$

5. $5(x - 1) - x(x - 1)$

6. $5n(2n - 7) + n(2n - 7)$

7. $6m(m + 3)^4 + m(m + 3)^3$

8. $3p(p + 2)^6 + 9p^2(p + 2)^7$

9. $ac + ad + bc + bd$

10. $xy + 3x + 2y + 6$

11. $x^3 - 3x^2 + 2x - 6$

12. $m^4 - 2m^3 - 3m + 6$

13. $6 - 3x - 2y + xy$

14. $15 + 6y - 20x - 8xy$

15. $15 + 2x^2y - 3x^2 - 10y$

16. $3x^3 - 20y - 15x^2 + 4xy$

17. $a^2 + 3a - 10$

18. $y^2 + 14y + 45$

19. $x^2 - 11xy + 18y^2$

20. $p^2 - pq - 12q^2$

21. $24 + 14x + x^2$

22. $24 - 2m - m^2$

23. $x^4 + 3x^2 - 18$

24. $x^6 + 2x^3 - 63$

25. $2x^2 + 5x + 3$

26. $3x^2 + 8x + 4$

27. $3a^2 + 11ab - 20b^2$

28. $5x^2 + 18xy - 8y^2$

29. $4x^2 - 11x - 3$

30. $6p^2 + 23p - 4$

31. $15 - 16t + 4t^2$

32. $12 - 17t + 6t^2$

33. $12x^2 - 23xy + 10y^2$

34. $18x^2 + 23xy - 6y^2$

35. $2n^4 + 7n^2 + 5$

36. $3y^4 - 13y^2 + 10$

37. $6x^6 - 17x^3y^2 + 10y^4$

38. $10x^8 - x^4y - 3y^2$

39. $8y^{12} + 13y^6 - 6$

40. $12a^8 + 8a^4 - 15$

41. $x^2 - 81$

42. $x^2 - 36$

43. $25x^2 - y^2$

44. $x^2 - 16y^{-4}$

45. $4(x^2 + 5)^2 - 25x^2$

46. $49 - (m + n)^2$

47. $t^3 + 8$

48. $125 + a^3$

49. $27x^{-3} + 8y^6$

50. $y^3 - 27$

51. $8 - 125n^3$

52. $64n^3 - 1$

53. $t^2 + 8t + 16$

54. $25 - 10x + x^2$

55. $4a^4 - 12a^2b^2 + 9b^4$

56. $9p^6 - 6p^3 + 1$

57. $x^2 - 4x + 4 - y^2$

58. $x^2 + 12x + 36 - 9y^2$

59. $4x^2 + 12x - 4y^2 + 9$

60. $9x^2 - 16y^2 - 30x + 25$

61. $x^4 + 4$

62. $64y^4 + 1$

63. $a^4 + a^2 + 25$

64. $n^4 - 23n^2 + 1$

65. $x^4 + x^2y^2 + y^4$

66. $a^4 - 7a^2b^2 + b^4$

67. $25m^4 + 61m^2n^2 + 49n^4$

68. $16x^4 + 20x^2y^2 + 9y^4$

69. $4x^3y - 28x^2y + 48xy$

70. $3x^4y^2 - 6x^3y^3 - 45x^2y^4$

71. $6x^3 - 8x^2 - 8x$

72. $12x^3 + 3x^2 - 9x$

73. $5a^3 - 45a$

74. $2m^3n - 32mn$

75. $16x^4y + 2xy$

76. $3x^5y^2 - 81x^2y^2$

77. $2n^4 - 32$

78. $81x - 16x^5$

79. $4x^4 - 13x^2 + 9$

80. $9x^4 - 37x^2y^2 + 4y^4$

81. $3a^7 + 6a^4 - 9a$

82. $2x^6y + 26x^3y + 80y$

83. $8x^3y + 8x^2y - 18xy - 18y$

84. $6x^4 + 3x^3 - 24x^2 - 12x$

85. $x^6 - 1$

86. $64 - x^6$

87. $a^9 - 16a$

88. $512 - 2y^8$

Critical Thinking

In Exercises 89–98, factor completely over the set of real numbers.

89. $x^2 - 10$

90. $3t^2 - 4y^2$

91. $a^3 + 5b^3$

92. $8x^3 - 2$

93. $x^4 + 9$

94. $n^4 + 36$

95. $x^2 + 6x + 6$

96. $x^4 - 2x^2 + 4$

97. $t^8 - 16$

98. $m^8 - 256$

In Exercises 99–102, factor completely over the set of complex numbers.

99. $x^2 + 25$

100. $y^2 + 18$

101. $t^4 - 36$

102. $t^4 - 256$

 The expressions in Exercises 103–106 occur in calculations involving derivatives in calculus. Each expression may be simplified by factoring out the common factors from each term. Simplify each expression.

103. $2x[3(x + 3)^2] + (x + 3)^3(2)$

104. $4x^3[5(1 - 2x)^4(-2)] + (1 - 2x)^5(12x^2)$

105. $(2x + 1)^2[4(3x - 1)^3(3)] + (3x - 1)^4[2(2x + 1)]$

106. $(x^2 + 1)^5[4(3x^3 + 4)^3(9x^2)] + (3x^3 + 4)^4[5(x^2 + 1)^4(2x)]$

107. The total amount S of aluminum used to make a beer can is given by the formula

$$S = 2\pi rht + 2\pi r^2 t,$$

where r is the radius, h is the height, and t is the thickness of the beer can. Factor the expression on the right-hand side of this formula.

108. Finding the number of households (in units of 1000) that use a certain product after it has been on the market for n months involves working with the expression

$$32n - \tfrac{4}{3}n^2.$$

Factor this expression so that the coefficient of n in the binomial factor is 1.

 Calculator Activities

In Exercises 109–114, evaluate each expression in two ways:

(i) *by using your calculator directly.*

(ii) *without the aid of a calculator, by using the factoring techniques discussed in this section.*

109. $(891 \cdot 240) + (891 \cdot 760)$

110. $(80 \cdot 27) + (80 \cdot 23) + (20 \cdot 27) + (20 \cdot 23)$

111. $65^2 - 55^2$

112. $1000^2 - 900^2$

113. $60^2 + 2(60)(40) + 40^2$

114. $10^3 - 9^3$

115. Finding the response time of mice to a certain stimulus involves working with the expression

$$10.24 + 9.24A^2 - A^4,$$

where A is the age of the subject receiving the stimulus. Factor this expression completely over the set of rational numbers.

116. The velocity V of sound in seawater may be approximated by the formula

$$V = 4800 + 14t - 0.12t^2,$$

where t is the temperature of the seawater. Factor the right-hand side of this formula completely over the set of rational numbers.

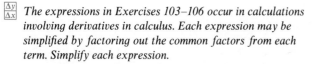

1.7   Algebraic Fractions

◆ **Introductory Comments**

The quotient of two algebraic expressions is called an **algebraic fraction**. If the numerator and denominator of an algebraic fraction are polynomials, then the algebraic fraction is referred to as a **rational expression**. Since division by zero is undefined, it is important that the variables in an algebraic fraction be restricted to those values that give a nonzero denominator. Thus, for the rational

expression

$$\frac{4y^2 + 5y + 3}{2y - 6},$$

we assume that its denominator $2y - 6 \neq 0$, which implies that $y \neq 3$. Throughout this section, we apply the properties of fractions (Section 1.1) when working with algebraic fractions.

◆ Generating Equivalent Algebraic Fractions

To generate equivalent algebraic fractions or to reduce an algebraic fraction to lowest terms, we apply the **fundamental property of fractions**.

Fundamental Property of Fractions

For all fractions $\dfrac{a}{b}$ ($b \neq 0$) and all real numbers k ($k \neq 0$),

$$\frac{a}{b} = \frac{ak}{bk}.$$

Suppose we wish to change $(2x + 3)/2x$ to an equivalent rational expression whose denominator is $8x^3y$. Since $2x(4x^2y) = 8x^3y$, we multiply both numerator and denominator of $(2x + 3)/2x$ by $4x^2y$ as follows:

$$\frac{2x + 3}{2x} = \frac{(2x + 3) \cdot (4x^2y)}{2x \cdot (4x^2y)} = \frac{8x^3y + 12x^2y}{8x^3y}$$

Equivalent fractions

Reversing the procedure, we can reduce $(8x^3y + 12x^2y)/8x^3y$ to lowest terms by factoring and reducing:

$$\frac{8x^3y + 12x^2y}{8x^3y} = \frac{\overset{1}{\cancel{4x^2y}}(2x + 3)}{\underset{2x}{\cancel{8x^3y}}} = \frac{2x + 3}{2x}.$$

Caution
To reduce the fraction as follows

$$\frac{\overset{1}{\cancel{8x^3y}} + 12x^2y}{\underset{1}{\cancel{8x^3y}}} \quad \text{is WRONG!}$$

Remember, we can cancel common *factors* of the numerator and denominator, but never common *terms*.

EXAMPLE 1   Change each algebraic fraction to an equivalent algebraic fraction with a denominator of $x(x + 1)^2$.

(a) $\dfrac{x - 1}{x^2 + x}$ (b) $\dfrac{2x}{x^2 + 2x + 1}$

SOLUTION

(a) $\dfrac{x - 1}{x^2 + x} = \dfrac{x - 1}{x(x + 1)} = \dfrac{(x - 1) \cdot (x + 1)}{x(x + 1) \cdot (x + 1)} = \dfrac{x^2 - 1}{x(x + 1)^2}$

> We must multiply $x(x + 1)$ by $(x + 1)$ to obtain $x(x + 1)^2$.

(b) $\dfrac{2x}{x^2 + 2x + 1} = \dfrac{2x}{(x + 1)^2} = \dfrac{2x \cdot (x)}{(x + 1)^2 \cdot (x)} = \dfrac{2x^2}{x(x + 1)^2}$

> We must multiply $(x + 1)^2$ by x to obtain $x(x + 1)^2$.

PROBLEM 1   Change $\dfrac{x - 2}{x + 5}$ to an equivalent algebraic fraction whose denominator is $3x^2 + 13x - 10$.

We say that an algebraic fraction is *reduced to lowest terms* when its numerator and denominator contain no common factor other than ± 1.

EXAMPLE 2   Reduce each fraction to lowest terms.

(a) $\dfrac{9 - x^2}{x^2 - 2x - 15}$ (b) $\dfrac{1 - x}{x^3 - 1}$

SOLUTION

(a) $\dfrac{9 - x^2}{x^2 - 2x - 15} = \dfrac{\cancel{(3 + x)}(3 - x)}{\cancel{(x + 3)}(x - 5)} = \dfrac{3 - x}{x - 5}$

> We use the fact that $a - b = -(b - a)$ to obtain common factors that cancel.

(b) $\dfrac{1 - x}{x^3 - 1} = \dfrac{1 - x}{(x - 1)(x^2 + x + 1)} = \dfrac{-\cancel{(x - 1)}}{\cancel{(x - 1)}(x^2 + x + 1)}$

$= \dfrac{-1}{(x^2 + x + 1)}$

PROBLEM 2   Repeat Example 2 for $\dfrac{16 - x^2}{2x^3 - 5x^2 - 12x}$.

◆ Multiplying and Dividing Algebraic Fractions

To multiply and divide algebraic fractions, we use the multiplication and division properties of fractions, first given in Section 1.1 and restated here:

Multiplication and Division Properties of Fractions

For all fractions $\dfrac{a}{b}$ and $\dfrac{c}{d}$ ($b \neq 0, d \neq 0$),

1. Multiplication Property $\qquad \dfrac{a}{b} \cdot \dfrac{c}{d} = \dfrac{ac}{bd}$

2. Division Property $\qquad \dfrac{a}{b} \div \dfrac{c}{d} = \dfrac{a}{b} \cdot \dfrac{d}{c} \quad (c \neq 0)$

A good strategy is to completely factor numerators and denominators, relative to the set of integers, before applying these properties. In this way, common factors can be cancelled first, yielding a product that is reduced to lowest terms.

E X A M P L E 3 Perform the indicated operations.

(a) $\dfrac{x-6}{2x^2} \cdot \dfrac{x^2-x}{36-x^2}$ $\qquad$ **(b)** $\dfrac{2x^3+16}{2x^2+x-6} \div \dfrac{2x^2y-4xy+8y}{4x^2-12x+9}$

S O L U T I O N

To develop further cancellations, change $a - b$ to $-(b - a)$.

(a) $\dfrac{x-6}{2x^2} \cdot \dfrac{x^2-x}{36-x^2} = \dfrac{(x-6)\cdot x(x-1)}{2x^2 \cdot (6+x)(6-x)} = \dfrac{-(6-x)\cdot x(x-1)}{2x^2 \cdot (6+x)(6-x)}$

$$= \dfrac{-(x-1)}{2x(6+x)}$$

(b) $\dfrac{2x^3+16}{2x^2+x-6} \div \dfrac{2x^2y-4xy+8y}{4x^2-12x+9}$

$$= \dfrac{2x^3+16}{2x^2+x-6} \cdot \dfrac{4x^2-12x+9}{2x^2y-4xy+8y}$$

$$= \dfrac{2(x+2)(x^2-2x+4)}{(2x-3)(x+2)} \cdot \dfrac{(2x-3)^2}{2y(x^2-2x+4)}$$

$$= \dfrac{2(x+2)(x^2-2x+4)\cdot(2x-3)^2}{(2x-3)(x+2)\cdot 2y(x^2-2x+4)}$$

$$= \dfrac{2x-3}{y}$$

PROBLEM 3   Repeat Example 3 for $\dfrac{x^2}{2x^2 - 5x + 2} \div \dfrac{4x}{4 - x^2}$.   ◆

◆ Adding and Subtracting Algebraic Fractions

To add and subtract algebraic fractions, we use the addition and subtraction properties of fractions, first given in Section 1.1 and repeated as follows:

Addition and Subtraction Properties of Fractions

For all fractions $\dfrac{a}{b}$ and $\dfrac{c}{b}$ ($b \neq 0$),

1. Addition Property $\dfrac{a}{b} + \dfrac{c}{b} = \dfrac{a + c}{b}$ **2.** Subtraction Property $\dfrac{a}{b} - \dfrac{c}{b} = \dfrac{a - c}{b}$

After applying these properties, always look for common factors to see if the sum or difference can be reduced to lower terms.

EXAMPLE 4   Perform the indicated operations.

(a) $\dfrac{x + y}{xy} - \dfrac{y - 6x}{xy}$ (b) $\dfrac{2x + 4}{x - 5} + \dfrac{x + 3}{5 - x}$

SOLUTION

> Remember the parentheses and subtract the entire expression.

(a) $\dfrac{x + y}{xy} - \dfrac{y - 6x}{xy} = \dfrac{(x + y) - (y - 6x)}{xy} = \dfrac{x + y - y + 6x}{xy}$

$= \dfrac{7x}{xy} = \dfrac{7}{y}$ **Reduce to lowest terms**

(b) $\dfrac{2x + 4}{x - 5} + \dfrac{x + 3}{5 - x} = \dfrac{2x + 4}{x - 5} + \dfrac{x + 3}{-(x - 5)}$

> To obtain like denominators, use the facts that $a - b = -(b - a)$ and $a/(-b) = -a/b$.

$= \dfrac{2x + 4}{x - 5} + \dfrac{-(x + 3)}{x - 5}$

$= \dfrac{(2x + 4) + (-x - 3)}{x - 5}$

$= \dfrac{x + 1}{x - 5}$ ◆

PROBLEM 4 Find the difference $\dfrac{2x^2 - 9}{2x^2 + 11x + 5} - \dfrac{x^2 + 16}{2x^2 + 11x + 5}$. ◆

To add or subtract algebraic fractions with *unlike denominators*, we first find the **least common denominator**, which is abbreviated as **LCD**. The following two-step procedure may be used to find the LCD.

Procedure for Determining the LCD	Example: $\dfrac{x-1}{x^2+x} - \dfrac{2x}{x^2+2x+1}$
1. *Completely factor* each denominator, and use exponential notation to represent repeated prime factors that occur in any one of the denominators.	1. $\dfrac{x-1}{\boxed{x}\,\boxed{(x+1)}} - \dfrac{2x}{\boxed{(x+1)^2}}$
2. The product of each *different prime factor* to the *highest power* it occurs in any one of the denominators is the LCD.	2. LCD is $x(x+1)^2$.

After determining the LCD, we apply the fundamental property of fractions and change the fractions to equivalent fractions, each having the LCD as its new denominator. Once the denominators are the same, we use the addition and subtraction properties of fractions to find the sum or difference of the fractions.

EXAMPLE 5 Perform the indicated operations.

(a) $\dfrac{x-1}{x^2+x} - \dfrac{2x}{x^2+2x+1}$ (b) $\dfrac{8}{4-y^2} + \dfrac{y}{2+y} - \dfrac{1}{y-2}$

SOLUTION

(a) $\dfrac{x-1}{x^2+x} - \dfrac{2x}{x^2+2x+1} = \dfrac{x-1}{x(x+1)} - \dfrac{2x}{(x+1)^2}$

> Factor the denominators.
> The LCD is $x(x+1)^2$.

$= \dfrac{(x-1)\cdot(x+1)}{x(x+1)\cdot(x+1)} - \dfrac{2x\cdot x}{(x+1)^2 \cdot x}$

$= \dfrac{x^2-1}{x(x+1)^2} - \dfrac{2x^2}{x(x+1)^2}$

$= \dfrac{-x^2-1}{x(x+1)^2}$

(b) $\dfrac{8}{4-y^2}+\dfrac{y}{2+y}-\dfrac{1}{y-2}$

$$=\dfrac{8}{(2+y)(2-y)}+\dfrac{y}{2+y}-\dfrac{-1}{2-y}$$

Factoring and rearranging the last fraction yields an LCD of $(2+y)(2-y)$.

$$=\dfrac{8}{(2+y)(2-y)}+\dfrac{y\cdot(2-y)}{(2+y)\cdot(2-y)}-\dfrac{-1\cdot(2+y)}{(2-y)\cdot(2+y)}$$

$$=\dfrac{8+(2y-y^2)+(2+y)}{(2+y)(2-y)}$$

$$=\dfrac{10+3y-y^2}{(2+y)(2-y)}$$

Always look for common factors in the numerator and denominator, and reduce to lowest terms

$$=\dfrac{(5-y)(2+y)}{(2+y)(2-y)}=\dfrac{5-y}{2-y}$$

PROBLEM 5 Repeat Example 5 for $\dfrac{x+1}{x^2-2x+1}-\dfrac{1}{x-1}$.

◆ Complex Fractions

A **complex fraction** is a fraction that contains at least one fraction in its numerator, its denominator, or in both numerator and denominator. To simplify a complex fraction, first find the LCD for *all* fractions appearing in its numerator and denominator. Then apply the fundamental property of fractions and multiply the numerator and denominator of the complex fraction by this LCD.

EXAMPLE 6 Simplify each complex fraction.

(a) $\dfrac{16-\dfrac{1}{x^2}}{4+\dfrac{1}{x}}$ **(b)** $\dfrac{1+\dfrac{2}{y-1}}{\dfrac{y^2+y}{y^2+y-2}}$

SOLUTION

(a) The LCD for $\dfrac{1}{x^2}$ and $\dfrac{1}{x}$ is x^2. Thus, we multiply numerator and denominator by x^2 as follows:

$$\frac{16 - \dfrac{1}{x^2}}{4 + \dfrac{1}{x}} = \frac{\left(16 - \dfrac{1}{x^2}\right) \cdot x^2}{\left(4 + \dfrac{1}{x}\right) \cdot x^2}$$

$$= \frac{16x^2 - 1}{4x^2 + x}$$

$$= \frac{(4x + 1)(4x - 1)}{x(4x + 1)} = \frac{4x - 1}{x} \qquad \textbf{Factor and reduce}$$

(b) The LCD for $\dfrac{2}{y - 1}$ and $\dfrac{y^2 + y}{(y + 2)(y - 1)}$ is $(y - 1)(y + 2)$. Thus, we multiply numerator and denominator by $(y - 1)(y + 2)$ as follows:

$$\frac{1 + \dfrac{2}{y - 1}}{\dfrac{y^2 + y}{y^2 + y - 2}} = \frac{\left(1 + \dfrac{2}{y - 1}\right) \cdot [(y - 1)(y + 2)]}{\dfrac{y^2 + y}{(y + 2)(y - 1)} \cdot [(y - 1)(y + 2)]}$$

$$= \frac{(y - 1)(y + 2) + 2(y + 2)}{y^2 + y}$$

$$= \frac{y^2 + 3y + 2}{y^2 + y}$$

$$= \frac{(y + 2)(y + 1)}{y(y + 1)} = \frac{y + 2}{y} \qquad \textbf{Factor and reduce} \qquad \blacklozenge$$

As an alternate method for simplifying a complex fraction, perform the indicated additions and subtractions in the numerator and denominator. After this is accomplished, apply the division property of fractions and simplify. Using Example 6(a), we have

$$\frac{16 - \dfrac{1}{x^2}}{4 + \dfrac{1}{x}} = \frac{\dfrac{16x^2}{x^2} - \dfrac{1}{x^2}}{\dfrac{4x}{x} + \dfrac{1}{x}} = \frac{\dfrac{16x^2 - 1}{x^2}}{\dfrac{4x + 1}{x}}$$

after performing the indicated addition and subtraction. Now apply the division property of fractions:

$$\frac{\dfrac{16x^2 - 1}{x^2}}{\dfrac{4x + 1}{x}} = \frac{16x^2 - 1}{x^2} \cdot \frac{x}{4x + 1} = \frac{(4x + 1)(4x - 1) \cdot x}{x^2 \cdot (4x + 1)} = \frac{4x - 1}{x}$$

PROBLEM 6 Simplify $\dfrac{\dfrac{3}{x+h}-\dfrac{3}{x}}{h}$. ◆

◆ **Some Algebra of Calculus**

In calculus, it is often necessary to rationalize an expression containing the form $\sqrt{a}-\sqrt{b}$ so that other evaluations can be performed. To rationalize such an expression, we apply the fundamental property of fractions and multiply both numerator and denominator by $\sqrt{a}+\sqrt{b}$. In computing the product $(\sqrt{a}-\sqrt{b})(\sqrt{a}+\sqrt{b})$, we use the fact that $(A-B)(A+B)=A^2-B^2$ as follows:

$$(\sqrt{a}-\sqrt{b})(\sqrt{a}+\sqrt{b})=(\sqrt{a})^2-(\sqrt{b})^2=a-b.$$

Each of the factors $\sqrt{a}-\sqrt{b}$ and $\sqrt{a}+\sqrt{b}$ is called the **conjugate** of the other factor.

EXAMPLE 7 Rationalize the numerator of $\dfrac{\sqrt{x+h}-\sqrt{x}}{h}$.

SOLUTION We multiply both numerator and denominator by the conjugate of the numerator as follows:

$$\begin{aligned}\frac{\sqrt{x+h}-\sqrt{x}}{h}&=\frac{(\sqrt{x+h}-\sqrt{x})\cdot(\sqrt{x+h}+\sqrt{x})}{h\cdot(\sqrt{x+h}+\sqrt{x})}\\&=\frac{(x+h)-(x)}{h\cdot(\sqrt{x+h}+\sqrt{x})}\\&=\frac{h}{h\cdot(\sqrt{x+h}+\sqrt{x})}\\&=\frac{1}{\sqrt{x+h}+\sqrt{x}}\end{aligned}$$ ◆

PROBLEM 7 Rationalize the denominator of $\dfrac{1}{\sqrt{x+h}+\sqrt{x}}$. ◆

Algebraic expressions containing negative exponents develop frequently in calculus. For such expressions, we use $a^{-n}=1/a^n$ to form algebraic fractions, and then simplify.

EXAMPLE 8 Simplify.

(a) $6x(1-3x)^{-3}+(1-3x)^{-2}$

(b) $\dfrac{2x(x^2+1)^{1/3}-\frac{2}{3}x^3(x^2+1)^{-2/3}}{(x^2+1)^{2/3}}$

SOLUTION

(a) We may use $a^{-n} = 1/a^n$ to form algebraic fractions, then add the fractions as follows:

$$6x(1 - 3x)^{-3} + (1 - 3x)^{-2} = \frac{6x}{(1 - 3x)^3} + \frac{1}{(1 - 3x)^2}$$

$$= \frac{6x}{(1 - 3x)^3} + \frac{1 \cdot (1 - 3x)}{(1 - 3x)^2 \cdot (1 - 3x)}$$

$$= \frac{6x + (1 - 3x)}{(1 - 3x)^3}$$

$$= \frac{3x + 1}{(1 - 3x)^3}$$

(b) We use $a^{-n} = 1/a^n$ to form the following complex fraction:

$$\frac{2x(x^2 + 1)^{1/3} - \dfrac{2x^3}{3(x^2 + 1)^{2/3}}}{(x^2 + 1)^{2/3}}$$

Now we multiply numerator and denominator by $3(x^2 + 1)^{2/3}$ as follows:

$$\frac{\left[2x(x^2 + 1)^{1/3} - \dfrac{2x^3}{3(x^2 + 1)^{2/3}} \right] \cdot 3(x^2 + 1)^{2/3}}{(x^2 + 1)^{2/3} \cdot 3(x^2 + 1)^{2/3}} = \frac{6x(x^2 + 1)^1 - 2x^3}{3(x^2 + 1)^{4/3}}$$

$$= \frac{4x^3 + 6x}{3(x^2 + 1)^{4/3}} \qquad \blacklozenge$$

An alternate method for simplifying Example 8(a) is to factor out $(1 - 3x)^{-3}$ as follows:

$$6x(1 - 3x)^{-3} + (1 - 3x)^{-2} = (1 - 3x)^{-3}[6x + (1 - 3x)^1]$$

$$= (1 - 3x)^{-3}(3x + 1)$$

$$= \frac{3x + 1}{(1 - 3x)^3}$$

PROBLEM 8   Simplify $(2 - x)^{-1/2} - x(2 - x)^{-3/2}$ by factoring out $(2 - x)^{-3/2}$. $\blacklozenge$

Exercises 1.7

 Basic Skills

In Exercises 1–10, find the missing numerator so that the algebraic fractions are equivalent.

1. $\dfrac{3m - 1}{5m} = \dfrac{?}{10m^2n}$

2. $\dfrac{x - 1}{3x^2} = \dfrac{?}{9x^3y^2}$

3. $\dfrac{1}{2x^2 - x} = \dfrac{?}{x(2x - 1)^2}$

4. $\dfrac{2}{a + b} = \dfrac{?}{a^2(a + b)^3}$

5. $\dfrac{x - 3}{2x - 3} = \dfrac{?}{2x^2 + 3x - 9}$

6. $\dfrac{2p - 5q}{2p + 5q} = \dfrac{?}{4p^2 - 25q^2}$

7. $\dfrac{4y}{x^2+4xy}=\dfrac{?}{3x^3+11x^2y-4xy^2}$

8. $\dfrac{3y}{2x^2+x}=\dfrac{?}{4x^3-8x^2-5x}$

9. $\dfrac{t+2}{t^2+2t+4}=\dfrac{?}{t^3-8}$ 10. $\dfrac{x-1}{x+1}=\dfrac{?}{2x^3+2}$

In Exercises 11–22, reduce the algebraic fraction to lowest terms.

11. $\dfrac{5xy}{5xy-10x^2y}$ 12. $\dfrac{2x-4x^2}{2x}$

13. $\dfrac{4m-4n}{12n-12m}$ 14. $\dfrac{9x-27y}{9y-3x}$

15. $\dfrac{2x^2+11x+5}{x^2+4x-5}$ 16. $\dfrac{16b^2-a^2}{a^2+ab-12b^2}$

17. $\dfrac{n^2-5n-6}{36-n^2}$ 18. $\dfrac{4x^2-5x-6}{8-x^3}$

19. $\dfrac{x^3-x}{x^4+3x^3-x^2-3x}$ 20. $\dfrac{k^4-1}{k^8-1}$

21. $\dfrac{9-(x+y)^2}{(x-3)^2-y^2}$ 22. $\dfrac{x^2-2xy+y^2}{x^3-x^2y-xy^2+y^3}$

In Exercises 23–48, perform the indicated operations and reduce to lowest terms.

23. $\dfrac{x^2+2x+1}{2xy}\cdot\dfrac{6xy}{x^2-1}$ 24. $\dfrac{a}{(a-b)^2}\cdot\dfrac{a^2-b^2}{a^2}$

25. $\dfrac{4-n^2}{(n-1)^2}\cdot\dfrac{n^2+3n-4}{n^2+2n-8}$ 26. $\dfrac{(x+2)^3}{x^3+8}\cdot\dfrac{x^3-2x^2+4x}{x^2+4x+4}$

27. $\dfrac{12x^2}{x^2+2xy+y^2}\div\dfrac{6x}{x+y}$ 28. $\dfrac{m+3}{5m}\div\dfrac{m^2-9}{10m^3}$

29. $\dfrac{3a^3-81b^3}{9b^2-a^2}\div(a^2+3ab+9b^2)$

30. $\dfrac{9-(x-y)^2}{3xy^2}\div\dfrac{x-y-3}{3xy}$

31. $\dfrac{3n^2+7n+3}{4n-1}+\dfrac{5-n^2}{1-4n}$

32. $\dfrac{6x^2-12x}{x^2+2x+4}+\dfrac{(x-2)^3}{x^2+2x+4}$

33. $\dfrac{(x+1)^2}{3x}-\dfrac{2x+1}{3x}$ 34. $\dfrac{b-a}{2a-b}-\dfrac{a-b}{b-2a}$

35. $\dfrac{3}{m^2-9}+\dfrac{1}{m+3}$ 36. $\dfrac{x+2}{x-2}+\dfrac{x-2}{x+2}$

37. $\dfrac{1}{(x-1)^3}+\dfrac{1}{1-2x+x^2}$ 38. $\dfrac{x+9}{x^2-6x+9}+\dfrac{2}{x^2-9}$

39. $\dfrac{2-x}{x^2+2x}-\dfrac{x}{x^2+4x+4}$ 40. $\dfrac{(x-4)^2}{x+4}-\dfrac{(x+4)^2}{x-4}$

41. $\dfrac{1}{a^2+a}-\dfrac{1}{a-a^2}$ 42. $\dfrac{1}{t-1}-\dfrac{t^2+t}{t^3-1}$

43. $\dfrac{a+1}{a^2+a+1}+\dfrac{1}{1-a}+\dfrac{a^2+2a+3}{a^3-1}$

44. $\dfrac{x-3}{3+x}+\dfrac{x+3}{3-x}+\dfrac{36}{x^2-9}$

45. $\left(\dfrac{x+y}{x-y}-\dfrac{x-y}{x+y}\right)\div\left(\dfrac{x^2+y^2}{x^2-y^2}-\dfrac{x^2-y^2}{x^2+y^2}\right)$

46. $\left(x+\dfrac{4}{x}\right)\left(\dfrac{x^2}{4}-1\right)\div\left(1-\dfrac{16}{x^4}\right)$

47. $\left(\dfrac{b}{ac}+\dfrac{2}{a}+\dfrac{c}{ab}-\dfrac{a}{bc}\right)\left(1+\dfrac{b+c}{a-b-c}\right)$

48. $\left(\dfrac{n^2}{25n^2-4}\right)\left(5n-8-\dfrac{4}{n}\right)\left(5n+8-\dfrac{4}{n}\right)$

In Exercises 49–54, simplify each complex fraction.

49. $\dfrac{\frac{1}{a}+1}{\frac{1}{a^2}-1}$ 50. $\dfrac{1-\frac{1}{4n^2}}{2-\frac{1}{n}}$

51. $\dfrac{\frac{9}{x}-x}{\frac{1}{x^2}-\frac{8}{x^3}+\frac{15}{x^4}}$ 52. $\dfrac{\frac{1}{x^2}-\frac{1}{y^2}}{\frac{1}{x^2}+\frac{2}{xy}+\frac{1}{y^2}}$

53. $\dfrac{\frac{x-2}{x+2}+\frac{x+2}{x-2}}{\frac{x^2-4}{(x-2)^2}-1}$ 54. $\dfrac{\frac{3-x}{3+x}+\frac{9+x^2}{9-x^2}}{\frac{27+x^3}{(3-x)^2}}$

In Exercises 55–60, rationalize the numerator and reduce to lowest terms.

55. $\dfrac{\sqrt{a}+\sqrt{b}}{\sqrt{a}-\sqrt{b}}$ 56. $\dfrac{x-\sqrt{x^2-1}}{x+\sqrt{x^2-1}}$

57. $\dfrac{3-\sqrt{t+3}}{t-6}$ 58. $\dfrac{\sqrt{2a}+\sqrt{2b}}{\sqrt{2ab}}$

59. $\dfrac{\sqrt{(x+h)^2+1}-\sqrt{x^2+1}}{h}$

60. $\dfrac{\sqrt{2+(x+h)}-\sqrt{2+x}}{h}$

In Exercises 61–66, rationalize the denominator and reduce to lowest terms.

61. $\dfrac{\sqrt{3}+\sqrt{x}}{\sqrt{3}-\sqrt{x}}$

62. $\dfrac{5-m}{\sqrt{5n}+\sqrt{mn}}$

63. $\dfrac{1}{x-\sqrt{x^2-1}}$

64. $\dfrac{a^2}{\sqrt{a^2+b^2}+b}$

65. $\dfrac{mn\sqrt{mn}}{m\sqrt{n}+n\sqrt{m}}$

66. $\dfrac{b\sqrt{a}+a\sqrt{b}}{b\sqrt{a}-a\sqrt{b}}$

 In Exercises 67–72, use $a^{-n}=1/a^n$ to form an algebraic fraction, then simplify.

67. $7t(2-3t)^{-4}-3(2-3t)^{-3}$

68. $\frac{1}{3}x(3-2x)^{-1}+(3-2x)^{-2}$

69. $\frac{4}{5}n(1-4n)^{-6/5}+(1-4n)^{-1/5}$

70. $x(3-x)^{-4/3}-(3-x)^{-1/3}$

71. $-t(3-4t)^{-3/4}+(3-4t)^{1/4}$

72. $\frac{3}{4}y^2(y^2+1)^{-2/3}-(y^2+1)^{1/3}$

In Exercises 73–82, use $a^{-n}=1/a^n$ to form a complex fraction, then simplify.

73. $\dfrac{a^{-1}+b^{-1}}{a^{-2}-b^{-2}}$

74. $\dfrac{2+mn^{-1}}{2-m^{-1}n}$

75. $\dfrac{x+8+12x^{-1}}{1+7x^{-1}+10x^{-2}}$

76. $\dfrac{xy^{-2}+x^{-2}y}{x^{-2}-x^{-1}y^{-1}+y^{-2}}$

77. $\dfrac{1+(n-1)^{-1}}{1-(n+1)^{-1}}$

78. $\dfrac{a}{1+(a-1)(a+1)^{-1}}$

79. $\dfrac{6t^{4/3}-(t^2+1)t^{-2/3}}{t^{2/3}}$

80. $\dfrac{(x^2+4)^{1/2}-x^2(x^2+4)^{-1/2}}{x^2+4}$

81. $\dfrac{\frac{1}{4}x^{-3/4}(x+1)^{1/4}-\frac{1}{4}x^{1/4}(x+1)^{-3/4}}{(x+1)^{1/2}}$

82. $\dfrac{\frac{1}{3}x^{-2/3}(1-2x)^{1/3}-\frac{1}{3}x^{1/3}(1-2x)^{-2/3}(-2)}{[(1-2x)^{1/3}]^2}$

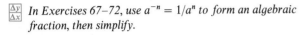

Critical Thinking

83. If we reduce $\dfrac{x-1}{x^2-1}$ to lowest terms, we obtain $\dfrac{1}{x+1}$. Are these two rational expressions equal for all real values of x? Explain.

84. Rationalize the numerator of

$$\frac{\sqrt[3]{x+h}-\sqrt[3]{x}}{h}$$

by using the rationalizing factor

$$\sqrt[3]{(x+h)^2}+\sqrt[3]{x+h}\cdot\sqrt[3]{x}+\sqrt[3]{x^2}.$$

What special product in Section 1.4 gives forth this rationalizing factor?

85. The sum or difference of a polynomial and a rational expression is called a *mixed algebraic expression*. Use the addition and subtraction properties for fractions (reading from right to left) to write each rational expression as a mixed algebraic expression.

(a) $\dfrac{y^2+7y-5}{y}$ (b) $\dfrac{14-(t+7)^2}{t+7}$

86. Use the addition and subtraction properties for frac-

tions (reading from right to left) to write each rational expression as a sum or difference of simpler rational expressions, each reduced to lowest terms.

(a) $\dfrac{12xy^2-7x^2y^3+6x-9}{3x^2y}$

(b) $\dfrac{3a(x+y)^2+(x+y)^3-2(x+y)}{2(x+y)^2}$

87. If two batteries are connected in parallel, as shown in the figure, and have the same internal resistance r and the same voltage E, then the current i supplied to the external load R is given by the formula

$$i=\frac{E}{R+\dfrac{r}{2}}.$$

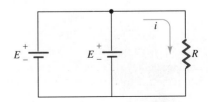

Simplify the complex fraction that appears on the right-hand side of this formula.

88. If a payment of R dollars is made at the end of each time period for n periods into an annuity that earns interest at the rate of r per period, then the present value A of the annuity is given by the formula

$$A = R\left[\frac{1 - (1 + r)^{-n}}{r}\right].$$

Rewrite the expression on the right-hand side of this formula as a single fraction that contains no negative exponent.

89. Finding the number of crimes committed in a certain part of the city involves working with the expression

$$\frac{60}{2 + \sqrt{n}} - n,$$

where n is the number of police that are assigned to the area. Write this expression as a single fraction with a rationalized denominator.

90. The number of items that a machine can produce during an 8-hour day is $t^2 + 16t$, where t is the number of hours that the machine operates. Suppose the total cost to produce these items is $8t + 128$. Find a simplified algebraic expression that represents the *average cost* of production.

 Calculator Activities

91. A single lens with a focal length of 11.1 cm projects an image of an object onto a screen, as shown in the sketch. The distance (in cm) from the lens to the screen may be found by evaluating

$$\frac{1}{\dfrac{1}{11.1} - \dfrac{1}{12.4}}.$$

Use the reciprocal key ($\boxed{1/\text{x}}$ or $\boxed{\text{x}^{-1}}$) on your calculator to find this distance.

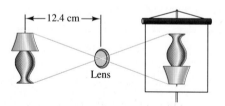

Lens

92. The total resistance (in ohms, Ω) between points A and B in the circuit shown may be found by evaluating

$$\frac{1}{\dfrac{1}{20.6} + \dfrac{1}{50.4} + \dfrac{1}{100.8}}.$$

Use the reciprocal key ($\boxed{1/\text{x}}$ or $\boxed{\text{x}^{-1}}$) on your calculator to determine the total resistance between points A and B.

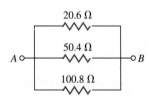

EXERCISE 92

In Exercises 93 and 94, evaluate each expression by each method:

(a) *Perform the indicated operations* without *the use of a calculator. Express the answer as a rational number a/b reduced to lowest terms.*

(b) *Use the reciprocal key ($\boxed{1/\text{x}}$ or $\boxed{\text{x}^{-1}}$) on your calculator. Express the answer as a terminating or repeating decimal. Compare your answer with part (a).*

93. $1 + \dfrac{1}{2 + \dfrac{1}{3 + \frac{1}{4}}}$

94. $1 + \dfrac{1}{1 + \dfrac{1}{1 + \dfrac{1}{1 + \dfrac{1}{1 + \frac{1}{1}}}}}$

Chapter 1 Review

Questions for Group Discussion

1. What is the difference between a *rational number* and an *irrational number*?

2. Is the expression $-x$ always negative? Explain.

3. From a given fraction, how can an *equivalent fraction* be generated?

4. For what values of a does $\dfrac{|a|}{a} = -1$?

5. If $a > 0$ and $b > 0$, discuss what happens to the fraction a/b in each case:
   (a) a remains the same and b gets larger.
   (b) b remains the same and a gets larger.
   (c) a remains the same and b gets smaller.
   (d) b remains the same and a gets smaller.

6. State the six *fundamental properties of the real numbers*.

7. List the *laws of exponents*. Illustrate each rule with an example.

8. Look up the meanings of the following prefixes in a dictionary: *pico-*, *deci-*, *nano-*, *giga-*, and *hecto-*. Using these definitions, explain the meanings of picosecond, decimeter, nanofarad, gigawatt, hectoliter.

9. What is meant by the *principal nth root* of a real number? Is the principal *n*th root of a real number always a real number? Explain.

10. Explain the procedure of *rationalizing a numerator or denominator*. Illustrate with an example.

11. Describe the process of *reducing* an algebraic fraction to lowest terms.

12. What is a *polynomial*?

13. What is meant by the *degree of a polynomial*? Illustrate with an example.

14. Explain the procedure for combining *like terms* into a single term.

15. State the procedures for adding, subtracting, and multiplying polynomials.

16. State the procedures for adding, subtracting, and multiplying complex numbers.

17. What is a *complex conjugate*? How is it used in finding the quotient of two complex numbers? Illustrate with an example.

18. What is an *imaginary number*? How does it differ from a *pure imaginary number*?

19. What are the two major subsets of the set of *complex numbers*? List some other subsets of the set of complex numbers.

20. Does every complex number have a reciprocal? Explain.

21. Name several types of factoring procedures. Illustrate each type with an example.

22. What is meant by *factoring completely* over the integers? over the reals?

23. What are the conditions for a trinomial to be a *perfect square*?

24. Explain the procedure of determining the *LCD* for a sum or difference of algebraic fractions.

25. What is a *complex fraction*? Explain two methods that can be used to simplify a complex fraction. Use one example to illustrate the two methods.

26. How are *conjugates* used to rationalize a denominator or numerator?

Review Exercises

In Exercises 1–12, use the definitions and properties of the real numbers to complete each statement.

1. $8m + 3 = 3 + \boxed{}$

2. $-(x + 9) + (x + 9) = \boxed{}$

3. $6 - (-3x) = \boxed{} + 6$

4. $-(8 - t) = t - \boxed{}$

5. $(-3)[-(x + \sqrt{5})] = \boxed{}(x + \sqrt{5})$

6. $(x + 2)(1 - t) = (x + 2) - (x + 2)\boxed{}$

7. $\dfrac{x}{3} - 2 = \dfrac{x}{3} - \dfrac{\boxed{}}{3} = \dfrac{\boxed{}}{3}$

8. $\dfrac{2x}{3} \div 3 = \dfrac{2x}{3} \cdot \boxed{} = \boxed{}$

9. $\dfrac{3x + 5}{2} = \boxed{}(3x + 5)$

10. $(n - 4)\boxed{} = 1$

11. $\dfrac{3}{8 - x} = \dfrac{\boxed{}}{x - 8}$

12. $(\sqrt{3x} + y) + \boxed{} = \sqrt{3x}$

In Exercises 13–16, rewrite each statement using inequality notation.

13. a is at least 7.

14. b is at most -9.

15. c is negative and greater than -10.

16. d is nonnegative and at most 6.

In Exercises 17 and 18, find the distance from A to B. Write the answer without absolute value bars.

17. The coordinate of point A is -9 and the coordinate of point B is 6.

18. The coordinate of point A is 3 and the coordinate of point B is π.

In Exercises 19–28, simplify each exponential expression. Write the answer without zero or negative exponents. Assume the variables represent positive real numbers.

19. $(-2x^4y^5)(5xy^4)^2$

20. $[-2(x + 3)^2]^3$

21. $(2x + 1)^{-3}(2x + 1)^{-1}(2x + 1)^0$

22. $\left(\dfrac{3x^3}{2}\right)^{-2}$

23. $\dfrac{(x - 1)^4}{(x - 1)^3}$

24. $\dfrac{(2m^2n)^{-4}}{8m^{-2}}$

25. $(x + 1)^{-2/3}(x + 1)^{5/3}$

26. $(36x^4y^{-6})^{3/2}y^8$

27. $\left(\dfrac{16x^2}{y^8}\right)^{-3/4}$

28. $\dfrac{[9(x + 4)^{-3}]^{-1/2}}{(x + 4)^{1/2}}$

In Exercises 29 and 30, perform the indicated operations using scientific notation.

29. $\dfrac{0.00000000018}{60,000,000,000,000}$

30. $(30,000,000,000)(0.0000000002)^3$

In Exercises 31 and 32, rewrite each expression in exponential form.

31. $\sqrt[3]{(x + 2)^2}$

32. $\sqrt{x} \cdot \sqrt[4]{x^3}$

In Exercises 33 and 34, rewrite each exponential expression in radical form.

33. $(x + 4)^{-1/6}$

34. $(2ab)^{3/4}$

In Exercises 35–38, write each radical expression in simplified radical form. Assume the variables represent positive real numbers.

35. $\sqrt{32(x + y)^3}$

36. $\sqrt{\sqrt[3]{36x^4}}$

37. $\sqrt[4]{32x^2y^2} \cdot \sqrt[4]{2xy^2}$

38. $\sqrt{\dfrac{8m^5}{(n + 2)^6}}$

In Exercises 39–40, simplify by assuming the variables represent any real number.

39. $(9x^2y^2)^{1/2}$

40. $[(-2)^4(y - 1)^8]^{1/4}$

In Exercises 41–60, perform the indicated operations. Be sure to combine any like terms.

41. $(a^2 - 3a) + (5a - 4a^2) - (a + 5a^2)$

42. $(x^2 - 2x\sqrt{27x}) - (\sqrt{3x^3} - 3x^2)$

43. $(3\sqrt{40x} - 4\sqrt{90x}) + 2\sqrt{160x}$

44. $3x^{1/2} + (2x^{-1} + 4x^{1/2} - 3x^{-1})$

45. $(x^3 + 2x^2 - 3)(2x^3 - 3x - 1)$

46. $(5n^{-1/2} - 3n^{-3/2})(2n^{3/2} - n^{-1/2})$

47. $(5x^2 + 2y)(6x^2 - 5y)$

48. $(3\sqrt{ab} - 2)(4\sqrt{ab} - 5)$

49. $(2t - 5)(2t + 5)$

50. $(\sqrt{2x} - \sqrt{3y})(\sqrt{2x} + \sqrt{3y})$

51. $(x - 4)(x^2 + 4x + 16)$

52. $(4m^{-2} - 6m^{-1} + 9)(2m^{-1} + 3)$

53. $(3x - 7y)^2$ **54.** $(\sqrt{2n} + 1)^2$

55. $(2x - 5)^3$ **56.** $(a^{-2} - 2b)^3$

57. $[(x + 3)(x - 3)]^3$ **58.** $2(1 - a)^2(-5a) + 10a$

59. $y^2 - [x(3x - y) - (3x - 4y)(x + y)]$

60. $x\{x^2 - [(2x + 3)2x - x(3x + 2)]\}$

In Exercises 61–62, find a polynomial that represents the area of the shaded region.

61.

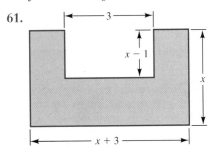

62.

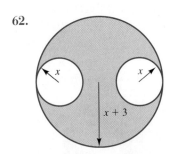

In Exercises 63–82, factor completely over the set of integers.

63. $3y^2(y - 2)^3 + 6y(y - 2)^4$

64. $x(1 - 3x)^6 + (1 - 3x)^5$

65. $m^3 - 3m^2 + m - 3$ **66.** $12 - 9y - 8x + 6xy$

67. $6x^2 - xy - 12y^2$ **68.** $8t^2 + 6t - 5$

69. $2x^4 - 13x^2 + 15$ **70.** $m^2 + 10mn + 16n^2$

71. $x^2 + 2x - 15$ **72.** $a^6 - 7a^3b + 12b^2$

73. $25x^2 - 9y^4$ **74.** $x^3 + 27y^3$

75. $8x^3 - 1$ **76.** $9m^4 - 12m^2n^2 + 4n^4$

77. $t^3 + 6t^2 + 12t + 8$ **78.** $x^4 - 4y^4$

79. $8x^3y + 2x^2y - 6xy$ **80.** $m^6 - 4m^3 - 32$

81. $2xy + 4x^2y - 18x^3y - 36x^4y$

82. $16t^8 - 1$

In Exercises 83 and 84, find the missing numerator so that the fractions are equivalent.

83. $\dfrac{3}{x + 3} = \dfrac{?}{x(x + 3)^3}$ **84.** $\dfrac{x - 2}{x^2 - 2x + 4} = \dfrac{?}{x^3 + 8}$

In Exercises 85 and 86, reduce the fraction to lowest terms.

85. $\dfrac{n^2 + 5n - 36}{3n^2 - 14n + 8}$ **86.** $\dfrac{16x - x^5}{x^2 - 2x}$

In Exercises 87–102, perform the indicated operations.

87. $\dfrac{-5}{2x} + \dfrac{7}{2x}$ **88.** $\dfrac{2}{\pi} - \dfrac{9}{\pi}$

89. $\dfrac{-9}{5x} \cdot \dfrac{10x^2}{3}$ **90.** $\dfrac{3x}{2} \div \dfrac{15x}{8}$

91. $\dfrac{7y}{8x} + \dfrac{-5x^2}{12y}$ **92.** $\dfrac{5x}{9} - \dfrac{y}{12}$

93. $\dfrac{2x^2 - 3x - 9}{6xy} \cdot \dfrac{2x^2}{9 - x^2}$

94. $\dfrac{2a^2 + ab - 3b^2}{4a^2 - 9b^2} \div \dfrac{b^3 - a^3}{3b - 2a}$

95. $\dfrac{2x + 3}{x^2 - 9} - \dfrac{5x - 6}{x^2 - 9}$

96. $\dfrac{3m(m - n)}{4m - n} + \dfrac{n^2 - m^2}{n - 4m}$

97. $\dfrac{3}{x^2 + 4x} - \dfrac{3}{x^2 + 8x + 16}$

98. $\dfrac{1}{5 + x} + \dfrac{1}{5 - x} - \dfrac{x^2 - 7x}{x^2 - 25}$

99. $\dfrac{1}{x^2 - 4xy + 4y^2} + \dfrac{1}{x^2 - 4y^2}$

100. $\dfrac{y^{-2} - x^{-2}}{x^{-1} - y^{-1}}$ **101.** $\dfrac{1 + (a - 1)^{-1}}{(a - 1)^{-1}}$

102. $\dfrac{1 - \dfrac{1}{25x^2}}{5 - \dfrac{1}{x}}$

In Exercises 103 and 104, rationalize the numerator and simplify.

103. $\sqrt[4]{\dfrac{(x-5)^3}{3x}}$ **104.** $\dfrac{\sqrt{3x}+\sqrt{3y}}{\sqrt{3xy}}$

In Exercises 105 and 106, rationalize the denominator and simplify.

105. $\dfrac{1}{2x-\sqrt{4x^2-1}}$ **106.** $\dfrac{1}{\sqrt[3]{(3-x)^2}}$

In Exercises 107–110, simplify and express the answer in the form b or bi, where b is a real number.

107. $4\sqrt{-45}+3\sqrt{-80}-\sqrt{-20}$

108. $\sqrt{-24}\cdot\sqrt{6}\cdot\sqrt{-16}$

109. $(2i)^6$ **110.** $5i^{-33}$

In Exercises 111–114, simplify and express the answer in the complex number form a + bi.

111. $(9+i)+(3-2i)$

112. $(4-3\sqrt{-9})-(9+3\sqrt{-25})$

113. $(3-5i)(4+7i)$ **114.** $\dfrac{(1-i)^2}{3+4i}$

115. The wavelength λ (the Greek letter *lambda*) of a radio wave of frequency f is given by the formula

$$\lambda=\frac{3\times10^8}{f},$$

where λ is in meters and f is in hertz. Determine λ if $f=7.5\times10^6$ hertz.

116. The change in length δ (the Greek letter *delta*) of a concrete slab in a sidewalk due to thermal expansion is given by the formula

$$\delta=5.5\times10^{-6}\,tl,$$

where t is the change in temperature and l is the original length of the slab. Suppose 30.0-ft concrete slabs are being laid with an expansion gap between the end of one slab and the beginning of the next. What is the minimum expansion gap, measured in *inches*, that should be left between the slabs if the temperature of the concrete in the summer can reach 170 °F and the concrete is being laid at a temperature of 44 °F?

117. The land area of China is approximately 3,700,000 square miles. In 1980, the population of China was approximately one billion. At that time, what was the amount of land area per person, measured in *square feet*? (*Hint:* 5280 ft = 1 mi)

118. The United States natural gas reserve is estimated to be about 180 trillion cubic feet at present, and consumption is averaging approximately 9×10^{11} cubic feet per year. Assuming consumption continues at this rate, how long will the present supply last?

CHAPTER 2

Techniques of Solving Equations and Inequalities

A contractor agrees to put in a swimming pool in a given amount of time. The contract specifies that he is to receive $250 per day for each day's work during the allotted time and he will forfeit $100 for each day taken beyond that time. Find the time allowed in the contract if he received $3900 for 24 days of work but did not complete the project in the allotted time.

(For the solution, see Example 1 in Section 2.2.)

2.1   An Introduction to Solving Equations

◆ **Introductory Comments**

An **equation** is a statement declaring that two algebraic expressions are equal. Two examples of equations in one variable are

$$3x + 4x = 7x \quad \text{and} \quad 2x + 1 = 7.$$

The equation $3x + 4x = 7x$ is true no matter what value we choose to replace x. An equation that becomes true when the variable is replaced by *any* permissible number is called an **identity**. The equation $2x + 1 = 7$ is true only when we replace x with 3 and is false for all other replacements. An equation that is true only for some values of the variable, but not for others (or is never true for any value of the variable), is called a **conditional equation**.

We refer to values of the variable that make the equation a true statement as the **roots** or **solutions** to the equation. Thus, we say 3 is a root or solution to the equation $2x + 1 = 7$, since 3 is the number that *satisfies* this equation. The equation $2x = 6$ also has a solution of 3. Equations that have the same solutions are said to be **equivalent equations**. We can generate equivalent equations from a given equation by using the following rules.

Rules for Generating Equivalent Equations

1. Add the same expression to both sides of a given equation, or subtract the same expression from both sides.
2. Multiply or divide both sides of a given equation by the same *nonzero* expression.
3. Interchange the left-hand and right-hand sides of a given equation.
4. Simplify algebraic expressions that appear on either side of a given equation.

In this section, we use these rules for generating equivalent equations to help *solve* an equation. To **solve** an equation means to find *all* values of the variable that make the equation a true statement.

Solving Linear Equations

Equations such as $2x + 1 = 7$ that are of the form

$$ax + b = c$$

where a, b, and c are real numbers and $a \neq 0$, are called **linear** or **first-degree equations** in one variable. To solve a linear equation, we generate a chain of equivalent equations until the variable is isolated on one side of the equation. The solution to the equation then appears on the other side. Thus, to solve the linear equation $2x + 1 = 7$, we proceed as follows:

$$2x + 1 = 7$$

$$2x = 6 \qquad \text{Subtract 1 from both sides}$$

$$x = 3 \qquad \text{Divide both sides by 2}$$

| Isolated variable | Solution |

Many equations not initially of the form $ax + b = c$ can be transformed into this form by using the rules for generating equivalent equations. Once we have generated an equivalent equation of the linear form $ax + b = c$, the solution can be obtained easily.

EXAMPLE 1   Solve each equation.

(a) $2y - 11 = 5y - 5$ (b) $3x + 9 - 2(4 - 7x) = 5(13 - 3x)$

SOLUTION

(a)   To solve this equation, we proceed as follows:

$$2y - 11 = 5y - 5$$

$$-11 = 3y - 5 \qquad \text{Subtract } 2y \text{ from both sides}$$

$$-6 = 3y \qquad \text{Add 5 to both sides}$$

$$y = -2 \qquad \text{Divide both sides by 3 and interchange the sides}$$

Since the solution to the last equation is obviously -2, the solution to the original equation must also be -2.

(b)   To solve this equation, we proceed as follows:

$$3x + 9 - 2(4 - 7x) = 5(13 - 3x)$$

$$3x + 9 - 8 + 14x = 65 - 15x \qquad \text{Apply the distributive property}$$

$$17x + 1 = 65 - 15x \qquad \text{Combine like terms on the left-hand side}$$

$$32x + 1 = 65 \qquad \text{Add } 15x \text{ to both sides}$$

$$32x = 64 \qquad \text{Subtract 1 from both sides}$$

$$x = 2 \qquad \text{Divide both sides by 32}$$

Since the solution to this equation is obviously 2, the solution to the original equation must also be 2.   ◆

To **check** the solution of an equation, we replace the variable in the *original* equation with the derived solution to see if the equation becomes true. For instance, we can check the solution of the equation in Example 1(a) by replacing y with -2 as follows:

$$2y - 11 = 5y - 5$$

$$\text{Check:} \quad 2(-2) - 11 = 5(-2) - 5?$$

$$-4 - 11 = -10 - 5?$$

$$-15 = -15 \quad \checkmark$$

You can check the solution to Example 1(b) by replacing x with 2 in the equation $3x + 9 - 2(4 - 7x) = 5(13 - 3x)$.

PROBLEM 1   Solve $(n - 1)^2 + 3(n - 3) = (n + 1)(n + 4)$ and check.   ◆

◆ Fractional Equations Reducible to Linear Form

An equation that contains one or more algebraic fractions is called a **fractional equation**. Many fractional equations are reducible to *linear form* once the denominators in the fractional equation have been eliminated. To eliminate the denominators, we multiply both sides of the fractional equation by the least common denominator (LCD) for all the fractions in the equation.

EXAMPLE 2   Solve each fractional equation.

(a) $\dfrac{5x}{6} - x = \dfrac{1}{8}$ (b) $\dfrac{2y + 3}{5} - \dfrac{y - 1}{15} = \dfrac{3y - 1}{3}$

SOLUTION

(a)   The LCD for the denominators 6 and 8 is 24. Thus, we multiply both sides of this equation by 24 and cancel the denominators as follows:

$$24\left(\dfrac{5x}{6} - x\right) = 24\left(\dfrac{1}{8}\right) \qquad \text{Multiply both sides by 24}$$

$$20x - 24x = 3 \qquad \text{Simplify}$$

$$-4x = 3 \qquad \text{Collect like terms on the left-hand side}$$

$$x = -\dfrac{3}{4} \qquad \text{Divide both sides by } -4$$

Thus, the solution of this fractional equation is $-\frac{3}{4}$. You can check the solution.

(b) The LCD for the denominators 5, 15, and 3 is 15. Thus, we multiply both sides of the equation by 15 as follows:

$$15\left(\frac{2y+3}{5} - \frac{y-1}{15}\right) = 15\left(\frac{3y-1}{3}\right)$$ Multiply both sides by 15

$$3(2y+3) - (y-1) = 5(3y-1)$$ Simplify

$$6y+9 - y+1 = 15y - 5$$ Apply the distributive property

$$5y+10 = 15y - 5$$ Collect like terms on the left-hand side

$$10 = 10y - 5$$ Subtract 5y from both sides

$$15 = 10y$$ Add 5 to both sides

$$y = \frac{3}{2}$$ Divide both sides by 10, interchange sides, and reduce

Thus, the solution of this fractional equation is $\frac{3}{2}$. You can check the solution.   ◆

PROBLEM 2 Solve $\dfrac{n-2}{12} - \dfrac{n-5}{8} = n - 1$ and check.   ◆

If we multiply both sides of a given equation by *zero*, the equation generated is *not* an equivalent equation. Thus, when we multiply both sides of a fractional equation by an LCD that contains a variable, we *must* check the solution to be certain that the LCD is indeed a *nonzero* quantity.

EXAMPLE 3 Solve each fractional equation.

(a) $\dfrac{9}{x-1} - 4 = \dfrac{1}{2}$ **(b)** $\dfrac{3}{2} - \dfrac{1}{m+2} = \dfrac{m}{2m+4}$

SOLUTION

(a) The LCD for the denominators $(x-1)$ and 2 is $2(x-1)$. We multiply both sides of this fractional equation by $2(x-1)$, assuming $2(x-1) \neq 0$:

$$2(x-1)\left[\frac{9}{x-1} - 4\right] = 2(x-1)\left[\frac{1}{2}\right]$$ Multiply both sides by $2(x-1)$

$$18 - 8(x-1) = x - 1$$ Simplify

EXAMPLE 3 (*continued*)

$$26 - 8x = x - 1$$ **Collect like terms on the left-hand side**

$$27 = 9x$$ **Add $8x$ and add 1 to both sides**

$$x = 3$$ **Divide both sides by 9 and interchange sides**

If $x = 3$, then the LCD

$$2(x - 1) = 2(3 - 1) = 4,$$

is a *nonzero* quantity. Thus, we have generated equivalent equations, and the solution of the original equation must also be 3. You can check this solution by replacing x with 3 in the original equation.

(b) The LCD for the denominators 2, $m + 2$, and $2m + 4$, which factors to $2(m + 2)$, is $2(m + 2)$. We multiply both sides of this fractional equation by $2(m + 2)$, assuming $2(m + 2) \neq 0$:

$$2(m + 2)\left[\frac{3}{2} - \frac{1}{m + 2}\right] = 2(m + 2)\left[\frac{m}{2(m + 2)}\right]$$ **Multiply both sides by $2(m + 2)$**

$$3(m + 2) - 2 = m$$ **Simplify**

$$3m + 4 = m$$ **Collect like terms on left-hand side**

$$4 = -2m$$ **Subtract $3m$ from both sides**

$$m = -2$$ **Divide both sides by -2 and interchange sides**

However, if $m = -2$, then the LCD

$$2(m + 2) = 2(-2 + 2) = 0.$$

Since multiplying both sides of an equation by zero does *not* generate equivalent equations, the apparent solution of -2 must be discarded. Therefore, we conclude this fractional equation has *no solution*.   ◆

PROBLEM 3   Suppose you replace m with -2 in the original equation in Example 3(b). What dilemma do you confront?   ◆

Note: A "solution," such as -2 in Example 3(b), that develops through the algebraic process but does not satisfy the original equation is called an **extraneous root**. An extraneous root may develop when we multiply both sides of an equation by a variable expression. Therefore, it is essential to check the solutions that develop, and discard any extraneous root(s).

◆ Literal Equations and Formulas

An equation that contains letters other than the variable for which we wish to solve is called a **literal equation**. The linear equation $ax + b = c$ is an example of a literal equation. To solve this equation for x, we proceed as follows:

$$ax + b = c$$

$$ax = c - b \qquad \text{Subtract } b \text{ from both sides}$$

$$x = \frac{c - b}{a} \qquad \text{Divide both sides by } a \ (a \neq 0)$$

To solve other literal equations in which the variable appears *more than once*, we use the rules for generating equivalent equations and apply the following steps:

1. *Group* all the variable terms for which we wish to solve on one side of the equation.
2. *Factor out* the variable for which we wish to solve.
3. *Divide* each side of the equation by the expression being multiplied by the variable.

The procedure is illustrated in the next example.

EXAMPLE 4   Solve the literal equation $(x + a)(x + b) = x(x + c)$ for x.

SOLUTION   To solve this equation for x, we proceed as follows:

$$(x + a)(x + b) = x(x + c)$$

$$x^2 + ax + bx + ab = x^2 + cx \qquad \text{Apply the distributive property}$$

$$ab = cx - ax - bx \qquad \text{Group the } x \text{ terms on one side}$$

$$ab = x(c - a - b) \qquad \text{Factor out the variable } x$$

$$x = \frac{ab}{c - a - b} \qquad \begin{array}{l}\textit{Divide both sides by } (c - a - b) \\ \text{with } c - a - b \neq 0\end{array} \qquad ◆$$

PROBLEM 4   In Example 4, we grouped the x terms on the *right-hand side* of the equation. Solve Example 4 by grouping the x terms on the *left-hand side* of the equation. Show that the answer you obtain is the same as that given in Example 4.   ◆

A **formula** is a mathematical or scientific rule in the form of a literal equation that describes a special relationship between two or more variables. Often, it is necessary to rearrange a formula and solve for one of its variables.

E X A M P L E 5   Using the lens formula $\dfrac{1}{p} + \dfrac{1}{q} = \dfrac{1}{f}$, solve for the focal length f.

S O L U T I O N   To solve this formula for f, we proceed as follows:

$$\frac{1}{p} + \frac{1}{q} = \frac{1}{f}$$

$$qf + pf = pq \qquad \text{Multiply both sides by the LCD, } pqf$$

$$f(q + p) = pq \qquad \text{Factor out } f$$

$$f = \frac{pq}{q + p} \qquad \text{Divide both sides by } (q + p) \text{ with } q + p \neq 0 \qquad \blacklozenge$$

P R O B L E M 5   Solve the area of a trapezoid formula $A = \dfrac{h}{2}(b_1 + b_2)$ for the base b_1. $\blacklozenge$

Exercises 2.1

Basic Skills

In Exercises 1–32, solve each equation.

1. $2x - 7 = -21$

2. $14 = 22 + 3y$

3. $26 = 6 - 5t$

4. $12 - 6x = -18$

5. $3n + 4 = n - 4$

6. $12m + 5 = 2m - 5$

7. $8x + 4 + 3x = x - 11$

8. $7x - 3 + 2x = 4x + 3 + x$

9. $9x - 2(4x + 1) = 3$

10. $16 - (3x + 7) = 3x$

11. $(m + 1)^2 = m^2 - 3$

12. $(3y + 2)^2 - 5y = 3(3y^2 - 1)$

13. $x(x + 2)(x + 4) - (x + 2)^3 = 4$

14. $(y - 1)^3 + (y + 1)^3 = 2(y^3 + 1)$

15. $\dfrac{5x}{6} - 1 = \dfrac{x}{3}$

16. $0.4x - 6 = 0.5x - 2$

17. $\dfrac{x}{2} + \dfrac{x}{4} = \dfrac{1}{6} + \dfrac{x}{3}$

18. $\dfrac{x}{2} + \dfrac{x}{3} + \dfrac{x}{5} = 1 - \dfrac{x}{6}$

19. $y + 0.5(y + 3) = 6$

20. $4 + \dfrac{n}{6} = \dfrac{n + 4}{2}$

21. $\dfrac{t - 1}{6} = \dfrac{2t + 3}{12} - \dfrac{7t - 4}{10}$

22. $\dfrac{x}{12} - \dfrac{x - 3}{8} = \dfrac{2x - 1}{3}$

23. $\dfrac{3}{x + 2} = \dfrac{4}{x - 1}$

24. $\dfrac{4}{2x - 1} - \dfrac{4}{x + 4} = 0$

25. $\dfrac{3}{x + 2} - \dfrac{1}{x - 2} = \dfrac{x}{x^2 - 4}$

26. $\dfrac{1}{x + 1} - \dfrac{1}{x - 1} = \dfrac{2x}{x^2 - 1}$

27. $\dfrac{2x - 1}{4x + 1} - \dfrac{3x + 2}{6x + 1} = 0$

28. $\dfrac{2x + 3}{3x} + \dfrac{1}{3} = \dfrac{x - 3}{x + 3}$

29. $\dfrac{3}{x - 2} - \dfrac{1}{2 - 3x} = \dfrac{2}{3x^2 - 8x + 4}$

30. $\dfrac{2}{x^2 + 5x + 4} + \dfrac{1}{x^2 + x} = \dfrac{2}{x^2 + 4x}$

31. $\dfrac{3}{x^3 + 1} + \dfrac{x}{x^2 - x + 1} = \dfrac{1}{x + 1}$

32. $\dfrac{1}{8x^3 - 1} = \dfrac{2x}{4x^2 + 2x + 1} - \dfrac{1}{2x - 1}$

In Exercises 33–44, solve each literal equation for x.

33. $mx = nx + k$

34. $ax - b(c + x) = kx$

35. $\dfrac{a + bx}{k} = 3x + a$

36. $\dfrac{a - x}{b} = \dfrac{c - x}{a}$

37. $\dfrac{1}{m} - \dfrac{1}{x} = \dfrac{1}{n} + \dfrac{1}{x}$

38. $ax + b = \dfrac{x}{c} + \dfrac{x}{d}$

39. $(a - b)(x - c) = (b - c)x$

40. $(x + m)(x - n) = (x - m)^2$

41. $\dfrac{2x + k}{x - k} - \dfrac{x - k}{x + k} = 1$

42. $\dfrac{x}{a} - \dfrac{a}{a + b} = \dfrac{x}{a - b}$

43. $\dfrac{a + x}{a - x} = \dfrac{x^2}{a^2 - x^2}$

44. $\dfrac{x}{m + n} + \dfrac{m}{n - m} = \dfrac{x}{m^2 - n^2}$

In Exercises 45–54, solve each formula for the indicated variable.

45. Electrical current: $I = \dfrac{E}{R}$

   (a) for the voltage E (b) for the resistance R

46. Tension in an elevator cable: $T = m(g + a)$

   (a) for the mass of the elevator m
   (b) for the acceleration a

47. Thermal expansion: $L = L_0(1 + \mu\,\Delta t)$

   (a) for the original length L_0
   (b) for the temperature change Δt

48. nth term of an arithmetic sequence: $a_n = a_1 + (n - 1)d$

   (a) for the common difference d
   (b) for the number of terms n

49. Amount of money accrued: $A = P + Prt$

   (a) for the interest rate r
   (b) for the principal invested P

50. Mechanical advantage of a differential hoist:
   $$M = \dfrac{2R}{R - r}$$

   (a) for the radius of the small pulley r
   (b) for the radius of the large pulley R

51. Thrust of a spaceship's engine: $F = \dfrac{mv - mv_0}{t}$

   (a) for the burn time t
   (b) for the mass of the spaceship m

52. Sum of a finite geometric series: $S = \dfrac{a_1 - ra_n}{1 - r}$

   (a) for the first term a_1 (b) for the common ratio r

53. Specific heat of a substance: $S = \dfrac{H}{mt_2 - mt_1}$

   (a) for the original temperature t_1
   (b) for the mass of the object m

54. Deflection of a steel beam: $D = \dfrac{3LPx^2 - Px^3}{8EI}$

   (a) for the length of the beam L
   (b) for the concentrated load P

Critical Thinking

55. Is the equation $A = B$, where A and B are algebraic expressions, equivalent to the equation $A - B = 0$? to $\dfrac{A}{B} = 1$? to $A^2 = B^2$? Support each answer with an example.

56. Find a value of a such that the given pair of equations are equivalent.

   (a) $a - 4x = 18$ and $2x + 3 = 13$
   (b) $2x - 3a = 5x + 3$ and $3(x + 2) = 6(x + 3)$

 The equations in Exercises 57 and 58 develop in calculus through a procedure called implicit differentiation. *The symbol dy/dx represents the derivative of y with respect to x. Solve each equation for dy/dx.*

57. $x\dfrac{dy}{dx} + y + \dfrac{dy}{dx} = 7$

58. $2x^2y\dfrac{dy}{dx} + 2xy^2 - 2 = 9y^2\dfrac{dy}{dx} + \dfrac{dy}{dx}$

59. Which of the following equations is an identity?

   (a) $(2x - 1)^2 + 4x = (2x + 1)(2x - 1)$
   (b) $(2x - 1)(3x + 4) - (5x + 4) = 6x^2$

EXERCISE 59 *(continued)*

(c) $\dfrac{x^3 - 8}{x - 2} = x^2 + 4$

(d) $\dfrac{2x^2 - 5x + 4}{x} = 2x - 5 + \dfrac{4}{x}$

60. Given that $x = 3$, find the fallacy in the following argument:

$$x = 3$$

$$x - 3 = 0 \qquad \text{Subtract 3 from both sides}$$

$$2x - 6 = 0 \qquad \text{Multiply both sides by 2}$$

$$x^2 + 2x - 6 = x^2 \qquad \text{Add } x^2 \text{ to both sides}$$

$$x^2 + 2x - 15 = x^2 - 9 \qquad \text{Subtract 9 from both sides}$$

$$(x + 5)(x - 3) = (x + 3)(x - 3) \qquad \text{Factor each side}$$

$$x + 5 = x + 3 \qquad \text{Divide both sides by } (x - 3)$$

$$5 = 3 \qquad \text{Subtract } x \text{ from both sides}$$

 Calculator Activities

In Exercises 61–66, use a calculator to help solve each equation. Write each solution as a decimal number rounded to three significant digits.

61. $19.3x - 14.2 = 10.8(13.3 - 23.6x)$

62. $(x + 3.25)^2 - 4.24(x - 9.62) = x^2$

63. $\dfrac{x - 22.6}{18.2} = \dfrac{x - 18.2}{22.6}$

64. $57.2 + 29.1 + \dfrac{x}{57.2} = \dfrac{x}{29.1}$

65. $\dfrac{6.32x^2}{8.32 - 4.75x} + 6.32 + \dfrac{6.32x}{4.75} = 0$

66. $\dfrac{x - (8.62 \times 10^{-7})}{2.72 \times 10^{-3}} = 2.85 \times 10^{-4}$

67. The relationship between the air temperature T_a (in °F), the dew point temperature T_d (in °F), and the height h (in feet) to the base of a cumulus cloud is given by the formula

$$h = \frac{T_a - T_d}{0.0045}.$$

What is the dew point temperature when the base of the cloud is 5200 ft above the ground and the air temperature is 72 °F?

68. Federal regulations suggest that if a bank has D dollars in deposits with a reserve ratio of r, then it can make loans in the amount of L dollars according to the formula

$$L = \frac{D(1 - r)}{r + 0.25(1 - r)}.$$

What is the reserve ratio of a bank with $1,275,000 in deposits that makes loans totaling $375,000?

2.2 An Introduction to Solving Word Problems

◆ **Basic Strategy**

An essential part of our study of algebra is to become skillful in reading word problems and changing the words into algebraic equations that can be solved for the *unknowns*. Unfortunately, no fixed, ironclad procedure can be given to solve *all* word problems. Setting up each problem requires some ingenuity and a skill that comes only with a great deal of practice. The following general guidelines are offered as a basic strategy for solving word problems.

Basic Strategy for Solving Word Problems

1. Read the problem carefully and determine what the *question* is asking you to find. Ask yourself what is known and what is unknown.

2. Assign one of the *unknowns* the variable x (or any other letter you wish), and express each of the other unknowns in terms of x. When appropriate, draw a picture of the situation being described.

3. Develop an *equation* that relates the known and unknown quantities. This relationship may be given by an established formula or may require some intuition.

4. Solve this equation for x. Use this *solution* to answer the *question* the problem asked.

5. *Check* your answer by making certain it satisfies the conditions of the problem.

Note the manner in which we apply this basic strategy to help solve our first example.

EXAMPLE 1 A contractor agrees to put in a swimming pool in a given amount of time. The contract specifies that he is to receive $250 per day for each day's work during the allotted time and he will forfeit $100 for each day taken beyond that time. Find the time allowed in the contract if he received $3900 for 24 days of work but did not complete the project in the allotted time.

SOLUTION

Question: What is the time allowed in the contract to complete the project?

Unknowns: Let

x = time (in days) allowed in the contract to complete the project, $x \le 24$.

Since the contractor took 24 days to complete the work, we have

$24 - x$ = time (in days) worked beyond that in the contract to complete the project.

Equation: $\boxed{\text{Contract amount}} - \boxed{\text{Forfeited amount}} = \boxed{\text{Received amount}}$

$$250x \quad - \quad 100(24 - x) \quad = \quad 3900$$

Solution:
$$250x - 2400 + 100x = 3900$$
$$350x - 2400 = 3900$$
$$350x = 6300$$
$$x = 18$$

Therefore, the time allowed in the contract is 18 days.

Check: Contract amount: 18 days @ $250 per day = $4500
Forfeited amount: 6 days @ $100 per day = $ 600
Difference: $3900   ◆

P R O B L E M 1   Repeat Example 1 if the contractor received $4200 for 28 days of work.   ◆

The problems in the remainder of this section are classified as *percent mixture problems, geometry problems, investment problems, business problems, uniform motion problems,* and *work problems.* Although each type is labeled differently, the basic strategy for solving each problem remains the same.

◆ Percent Mixture Problems

Recall from arithmetic that *r percent* ($r\%$) is equivalent to the fraction $\dfrac{r}{100}$. An interesting type of word problem that involves percents is the *percent mixture problem.* If a mixture of b liters contains $r\%$ of a certain ingredient, then the amount a of that ingredient (in liters) in the mixture is given by

$$a = \frac{r}{100} \cdot b$$

We apply this formula in our next example.

E X A M P L E 2   In order to hatch brine shrimp, a 5% salt solution is needed. Presently, the hatchery has a 25-liter solution that is 8% salt. How many liters of water must be added to this solution to obtain the required 5% salt solution?

SOLUTION

Question: How much water must be added to obtain a 5% salt solution?
Unknowns: Let

x = the amount of water (in liters) to be added to the present
8% salt solution in order to obtain the required 5% solution.

Since the present 8% salt solution is already 25 liters, we can state that

$25 + x$ = *total amount* (in liters) in the required 5% salt solution.

The amount of salt (in liters) in the present 8% salt solution is

$$8\% \text{ of } 25 = \tfrac{8}{100}(25) = 2.$$

The amount of salt (in liters) in the required 5% solution is given by

$$5\% \text{ of } (25 + x) = \tfrac{5}{100}(25 + x).$$

Equation: Since the amount of salt in both solutions *remains the same,* we have

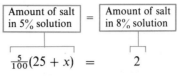

$$\frac{5}{100}(25 + x) \quad = \qquad 2$$

Solution: $\qquad$ $5(25 + x) = 200$ $\qquad$ **Multiply both sides by 100, the LCD**

$$125 + 5x = 200$$

$$5x = 75$$

$$x = 15$$

Thus, 15 liters of water should be added to the present 8% solution in order to obtain the required 5% salt solution.

Check: You can check this answer yourself.   ◆

PROBLEM 2   Referring to Example 2, how much water must be evaporated from the present 25-liter solution, which is 8% salt, in order to obtain a 10% salt solution for the brine shrimp?   ◆

◆ **Geometry Problems**

To work some geometry problems you may need to recall some basic geometric facts that you have learned in previous mathematics courses. The following facts will be useful in our discussion:

1. *Complementary angles:*   Two positive angles whose sum is 90°.
2. *Supplementary angles:*   Two positive angles whose sum is 180°.
3. *Interior angles of a triangle:*   Their sum is 180°.
4. *Pythagorean theorem:*   In a right triangle, the square of the length of the hypotenuse is the sum of the squares of the lengths of the legs.
5. *Perimeter, area, and volume formulas:*   See inside cover of text.

EXAMPLE 3   A length of rope hangs from the top of a flagpole with 6 feet coiled up on the ground at the base of the pole. When the free end of the rope is brought out 18 feet from the base of the pole, it just touches the ground. Assuming the flagpole is on level ground, what is the height of the flagpole?

SOLUTION

Question: What is the height of the flagpole?

Unknowns: Referring to Figure 2.1, let

$$x = \text{height of the flagpole (in feet).}$$

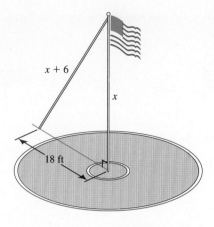

FIGURE 2.1

The quantities x, $x + 6$, and 18 are related by the Pythagorean theorem.

Since the rope is 6 feet longer than the height of the flagpole, we have

$$x + 6 = \text{length of the rope (in feet)}.$$

Equation: From the sketch in Figure 2.1, we can see that x, $x + 6$, and 18 are related by the *Pythagorean theorem.* Thus,

$$\boxed{(\text{Hypotenuse})^2} = \boxed{(\text{Leg})^2} + \boxed{(\text{Leg})^2}$$

$$(x + 6)^2 = x^2 + 18^2$$

Solution:
$$x^2 + 12x + 36 = x^2 + 324$$
$$12x = 288$$
$$x = 24$$

Therefore, the height of the pole is 24 feet.

Check: You can check this answer yourself.   ◆

PROBLEM 3   Rework Example 3 if the coil of rope is 4 feet and the free end must be brought out 20 feet before it just touches the ground.   ◆

◆ **Investment Problems**

The amount of simple interest i earned on a principal P invested at a certain rate of interest r per year over a time t in years is given by the formula

$$\boxed{i = Prt}$$

For example, the simple interest earned on $1000 invested at 9% per year for one year is

$$i = (\$1000)\left(\frac{9}{100}\right)(1) = \$90.$$

Many *investment problems* can be solved with this basic formula.

EXAMPLE 4   Norma withdrew all her money from her regular savings account, which paid 5% per year. She reinvested $5000 of the money in a term certificate that paid 7% per year and the remainder of the money in treasury bills that paid 9% per year. As a result, she earned $1000 more in interest in one year than she would have by leaving her money in the regular savings account. How much money did she invest in treasury bills?

SOLUTION

Question: How much money is invested in treasury bills?

Unknowns: Let

x = amount of money (in dollars) invested in treasury bills.

Then

$x + 5000$ = amount of money (in dollars) in the regular savings account.

Using $i = Prt$, we have

Interest from term certificate: $i = Prt = 5000(\frac{7}{100})(1) = 350.$

Interest from treasury bills: $i = Prt = x(\frac{9}{100})(1) = \dfrac{9x}{100}.$

Interest from savings account: $i = Prt = (x + 5000)(\frac{5}{100})(1) = \dfrac{5x}{100} + 250.$

Equation: Since she made $1000 more interest from the term certificate and treasury bills than from the regular savings account, we know that the *difference* in the interests must be $1000. Hence,

$$\left[\begin{array}{c}\text{Interest from}\\\text{term certificate}\end{array} + \begin{array}{c}\text{Interest from}\\\text{treasury bills}\end{array}\right] - \left[\begin{array}{c}\text{Interest from}\\\text{savings account}\end{array}\right] = \boxed{\$1000}$$

$$\left(350 + \frac{9x}{100}\right) \quad - \quad \left(\frac{5x}{100} + 250\right) = \quad 1000$$

Solution: $(35{,}000 + 9x) - (5x + 25{,}000) = 100{,}000$ **Multiply both sides by 100**

$$4x + 10{,}000 = 100{,}000$$

$$4x = 90{,}000$$

$$x = 22{,}500$$

Thus, she invested $22,500 in treasury bills.

Check:	Interest from term certificate:	$ 5,000(\frac{7}{100})(1) = $ 350
	Interest from treasury bills:	$22,500(\frac{9}{100})(1) = $2025
	Total interest:	$2375
	Interest from savings account:	$27,500(\frac{5}{100})(1) = $1375
	Difference:	$1000

PROBLEM 4   Norma invested $27,500 in two accounts. Part of the money earns 7% and part earns 9%. If the yearly return on the 7% and 9% investments is $2325, how much did she invest at 7%?   ◆

◆ **Business Problems**

The **total cost** to run a business is the sum of the **fixed cost** (expenses that do not depend on the level of production, such as mortgage payment, insurance, and utilities) and the **variable cost** (expenses that vary with the level of production, such as maintenance of equipment, cost of materials, and labor):

> Total cost = Variable costs + fixed costs

The **total revenue** that a business earns is the product of the price it charges per unit and the number of units sold:

> Total revenue = (Price per unit)(number of units sold)

The **profit** that a business makes is the difference between the total revenue and the total cost:

> Profit = Total revenue − total cost

E X A M P L E 5   The fixed cost to run a wood stove company is $1500 per month and the variable cost to produce a stove is $300 per unit. If each stove sells for $400, determine the number of units that must be sold each month for the company to make a profit of $5000 per month.

SOLUTION

Question: How many units must be sold to make a profit of $5000 per month?

Unknowns: Let

$$x = \text{number of units that must be sold.}$$

Then

$$300x = \text{variable cost (in dollars)}$$

$$300x + 1500 = \text{total cost (in dollars)}$$

Variable cost   Fixed cost

$$400x = \text{total revenue (in dollars)}$$

Price per unit   Number of units sold

Equation:

| Profit | = | Total revenue | − | Total cost |

$$5000 \quad = \quad 400x \quad - (300x + 1500)$$

Solution:

$$5000 = 100x - 1500$$
$$6500 = 100x$$
$$x = 65$$

Hence 65 units must be sold each month for a profit of $5000 per month.

Check: You can check this answer yourself.   ◆

PROBLEM 5   How many units must be sold each month by the company described in
Example 5 in order to break even (make no profit)?   ◆

◆ **Uniform Motion Problems**

Objects that move at a constant rate of speed are in *uniform motion.* The formula

$$d = rt$$

is used to find the distance *d* traveled when given the constant rate of speed *r*
over a given time *t*. Rearranging this formula into the forms

$$r = \frac{d}{t} \quad \text{and} \quad t = \frac{d}{r}$$

is often helpful when we are solving problems concerning uniform motion.

EXAMPLE 6   A college crew team practices on the Charles River in Boston. In order to
qualify for the NCAA championship trials, the team must be able to row at a
constant rate of 15 miles per hour (mph) in still water. Their coach observes
that while rowing at a constant rate of speed, they travel 3 miles downstream
in the same time it takes them to row 2 miles upstream. He knows that the
Charles River flows at a constant rate of 3 mph. Will his team qualify for the
NCAA championship trials?

SOLUTION

Question: Will the team qualify for the trials?
Unknowns: Let

$$x = \text{the crew's rate of rowing (in mph) in still water.}$$

Since the river flows at a constant rate of 3 mph, we can state that

$x + 3 =$ the crew's rate of rowing (in mph) *downstream*, and

$x - 3 =$ the crew's rate of rowing (in mph) *upstream*.

We can arrange these facts as a table:

	Rate, r	Distance, d	Time, $t = \dfrac{d}{r}$
Downstream	$x + 3$	3	$\dfrac{3}{x + 3}$
Upstream	$x - 3$	2	$\dfrac{2}{x - 3}$

Equation: Since it is given that the time to travel 3 miles downstream *is the same as* the time to travel 2 miles upstream, we have

$$\boxed{\text{Time downstream}} = \boxed{\text{Time upstream}}$$

$$\frac{3}{x + 3} = \frac{2}{x - 3}$$

Solution:

$$3(x - 3) = 2(x + 3) \qquad \text{Multiply each side by the LCD, } (x + 3)(x - 3)$$

$$3x - 9 = 2x + 6$$

$$x = 15$$

Thus, the crew's rate in still water is 15 mph. At this rate they will just qualify for the NCAA championship trials.

Check: You should check this answer yourself.   ◆

PROBLEM 6   Rework Example 6 if the speed of the river is 2 mph and the crew can row 4 miles downstream in the same time it takes them to row 3 miles upstream. ◆

◆ Work Problems

Suppose it takes t hours to complete a certain job when working at a constant rate of speed. Then $1/t$ of the job is done in 1 hour. For example, if it takes 3 hours to mow a lawn, then $1/3$ of the job is done in 1 hour. We can use this idea to solve many different types of *work problems*.

EXAMPLE 7

Two spillways from a reservoir are used to flood a cranberry bog. When the larger spillway is used alone, the bog will flood in 36 hours. If both spillways are used together, the bog will flood in 24 hours. How long would it take the smaller spillway alone to flood the cranberry bog?

SOLUTION

Question: How long does it take to flood the bog using the smaller spillway alone?

Unknowns: Let

$$x = \text{the amount of time (in hours) for the smaller spillway alone to flood the bog.}$$

Equation:

Part of job done by smaller spillway in 1 hour	+	Part of job done by larger spillway in 1 hour	=	Part of job done by both together in 1 hour
$\dfrac{1}{x}$	+	$\dfrac{1}{36}$	=	$\dfrac{1}{24}$

Solution:

$$72x\left(\frac{1}{x} + \frac{1}{36}\right) = 72x\left(\frac{1}{24}\right) \qquad \textbf{Multiply both sides by } 72x, \textbf{ the LCD}$$

$$72 + 2x = 3x$$

$$x = 72$$

Thus, it takes 72 hours to flood the cranberry bog when the smaller spillway is working alone.

Check: You can check this answer yourself.   ◆

PROBLEM 7

Suppose a smaller spillway takes 48 hours to flood the bog and a larger spillway takes 36 hours to flood the bog if each spillway operates independently. How long does it take to flood the cranberry bog if both spillways operate simultaneously?   ◆

Exercises 2.2

 Basic Skills

In Exercises 1–38, set up an equation, and then solve.

1. The sum of three consecutive integers is -147. Find the three numbers.

2. Find two consecutive, odd integers whose sum is 152.

3. Find two consecutive, even integers such that the difference of their squares is 60.

4. One number is three times another, and the sum of their reciprocals is $\frac{1}{3}$. Find the two numbers.

5. A rectangle is twice as long as it is wide. If its perimeter is 98 ft, what are its dimensions?

6. Two of the interior angles of a triangle are equal. The third interior angle is 16° less than one of the equal angles. Find the three interior angles of the triangle.

7. Find the measure of an angle if the sum of its complement and supplement is 200°.

8. Find the measure of an angle whose supplement is 20° more than twice its complement.

9. Four years ago, a father was four times as old as his son. Today, the father is only three times as old as his son. What are their present ages?

10. At present, a mother's age is four times her daughter's age. In 20 years, the mother will be twice as old as her daughter. Find their present ages.

11. The hypotenuse of a right triangle is 1 inch longer than one of its legs. The other leg is 5 inches long. What is the length of the hypotenuse?

12. The length of a rectangle is twice its width. If each side of the rectangle were increased by 3 centimeters, the new perimeter would be 48 centimeters. What are the dimensions of the original rectangle?

13. Dinner for eight people amounted to $240. The cashier was paid with an equal number of $1, $5, and $10 bills. How was this accomplished?

14. Two daughters and a son inherit one million dollars from their father's estate with the stipulation that the oldest daughter receive three times as much as the youngest daughter and half as much as the son. Find the amount each inherits.

15. At a Red Sox baseball game, box seats are $9 and grandstand seats are $6. A total of 24,320 seats is sold for a particular game, and the receipts from the box seats equal those from the grandstand seats. How many box seats are sold?

16. A developer purchased 22 acres of land for $900,000. He agreed to pay $50,000 per acre for buildable land and $10,000 per acre for marshland. How many acres of buildable land were purchased?

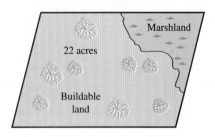

17. A woman invests $20,000, part at 8% per year and the rest at 12% per year. The interest she earns from these

two investments is the same as if she invested the $20,000 at a single interest rate of $8\frac{3}{4}\%$ per year. How much money is invested at each rate?

18. A man can paint his house in 8 days. His son could do it alone in 12 days. How long would it take to paint the house if they work together?

19. How much water must be added to one liter of a solution that is 75% alcohol in order to make the mixture 60% alcohol?

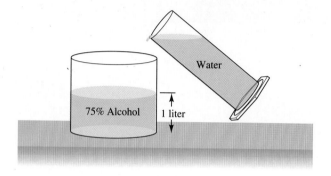

20. A textbook publishing company sells an algebra book for $28.95. The manufacturing cost of each book is $10.55 and the monthly overhead cost for the company is $5520. How many algebra books must the company sell each month to break even?

21. A furniture manufacturer produces brass beds. The variable cost per unit is $250 and the manufacturer's fixed cost is $4000 per month. If the company can produce 20 beds per month, what price should it charge for each bed in order to make a profit of $6000 per month?

22. Two cyclists start from towns 81 miles apart and travel toward each other. One travels at a constant rate of 12 miles per hour (mph) while the other travels at a constant rate of 15 mph. In how many hours will they meet?

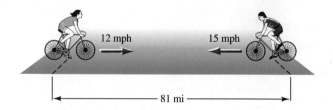

23. The IQ (intelligence quotient) of a child is determined by taking the ratio of the child's mental age to chronological age and then multiplying by 100. What is the

chronological age of a child whose IQ is 150 and mental age is 12?

24. Last year a group of airline mechanics made $12.50 per hour and worked 260 eight-hour days. This year, after being on strike, they accepted a new contract that increased their hourly wage to $13.00 per hour. How many days were the mechanics on strike this year if their annual earnings equal that of last year?

25. One batch of insecticide contains 0.010% of an active ingredient and another batch contains 0.025% of the same active ingredient. How many pounds of each batch must be used to produce 120 pounds of an insecticide that contains 0.015% of the active ingredient?

26. A professor receives an 8% pay increase plus a merit increase of $1000. His salary is now $40,312. What was his salary before these pay increases?

27. An amount of $8000 is split so that the interest from one part, which is invested at 6% per year, is double that of the other part, which is invested at 9% per year. How is this accomplished?

28. All items in a store are marked 30% off. A microwave oven is reduced to a new price of $126. What is its original price?

29. Oil is being transferred from one barge to another. The oil in one barge is 6 ft deep and is decreasing at a constant rate of 1 inch per hour. The oil in the other barge is $2\frac{1}{2}$ ft deep and is increasing at a constant rate of 3 inches per hour. When will the depth of the oil in each barge be the same, and what is this depth?

30. Two hours after a small plane left the airport a jet fighter takes off and overtakes the small plane in one half hour. The jet fighter travels 600 mph faster than the smaller plane. What is the speed of both aircraft?

31. A man jogged from his home to the coffee shop at a rate of 7 miles per hour (mph) and walked back home at the rate of 3 mph. He spent 20 minutes in the coffee shop and the total trip took 2 hours. How far is the coffee shop from his home?

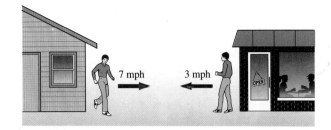

EXERCISE 31

32. A swimming pool can be filled with two hoses in 8 hours. Using the smaller hose alone requires 20 hours to fill the pool. How long would it take to fill the pool using only the larger hose?

33. An inlet pipe can fill a tank in 15 hours, while an outlet pipe can empty it in 25 hours. The tank is empty and both pipes are opened. How long will it take to fill the tank?

34. The perimeter of an isosceles triangle is 48 inches and the altitude to the nonequal side is 12 inches. What is the length of one of the equal sides?

35. A house is insured for 80% of its market value. If destroyed by fire, the owner would receive $56,000 less than the market value of the house. What is the market value of the house?

36. A student's grade in a math course depends on three hourly exams and one final exam. The final exam counts twice as much as an hourly exam. A student has scores of 72%, 76%, and 82% on the hourly exams. What score does the student need on the final exam for an 80% average?

37. A homeowner uses 1200 gallons of heating oil per year to heat his home. By installing storm windows that cost $1500, he will decrease yearly oil consumption by 25%. If he pays $0.80 per gallon for heating oil, in how many years will he recover the cost of the storm windows?

38. The denominator of a fraction exceeds its numerator by 18. The reduced value of the fraction is $\frac{4}{7}$. What is the fraction?

Critical Thinking

In Exercises 39–42, set up an equation, then solve.

39. Oak flooring is installed in a square room and a rectangular room of a new home. The square room requires 18 square feet (sq ft) less flooring than the rectangular room. The length of the rectangular room is 6 feet more than a side of the square room, and the width of the rect-

angular room is 4 feet less than a side of the square room. Find the amount of oak flooring used in each room.

40. How many pounds of cashews, worth $6.40 per pound, must be added to 12 pounds of peanuts, worth $2.40 per pound, to form a mixture that is worth $4.00 per pound?

41. A retailer buys men's suits wholesale and sells them retail at a 40% markup. During a sale she reduces the retail price by 40% and sells the suits for $168. What is the wholesale price she pays for each suit? (*Hint:* The answer is not $168.)

42. An automobile, traveling west on the Massachusetts Turnpike, is 30 miles from Boston at 9:00 A.M. and 96 miles from Boston at 10:00 A.M. Another car, traveling east on the turnpike, is 124 miles from Boston at 9:00 A.M. and 70 miles from Boston at 10:00 A.M. What is the exact time when the vehicles pass each other?

*A **proportion** is an equation that states two ratios are equal to each other. In Exercises 43–46, set up a proportion and solve by using the cross product property (Section 1.1).*

43. A man 6 ft tall casts a shadow of 4 ft. The flagpole next to the man casts a shadow of $25\frac{1}{2}$ ft. Determine the height of the flagpole.

44. A wildlife management team catches 40 red-spotted turtles, tags them, and then returns them to the lake. After one month, the team catches 26 red-spotted turtles and finds that 8 of these are tagged. Use a proportion to estimate the number of red-spotted turtles in the lake.

45. A musical chord is composed of three notes with frequencies in the ratio 4:5:6. If the first note in the chord has a frequency of 304 hertz, find the frequencies of the other two notes.

46. In a step-up transformer, the ratio of the number of turns of wire in the primary coil to the number of turns of wire in the secondary coil is 1:240. There are 14,460 turns of wire altogether. How many turns of wire are in each coil?

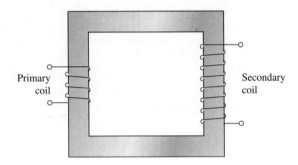

Calculator Activities

In Exercises 47–50, use your calculator to solve each problem. Record each answer to three significant digits.

47. A four-foot-wide brick walkway surrounds a circular pond as shown in the sketch. If 1600 square feet of brick were used to construct the walkway, find the radius of the pond.

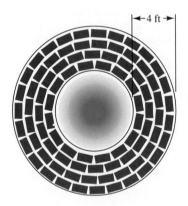

48. What is the radius of a circle when its circumference exceeds its diameter by 1 centimeter?

49. A racetrack used for horse racing has straight sides and semicircular ends. One lap is 1 mile, and the length of each straight part is twice the diameter of a semicircular end. Find the length of the straight part of the racetrack.

50. Technicians on a ship in the Persian Gulf hear the explosion of a floating mine 10 seconds sooner through the water than through the air. How far away is the explosion if the speed of sound is 1531 meters per second through seawater and 331 meters per second through air?

<table>
</table>

2.3   Quadratic Equations

Any equation of the form

$$ax^2 + bx + c = 0$$

where x is the variable and a, b, and c are real numbers with $a \neq 0$, is a **quadratic**, or **second degree equation** in **standard form**. In this section, we discuss several methods for solving quadratic equations.

◆ **The Zero Product Property**

Some quadratic equations can easily be solved by factoring. This method relies on the **zero product property**.

Zero Product Property

For real numbers p and q,

if $pq = 0$,   then either $p = 0$ or $q = 0$ (or both p and q are zero).

For example, to solve the quadratic equation $x^2 - x = 12$, we proceed as follows:

$$x^2 - x = 12$$

$$x^2 - x - 12 = 0 \qquad \text{Write in standard form}$$

$$(x - 4)(x + 3) = 0 \qquad \text{Factor the trinomial}$$

$$x - 4 = 0 \quad \text{or} \quad x + 3 = 0 \qquad \text{Apply the zero product property}$$

$$x = 4 \quad \text{or} \qquad x = -3 \qquad \text{Solve for } x$$

Thus, the roots of $x^2 - x = 12$ are 4 and -3. Checking the solutions, we have

Check I:	*Check II:*
$x^2 - x = 12$	$x^2 - x = 12$
$(4)^2 - (4) = 12 \quad ?$	$(-3)^2 - (-3) = 12 \quad ?$
$16 - 4 = 12 \quad ?$	$9 + 3 = 12 \quad ?$
$12 = 12 \quad \checkmark$	$12 = 12 \quad \checkmark$

Do not begin the factoring process until the quadratic equation has been written in standard form. To write

$$x^2 - x = 12$$

$$x(x - 1) = 12$$

$$x = 12 \quad \text{or} \quad x - 1 = 12 \text{ is WRONG!}$$

Remember, we must have *factors* on one side of the equation and *zero* on the other side of the equation in order to apply the zero product property.

EXAMPLE 1   Write each equation in the standard quadratic form $ax^2 + bx + c = 0$. Then factor and solve by using the zero product property.

(a) $(x - 3)(x - 10) = x - 10$ (b) $\dfrac{1}{n - 3} - \dfrac{6}{n^2 - 9} = 2$

SOLUTION

(a)   To solve this equation, we proceed as follows:

$$(x - 3)(x - 10) = x - 10$$

$$x^2 - 13x + 30 = x - 10 \qquad \textbf{Multiply}$$

$$x^2 - 14x + 40 = 0 \qquad \textbf{Write in standard form}$$

$$(x - 4)(x - 10) = 0 \qquad \textbf{Factor the trinomial}$$

$$x - 4 = 0 \quad \text{or} \quad x - 10 = 0 \qquad \textbf{Apply the zero product property}$$

$$x = 4 \quad \text{or} \qquad x = 10$$

Thus, the roots of this equation are 4 and 10. You should check both roots.

(b)   The LCD for $n - 3$ and $n^2 - 9$, which factors to $(n - 3)(n + 3)$, is $(n - 3)(n + 3)$. We can eliminate fractions by multiplying both sides of this fractional equation by the LCD, provided $(n - 3)(n + 3) \neq 0$. Thus,

$$\frac{1}{n - 3} - \frac{6}{n^2 - 9} = 2$$

$$(n + 3) - 6 = 2(n - 3)(n + 3) \qquad \textbf{Multiply both sides by } (n - 3)(n + 3)$$

$$n - 3 = 2n^2 - 18 \qquad \textbf{Simplify both sides}$$

$$2n^2 - n - 15 = 0 \qquad \textbf{Write in standard form}$$

$$(2n + 5)(n - 3) = 0 \qquad \textbf{Factor}$$

$$2n + 5 = 0 \quad \text{or} \quad n - 3 = 0 \qquad \text{Apply zero product property}$$

$$n = -\frac{5}{2} \quad \text{or} \qquad n = 3$$

Extraneous root, so discard.

Thus, the only root of this equation is $-\frac{5}{2}$.   ◆

Do not divide both sides of an equation by a variable expression. You may lose one or more of the roots. In Example 1(a), it is tempting to divide both sides by $(x - 10)$ and simplify:

$$(x - 3)(x - 10) = x - 10$$

$$x - 3 = 1$$

$$x = 4 \text{ is WRONG!}$$

If we take this route, we will lose the root $x = 10$.

PROBLEM 1   Repeat Example 1 for $x^2 - x - 12 = 3(x + 2)(x - 2)$.   ◆

◆ **The Square Root Property**

Consider the equation $x^2 = k$, where k is any real number. One method of solving this equation is to factor completely over the set of complex numbers and apply the zero product property as follows:

$$x^2 = k$$

$$x^2 - k = 0 \qquad \text{Write in standard form}$$

$$(x - \sqrt{k})(x + \sqrt{k}) = 0 \qquad \text{Factor as the difference of squares}$$

$$x - \sqrt{k} = 0 \quad \text{or} \quad x + \sqrt{k} = 0 \qquad \text{Apply zero product property}$$

$$x = \sqrt{k} \quad \text{or} \qquad x = -\sqrt{k}$$

Thus, the two roots of $x^2 = k$ are $\sqrt{k}$ and $-\sqrt{k}$ or, written more compactly, $\pm\sqrt{k}$. It is important to note the following situations:

1. If $k > 0$, then both roots are real numbers.
2. If $k < 0$, then both roots are pure imaginary numbers.
3. If $k = 0$, then there is one root, which is 0.

◆ **Square Root Property**

> If $x^2 = k$, where k is any real number, then $x = \pm\sqrt{k}$.

The square root property may be used to solve any quadratic equation of the form $ax^2 + c = 0$.

EXAMPLE 2 Solve each quadratic equation by using the square root property.

(a) $4x^2 + 9 = 0$ **(b)** $(n + 3)^2 = 8$

SOLUTION

(a) To solve this equation, we proceed as follows:

$$4x^2 + 9 = 0$$

$$4x^2 = -9 \qquad \text{Subtract 9 from both sides}$$

$$x^2 = -\frac{9}{4} \qquad \text{Divide both sides by 4}$$

$$x = \pm\sqrt{-\frac{9}{4}} = \pm\frac{3}{2}i \qquad \text{Apply square root property}$$

Thus, the equation has two pure imaginary roots, $\frac{3}{2}i$ and $-\frac{3}{2}i$.

(b) To solve this equation, we proceed as follows:

$$(n + 3)^2 = 8$$

$$n + 3 = \pm\sqrt{8} \qquad \text{Apply square root property}$$

$$n = -3 \pm \sqrt{8} \qquad \text{Add } -3 \text{ to both sides}$$

$$n = -3 \pm 2\sqrt{2} \qquad \text{Simplify the radical}$$

Thus, the equation has two real roots, $-3 + 2\sqrt{2}$ and $-3 - 2\sqrt{2}$. Using a calculator, we find that $-3 + 2\sqrt{2} \approx -0.172$ and $-3 - 2\sqrt{2} \approx -5.83$. ◆

PROBLEM 2 Solve the quadratic equation $(x + 2)^2 - 9 = 0$. ◆

◆ **Completing the Square**

We now develop a general technique by which all quadratic equations of the form $x^2 + bx = c$ may be solved. This method—referred to as **completing the square**—requires that we add a constant to both sides of the equation $x^2 + bx = c$ so that the left-hand side becomes a perfect square trinomial. Note that if we add the *square of half the coefficient of* x, or $(b/2)^2$, to the left-hand side of this equation, we obtain

$$x^2 + bx + \left(\frac{b}{2}\right)^2$$

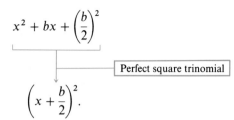

Perfect square trinomial

which factors to

$$\left(x + \frac{b}{2}\right)^2.$$

After writing the left-hand side of the equation as a perfect square, we then apply the square root property to obtain the solutions. The procedure is illustrated in the next example.

EXAMPLE 3   Solve each quadratic equation by completing the square.

(a) $x^2 + 4 = 8x$ (b) $3t^2 + 4t = 2$

SOLUTION

(a) Before we start to complete the square, we must write the equation in the form $x^2 + bx = c$. Thus, we proceed as follows:

$$x^2 + 4 = 8x$$

$$x^2 - 8x = -4 \qquad \text{Write in the form } x^2 + bx = c$$

$$x^2 - 8x + 16 = -4 + 16 \qquad \text{Complete the square by adding 16 on \textit{both} sides}$$

(Half of $-8)^2$

$$(x - 4)^2 = 12 \qquad \text{Factor the perfect square trinomial}$$

$$x - 4 = \pm\sqrt{12} \qquad \text{Apply square root property}$$

$$x = 4 \pm 2\sqrt{3} \qquad \text{Add 4 to both sides and simplify the radical}$$

Thus, the roots are $4 + 2\sqrt{3} \approx 7.46$ and $4 - 2\sqrt{3} \approx 0.536$.

(b) To write $3t^2 + 4t = 2$ in the form $x^2 + bx = c$, we divide both sides by 3, then complete the square as follows:

$$3t^2 + 4t = 2$$

$$t^2 + \frac{4}{3}t = \frac{2}{3} \qquad \text{Divide both sides by 3}$$

$$t^2 + \frac{4}{3}t + \frac{4}{9} = \frac{2}{3} + \frac{4}{9} \qquad \text{Complete the square by adding } \frac{4}{9} \text{ to both sides}$$

(Half of $\frac{4}{3})^2$

EXAMPLE 3(b) (*continued*)

$$\left(t + \frac{2}{3}\right)^2 = \frac{10}{9} \qquad \text{Factor the perfect square trinomial}$$

$$t + \frac{2}{3} = \pm\sqrt{\frac{10}{9}} \qquad \text{Apply square root property}$$

$$t = -\frac{2}{3} \pm \frac{\sqrt{10}}{3} \qquad \begin{array}{l}\text{Add } -\frac{2}{3} \text{ to both sides} \\ \text{and simplify the radical}\end{array}$$

$$t = \frac{-2 \pm \sqrt{10}}{3} \qquad \text{Combine the fractions}$$

Thus, the roots are $\dfrac{-2 + \sqrt{10}}{3} \approx 0.387$ and $\dfrac{-2 - \sqrt{10}}{3} \approx -1.72.$ ◆

PROBLEM 3   Solve the quadratic equation $4y^2 = 3y + 27$ by each method:

(a)   completing the square.

(b)   factoring and using the zero product property.   ◆

◆ ## Quadratic Formula

The method of completing the square enables us to derive a formula that can be used to solve any quadratic equation that is written in standard form. Beginning with $ax^2 + bx + c = 0$, we proceed as follows:

$$ax^2 + bx + c = 0$$

$$ax^2 + bx = -c \qquad \text{Subtract } c \text{ from both sides}$$

$$x^2 + \frac{b}{a}x = -\frac{c}{a} \qquad \text{Divide both sides by } a$$

$$x^2 + \frac{b}{a}x + \frac{b^2}{4a^2} = \frac{b^2}{4a^2} - \frac{c}{a} \qquad \begin{array}{l}\text{Complete the square by adding} \\ b^2/(4a^2) \text{ to both sides}\end{array}$$

$$\uparrow$$
$$\boxed{(\text{Half of } b/a)^2}$$

$$\left(x + \frac{b}{2a}\right)^2 = \frac{b^2 - 4ac}{4a^2} \qquad \begin{array}{l}\text{Factor the perfect square} \\ \text{trinomial and add fractions} \\ \text{on the right-hand side}\end{array}$$

$$x + \frac{b}{2a} = \pm\sqrt{\frac{b^2 - 4ac}{4a^2}} \qquad \text{Apply square root property}$$

$$x + \frac{b}{2a} = \pm\frac{\sqrt{b^2 - 4ac}}{2a} \qquad \text{Simplify the radical expression}$$

$$x = \frac{-b \pm \sqrt{b^2 - 4ac}}{2a} \qquad \text{Add } -b/(2a) \text{ to both sides}$$

This last equation represents a formula that can be used to find the roots of any quadratic equation that is written in standard form. We refer to this formula as the **quadratic formula**.

◆ Quadratic Formula

If $ax^2 + bx + c = 0\,(a \neq 0)$, then

$$x = \frac{-b \pm \sqrt{b^2 - 4ac}}{2a}.$$

In the quadratic formula, the quantity $b^2 - 4ac$ is called the **discriminant**. It indicates the *nature of the roots* of the quadratic equation $ax^2 + bx + c = 0$, when a, b, and c are real numbers. We consider three cases.

Case 1

If $b^2 - 4ac > 0$, then $\sqrt{b^2 - 4ac}$ is a *real number* and the quadratic formula gives two distinct real roots:

$$\frac{-b + \sqrt{b^2 - 4ac}}{2a} \qquad \text{and} \qquad \frac{-b - \sqrt{b^2 - 4ac}}{2a}$$

If a, b, and c are rational numbers and $b^2 - 4ac$ is the square of a rational number, then these two distinct real roots are rational numbers. Otherwise, the roots are irrational numbers.

Case 2

If $b^2 - 4ac < 0$, then $\sqrt{b^2 - 4ac}$ is a *pure imaginary number* and the quadratic formula gives two complex conjugate roots:

$$-\frac{b}{2a} + \frac{\sqrt{|b^2 - 4ac|}}{2a}\,i \qquad \text{and} \qquad -\frac{b}{2a} - \frac{\sqrt{|b^2 - 4ac|}}{2a}\,i$$

Case 3

If $b^2 - 4ac = 0$, then $\sqrt{b^2 - 4ac}$ is *zero* and the quadratic formula gives only one real root:

$$-\frac{b}{2a}$$

The quantity $-\dfrac{b}{2a}$ is called a **double root** or **repeated root of multiplicity two**.

We now summarize the nature of the roots of the quadratic equation $ax^2 + bx + c = 0$, where a, b, and c are real numbers:

Discriminant	Nature of the roots
$b^2 - 4ac > 0$	Two distinct real roots
$b^2 - 4ac < 0$	Two complex conjugate roots
$b^2 - 4ac = 0$	One real root of multiplicity two

E X A M P L E 4   Solve each quadratic equation by using the quadratic formula.

(a) $x^2 + 6x = 18$ (b) $y^2 + 5 = 2y$ (c) $2x^2 - 2\sqrt{10}\,x + 5 = 0$

S O L U T I O N

(a)   Before identifying a, b, and c, we write $x^2 + 6x = 18$ in standard form,

$$x^2 + 6x - 18 = 0.$$

Now, $a = 1$, $b = 6$, and $c = -18$. Using the quadratic formula, we have

$$x = \frac{-6 \pm \sqrt{6^2 - 4(1)(-18)}}{2(1)} = \frac{-6 \pm \sqrt{108}}{2}$$

$$= \frac{-6 \pm 6\sqrt{3}}{2} \qquad \text{Simplify the radical}$$

$$= \frac{2(-3 \pm 3\sqrt{3})}{2} \qquad \text{Factor}$$

$$= -3 \pm 3\sqrt{3} \qquad \text{Reduce}$$

Thus, the two real roots are $-3 + 3\sqrt{3} \approx 2.20$ and $-3 - 3\sqrt{3} \approx -8.20$.

(b)   We begin by writing $y^2 + 5 = 2y$ in standard form,

$$y^2 - 2y + 5 = 0.$$

Therefore, $a = 1$, $b = -2$, and $c = 5$. Using the quadratic formula, we have

$$y = \frac{-(-2) \pm \sqrt{(-2)^2 - 4(1)(5)}}{2(1)} = \frac{2 \pm \sqrt{-16}}{2}$$

$$= \frac{2 \pm 4i}{2} \qquad \text{Simplify the radical}$$

$$= \frac{2(1 \pm 2i)}{2} \qquad \text{Factor}$$

$$= 1 \pm 2i \qquad \text{Reduce}$$

Thus, the roots are the complex conjugates $1 + 2i$ and $1 - 2i$.

(c) For $2x^2 - 2\sqrt{10}\,x + 5 = 0$, we have $a = 2$, $b = -2\sqrt{10}$, and $c = 5$. Using the quadratic formula, we find

$$x = \frac{-(-2\sqrt{10}) \pm \sqrt{(-2\sqrt{10})^2 - 4(2)(5)}}{2(2)} = \frac{2\sqrt{10} \pm \sqrt{0}}{4}$$

$$= \frac{2\sqrt{10}}{4}$$

$$= \frac{\sqrt{10}}{2} \qquad \text{Reduce}$$

We refer to $\dfrac{\sqrt{10}}{2}$ as a *repeated root of multiplicity two*.   ◆

We can check the solutions to every quadratic equation, whether the roots are irrational numbers or imaginary numbers. Checking the irrational roots $-3 \pm 3\sqrt{3}$ for the quadratic equation $x^2 + 6x = 18$ in Example 4(a), we have

Check I:

$$(-3 + 3\sqrt{3})^2 + 6(-3 + 3\sqrt{3}) = 18 \quad ?$$
$$(9 - 18\sqrt{3} + 27) + (-18 + 18\sqrt{3}) = 18 \quad ?$$
$$18 = 18 \quad \checkmark$$

Check II:

$$(-3 - 3\sqrt{3})^2 + 6(-3 - 3\sqrt{3}) = 18 \quad ?$$
$$(9 + 18\sqrt{3} + 27) + (-18 - 18\sqrt{3}) = 18 \quad ?$$
$$18 = 18 \quad \checkmark$$

When checking complex conjugate roots, like those in Example 4(b), remember that $i^2 = -1$.

PROBLEM 4 Check the solutions to the quadratic equations in parts (b) and (c) of Example 4.   ◆

◆ **Application: Parking Lot Addition**

Many of the applications discussed in Section 2.2, such as the geometric problems, uniform motion problems, work problems, and so on, lead to quadratic equations.

EXAMPLE 5 An existing rectangular parking lot is 100 feet by 220 feet. Because of increased business, the store owner wants to double this parking area by adding an area of uniform width, as shown in Figure 2.2. What are the dimensions of the new parking area?

Additional parking area of uniform width

Existing parking area

100 ft

220 ft

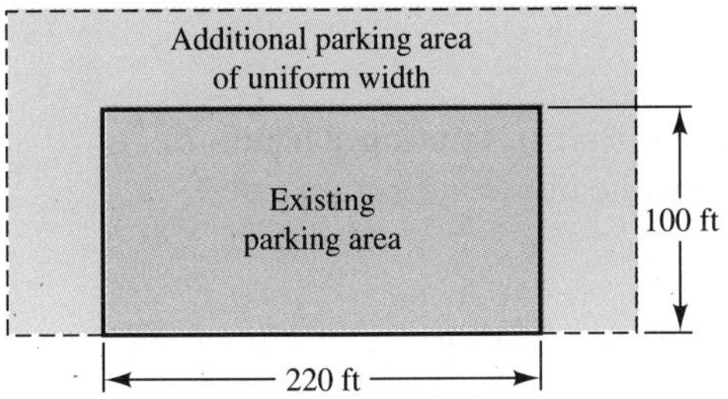

FIGURE 2.2

SOLUTION Using the basic strategy discussed in Section 2.2, we proceed as follows:

Question: What are the dimensions of the new parking area?

Unknowns: Referring to Figure 2.3, we let

$x =$ the uniform width (in feet) of the additional parking area.

FIGURE 2.3

Redrawing the sketch and labeling the
dimensions of the new parking lot.

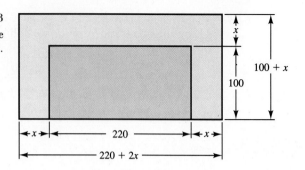

Then

$$220 + 2x = \text{the overall length (in feet) of the new parking area}$$

and

$$100 + x = \text{the overall width (in feet) of the new parking area.}$$

Equation: Since the area of a rectangle is the product of its length and width,
we have

| New length | · | New width | = | *Double* the existing area |

$$(220 + 2x) \quad \cdot \quad (100 + x) \quad = \quad 2(220)(100)$$

Solution: $22,000 + 420x + 2x^2 = 44,000$

$2x^2 + 420x - 22,000 = 0$ **Write in standard form**

$x^2 + 210x - 11,000 = 0$ **Divide both sides by 2**

Now, applying the quadratic formula, we have

$$x = \frac{-210 \pm \sqrt{(210)^2 - 4(1)(-11,000)}}{2(1)} = \frac{-210 \pm \sqrt{88,100}}{2}.$$

Using a calculator, we find the approximate roots to be

$$\frac{-210 + \sqrt{88,100}}{2} \approx 43.4 \qquad \text{and} \qquad \frac{-210 - \sqrt{88,100}}{2} \approx -253.4.$$

Since width cannot be negative, we discard the negative solution. Therefore,
the only solution is 43.4, and we can determine the dimensions of the new
parking lot to the nearest tenth of a foot:

Width: $100 + x = 100 + 43.4 = 143.4$ ft

Length: $220 + 2x = 220 + 2(43.4) = 306.8$ ft

Check: The new area is

$$(143.4 \text{ ft})(306.8 \text{ ft}) = 43{,}995.12 \text{ sq ft} \approx 44{,}000 \text{ sq ft,}$$

which is double the original area.   ◆

PROBLEM 5   Repeat Example 5 if the owner wants to double the existing area by *uniformly* increasing the length and width as shown in Figure 2.4.   ◆

FIGURE 2.4

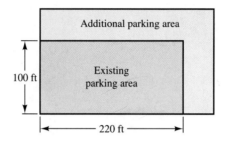

Exercises 2.3

 Basic Skills

In Exercises 1–20, write each equation in the standard quadratic form $ax^2 + bx + c = 0$. Then factor and solve by using the zero product property.

1. $x^2 + 2x = 8$
2. $x^2 - 48 = 13x$
3. $6y^2 = 12y$
4. $3t^2 = t$
5. $n(2n - 3) = 5$
6. $(x - 1)(3x + 2) = 2$
7. $(x + 4)(x + 6) = x + 6$
8. $(m - 7)(m + 4) = 2(m - 7)$
9. $y^2 + (y - 2)(y + 8) = 2y$
10. $(x + 3)^2 = x(3x - 1)$
11. $\dfrac{n - 1}{5} = \dfrac{4}{n}$
12. $\dfrac{x}{27} - \dfrac{1}{x} = \dfrac{2}{9}$
13. $\dfrac{1}{x} - \dfrac{1}{6} = \dfrac{x - 1}{x + 1}$
14. $\dfrac{1}{p} + \dfrac{3}{4} = \dfrac{2p + 3}{3p - 1}$
15. $\dfrac{5 - x}{x - 3} = \dfrac{3}{x^2 - 4x + 3}$
16. $\dfrac{7}{x^2 - 4} - \dfrac{2}{x + 2} = 1$
17. $\dfrac{x + 2}{x + 1} - \dfrac{x - 3}{x - 1} = \dfrac{x^2 + 2x - 1}{x^2 - 1}$
18. $\dfrac{x}{2x + 1} - \dfrac{3}{3 - x} = \dfrac{13}{2x^2 - 5x - 3}$
19. $\dfrac{n - 1}{n - 3} - 2 = \dfrac{1}{2} - \dfrac{n - 5}{n^2 - 5n + 6}$
20. $\dfrac{2}{m - 1} + \dfrac{1}{m^2 - 4m + 3} + 2 = \dfrac{3}{m - 3}$

In Exercises 21–30, solve each equation by using the square root property.

21. $x^2 - 36 = 0$
22. $100 - 9n^2 = 0$
23. $4p^2 - 15 = 0$
24. $3x^2 - 25 = 0$
25. $4x^2 + 49 = 0$
26. $16x^2 + 75 = 0$
27. $(4y + 3)^2 = 32$
28. $(2m - 1)^2 + 24 = 0$
29. $n^2 - 10n + 25 = -18$
30. $4t^2 + 4t + 1 = 28$

In Exercises 31–44, solve each equation by completing the square.

31. $x^2 + 12x = 12$
32. $x^2 = 10x + 15$
33. $n^2 - 4n + 8 = 0$
34. $y^2 + 14y + 58 = 0$
35. $x^2 = 3x + 9$
36. $m^2 + 5m = 5$
37. $p^2 - \frac{4}{5}p = \frac{21}{25}$
38. $x^2 + \frac{3}{2}x = \frac{7}{16}$
39. $16y^2 + 24y = 27$
40. $4x^2 - 12x = 7$
41. $3x^2 - 9 = 2x$
42. $5t^2 + 2t + 1 = 0$
43. $-x^2 + 5x - 10 = 0$
44. $-2x^2 + 7x = 9$

In Exercises 45–60, write each equation in the standard quadratic form $ax^2 + bx + c = 0$. Then solve by using the quadratic formula.

45. $x^2 = 11x + 1$

46. $m^2 - 11 = 2m$

47. $20y^2 + y = 12$

48. $18x^2 + 3x = 10$

49. $5x^2 + 1 = 4x$

50. $2x^2 + 3 = 4x$

51. $(2x - 1)(x + 4) = (x + 1)^2$

52. $y(y - 3) = (2y + 3)(y - 1)$

53. $2(x^2 + 2) = \sqrt{7}\,x$

54. $3x^2 + 2(1 + \sqrt{6}\,x) = 0$

55. $\dfrac{1}{n} - \dfrac{2n}{5} = \dfrac{1}{2n}$

56. $\dfrac{2x}{3} = x^2 + \dfrac{1}{2}$

57. $\dfrac{1}{p} - \dfrac{1}{3} = \dfrac{4}{p - 3}$

58. $\dfrac{x}{x + 1} - \dfrac{x + 1}{x} = \dfrac{1}{2}$

59. $\dfrac{x}{x^2 - 4} + \dfrac{1}{x + 2} = 1$

60. $\dfrac{2}{3t + 2} + 1 = \dfrac{t}{3t - 2} + \dfrac{8}{4 - 9t^2}$

In Exercises 61–70, set up an equation, then solve.

61. Find two integers whose sum is 32 and whose product is 252.

62. Find two consecutive, positive even integers such that the sum of their squares is 340.

63. The area of a triangle is 54 square inches and its base is 3 inches shorter than its height. Find the base and height of the triangle.

64. The perimeter of a rectangle is 84 feet and the diagonal is 30 feet. Find the length and width of the rectangle.

65. Having just purchased a coal-burning stove, a homeowner constructs a coal bin with rectangular sides to hold four tons of coal. The height of the coal bin is 4 feet and the length is 2 feet more than the width. Allowing 35 cubic feet for each ton of coal, find the dimensions of the coal bin.

66. A picture is encased in a wood frame of uniform width as shown in the sketch. The dimensions of the picture are 8 inches by 10 inches and the total area of both picture and frame is $131\tfrac{1}{4}$ square inches. Find the width of the frame.

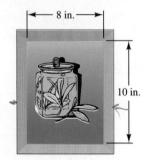

67. When the price of a product is x dollars, a manufacturer will *supply* the market with $20x - 10$ units of their product and consumers will *demand* to buy $\dfrac{280}{x}$ units. At what price will supply equal demand?

68. A company determines that when each unit that they manufacture is priced to sell at x dollars, where $x < \$10$, the number of units sold is $360 - 9x$. What is the price of each unit if the total revenue from sales is $1575?

69. Using a large and small pump together, a homeowner can remove all the water from a cellar in 24 minutes. Using only one pump, it takes 20 minutes longer to remove all the water with the smaller pump than with the larger one. Find the times required to remove this amount of water using each pump alone.

70. A college crew team can row 6 miles down a river and back again in 1 hour 15 minutes. If the river flows at 2 mph, find the crew's rate of speed in still water.

Critical Thinking

In Exercises 71–80, solve each literal equation for x.

71. $ax^2 + bx = 0$

72. $a^2x^2 - b^2 = 0$

73. $x^2 + 2ax + a^2 = b^2$

74. $a^2x^2 + b^2 = 2abx$

75. $x^2 + 2bx - a^2 = 2ab$

76. $x^2 + x = a^2 + a$

77. $\dfrac{a}{x - 1} + a = bx$

78. $\dfrac{a}{x} + x = a + \dfrac{1}{x}$

79. $x + \dfrac{1}{a} = a + \dfrac{1}{x}$

80. $\dfrac{x}{a} + \dfrac{a}{x} = \dfrac{a}{b} + \dfrac{b}{a}$

In Exercises 81–86, find the value(s) of k such that each quadratic equation has exactly one root of multiplicity two.

81. $x^2 + 16x + k = 0$ **82.** $kx^2 - 4x + 2 = 0$

83. $2x^2 - kx + 3 = 0$ **84.** $5x^2 - (\sqrt{3}\,k)x + 3 = 0$

85. $kx^2 - kx + 2 = 0$ **86.** $3x^2 - kx + k = 0$

87. An apartment complex consists of 120 units. When the rent is $600 per month, all apartments are rented. However, for each $20 increase per month in rent, three vacancies occur. What should the owners of the apartment complex charge for rent if they wish to receive $73,500 per month from rents?

88. A campus bookstore can order sweatshirts with the university emblem at $12 per sweatshirt when less than 20 shirts are ordered from the supplier. If more than 20 but less than 200 sweatshirts are ordered, the price per shirt is reduced by 3 cents times the number of shirts ordered. How many sweatshirts can the bookstore order for $900?

89. A group of college students rent a bus for a ski trip. The cost for the bus is $480, and they divide this expense evenly. However, because of sickness, four students are unable to attend. This increases the share that each person must pay by $10. How many students go on the ski trip?

90. A real estate developer bought two parcels of land for $600,000 each. The larger parcel contains 10 acres more than the smaller parcel, yet cost $10,000 per acre less. How many acres are in each parcel?

 Calculator Activities

In Exercises 91–94, use a calculator to help solve each equation. Write each solution as a decimal number rounded to three significant digits.

91. $3.81x^2 - 8.94x = 12.6$

92. $0.238x^2 + 0.342x - 1.14 = 0$

93. $872x^2 = 169x + 906$

94. $-968x^2 + 543 = 272x$

95. A rectangular flower garden, 16.6 feet by 32.2 feet, is to be surrounded by a brick walkway of uniform width, as shown in the sketch. The contractor has enough brick to cover an area of 400 square feet. How wide should the walk be if all bricks are to be used?

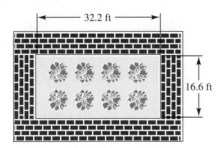

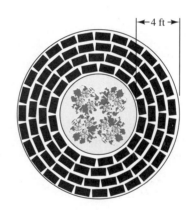

EXERCISE 96

97. Two ferries leave Woods Hole for Nantucket. One ferry travels 2.2 knots faster than the other and requires 24 minutes less time to make the trip. Find the speed (in knots) of each ferry if the distance between ports is 24 nautical miles.

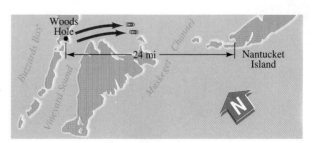

96. A four-foot walkway surrounds a circular flower garden, as shown in the sketch. The area of the walk is 44% of the area of the garden. Find the radius of the garden.

98. An investor wants to purchase a certain number of shares of stock for $2400. Her broker suggests she wait one week before doing so. At the end of the week the stock has declined $10 per share. By waiting, the investor is able to purchase 12 more shares of this stock with her $2400. How many shares of stock did she buy?

2.4   Other Types of Equations

Throughout this section, we consider several miscellaneous types of equations— *power equations*, *radical equations*, *factorable-type equations*, and *quadratic-type equations*. Since the algebraic methods we have discussed so far do not enable us to find *all* the solutions to these types of equations, we will be concerned with finding only the *real roots*. In Chapter 5, we discuss in detail the real as well as the imaginary solutions to similar types of equations.

◆ Power Equations

We refer to any equation of the form $x^{m/n} = k$, where m and n are integers ($n \neq 0$) and k is a real number, as a **power equation**. First, we consider power equations of the form $x^m = k$, where m is a positive integer. Some observations about the solutions of a power equation:

1. For $x^m = k$, if $k > 0$ and m a positive *even* integer, the equation has two real solutions, $\pm\sqrt[m]{k}$. For example, if $x^4 = 10$, then the two real solutions are $\pm\sqrt[4]{10}$.

2. For $x^m = k$, if $k < 0$ and m a positive *even* integer, the equation has no real solution. For example, $x^4 = -10$ has no real solution. No real number raised to an even power is negative.

3. For $x^m = k$, if k is any real number and m a positive *odd* integer, the equation has only one real solution, $\sqrt[m]{k}$. For example, if $x^5 = -18$, then the only real solution is $\sqrt[5]{-18} = -\sqrt[5]{18}$.

EXAMPLE 1   Find all the real solutions for each power equation.

(a) $(x - 2)^6 = 128$ (b) $x^{-3} = -64$

SOLUTION

(a) To solve this equation, we proceed as follows:

$$(x - 2)^6 = 128$$
$$x - 2 = \pm\sqrt[6]{128}$$
$$x = 2 \pm \sqrt[6]{128} = 2 \pm 2\sqrt[6]{2} \qquad \text{Simplify the radical}$$

Thus, the two real roots are $2 + 2\sqrt[6]{2} \approx 4.24$ and $2 - 2\sqrt[6]{2} \approx -0.245$.

(b) Writing $x^{-3} = -64$ in terms of a positive integer exponent, we have

$$\frac{1}{x^3} = -64 \qquad \text{or} \qquad x^3 = -\frac{1}{64},$$

provided $x \neq 0$. Therefore,

$$x = \sqrt[3]{-\dfrac{1}{64}} = -\dfrac{1}{4}.$$

Thus, the only real solution is $-\frac{1}{4}$.   ◆

PROBLEM 1   Repeat Example 1 for $(2x - 3)^4 = -16$.   ◆

To solve power equations of the form $x^{m/n} = k$, where m and n are positive integers that are relatively prime and k is a real number, it is useful to raise both sides to the nth power to obtain $x^m = k^n$. The equation $x^m = k^n$ may then be solved by the methods suggested in Example 1.

But, does raising both sides of an equation to a positive integer power generate equivalent equations? Consider the equation

$$x = k, \quad \text{where } k \text{ is a real number.}$$

Of course, the solution to this equation is k. If we raise both sides of this equation to the nth power, where n is a positive integer, we obtain

$$x^n = k^n.$$

To solve this power equation for x we must consider two possibilities:

1. If n is a positive *odd* integer,

$$x = \sqrt[n]{k^n}$$

$$x = k$$

Equivalent equations are generated.

2. If n is a positive *even* integer,

$$x = \pm\sqrt[n]{k^n}$$

$$x = \pm k$$

Equivalent equations are *not* generated.

Note that for a positive *even* integer power n, the solution of the original equation, k, appears as one of the solutions of the power equation $x^n = k^n$. However, the power equation also has an *extraneous root*, $-k$, that is *not* a solution of the original equation. Therefore, when raising both sides of an equation to a positive *even* integer power, it is essential to *check all solutions* that develop in order to make certain that no extraneous root is recorded as a solution to the original equation.

E X A M P L E   2   Find all the real solutions of each power equation.

(a) $(2x - 1)^{2/3} = 9$ **(b)** $x^{-5/2} = -\dfrac{1}{32}$

S O L U T I O N

(a)   To solve this equation, we proceed as follows:

$$(2x - 1)^{2/3} = 9$$

$$(2x - 1)^2 = 9^3 \qquad \text{\textbf{Cube both sides}}$$

$$2x - 1 = \pm\sqrt{9^3}$$

$$x = \frac{1 \pm \sqrt{9^3}}{2}$$

$$= \frac{1 \pm 27}{2}$$

This simplifies to $x = 14$ or $x = -13$. Since raising both sides of an equation to a positive *odd* integer power does generate equivalent equations, we know that both 14 and -13 are also roots of the original equation. You can check both solutions.

(b)   Writing $x^{-5/2} = -\dfrac{1}{32}$ in terms of a *positive* exponent, we have

$$\frac{1}{x^{5/2}} = -\frac{1}{32} \qquad \text{or} \qquad x^{5/2} = -32,$$

provided $x \neq 0$. Now, squaring both sides gives us

$$x^5 = (-32)^2$$

$$x = \sqrt[5]{(-32)^2} = \left(\sqrt[5]{-32}\right)^2 = (-2)^2 = 4$$

Since we raised both sides of an equation to a positive *even* integer power, we must check for extraneous roots. Replacing x with 4 in the original equation, we find

$$(4)^{-5/2} = \frac{1}{(4)^{5/2}} = \frac{1}{2^5} = \frac{1}{32} \neq -\frac{1}{32}.$$

Hence, 4 is an extraneous root, and we conclude that the original equation has no real solution.   ◆

Note: In Example 2(b), the equation $x^{5/2} = -32$ is equivalent to $(x^5)^{1/2} = -32$. The solution to this equation must be a real number whose principal square root is negative. Since the principal square root of a real number is never negative, we know that the original equation must have no real solution.

PROBLEM 2   Find all the real solutions of $(x - 3)^{3/4} = 2$.   ◆

◆ **Radical Equations**

Equations that contain the unknown in a radicand are called **radical equations**. To solve a radical equation containing one radical expression, we begin by isolating the radical expression on one side of the equation. Next, we eliminate the radical by raising both sides of the equation to the power that is equal to the index of the radical. Since raising both sides of an equation to a positive even integer power may introduce extraneous roots, be sure to check *all* solutions.

EXAMPLE 3   Solve each radical equation.

(a) $\sqrt{2x - 1} + 2 = x$ **(b)** $x - \sqrt[3]{x^3 - 4x + 8} = 0$

SOLUTION

(a) To solve this equation, we proceed as follows:

$$\sqrt{2x - 1} + 2 = x$$

$$\sqrt{2x - 1} = x - 2$$ Isolate the radical by subtracting 2 from both sides

$$(\sqrt{2x - 1})^2 = (x - 2)^2$$ Square both sides

$$2x - 1 = x^2 - 4x + 4$$ Expand and simplify

$$x^2 - 6x + 5 = 0$$ Write the quadratic equation in standard form

$$(x - 5)(x - 1) = 0$$ Solve the quadratic equation

$$x - 5 = 0 \quad \text{or} \quad x - 1 = 0$$

$$x = 5 \quad \text{or} \quad x = 1$$

Since we raised both sides of an equation to a positive *even* integer power, we must check for extraneous roots. Substituting $x = 5$ and $x = 1$ in the original equation, we find

Check I: *Check II:*

$$\sqrt{2(5) - 1} + 2 = (5) \quad ?$$ $$\sqrt{2(1) - 1} + 2 = (1) \quad ?$$

$$\sqrt{9} + 2 = 5 \quad ?$$ $$\sqrt{1} + 2 = 1 \quad ?$$

$$5 = 5 \quad \checkmark$$ $$3 \neq 1.$$

Hence, 5 is the only root of the original equation.

(b)   To solve this equation, we proceed as follows:

$$x - \sqrt[3]{x^3 - 4x + 8} = 0$$

$$x = \sqrt[3]{x^3 - 4x + 8} \qquad \text{**Isolate the radical**}$$

$$(x)^3 = \left(\sqrt[3]{x^3 - 4x + 8}\right)^3 \qquad \text{**Cube both sides**}$$

$$x^3 = x^3 - 4x + 8 \qquad \text{**Simplify**}$$

$$4x = 8 \qquad \text{**Solve for the unknown**}$$

$$x = 2$$

Since raising both sides of an equation to a positive *odd* integer power does generate equivalent equations, we know that 2 must also be a solution of the original equation. You can check this solution.   ◆

P R O B L E M   3   Solve the radical equation $\sqrt{3y} + 6 = y$.   ◆

If a radical equation contains two or more square root expressions, isolate the most complicated square root expression on one side of the equation and then square both sides. Continue this procedure until all radicals have been eliminated. Then solve the resulting equation for the unknown.

E X A M P L E   4   Solve the equation $\sqrt{9n - 2} - \sqrt{3n + 7} = 1$.

S O L U T I O N   To solve this equation, we proceed as follows:

$$\sqrt{9n - 2} - \sqrt{3n + 7} = 1$$

$$\sqrt{9n - 2} = 1 + \sqrt{3n + 7} \qquad \text{**Isolate one of the radicals**}$$

$$\left(\sqrt{9n - 2}\right)^2 = \left(1 + \sqrt{3n + 7}\right)^2 \qquad \text{**Square both sides**}$$

$$9n - 2 = 1 + 2\sqrt{3n + 7} + (3n + 7) \qquad \text{**Expand and simplify**}$$

$$6n - 10 = 2\sqrt{3n + 7} \qquad \text{**Isolate the radical**}$$

$$3n - 5 = \sqrt{3n + 7} \qquad \text{**Divide both sides by 2**}$$

$$(3n - 5)^2 = \left(\sqrt{3n + 7}\right)^2 \qquad \text{**Square both sides again**}$$

$$9n^2 - 30n + 25 = 3n + 7 \qquad \text{**Expand and simplify**}$$

$$9n^2 - 33n + 18 = 0 \qquad \text{**Write the quadratic equation in standard form**}$$

$$3n^2 - 11n + 6 = 0 \qquad\qquad \textbf{Divide both sides by 3}$$

$$(3n - 2)(n - 3) = 0 \qquad\qquad \textbf{Solve for the unknown}$$

$$3n - 2 = 0 \quad \text{or} \quad n - 3 = 0$$

$$n = \frac{2}{3} \quad \text{or} \qquad n = 3$$

Since we raised both sides of an equation to a positive *even* integer power, we must check for extraneous roots. Substituting $n = \frac{2}{3}$ and $n = 3$ in the original equation, we find

Check I: $\quad \sqrt{9(\frac{2}{3}) - 2} - \sqrt{3(\frac{2}{3}) + 7} = \sqrt{4} - \sqrt{9} = -1 \neq 1.$

Check II: $\quad \sqrt{9(3) - 2} - \sqrt{3(3) + 7} = \sqrt{25} - \sqrt{16} = 1.$ ✓

Thus, the only solution of the original equation is 3. ◆

PROBLEM 4   Solve the radical equation $\sqrt{x^2 + 7} + \sqrt{x^2 + 16} = 9.$ ◆

◆ **Factorable-Type Equations**

If one side of an equation can be factored into a product of algebraic expressions and the other side of the equation equals zero, then we may use the zero product property (Section 2.3) to help solve these **factorable-type equations**.

EXAMPLE 5   Find the real roots of each equation by factoring and applying the zero product property.

(a) $\quad 3y^4 - 6y^3 + 12y^2 = 24y$ **(b)** $\quad 4x^2(x - 3)^{1/3} + 8x(x - 3)^{4/3} = 0$

SOLUTION

(a) First, we develop a zero on one side of the equation by bringing all terms to the other side.

$$3y^4 - 6y^3 + 12y^2 - 24y = 0 \qquad \textbf{Subtract } 24y \textbf{ from both sides}$$

$$3y(y^3 - 2y^2 + 4y - 8) = 0 \qquad \textbf{Factor out } 3y$$

$$3y[y^2(y - 2) + 4(y - 2)] = 0 \qquad \textbf{Factor by grouping terms}$$

$$3y(y - 2)(y^2 + 4) = 0$$

$$3y = 0 \quad \text{or} \quad y - 2 = 0 \quad \text{or} \quad y^2 + 4 = 0 \qquad \begin{array}{l}\textbf{Apply the zero}\\\textbf{product property}\\\textbf{and solve for } y\end{array}$$

$$y = 0 \qquad\qquad y = 2 \qquad\qquad \text{No real solution}$$

Thus, the real roots are 0 and 2.

(b) This type of equation occurs frequently in calculus. We begin by removing common factors so that the remaining factor is a polynomial with integer coefficients as follows:

$$4x^2(x-3)^{1/3} + 8x(x-3)^{4/3} = 0$$

$$4x(x-3)^{1/3}[x + 2(x-3)] = 0 \qquad \text{Factor out } 4x(x-3)^{1/3}$$

$$4x(x-3)^{1/3}(3x-6) = 0 \qquad \text{Simplify the remaining factor}$$

$$4x = 0 \quad \text{or} \quad (x-3)^{1/3} = 0 \quad \text{or} \quad 3x - 6 = 0 \qquad \text{Apply the zero product property and solve for } x$$

$$x = 0 \qquad\qquad x = 3 \qquad\qquad x = 2$$

Thus, the roots are 0, 2, and 3. ◆

PROBLEM 5 Repeat Example 5 for $3t^2(t-3)^{1/2} - 4t(t-3)^{3/2} = 0$. ◆

◆ Quadratic-Type Equations

Any equation of the form $au^2 + bu + c = 0$, where a, b, and c are real numbers with $a \neq 0$ and u is an algebraic expression, is called a **quadratic-type equation**. For example,

$$x^6 - 9x^3 + 8 = 0$$

is an equation of quadratic type, where $u = x^3$ and $u^2 = x^6$. Note that the ratio of the exponents for x^6 and x^3 is 2 to 1. This 2 to 1 ratio of the exponents is a characteristic of all quadratic-type equations. To solve the equation $x^6 - 9x^3 + 8 = 0$, we use factoring and the zero product property as follows:

$$x^6 - 9x^3 + 8 = 0$$

$$(x^3 - 8)(x^3 - 1) = 0 \qquad \text{Factor}$$

$$x^3 - 8 = 0 \quad \text{or} \quad x^3 - 1 = 0 \qquad \text{Apply the zero product property and solve for } x$$

$$x^3 = 8 \qquad\qquad x^3 = 1$$

$$x = 2 \qquad\qquad x = 1$$

Thus, the real roots of the equation $x^6 - 9x^3 + 8 = 0$ are 2 and 1. You can check both solutions.

To solve an equation of quadratic type, it may be necessary to apply the quadratic formula. The procedure is illustrated in our next example.

EXAMPLE 6   Find the real roots of the equation $y - 2\sqrt{y} - 1 = 0$.

SOLUTION   Rewriting the radical expression in exponential form, we obtain

$$y - 2y^{1/2} - 1 = 0.$$

Note that the ratio of the exponents is 2 to 1. Thus, $y - 2y^{1/2} - 1 = 0$ appears to be an equation of quadratic type. By making the substitutions

$$u = y^{1/2} \qquad \text{and} \qquad u^2 = y$$

we can write this equation as

$$u^2 - 2u - 1 = 0$$

A quadratic equation in standard form

Since this equation is not factorable over the set of integers, we apply the quadratic formula as follows:

$$u = \frac{-(-2) \pm \sqrt{(-2)^2 - 4(1)(-1)}}{2(1)}$$

$$= \frac{2 \pm \sqrt{8}}{2} = 1 \pm \sqrt{2}$$

We now use the fact that $u = y^{1/2}$ and solve for y as follows:

$$u = 1 \pm \sqrt{2}$$

$$y^{1/2} = 1 \pm \sqrt{2} \qquad \text{Substitute } y^{1/2} \text{ for } u$$

$$y = (1 \pm \sqrt{2})^2 \qquad \text{Square both sides}$$

$$y = 3 \pm 2\sqrt{2} \qquad \text{Expand the binomial}$$

Since we raised both sides of an equation to a positive even integer power, we must check for extraneous roots. Using a calculator and substituting

$$y = 3 + 2\sqrt{2} \approx 5.82843 \qquad \text{and} \qquad y = 3 - 2\sqrt{2} \approx 0.17157$$

into the original equation, we find that $3 - 2\sqrt{2}$ is an extraneous root. Hence, we conclude that the only solution is $3 + 2\sqrt{2}$. ◆

PROBLEM 6   We can also think of $y - 2\sqrt{y} - 1 = 0$ as a radical equation. Solve this radical equation by isolating the square root and squaring both sides. You should obtain the same solution as in Example 6. ◆

Exercises 2.4

Basic Skills

In Exercises 1–24, find the real roots of each equation.

1. $x^4 = 81$

2. $n^6 = 128$

3. $y^3 = 54$

4. $x^5 = -32$

5. $(x-1)^4 = 48$

6. $(2h+2)^5 = -64$

7. $x^{-2} = 25$

8. $x^{-8} = -256$

9. $x^{-5} = -\frac{1}{32}$

10. $k^{-4} = \frac{81}{16}$

11. $(2t+1)^{-3} = -27$

12. $(3x-2)^{-2} = 9$

13. $n^{1/2} = 11$

14. $x^{4/3} = 16$

15. $x^{2/3} = 2$

16. $m^{3/2} = 4$

17. $x^{7/4} = -8$

18. $x^{4/5} = -16$

19. $p^{-3/2} = \frac{27}{8}$

20. $x^{-5/3} = -\frac{1}{32}$

21. $(2y)^{-5/6} = -8$

22. $(4y)^{-2/3} = 9$

23. $(x^2 - 2x - 7)^{4/3} = 16$

24. $(x^2 + 15x)^{-1/2} = \frac{1}{10}$

In Exercises 25–40, solve each radical equation. Be sure to check for extraneous roots.

25. $3\sqrt{x+1} = 4$

26. $4\sqrt{3x+1} = 9$

27. $\sqrt{n^4 + 9} - 5 = 0$

28. $\sqrt[3]{m^3 + 8} + 3 = 0$

29. $\sqrt[5]{3x^5 - 2} = x$

30. $\sqrt[4]{x^4 - x + 3} = x$

31. $\sqrt{1-t} - t = 5$

32. $\sqrt{5x-1} + 3 = x$

33. $x - \sqrt[3]{x^3 - 4x^2 - 3x - 1} = 1$

34. $1 + \sqrt[4]{x^4 - 4x^3 + 3} = x$

35. $\sqrt{2y} = \sqrt{y+1} + 1$

36. $\sqrt{t-3} = 2\sqrt{t} - 3$

37. $\sqrt{5x-4} - \sqrt{4x-7} = 1$

38. $\sqrt{x} - \sqrt{3x-2} + \sqrt{x-5} = 0$

39. $\sqrt{3x} + \sqrt{4x-3} = \sqrt{11x+3}$

40. $\sqrt{y-5} + \sqrt{2y-3} = \sqrt{2y+4}$

In Exercises 41–50, find the real solutions of each equation by factoring and applying the zero product property.

41. $y^3 - y = 0$

42. $x^4 + 27x = 0$

43. $2x^4 = 16x^2$

44. $y^3 = 8y^6$

45. $x^4 - 3x^3 - 2x^2 = 0$

46. $10x^2 + 3x = 8x^3$

47. $2x^5 - 3x^4 = x^3$

48. $5x^3 - 8x^2 = 2x$

49. $2x^3 - 3x^2 + 4x = 6$

50. $2n^6 - 6n^2 + 12n - 4n^5 = 0$

$\boxed{\frac{\Delta y}{\Delta x}}$ *Equations like those given in Exercises 51–60 develop frequently in calculus. Find the real solutions of each equation by factoring and applying the zero product property.*

51. $3x^2(x+5)^2 + 2x(x+5)^3 = 0$

52. $4x^2(x+1)^3 + 2x(x+1)^4 = 0$

53. $x^2(1-x)^{-2} + 2x(1-x)^{-1} = 0$

54. $-6x^2(x^2+5)^{-4} + (x^2+5)^{-3} = 0$

55. $6t^{4/3} + 4t^{1/3}(2t-28) = 0$

56. $2x(x-1)^{1/2} + 3(x-1)^{3/2} = 0$

57. $\frac{4}{3}x(x-2)^{-2/3} + 4(x-2)^{1/3} = 0$

58. $\frac{3}{4}y(2y+3)^{-1/4} + (2y+3)^{3/4} = 0$

59. $-\frac{1}{2}x^2(x-3)^{-3/2} + 2x(x-3)^{-1/2} = 0$

60. $-\frac{4}{5}t^3(t^2-1)^{-6/5} + 4t(t^2-1)^{-1/5} = 0$

Each equation in Exercises 61–76 is of quadratic type. Find the real solutions of each equation.

61. $y^6 - 7y^3 - 8 = 0$

62. $2y^8 - 7y^4 - 4 = 0$

63. $x^4 + 2x^2 = 5$

64. $x^6 - 4x^3 = 3$

65. $4x^{-4} - 3x^{-2} - 1 = 0$

66. $x^{-2} - x^{-1} - 6 = 0$

67. $2x^{2/3} + 7x^{1/3} = 4$

68. $x + \sqrt{x} = 12$

69. $x^{-3/2} - 16 = 6x^{-3/4}$

70. $x^{-4/3} = 3x^{-2/3} + 4$

71. $(x-1)^4 - 8(x-1)^2 - 9 = 0$

72. $4(x+1)^{-6} + 4(x+1)^{-3} = 3$

73. $2\sqrt{2x+1} + \sqrt[4]{2x+1} = 6$

74. $\sqrt[3]{x+1} = \sqrt[6]{x+1} + 12$

75. $x^2 + 2x + \sqrt{x^2 + 2x} - 5 = 15$

76. $x^2 - 3x + 2\sqrt{x^2 - 3x} + 2 = 10$

Critical Thinking

In Exercises 77–78, solve each fractional equation.

77. $2\sqrt{2x} + \sqrt{2x + 5} = \dfrac{21}{\sqrt{2x + 5}}$

78. $\dfrac{5}{\sqrt{x - 1}} = 2\sqrt{x - 1} - \sqrt{3x + 3}$

In Exercises 79–80, solve each literal equation for x.

79. $2a + \sqrt{3x^2 - 4ax + 1} = 2x$

80. $\sqrt{a - \sqrt{ax + x^2}} = \sqrt{a} - \sqrt{x}$

81. The hypotenuse of a right triangle is 2 centimeters more than its longer leg. The perimeter of the triangle is 40 cm. What is the length of the hypotenuse?

82. The area of a rectangle is 192 square inches and a diagonal of the rectangle is 20 inches. Find the length and width of the rectangle.

83. A company determines that when it produces q units of a product, its total revenue from sales is $200q^{1/2}$ dollars. Suppose the company can produce no more than 500 units and that the variable cost per unit is $4 and the fixed cost is $1600. How many units must the company produce to break even (no profit).

84. Biologists estimate that the population P of the endangered red-spotted turtle is decreasing according to the formula

$$P = 90 - 10(34t + 22)^{2/5},$$

where t is the time in years. If the population continues to decrease at this rate, in how many years will this species become extinct?

Calculator Activities

In Exercises 85–90, use a calculator to find the real roots of the equation. Write each solution as a decimal number rounded to three significant digits.

85. $y^{3/2} = 2.719$

86. $(2.25t - 4.06)^{-4} = 2.54$

87. $\sqrt{21.7 - 1.23x} = 29.6$

88. $\sqrt[3]{54.8x - 16.1x^2} = x$

89. $2.1x^4 - 5.6x^2 - 8.9 = 0$

90. $(5.9x^2 - 6.4)(2.7x + 5.8) = 5.9x^2 - 6.4$

91. A campsite is set up on the bank of a river that is 40 ft wide. On the other side of the river and 200 ft downstream is another campsite. A newly constructed footbridge and walkway join the two campsites, as shown in the sketch. The walkway, which runs along the edge of the straight river, cost $2 per foot to construct. The footbridge, which is built diagonally over the river, cost $10 per foot. If the total cost of this project was $840, find the length of the walkway.

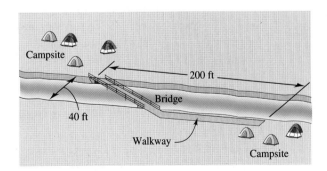

92. The circumference C of an oval football stadium may be approximated by the formula

$$C \approx 2\pi \sqrt{\frac{a^2 + b^2}{2}},$$

where a and b are the distances measured from the center of the stadium to the extreme points as shown in the sketch. Approximate the length of the stadium if its circumference is 1727 meters and its width is 351.2 meters.

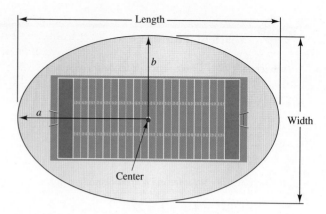

EXERCISE 92

2.5 An Introduction to Solving Inequalities

◆ **Introductory Comments**

An **inequality** is a statement declaring one algebraic expression is *less than* ($<$), *greater than* ($>$), *less than or equal to* ($\leq$), or *greater than or equal to* ($\geq$) another algebraic expression. Two examples of an inequality in one variable are

$$x + 1 > x \qquad \text{and} \qquad -2x \leq 6.$$

The inequality $x + 1 > x$ is true no matter what real number is chosen to replace x. An inequality that becomes true when the variable is replaced by *any* permissible number is called an **absolute inequality**. The inequality $-2x \leq 6$ is true only when x is replaced by real numbers greater than or equal to -3. An inequality that is true only for some values of the variable but not for others (or is never true for any value of the variable) is called a **conditional inequality**.

In this section, we introduce methods that enable us to solve an inequality. To **solve an inequality** means to find *all* values of the variable that make the inequality a true statement. We refer to these special values as the **solution set** of the inequality. Thus, we say that the solution set of the inequality $-2x \leq 6$ is all real numbers greater than or equal to -3.

◆ **Interval Notation**

Roughly speaking, any portion of the real number line that corresponds geometrically to a line segment is called a **bounded interval**, and any portion of the real number line that corresponds geometrically to a ray (or the entire real number line) is called an **unbounded interval**. In this text, we use **interval notation** to describe the solution set of an inequality. For example, the solution set of the inequality $-2x \leq 6$ may be expressed by the unbounded interval $[-3, \infty)$. The bracket preceding -3 indicates that the endpoint -3 is included in this *solution interval*. The symbol ∞, read infinity, is not a real number. It is used to indicate that the solution interval has no right-hand boundary.

Next, we list some basic inequalities along with their solution intervals and graphs on the real number line. In each case, a and b are real numbers and x is the variable. A closed dot on the graph indicates that the point is included as part of the solution set, whereas an open circle indicates that the point is not part of the solution set. We have broken bounded intervals into three classes: *open*, *closed*, and *half-open intervals*.

Inequality	Solution Interval	Graph
Unbounded intervals:		
$x > a$	(a, ∞)	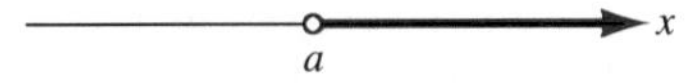
$x \geq a$	$[a, \infty)$	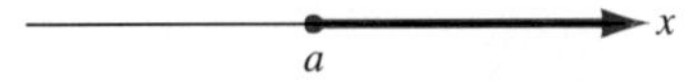
$x < a$	$(-\infty, a)$	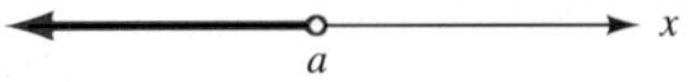
$x \leq a$	$(-\infty, a]$	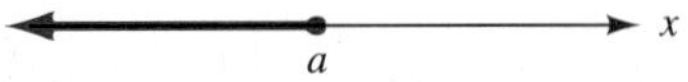
$-\infty < x < \infty$	$(-\infty, \infty)$	
Open interval:		
$a < x < b$	(a, b)	
Closed interval:		
$a \leq x \leq b$	$[a, b]$	
Half-open intervals:		
$a < x \leq b$	$(a, b]$	
$a \leq x < b$	$[a, b)$	

Note: For unbounded intervals, $-\infty$ is always preceded by a parenthesis, and ∞ is always followed by a parenthesis. Also, in interval notation the smaller of the two numbers is always written first.

EXAMPLE 1

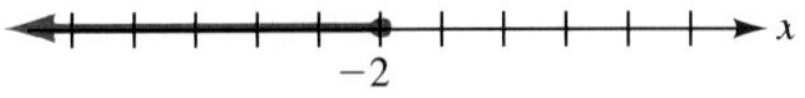

FIGURE 2.5

Graph of the solution set $(-\infty, -2]$ for the inequality $x \leq -2$.

FIGURE 2.6

Graph of the solution set $(-2, 5)$ for the inequality $-2 < t < 5$.

Using interval notation, write the solution set of each inequality, then graph the solution set.

(a) $x \leq -2$ **(b)** $-2 < t < 5$

SOLUTION

(a) The solution set for $x \leq -2$ is written $(-\infty, -2]$. The graph of the solution set is shown in Figure 2.5. The solid dot at -2 means that -2 is *included* in the solution set.

(b) The solution set for $-2 < t < 5$ is written $(-2, 5)$. The graph of the solution set is shown in Figure 2.6. The open circles at -2 and 5 mean that -2 and 5 are *excluded* from the solution set. ◆

PROBLEM 1 Repeat Example 1 for each inequality:

(a) $t < -3$ **(b)** $8 \le x < 12$ ◆

◆ **Solving Linear Inequalities**

Inequalities that have the same solution set are said to be **equivalent inequalities**. We can generate equivalent inequalities from a given inequality by using the following rules.

Rules for Generating Equivalent Inequalities

1. Add or subtract the same quantity on both sides of the given inequality.

2. Multiply or divide both sides of a given inequality by the same *positive* quantity.

3. Multiply or divide both sides of a given inequality by the same *negative* quantity and *reverse the inequality* (that is, change $<$ to $>$, $>$ to $<$, $\le$ to $\ge$, or $\ge$ to $\le$).

4. Interchange the left-hand and right-hand sides of a given inequality, and *reverse the inequality*.

5. Simplify algebraic expressions that appear on either side of the given inequality.

Caution   Although the rules for generating equivalent inequalities are similar to those for generating equivalent equations, there are two major differences:

1. When multiplying or dividing both sides of an inequality by the same *negative* number, we must *reverse the inequality*. For example, the solution set of $-2x \le 6$ is $[-3, \infty)$. Dividing both sides of $-2x \le 6$ by -2 and reversing the inequality gives $x \ge -3$, whose solution set is also $[-3, \infty)$.

2. When interchanging the left-hand and right-hand sides of an inequality, we must also reverse the inequality. For example, the solution set of $x > 3$ is $(3, \infty)$. Interchanging the sides and reversing the inequality gives us $3 < x$, whose solution set is also $(3, \infty)$.

Inequalities of the form

$$ax + b < c, \qquad ax + b \le c, \qquad ax + b > c, \quad \text{or} \quad ax + b \ge c$$

where a, b, and c are real numbers ($a \ne 0$), are referred to as **linear inequalities**. To solve a linear inequality we apply the rules for generating equivalent inequalities, and transform the given inequality into an equivalent one whose solution set is obvious. For example, to solve the linear inequality $2t - 1 \le 7$, we generate equivalent inequalities as follows:

$$2t - 1 \leq 7$$

$$2t \leq 8 \qquad \text{Add 1 to each side}$$

$$t \leq 4 \qquad \text{Divide both sides by 2}$$

Since the solution set for $t \leq 4$ is obviously $(-\infty, 4]$, and $t \leq 4$ is equivalent to $2t - 1 \leq 7$, the solution set for $2t - 1 \leq 7$ is also $(-\infty, 4]$.

EXAMPLE 2 Solve each inequality. Use interval notation to describe the solution set.

(a) $5(x + 1) < 2(4 - 3x) + 2x$ (b) $\dfrac{2x}{3} - \dfrac{6x - 1}{6} \leq -\dfrac{1}{2}$

SOLUTION

(a) To solve this inequality, we proceed as follows:

$$5(x + 1) < 2(4 - 3x) + 2x$$

$$5x + 5 < 8 - 4x \qquad \text{Simplify each side}$$

$$9x + 5 < 8 \qquad \text{Add } 4x \text{ to both sides}$$

$$9x < 3 \qquad \text{Subtract 5 from both sides}$$

$$x < \frac{1}{3} \qquad \text{Divide both sides by 9 and reduce}$$

Since the solution set of this last inequality is $(-\infty, \frac{1}{3})$, the solution set of the original inequality is also $(-\infty, \frac{1}{3})$.

(b) To solve this inequality, we begin by eliminating fractions as follows:

$$\frac{2x}{3} - \frac{6x - 1}{6} \leq -\frac{1}{2}$$

$$4x - (6x - 1) \leq -3 \qquad \text{Multiply both sides by 6, the LCD}$$

$$-2x + 1 \leq -3 \qquad \text{Simplify the left-hand side}$$

$$-2x \leq -4 \qquad \text{Subtract 1 from both sides}$$

$$x \geq 2 \qquad \text{Divide both sides by } -2 \text{ and } \textit{reverse the inequality}$$

Since the solution set of this last inequality is $[2, \infty)$, the solution set of the original inequality is also $[2, \infty)$. ◆

PROBLEM 2 Repeat Example 2 for $\dfrac{y+1}{2} - \dfrac{5(y-1)}{8} > 1$.   ◆

◆ **Solving Double Inequalities**

To solve a *double inequality*, such as

$$-3 < 2x - 1 \le 5,$$

in which the unknown appears only in the middle member, we generate equivalent double inequalities by performing operations on all *three* members as follows:

$$-3 < 2x - 1 \le 5$$

$$-2 < \quad 2x \quad \le 6 \qquad \textbf{Add 1 to each member}$$

$$-1 < \quad x \quad \le 3 \qquad \textbf{Divide each member by 2}$$

Thus, the solution set is $(-1, 3]$.

EXAMPLE 3 Solve the double inequality $-1 \le \dfrac{1-2t}{3} \le 1$. Use interval notation to describe the solution set.

SOLUTION To solve this inequality, we proceed as follows:

$$-1 \le \dfrac{1-2t}{3} \le 1$$

$$-3 \le 1 - 2t \le 3 \qquad \textbf{Multiply each member by 3}$$

$$-4 \le \quad -2t \quad \le 2 \qquad \textbf{Subtract 1 from each member}$$

$$2 \ge \quad t \quad \ge -1 \qquad \textbf{Divide each member by -2 and} \\ \textit{reverse the inequalities}$$

$$-1 \le \quad t \quad \le 2 \qquad \textbf{Interchange left-hand and right-hand members} \\ \textbf{and again reverse the inequalities}$$

Thus, the solution set is $[-1, 2]$.   ◆

PROBLEM 3 Repeat Example 3 for $5 < 3(3 - 2y) \le 6$.   ◆

To solve a double inequality in which the unknown appears in more than one member, we write the double inequality $a < b < c$ as two inequalities,

$$a < b \qquad \textit{and} \qquad b < c.$$

The solution set for such a double inequality is all real numbers common to the solution intervals of both $a < b$ and $b < c$.

EXAMPLE 4   Solve the double inequality $x - 4 < 2(x - 1) < x + 5$. Use interval notation to describe the solution set.

SOLUTION   Write the double inequality as an equivalent pair:

$$x - 4 < 2(x - 1) \qquad \text{and} \qquad 2(x - 1) < x + 5$$

Solving individually, we have

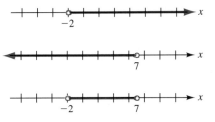

$$
\begin{array}{ll}
x - 4 < 2(x - 1) & \qquad 2(x - 1) < x + 5 \\
x - 4 < 2x - 2 & \qquad 2x - 2 < x + 5 \\
-x < 2 & \qquad x < 7 \\
x > -2 &
\end{array}
$$

Solution set: $(-2, \infty)$ *Solution set:* $(-\infty, 7)$

FIGURE 2.7
The solution set for the double inequality $x - 4 < 2(x - 1) < x + 5$ is $(-2, \infty) \cap (-\infty, 7) = (-2, 7)$.

Now, the solution set of the original inequality is all real numbers common to both solution intervals $(-2, \infty)$ *and* $(-\infty, 7)$. From the graphs in Figure 2.7, we see that the solution set is $(-2, 7)$.   ◆

Note: The intersection of two sets A and B is denoted $A \cap B$, which we read as "A intersection B." This is the set of elements that are members of both set A and set B. Referring to Example 4, the solution set can be expressed as

$$(-2, \infty) \cap (-\infty, 7) = (-2, 7).$$

PROBLEM 4   Repeat Example 4 for $2x - 3 < 7 \le 5 - x$.   ◆

◆ **Application: Telephone Calls to Beijing**

The basic strategy for solving word problems that we discussed in Section 2.2, may be extended to inequalities. We conclude this section with an applied problem in which we use a double inequality to obtain the solution.

EXAMPLE 5   During a recent pro-democracy demonstration in Beijing, Chinese students at a college in the Boston area called their families back home. Each call cost $12.50 plus $1.55 for each additional minute after the first 3 minutes. If the price of the students' calls ranged from $24.90 to $46.60, find the range of the time of the calls (in minutes).

SOLUTION   We proceed as follows:

Question: What is the time interval for the calls?
Unknowns: Let

$$x = \text{time (in minutes) of a call to Beijing, } x \ge 3.$$

Then

$$12.50 + 1.55(x - 3) = \text{the cost (in \$) of a call.}$$

Inequality: Since the price of the calls ranged from \$24.90 to \$46.60, we have

$$24.90 \leq 12.50 + 1.55(x - 3) \leq 46.60$$

Solution: $12.40 \leq 1.55(x - 3) \leq 34.10$ Subtract 12.50 from each
member

$$8 \leq \quad x - 3 \quad \leq 22$$ Divide each member by 1.55

$$11 \leq \quad x \quad \leq 25$$ Add 3 to each member

Hence, the time of the calls ranged from 11 minutes to 25 minutes.

Check: You can check the solution. ◆

P R O B L E M 5   Repeat Example 5 if the price of the calls ranged from \$31.10 to \$69.85. ◆

Exercises 2.5

Basic Skills

In Exercises 1–8, use interval notation to write the solution set of each inequality. Then graph the solution set.

1. $x > 4$

2. $m \geq -2$

3. $t \leq -5$

4. $x < 6$

5. $-4 \leq x \leq 1$

6. $-3 < t < -1$

7. $3 < n \leq 5$

8. $6 \geq x > -2$

In Exercises 9–24, solve each inequality. Use interval notation to describe the solution set.

9. $3x - 2 < 7$

10. $7n - 4 > 3$

11. $3 - 2n > 13$

12. $15 - 4x < -3$

13. $2x + 6 \geq 3x + 4$

14. $3m - 4 \leq m - 8$

15. $2(y + 3) > 3 - (4 - y)$

16. $4x - (x - 2) < x - 3$

17. $(x + 3)(x - 2) \leq (x + 4)(x - 2)$

18. $(x - 2)^2 - 2x \geq (x + 4)(x - 5) - 1$

19. $\dfrac{2p}{3} > \dfrac{p}{9} - \dfrac{5}{6}$

20. $\dfrac{x}{3} + \dfrac{x}{5} < 1 + \dfrac{11x}{15}$

21. $\dfrac{x + 5}{2} - 3 \leq \dfrac{3x}{5} - \dfrac{x - 4}{4}$

22. $\dfrac{h + 1}{2} + \dfrac{2h + 3}{4} \geq 2$

23. $\dfrac{x(3x - 1)}{3} + 4 \geq \dfrac{x(2x - 3)}{2}$

24. $2\left(\dfrac{x}{3} + 1\right) - 3\left(\dfrac{x}{2} - 1\right) < \dfrac{5}{4}$

In Exercises 25–36, solve each double inequality. Use interval notation to describe the solution set.

25. $-2 < 2h - 3 < 5$

26. $1 \leq 4 - 3x < 10$

27. $-3 < \dfrac{4 - x}{4} \leq 2$

28. $-8 \leq \dfrac{2(t - 4)}{5} \leq -1$

29. $\dfrac{1}{4} \leq \dfrac{6x - 2}{3} \leq \dfrac{5}{6}$

30. $-\dfrac{1}{6} < \dfrac{2y - 1}{-3} < \dfrac{9}{8}$

31. $n < 3 - 2n < 9$

32. $2x - 5 \leq x < 18$

33. $2(2x - 3) < x + 3 \leq 3x$

34. $2(x - 3) < 6 \leq 3x - 3$

35. $\dfrac{8y}{3} \leq \dfrac{2y + 5}{4} < 1$

36. $\dfrac{2z + 1}{5} < \dfrac{2 - z}{2} < \dfrac{9z}{10}$

In Exercises 37–46, set up an inequality, then solve.

37. Company A rents cars for $25 per day plus 30 cents for each mile driven. Company B charges $60 per day plus 10 cents for each mile driven. For what range of mileage does it cost less to rent a car from Company B than from Company A?

38. A salesperson earns a monthly salary of $800 plus 5% commission on sales. How much merchandise must this person sell in one month to make at least $2160 but not more than $2400?

39. A student has scores of 84%, 72%, and 62% on her first three math tests. What range of scores on her fourth test will give her an average greater than or equal to 70% but less than 80%?

40. The legal speed limit along a certain highway is between 45 mph and 60 mph, inclusive. If a driver travels for four hours on this highway, what is the range in miles that can be covered?

41. The perimeter of an isosceles triangle is greater than 20 inches but less than 25 inches. The base is one half as long as one of the equal sides. What is the range of possible values for the base?

42. The relationship between Fahrenheit (F) and Celsius (C)

temperatures is given by

$$C = \tfrac{5}{9}(F - 32).$$

A welder's torch heats to temperatures between 2000 °C and 2200 °C, inclusive. What is the corresponding range of Fahrenheit temperatures?

43. Find all sets of three consecutive positive odd integers such that the largest is greater than twice the smallest.

44. Find all sets of four consecutive positive even integers such that the largest is greater than twice the smallest.

45. An author requests that a publishing company pay a lump sum of $100,000 for the exclusive rights to his book. The company is willing to pay the author a lump sum of $10,000 plus a royalty of $3 for each book sold. Under what condition is the publisher's offer more economically attractive to the author than his initial request?

46. An architect's plan for a new school suggests that each student sitting in a closed classroom be allowed between 320 cubic feet and 400 cubic feet of air space. What is the allowable range for the height of the ceiling of a classroom that seats 27 students if the length and width of the room are 36 feet and 24 feet, respectively?

 Critical Thinking

In Exercises 47–50, solve the literal inequality for x subject to the given condition(s).

47. (a) $ax + b > c$ given $a > 0$
    (b) $ax + b > c$, given that $a < 0$

48. (a) $ax + b < cx + d$ given $a > c$
    (b) $ax + b < cx + d$ given $a < c$

49. (a) $\dfrac{x}{a} + \dfrac{y}{a} \geq c$ given $a > 0$
    (b) $\dfrac{x}{a} + \dfrac{y}{a} \geq c$ given $a < 0$

50. (a) $\dfrac{x}{a} + \dfrac{y}{b} \leq c$ given $a > 0, b \neq 0$
    (b) $\dfrac{x}{a} + \dfrac{y}{b} \leq c$ given $a < 0, b \neq 0$

51. Determine the interval for which each expression is undefined in the real number system.

    (a) $\sqrt{3x + 8}$ (b) $\sqrt[4]{2 - 5x}$

52. Explain what is *wrong* with each of the given interval notations.

    (a) $(0, -\infty)$ (b) $(1, \infty]$

53. Given that $x > 5$, find the fallacy in the following argument.

$x > 5$	
$5x > 25$	Multiply both sides by 5
$5x - x^2 > 25 - x^2$	Subtract x^2 from both sides
$x(5 - x) > (5 + x)(5 - x)$	Factor each side
$x > 5 + x$	Divide both sides by $5 - x$
$0 > 5$	Subtract x from both sides

54. Is the inequality $A < B$, where A and B are algebraic expressions, equivalent to the inequality $A - B < 0$? to $\dfrac{A}{B} < 1$? to $A^2 < B^2$? Explain.

Calculator Activities

In Exercises 55–60, use a calculator to solve each inequality. Describe each solution set using interval notation with decimal numbers rounded to three significant digits.

55. $-9.62x + 5.34 \geq 6.75$

56. $42.91x - 18.63 \leq 52.72x + 72.35$

57. $0.981 < 0.732x + 1.745 < 1.624$

58. $15.5 > 27.5 - 18.3x > 11.2$

59. $2.75x \leq 5.62 - 8.98x < 9.62$

60. $2.7(5.4x - 2.2) > 18.3 \geq 1.6(4.3 - 1.7x)$

61. In order to widen a highway, the state takes by eminent domain a strip of land from Lot A, which is shown in the sketch. Find the range of widths for this strip of land if Lot A must remain at least 1 acre after removing the strip.   (*Hint:* 1 acre = 43,560 square feet)

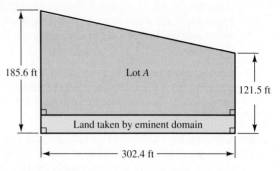

EXERCISE 61

62. For a certain gas, the relationship between pressure P (measured in pounds per square inch) and volume V (measured in cubic inches) is $PV = 804.6$. If $25.2 \leq P \leq 40.1$, what are the corresponding values of V?

2.6 Polynomial and Rational Inequalities

We now turn our attention to solving some *nonlinear* types of inequalities, such as those containing a polynomial of degree greater than 1 or a rational expression with a variable denominator. As we illustrate in this section, solving these types of inequalities involves working with the *critical values* of the inequality.

◆ **Polynomial Inequalities**

An inequality that can be written with a polynomial on the left-hand side and zero on the right-hand side is called a **polynomial inequality in standard form**. The inequality

$$x^2 - x - 6 > 0$$

is an example of a polynomial inequality in standard form. The solution set of this inequality is all real numbers x that make the trinomial $x^2 - x - 6$ a positive value. The following table shows values of the trinomial $x^2 - x - 6$ for integer values of x from -6 to 6.

x	-6	-5	-4	-3	-2	-1	0	1	2	3	4	5	6
$x^2 - x - 6$	36	24	14	6	0	-4	-6	-6	-4	0	6	14	24

We refer to -2 and 3 as the **critical values** of the inequality $x^2 - x - 6 > 0$, since these are the values of x for which $x^2 - x - 6$ is *neither positive nor nega-*

tive. Of course, -2 and 3 are also the roots of the quadratic equation

$$x^2 - x - 6 = 0,$$

and may be found by factoring and applying the zero product property as follows:

$$x^2 - x - 6 = 0$$

$$(x + 2)(x - 3) = 0$$

$$x + 2 = 0 \quad \text{or} \quad x - 3 = 0$$

$$x = -2 \quad \text{or} \quad x = 3$$

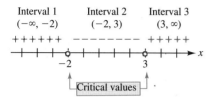

FIGURE 2.8

The sign of $x^2 - x - 6$ is *positive* on the intervals $(-\infty, -2)$ and $(3, \infty)$, and *negative* on the interval $(-2, 3)$.

Referring to the preceding table, it appears that $x^2 - x - 6$ is always *negative* for values of x between the critical values of -2 and 3. However, it appears that for values of x less than the critical value of -2 or greater than the critical value of 3, $x^2 - x - 6$ is always *positive*. As illustrated in Figure 2.8, these two critical values seem to divide the real number line into three intervals, and the sign of $x^2 - x - 6$ is constant on each of these intervals. That is, *critical values form intervals and the algebraic sign of the polynomial is constant in each interval.* Thus, referring to Figure 2.8, we state that the solution set for the polynomial inequality $x^2 - x - 6 > 0$ is all real numbers in the intervals

$$(-\infty, -2) \quad or \quad (3, \infty).$$

The **union** of two sets A and B is denoted $A \cup B$, which we read as "*A* union *B*." This is the set of elements that are either members of set A *or* members of set B, or are members of both sets A and B. Thus, more formally, we express the solution set for the polynomial inequality $x^2 - x - 6 > 0$ as

$$(-\infty, -2) \cup (3, \infty).$$

We now summarize the steps for solving a polynomial inequality. We refer to this procedure as the **critical value method**.

Critical Value Method for Solving an Inequality

1. Write the polynomial inequality in *standard form* (polynomial on the left-hand side and zero on the right-hand side).

2. Find the *critical values* by setting the polynomial equal to zero and solving for the unknown.

3. Place the critical values on a real number line and label the *intervals* into which these points divide the line.

4. In each interval, select an arbitrary test number. Use these test numbers to determine the *algebraic sign* of the polynomial throughout each interval.

5. Observe the algebraic signs in Step 4 to find the *solution set* for the polynomial inequality in standard form.

EXAMPLE 1   Use the critical value method to solve each polynomial inequality.

(a) $2x^2 + 3x < 20$ **(b)** $x^3 \geq 2x^2 + 9x$

SOLUTION

(a)   We proceed as follows:

Standard form: $2x^2 + 3x - 20 < 0$

Critical values: $2x^2 + 3x - 20 = 0$

$$(2x - 5)(x + 4) = 0$$

$$2x - 5 = 0 \quad \text{or} \quad x + 4 = 0$$

$$x = \tfrac{5}{2} \qquad\qquad x = -4$$

Hence, the two critical values are -4 and $\tfrac{5}{2}$.

Intervals: The two critical values divide the real number line into three intervals as shown:

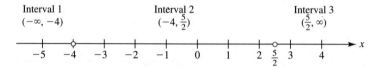

```
Interval 1                 Interval 2                 Interval 3
(-∞, -4)                   (-4, 5/2)                  (5/2, ∞)
```

Algebraic signs: Choose a test number in each interval and determine the algebraic sign of $2x^2 + 3x - 20$. For convenience, we use the *factored form* of this polynomial.

Interval	Test Number	Sign of $(2x - 5)(x + 4)$	
$(-\infty, -4)$	-5	$[2(-5) - 5][(-5) + 4] = (-)(-) = +$	(> 0)
$(-4, \tfrac{5}{2})$	0	$[2(0) - 5][(0) + 4] = (-)(+) = -$	(< 0)
$(\tfrac{5}{2}, \infty)$	3	$[2(3) - 5][(3) + 4] = (+)(+) = +$	(> 0)

Solution set: Since $2x^2 + 3x - 20 < 0$ in the interval $(-4, \tfrac{5}{2})$, the solution set for $2x^2 + 3x < 20$ is $(-4, \tfrac{5}{2})$.

(b)   We proceed as follows:

Standard form: $x^3 - 2x^2 - 9x \geq 0$

Critical values: $x^3 - 2x^2 - 9x = 0$

$$x(x^2 - 2x - 9) = 0$$

$$x = 0 \quad \text{or} \quad x^2 - 2x - 9 = 0$$

To find x, apply the quadratic formula.

$$x = \frac{-(-2) \pm \sqrt{(-2)^2 - 4(1)(-9)}}{2(1)} = 1 \pm \sqrt{10}$$

Hence, there are three critical values: 0, $1 + \sqrt{10} \approx 4.2$, and $1 - \sqrt{10} \approx -2.2$.

Intervals: These three critical values form four intervals, as shown:

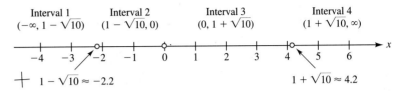

Algebraic signs: We choose a test number in each interval and determine the algebraic sign of the polynomial:

Interval	Test Number	Sign of $x^3 - 2x^2 - 9x$
$(-\infty, 1 - \sqrt{10})$	-3	$(-3)^3 - 2(-3)^2 - 9(-3) = -18$ (<0)
$(1 - \sqrt{10}, 0)$	-1	$(-1)^3 - 2(-1)^2 - 9(-1) = +6$ (>0)
$(0, 1 + \sqrt{10})$	1	$(1)^3 - 2(1)^2 - 9(1) = -10$ (<0)
$(1 + \sqrt{10}, \infty)$	5	$(5)^3 - 2(5)^2 - 9(5) = +30$ (>0)

Solution set: Since $x^3 - 2x^2 - 9x > 0$ in the interval $\left(1 - \sqrt{10}, 0\right)$ *or* the interval $\left(1 + \sqrt{10}, \infty\right)$, and $x^3 - 2x^2 - 9x = 0$ for $x = 0$ and $1 \pm \sqrt{10}$, we conclude that the solution set for $x^3 \geq 2x^2 + 9x$ is

$$[1 - \sqrt{10}, 0] \cup [1 + \sqrt{10}, \infty).\qquad\qquad\blacklozenge$$

PROBLEM 1   Repeat Example 1 for $x^5 < 16x$. $\blacklozenge$

◆ **Rational Inequalities**

An inequality that can be written with a rational expression on the left-hand side and zero on the right-hand side is called a **rational inequality in standard form**. The rational inequality

$$\frac{5x - 2}{x - 1} \leq 4$$

can be written in standard form by proceeding as follows:

$$\frac{5x - 2}{x - 1} - 4 \leq 0 \qquad \text{Subtract 4 from both sides}$$

$$\frac{(5x - 2) - 4(x - 1)}{x - 1} \leq 0 \qquad \text{Write as a single fraction}$$

$$\frac{x + 2}{x - 1} \leq 0 \qquad \text{Simplify}$$

$$\boxed{\text{Standard form}}$$

When writing a rational inequality in standard form, we cannot multiply both sides of the inequality by an expression that contains the unknown. Since we don't know whether the unknown quantity is positive or negative, we don't know whether the direction of the inequality is changed. For example, to eliminate the fraction by writing

$$\frac{5x - 2}{x - 1} \le 4 \qquad \text{as} \qquad 5x - 2 \le 4(x - 1) \text{ is WRONG!}$$

A rational expression changes sign only when its numerator or denominator changes its algebraic sign. Thus, to solve a rational inequality in standard form, we begin by finding the values of the unknown that make its numerator or denominator zero. Again, we refer to these values as critical values. Once we have determined the critical values, we can find the solution set for the rational inequality by using a procedure similar to the one we outlined for solving polynomial inequalities.

Note: Remember that a rational expression is undefined when its denominator is zero. Therefore, the solution set of a rational inequality cannot include any critical value that yields a denominator of zero.

E X A M P L E 2 Solve each rational inequality.

(a) $\dfrac{5x - 2}{x - 1} \le 4$ **(b)** $\dfrac{2x + 3}{x - 3} > \dfrac{x + 1}{x - 1}$

S O L U T I O N

(a) *Standard form:* As we previously found the rational inequality $\dfrac{5x - 2}{x - 1} \le 4$ is equivalent to

$$\frac{x + 2}{x - 1} \le 0.$$

Critical values: We find the critical values of a rational inequality in standard form by setting its numerator and denominator equal to zero.

Critical value of numerator:	*Critical value of denominator:*
$x + 2 = 0$	$x - 1 = 0$
$x = -2$	$x = 1$

Intervals: These two critical values divide the real number line into three intervals as shown:

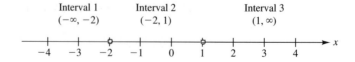

Algebraic signs: We choose a test number in each interval and check the algebraic sign of the rational expression.

Interval	Test Number	Sign of $\dfrac{x+2}{x-1}$
$(-\infty, -2)$	-3	$\dfrac{(-3)+2}{(-3)-1} = \dfrac{(-)}{(-)} = +$ (>0)
$(-2, 1)$	0	$\dfrac{(0)+2}{(0)-1} = \dfrac{(+)}{(-)} = -$ (<0)
$(1, \infty)$	2	$\dfrac{(2)+2}{(2)-1} = \dfrac{(+)}{(+)} = +$ (>0)

Solution set: Since $\dfrac{x+2}{x-1} < 0$ in the interval $(-2, 1)$ and $\dfrac{x+2}{x-1} = 0$ for $x = -2$, the solution set of $\dfrac{x+2}{x-1} \leq 0$ is $[-2, 1)$. Thus, the solution set for $\dfrac{5x-2}{x-1} \leq 4$ is also $[-2, 1)$.

(b) *Standard form:*

$$\frac{2x+3}{x-3} > \frac{x+1}{x-1}$$

$$\frac{2x+3}{x-3} - \frac{x+1}{x-1} > 0 \qquad \text{Subtract } \frac{x+1}{x-1} \text{ from both sides}$$

$$\frac{(2x+3)(x-1) - (x+1)(x-3)}{(x-3)(x-1)} > 0 \qquad \text{Subtract fractions}$$

$$\frac{x^2 + 3x}{(x-3)(x-1)} > 0 \qquad \text{Simplify the numerator}$$

Critical values of numerator: *Critical values of denominator:*

$$x^2 + 3x = 0 \qquad\qquad\qquad (x-3)(x-1) = 0$$

$$x(x+3) = 0 \qquad\qquad\quad x-3 = 0 \quad \text{or} \quad x-1 = 0$$

$$x = 0 \quad \text{or} \quad x+3 = 0 \qquad\quad x = 3 \qquad\qquad x = 1$$

$$x = -3$$

Thus, the four critical values are -3, 0, 1, and 3.

Intervals: These four critical values divide the real number line into five intervals as shown.

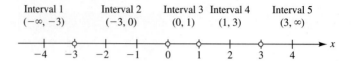

Algebraic signs: For convenience, we use the factored form of the rational expression $\dfrac{x^2 + 3x}{(x - 3)(x - 1)}$.

Interval	Test Number	Sign of $\dfrac{x(x + 3)}{(x - 3)(x - 1)}$	
$(-\infty, -3)$	-4	$\dfrac{(-4)[(-4) + 3]}{[(-4) - 3][(-4) - 1]} = \dfrac{(-)(-)}{(-)(-)} = +$	(>0)
$(-3, 0)$	-1	$\dfrac{(-1)[(-1) + 3]}{[(-1) - 3][(-1) - 1]} = \dfrac{(-)(+)}{(-)(-)} = -$	(<0)
$(0, 1)$	$\tfrac{1}{2}$	$\dfrac{(1/2)[(1/2) + 3]}{[(1/2) - 3][(1/2) - 1]} = \dfrac{(+)(+)}{(-)(-)} = +$	(>0)
$(1, 3)$	2	$\dfrac{(2)[(2) + 3]}{[(2) - 3][(2) - 1]} = \dfrac{(+)(+)}{(-)(+)} = -$	(<0)
$(3, \infty)$	4	$\dfrac{(4)[(4) + 3]}{[(4) - 3][(4) - 1]} = \dfrac{(+)(+)}{(+)(+)} = +$	(>0)

Solution set: Since $\dfrac{x^2 + 3x}{(x - 3)(x - 1)} > 0$ in the intervals $(-\infty, -3)$ *or* $(0, 1)$ *or* $(3, \infty)$, the solution set for $\dfrac{2x + 3}{x - 3} > \dfrac{x + 1}{x - 1}$ is

$$(-\infty, -3) \cup (0, 1) \cup (3, \infty).$$ ◆

P R O B L E M   2   Repeat Example 2 for $\dfrac{x^2 + 2x - 4}{x + 2} \geq 1$. ◆

◆ **Application: Hitting a Pop Fly**

Nearly four hundred years ago, Galileo discovered the laws of motion for freely falling objects. He discovered that when a projectile is fired straight upward from the ground with an initial velocity v (in feet per second), its distance d (in feet) above the ground after time t (in seconds) is given by

$$d = vt - 16t^2$$

EXAMPLE 3   A baseball player hits a pop fly straight upward with an initial velocity of 64 feet per second (ft/s). What is the time interval when the baseball is more than 48 ft above the ground?

SOLUTION   Since $v = 64$ ft/s, the distance d (in feet) above the ground after time t (in seconds) is given by

$$d = 64t - 16t^2.$$

Now, to determine the time when the baseball is more than 48 ft above the ground, we must solve the polynomial inequality

$$64t - 16t^2 > 48$$

$$-16t^2 + 64t - 48 > 0 \qquad \text{Write in standard form}$$

$$t^2 - 4t + 3 < 0 \qquad \begin{array}{l}\text{Divide both sides by } -16 \\ \text{and } \textit{reverse the inequality}\end{array}$$

$$(t - 3)(t - 1) < 0 \qquad \text{Factor}$$

For $(t - 3)(t - 1) < 0$, we have

Critical values: 1, 3

Algebraic signs:

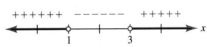

Solution set: (1, 3).

Thus, the baseball is more than 48 ft above the ground during the time interval from 1 to 3 seconds.   ◆

PROBLEM 3   Referring to Example 3, how long does it take the baseball to hit the ground?
◆

Exercises 2.6

Basic Skills

In Exercises 1–20, solve each polynomial inequality. Write the solution set using interval notation.

1. $x^2 - 9 < 0$

2. $p^2 > 25$

3. $y^2 \geq 12y$

4. $x^2 + 20x \leq 0$

5. $x^2 - 8x \leq 9$

6. $y(y + 10) \geq 24$

7. $4x(x + 3) > 7$

8. $4x^2 + 9 < 20x$

9. $n^2 < 11n + 1$

10. $x^2 + 3x > 8 - 3x$

11. $(x - 1)^2 \geq (2x + 3)^2$

12. $(4n - 1)(2n + 5) \geq (3n + 1)^2$

13. $x^3 \leq 4x^2 + 8x$

14. $2x^4 + 3x^3 > 2x^2$

15. $t^4 + 3t^3 > 2t^2 + 6t$

16. $2h^3 + 6h \leq 3h^2 + 9$

17. $m^4 + 4m^2 < 45$

18. $x^6 \geq x^3 + 2$

19. $2x^2(3x - 2)^3 + (3x - 2)^4 \geq 0$

20. $t(2t + 1)^2(t - 3)^3(t + 2) < 0$

In Exercises 21–36, solve each rational inequality. Write the solution set using interval notation.

21. $\dfrac{1}{x} > 2$

22. $\dfrac{2}{3x} < 1$

23. $\dfrac{x+3}{x-1} < 0$

24. $\dfrac{3-2x}{5x} > 0$

25. $\dfrac{x-4}{2x+1} \geq 3$

26. $\dfrac{3x+2}{x-1} \leq 1$

27. $\dfrac{x^2+3x-4}{x-2} \leq 0$

28. $\dfrac{x^2+5x-14}{x^2+5x-24} \geq 0$

29. $\dfrac{x^2+6x+2}{x^2-25} < 0$

30. $\dfrac{3x-4}{x^2+4x+1} < 0$

31. $\dfrac{1}{x} > \dfrac{2x+1}{x+8}$

32. $\dfrac{2x-3}{x+4} < \dfrac{x-1}{x+3}$

33. $\dfrac{1}{x-2} + \dfrac{x}{x+3} < \dfrac{1}{2}$

34. $x + \dfrac{x-3}{2x+5} < 3$

35. $\dfrac{7}{x^2+7x+12} \geq \dfrac{x-3}{x+4}$

36. $\dfrac{x}{x^2+2x-15} \geq \dfrac{2}{x^2-25}$

In Exercises 37–44 set up an inequality, then solve.

37. The length of a rectangle is 2 meters more than its width, and its area is less than 99 square meters. What is the range of values for the length of the rectangle?

38. One leg of a right triangle is 3 inches longer than the other leg. If the hypotenuse is less than $\sqrt{39}$ inches, what is the range of values for the shorter leg?

39. Find all sets of three consecutive even integers such that the square of the largest integer is greater than the sum of the squares of the other two integers.

40. Find all sets of two consecutive integers such that the square of the smaller integer is less than four times the larger integer.

41. When the price of a product is p dollars, a manufacturer will *supply* the market with $4p^2 - 5p$ units of the product and consumers will *demand* to buy $30 - p^2$ units. Under what conditions is supply greater than demand?

42. In an ant colony, the relationship between the number of adult ants, A, and the number of larvae, L, that survive to become adults is given by

$$L = \dfrac{6000A}{A+1200}$$

Under what conditions is the number of larvae that survive to become adults less than the number of adult ants in the colony?

43. If a rocket is fired straight upward from the ground with an initial velocity of 320 ft/s, what is the time interval when the rocket is more than 1200 ft above the ground? (See Example 3.)

44. If an arrow is shot straight upward from the ground with an initial velocity of 96 ft/s, what are the time intervals when the arrow is less than 80 ft above the ground? (See Example 3.)

    Critical Thinking

In Exercises 45–48, solve each double inequality. Use interval notation to describe the solution set.

45. $6 < x^2 - x < 20$

46. $10 - x < x^2 - 4x \leq 2x + 16$

47. $-3 \leq \dfrac{2x-1}{x+2} \leq 7$

48. $\dfrac{1}{x} \leq \dfrac{2-6x}{x-5} < 10$

49. Find the range of values for x if the area of the region shaded in the sketch is from 42 to 127, inclusive.

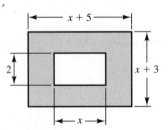

EXERCISE 49

50. Find the range of values for x if the area of the region shaded in the sketch is from 21π to 33π, inclusive.

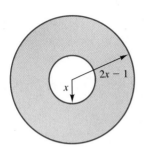

In Exercises 51–54, determine the interval(s) for which each expression is undefined in the real number system.

51. $\sqrt{x^2 - 2x - 8}$ 52. $\sqrt[4]{x^3 + 7x^2 + 10x}$

53. $\sqrt{\dfrac{2x + 3}{x - 1}}$ 54. $\sqrt{\dfrac{x^2 - 4}{x + 1}}$

In Exercises 55–58, determine the values of k for which each quadratic equation has real roots.

55. $x^2 + kx + 6 = 0$ 56. $2x^2 + kx + 3 = 0$

57. $x^2 + kx + k = 0$ 58. $kx^2 + x + k = 0$

 Calculator Activities

In Exercises 59–64, use a calculator to solve each inequality. Use interval notation to describe the solution set with decimal numbers rounded to three significant digits.

59. $27.93x^2 - 92.24 > 0$ 60. $2.6x^2 - 5.4x \le 9.2$

61. $42.6x^4 \le 18.3x^2 + 92.4$ 62. $\dfrac{x + 3.74}{x - 9.24} < 8.25$

63. $\dfrac{39.4x}{14.5x + 12.8} < 10.0x$ 64. $\dfrac{1.9x^2 - 8.7x + 2.6}{5.5x + 3.5} > 0$

65. If two resistances, R_1 and R_2, are connected in parallel in an electric circuit, then the total effective resistance R can be found by using the formula

$$R = \frac{R_1 R_2}{R_1 + R_2}.$$

If R_1 is 10.6 ohms (Ω), what is the range of values for R_2 if the total effective resistance is positive but not more than 8.2 Ω?

66. A highway safety group recommends the formula

$$d = 0.045v^2 + v$$

for determining the minimum safe distance d (in feet) that should be maintained between two cars traveling along a highway at a speed of v miles per hour (mph). If the distance between two cars does not exceed 200 ft, find the interval of safe speeds.

2.7 Equations and Inequalities Involving Absolute Value

◆ **Introductory Comments**

Recall from Section 1.1, the *absolute value of a*, or $|a|$, is the distance on the real number line between zero and the number a, without regard to direction. More formally,

$$|a| = \begin{cases} a & \text{if } a \ge 0 \\ -a & \text{if } a < 0. \end{cases}$$

In this section we solve equations and inequalities involving absolute value. The key to solving these types of equations and inequalities is to develop ways of *removing the absolute value signs*.

◆ Absolute Value in Equations

Consider the equation $|x| = k$, where k is a positive real number. By the definition of absolute value, we have

$$|x| = x \quad \text{if } x \geq 0, \qquad |x| = -x \quad \text{if } x < 0.$$

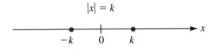

Thus, solving the equation $|x| = k$, where k is a positive real number, is equivalent to solving the pair of equations $x = k$ and $-x = k$ individually. Hence, the equation $|x| = k$, where k is a positive real number, is equivalent to $x = \pm k$. Geometrically, $\pm k$ are the coordinates of two points on the real number line that are exactly k units from 0 (see Figure 2.9). We can extend this idea to equations of the form $|u| = k$, where u is an algebraic expression.

FIGURE   2.9
Graph of $|x| = k$, which is equivalent to $x = \pm k$.

 Absolute Value in Equations

If u is an algebraic expression, and

1. If k is a *positive* real number, then the equation $|u| = k$ is equivalent to $u = \pm k$.

2. If k is a *negative* real number, then the equation $|u| = k$ has no solution.

3. If k is *zero*, then the equation $|u| = k$ is equivalent to $u = 0$.

E X A M P L E   1   Solve each equation.

(a) $|4x - 3| = 5$ **(b)** $|x^2 - 6x| = 12 - 7x$

S O L U T I O N

(a)   Since $5 > 0$, solving $|4x - 3| = 5$ is equivalent to solving the equations $4x - 3 = \pm 5$. Thus,

$$
\begin{array}{ccc}
4x - 3 = 5 & \text{or} & 4x - 3 = -5 \\
4x = 8 & & 4x = -2 \\
x = 2 & & x = -\frac{1}{2}
\end{array}
$$

Thus, the solutions of the original equation are 2 and $-\frac{1}{2}$. You can check both solutions.

(b)   For $|x^2 - 6x| = 12 - 7x$, we don't know whether $12 - 7x > 0$. We proceed as in part (a), but we *must* check our solutions to be certain that

$12 - 7x$ is not a negative quantity. Thus,

$$x^2 - 6x = 12 - 7x \qquad \text{or} \qquad x^2 - 6x = -(12 - 7x)$$
$$x^2 + x - 12 = 0 \qquad\qquad\qquad x^2 - 13x + 12 = 0$$
$$(x + 4)(x - 3) = 0 \qquad\qquad\qquad (x - 12)(x - 1) = 0$$
$$x = -4 \ \text{ or } \ x = 3 \qquad\qquad\qquad x = 12 \ \text{ or } \ x = 1$$

Checking these solutions, we find

Check I: $|(-4)^2 - 6(-4)| = 12 - 7(-4)$
$$|40| = 40$$
$$40 = 40 \ \checkmark$$

Check II: $|(3)^2 - 6(3)| = 12 - 7(3)$
$$|-9| = -9$$
$$9 \neq -9$$

Check III: $|(12)^2 - 6(12)| = 12 - 7(12)$
$$|72| = -72$$
$$72 \neq -72$$

Check IV: $|(1)^2 - 6(1)| = 12 - 7(1)$
$$|-5| = 5$$
$$5 = 5 \ \checkmark$$

Thus, the only solutions are -4 and 1.   ◆

PROBLEM 1   Solve $|1 - 2x| = 3x + 6$.   ◆

◆ **Properties of Absolute Value**

Next, we list four basic **properties of absolute value** for products, quotients, sums, and differences.

Properties of Absolute Value

For all real numbers a and b,

1. $|ab| = |a||b|$

2. $\left|\dfrac{a}{b}\right| = \dfrac{|a|}{|b|}$ $(b \neq 0)$

3. $|a + b| \leq |a| + |b|$ *(the triangle inequality)*

4. $|a - b| \geq |a| - |b|$

Some illustrations of these properties, with $a = -8$ and $b = 4$:

1. $\quad |-8 \cdot 4| = |-8||4|$

$\qquad |-32| = 8 \cdot 4$

$\qquad\quad 32 = 32$

2. $\quad \left|\dfrac{-8}{4}\right| = \dfrac{|-8|}{|4|}$

$\qquad |-2| = \dfrac{8}{4}$

$\qquad\quad 2 = 2$

3. $\quad |-8+4| \le |-8| + |4|$

$\qquad |-4| \le 8 + 4$

$\qquad\quad 4 \le 12$

4. $\quad |-8-4| \ge |-8| - |4|$

$\qquad |-12| \ge 8 - 4$

$\qquad\quad 12 \ge 4$

These properties may be used to help solve certain equations and inequalities. For instance, consider the equation

$$|u| = |v|,$$

where u and v are algebraic expressions and $v \ne 0$. Dividing both sides by $|v|$, $v \ne 0$, we obtain

$$\frac{|u|}{|v|} = 1$$

$$\left|\frac{u}{v}\right| = 1 \qquad \textbf{Property of absolute value}$$

$$\frac{u}{v} = 1 \quad \text{or} \quad \frac{u}{v} = -1$$

$$u = v \quad \text{or} \quad u = -v$$

We can summarize this information as follows.

Equations of the Form $|u| = |v|$

The equation

$$|u| = |v|$$

is equivalent to

$$u = \pm v.$$

EXAMPLE 2   Solve each equation.

(a) $|2x - 3| = |x - 6|$ (b) $|x^2 - 2x| = |8x|$

SOLUTION

(a) Solving the equation $|2x - 3| = |x - 6|$ is equivalent to solving
$2x - 3 = \pm(x - 6)$. Thus,

$$2x - 3 = x - 6 \quad \text{or} \quad 2x - 3 = -(x - 6)$$
$$x = -3 \qquad\qquad 2x - 3 = -x + 6$$
$$3x = 9$$
$$x = 3$$

Thus, the original equation has two solutions, ± 3.

(b) Solving the equation $|x^2 - 2x| = |8x|$ is equivalent to solving
$x^2 - 2x = \pm 8x$. Thus,

$$x^2 - 2x = 8x \quad \text{or} \quad x^2 - 2x = -8x$$
$$x^2 - 10x = 0 \qquad\qquad x^2 + 6x = 0$$
$$x(x - 10) = 0 \qquad\qquad x(x + 6) = 0$$
$$x = 0 \quad \text{or} \quad x = 10 \qquad x = 0 \quad \text{or} \quad x = -6$$

Thus, the original equation has three solutions: 0, 10, and -6.   ◆

PROBLEM 2   Solve $|4x - 3| = |2 - x|$.   ◆

◆ **Absolute Value in Inequalities**

We now turn our attention to solving inequalities that involve absolute value.
The inequality $|x| < k$, where k is a positive real number, can be interpreted as
the set of all real numbers that are *less than k* units from 0. Thus, as shown in
Figure 2.10, $|x| < k$ is equivalent to $-k < x < k$.

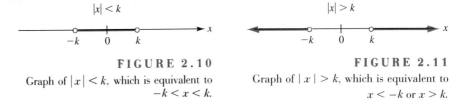

FIGURE 2.10   FIGURE 2.11
Graph of $|x| < k$, which is equivalent to   Graph of $|x| > k$, which is equivalent to
$-k < x < k$. $x < -k$ or $x > k$.

The inequality $|x| > k$ can be interpreted as the set of all real numbers that
are *greater than k* units from 0. Thus, as shown in Figure 2.11, $|x| > k$ is equiva-
lent to $x < -k$ or $x > k$. We now extend these ideas to inequalities of the forms
$|u| < k$ and $|u| > k$, where u is an algebraic expression.

Absolute Value in Inequalities

If k is a positive real number and u is an algebraic expression, then

1. $|u| < k$ is equivalent to $-k < u < k$.   **2.** $|u| > k$ is equivalent to $u < -k$ or $u > k$.

E X A M P L E 3   Solve each inequality. Use interval notation to describe the solution set.

(a) $|2x - 3| < 5$ (b) $|2 - t| - 3 \geq 1$

S O L U T I O N

(a) The inequality $|2x - 3| < 5$ is equivalent to

$$-5 < 2x - 3 < 5$$

$$-2 < \quad 2x \quad < 8 \qquad \text{Add 3 to each member}$$

$$-1 < \quad x \quad < 4 \qquad \text{Divide each member by 2}$$

Thus, the solution set of $|2x - 3| < 5$ is $(-1, 4)$.

(b) Adding 3 to both sides of the inequality $|2 - t| - 3 \geq 1$ gives us $|2 - t| \geq 4$. Now, the inequality $|2 - t| \geq 4$ is equivalent to

$$2 - t \leq -4 \qquad \text{or} \qquad 2 - t \geq 4$$

$$-t \leq -6 \qquad\qquad\qquad -t \geq 2$$

$$t \geq 6 \qquad\qquad\qquad\quad t \leq -2$$

Thus, the solution set of the original inequality consists of all real numbers belonging to either of the intervals $(-\infty, -2]$ *or* $[6, \infty)$. Hence, the solution set of $|2 - t| - 3 \geq 1$ is $(-\infty, -2] \cup [6, \infty)$. ◆

P R O B L E M 3   Repeat Example 3 for $|-6x| - 4 > 5$. ◆

In the next example, we illustrate the general technique for solving polynomial and rational inequalities involving absolute value.

E X A M P L E 4   Solve each inequality. Use interval notation to describe the solution set.

(a) $|x^2 - 3| < 1$ (b) $\left|\dfrac{x - 4}{x - 1}\right| \geq 2$

S O L U T I O N

(a) The inequality $|x^2 - 3| < 1$ is equivalent to

$$-1 < x^2 - 3 < 1,$$

which in turn is equivalent to

$$-1 < x^2 - 3 \quad \text{and} \quad x^2 - 3 < 1$$
$$x^2 - 2 > 0 \qquad\qquad x^2 - 4 < 0$$

We now use the critical value method (from Section 2.6) for solving a polynomial inequality.

For $x^2 - 2 > 0$:	For $x^2 - 4 < 0$:
Critical values: $\pm\sqrt{2}$	*Critical values:* ± 2
Algebraic signs:	*Algebraic signs:*

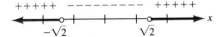 

Solution set: $(-\infty, -\sqrt{2}) \cup (\sqrt{2}, \infty)$ *Solution set:* $(-2, 2)$

Now the solution set of the original inequality is all real numbers common to both solution intervals $(-\infty, -\sqrt{2}) \cup (\sqrt{2}, \infty)$ *and* $(-2, 2)$. Hence, we conclude that the solution set of $|x^2 - 3| < 1$ is $(-2, -\sqrt{2}) \cup (\sqrt{2}, 2)$.

(b) The inequality $\left|\dfrac{x-4}{x-1}\right| \geq 2$ is equivalent to

$$\frac{x-4}{x-1} \leq -2 \quad \text{or} \quad \frac{x-4}{x-1} \geq 2.$$

Using the procedure we discussed in Section 2.6 for solving a rational inequality, we first change to standard form as follows:

$$\frac{x-4}{x-1} \leq -2 \qquad\qquad \frac{x-4}{x-1} \geq 2$$

$$\frac{x-4}{x-1} + 2 \leq 0 \qquad\qquad \frac{x-4}{x-1} - 2 \geq 0$$

$$\frac{3x-6}{x-1} \leq 0 \qquad\qquad \frac{-x-2}{x-1} \geq 0$$

For $\dfrac{3x-6}{x-1} \leq 0$:	For $\dfrac{-x-2}{x-1} \geq 0$:
Critical values: 1, 2	*Critical values:* -2, 1
Algebraic signs:	*Algebraic signs:*

 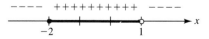

Solution set: $(1, 2]$ *Solution set:* $[-2, 1)$

The solution set of the original inequality is all real numbers belonging to either of the intervals (1, 2] *or* [−2, 1). Hence, we conclude that the solution set of $\left|\dfrac{x-4}{x-1}\right| \geq 2$ is $[-2, 1) \cup (1, 2]$.   ◆

PROBLEM 4   Repeat Example 4 for $\left|\dfrac{y-2}{y}\right| < 3$.   ◆

Exercises 2.7

Basic Skills

In Exercises 1–28, solve each equation.

1. $|3x| = 12$

2. $3|-2t| = 15$

3. $6|5n - 3| = 12$

4. $|4y - 3| - 5 = 4$

5. $|9 - 3y| = 0$

6. $|3x - 11| = -4$

7. $|x^2 - 4x + 2| - 2 = 0$

8. $|n^2 - 7| - 9 = 0$

9. $|3x^3 - 4x^2 + 4x - 10| = -5$

10. $|2x^3 - 4x^2 + 6x| = 0$

11. $\left|\dfrac{3}{x+4}\right| = 7$

12. $\dfrac{|5n-1|}{|2-3n|} = \dfrac{2}{3}$

13. $|3x|\left|\dfrac{1}{x+1}\right| = 4$

14. $|x+1|\left|\dfrac{3}{x+2}\right| = 2$

15. $|3 - 2t| = 6t$

16. $|4m - 1| = 2m + 5$

17. $|x^2 - 3x| = 4x$

18. $|x^2 - 4x + 3| + 6 = 2x$

19. $|x| + 9 = x^2 + x$

20. $|x^2 - 3x| = 6 - 4x$

21. $|3h - 5| = |2h + 10|$

22. $|2x + 7| = |6x - 5|$

23. $|x^2 + 9| - |10x| = 0$

24. $|x^2 + 3x| = |5x + 15|$

25. $|x||x - 4| = |3x - 12|$

26. $|y|^2 - 3|y| = 28$

27. $\left|\dfrac{2|x|}{3} + 1\right| = 0$

28. $\left|\dfrac{3|x-1|}{4} - 1\right| = 2$

In Exercises 29–46, solve each inequality. Use interval notation to describe the solution set.

29. $|x| < 3$

30. $|2x| - 2 \geq 8$

31. $\left|\dfrac{2x}{3} - 1\right| > 0$

32. $5\left|\dfrac{-x}{10}\right| + 3 < 13$

33. $|5 - 2x| > 3$

34. $|4x + 3| + 2 < 17$

35. $\left|\dfrac{4-x}{3}\right| \leq \dfrac{1}{2}$

36. $\left|\dfrac{2(4-3x)}{3}\right| > 8$

37. $\left|\dfrac{3x}{4} + \dfrac{1-3x}{6}\right| \geq 1$

38. $\left|\dfrac{1-2x}{4} + \dfrac{1+2x}{6}\right| \leq \dfrac{5}{9}$

39. $|y^2 - 6| \geq 3$

40. $|2 - x^3| < 6$

41. $|x^2 - 4x| - 4 < 0$

42. $|-n^2 - 2n + 1| \geq 1$

43. $\left|\dfrac{2x-1}{x}\right| > 1$

44. $\left|\dfrac{4}{x-3}\right| - 2 \leq 0$

45. $\left|\dfrac{4x+3}{x-2}\right| \leq 6$

46. $\left|\dfrac{x^2-3}{x}\right| > 2$

Critical Thinking

In Exercises 47 and 48, solve each literal equation for x.

47. $|ax + b| = 1$

48. $|ax + b| = |cx + d|$

In Exercises 49 and 50, solve each inequality.

49. $|2x + 3| \leq |4x - 1|$

50. $|6x + 5| \geq |2 - x|$

$\boxed{\frac{\Delta y}{\Delta x}}$ *In calculus, the epsilon (ε), delta (δ) definition of a limit involves working with inequalities that contain absolute values. In Exercises 51–54, find δ (in terms of ε) so that the given implication is true.*

51. $|x - 5| < \delta$ implies $|(2x - 3) - 7| < \varepsilon$

52. $|x - 3| < \delta$ implies $|(3x - 5) - 4| < \varepsilon$

53. $|x + 3| < \delta$ implies $|\frac{1}{2}(x - 4) + \frac{7}{2}| < \varepsilon$

54. $|x + 2| < \delta$ implies $|\frac{1}{4}(5x + 2) + 2| < \varepsilon$

55. Matchsticks in quantities of 1000 are packaged into boxes by a machine that makes an error of at most five matchsticks per box.

(a) If x is the number of matchsticks in a box, write an inequality using absolute value that describes this situation.

(b) Solve the inequality in part (a) for x.

56. A pediatrician determines that the weight x in pounds of 95% of the babies born at a certain hospital satisfies the inequality $|x - 7.2| \leq 2.3$. What is the weight of the other 5% of the babies?

 Calculator Activities

In Exercises 57–60, use a calculator to solve each equation. Round the answers to three significant digits.

57. $|2.543x - 5.675| = 9.327$

58. $|25.3x - 52.6| = |72.3 - 19.4x|$

59. $\left| \dfrac{0.324}{x + 0.798} \right| = 5.36$

60. $|0.34x^2 - 6.24x - 2.23| = 1.10$

In Exercises 61–63, use a calculator to solve each inequality. Use interval notation to describe each solution set with decimal numbers rounded to three significant digits.

61. $|19.24x + 15.27| < 12.75$

62. $|-15.8x| - 91.3 > 18.6$

63. $\left| \dfrac{9.21 - 4.34x}{6.89} \right| \geq 8.33$

64. Precision parts must be manufactured to certain standards. The allowable deviation from the standard is called the *tolerance*. Suppose a resistor in a computer has a specified value of 40.2 ohms (Ω) with tolerance $\pm 5\%$.

(a) Write an inequality using absolute value that describes the permissible range of values for this resistor R.

(b) Solve the inequality in part (a) for R.

Chapter 2 Review

 Questions for Group Discussion

1. Explain the difference between an *identity* and a *conditional equation*.

2. What is the basic form of a *linear equation* in one variable? What is meant by the *root* of such an equation?

3. What is an *extraneous root*? Explain how such a root may develop.

4. What is meant by a *literal equation*? Give an example of a literal equation that is quadratic.

5. Explain what can happen to one or more of the roots of an equation if both sides of the equation are divided by a variable expression. Illustrate with an example.

6. Describe the method of *completing the square*.

7. By using the discriminant, explain how you can determine whether the roots of a quadratic equation are (i) real, (ii) imaginary, and (iii) equal.

8. Explain the difference between a *quadratic equation* and an *equation of quadratic type*.

9. What is meant by an *unbounded interval*?

10. Explain the difference between an *open interval* and a *closed interval*.

11. State some conditions for which it is necessary to *reverse the inequality* when generating equivalent inequalities. Illustrate by examples.

12. What is the importance of the *critical values* of a polynomial inequality? of a rational inequality?

13. If k is a positive real number, give a geometrical interpretation for each of the following:
    (a) $|x| = k$ (b) $|x| < k$ (c) $|x| > k$

14. What is the difference between the *union* and *intersection* of two sets? When is the union symbol needed to describe the solution set of an inequality?

15. Given that a is a real number, are the roots of the equation $5x^2 + 2ax + a^2 = 0$ real or imaginary? Explain your reasoning.

16. Is it possible to find three consecutive integers whose sum equals the product of the smallest and largest? Explain.

Review Exercises

In Exercises 1–40, solve for x.

1. $18 = 24 - 3x$

2. $3x + 4 = 7x - 8$

3. $6x - (3 - 2x) = 2(x - 3)$

4. $4x^2 - (2x - 3)^2 = 9$

5. $\dfrac{1}{x} - \dfrac{1}{a} = \dfrac{1}{b} - \dfrac{1}{x}$

6. $\dfrac{x}{8} - \dfrac{3x + 1}{10} = \dfrac{3 - x}{20}$

7. $\dfrac{x}{2x - 1} + \dfrac{1}{2} = \dfrac{3x - 1}{6x - 3}$

8. $\dfrac{3}{x + 3} - \dfrac{1}{x - 3} = \dfrac{x}{x^2 - 9}$

9. $|4 - 3x| = 16$

10. $|2x - 5| = 4x + 1$

11. $x^2 - 6x = 16$

12. $x(3x - 4) = 4$

13. $\dfrac{x}{x + 4} - \dfrac{2}{x - 1} = \dfrac{20}{x^2 + 3x - 4}$

14. $\dfrac{x - 1}{x} = \dfrac{5}{2} - \dfrac{x}{x - 1}$

15. $x^2 - 16 = 0$

16. $(2x - 1)^2 = 18$

17. $3x^2 + 48 = 0$

18. $-2x^2 = 16$

19. $x^2 = 4x + 10$
    (Use the method of completing the square.)

20. $2x^2 + 3x = 1$
    (Use the method of completing the square.)

21. $x^2 + 5 = 2x$

22. $x^2 + 8x + 20 = 0$

23. $2(x^2 - 3) = \sqrt{6}\,x$

24. $2x^2 = 2\sqrt{2}\,x - 3$

25. $\dfrac{1}{2} - \dfrac{1}{x} = \dfrac{1}{x - 2}$

26. $x + \dfrac{1}{x} = \dfrac{1}{a} + a$

27. $x^{2/3} = \frac{3}{4}$

28. $(x^2 + 6x)^{-1/2} = \frac{1}{4}$

29. $\sqrt{2 - 2x} = 3x + 1$

30. $\sqrt[3]{x^3 + 7} = x + 1$

31. $\sqrt{5x - 3} = \sqrt{2x - 6}$

32. $\sqrt{6x + 4} - \sqrt{x - 1} = \sqrt{x + 7}$

33. $\sqrt{a + x} + \sqrt{a} = \sqrt{a - x}$

34. $\sqrt{a - x} + \sqrt{b - x} = \sqrt{a + b - 2x}$

35. $x^4 - 8x^3 = 33x^2$

36. $2x^3 + 3x^2 + 18x + 27 = 0$

37. $6x^2(x - 4)^3 = 2x(x - 4)^4$

38. $x(x + 1)^{-1} - 6(x + 1)^{-2} = 0$

39. $x^4 + 36 = 13x^2$

40. $x^2(x^2 + 40) = -144$

In Exercises 41–50, find the real roots of each equation.

41. $x^{-3} = -125$

42. $(p - 2)^4 = 81$

43. $y^{1/4} + 5y^{1/8} = 14$

44. $3x^{-4} + 4x^{-2} - 7 = 0$

45. $2x^{-4/3} + 5x^{-2/3} = 12$

46. $2\sqrt{3m+4} + \sqrt[4]{3m+4} = 10$

47. $n^2(3n-2)^{1/2} + n(3n-2)^{3/2} = 0$

48. $6x^6 + 27x^2 = 4x^4 + 18$

49. $x^4 - 6x^2 = 3$

50. $x^6 + 10x^3 + 7 = 0$

In Exercises 51–54, solve each formula for the indicated variable.

51. Resistance of a conductor: $R = \dfrac{\mu L + \mu Lt}{d^2}$. Solve for the length of the conductor L.

52. Energy output of a battery: $W = I^2 R_1 t + I^2 R_2 t$. Solve for the current I.

53. Surface area of a right circular cylinder: $A = 2\pi r^2 + 2\pi rh$. Solve for the radius r.

54. Bending moment of a simply supported beam: $M = \frac{1}{2}Lwx - \frac{1}{2}wx^2$. Solve for the distance from the end of the beam x.

In Exercises 55–74, solve each inequality. Use interval notation to describe the solution set.

55. $4 - x > 7$

56. $3x - 2 < 5x - 4$

57. $\dfrac{x}{4} \le \dfrac{2x}{5} - \dfrac{3}{10}$

58. $\dfrac{1}{6} - \dfrac{t-2}{12} \ge \dfrac{t-1}{4}$

59. $-2 \le \dfrac{2-y}{3} \le 4$

60. $x \le 3(x-2) \le 6$

61. $|5 - 3n| > 8$

62. $\left|\dfrac{2x-3}{5}\right| < \dfrac{2}{3}$

63. $3x^2 + 7x \le 6$

64. $(2y+1)(y-3) \ge (2y+1)^2$

65. $x^3 + 2x^2 \ge 4x$

66. $x^4 - 5x^2 + 4 < 0$

67. $\dfrac{p-4}{p+2} \ge 0$

68. $\dfrac{x^2 - 2x + 3}{x+1} \le 0$

69. $\dfrac{1}{x} < \dfrac{3(x-1)}{x+4}$

70. $m + \dfrac{3m}{2-m} > 4$

71. $\left|\dfrac{x+6}{x-3}\right| > 4$

72. $\left|\dfrac{3x-5}{2x-1}\right| < 5$

73. $\left|\dfrac{w-3}{w+2}\right| \le 2$

74. $|3x+1| \ge |1-x|$

In Exercises 75–96, set up an equation or inequality, and then solve.

75. In his will, a man decides to leave two thirds of his estate to his wife, one eighth to his son, and the remainder, which is $50,000, to his daughter. What is the value of his estate?

76. When a fringe of uniform width is added to each of the shorter ends of a 6-foot by 9-foot rug, as shown in the sketch, the area of the rug is increased by 4 square feet. Find the width of the fringe.

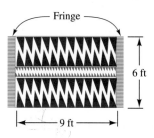

77. When $12,500 is invested, part at 8% and the remainder at 9%, the yearly interest is $41 more than if all the funds had been invested at 8%. How much money is invested at each rate?

78. A goldsmith has two alloys of gold. One is 80% pure gold and the other is 60% pure gold. How many ounces of each alloy must be used to make 60 ounces of an alloy that is 65% gold?

79. If each side of a square is increased by 4 inches, the area of the square is increased by 80 square inches. Find the side of the square.

80. If the radius of a circle is increased by 1 meter, the area of the circle is increased by 19π square meters. Find the radius of the circle.

81. An open box is made from a 6-inch by 9-inch piece of metal by cutting out a square from each corner and then turning up the sides along the dashed lines, as shown in the figure. If the area of the bottom of the box is 18 square inches, what is the height of the box?

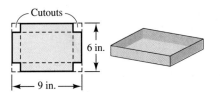

82. The dimensions of a picture are 12 inches by 15 inches. Encasing the picture in a frame of uniform width increases its area by 50%. What is the width of the frame?

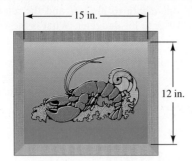

15 in.

12 in.

83. Preparing for a triathlon, a man runs 2 miles at the rate of 8 mph and returns home along the same route by riding a bicycle. The entire trip takes 20 minutes. What is the rate of speed at which he pedals?

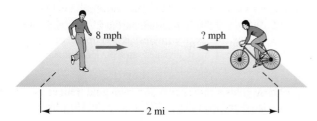

8 mph   ? mph

2 mi

84. The difference of the cubes of two consecutive positive integers is 169. What are the two numbers?

85. Find k such that one root of $25x^2 + kx + 1 = 0$ is two more than the other root.

86. The sum of two numbers is 6 and the sum of their reciprocals is $\frac{8}{9}$. What are the numbers?

87. A gas tank is filled in $3\frac{3}{4}$ hours by two pipelines together. Working alone, the larger pipeline can fill the tank in four hours less time than the smaller one. What time is required by each pipeline alone to fill the tank?

88. The fixed cost to run a porcelain-doll company is $2200 per month and the variable cost to produce a doll is $12 per unit. If each doll sells for $30, determine the number of units that must be sold each month for the company to make a profit of $5000 per month.

89. If a number is increased by 57, its positive square root is increased by 3. Find the number.

90. Find the length and width of a rectangle if its diagonal is 2 centimeters more than the longer side, which in turn is twice the shorter side.

91. You have a choice of two methods of determining your yearly salary: One method pays a base salary of $24,000 plus 2% commission on the total amount of sales. The other has no base salary but pays 12% commission on sales. When is it financially better to choose the second method?

92. A student's hourly exam scores are 61%, 84%, 75%, and 86%. The final exam counts twice as much as an hourly exam. What is the possible range of scores on the final exam that will give the student an average of at least 70% but less than 80%?

93. Find the interval(s) for which real numbers have the following property: The cube of the number is less than the number itself.

94. The base of a triangle is 4 feet more than its height, and the area of the triangle is less than 96 square feet. What is the range of values for the base of the triangle?

95. The director of a college library is trying to decide which of two copying machines to purchase. Model A costs $5000 with a maintenance agreement of $20 per month. Model B costs $4000 with a maintenance agreement of $25 per month. Under what conditions is model A more economically attractive than model B?

96. A pediatrician determines that the length x in inches of 90% of the babies born at a certain hospital satisfies the inequality

$$|x - 20.6| \le 2.1.$$

What is the length of the other 10% of the babies?

Cumulative Review Exercises

Chapters 1 & 2

1. Factor each expression completely.
   (a) $x(x + 1)(2x - 5) - 12(x + 1)$
   (b) $x^2 - 4xy - 4a^2 + 8ay$

2. Simplify each fractional expression.
   (a) $\dfrac{(x - 3)^{1/2} - 4(x - 3)^{-1/2}}{x - 7}$
   (b) $\dfrac{x^{-1} + y^{-1}}{(x + y)^{-1}}$

3. If 17 is 5 more than $3x$, what is the value of $2x - 3$?

4. Express $\left(x^{-2}\sqrt{y}\right)^{-6}$ using only positive exponents.

5. Solve each equation for x.
   (a) $(a + x)^{1/2} - (a - x)^{1/2} = x$
   (b) $\dfrac{x + 2}{a} + \dfrac{a}{x - 2a} = 2$

6. Simplify each expression. Assume $x > y > 0$.
   (a) $(x + y)^2(x^{-1} + y^{-1})^{-2}$
   (b) $\sqrt{x^4 - x^3y - x^2y^2 + xy^3}$

7. Express $8^0 + 3\tfrac{1}{2} - 6^{-1} + \dfrac{1}{\sqrt[3]{-27}} + 8^{2/3}$ as an integer.

8. If $\dfrac{\sqrt{19} - \sqrt{10}}{x} = \dfrac{x}{\sqrt{19} + \sqrt{10}}$, find x.

9. For what values of k does the equation $4x^2 - 12x + k = 0$ have the given roots?
   (a) one root that is zero   (b) equal roots
   (c) real roots   (d) imaginary roots

10. For the equation $ax^2 - 5x + 2 = 0$, find the value of a if one root is to be the reciprocal of the other.

11. Three hundred children are seated in an auditorium in a certain number of rows with the same number of children in each row. Placing two more children in each row would result in five fewer rows being needed. In how many rows are the children seated?

12. Into what two amounts can $22,400 be split so that the annual interest from one part, which is invested at 8%, is double that from the other part, which is invested at 10%?

13. Expand and simplify $(1 + i)^3$, where $i = \sqrt{-1}$.

14. Solve the inequality $x^2 - 4 \le \dfrac{1}{x^2}$ for x.

15. Perform the indicated operations and reduce to lowest terms.
    (a) $\left(\dfrac{x^3 + y^3}{x^2 - y^2}\right) \div \left(\dfrac{x^2}{x - y} - y\right)$
    (b) $\dfrac{x^{1/2} + 1}{x^{1/2} - 1} - \dfrac{x^{1/2} - 1}{x^{1/2} + 1} - \dfrac{4}{x - 1}$

16. Arrange the given numbers in order from smallest to largest:

$$4^{-3/2} \qquad \left(\tfrac{1}{4}\right)^{-2} \qquad (4^{-1/3})^6 \qquad \left(\tfrac{1}{4}\right)^{-1/2} \qquad \dfrac{1}{4^0}$$

17. The circumference of a rear wheel on a dragster is 4 feet more than the circumference of a front wheel. A front wheel makes 88 more revolutions than a rear wheel in traveling a quarter-mile track. What is the circumference of each wheel?

18. A business determines that consumers will purchase x units of their product when the price in dollars of each unit is $\dfrac{60 - x}{3}$. How many units are sold if the total revenue from sales is $225?

19. Perform the indicated operations and reduce to lowest terms.
    (a) $\dfrac{a}{(a - b)(a - c)} + \dfrac{b}{(b - c)(b - a)} + \dfrac{a}{(c - a)(c - b)}$
    (b) $\dfrac{1 + \dfrac{1}{x^2} + \dfrac{1}{x^4}}{1 + \dfrac{1}{x} + \dfrac{1}{x^2}}$

20. Simplify $(x^2 + 10x + 25)^{1/2} - (x^2 - 10x + 25)^{1/2}$ given that:
    (a) x is any real number.   (b) $x \ge 5$.

21. Simplify $\dfrac{\sqrt{54} + \sqrt{27}}{\sqrt{6} - \sqrt{3}}$ and rationalize the denominator.

22. Solve the equation

$$\dfrac{2x + \sqrt{4x^2 - 1}}{2x - \sqrt{4x^2 - 1}} = \dfrac{1}{4}$$

by rationalizing the numerator and then applying the square root property.

23. A square piece of sheet metal is to be made into a rectangular box with an open top by cutting out a 2-inch by 2-inch square from each corner and bending along the dotted lines as shown in the sketch. What is the range of dimensions of the original piece of sheet metal if the volume of the box is from 50 to 98 cubic inches?

24. A car radiator contains 15 quarts of a 20% antifreeze solution. How many quarts of 20% solution must be drained and replaced with pure antifreeze (100% solution) to achieve a solution of at least 60% antifreeze?

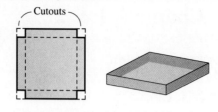

E X E R C I S E   2 3

3

Graphs and Functions

The real estate tax T on a property varies *directly* as its assessed value V.

(a) Express T as a function of V if $T = \$2800$ when $V = \$112,000$.

(b) State the domain of the function defined in part (a), then sketch its graph.

(For the solution, see Example 3 in Section 3.7.)

3.1   Working in the Cartesian Plane

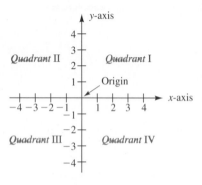

FIGURE 3.1
The coordinate plane, or Cartesian plane.

◆ Introductory Comments

Recall from Section 1.1 that for each point on the real number line, there corresponds a real number called its *coordinate*. In this section, we extend this idea by assigning to each point in a plane a pair of real numbers. To do this, we construct horizontal and vertical real number lines that intersect at the zero points of the two lines (see Figure 3.1). The two lines are called *coordinate axes* and the *plane* in which they lie is called the **coordinate plane** or **Cartesian plane**, after the French mathematician René Descartes (1596–1650). The point where the axes intersect is called the **origin**. The horizontal number line has its positive direction to the right and is usually called the **x-axis**. The vertical number line has its positive direction upward and is usually called the **y-axis**. The coordinate axes divide the plane into four regions, or **quadrants**, which are labeled with Roman numerals, as shown in Figure 3.1.

◆ Plotting Points

We can now assign to each point P in the coordinate plane a unique pair of numbers, called its **rectangular coordinates** or **Cartesian coordinates**. By drawing horizontal and vertical lines through P, we find that the vertical line intersects the x-axis at some point with coordinate a, and the horizontal line intersects the y-axis at some point with coordinate b, as shown in Figure 3.2. Thus, we assign the pair (a, b) to the point P. In the pair (a, b), the first number a is called the **x-coordinate** or **abscissa** of P, and the second number b is called the **y-coordinate** or **ordinate** of P. Since the x-coordinate is always written first, we refer to (a, b) as an **ordered pair** of numbers.

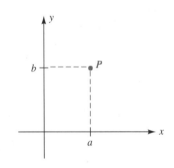

FIGURE 3.2
The coordinates of point P are (a, b).

Conversely, we can locate any ordered pair (a, b) in the coordinate plane by constructing a vertical line through a on the x-axis and a horizontal line through b on the y-axis. The intersection of these lines determines a unique point P, which we designate as $P(a, b)$. Thus, for each point in the coordinate plane there corresponds a unique ordered pair of real numbers, and for each ordered pair of real numbers there is a unique point in the coordinate plane. Hence, we have a *one-to-one correspondence* between pairs of real numbers and points in a coordinate plane.

EXAMPLE 1   Plot the points $A(2, 3)$, $B(-3, 2)$, $C(-4, -3)$, and $D(\frac{3}{2}, -4)$ on a coordinate plane. Specify the quadrant in which each point lies.

SOLUTION   The point $A(2, 3)$ is in quadrant I, $B(-3, 2)$ is in quadrant II, $C(-4, -3)$ is in quadrant III, and $D(\frac{3}{2}, -4)$ is in quadrant IV. The points are plotted in Figure 3.3.

FIGURE 3.3

Plotting points associated with ordered pairs.

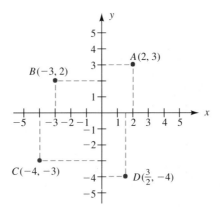

PROBLEM 1 Plot the points $A(-3, 1)$, $B(0, -4)$, $C(4, 0)$, and $D(1\frac{1}{4}, 2)$ on a coordinate plane.

The Distance Formula

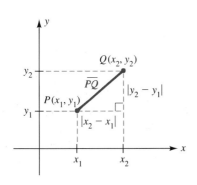

FIGURE 3.4

The length of $\overline{PQ}$, denoted as PQ, may be determined by the Pythagorean theorem.

By convention, the *line segment* that joins the points $P(x_1, y_1)$ and $Q(x_2, y_2)$ in the coordinate plane is designated as $\overline{PQ}$, and the *length* of $\overline{PQ}$ is designated as PQ. To determine PQ, we begin by constructing a right triangle, as shown in Figure 3.4. The length of the horizontal leg of the triangle is $|x_2 - x_1|$ and the length of the vertical leg is $|y_2 - y_1|$. By the Pythagorean theorem, we can find the length of $\overline{PQ}$:

$$(PQ)^2 = |x_2 - x_1|^2 + |y_2 - y_1|^2$$
$$PQ = \sqrt{|x_2 - x_1|^2 + |y_2 - y_1|^2}$$
$$PQ = \sqrt{(x_2 - x_1)^2 + (y_2 - y_1)^2}$$

We refer to this last equation as the **distance formula**.

Distance Formula

The **distance** between two points $P(x_1, y_1)$ and $Q(x_2, y_2)$ in the coordinate plane is given by

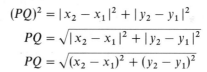

$$PQ = \sqrt{(x_2 - x_1)^2 + (y_2 - y_1)^2}.$$

Note: When we use the distance formula, it does not matter which point is called (x_1, y_1) and which is called (x_2, y_2). This is because $(x_2 - x_1)$ and $(x_1 - x_2)$ are negatives of each other, as are $(y_2 - y_1)$ and $(y_1 - y_2)$, and squaring a number or its negative gives us the same numerical result.

EXAMPLE 2   Find the length of the line segment joining the points $A(-2, 3)$ and $B(4, 1)$.

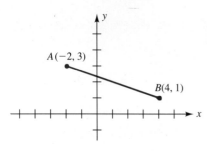

FIGURE 3.5
The length of the line segment joining the points A and B can be found by using the distance formula.

SOLUTION   Figure 3.5 shows the points A and B in the coordinate plane. We designate the length of the line segment that joins the points A and B as AB. We use the distance formula with $(x_1, y_1) = (-2, 3)$ and $(x_2, y_2) = (4, 1)$:

$$AB = \sqrt{[4 - (-2)]^2 + (1 - 3)^2}$$
$$= \sqrt{(6)^2 + (-2)^2}$$
$$= \sqrt{40}$$
$$= 2\sqrt{10}$$
$$\approx 6.32 \qquad \blacklozenge$$

PROBLEM 2   Find the length of the line segment joining the points $P(-3, 1)$ and $Q(1, -2)$.
$\blacklozenge$

◆ The Midpoint Formula

Given two points $P(x_1, y_1)$ and $Q(x_2, y_2)$ in the coordinate plane, we can find the coordinates of the midpoint M of $\overline{PQ}$ by finding the average value of the x-coordinates and the average value of the y-coordinates of the endpoints P and Q. We refer to this fact as the **midpoint formula**.

◢ Midpoint Formula

> The coordinates of the **midpoint** M of a line segment joining the points $P(x_1, y_1)$ and $Q(x_2, y_2)$ are
>
> $$\left(\frac{x_1 + x_2}{2}, \frac{y_1 + y_2}{2}\right)$$

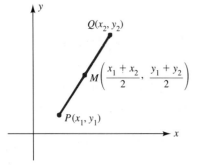

FIGURE 3.6
The midpoint of $\overline{PQ}$ is M, provided $PM = MQ$ and $PM + MQ = PQ$.

To verify the midpoint formula, we use Figure 3.6 and the distance formula to show that $PM = MQ$ and $PM + MQ = PQ$. Note that

$$PM = \sqrt{\left(\frac{x_1 + x_2}{2} - x_1\right)^2 + \left(\frac{y_1 + y_2}{2} - y_1\right)^2}$$
$$= \sqrt{\left(\frac{x_2 - x_1}{2}\right)^2 + \left(\frac{y_2 - y_1}{2}\right)^2}$$
$$= \tfrac{1}{2}\sqrt{(x_2 - x_1)^2 + (y_2 - y_1)^2}$$

and

$$MQ = \sqrt{\left(x_2 - \frac{x_1 + x_2}{2}\right)^2 + \left(y_2 - \frac{y_1 + y_2}{2}\right)^2}$$
$$= \sqrt{\left(\frac{x_2 - x_1}{2}\right)^2 + \left(\frac{y_2 - y_1}{2}\right)^2}$$
$$= \tfrac{1}{2}\sqrt{(x_2 - x_1)^2 + (y_2 - y_1)^2}.$$

Thus $$PM + MQ = \sqrt{(x_2 - x_1)^2 + (y_2 - y_1)^2} = PQ.$$

Since $PM = MQ$ and the points P, M, and Q are *collinear* $(PM + MQ = PQ)$, we conclude that M is the midpoint of $\overline{PQ}$.

EXAMPLE 3   Find the coordinates of the midpoint of the line segment joining the points $A(-2, 3)$ and $B(4, 1)$.

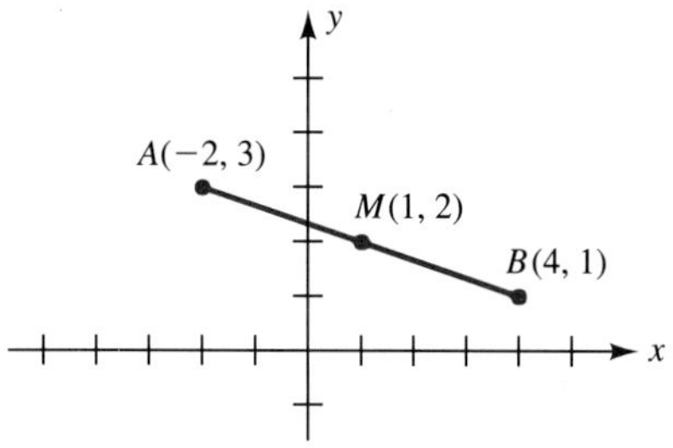

SOLUTION   Using the midpoint formula with $(x_1, y_1) = (-2, 3)$ and $(x_2, y_2) = (4, 1)$, we have

$$\left(\frac{-2 + 4}{2}, \frac{3 + 1}{2} \right) = (1, 2).$$

Thus, as shown in Figure 3.7, the coordinates of the midpoint M are $(1, 2)$.   ◆

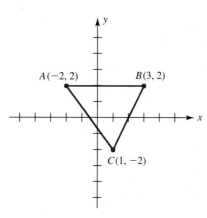

FIGURE 3.7
The coordinates of the midpoint of $\overline{AB}$ are $(1, 2)$.

PROBLEM 3   Find the coordinates of the midpoint of the line segment joining the points $A(-4, -3)$ and $B(5, -1)$.   ◆

◆ **Analytic Geometry**

As illustrated in the next example, we can place geometric figures on the coordinate axes and solve related problems by using the distance and midpoint formulas. We refer to this branch of mathematics, which uses algebra to solve geometric problems, as **analytic geometry**.

EXAMPLE 4   Show that the triangle with vertices at the points $A(-2, 2)$, $B(3, 2)$, and $C(1, -2)$ is an isosceles triangle (two sides of equal length). Then find the area of triangle ABC.

SOLUTION   We begin by plotting the points and constructing the triangle, as shown in Figure 3.8. Now, using the distance formula, we have

$$AB = \sqrt{[3 - (-2)]^2 + (2 - 2)^2} = \sqrt{(5)^2 + (0)^2} = 5,$$
$$BC = \sqrt{(1 - 3)^2 + (-2 - 2)^2} = \sqrt{(-2)^2 + (-4)^2} = \sqrt{20} = 2\sqrt{5},$$
$$AC = \sqrt{[1 - (-2)]^2 + (-2 - 2)^2} = \sqrt{(3)^2 + (-4)^2} = \sqrt{25} = 5.$$

Since $AB = AC$, we conclude the triangle is isosceles with BC the nonequal side.

To find the area of triangle ABC, recall from geometry that the altitude to the nonequal side of an isosceles triangle bisects that side. Thus, if $\overline{AM}$ is the altitude to the base $\overline{BC}$, then M is the midpoint of $\overline{BC}$, as shown in Figure 3.9.

FIGURE 3.8
The triangle formed by the points $A(-2, 2)$, $B(3, 2)$, and $C(1, -2)$.

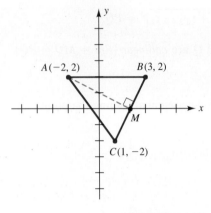

FIGURE 3.9
The altitude to the nonequal side of an isosceles triangle bisects that side. Thus, M is the midpoint of $\overline{BC}$.

By the midpoint formula, the coordinates of the midpoint M are

$$\left(\frac{1+3}{2}, \frac{-2+2}{2}\right) = (2, 0)$$

and, by the distance formula, the length of the altitude $\overline{AM}$ is

$$AM = \sqrt{[(2-(-2)]^2 + (0-2)^2} = \sqrt{(4)^2 + (-2)^2} = \sqrt{20} = 2\sqrt{5}.$$

Hence, the area of triangle ABC is

$$\begin{aligned}\text{Area} &= \tfrac{1}{2}(\text{base})(\text{height}) = \tfrac{1}{2}(BC)(AM) \\ &= \tfrac{1}{2}(2\sqrt{5})(2\sqrt{5}) \\ &= 10 \text{ square units.}\end{aligned}$$

As an alternative method for finding the area of triangle ABC, let $\overline{CN}$ be the altitude to base $\overline{AB}$. Since $\overline{AB}$ is a horizontal line segment, $CN = 4$. Hence,

$$\text{Area} = \tfrac{1}{2}(AB)(CN) = \tfrac{1}{2}(5)(4) = 10 \text{ square units.} \qquad \blacklozenge$$

The *converse* of the Pythagorean theorem is also true. If the square of the longest side of a triangle is equal to the sum of the squares of the two other sides, then the triangle is a right triangle, with the right angle opposite the longest side.

PROBLEM 4   Use the distance formula and the converse of the Pythagorean theorem to show that the triangle with vertices at the points $A(2, 5)$, $B(-2, 3)$, and $C(4, 1)$ is an isosceles right triangle. Then find the area of triangle ABC. $\qquad \blacklozenge$

Exercises 3.1

 Basic Skills

In Exercises 1–6, fill in the blank to complete the statement.

1. If both coordinates are negative, the point is located in quadrant _____.

2. If the x-coordinate is _____ and the y-coordinate is _____, the point is located in quadrant II.

3. If the x-coordinate is _____ and the y-coordinate is _____, the point is located in quadrant IV.

4. If the x-coordinate is zero, the point is located on the _____.

5. If the y-coordinate is zero, the point is located on the _____.

6. If both coordinates are _____, the point is located at the origin.

In Exercises 7 and 8, name the ordered pair associated with each point.

7.

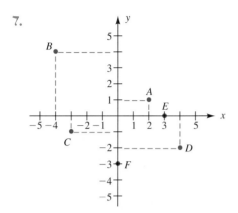

8.

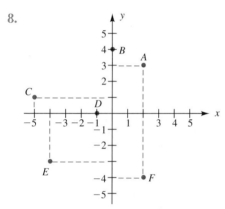

In Exercises 9–18, find

(a) *the length of the line segment joining the points A and B, and*

(b) *the coordinates of the midpoint M of the line segment joining the points A and B.*

9. $A(1, 2)$, $B(4, 6)$ **10.** $A(3, -6)$, $B(-2, 6)$

11. $A(-2, 3)$, $B(1, 8)$ **12.** $A(-1, -2)$, $B(-3, 2)$

13. $A(-5, 2)$, $B(3, -5)$

14. $A(-2, -1)$, $B(-6, -5)$

15. $A(-\frac{1}{2}, \frac{2}{3})$, $B(-\frac{3}{4}, 1)$ **16.** $A(\frac{1}{2}, \frac{1}{2})$, $B(\frac{2}{3}, \frac{5}{8})$

17. $A(0, \sqrt{2})$, $B(-4, 0)$ **18.** $A(\sqrt{3}, -1)$, $B(0, 2)$

19. Find a formula for the distance d between the origin and the point (x, y) in the coordinate plane.

20. Find the coordinates of the midpoint M of a line segment joining the origin to the point (x, y) in the coordinate plane.

21. Find the perimeter of a triangle whose vertices are the points $A(-2, 1)$, $B(1, 3)$, and $C(4, -3)$.

22. Find the perimeter of a quadrilateral whose vertices are the points $A(-1, 0)$, $B(2, 4)$, $C(8, -4)$, and $D(4, -12)$.

23. Show that the points $A(1, 4)$, $B(2, -3)$, and $C(-1, -2)$ are the vertices of a right triangle. Then find the area of triangle ABC.

24. Show that the triangle whose vertices are the points $A(-5, 14)$, $B(1, 4)$, and $C(11, 10)$ is isosceles. Then find its area.

25. Show that the triangle whose vertices are $A(1, 1)$, $B(-1, -1)$, and $C(\sqrt{3}, -\sqrt{3})$ is an equilateral triangle. Then find its area.

26. Show that the triangle whose vertices are $A(0, 6)$, $B(2, 0)$, and $C(8, 2)$ is an isosceles right triangle. Then find its area.

27. A line segment joins the points $A(8, -12)$ and $B(-4, 6)$. Use the midpoint formula to find the coordinates of the three points that divide this line segment into four equal parts.

28. Use the distance formula to determine if the points $A(-2, -3)$, $B(1, 3)$, and $C(2, 5)$ are collinear.

29. A *median* of a triangle is a line segment that joins a vertex of the triangle to the midpoint of the opposite side. Find the lengths of the medians of a triangle whose vertices are $A(-5, 4)$, $B(5, 2)$, and $C(-1, -4)$.

30. The midpoint of a line segment is $M(3, -2)$. One endpoint of the segment has coordinates $(6, 3)$. Find the coordinates of the other endpoint.

Critical Thinking

31. Find the coordinates of a point on the *y*-axis that are equidistant from the points $A(-2, -4)$ and $B(3, 5)$.

32. Find the coordinates of a point on the *x*-axis that are equidistant from the points $A(-2, -4)$ and $B(3, 5)$.

For Exercises 33–36, refer to the right triangle with vertices A, O, and B as shown:

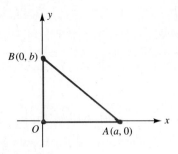

33. Find the coordinates of the midpoint M of the hypotenuse $\overline{AB}$.

34. Find AM and BM.

35. Find OM.

36. What conclusion can you make about the midpoint of the hypotenuse in regard to the three vertices of the right triangle?

For Exercises 37–40, refer to the parallelogram with vertices A, B, C, and O as shown:

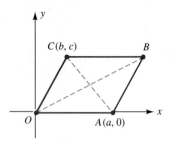

37. Find the coordinates of point B.

38. Find the midpoint of the diagonal $\overline{AC}$.

39. Find the midpoint of the diagonal $\overline{OB}$.

40. What conclusion can you make about the diagonals of a parallelogram?

A point P lies on the line segment joining the points $A(x_1, y_1)$ and $B(x_2, y_2)$ as shown:

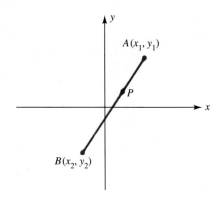

If the ratio $AP/AB = r$, then the point P has the coordinates

$$P(x_1 + r(x_2 - x_1), y_1 + r(y_2 - y_1)).$$

Use this formula for Exercises 41 and 42.

41. A line segment joins the points $A(6, 4)$ and $B(2, -2)$. Find the coordinates of the two points P_1 and P_2 that divide this line segment into three equal parts.

42. A line segment joins the points $A(5, 6)$ and $B(-3, -4)$. Find the coordinates of the four points P_1, P_2, P_3, and P_4 that divide the line segment into five equal parts.

 Calculator Activities

In Exercises 43–46, find

(a) *the length of the line segment that joins the given point to the origin.*

(b) *the coordinates of the midpoint M of the line segment that joins the given point to the origin.*

43. $(2.56, 3.20)$ 44. $(-10.6, 14.8)$

45. $(-46.9, -76.8)$ 46. $(123.6, -457.9)$

47. Find the radius and area of a circle that passes through the point $A(2.24, 3.71)$ and has its center at the origin.

48. The line segment that joins the point $A(4.3, 2.5)$ to $B(-2.7, -4.9)$ is the diameter of a circle. Find the center, radius, and area of the circle.

3.2 ◆ Graphs of Equations

◆ **Introductory Comments**

In Chapter 2, we solved various types of equations containing one unknown. We now turn our attention to equations in *two unknowns*. A solution of an equation in two unknowns x and y is an ordered pair (x_1, y_1) such that when x is replaced by x_1 and y by y_1 the equation becomes true. Consider the equation

$$y = 2x + 1$$

with the two unknowns x and y. We say that the ordered pair $(0, 1)$ is a solution of this equation, since

$$y = 2x + 1$$
$$1 = 2(0) + 1$$
$$1 = 1 \quad \text{is true.}$$

To find other ordered pairs that are solutions of this equation, we arbitrarily choose values of x and then determine the corresponding values of y. Under these conditions we say that x is the **independent variable** and y the **dependent variable** in the equation. In order to determine other solutions to this equation, it is convenient to set up a *table of values* such as the one shown in Table 3.1.

Table 3.1
Table of values for $y = 2x + 1$

x	−2	−1	0	1	2
$y = 2x + 1$	−3	−1	1	3	5

Thus, along with $(0, 1)$ we have $(-2, -3), (-1, -1), (1, 3)$, and $(2, 5)$ as four other solutions of the equation $y = 2x + 1$. By continuing this process, we can generate many other ordered pairs that satisfy this equation. The **graph** of an equation is the set of *all* ordered pairs (x, y) in the coordinate plane that are solutions of the given equation. The graph of the equation $y = 2x + 1$ is a straight line, as shown in Figure 3.10. The arrowhead at each end of the line indicates that the graph continues indefinitely in that direction. Note that every point on this line is a solution of the equation $y = 2x + 1$.

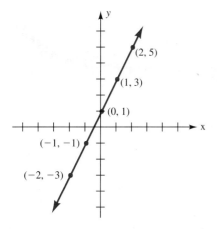

FIGURE 3.10
The graph of the equation $y = 2x + 1$
is a straight line.

◆   **Point-Plotting Method**

The *point-plotting method* is a procedure used in elementary algebra courses to *sketch the graph* of an equation.

Point-Plotting Method

To sketch the graph of an equation by the **point-plotting method,**

1. set up a *table of values* and find a few ordered pairs that satisfy the equation.

2. *Plot* and label the corresponding points in the coordinate plane.

3. Look for a pattern, and *connect the plotted points* to form a smooth curve.

Note: Although this method works well for some simple equations in two unknowns, it is inadequate for sketching the graph of more complicated equations. Throughout this chapter, we will introduce various graphical aids that are useful for sketching an accurate graph by plotting as few points as possible.

E X A M P L E 1   Sketch the graph of each equation by using the point-plotting method.

   (a) $y = |x|$ **(b)** $x = y^2$ **(c)** $2y - x^3 = 0$

S O L U T I O N

(a) We begin by selecting arbitrary values of x, and then finding their corresponding values of y. The following table of values organizes our work.

x	-4	-2	-1	0	1	2	4
$y = |x|$	4	2	1	0	1	2	4

We now plot and label the points given by this table and connect them according to the suggested pattern. The graph of $y = |x|$ is the V-shaped curve shown in Figure 3.11. The arrowheads on the curve indicate that the graph continues upward to the left and upward to the right, according to the suggested pattern.

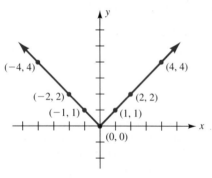

FIGURE 3.11
Graph of $y = |x|$.

(b) For the equation $x = y^2$, it is easier to choose arbitrary values of y and then determine the corresponding values of x. In doing so, we are treating y as the independent variable and x as the dependent variable. The following table of values shows several values of y and their corresponding x values.

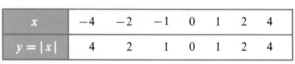

y	-3	-2	-1	0	1	2	3
$x = y^2$	9	4	1	0	1	4	9

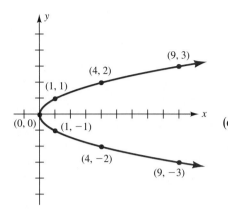

FIGURE 3.12
Graph of $x = y^2$.

We plot and label the points given by this table, and connect them according to the suggested pattern. *Remember that the x-coordinate is always written first in an ordered pair.* The graph of $x = y^2$ is shown in Figure 3.12. It is a cup-shaped curve called a *parabola*. The arrowheads on the curve indicate that the curve continues indefinitely, according to the suggested pattern. This type of curve is studied in detail in Chapter 4.

(c) It is best to solve the equation $2y - x^3 = 0$ for either x or y before setting up a table of values. For this equation, solving for y is easier than solving for x, and we obtain

$$y = \tfrac{1}{2}x^3.$$

The following table of values shows several values of x and their corresponding y-values.

x	-2	-1	0	1	2
$y = \tfrac{1}{2}x^3$	-4	$-\tfrac{1}{2}$	0	$\tfrac{1}{2}$	4

We now plot and label the points given by this table and draw a smooth curve through them. The graph of $2y - x^3 = 0$ is shown in Figure 3.13. The arrowheads on the curve indicate that the graph continues indefinitely, downward to the left and upward to the right. ◆

FIGURE 3.13
Graph of $2y - x^3 = 0$.

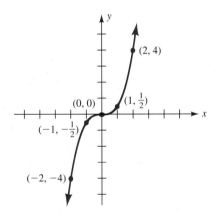

PROBLEM 1 Sketch the graph of the equation $4x + y = 8$. ◆

◆ Symmetry and Intercepts

Referring to Figure 3.11, if we fold the coordinate plane along the y-axis, the right-hand and left-hand portions of the graph will coincide. Thus, we say the graph of $y = |x|$ is **symmetric with respect to the y-axis**. This means that for each point (x, y) on the graph there corresponds a point $(-x, y)$ that is also on the graph. Referring to Figure 3.12, if we fold the coordinate plane along the x-axis, the upper and lower portions of the graph will coincide. Thus, we say that the graph of $x = y^2$ is **symmetric with respect to the x-axis**. This means that for each point (x, y) on the graph there corresponds a point $(x, -y)$ that is also on the graph. Finally, referring to Figure 3.13, if we fold the coordinate plane along the x-axis, and then along the y-axis, the upper and lower portions of the graph will coincide. Thus, we say the graph of $y = \frac{1}{2}x^3$ is **symmetric with respect to the origin**. This means that for each point (x, y) on the graph, there corresponds a point $(-x, -y)$ that is also on the graph. Knowing the symmetry of the graph of an equation enables us to plot only half as many points as we would otherwise need. The following tests can be used to tell whether an equation has any of these three types of symmetry.

 Tests for Symmetry

The graph of an equation in two unknowns x and y is

1. **symmetric with respect to the y-axis** if replacing x with $-x$ yields an equivalent equation.

2. **symmetric with respect to the x-axis** if replacing y with $-y$ yields an equivalent equation.

3. **symmetric with respect to the origin** if replacing x and y with $-x$ and $-y$, respectively, yields an equivalent equation.

When we sketch the graph of an equation, it is helpful to find where the curve crosses the x-axis and y-axis. Such points are called the **intercepts** of the graph. The **x-intercepts** are the x-coordinates of the points where the curve crosses the x-axis. Since y is zero when the curve crosses the x-axis, the x-intercepts can be found by letting y be zero and solving the equation for x. The **y-intercepts** are the y-coordinates of the points where the curve crosses the y-axis. Since x is zero when the curve crosses the y-axis, the y-intercepts can be found by letting x be zero and solving the equation for y. Intercepts and symmetry are useful aids in graphing an equation.

EXAMPLE 2   Use the ideas of symmetry and intercepts to sketch the graph of each equation.

(a) $x^2 + y + 2 = 0$ **(b)** $xy^2 = 1$ **(c)** $y = x^3 - 3x$

SOLUTION

(a) *Symmetry:* Replacing x with $-x$ in the equation $x^2 + y + 2 = 0$ gives us

$$(-x)^2 + y + 2 = 0.$$

Since $(-x)^2 = x^2$ for all real x, we have an equivalent equation. Hence, the graph of $x^2 + y + 2 = 0$ is symmetric with respect to the y-axis.

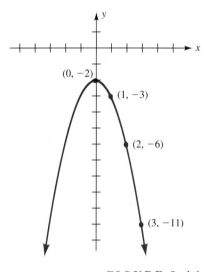

FIGURE 3.14
The graph of $x^2 + y + 2 = 0$ is symmetric with respect to the y-axis.

x-intercepts: Letting $y = 0$ in the equation $x^2 + y + 2 = 0$, we obtain

$$x^2 + (0) + 2 = 0 \qquad \text{or} \qquad x^2 = -2,$$

which has no real solution. Thus, we conclude the graph has no x-intercept.

y-intercepts: Letting $x = 0$ in the equation $x^2 + y + 2 = 0$, we obtain

$$(0)^2 + y + 2 = 0 \qquad \text{or} \qquad y = -2.$$

Hence the y-intercept is -2.

To sketch the graph of $x^2 + y + 2 = 0$, we draw a smooth curve through the y-intercept and the points in the following table of values.

x	1	2	3
$y = -x^2 - 2$	-3	-6	-11

We can then use symmetry to draw the mirror image on the left-hand side of the y-axis. The graph of $x^2 + y + 2 = 0$ is shown in Figure 3.14.

(b)   *Symmetry:*   Replacing y with $-y$ in the equation $xy^2 = 1$, gives us

$$x(-y)^2 = 1.$$

Since $(-y)^2 = y^2$ for all real y, we have an equivalent equation. Hence, the graph of $xy^2 = 1$ is symmetric with respect to the x-axis.

x-intercepts: Letting $y = 0$ in the equation $xy^2 = 1$, we obtain

$$x(0)^2 = 1 \qquad \text{or} \qquad 0x = 1.$$

However, since no number times zero is one, we conclude that the graph has no x-intercept.

y-intercepts: Letting $x = 0$ in the equation $xy^2 = 1$, we obtain

$$(0)y^2 = 1.$$

However, since no number squared times zero is one, we conclude that the graph has no y-intercept.

To sketch the graph of $xy^2 = 1$, we draw a smooth curve through the points in the following table of values.

y	$\frac{1}{3}$	$\frac{1}{2}$	1	2	3
$x = 1/y^2$	9	4	1	$\frac{1}{4}$	$\frac{1}{9}$

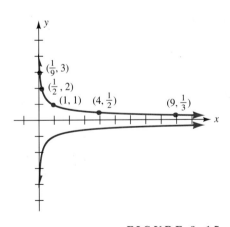

FIGURE 3.15
The graph of $xy^2 = 1$ is symmetric with respect to the x-axis.

We can then use symmetry to draw the mirror image below the x-axis. The graph of $xy^2 = 1$ is shown in Figure 3.15. Since the graph has no x- or y-intercept, it does not cross either axis.

(c) *Symmetry:* Replacing x with $-x$ and y with $-y$ in the equation $y = x^3 - 3x$ gives us an equivalent equation:

$$-y = (-x)^3 - 3(-x)$$

$$-y = -x^3 + 3x \qquad \textbf{Simplify}$$

$$y = x^3 - 3x \qquad \textbf{Multiply both sides by } -1$$

Hence, the graph of $y = x^3 - 3x$ is symmetric with respect to the origin.

x-intercepts: Letting $y = 0$ in the equation $y = x^3 - 3x$, we obtain

$$0 = x^3 - 3x$$

$$0 = x(x^2 - 3)$$

$$x = 0 \quad \text{or} \quad x = \pm\sqrt{3}$$

Hence, we conclude the x-intercepts are 0, $\sqrt{3}$, and $-\sqrt{3}$.

y-intercepts: Letting $x = 0$ in the equation $y = x^3 - 3x$, we obtain

$$y = (0)^3 - 3(0) \qquad \text{or} \qquad y = 0.$$

Since the y-intercept is 0 and an x-intercept is also 0, the graph must pass through the origin $(0, 0)$.

We can use the nonnegative intercepts and the points in the following table of values to sketch the graph of $y = x^3 - 3x$ on the right-hand side of the y-axis.

x	$\frac{1}{2}$	1	2	3
$y = x^3 - 3x$	$-\frac{11}{8}$	-2	2	18

We can then use symmetry to reflect this portion of the graph about the y-axis and then about the x-axis to obtain the other portion of the graph. The graph of $y = x^3 - 3x$ is shown in Figure 3.16. Note the selection of different scales for the x-axis and y-axis. ◆

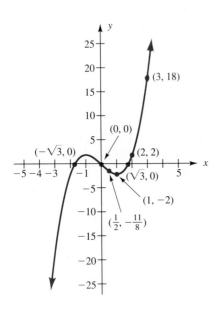

FIGURE 3.16
The graph of $y = x^3 - 3x$ is symmetric with respect to the origin.

PROBLEM 2   Repeat Example 2 for $y^2 + x = 4$. ◆

◆ **Circles**

We can sketch the graph of a circle simply by recognizing and inspecting its equation. A **circle** is the set of all points in a plane that lie a fixed distance from a given point. We call the fixed distance the **radius** and the given point the **center** of the circle. Shown in Figure 3.17 is a circle with radius r and center (h, k). The point (x, y) lies on this circle if and only if the distance from the center (h, k) to (x, y) is r. By using the distance formula (Section 3.1), we have

$$r = \sqrt{(x - h)^2 + (y - k)^2}.$$

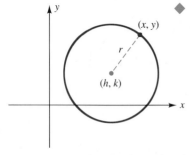

FIGURE 3.17
Circle with center (h, k) and radius r.

If we square both sides of this equation, we obtain the equation of a circle in standard form.

Equation of a Circle in Standard Form

The **equation of a circle in standard form** with center (h, k) and radius r is

$$(x - h)^2 + (y - k)^2 = r^2.$$

EXAMPLE 3   Determine the equation of a circle in standard form with center $(2, -1)$ and radius 3.

SOLUTION   Using the equation of a circle in standard form with $h = 2$, $k = -1$, and $r = 3$, we have

$$(x - 2)^2 + [y - (-1)]^2 = 3^2$$
$$(x - 2)^2 + (y + 1)^2 = 9.$$

The graph of this equation is given in Figure 3.18.

FIGURE 3.18

A circle with center $(2, -1)$ and radius 3 has the equation $(x - 2)^2 + (y + 1)^2 = 9$.

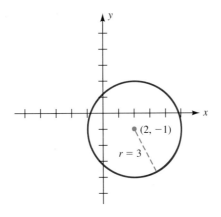

Conversely, we can state that the graph of any equation of the form

$$(x - h)^2 + (y - k)^2 = r^2 \quad \text{with } r > 0$$

is a circle with center (h, k) and radius r.

PROBLEM 3   Sketch the graph of the equation $(x + 2)^2 + (y - 1)^2 = 25$.

Squaring the expressions $(x - h)$ and $(y - k)$ in the equation

$$(x - h)^2 + (y - k)^2 = r^2$$

gives us

$$x^2 - 2hx + h^2 + y^2 - 2ky + k^2 = r^2$$
$$x^2 + y^2 - 2hx - 2ky + (h^2 + k^2 - r^2) = 0.$$

Replacing the constants $-2h$ with D, $-2k$ with E, and $(h^2 + k^2 - r^2)$ with F, we have

$$x^2 + y^2 + Dx + Ey + F = 0$$

This equation is called the **equation of a circle in general form** and is characterized by the presence of x^2 and y^2 terms, each having a coefficient of 1. As illustrated in the next example, *intercepts, symmetry, and equation recognition are all important aids in graphing an equation.*

EXAMPLE 4   Sketch the graph of the equation $x^2 + y^2 = 25$.

SOLUTION   Using the tests for symmetry, we find that the graph of this equation is symmetric with respect to *both axes* as well as being symmetric with respect to the origin. The x-intercepts are ± 5 and the y-intercepts are ± 5. Using these facts, and plotting a few additional points, we can graph the equation. However, it is simpler to recognize that $x^2 + y^2 = 25$ is an equation of the form

$$x^2 + y^2 + Dx + Ey + F = 0, \quad \text{where } D = E = 0 \text{ and } F = -25.$$

Writing the equation $x^2 + y^2 = 25$ in the form

$$(x - 0)^2 + (y - 0)^2 = 5^2$$

Standard form of a circle

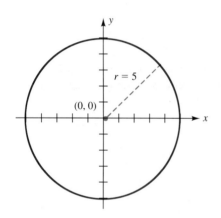

FIGURE 3.19
The graph of $x^2 + y^2 = 25$ is a circle with center $(0, 0)$ and radius 5.

allows us to identify the graph of this equation as a circle with center $(0, 0)$ and radius 5. The graph of the circle is now immediate and is shown in Figure 3.19. ◆

PROBLEM 4   Sketch the graph of the equation $x^2 + y^2 = 4$. ◆

To graph an equation of the form $x^2 + y^2 + Dx + Ey + F = 0$, it is best to write the equation in the standard form $(x - h)^2 + (y - k)^2 = r^2$. Once the equation is in standard form, we can easily draw its graph. As illustrated in the next example, we sometimes use the process of completing the square (Section 2.3) to write the equation $x^2 + y^2 + Dx + Ey + F = 0$ in the form $(x - h)^2 + (y - k)^2 = r^2$.

EXAMPLE 5   Sketch the graph of each equation.

(a) $x^2 + y^2 + 6x = 0$ (b) $4x^2 + 4y^2 - 8x + 20y + 13 = 0$

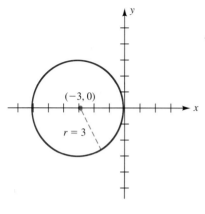

FIGURE 3.20

The graph of $x^2 + y^2 + 6x = 0$ is a circle with center $(-3, 0)$ and radius 3.

SOLUTION

(a) The equation $x^2 + y^2 + 6x = 0$ appears to be the equation of a circle in general form with $D = 6$, $E = 0$, and $F = 0$. Using the process of completing the square, we have

$$(x^2 + 6x \quad) + y^2 = 0 \qquad \text{Regroup}$$

$$(x^2 + 6x + 9) + y^2 = 9 \qquad \text{Complete the square by adding 9 to both sides}$$

$$(x + 3)^2 + y^2 = 9 \qquad \text{Factor the perfect square trinomial}$$

$$[x - (-3)]^2 + (y - 0)^2 = 3^2 \qquad \text{Write in standard form}$$

Thus, the center of the circle is $(-3, 0)$ and the radius is 3. The graph of this circle is shown in Figure 3.20.

(b) Dividing both sides by 4, we obtain

$$x^2 + y^2 - 2x + 5y + \tfrac{13}{4} = 0,$$

which appears to be the equation of a circle in general form with $D = -2$, $E = 5$, and $F = \tfrac{13}{4}$. Using the process of completing the square, we have

$$(x^2 - 2x \quad) + (y^2 + 5y \quad) = -\frac{13}{4} \qquad \text{Regroup}$$

Add 1 to both sides.

$$(x^2 - 2x + 1) + \left(y^2 + 5y + \frac{25}{4}\right) = -\frac{13}{4} + \left(1 + \frac{25}{4}\right) \qquad \text{Complete the squares}$$

Add $\tfrac{25}{4}$ to both sides.

$$(x - 1)^2 + \left(y + \frac{5}{2}\right)^2 = 4 \qquad \text{Factor}$$

$$(x - 1)^2 + \left[y - \left(-\frac{5}{2}\right)\right]^2 = 2^2 \qquad \text{Write in standard form}$$

FIGURE 3.21

The graph of
$4x^2 + 4y^2 - 8x + 20y + 13 = 0$
is a circle with center $(1, -\frac{5}{2})$
and radius 2.

Thus, the center of this circle is $(1, -\frac{5}{2})$ and the radius is 2. The graph of this equation is shown in Figure 3.21.   ◆

Not every equation of the form $x^2 + y^2 + Dx + Ey + F = 0$ is a circle. If this equation is written in the standard form $(x - h)^2 + (y - k)^2 = r^2$ and $r^2 = 0$, then its graph is the single point (h, k). If $r^2 < 0$ when the equation is written in standard form, then it has no graph, since no point (x, y) with real coordinates can satisfy the equation.

PROBLEM 5   Is $x^2 + y^2 + 4x + 6y + 13 = 0$ the equation of a circle? Explain.   ◆

Exercises 3.2

 Basic Skills

In Exercises 1–12, determine the x-intercept(s) and y-intercept(s) for the graph of each equation.

1. $6x + 5y = 10$

2. $y = 2x - 7$

3. $x^2 + 3y - 36 = 0$

4. $y = 2x^2 + 5x - 3$

5. $y = 8x - x^4$

6. $xy = 4$

7. $x^2 + y^2 - 6x - 16 = 0$

8. $4x^2 + y^2 + 4x - 2y + 1 = 0$

9. $y = \dfrac{\sqrt{x^2 - 2}}{x}$

10. $y = \dfrac{x - 4}{2x - 3}$

11. $3y = x^2 + 2x + 9$

12. $x^2 + 4x - 6y = 3$

In Exercises 13–24, determine if the graph of each equation has symmetry with respect to the x-axis, y-axis, or origin.

13. $x^2 + 4y = 9$

14. $x - y^4 = 4$

15. $xy = 2$

16. $y = x^{2/3}$

17. $x = 3 - |y|$

18. $x^2 + 4y^2 = 16$

19. $x^2 + 1 = xy$

20. $x^2y + y = 1$

21. $y = \sqrt{x - 2}$

22. $y = 3x - 2x^3 + x^5$

23. $|x| + |y| = 1$

24. $y = 3x^3 - 2x^2$

In Exercises 25–44, use the point-plotting method and the ideas of symmetry and intercepts to sketch the graph of each equation.

25. $y = x$

26. $y = x^2$

27. $y = \sqrt{x}$

28. $y = x^{1/3}$

29. $xy = 1$

30. $y - 3x = 0$

31. $3x + 2y = 12$

32. $4x - 3y = 12$

33. $x = y^2 + 2y$

34. $y = x^2 - 4x$

35. $y = 12x - x^3$

36. $y = x^4 - 4x^2$

37. $y = |2x - 5|$

38. $x = -|4 - y^2|$

39. $x = -\sqrt{4 - y^2}$

40. $y = \sqrt{x^2 - 9}$

41. $|x| + |y| = 2$

42. $|x| - |y| = 2$

43. $y = \dfrac{1}{x^2 + 1}$

44. $y = \dfrac{x^2 + 1}{x^2}$

In Exercises 45–56, give the center and radius for each equation that defines a circle. Then sketch the graph.

45. $x^2 + (y - 2)^2 = 16$

46. $(x + 5)^2 + (y + 2)^2 = 4$

47. $x^2 + y^2 = 9$

48. $x^2 + y^2 - 36 = 0$

49. $x^2 + y^2 - 2y = 3$

50. $x^2 + y^2 + 5x = 0$

51. $x^2 + y^2 - 6x + 8y + 9 = 0$

52. $x^2 + y^2 - x - 4y = 2$

53. $4x^2 + 4y^2 - 8x + 12y = 3$

54. $4x^2 + 4y^2 + 20x - 16y + 41 = 0$

55. $3x^2 + 3y^2 - 4x - 8y + 24 = 0$

56. $2x^2 + 2y^2 - x + 20y + 40 = 0$

In Exercises 57–64, write the equation of a circle that has the given characteristics.

57. Center at $(0, 0)$; radius of 2

58. Center at $(-3, 0)$; radius of 3

59. Center at $(2, -3)$; radius of $\sqrt{2}$

60. Center at $(-1, 2)$; radius of $\frac{2}{3}$

61. Center at the origin; passes through the point $(-3, 4)$

62. Center at $(-1, 3)$; passes through the point $(1, -1)$

63. Line segment from $(2, 3)$ to $(-2, 5)$ is the diameter

64. Line segment from $(-1, -1)$ to $(3, 1)$ is the radius (*Hint:* There are two possible answers.)

 Critical Thinking

65. Graph the equation $y = |x|$. Then on the same set of axes, sketch the graphs of the equations $y = |x| + 1$, $y = |x| + 2$, and $y = |x| + 3$. Compare the graphs of these equations. What effect does the constant c have on the graph of $y = |x| + c$ when $c > 0$?

66. Graph the equation $y = x^3$. Then on the same set of axes, sketch the graphs of the equations $y = x^3 - 1$, $y = x^3 - 2$, and $y = x^3 - 3$. Compare the graphs of these equations. What effect does the constant c have on the graph of $y = x^3 + c$ when $c < 0$?

67. Graph the equation $y = \sqrt{x}$. Then on the same set of axes, sketch the graphs of the equations $y = \sqrt{x-1}$, $y = \sqrt{x-2}$, and $y = \sqrt{x-3}$. Compare the graphs of these equations. What effect does the constant c have on the graph of $y = \sqrt{x+c}$ when $c < 0$?

68. Graph the equation $y = x^2$. Then on the same set of axes, sketch the graphs of the equations $y = (x+1)^2$, $y = (x+2)^2$, and $y = (x+3)^2$. Compare the graphs of these equations. What effect does the constant c have on the graph of $y = (x+c)^2$ when $c > 0$?

69. Two circles having the same center but different radii are called *concentric circles*.

   (a) If the ratio of the radii of two concentric circles is 2:1 and the equation of the circle with the larger radii is $x^2 + y^2 - 10x + 6y = 2$, determine the equation of the other circle.

(b) Find the area between the two circles described in part (a).

70. Given the circles $x^2 + y^2 + 2x + 4y + 1 = 0$ and $x^2 + y^2 - 8y = 0$,

   (a) find the distance between their centers.
   (b) find the coordinates of the midpoint of the line segment that joins their centers.

71. A *tangent* to a circle is a line that intersects the circle in one and only one point. Find the equation of a circle with diameter 10 if both the x-axis and y-axis are tangent to the circle. (*Hint:* There are four possible answers.)

72. For the circle $x^2 + y^2 + Dx + Ey + F = 0$, use the method of completing the square to determine a formula in terms of D, E, and F for the radius of the circle.

 Calculator Activities

In Exercises 73–76, find the x-intercept(s) and y-intercept(s) for the graph of each equation. Round off each intercept to three significant digits.

73. $8.75x - 2.95y = 7.84$

74. $21.5x^3 + 39.7y - 46.8 = 0$

75. $(x + 2.41)^2 + (y - 7.93)^2 = 9.22$

76. $2.7x^2 + 5.9x - 4.3y = 8.9$

In Exercises 77 and 78, determine the center, radius, and x-intercept(s) and y-intercept(s) of the circle with the given equation. Round off each answer to three significant digits.

77. $x^2 + y^2 - 7.22x - 4.84y + 12.34 = 0$

78. $4.1x^2 + 4.1y^2 - 24.1x - 16.3y + 25.2 = 0$

79. A castle entrance has the form of a Gothic arch, as shown in the figure.

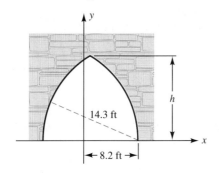

(a) Taking the axes as shown, find the equation of the circle that forms the left-hand side of the arch.
(b) Use the equation in part (a) to find the height h of the arch, rounded to three significant digits.

80. The radius of a Ferris wheel is 10.1 meters, as shown in the figure.

   (a) Taking the axes as shown, find the equation of the circle that forms the Ferris wheel.
   (b) Use the equation in part (a) to find the distance x from a chair to the vertical axis of the wheel when the height of the chair is 4.2 meters above the ground. Round the answer to three significant digits.

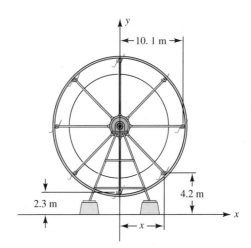

3.3 ◆ Functions

◆ Introductory Comments

Given an equation in two unknowns x and y, suppose we *choose* values for x and *obtain* corresponding values for y. Hence, we are using x as the *independent variable* and y as the *dependent variable*. The set of all numbers from which we may choose is called the **domain** and the set of all numbers that we obtain is called the **range**. If for each value of x in the domain there corresponds one and only one value of y in the range, we say that y is a **function** of x.

◆◆ Definition of a Function

Input: $x = 3$

Rule:
"Square the input."

Output: $y = 9$

FIGURE 3.22
Function machine for $y = x^2$

> A **function** from a set X to a set Y is a rule of correspondence that assigns to each element x in X *exactly one* element y in Y.

A function can be represented by a list or table, by a graph, or by a formula or equation. Many functions in this text are specified by equations. For example, consider the equation $y = x^2$. It is useful to think of the x-values as **inputs** and their corresponding y-values as **outputs**. We can think of this equation as a rule that says "Square the input." A *function machine* for this equation is shown in Figure 3.22. If we place $x = 3$ in the input hopper, the machine follows the rule "square the input" and gives us an output of $y = 9$. The rule "square the input" assigns to each input value x one and only one output value y. Thus we say the equation $y = x^2$ *defines y as a function of x.*

Not every equation in two unknowns x and y defines y as a function of x. Consider the equation $y^2 = x$. If we choose the input $x = 9$, the equation becomes $y^2 = 9$ and, consequently, $y = \pm 3$. In this equation we obtain two outputs for the input $x = 9$. Since our definition of a function requires that there be exactly one output for each input, we know that the equation $y^2 = x$ does not define y as a function of x.

◆ Vertical Line Test

Suppose we are given the graph of an equation in two unknowns x and y. If every vertical line that can be drawn in the coordinate plane intersects the graph *at most once*, then we can say the equation has exactly one output y for each input value x that we can assign. Hence, the equation defines y as a function of x. We refer to this graphical method of determining whether y is a function of x as the **vertical line test**.

◆◆ Vertical Line Test

> An equation defines y as a function of x if and only if every vertical line in the coordinate plane intersects the graph of the equation at most once.

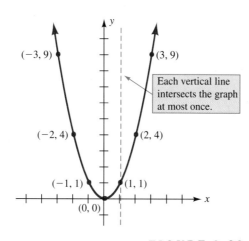

FIGURE 3.23

The graph of $y = x^2$ passes the vertical
line test.

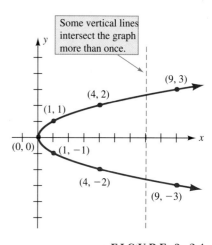

FIGURE 3.24

The graph of $y^2 = x$ fails the vertical
line test.

Notice the graph of $y = x^2$ (see Figure 3.23) passes the vertical line test, while the graph of $y^2 = x$ (see Figure 3.24) fails this test.

E X A M P L E   1   Determine if the equation defines y as a function of x.

(a) $x^2 + y - 4 = 0$ (b) $x^2 + y^2 - 4 = 0$

S O L U T I O N

(a) Using the tests for symmetry in Section 3.2, we know that the graph of the equation $x^2 + y - 4 = 0$ is symmetric with respect to the y-axis. The x-intercepts are ± 2 and the y-intercept is 4. The graph of this equation is shown in Figure 3.25. Note that every vertical line intersects this graph *at most once*. Thus, by the vertical line test, the equation $x^2 + y - 4 = 0$ defines y as a function of x. Also, if we solve the equation for y, we obtain

$$y = 4 - x^2.$$

From this form of the equation we can see that for each input value x we choose, one, and only one, output value y corresponds to it. The equation $x^2 + y - 4 = 0$ defines a *quadratic function*. (Quadratic functions are discussed in detail in Chapter 4.)

(b) From Section 3.2, we know that the graph of the equation $x^2 + y^2 - 4 = 0$ is a circle with center (0, 0) and radius 2. The graph of this equation is shown in Figure 3.26. Since at least one vertical line intersects this graph twice, we know that the equation $x^2 + y^2 - 4 = 0$ does not define y as a function of x. Also, if we solve this equation for y, we obtain

$$y = \pm\sqrt{4 - x^2}.$$

FIGURE 3.25

The graph of $x^2 + y - 4 = 0$ passes the
vertical line test.

FIGURE 3.26
The graph of $x^2 + y^2 - 4 = 0$ fails the vertical line test.

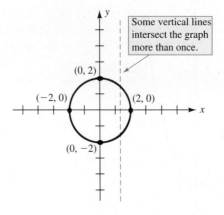

This form of the equation makes it obvious that we have *two* outputs for each input value x in the interval $(-2, 2)$. ◆

Note: Taken separately, the equations $y = \sqrt{4 - x^2}$ and $y = -\sqrt{4 - x^2}$ define y as a function of x, and the graph of each function is half a circle. Referring to Figure 3.26 we see that the graph of $y = \sqrt{4 - x^2}$ is the semicircle above the x-axis and the graph of $y = -\sqrt{4 - x^2}$ is the semicircle below the x-axis. Note that each semicircle passes the vertical line test.

PROBLEM 1   Determine if the equation $2x + y^2 = 4$ defines y as a function of x. ◆

◆ **Functional Notation**

The letters f, F, g, G, h, and H are often used to represent functions. The *functional notation* $f(x)$ is read "f of x" and is defined as follows.

◆ Functional Notation
_____

> If f is a function and x is an input for the function, then the **functional notation**
>
> $$f(x), \text{ read "}f\text{ of }x\text{,"}$$
>
> denotes the corresponding output of the function.

The function f defined by the equation $y = x^2$ may be written as

$$f(x) = x^2,$$

where f specifies a rule for determining the value of an output $f(x)$ from a given input x. For instance, $f(-3)$ denotes the output for a given input of -3 using the rule "square the input." Hence,

$$f(-3) = (-3)^2 = 9.$$

Table 3.2 shows several other inputs x and outputs $f(x)$ for the function f defined by $f(x) = x^2$. We refer to the process of determining the value of $f(x)$ from a given input x as *computing the functional value*.

Table 3.2
Table of values for $f(x) = x^2$.

x	-3	-2	-1	0	1	2	3
$f(x)$	9	4	1	0	1	4	9

EXAMPLE 2   Let G be a function defined by $G(x) = x^2 - 4x + 3$. Compute each functional value:

(a) $G(-2)$ (b) $G(2a)$ (c) $G(x-1)$

SOLUTION

(a)   The notation $G(-2)$ denotes the output obtained from an input of -2. Replacing x with -2 in the rule for G, we obtain

$$G(-2) = (-2)^2 - 4(-2) + 3$$
$$= 4 + 8 + 3$$
$$= 15.$$

(b)   The notation $G(2a)$ denotes the output obtained from an input of $2a$. Replacing x with $2a$ in the rule for G, we obtain

$$G(2a) = (2a)^2 - 4(2a) + 3$$
$$= 4a^2 - 8a + 3.$$

(c)   The notation $G(x-1)$ denotes the output obtained from an input of $x - 1$. Replacing x with $x - 1$ in the rule for G, we obtain

$$G(x-1) = (x-1)^2 - 4(x-1) + 3$$
$$= x^2 - 2x + 1 - 4x + 4 + 3$$
$$= x^2 - 6x + 8. \qquad \blacklozenge$$

PROBLEM 2   For the function G defined in Example 2, find $G(x-h)$. $\blacklozenge$

As illustrated in the next example, a function can be defined by a multipart rule. We refer to such a function as a **piecewise-defined function**. When we work with a piecewise-defined function, it is important to observe how the rule changes over different subsets of the domain.

EXAMPLE 3   Let h be a function defined by

$$h(x) = \begin{cases} 1 - x^2 & \text{if } x < 1 \\ x - 1 & \text{if } 1 \le x \le 3 \\ 4 & \text{if } x > 3. \end{cases}$$

Compute each functional value:

(a) $h(-2)$ (b) $h(3)$ (c) $h(\pi)$

SOLUTION

(a)   If $x < 1$, then the function h is defined by $h(x) = 1 - x^2$. Thus, for $x = -2$, we have

$$h(-2) = 1 - (-2)^2 = 1 - 4 = -3.$$

(b)   If $1 \le x \le 3$, then the function h is defined by $h(x) = x - 1$. Thus, for $x = 3$, we have

$$h(3) = (3) - 1 = 2.$$

(c)   If $x > 3$, then the function h is defined by $h(x) = 4$. Thus, for $x = \pi \approx 3.14$, we have

$$h(\pi) = 4.$$ ◆

PROBLEM 3   For the function h defined in Example 3, find $h(1)$.   ◆

◆ **Finding the Domain of a Function**

We can think of the domain of a function as the set of all possible inputs. When a function is defined by an equation, the domain is assumed to be the set of all real numbers that give real number outputs. For the function f defined by $f(x) = x^2$, every real number assigned to x leads to a corresponding real number for $f(x)$. Thus, the domain of the function f is the set of all real numbers. In this text, we express the domain of a function by using *interval notation* (see Section 2.5). Thus, we express the domain of f as $(-\infty, \infty)$.

The *range* of a function is the set of all outputs that we obtain from the elements in the domain. Consider the function f defined by $f(x) = x^2$. When real numbers are squared, we obtain *nonnegative* real numbers. Thus, using interval notation, we conclude that the range of f is $[0, \infty)$. For most functions, however, the range is quite difficult to find and is best obtained from the graph of the function, which we will discuss later in this section. As illustrated in the next example, we must be aware of two properties when determining the domain of a function:

1. Division by zero   and   2. Even roots of negative numbers.

EXAMPLE 4   Determine the domain of each function.

(a) $F(x) = \dfrac{2x + 5}{x^3 - 3x^2 + 2x - 6}$

(b) $g(x) = \dfrac{\sqrt{2 - x}}{x + 5}$

SOLUTION

(a)   Remember that division by zero is undefined. Thus, the domain of this function is all real numbers x except those for which the denominator equals zero:

$$x^3 - 3x^2 + 2x - 6 = 0$$

$$x^2(x - 3) + 2(x - 3) = 0 \qquad \textbf{Factor by grouping terms}$$

$$(x - 3)(x^2 + 2) = 0$$

$$x = 3 \quad \text{or} \quad x^2 + 2 = 0$$

$$\boxed{\text{No real solution}}$$

Hence, the domain of this function is all real numbers *except* $x = 3$. Using interval notation, we express the domain of this function as

$$(-\infty, 3) \cup (3, \infty).$$

(b)   Remember, even roots of negative numbers are *not* real numbers. Thus, the radicand $2 - x$ must be *nonnegative*, that is,

$$2 - x \geq 0$$

$$-x \geq -2$$

$$x \leq 2.$$

Also, we must exclude $x = -5$ from the domain, since this value of x makes the denominator zero. Hence, the domain of this function is all real numbers less than or equal to 2 except $x = -5$. Using interval notation, we express the domain of this function as

$$(-\infty, -5) \cup (-5, 2].$$ ◆

PROBLEM 4   Determine the domain of the function H defined by $H(x) = \dfrac{x + 5}{\sqrt{2 - x}}$.   ◆

◆ **The Graph of a Function**

The graph of a function f is the same as the graph of the equation $y = f(x)$ and is defined as follows.

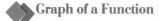 Graph of a Function

> The **graph of a function** f is the set of all points (x, y) in the coordinate plane such that x is in the domain of f and $y = f(x)$.

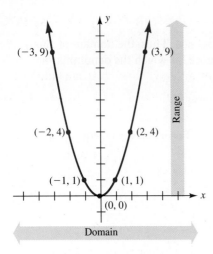

FIGURE 3.27
Graph of $f(x) = x^2$.
Domain: $(-\infty, \infty)$.   Range: $[0, \infty)$.

The point-plotting method we discussed in Section 3.2 may be used to sketch the graph of a simple function. We select various x-values from the domain of f and find the corresponding outputs $f(x)$. Using the inputs as the x-coordinates and the outputs as the y-coordinates, we plot the points $(x, f(x))$ in the coordinate plane. Connecting these points to form a smooth curve gives us the graph of f.

The graph of the function f defined by $f(x) = x^2$ is shown in Figure 3.27. The graph of this function is the same as the graph of the equation $y = x^2$ (see Figure 3.23). Observe that the domain and range of the function f are evident from its graph.

Note that the graph of $f(x) = x^2$ (Figure 3.27) is symmetric with respect to the y-axis. We say that a function is an *even function* if its graph is symmetric with respect to the y-axis or that it is an *odd function* if its graph is symmetric with respect to the origin. The following tests can be used to determine if a function is even or odd.

Tests for Even and Odd Functions

1. A function f is an **even function** if $f(-x) = f(x)$ for every x in the domain of f.

2. A function f is an **odd function** if $f(-x) = -f(x)$ for every x in the domain of f.

EXAMPLE 5   Determine if the function is even, odd, or neither.

   (a) $g(x) = \sqrt{x^2 - 4}$ **(b)** $h(x) = x - x^3$

SOLUTION

(a)   The function g is *even*, since

$$g(-x) = \sqrt{(-x)^2 - 4} = \sqrt{x^2 - 4} = g(x).$$

Thus, we know the graph of the function g is symmetric with respect to the y-axis.

(b)   The function h is *odd*, since

$$h(-x) = (-x) - (-x)^3 = -x + x^3 = -(x - x^3) = -h(x).$$

Thus, we know the graph of the function h is symmetric with respect to the origin.   ◆

PROBLEM 5   Repeat Example 5 for the function h defined by $h(x) = x - x^2$.   ◆

The **zeros** of a function f are the values of x for which $f(x) = 0$. The *real zeros* of a function f are the x-intercepts of the graph of f.

EXAMPLE 6   Find the real zeros of each function and sketch the graph of each function:

(a) $g(x) = \sqrt{x^2 - 4}$ **(b)** $h(x) = x - x^3$

SOLUTION

(a)   The zeros of g are the roots of the equation $g(x) = \sqrt{x^2 - 4} = 0$. Squaring both sides, we have

$$x^2 - 4 = 0 \quad \text{or} \quad x = \pm 2.$$

Thus, the zeros of the function g are ± 2, and the x-intercepts of the graph of g are also ± 2. Using the point-plotting method along with the fact that g is an even function with real zeros ± 2, we can sketch the graph of g. The graph of the function g is shown in Figure 3.28.

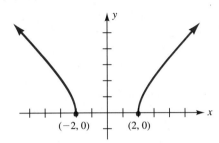

FIGURE 3.28
Graph of $g(x) = \sqrt{x^2 - 4}$, an *even* function with *zeros* ± 2.

(b)   The zeros of h are the roots of the equation $h(x) = x - x^3 = 0$. Factoring and applying the zero product property, we have

$$x - x^3 = 0$$
$$x(1 - x^2) = 0$$
$$x = 0 \quad \text{or} \quad 1 - x^2 = 0$$
$$x = \pm 1$$

Thus, the zeros of the function h are 0 and ± 1, and the x-intercepts of the graph of h are also 0 and ± 1. Using the point-plotting method along with the fact that h is an odd function with real zeros 0 and ± 1, we can sketch the graph of h. The graph of the function h is shown in Figure 3.29.   ◆

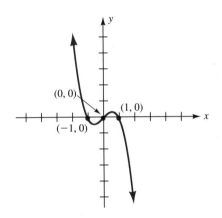

FIGURE 3.29
Graph of $h(x) = x - x^3$, an *odd* function with *zeros* 0 and ± 1.

The domain and range of a function are apparent from its graph. From the graph of the function g in Figure 3.28, we can state that the domain of g is $(-\infty, -2] \cup [2, \infty)$ and the range is $[0, \infty)$.

PROBLEM 6   Use the graph of the function h in Figure 3.29 to find the domain and range of h.   ◆

◆ **The Graphing Calculator**

Recent technology enables us to generate the graph of a function by using a personal computer with graphing software or a hand-held graphing calculator. Although you may not have access to a personal computer with graphing software, you can purchase a hand-held graphing calculator at a cost that is slightly more than a regular scientific calculator.

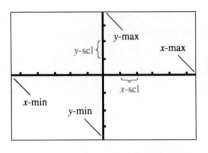

FIGURE 3.30

Viewing rectangle on a graphing calculator.

The key to using a graphing calculator is to select a viewing screen that shows the portion of the graph you wish to display. The viewing screen on a graphing calculator, which is usually called the *viewing rectangle*, represents a portion of the coordinate plane. On most graphing calculators, the [RANGE] key is used to select a viewing rectangle. After pressing the [RANGE] key, you enter several values:

x-min	x-axis minimum value	**y-min**	y-axis minimum value
x-max	x-axis maximum value	**y-max**	y-axis maximum value
x-scl	distance between scale marks on x-axis	**y-scl**	distance between scale marks on y-axis

These are shown in Figure 3.30.

For instance, to generate the graph of $g(x) = \sqrt{x^2 - 4}$ (see Figure 3.28) on a graphing calculator, we begin by pressing the [RANGE] key to choose a viewing rectangle. Although the choice of a viewing rectangle is arbitrary, Figure 3.28 suggests the viewing rectangle

$$[-6, 6] \qquad \text{by} \qquad [-2, 6]$$

x-min　x-max　　　　y-min　y-max

with x- and y-scales of 1 unit. After entering these values, we may now obtain the graph of $g(x) = \sqrt{x^2 - 4}$ in the viewing rectangle. For instance, on a *Texas Instrument TI-81* we enter the keying sequence

[Y=] [√] [(] [ALPHA] [X] [X²] [−] [4] [)] [GRAPH]

or in the viewing rectangle of a *CASIO fx-7000 G* we enter the keying sequence

[GRAPH] [√] [(] [ALPHA] [X] [X²] [−] [4] [)] [EXE].

The procedure with other graphing calculators is similar. If you have access to a graphing calculator, you may wish to consult the Graphing Calculator Supplement that accompanies this text for more detailed explanations.

The [TRACE] key on a graphing calculator can be used to display the x- and y-coordinates associated with each dot, or *pixel*, along the graph. With the [ZOOM] key we can magnify the graph n times. Look for the logo ▦ in the *Calculator Activities* section of each exercise set. The problems with this logo are designed for students who have access to a graphing calculator.

Do not use a graphing calculator to obtain a quick solution with no understanding of the fundamental mathematical ideas being presented. To become dependent upon this machine for solutions can produce disastrous effects. Instead, use it to explore, investigate, and check your mathematical problems. In this way, you will enhance your learning and understanding of college mathematics.

Exercises 3.3

 Basic Skills

In Exercises 1–14, determine if the given equation defines y as a function of x.

1. $x^2 + y - 1 = 0$ 2. $3x + 2y = 6$

3. $x + y^2 = 1$ 4. $x^2 + y - 2x = 0$

5. $y = \sqrt{x^2 - 4}$ 6. $y = |x - 5|$

7. $x^2 + y^2 - 4x = 0$ 8. $x^2 - y^2 - 4 = 0$

9. $|x|y = 1$ 10. $|xy| = 1$

11. $y^3 + x - 1 = 0$ 12. $x^3 + y^3 = 1$

13. $xy - x^2 = 1$ 14. $x^2 y + y = 1$

Given the functions f, g, and h, defined by

$$f(x) = 4x^2 - 2x + 1$$

$$g(x) = \sqrt{x} - 4 \quad \text{and} \quad h(x) = |2x + 3|,$$

respectively, compute the functional values given in Exercises 15–34.

15. $f(2)$ 16. $g(9)$

17. $h(-4)$ 18. $g(\frac{1}{4}) + g(1)$

19. $f(\sqrt{2})$ 20. $-h(\frac{3}{2})$

21. $h(ab)$ 22. $f(3p)$

23. $g(t^2), \quad t \ge 0$ 24. $f(-x)$

25. $f\left(\dfrac{n}{2}\right)$ 26. $h(x^2 - 3)$

27. $g(x) - g(0)$ 28. $f(x + 2)$

29. $g(1 + x^2)$ 30. $f(x + h) - f(x)$

31. $f(\sqrt{x - 2})$ 32. $h\left(-\dfrac{x}{2}\right)$

33. $f(g(x))$ 34. $g(f(x))$

In Exercises 35–38, for the given piecewise-defined function, compute each functional value.

35. $f(x) = \begin{cases} 2x - 5 & \text{if } x < 0 \\ x^2 & \text{if } x \ge 0 \end{cases}$

   (a) $f(3)$ (b) $f(-3)$
   (c) $f(2)$ (d) $f(0)$

36. $g(x) = \begin{cases} \sqrt{3 - x} & \text{if } x \le 3 \\ 2 - x & \text{if } 3 < x < 6 \\ 1 & \text{if } x \ge 6 \end{cases}$

   (a) $g(0)$ (b) $g(4)$
   (c) $g(3)$ (d) $g(7)$

37. $h(x) = \begin{cases} x & \text{if } x > 0 \\ 0 & \text{if } x = 0 \\ -x & \text{if } x < 0 \end{cases}$

   (a) $h(7)$ (b) $h(-3)$
   (c) $h(5) - h(0)$ (d) $h(x + 1)$

38. $f(x) = \begin{cases} 1 & \text{if } x \text{ is a rational number} \\ -1 & \text{if } x \text{ is an irrational number} \end{cases}$

   (a) $f(0)$ (b) $f(6)$
   (c) $f(\pi)$ (d) $f(\sqrt{2})$

In Exercises 39–54, state the domain of each function using interval notation.

39. $f(x) = 3x + 2$ 40. $g(x) = 1 - x^2$

41. $f(x) = x^3 + 2$ 42. $f(x) = \sqrt[3]{2x - 1}$

43. $H(x) = \sqrt{4 - x}$ 44. $h(x) = \sqrt{2x - 3}$

45. $g(x) = -\sqrt{16 - x^2}$ 46. $F(x) = \sqrt{x^2 - 4}$

47. $f(x) = \dfrac{x}{x + 2}$ 48. $f(x) = \dfrac{-3}{\sqrt{2 - x}}$

49. $G(x) = \dfrac{1}{x^2 - 4}$ 50. $f(x) = \dfrac{2x}{x^2 + 3}$

51. $f(x) = \dfrac{-3}{x^2 + 3x - 10}$

52. $f(x) = \dfrac{2x - 3}{x^3 - 4x^2 + x - 4}$

53. $F(x) = \dfrac{\sqrt{2x - 1}}{4 - x}$

54. $H(x) = \sqrt{x^3 - 2x^2 - 8x}$

In Exercises 55–64, determine if the function is even, odd, or neither. If you have access to a graphing calculator, verify your answer by looking at the symmetry of the graph.

55. $F(x) = x$ 56. $g(x) = x^3$

57. $f(x) = 3 - |x|$

58. $f(x) = 2x^2 - 4$

59. $f(x) = 4x - x^5$

60. $f(x) = x^4 + 2x^2 + 3$

61. $g(x) = \dfrac{3 - x^2}{2 + x^6}$

62. $H(x) = \dfrac{x + x^3}{2|x|}$

63. $h(x) = x^3 - 3x + 1$

64. $g(x) = (3x^2 - 2x)^2$

In Exercises 65–74, find the real zeros of each function. If you have access to a graphing calculator, verify your answer by tracing to the x-intercepts of the graph.

65. $g(x) = 3x - 5$

66. $f(x) = 4 + 5x$

67. $F(x) = x^2 - 3x - 40$

68. $g(x) = 2x^2 - 7x + 3$

69. $h(x) = x^2 - 8x + 4$

70. $f(x) = 2x^4 - 5x^2 - 12$

71. $g(x) = \dfrac{2x - 3}{x^2 + 4}$

72. $G(x) = \dfrac{9 - x^2}{2x + 3}$

73. $H(x) = 5 - \sqrt{x^4 + 9}$

74. $h(x) = x - \sqrt{4x^2 - x - 2}$

In Exercises 75–84, a function and its graph are given.

(a) *Find the domain of the function.*
(b) *Find the range of the function.*
(c) *Find any real zeros of the function.*
(d) *Determine if the function is even, odd, or neither.*

75. $f(x) = |x| - 2$

76. $g(x) = 3x - 2$

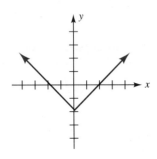

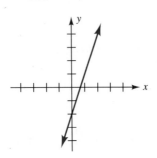

77. $g(x) = x^2 + 2x$

78. $H(x) = 2x^2 - x^4$

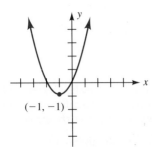

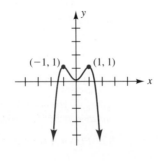

79. $f(x) = \sqrt{5 - x^2}$

80. $g(x) = -\sqrt{x^2 - 3}$

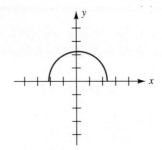

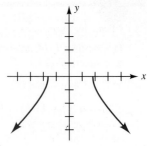

81. $f(x) = \dfrac{-2}{x^2 + 1}$

82. $g(x) = \dfrac{x^2 + 1}{x}$

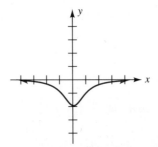

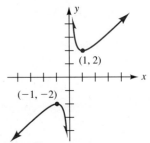

83. $g(x) = x^5 - 5x, \quad -2 \le x \le 2$

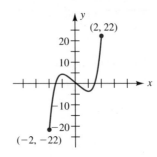

84. $f(x) = \begin{cases} x^2 & \text{if } -2 \le x < 1 \\ 3 - 2x & \text{if } \ \ 1 \le x \le 3 \end{cases}$

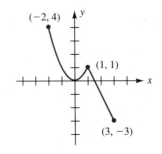

Critical Thinking

 *In calculus, it is necessary to evaluate the **difference quotient***

$$\frac{f(x + \Delta x) - f(x)}{\Delta x}$$

for a given function f. The symbol Δx is read "delta x" and is commonly used to denote a small change in x. In Exercises 85–92, find the difference quotient for the given function f, and simplify the result.

85. $f(x) = 2x + 1$

86. $f(x) = 2 - 3x$

87. $f(x) = x^2$

88. $f(x) = x^3$

89. $f(x) = x^2 + 2x - 3$

90. $f(x) = 2x^2 - 3x$

91. $f(x) = \dfrac{1}{x}$

[*Hint:* Subtract fractions in the numerator and eliminate Δx.]

92. $f(x) = \sqrt{x}$

[*Hint:* Rationalize the numerator and eliminate Δx.]

93. The radius r (in meters) of an oil spill is a function of the time t in minutes that the oil has been leaking. This function is given by

$$r(t) = \sqrt{2t - 1}, \qquad t \geq 1.$$

(a) Evaluate and interpret $r(25)$.
(b) Find the time t when $r(t) = 5$ meters.

94. The bending moment M in pound-feet (lb-ft) along a simply supported 9-ft steel beam that carries a uniformly distributed load of 800 lb/ft is a function of the distance x (in feet) from one end of the beam and is given by

$$M(x) = \tfrac{1}{2}(9)(800)x - \tfrac{1}{2}(800)x^2$$
$$= 3600x - 400x^2, \qquad 0 \leq x \leq 9.$$

(a) Evaluate and interpret $M(1)$.
(b) Find the distance when $M(x) = 8100$ lb/ft.

95. The population size P of a certain organism is a function of the temperature t (in degrees Celsius) of the medium in which the organism exists and is given by

$$P(t) = 3t^2, \qquad 0 \leq t \leq 40.$$

If t_1 is the present temperature of the medium, compare $P(t_1)$ with $P(\tfrac{1}{2}t_1)$, and state the effect that halving the temperature has on the population size.

96. In a psychological experiment, a student is given electrical shocks of varying intensities and asked to rate the intensity of each shock in relation to an initial shock s_0, which is given a rating of 10. It is found that the response number R given by the student is a function of the magnitude (in milliamps) of the shock s and is defined by

$$R(s) = \frac{s^{3/2}}{40}.$$

Compare $R(s_0)$ with $R(4s_0)$, and state the effect that quadrupling the intensity has on the response number.

Calculator Activities

For Exercises 97–100, use the functions f and g defined by

$$f(x) = 14.6x^2 - 12.9x + 25.4 \qquad \text{and}$$

$$g(x) = \frac{26.5x - 92.4}{86.3 - 29.6x}$$

to compute the functional values to three significant digits.

97. $f(2.6)$

98. $f(-1.31)$

99. $g(-23.5)$

100. $g(0.035)$

In Exercises 101–104, find the real zeros of each function to three significant digits. If you have access to a graphing calculator, verify your answer by tracing to the x-intercept(s) of the graph.

101. $f(x) = 2.478x - 5.937$

102. $g(x) = 2.6x^2 - 3.7x - 8.9$

103. $h(x) = \dfrac{27.9x^2 - 93.2}{81.5x + 42.3}$

104. $F(x) = \sqrt{0.532x - 0.273} - 0.632$

105. The cost C (in dollars) for a taxicab fare is a function of the miles x driven and is given by

$$C(x) = 1.50 + 1.45x, \quad x \geq 0.$$

(a) Evaluate and interpret $C(12.4)$.
(b) Determine the miles driven when $C(x) = \$33.98$.

106. If \$5000 is borrowed at $10\frac{1}{2}\%$ simple interest per year,

the amount A (in dollars) that must be paid back is a function of the time t (in years) over which it is borrowed and is given by

$$A(t) = 5000(1 + 0.105t), \quad t \geq 0.$$

(a) Evaluate and interpret $A(12)$.
(b) Find the time t if $A(t) = \$8150$.

3.4   Techniques of Graphing Functions

◆ Introductory Comments

Figure 3.31 shows the graphs of eight basic functions that occur frequently in mathematics: **constant function, identity function, absolute value function, squaring function, cubing function, reciprocal function, square root function**, and **cube root function**. By plotting points or using a graphing calculator, we can verify the graph of each of these functions.

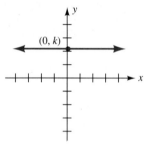

(a) Constant function:
   $f(x) = k$
   Domain: $(-\infty, \infty)$
   Range: $\{k\}$

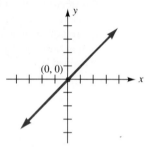

(b) Identity function:
   $f(x) = x$
   Domain: $(-\infty, \infty)$
   Range: $(-\infty, \infty)$

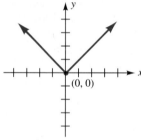

(c) Absolute value
   function: $f(x) = |x|$
   Domain: $(-\infty, \infty)$
   Range: $[0, \infty)$

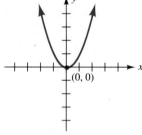

(d) Squaring function:
   $f(x) = x^2$
   Domain: $(-\infty, \infty)$
   Range: $[0, \infty)$

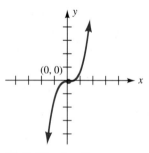

(e) Cubing function:
   $f(x) = x^3$
   Domain: $(-\infty, \infty)$
   Range: $(-\infty, \infty)$

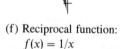

(f) Reciprocal function:
   $f(x) = 1/x$
   Domain: $(-\infty, 0) \cup (0, \infty)$
   Range: $(-\infty, 0) \cup (0, \infty)$

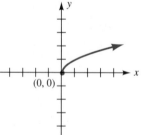

(g) Square root function:
   $f(x) = \sqrt{x}$
   Domain: $[0, \infty)$
   Range: $[0, \infty)$

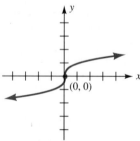

(h) Cube root function:
   $f(x) = \sqrt[3]{x}$
   Domain: $(-\infty, \infty)$
   Range: $(-\infty, \infty)$

FIGURE 3.31   Graphs of eight basic functions with domains and ranges specified.

In this section, we use the graphs of these eight basic functions to sketch the graphs of many other related functions by

1. shifting a graph vertically or horizontally,
2. reflecting a graph about the *x*-axis or *y*-axis, and
3. stretching or shrinking a graph.

◆ **Vertical Shift Rule**

Suppose we wish to sketch the graph of the function *F* defined by

$$F(x) = |x| + 2.$$

We can set up a table of values and plot points as shown in Figure 3.32. However, notice the graph of this function is the same as the graph of $f(x) = |x|$ [the absolute value function shown in Figure 3.31(c)], but *shifted vertically upward* 2 units. This is because when we substitute the same input into the functions *F* and *f*, the output $F(x)$ is always 2 more than the output $f(x)$. Thus, *without plotting points*, we can sketch the graph of $F(x) = |x| + 2$ by shifting the graph of $f(x) = |x|$ vertically upward 2 units. In summary, we state the **vertical shift rule**.

FIGURE 3.32
Comparison of the graphs of
$F(x) = |x| + 2$ and $f(x) = |x|$.

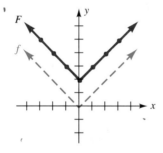

x	F(x)
−3	5
−2	4
−1	3
0	2
1	3
2	4
3	5

◆◆ **Vertical Shift Rule**

> If *f* is a function and *c* is a constant, then the graph of the function *F* defined by
>
> $$F(x) = f(x) + c$$
>
> is the same as the graph of *f* shifted *vertically upward* $|c|$ units if $c > 0$ or shifted *vertically downward* $|c|$ units if $c < 0$.

To obtain an accurate sketch, it is a good practice to determine and label the *axis intercepts* of the graph. The ***x*-intercepts** of the graph of a function *f* are the real zeros of the function and may be found by solving the equation $f(x) = 0$. The ***y*-intercept** of the graph of a function *f* may be found by evaluating $f(0)$.

EXAMPLE 1   Sketch the graph of each function and label the x-intercept and y-intercept.

(a) $F(x) = x^2 + 1$ **(b)** $G(x) = \sqrt{x} - 2$

SOLUTION

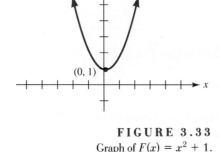

FIGURE 3.33
Graph of $F(x) = x^2 + 1$.

(a) We can think of this function as $F(x) = f(x) + 1$, where $f(x) = x^2$ [the squaring function shown in Figure 3.31(d)]. Thus, by the vertical shift rule, the graph of $F(x) = x^2 + 1$ is the same as the graph of $f(x) = x^2$ shifted vertically *upward* 1 unit. The graph of the function F is shown in Figure 3.33.

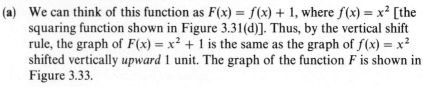

x-intercept: $F(x) = 0$ y-intercept: $F(0) = (0)^2 + 1 = 1$

$x^2 + 1 = 0$

$x^2 = -1$

No real solution.
No x-intercept.

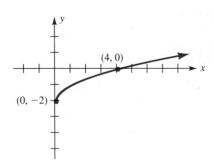

FIGURE 3.34
Graph of $G(x) = \sqrt{x} - 2$.

(b) We can think of this function as $G(x) = f(x) - 2$, where $f(x) = \sqrt{x}$ [the square root function shown in Figure 3.31(g)]. Thus, by the vertical shift rule, the graph of $G(x) = \sqrt{x} - 2$ is the same as the graph of $f(x) = \sqrt{x}$ shifted vertically *downward* 2 units. The graph of the function G is shown in Figure 3.34.

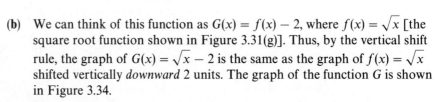

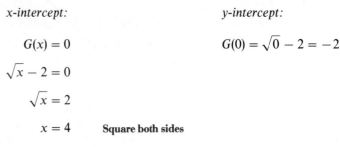

x-intercept: y-intercept:

$G(x) = 0$ $G(0) = \sqrt{0} - 2 = -2$

$\sqrt{x} - 2 = 0$

$\sqrt{x} = 2$

$x = 4$ **Square both sides**   ◆

PROBLEM 1   Sketch the graph of $h(x) = x^3 - 1$ and label the x-intercept and y-intercept.
                                                                                      ◆

◆ **Horizontal Shift Rule**

Suppose we wish to sketch the graph of the function F defined by

$$F(x) = |x + 2|.$$

We can set up a table of values and plot points as shown in Figure 3.35. However, notice that the graph of this function is the same as the graph of $f(x) = |x|$ [Figure 3.31(c)] *shifted horizontally to the left* 2 units. This is because when we substitute an input into F that is 2 *less than* the input we use in f, the values of the outputs $F(x)$ and $f(x)$ are equal. Thus, *without plotting points*, we can sketch the graph of $F(x) = |x + 2|$ by shifting the graph of $f(x) = |x|$ horizontally to the left 2 units. In summary, we state the **horizontal shift rule**.

FIGURE 3.35

Comparison of the graphs of
$F(x) = |x + 2|$ and $f(x) = |x|$.

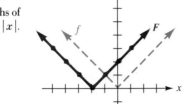

x	F(x)
−5	3
−4	2
−3	1
−2	0
−1	1
0	2
1	3

Horizontal Shift Rule

> If f is a function and c is a constant, then the graph of the function F defined by
>
> $$F(x) = f(x + c)$$
>
> is the same as the graph of f shifted *horizontally to the left* $|c|$ units if $c > 0$ or shifted *horizontally to the right* $|c|$ units if $c < 0$.

EXAMPLE 2   Sketch the graph of each function and label the *x*-intercept and *y*-intercept.

(a) $F(x) = (x + 1)^2$ (b) $G(x) = \sqrt[3]{x - 2}$

SOLUTION

(a)   We can think of this function as $F(x) = f(x + 1)$, where $f(x) = x^2$ [see Figure 3.31(d)]. Thus, by the horizontal shift rule, the graph of $F(x) = (x + 1)^2$ is the same as the graph of $f(x) = x^2$ shifted horizontally *to the left* 1 unit. The graph of the function F is shown in Figure 3.36.

x-intercept:

$$F(x) = 0$$

$$(x + 1)^2 = 0$$

$$x + 1 = 0 \qquad \text{Take the square root of both sides}$$

$$x = -1$$

y-intercept:

$$F(0) = (0 + 1)^2 = 1$$

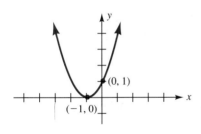

FIGURE 3.36

Graph of $F(x) = (x + 1)^2$

(b)   We can think of this function as $G(x) = f(x - 2)$, where $f(x) = \sqrt[3]{x}$ [the cube root function shown in Figure 3.31(h)]. Thus, by the horizontal shift rule, the graph of $G(x) = \sqrt[3]{x - 2}$ is the same as the graph of $f(x) = \sqrt[3]{x}$ shifted horizontally *to the right* 2 units.

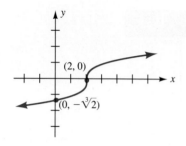

FIGURE 3.37
Graph of $G(x) = \sqrt[3]{x-2}$.

x-intercept:

$$G(x) = 0$$

$$\sqrt[3]{x-2} = 0$$

$$x - 2 = 0 \qquad \text{Cube both sides}$$

$$x = 2$$

y-intercept:

$$G(0) = \sqrt[3]{(0)-2} = \sqrt[3]{-2} \approx -1.26$$

The graph of the function G is shown in Figure 3.37.   ◆

PROBLEM 2   Sketch the graph of $h(x) = (x-1)^3$ and label the *x*-intercept and *y*-intercept.
                                                                          ◆

As illustrated in the next example, we can use both the vertical and horizontal shift rules to help sketch the graph of a function.

EXAMPLE 3   Sketch the graph of $F(x) = \dfrac{1}{x-2} + 1$ and label the *x*-intercept and *y*-intercept.

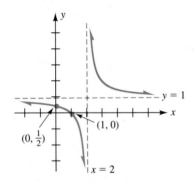

FIGURE 3.38
Graph of $F(x) = \dfrac{1}{x-2} + 1$.

SOLUTION We can think of this function as $F(x) = f(x-2) + 1$, where $f(x) = 1/x$ [the reciprocal function shown in Figure 3.31(f)]. Thus, by the horizontal and vertical shift rules, we obtain the graph of F by shifting the graph of $f(x) = 1/x$ to the right 2 units and upward 1 unit as shown in Figure 3.38.

x-intercept:

$$F(x) = 0$$

$$\frac{1}{x-2} + 1 = 0$$

$$1 + (x-2) = 0 \qquad \text{Multiply both sides} \atop \text{by } (x-2)$$

$$x = 1$$

y-intercept:

$$F(0) = \frac{1}{0-2} + 1 = \frac{1}{2}$$

                                                                          ◆

PROBLEM 3   Sketch the graph of $F(x) = (x+1)^2 - 4$ and label the *x*-intercepts and *y*-intercept.
                                                                          ◆

◆ ***x*-Axis and *y*-Axis Reflection Rules**

Consider the functions f and F defined by

$$f(x) = \sqrt{x} \qquad \text{and} \qquad F(x) = -\sqrt{x},$$

FIGURE 3.39
Comparison of the graphs of $f(x) = \sqrt{x}$
and $F(x) = -\sqrt{x}$.

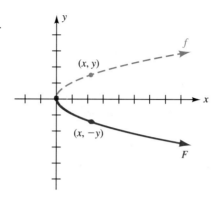

respectively. [As shown in Figure 3.31(g), f is the square root function.] If we substitute the same input into the functions f and F, the output $F(x)$ is always the *negative* of the output $f(x)$. Hence, as shown in Figure 3.39, the graph of F is the same as the graph of f *reflected about the x-axis*. In summary, we state the **x-axis reflection rule**.

x-Axis Reflection Rule

> If f is a function, then the graph of the function F defined by
>
> $$F(x) = -f(x)$$
>
> is the same as the graph of f *reflected about the x-axis*.

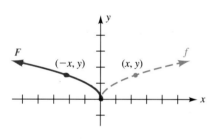

Consider the functions f and F defined by

$$f(x) = \sqrt{x} \qquad \text{and} \qquad F(x) = \sqrt{-x}$$

respectively. If we substitute an input into the function F that is the *opposite* of the input we use for f, the values of the outputs $F(x)$ and $f(x)$ are equal. Hence, as shown in Figure 3.40, the graph of F is the same as the graph of f *reflected about the y-axis*. In summary, we state the **y-axis reflection rule**.

FIGURE 3.40
Comparison of the graphs of $f(x) = \sqrt{x}$
and $F(x) = \sqrt{-x}$.

y-Axis Reflection Rule

> If f is a function, then the graph of the function F defined by
>
> $$F(x) = f(-x)$$
>
> is the same as the graph of f *reflected about the y-axis*.

EXAMPLE 4   Sketch the graph of each function and label the x-intercept and y-intercept.

(a) $F(x) = 2 - x^3$ (b) $h(x) = \sqrt{2 - x}$

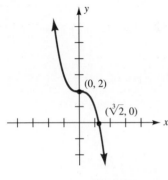

FIGURE 3.41
Graph of $F(x) = 2 - x^3$.

SOLUTION

(a) We can think of this function as $F(x) = -f(x) + 2$, where $f(x) = x^3$ [the cubing function shown in Figure 3.31(e)]. Thus, by the x-axis reflection rule and the vertical shift rule, the graph of $F(x) = 2 - x^3$ is obtained by reflecting the graph of $y = x^3$ about the x-axis and then shifting this graph upward 2 units as shown in Figure 3.41.

x-intercept:

$$F(x) = 0$$
$$2 - x^3 = 0$$
$$x^3 = 2$$
$$x = \sqrt[3]{2} \approx 1.26$$

y-intercept:

$$F(0) = 2 - (0)^3 = 2$$

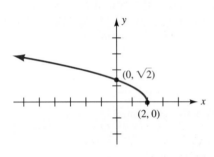

FIGURE 3.42
Graph of $h(x) = \sqrt{2 - x}$.

(b) We can think of this function as $h(x) = f(x - 2)$, where $f(x) = \sqrt{-x}$. Thus, by the y-axis reflection rule and the horizontal shift rule, the graph of $h(x) = \sqrt{2 - x}$ is obtained by reflecting the graph of $y = \sqrt{x}$ about the y-axis and then shifting this graph to the right 2 units as shown in Figure 3.42.

x-intercept:

$$h(x) = 0$$
$$\sqrt{2 - x} = 0$$
$$2 - x = 0 \quad \textbf{Square both sides}$$
$$x = 2$$

y-intercept:

$$h(0) = \sqrt{2 + 0} = \sqrt{2} \approx 1.41$$

◆

PROBLEM 4   Sketch the graph of $F(x) = -|x + 2|$ and label the x-intercept and y-intercept.

◆

◆ **Vertical Stretch and Shrink Rule**

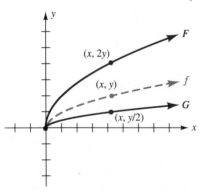

FIGURE 3.43
Comparison of the graphs of $f(x) = \sqrt{x}$,
$F(x) = 2\sqrt{x}$, and $G(x) = (1/2)\sqrt{x}$.

Consider the functions f, F, and G defined by

$$f(x) = \sqrt{x}, \qquad F(x) = 2\sqrt{x}, \qquad \text{and} \qquad G(x) = \tfrac{1}{2}\sqrt{x},$$

respectively. If we substitute the same input into the functions f, F, and G, the output $F(x)$ is *twice* the output $f(x)$, and the output $G(x)$ is *half* the output $f(x)$. As illustrated in Figure 3.43, the graphs of F and G are similar to the graph of f. The graph of F is *stretched vertically* by a factor of 2, and the graph of G is *shrunk vertically* by a factor of 2. In summary, we state the **vertical stretch and shrink rule**.

Vertical Stretch and Shrink Rule

If f is a function and c is a real number with $c > 1$, then the graph of the function F defined by

$$F(x) = cf(x)$$

is similar to the graph of f *stretched vertically* by a factor of c, and the graph of the function G defined by

$$G(x) = \frac{1}{c} f(x)$$

is similar to the graph of f *shrunk vertically* by a factor of c.

EXAMPLE 5 Sketch the graph of each function and label the x-intercept and y-intercept.

(a) $G(x) = -\dfrac{|x|}{4}$ **(b)** $F(x) = 4(x - 1)$

SOLUTION

(a) We can think of this function as $G(x) = \frac{1}{4}f(x)$, where $f(x) = -|x|$. Thus, by the x-axis reflection rule and the vertical stretch and shrink rule, we obtain the graph of $G(x) = -|x|/4$ by reflecting the graph of $y = |x|$ about the x-axis and then shrinking this graph by a factor of 4, as shown in Figure 3.44.

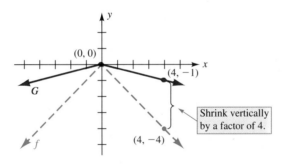

FIGURE 3.44

Graph of $G(x) = -\dfrac{|x|}{4}$

is formed from the graph of $f(x) = -|x|$.

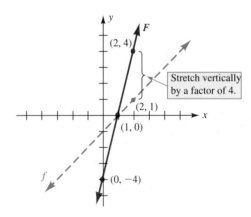

FIGURE 3.45

Graph of $F(x) = 4(x - 1)$ is formed from the graph of $f(x) = x - 1$.

(b) We can think of this function as $F(x) = 4f(x)$, where $f(x) = x - 1$. By the vertical shift rule, the graph of f is the same as the graph of $y = x$ [the identity function shown in Figure 3.31(b)], but shifted vertically downward 1 unit. The graph of $F(x) = 4(x - 1)$ is then obtained by stretching the graph of f by a factor of 4 as shown in Figure 3.45.

$$x\text{-intercept:} \qquad y\text{-intercept:}$$

$$F(x) = 0 \qquad F(0) = 4(0 - 1) = -4$$

$$4(x - 1) = 0$$

$$x - 1 = 0$$

$$x = 1 \qquad\qquad\qquad\qquad\qquad ◆$$

PROBLEM 5   Sketch the graph of $F(x) = 4x - 1$ and label the x-intercept and y-intercept.

◆

◆ **Increasing, Decreasing, and Constant Functions**

A function f is said to be an *increasing function* if as x increases the value of $f(x)$ also *increases*. A function f is said to be a *decreasing function* if as x increases $f(x)$ also *decreases*. A function f is said to be a *constant function* if as x increases, $f(x)$ *remains the same*. More precisely, we state the following definitions.

Increasing, Decreasing, and Constant Functions

1.   A function f is an **increasing function** if for all a and b in the domain of f,

$$f(a) < f(b) \text{ whenever } a < b.$$

2.   A function f is a **decreasing function** if for all a and b in the domain of f,

$$f(a) > f(b) \text{ whenever } a < b.$$

3.   A function f is a **constant function** if for all a and b in the domain of f,

$$f(a) = f(b).$$

From the graph of a function, we may determine if the function is increasing or decreasing. The graph of $F(x) = 4(x - 1)$, shown in Figure 3.45, always *rises* as we move from left to right along the x-axis. Thus, as x increases, $F(x)$ also increases, and we conclude the function F is an *increasing function*. The graph of $h(x) = \sqrt{2 - x}$, shown in Figure 3.42, always *falls* as we move from left to right along the x-axis. Hence, as x increases, $h(x)$ decreases, and we conclude the function h is a *decreasing function*.

The graph of $G(x) = -|x|/4$, shown in Figure 3.44, rises as we move from left to right until we reach the y-axis. The graph then falls as we continue to move along the x-axis from the y-axis toward the right. Thus, this function is *neither increasing nor decreasing* over its entire domain. This situation is typical of many functions that increase on some intervals in their domains and decrease on other intervals. For $G(x) = -|x|/4$, we say that G is *increasing on the interval* $(-\infty, 0)$ and *decreasing on the interval* $(0, \infty)$.

EXAMPLE 6 **(a)** Sketch the graph of the piecewise-defined function g defined by

$$g(x) = \begin{cases} 1 - x^2 & \text{if } x < 1 \\ x - 1 & \text{if } 1 \leq x \leq 3 \\ 4 & \text{if } x > 3 \end{cases}$$

(b) Use the graph to determine the intervals where the function is increasing, decreasing, or constant.

SOLUTION

(a) First, we apply the graphing techniques from this section to sketch the graphs of $y = 1 - x^2$, $y = x - 1$, and $y = 4$ with dashed lines, as shown in Figure 3.46. To form the graph of g, we darken the graph of

$$y = 1 - x^2 \quad \text{for } x < 1,$$
$$y = x - 1 \quad \text{for } 1 \leq x \leq 3,$$
and $$y = 4 \quad \text{for } x > 3,$$

as shown in Figure 3.47. We place a *solid dot* at the point (3, 2) to indicate that this point is part of the graph of g and we place an *open circle* at the point (3, 4) to indicate that this point is *not* part of the graph of g.

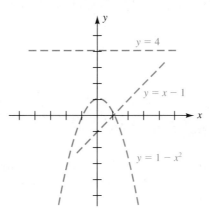

FIGURE 3.46
Graphs of $y = 1 - x^2$, $y = x - 1$, and $y = 4$, sketched with dashed lines.

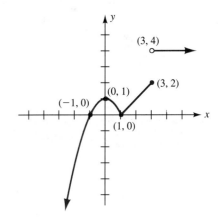

FIGURE 3.47
Graph of $g(x) = \begin{cases} 1 - x^2 & \text{if } x < 1 \\ x - 1 & \text{if } 1 \leq x \leq 3 \\ 4 & \text{if } x > 3. \end{cases}$

Increasing on intervals $(-\infty, 0)$ and $[1, 3]$.
Decreasing on the interval $(0, 1)$.
Constant on the interval $(3, \infty)$

(b) As we move along the *x*-axis from left to right, the graph *rises* until we reach the *y*-axis. From the *y*-axis to $x = 1$, the graph *falls*. From $x = 1$ to $x = 3$, the graph again *rises*. Finally, to the right of $x = 3$ the graph remains the same. Thus, we conclude the function is *increasing* on the intervals $(-\infty, 0)$ and $[1, 3]$, *decreasing* on the interval $(0, 1)$, and *constant* on the interval $(3, \infty)$. ◆

Note: Although the notion of *continuity* is studied in calculus, we briefly mention the idea at this time. A function is said to be *continuous* if no gap, break, or jump occurs in its graph. The graph of the function *g* in Figure 3.47 has a break at $x = 3$. Hence, we say that the function *g* is *discontinuous* at $x = 3$.

P R O B L E M 6   Determine whether the function *g* defined in Example 6 is continuous over each of the given intervals.

(a) $(-\infty, 3]$ **(b)** $(3, \infty)$ **(c)** $(-\infty, 4]$ **(d)** $(0, 9)$ ◆

Exercises 3.4

Basic Skills

In Exercises 1–42, sketch the graph of each function and label the x-intercept(s) and y-intercept. If you have access to a graphing calculator, verify each graph.

1. $f(x) = x + 3$

2. $g(x) = x^2 - 1$

3. $G(x) = x^3 - 2$

4. $H(x) = \dfrac{1}{x} + 2$

5. $F(x) = 3 + \sqrt{x}$

6. $f(x) = -2 + |x|$

7. $f(x) = |x - 1|$

8. $g(x) = \sqrt{x + 3}$

9. $H(x) = (x + 2)^2$

10. $F(x) = (x + 1)^3$

11. $F(x) = \dfrac{1}{x + 2}$

12. $f(x) = \sqrt[3]{x - 1}$

13. $h(x) = -\dfrac{1}{x}$

14. $g(x) = -x^3$

15. $f(x) = -x^2$

16. $H(x) = \sqrt[3]{-x}$

17. $g(x) = \sqrt{1 - x}$

18. $f(x) = -\sqrt{3 - x}$

19. $F(x) = (x - 1)^2 + 3$

20. $G(x) = |x - 4| - 1$

21. $f(x) = \sqrt{x + 2} - 1$

22. $h(x) = \dfrac{1}{x + 1} - 2$

23. $h(x) = -|x - 3|$

24. $f(x) = -(x + 1)$

25. $G(x) = 3 - x^2$

26. $H(x) = 2 - \sqrt{x}$

27. $F(x) = 2 - (x + 4)^3$

28. $h(x) = 1 - |x - 1|$

29. $f(x) = \sqrt[3]{x + 2} - 3$

30. $f(x) = 2 - \sqrt{4 - x}$

31. $H(x) = \dfrac{1}{2 - x}$

32. $g(x) = 3 + (2 - x)^2$

33. $g(x) = 3|x|$

34. $f(x) = -2x$

35. $f(x) = -\tfrac{1}{3}x^2$

36. $G(x) = \dfrac{x^3}{4}$

37. $G(x) = 2x - 1$

38. $F(x) = 3 - \tfrac{1}{2}x$

39. $F(x) = 2\sqrt{x - 4}$

40. $g(x) = 3 - 2|x|$

41. $F(x) = \dfrac{5}{x + 2}$

42. $h(x) = \dfrac{(x - 1)^2}{4}$

In Exercises 43–50, determine if the function is increasing, decreasing, or neither.

43. The function *G* in Exercise 3.

44. The function *g* in Exercise 8.

45. The function *g* in Exercise 17.

46. The function *g* in Exercise 14.

47. The function *G* in Exercise 25.

48. The function *G* in Exercise 20.

49. The function *f* in Exercise 29.

50. The function *f* in Exercise 34.

In Exercises 51–56, sketch the graph of each piecewise-defined function. Use the graph to determine the intervals where the function is increasing, decreasing, or constant.

51. $g(x) = \begin{cases} x + 1 & \text{if } x < 2 \\ 3 & \text{if } x \geq 2 \end{cases}$

52. $f(x) = \begin{cases} -x & \text{if } x \leq 4 \\ \sqrt{x} & \text{if } x > 4 \end{cases}$

53. $f(x) = \begin{cases} x + 2 & \text{if } x < 0 \\ 2 & \text{if } 0 \leq x < 3 \\ (x - 3)^2 & \text{if } x \geq 3 \end{cases}$

54. $h(x) = \begin{cases} x^3 & \text{if } x < 1 \\ 2 - x^2 & \text{if } 1 \leq x < 3 \\ -7 & \text{if } x \geq 3 \end{cases}$

55. $h(x) = \begin{cases} |x| & \text{if } x < 1 \\ -x^2 & \text{if } 1 \leq x \leq 3 \\ 1/x & \text{if } x > 3 \end{cases}$

56. $g(x) = \begin{cases} x^2 - 1 & \text{if } x \leq 1 \\ x - 1 & \text{if } 1 < x < 3 \\ 3 & \text{if } 3 \leq x \leq 5 \\ 13 - 2x & \text{if } x > 5 \end{cases}$

57. Determine if the function f, defined in Exercise 53, is continuous on each of the following intervals.

    (a) $(-\infty, 3)$ (b) $(-\infty, 5)$
    (c) $(-1, 3]$ (d) $[-6, 2)$

58. Determine if the function g, defined in Exercise 56, is continuous on each of the following intervals.

    (a) $(-\infty, 3)$ (b) $(-\infty, 5)$
    (c) $[3, \infty)$ (d) $[0, 8]$

 Critical Thinking

The graph of $y = f(x)$ is shown in the figure. Use this graph and the techniques of shifting and reflecting to sketch the graph of each of the equations in Exercises 59–62.

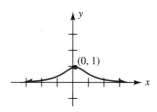

59. $y = -f(x)$

60. $y = f(x) - 3$

61. $y = f(x - 2)$

62. $y = 2 - f(x + 1)$

The graph of $y = \sqrt{4 - x^2}$ is shown in the figure. Use this graph to write an equation for each of the functions whose graphs are shown in Exercises 63–66.

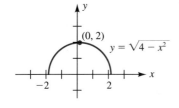

63.
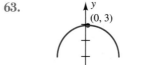

64.
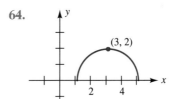

65.
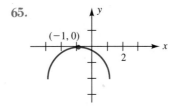

66.
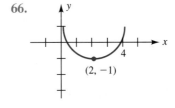

67. The cost C (in cents) of first-class postage is a function of a letter's weight w (in ounces) and is given by

$$C(w) = \begin{cases} 29 & \text{if } 0 < w \le 1 \\ 52 & \text{if } 1 < w \le 2 \\ 75 & \text{if } 2 < w \le 3 \\ 98 & \text{if } 3 < w \le 4 \\ 121 & \text{if } 4 < w \le 5 \end{cases}$$

This function is called a *step function*, because its graph looks like a step of stairs.

(a) Sketch the graph of this function.
(b) Evaluate and interpret $C(2.4)$.

68. In mathematics, the notation $[\![x]\!]$ denotes the largest integer n such that $n \le x$. For example,

$$[\![8.9]\!] = 8, \quad [\![5]\!] = 5, \quad [\![\pi]\!] = 3,$$

$$[\![-8.9]\!] = -9, \quad [\![-\sqrt{3}]\!] = -2,$$

and so on. The *greatest integer function* f defined by

$f(x) = [\![x]\!]$ is another example of a step function (see Exercise 67). Sketch the graph of the function f.

69. The volume V of air in a circular balloon is a function of its diameter d and is given by

$$V(d) = \frac{\pi}{6} d^3, \quad 0 \le d \le 10.$$

(a) Sketch the graph of this function.
(b) Evaluate and interpret $V(6)$.

70. The voltage v (in volts) that is applied to an electrical circuit is a function of time t (in milliseconds) and is given by

$$v(t) = \begin{cases} t^2 & \text{if } 0 \le t \le 2 \\ 4 & \text{if } 2 < t \le 6 \\ (t-8)^2 & \text{if } 6 < t \le 8 \end{cases}$$

(a) Sketch the graph of this function.
(b) Evaluate and interpret $v(7)$.

Calculator Activities

In Exercises 71–75, sketch the graph of each function and determine to three significant digits the x-intercept(s) and y-intercept. If you have access to a graphing calculator, verify each answer by tracing to the x-intercept(s) and y-intercept of the graph.

71. $f(x) = 2.35x + 9.45$ 72. $g(x) = 5.62 - 1.51x^3$

73. $h(x) = 0.85(x - 1.67)^2$

74. $F(x) = 1.34|x + 2.65| - 4.46$

75. $G(x) = 1.70 + \sqrt{2.32 - x}$

76. Use a graphing calculator to generate the graphs of the following equations in the same viewing rectangle:

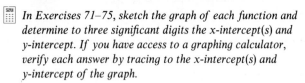

What rule in this section verifies the picture you observe?

77. Use a graphing calculator to generate the graphs of the following equations in the same viewing rectangle:

$$g(x) = (x + 2)^3 \qquad h(x) = (x + 1)^3 \qquad f(x) = x^3$$
$$G(x) = (x - 1)^3 \quad \text{and} \quad H(x) = (x - 2)^3$$

What rule in this section verifies the picture you observe?

78. Use a graphing calculator to generate the graphs of the following equations in the same viewing rectangle:

$$g(x) = 0.25x^2 \qquad h(x) = 0.5x^2 \qquad f(x) = x^2$$
$$G(x) = 2x^2 \quad \text{and} \quad H(x) = 4x^2$$

What rule in this section verifies the picture you observe?

3.5   Methods of Combining Functions

In arithmetic, we often use two real numbers to form a third number by the operations of addition, subtraction, multiplication, or division. In this section, we discuss ways in which two functions may be combined to form a third function. We begin by defining the algebraic operations of addition, subtraction, multiplication, and division of functions.

◆ Algebraic Operations

The *sum, difference, product,* and *quotient* of two functions f and g are defined as follows.

Operations on Functions

If f and g are two functions, then

1. their **sum** is the function $f + g$ defined by

$$(f + g)(x) = f(x) + g(x).$$

2. their **difference** is the function $f - g$ defined by

$$(f - g)(x) = f(x) - g(x).$$

3. their **product** is the function $f \cdot g$ defined by

$$(f \cdot g)(x) = f(x) \cdot g(x).$$

4. their **quotient** is the function f/g defined by

$$\left(\frac{f}{g}\right)(x) = \frac{f(x)}{g(x)}.$$

EXAMPLE 1   Use the graphs of the functions f and g in Figure 3.48 to compute each functional value:

(a) $(f + g)(-2)$ **(b)** $(f - g)(0)$ **(c)** $(f \cdot g)(2)$

(d) $\left(\dfrac{f}{g}\right)(1)$ **(e)** $(f + g)(3)$

FIGURE 3.48
The graphs of f and g for Example 1.

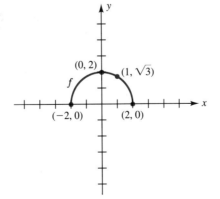

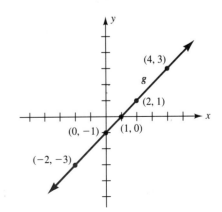

SOLUTION

(a) From the graph of f we find that $f(-2) = 0$, and from the graph of g we find that $g(-2) = -3$. Hence,

$$(f + g)(-2) = f(-2) + g(-2) = 0 + (-3) = -3.$$

(b) From the graph of f we find that $f(0) = 2$, and from the graph of g we find that $g(0) = -1$. Hence,

$$(f - g)(0) = f(0) - g(0) = 2 - (-1) = 3.$$

(c) From the graph of f we find that $f(2) = 0$, and from the graph of g we find that $g(2) = 1$. Hence,

$$(f \cdot g)(2) = f(2) \cdot g(2) = 0 \cdot 1 = 0.$$

(d) From the graph of f we find that $f(1) = \sqrt{3}$, and from the graph of g we find that $g(1) = 0$. Hence,

$$\left(\frac{f}{g}\right)(1) = \frac{f(1)}{g(1)} = \frac{\sqrt{3}}{0}, \quad \text{which is undefined.}$$

(e) Since 3 is not in the domain of f, $f(3)$ is undefined. Hence, the function $f + g$ is also undefined for the input $x = 3$.   ◆

PROBLEM 1   Use the graphs of the functions f and g in Example 1 to compute the functional values (a) $(g - f)(0)$ and (b) $\left(\dfrac{g}{f}\right)(1)$.   ◆

Observe from Example 1 that for $(f + g)(x)$, $(f - g)(x)$, $(f \cdot g)(x)$, and $(f/g)(x)$ to make sense, the value of x must be an acceptable input for both functions, f and g; that is, we must select an input x that is common to both the domain of f and the domain of g. Thus, *the domain of the function $f + g$, $f - g$, $f \cdot g$, or f/g is the intersection of the domains of f and g. Since division by zero is undefined, the domain of f/g must also exclude all inputs x for which $g(x) = 0$.*

EXAMPLE 2   Given the functions f and g defined by $f(x) = 3x^2$ and $g(x) = x^2 - \dfrac{1}{x}$, find each of the following functions. Then determine the domain of each function.

(a) $f + g$ (b) $g - f$ (c) $g \cdot f$ (d) f/g

SOLUTION

(a) The function $f + g$ is defined by

$$(f + g)(x) = f(x) + g(x) = 3x^2 + \left(x^2 - \frac{1}{x}\right) = 4x^2 - \frac{1}{x}.$$

(b)   The function $g - f$ is defined by

$$(g - f)(x) = g(x) - f(x) = \left(x^2 - \frac{1}{x} \right) - 3x^2 = -2x^2 - \frac{1}{x}.$$

(c)   The function $g \cdot f$ is defined by

$$(g \cdot f)(x) = g(x) \cdot f(x) = \left(x^2 - \frac{1}{x} \right) \cdot 3x^2 = 3x^4 - 3x.$$

(d)   The function f/g is defined by

$$\left(\frac{f}{g} \right)(x) = \frac{f(x)}{g(x)} = \frac{3x^2}{x^2 - \dfrac{1}{x}} = \frac{3x^3}{x^3 - 1}.$$

The domain of f is $(-\infty, \infty)$ and the domain of g is $(-\infty, 0) \cup (0, \infty)$. Thus, the domain of the functions $f + g$, $g - f$, and $g \cdot f$ is $(-\infty, 0) \cup (0, \infty)$. Since division by zero is undefined, the domain of f/g includes a further restriction:

$$x^2 - \frac{1}{x} \neq 0$$

$$x^3 - 1 \neq 0$$

$$x \neq 1$$

Thus, the domain of f/g is $(-\infty, 0) \cup (0, 1) \cup (1, \infty)$.   ◆

PROBLEM 2   Use the results of Example 2 to compute each functional value:

(a) $(g - f)(2)$ **(b)** $\left(\dfrac{f}{g} \right)(-1)$ ◆

◆   **Composite Functions**

We may also combine two functions by another method, the **composition of functions**. If f and g are functions, then f composed with g, denoted $f \circ g$, is formed by using the output of g as the input of f. Similarly, g composed with f, denoted $g \circ f$, is formed by using the output of f as the input of g.

◆ **Composite Function**

If f and g are functions, then the **composite function $f \circ g$** is defined by

$$(f \circ g)(x) = f(g(x))$$

and the **composite function $g \circ f$** is defined by

$$(g \circ f)(x) = g(f(x)).$$

Remember that the rule $f \circ g$ tells us to apply g first, then f, whereas the rule $g \circ f$ tells us to apply f first, then g.

EXAMPLE 3   Use the graphs of f and g in Figure 3.49 to compute each functional value.

(a) $(f \circ g)(4)$ (b) $(g \circ f)(0)$

FIGURE 3.49
The graphs of f and g for Example 3.

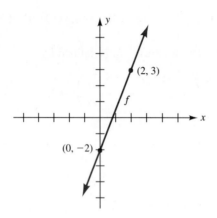
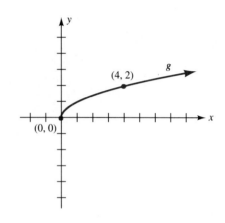

SOLUTION

(a) From the graph of g we observe that $g(4) = 2$. Thus,

$$(f \circ g)(4) = f(g(4)) = f(2).$$

Now, from the graph of f, we observe that $f(2) = 3$. Hence,

$$(f \circ g)(4) = 3.$$

(b) From the graph of f we observe that $f(0) = -2$. Thus,

$$(g \circ f)(0) = g(f(0)) = g(-2).$$

However, -2 is not in the domain of g. Hence, we conclude that

$$(g \circ f)(0) \text{ is undefined.} \qquad \blacklozenge$$

PROBLEM 3   Repeat Example 3 for $(f \circ g)(0)$. $\qquad\qquad\qquad\qquad\qquad\qquad\qquad\qquad$ $\blacklozenge$

Observe in Example 3(b) that for $g(f(x))$ to make sense, the values of $f(x)$ must be acceptable inputs for the function g. Remember, *the domain of $g \circ f$ is the set of all inputs x in the domain of f such that $f(x)$ is in the domain of g.* Likewise, *the domain of $f \circ g$ is the set of all inputs in the domain of g such that $g(x)$ is in the domain of f.*

EXAMPLE 4   Given the functions f and g defined by $f(x) = \sqrt{x}$ and $g(x) = x^2 - 4$, find the following functions. Then determine the domain of each function.

(a) $f \circ g$ (b) $g \circ f$

SOLUTION

(a) The function $f \circ g$ is defined by

$$(f \circ g)(x) = f(g(x)) = f(x^2 - 4) = \sqrt{x^2 - 4}.$$

The domain of $f \circ g$ is all real numbers in the domain of g such that $g(x)$ is in the domain of f. The domain of g is $(-\infty, \infty)$ and the domain of f is $[0, \infty)$. Thus, from the interval $(-\infty, \infty)$ we can select only inputs x such that the output $g(x)$ is *nonnegative*:

$$g(x) \geq 0$$
$$x^2 - 4 \geq 0$$
$$x \leq -2 \quad \text{or} \quad x \geq 2$$

Hence, the domain of $f \circ g$ is $(-\infty, -2] \cup [2, \infty)$.

(b) The function $g \circ f$ is defined by

$$(g \circ f)(x) = g(f(x)) = g(\sqrt{x}) = (\sqrt{x})^2 - 4 = x - 4.$$

The domain of $g \circ f$ is all real numbers in the domain of f such that $f(x)$ is in the domain of g. Since the domain of f is $[0, \infty)$, we can choose inputs only from this interval. Since each *nonnegative* input x gives us an output $f(x)$ that is in the domain of g, the domain of $g \circ f$ is also $[0, \infty)$. Hence, we write

$$(g \circ f)(x) = x - 4, \quad x \geq 0. \qquad \blacklozenge$$

Note: As illustrated in Example 4, the composite functions $f \circ g$ and $g \circ f$ are *not* necessarily the same. In Section 3.6, we will discuss a special class of functions in which $(f \circ g)(x) = (g \circ f)(x) = x$.

PROBLEM 4   For the functions f and g given in Example 4, find $(g \circ f)(9)$ by

(a)   first finding $f(9)$ and then evaluating $g(f(9))$.

(b)   using the result of Example 4(b), $(g \circ f)(x) = x - 4$. $\qquad \blacklozenge$

◆ **Decomposition of Functions**

To **decompose** a given function h, we try to express h as the composition of two simpler functions f and g; that is, we look for an *inner function g* and an *outer* *function f* such that

$$h(x) = f(g(x)) = (f \circ g)(x)$$

for all x in the domain of h. This procedure is used frequently in calculus and is illustrated in the next example.

E X A M P L E 5   Express each function as the composition of two functions f and g.

(a) $h(x) = (x^2 + 4)^3$ **(b)** $H(x) = \dfrac{1}{\sqrt{x+1}}$

S O L U T I O N

(a) One way to decompose the function h is to let the *inner function g* be defined by the quantity *inside the parentheses*. Thus, we let g be defined as

$$g(x) = x^2 + 4.$$

Now, we may define the *outer function f* as the cubing function

$$f(x) = x^3.$$

We can recombine our choices for f and g to verify that their composition is our original function h:

$$\begin{aligned}
(f \circ g)(x) &= f(g(x)) \\
&= f(x^2 + 4) \\
&= (x^2 + 4)^3 \\
&= h(x)
\end{aligned}$$

(b) One way to decompose the function H is to let the *inner function g* be defined by the quantity *inside the radical*. Thus, we let g be defined as

$$g(x) = x + 1, \quad x > -1.$$

Now, we may define the *outer function f* by

$$f(x) = \frac{1}{\sqrt{x}}.$$

We can verify our choices for f and g by finding the composite function $f \circ g$:

$$\begin{aligned}
(f \circ g)(x) &= f(g(x)) \\
&= f(x + 1) \\
&= \frac{1}{\sqrt{x+1}} \\
&= H(x)
\end{aligned}$$

◆

Note: In Example 5, other answers are possible. For instance, in part (a) we could let $f(x) = (x + 4)^3$ and $g(x) = x^2$. Then we have

$$(f \circ g)(x) = f(g(x))$$
$$= f(x^2)$$
$$= (x^2 + 4)^3$$
$$= h(x)$$

Remember that when we decompose a function h, the inner function g and the outer function f are not necessarily unique.

PROBLEM 5   Express the function H defined in Example 5(b) as the composition of two *other* functions f and g.   ◆

Exercises 3.5

 Basic Skills

In Exercises 1–16, use the graphs of f, g, and h shown below to compute (if possible) the given functional value.

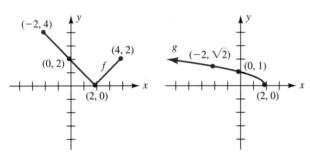

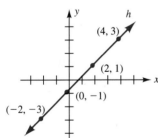

1. $(f + g)(-2)$
2. $(h + f)(0)$
3. $(g - h)(0)$
4. $(f - g)(2)$
5. $(f \cdot g)(4)$
6. $(h \cdot g)(-2)$
7. $\left(\dfrac{f}{g}\right)(-2)$
8. $\left(\dfrac{h}{f}\right)(0)$
9. $(g \circ f)(0)$
10. $(h \circ f)(-2)$

11. $(f \circ h)(-2)$
12. $(h \circ g)(3)$
13. $(f \circ g)(2)$
14. $(f \circ f)(-2)$
15. $[h \circ (f + g)](2)$
16. $[g - (h \circ f)](0)$

In Exercises 17–44, (a) find the indicated functions given that f, g, h, F, G, and H are defined as follows:

$$f(x) = 2x + 1 \qquad F(x) = \sqrt{x^2 - 4}$$

$$g(x) = x^2 + 1 \qquad G(x) = \frac{1}{x}$$

$$h(x) = \sqrt{x} \qquad H(x) = \frac{2}{x - 1}$$

and (b) find the domain of the indicated function.

17. $f + g$
18. $G + H$
19. $h + f$
20. $g - f$
21. $H - G$
22. $G - f$
23. $h \cdot F$
24. $f \cdot g$
25. $g \cdot G$
26. $\dfrac{G}{h}$
27. $\dfrac{f}{H}$
28. $\dfrac{F}{h}$
29. $f \circ g$
30. $g \circ f$
31. $G \circ f$
32. $G \circ F$
33. $F \circ h$
34. $g \circ h$

35. $G \circ H$ **36.** $H \circ G$

37. $H \circ f$ **38.** $G \circ G$

39. $f \circ f$ **40.** $H \circ H$

41. $f \cdot (g + h)$ **42.** $f \cdot g + f \cdot h$

43. $f \circ (g \circ h)$ **44.** $(f \circ g) \circ h$

45. Use the results of Exercise 21 to evaluate $(H - G)(3)$.

46. Use the results of Exercise 28 to evaluate $(F/h)(4)$.

47. Use the results of Exercise 35 to evaluate $(G \circ H)(2)$.

48. Use the results of Exercise 36 to evaluate $(H \circ G)(2)$.

 In Exercises 49–54, express the given function as a composition of the functions f, g, and h if

$$f(x) = x + 2 \quad g(x) = \sqrt{x} \quad \text{and} \quad h(x) = x^2 - 1.$$

49. $F(x) = \sqrt{x + 2}$ **50.** $G(x) = (x + 2)^2 - 1$

51. $H(x) = x - 1,\ x \geq 0$ **52.** $J(x) = x^2 + 1$

53. $Q(x) = \sqrt[4]{x}$ **54.** $P(x) = x^4 - 2x^2$

 In Exercises 55–62, express each function as the composition of two functions f and g. (Each exercise may have several correct answers.)

55. $h(x) = \dfrac{1}{\sqrt{x}}$ **56.** $H(x) = 4\sqrt{x - 2}$

57. $P(x) = (2x^3 - 1)^2$ **58.** $h(x) = \sqrt[3]{16 - x}$

59. $h(x) = |3x + 2|$ **60.** $Q(x) = \dfrac{1}{(x + 1)^4}$

61. $H(x) = 2(x - 1)^2 + 3(x - 1)$

62. $r(x) = \sqrt{4 - \sqrt{4 - x}}$

Critical Thinking

For Exercises 63–68, refer to the following table of values.

x	$f(x)$	$g(x)$
-1	4	3
0	7	2
2	3	-1
3	-1	-5

63. Given that $h(x) = (f \circ g)(x)$, find $h(2)$.

64. Given that $h(x) = (g \circ f)(x)$, find $h(2)$.

65. Given that $g(x) = (f \circ h)(x)$, find $h(2)$.

66. Given that $f(x) = (g \circ h)(x)$, find $h(2)$.

67. Given that $f(x) = (g \circ h)(x)$, find $h(3)$.

68. Given that $g(x) = (f \circ h)(x)$, find $h(-1)$.

69. Find $f(x)$ if $(g \circ f)(x) = 2x - 8$ and $g(x) = 6x + 4$.

70. A spherical balloon is being inflated with air. The volume V of the balloon is a function of its radius r and is given by $V(r) = \frac{4}{3}\pi r^3$. In turn, the radius r (in inches) is a function of the time t (in seconds) after the inflation process begins and is given by $r(t) = 3(t + 1)$.

(a) Determine $(V \circ r)(t)$.

(b) What does the composition function $V \circ r$ describe?

(c) Find the volume of the balloon after inflating it for 3 seconds.

(d) Find the time at which the volume is $36,000\pi$ cubic inches.

71. The impedance Z of an electrical circuit is a function of the inductance L in the circuit and is given by

$$Z(L) = \sqrt{20 + (12L)^2}.$$

(a) Decompose this function into two simpler functions f and g.

(b) Decompose this function into three simpler functions f, g, and h.

72. The mass m of a particle is a function of its velocity v and is given by

$$m(v) = \dfrac{2}{\sqrt{1 - \left(\dfrac{v}{300,000}\right)^2}}$$

(a) Decompose this function into two simpler functions f and g.

(b) Decompose this function into three simpler functions f, g, and h.

Calculator Activities

Given the functions f and g defined by

$$f(x) = 2x - 1 \quad \text{and} \quad g(x) = x^2 + 2x,$$

compute to three significant digits the functional values given in Exercises 73–78.

73. $(f + g)(2.02)$ 74. $(f - g)(19.97)$

75. $(f \cdot g)(32.8)$ 76. $(f/g)(0.672)$

77. $(f \circ g)(0.076)$ 78. $(g \circ f)(4.57)$

79. In a ski factory, the daily cost C (in dollars) of producing n pairs of skis is given by

$$C(n) = 225n - 0.8n^2, \quad 0 \le n \le 40$$

and the number of pairs produced in t hours is given by

$$n(t) = 4.5t, \quad t \le 10.$$

(a) Determine $(C \circ n)(t)$.
(b) What does the composition function $C \circ n$ describe?
(c) Find the daily production cost to the nearest dollar if the factory runs for 8.2 hours per day.
(d) How many hours does the factory operate if the daily production cost is $4228?

80. A bicycle manufacturer produces an all-terrain model at a cost of $55 per bicycle. Research has shown that when the selling price of the bike is x dollars, with $120 \le x \le 380$, the number of bikes sold per month is $836 - 2.2x$.

(a) Express the total revenue R per month as a function of x.
(b) Express the total cost C of the bikes that are sold per month as a function of x.
(c) Find and interpret $(R - C)(179)$.

3.6 Inverse Functions

If we select a number, cube it, and then take the cube root of the result, we obtain the original number with which we started. Reversing the procedure, taking the cube root first and then cubing the result, also returns us to the original number. Since one operation "undoes" the other, we say that cubing and taking a cube root are *inverse operations*. In this section, we discuss functions that behave in a similar manner and refer to them as *inverse functions*.

◆ **Verifying Inverse Functions**

Consider the cubing function f defined by $f(x) = x^3$ and the cube root function g defined by $g(x) = \sqrt[3]{x}$. The composite function $g \circ f$ represents cubing a number and then taking the cube root of the result, whereas the composite function $f \circ g$ represents taking the cube root of a number and then cubing the result. Note that

$$(g \circ f)(x) = g(f(x)) = g(x^3) = \sqrt[3]{x^3} = x$$
and
$$(f \circ g)(x) = f(g(x)) = f(\sqrt[3]{x}) = (\sqrt[3]{x})^3 = x.$$

In summary, composing f with g in either order is the *identity function*, the function that assigns each input to itself. Since one function "undoes" the other, we say that the cubing function and the cube root function are **inverse functions**.

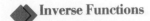**Inverse Functions**

> The functions f and g are **inverses** of each other if
>
> $$f(g(x)) = x \quad \text{for all } x \text{ in the domain of } g$$
>
> and $\qquad g(f(x)) = x \quad \text{for all } x \text{ in the domain of } f.$

If f is the inverse of g, then g is the inverse of f, that is, inverses come in pairs. Also, since inverse functions deal with the composition of functions, it is essential to check the domains and ranges to be certain that the domain of f equals the range of g and the domain of g equals the range of f.

E X A M P L E 1   Show that the functions f and g defined by

$$f(x) = 2x - 8 \qquad \text{and} \qquad g(x) = \tfrac{1}{2}x + 4$$

are inverses of each other.

S O L U T I O N Since the domain and range of both f and g are the set of all real numbers, we know the compositions of f and g exist. Thus, we proceed to show that $f(g(x)) = g(f(x)) = x$:

$$
\begin{aligned}
f(g(x)) &= f(\tfrac{1}{2}x + 4) & g(f(x)) &= g(2x - 8) \\
&= 2(\tfrac{1}{2}x + 4) - 8 & &= \tfrac{1}{2}(2x - 8) + 4 \\
&= x + 8 - 8 & &= x - 4 + 4 \\
&= x & &= x
\end{aligned}
$$

Since $f(g(x)) = g(f(x)) = x$, we conclude that the functions f and g are inverses of each other.   ◆

P R O B L E M 1   Repeat Example 1 if $f(x) = (x - 2)^5$ and $g(x) = \sqrt[5]{x} + 2$.   ◆

◆ **One-to-One Functions**

There exists an inverse function for a given function f only if f is a *one-to-one function*. A function f is said to be *one-to-one* if each element in the range of f is associated with only one element in its domain. More precisely, we state the following definition.

One-to-One Function

> A function f is a **one-to-one function** if, for a and b in the domain of f,
>
> $$f(a) = f(b) \quad \text{implies} \quad a = b.$$

A simple method for determining whether a function is one-to-one is to look at its graph. If every horizontal line that can be drawn in the coordinate plane intersects the graph at most once, then each output of the function is associated with only one input. Hence, the function is one-to-one. We refer to this graphical method of determining whether a function is one-to-one as the **horizontal line test**.

◆ Horizontal Line Test
_____

| A function f is one-to-one if no horizontal line in the coordinate plane intersects the graph of the function in more than one point. |

EXAMPLE 2   Determine if the given function is one-to-one.

(a) $f(x) = (x - 1)^3$ (b) $h(x) = x^2 - 1$

SOLUTION

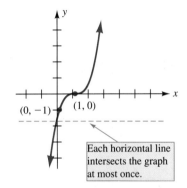

(0, −1)   (1, 0)

Each horizontal line intersects the graph at most once.

FIGURE 3.50

The graph of $f(x) = (x - 1)^3$ passes the *horizontal line test*. Hence, f is a *one-to-one* function.

(a) By the horizontal shift rule (Section 3.4), the graph of $f(x) = (x - 1)^3$ is the same as the graph of $y = x^3$ shifted to the right 1 unit, as shown in Figure 3.50. No horizontal line drawn in the coordinate plane intersects the graph more than once. Thus, by the horizontal line test, the function f is a one-to-one function. Also, note that

$$f(a) = f(b)$$

implies

$$(a - 1)^3 = (b - 1)^3$$

$$a - 1 = b - 1 \qquad \textbf{Take the cube root of both sides}$$

$$a = b \qquad \textbf{Add 1 to both sides}$$

Since $f(a) = f(b)$ implies $a = b$, we conclude that f is one-to-one.

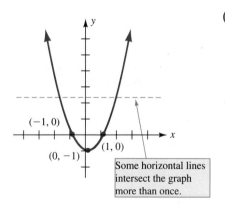

(−1, 0)

(0, −1)   (1, 0)

Some horizontal lines intersect the graph more than once.

FIGURE 3.51

The graph of $h(x) = x^2 - 1$ fails the *horizontal line test*. Hence, h is *not* one-to-one.

(b) By the vertical shift rule (Section 3.4), the graph of $h(x) = x^2 - 1$ is the same as the graph of $y = x^2$ shifted downward 1 unit, as shown in Figure 3.51. Notice the x-axis is one horizontal line that intersects the graph more than once. Thus, by the horizontal line test, the function h is not a one-to-one function. Also, note that

$$h(a) = h(b)$$

implies

$$a^2 - 1 = b^2 - 1$$

$$a^2 = b^2 \qquad \textbf{Add 1 to both sides}$$

$$a = \pm b \qquad \textbf{Take the square root of both sides}$$

Since $h(a) = h(b)$ does *not* imply $a = b$, we may also conclude that h is *not* one-to-one.   ◆

Note: The function f defined in Example 2(a) is an increasing function (see Section 3.4). *Every function that is either an increasing function or a decreasing function is also one-to-one.* Also, if a function is not one-to-one, we can often find a suitable restriction on its domain in order to form a new function that is one-to-one. For example, if we restrict the domain of the function h defined in Example 2(b) to *nonnegative numbers*, the new function that is formed is one-to-one.

PROBLEM 2   Show that the function H defined by $H(x) = x^2 - 1$ with $x \geq 0$ is a one-to-one function.   ◆

◆   **Finding the Inverse of a Function**

As we previously stated, there exists an inverse function for a given function f only if f is a one-to-one function. This is because each output of a one-to-one function comes from just one input. Thus, when we reverse the roles, we can assign to each output the input from which it came. We usually denote the inverse function of the function f by using the notation f^{-1} (read "f inverse") and write

$$f(f^{-1}(x)) = f^{-1}(f(x)) = x$$

In the notation f^{-1}, the superscript -1 denotes an inverse and is not to be confused with an exponent. That is, f^{-1} does not mean $1/f$.

The following procedure may be used to find the inverse of a one-to-one function.

Finding the Inverse Function of a One-to-One Function

If f is a one-to-one function, then the inverse function f^{-1} may be found by

1. replacing each x with $f^{-1}(x)$ in the equation that defines f, and then solving for $f^{-1}(x)$.
2. adjusting the domain of f^{-1} (if necessary) so that it equals the range of f.

This procedure is illustrated in the next example.

EXAMPLE 3   Find the inverse of the one-to-one function defined by

(a) $f(x) = x^3 - 1$ (b) $g(x) = \sqrt{x - 1}$

SOLUTION

(a) We can find the inverse of this one-to-one function as follows:

$$f(x) = x^3 - 1$$

$$f(f^{-1}(x)) = [f^{-1}(x)]^3 - 1 \qquad \textbf{Replace each } x \textbf{ with } f^{-1}(x)$$

$$x = [f^{-1}(x)]^3 - 1 \qquad \textbf{By definition, } f(f^{-1}(x)) = x$$

$$x + 1 = [f^{-1}(x)]^3 \qquad \textbf{Solve for } f^{-1}(x)$$

$$f^{-1}(x) = \sqrt[3]{x + 1}$$

Since the domain and range of both f and f^{-1} are the set of all real numbers, we conclude that the inverse function f^{-1} is defined by

$$f^{-1}(x) = \sqrt[3]{x + 1}.$$

(b) As shown in Figure 3.52, the function g is a one-to-one function with domain $[1, \infty)$ and range $[0, \infty)$. To find g^{-1}, we proceed as follows:

$$g(x) = \sqrt{x - 1}$$

$$g(g^{-1}(x)) = \sqrt{g^{-1}(x) - 1} \qquad \textbf{Replace each } x \textbf{ with } g^{-1}(x)$$

$$x = \sqrt{g^{-1}(x) - 1} \qquad \textbf{By definition, } g(g^{-1}(x)) = x$$

$$x^2 = g^{-1}(x) - 1 \qquad \textbf{Square both sides and solve for } g^{-1}(x)$$

$$g^{-1}(x) = x^2 + 1$$

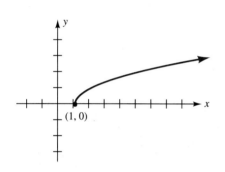

FIGURE 3.52
Graph of $g(x) = \sqrt{x - 1}$.

Since the domain of g^{-1} must be equal to the range of g, we must restrict the domain of g^{-1} to the interval $[0, \infty)$. Hence, we conclude that g^{-1} is defined by

$$g^{-1}(x) = x^2 + 1, \quad x \geq 0. \qquad \blacklozenge$$

Referring to Example 3(b), the result of composing g and g^{-1} in either order must be the identity function. Thus, to check Example 3(b), we show that $g(g^{-1}(x)) = g^{-1}(g(x)) = x$:

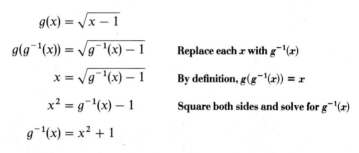

$$g(g^{-1}(x)) = g(x^2 + 1) \qquad\qquad g^{-1}(g(x)) = g^{-1}(\sqrt{x - 1})$$
$$= \sqrt{(x^2 + 1) - 1} \qquad\qquad = (\sqrt{x - 1})^2 + 1$$
$$= \sqrt{x^2} \qquad\qquad = (x - 1) + 1$$
$$= x \quad \text{since } x \geq 0. \qquad\qquad = x$$

PROBLEM 3   Check Example 3(a) by showing that $f(f^{-1}(x)) = f^{-1}(f(x)) = x$. $\qquad \blacklozenge$

Alternatively, the inverse of a one-to-one function f may be found by interchanging x and y in the equation $y = f(x)$ and solving for y. The procedure is illustrated in the next example.

EXAMPLE 4   Find the inverse of the one-to-one function f defined by $f(x) = 2x - 3$.

SOLUTION   To find the inverse, we can replace $f(x)$ with y and proceed as follows:

$$y = 2x - 3$$

$$x = 2y - 3 \qquad \text{Interchange } x \text{ and } y$$

$$x + 3 = 2y \qquad \text{Solve for } y$$

$$y = \frac{x + 3}{2}$$

Hence, the inverse of $y = f(x) = 2x - 3$ is $y = f^{-1}(x) = \dfrac{x + 3}{2}$.   ◆

PROBLEM 4   Find the inverse of the one-to-one function f defined by $f(x) = \dfrac{1}{x + 1}$.   ◆

◆ **Graphs of Inverse Functions**

The graphs of g and g^{-1}, the functions defined in Example 3(b), are shown in Figure 3.53. Notice that for every point (a, b) on the graph of g there corresponds a point (b, a) on the graph of g^{-1}. Thus, if we were to fold the coordinate plane along the dotted line $y = x$, the graphs of g and g^{-1} would coincide. In other words, the graphs of g and g^{-1} are *reflections of one another in the line* $y = x$. This special relationship between the graphs of g and g^{-1} is true for any function and its inverse.

FIGURE 3.53
The graphs of $g(x) = \sqrt{x - 1}$ and $g^{-1}(x) = x^2 + 1$, $x \geq 0$, are reflections of one another in the line $y = x$.

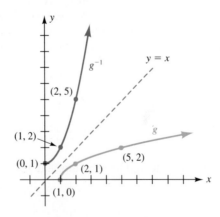

EXAMPLE 5   Suppose the function f is defined by $f(x) = 4 - x^2$ for $x \geq 0$. Sketch the graph of f^{-1}.

SOLUTION   We first sketch the graph of $f(x) = 4 - x^2$ for $x \geq 0$ by using the graphing techniques discussed in Section 3.4. The graph of f^{-1} is the same as this graph reflected in the line $y = x$, as shown in Figure 3.54. From the graphs of f and f^{-1}, we can observe that the domain of f^{-1} and the range of f is $(-\infty, 4]$. Also the range of f^{-1} and the domain of f is $[0, \infty)$.   ◆

FIGURE 3.54
Graph of f^{-1} formed from the graph of $f(x) = 4 - x^2$, $x \geq 0$ by reflecting about the line $y = x$.

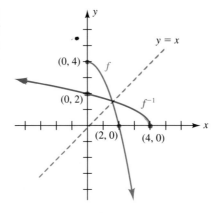

PROBLEM 5   Referring to Example 5, find the equation that defines f^{-1}.   ◆

Exercises 3.6

Basic Skills

In Exercises 1–10, show that the functions f and g are inverses of each other by showing that $f(g(x)) = g(f(x)) = x$.

1. $f(x) = 3x - 1$; $g(x) = \dfrac{x + 1}{3}$

2. $f(x) = 5 - 2x$; $g(x) = \dfrac{5 - x}{2}$

3. $f(x) = \dfrac{1}{x}$; $g(x) = \dfrac{1}{x}$

4. $f(x) = \sqrt[5]{x - 6}$; $g(x) = x^5 + 6$

5. $f(x) = x^3 - 8$; $g(x) = \sqrt[3]{x + 8}$

6. $f(x) = (x - 2)^3$; $g(x) = \sqrt[3]{x} + 2$

7. $f(x) = \dfrac{1}{x - 1}$; $g(x) = \dfrac{1}{x} + 1$

8. $f(x) = \dfrac{x}{x + 3}$; $g(x) = \dfrac{-3x}{x - 1}$

9. $f(x) = (x - 1)^2$, $x \geq 1$; $g(x) = \sqrt{x} + 1$

10. $f(x) = x^2 + 2$, $x \geq 0$; $g(x) = \sqrt{x - 2}$

In Exercises 11–20,

(a) *Determine if the function is one-to-one.*

(b) *If the function is not one-to-one, find a suitable restriction on its domain in order to form a new function that is one-to-one. There is no unique choice for this new function.*

11. $f(x) = 2x$

12. $f(x) = \dfrac{x^3}{3}$

13. $f(x) = x^2$

14. $f(x) = |x| - 3$

15. $f(x) = \sqrt{9 - x}$

16. $f(x) = \dfrac{1}{x - 3}$

17. $f(x) = |x + 2|$

18. $f(x) = 1 - \sqrt[3]{x}$

19. $f(x) = (x - 2)^2 - 1$

20. $f(x) = (x - 1)^{2/3}$

Each function f in Exercises 21–40 is a one-to-one function.

(a) *Find the inverse function f^{-1}.*

(b) *Sketch the graph of f and f^{-1} on the same coordinate plane and verify that the graphs of f and f^{-1} are symmetric with respect to the line $y = x$. If you have access to a graphing calculator, check each graph.*

21. $f(x) = 2x$

22. $f(x) = \dfrac{-x}{4}$

23. $f(x) = 3 - x$

24. $f(x) = x + 1$

25. $f(x) = 2 - x^3$

26. $f(x) = (x + 1)^3$

27. $f(x) = \sqrt[3]{x - 4}$

28. $f(x) = 3 + \sqrt[3]{x}$

29. $f(x) = \dfrac{1}{x + 1}$

30. $f(x) = \dfrac{1}{x} - 2$

31. $f(x) = \sqrt{x - 4} + 1$

32. $f(x) = \sqrt{x} + 3$

33. $f(x) = (x - 1)^2, \quad x \ge 1$

34. $f(x) = (x - 3)^2 + 2, \quad x \ge 3$

35. $f(x) = \sqrt{9 - x^2}, \quad 0 \le x \le 3$

36. $f(x) = \sqrt{x^2 - 9}, \quad x \ge 3$

37. $f(x) = x^2 + 3, \quad x \le 0$

38. $f(x) = 2 - x^2, \quad x \le 0$

39. $f(x) = \begin{cases} x^2 & \text{if } x \ge 0 \\ 2x & \text{if } x < 0 \end{cases}$

40. $f(x) = \begin{cases} x^3 & \text{if } x \ge 0 \\ -x^2 & \text{if } x < 0 \end{cases}$

Critical Thinking

In Exercises 41–44, the graph of a function f is shown. Determine whether the function f has an inverse. If so, sketch the graph of f^{-1} and complete this table of values:

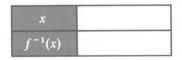

x	
$f^{-1}(x)$	

41.

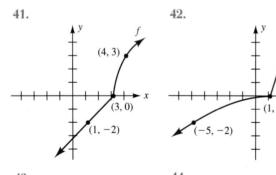

42.

43.

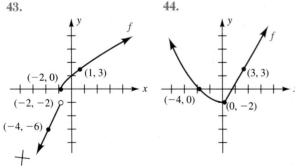

44.

For Exercises 45–50, assume the functions f and g are defined as follows:

$$f(x) = \dfrac{x - 2}{3} \quad \text{and} \quad g(x) = 3x + 1$$

45. Find $f^{-1}(4)$.

46. Find $g^{-1}(7)$.

47. Find $(f^{-1} \circ g)(-2)$.

48. Find $(f \circ g^{-1})(25)$.

49. Compare $(f \circ g)^{-1}(x)$ with $(g^{-1} \circ f^{-1})(x)$.

50. Compare $(g \circ f)^{-1}(x)$ with $(f^{-1} \circ g^{-1})(x)$.

In Exercises 51–54, determine the range of the given function by examining the domain of its inverse function.

51. $f(x) = \dfrac{1}{2x - 1}$

52. $f(x) = \dfrac{x}{6 - 3x}$

53. $f(x) = \dfrac{2x + 1}{3x - 2}$

54. $f(x) = \dfrac{9x - 3}{3x + 1}$

For Exercises 55–58, use the following facts:

(i) *The volume V of a sphere is a function of its radius r:*
$V = f(r) = \frac{4}{3}\pi r^3$.

(ii) *The surface area S of a sphere is also a function of its radius r:*
$S = g(r) = 4\pi r^2$.

55. Find the inverse function g^{-1}.

56. Find the inverse function f^{-1}.

57. Find and interpret $(f \circ g^{-1})(S)$.

58. Find and interpret $(g \circ f^{-1})(V)$.

Given the functions f and g defined by

$$f(x) = 3.24x - 3.46 \quad \text{and} \quad g(x) = \frac{2.72}{x - 8.65},$$

compute to three significant digits the functional values indicated in Exercises 59–64.

59. $f^{-1}(2.75)$ **60.** $g^{-1}(9.87)$

61. $(f^{-1} \cdot g)(35.9)$ **62.** $(f + g^{-1})(0.729)$

63. $(f \circ f^{-1})(0.173)$ **64.** $(g \circ g^{-1})(4.33)$

65. Use a graphing calculator to generate the graph of $y = x$ and the graphs of the given pair of functions in the same viewing rectangle. From the picture, determine if the given functions appear to be inverses of each other. Verify your answer.

    (a) $f(x) = \dfrac{3x + 1}{x}$ and $g(x) = \dfrac{1}{x - 3}$

    (b) $f(x) = \dfrac{x}{x - 2}$ and $g(x) = \dfrac{2x}{x - 1}$

66. In Chapter 6, we will define the *exponential function* $f(x) = e^x$ and the *logarithmic function* $g(x) = \ln x$. However, even at this point, we can observe a special relationship between these functions by using the calculator keys ⎡LN X⎤ and ⎡e^x⎤. Complete the following table and state the special relationship between $f(x) = e^x$ and $g(x) = \ln x$.

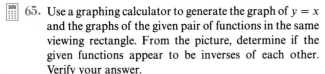

x	$f(x)$	$g(x)$	$f(g(x))$	$g(f(x))$
1				
2				
3				
4				
5				

3.7   Applied Functions and Variation

To solve many types of word problems—especially those that appear in calculus—we must begin by setting up an equation that defines a function. We may then analyze the function, draw its graph, and answer questions concerning the functional relationship between the quantities. In this section we practice setting up functions from words.

◆ Applied Functions

For many applied problems, we begin with an established formula and then use *substitution* to obtain a functional relationship between the desired variables. The procedure is illustrated in the next two examples.

EXAMPLE 1   Suppose 200 feet of fencing is needed to enclose a rectangular garden.

    (a)   Express the area A of the rectangular garden as a function of its length l.

    (b)   Give the domain of this function.

SOLUTION

    (a)   Drawing a rectangular garden, we let

$$l = \text{the length} \quad \text{and} \quad w = \text{the width}$$

as shown in Figure 3.55.

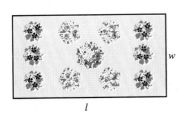

FIGURE 3.55

Rectangular garden of length l and width w

Recall that the area A of a rectangle is the product of its length and width. So

$$A = lw.$$

For the area A to be a function of its length l, we need to write A in terms of just l. Since 200 feet of fencing is needed to enclose the garden, we have

$$200 = 2l + 2w.$$

Solving this equation for w gives us

$$w = 100 - l.$$

Now, we substitute $100 - l$ for w in the area formula, $A = lw$, and obtain

$$A = lw = l(100 - l) = 100l - l^2.$$

The equation $A = 100l - l^2$ defines the area A of this rectangle as a function of its length l. If we wish to emphasize this functional relationship, we may use functional notation to write

$$A(l) = 100l - l^2.$$

(b) Algebraically, the domain of this function is the set of all real numbers but, geometrically, this domain does not make sense. This is because l represents the length of the rectangle, and lengths can only be *positive*. Thus, we must have $l > 0$. Also, since the perimeter is 200 feet, we must have $l < 100$, otherwise, the width w would be zero or negative. Because of the geometric restrictions placed on this function, the domain is $(0, 100)$.   ◆

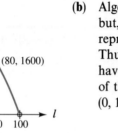

FIGURE 3.56

Graph of the function defined by the equation $A = 100l - l^2$ with $0 < l < 100$

To sketch the graph of $A = 100l - l^2 (0 < l < 100)$, we can choose a convenient scale for each axis and plot some points. The graph is shown in Figure 3.56. The vertical axis is labeled as A instead of y and the horizontal axis is labeled as l instead of x. From the graph, we can determine that the range of the function is $(0, 2500]$.

PROBLEM 1 Use the graph of $A = 100l - l^2$ in Figure 3.56 to determine the dimensions of the rectangle whose area is the *largest* possible.   ◆

EXAMPLE 2 Water is flowing into a conical funnel. The diameter of the base of the funnel is 8 inches and its height is 12 inches.

(a) Find the volume V of water in the funnel as a function of the height h of water in the funnel.

(b) Give the domain of this function, then sketch its graph.

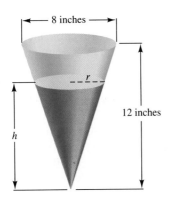

FIGURE 3.57
Conical funnel 8 inches wide and
12 inches deep.

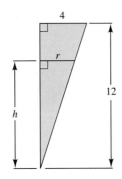

FIGURE 3.58
Similar triangles formed from the cross
section of the funnel.

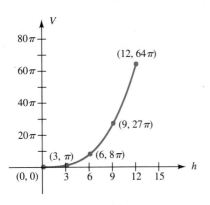

FIGURE 3.59
Graph of the function defined by the
equation $V = \frac{1}{27}\pi h^3$ with $0 \le h \le 12$.

SOLUTION

(a) We draw a conical funnel as in Figure 3.57 showing that

$$h = \text{the height of water in the funnel}$$

and

$$r = \text{the radius of the surface of the water.}$$

From elementary geometry, we know that the volume of water in the funnel is given by

$$V = \tfrac{1}{3}\pi r^2 h.$$

For the volume V to be a function of h, we must write the volume in terms of just h. To do this, we use similar triangles from the cross section of the funnel, as shown in Figure 3.58. Since corresponding sides of similar triangles are proportional, we have

$$\frac{4}{12} = \frac{r}{h}$$

or

$$r = \frac{h}{3}.$$

Now, we substitute $h/3$ for r in the volume formula $V = \tfrac{1}{3}\pi r^2 h$ and obtain

$$V = \tfrac{1}{3}\pi\left(\frac{h}{3}\right)^2 h$$

or

$$V = \tfrac{1}{27}\pi h^3.$$

If we wish to emphasize this functional relationship, we use functional notation to write

$$V(h) = \tfrac{1}{27}\pi h^3.$$

(b) Since the height of the funnel is 12 inches, the water level must be between 0 and 12 inches. Thus, the domain of the function defined by the equation $V = \tfrac{1}{27}\pi h^3$ is [0, 12]. Choosing convenient scales for the h-axis and V-axis, we sketch the graph by plotting some points. The graph of the function defined by the equation

$$V = \tfrac{1}{27}\pi h^3, \quad 0 \le h \le 12,$$

is shown in Figure 3.59. Note that the range of the function is [0, 64π]. ◆

PROBLEM 2

Suppose water is flowing into the funnel described in Example 2 at the rate of 8 cubic inches per minute. Find the volume V of water (in cubic inches) in the funnel as a function of the time t (in minutes), and state the domain of this function.   ◆

◆ Variation

In business, engineering, and science, the functional relationship between two quantities is often given in terms of *variation*. When we state that **y varies directly as x** or that **y is directly proportional to x**, we mean that the ratio of y to x is always the same. In other words,

$$\frac{y}{x} = k \qquad \text{or} \qquad y = kx$$

where k is called the *variation constant*.

EXAMPLE 3

The real estate tax T on a property varies directly as its assessed value V.

(a) Express T as a function of V if $T = \$2800$ when $V = \$112,000$.

(b) State the domain of the function defined in part (a), then sketch its graph.

SOLUTION

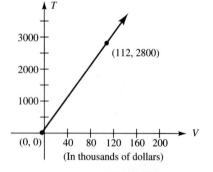

(a) Since T varies directly as V, we write

$$T = kV,$$

where k is the variation constant to be determined. To find k, we replace T with 2800 and V with 112,000 as follows:

$$T = kV$$
$$2800 = k(112,000)$$
$$k = \frac{2800}{112,000}$$
$$= \frac{1}{40}$$

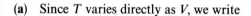

Thus, $T = \frac{1}{40}V$ or $T = \dfrac{V}{40}$.

(b) Since the assessed value V must be greater than or equal to zero, the domain is $[0, \infty)$. By the vertical stretch and shrink rule (Section 3.4), we know that the graph of $T = V/40$ is part of a straight line. Choosing convenient scales for the V-axis and T-axis, we show the graph in Figure 3.60. From the graph, we can see that the range is also $[0, \infty)$.   ◆

FIGURE 3.60

Graph of the function defined by the equation $T = V/40$, with $V \geq 0$.

PROBLEM 3   Using the results from Example 3(a), find the tax T on a piece of property with an assessed value of $168,000. Check your answer by using the graph in Figure 3.60.   ◆

When we state that *y* **varies inversely as** *x* or that *y* **is inversely proportional to** *x*, we mean that the product of y and x is always the same. In other words,

$$yx = k \qquad \text{or} \qquad y = \frac{k}{x}$$

where k is the *variation constant.*

EXAMPLE 4   The time t it takes a person to travel a fixed distance varies inversely as the rate of speed r at which the person travels.

(a) Express t as a function of r if $t = 4$ hours when $r = 60$ miles per hour.

(b) State the domain of the function defined in part (a), then sketch its graph.

SOLUTION

(a) Since t varies inversely as r, we write

$$t = \frac{k}{r},$$

where k is the variation constant to be determined. To find k, we replace t with 4 and r with 60 as follows:

$$t = \frac{k}{r}$$

$$4 = \frac{k}{60}$$

$$k = 240$$

Thus, $t = \dfrac{240}{r}$.

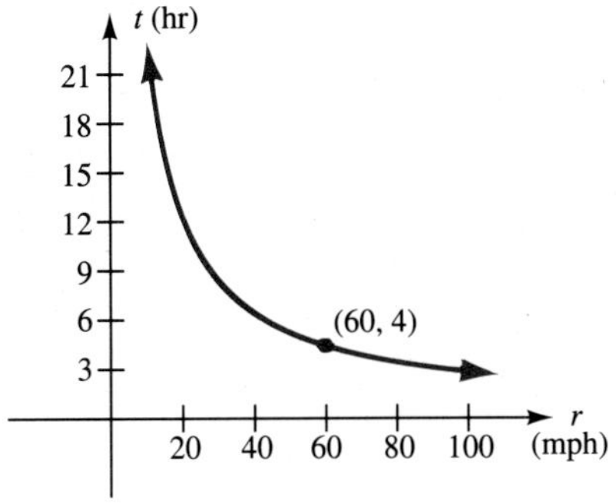

FIGURE 3.61

Graph of the function defined by the equation $t = 240/r$, with $r > 0$.

(b) Since rates of speed are assumed to be positive, the domain of the function defined by $t = 240/r$ is $(0, \infty)$. Choosing convenient scales for the r-axis and t-axis, we show the graph in Figure 3.61.   ◆

PROBLEM 4   Using the results of Example 4(a), find the time t it takes to travel a fixed distance if the rate of speed is 50 miles per hour. Check your answer by using the graph in Figure 3.61.   ◆

The following table gives several other types of variation that occur frequently. In each case, k denotes the variation constant.

Statement	Formula
y varies directly as the nth power of x.	$y = kx^n$
y varies inversely as the nth power of x.	$y = \dfrac{k}{x^n}$
y varies directly as x and inversely as z.	$y = \dfrac{kx}{z}$
y varies jointly as x and z.	$y = kxz$

EXAMPLE 5   The volume V of a right circular cylinder (see Figure 3.62) varies jointly as its height h and the square of its radius r. Describe what happens to the volume of a cylinder if its height and radius are doubled.

SOLUTION   Since V varies jointly as h and the square of r, we write

$$V = kr^2h$$

FIGURE 3.62
A right circular cylinder.

where k is the variation constant. If we double both the height and the radius we obtain

$$V = k(2r)^2(2h) = 8kr^2h.$$

From this equation we can see that the volume of the cylinder is *8 times the original volume*. Thus, when the height and radius of a cylinder are doubled, its volume becomes 8 times as large.   ◆

PROBLEM 5   The electrical resistance R of a wire varies directly as its length l and inversely as the square of its radius r. Describe what happens to the resistance of a wire if its length and radius are doubled.   ◆

Exercises 3.7

    Basic Skills

1. Express the diameter d of a circle as a function of its circumference C.

2. The length of a rectangle is 5 cm. Express its width w as a function of its perimeter P.

3. Suppose a leasing company charges $40 per day plus $0.20 per mile to rent a car. Express in dollars the daily cost C of renting a car as a function of the number n of miles driven.

4. Suppose a salesperson earns $200 per week plus 25% commission on all sales. Express in dollars the weekly earnings E of a salesperson as a function of the amount A of merchandise that she sells.

5. A square with side s and diagonal d is shown in the figure.

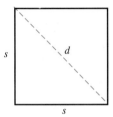

(a) Express s as a function of d.
(b) Express the area A of the square as a function of d.
(c) Express the perimeter P of the square as a function of d.

6. An equilateral triangle with side s and height h is shown in the figure.

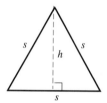

(a) Express h as a function of s.
(b) Express the area A of the triangle as a function of s.
(c) Express the perimeter P of the triangle as a function of h.

7. A small computer company has a present net worth of $125,000. It is estimated that the future weekly income and expenses for the company will be $30,000 and $26,000, respectively. Express the net worth W of the company at the end of t weeks as a function of t.

8. A $15,000 automobile depreciates 20% of its original value each year. Express the value V of the automobile at the end of t years as a function of t.

9. The radius r of a pile of sand in the shape of a cone is twice its height h.

(a) Express the volume V of sand as a function of h.
(b) Express the volume V of sand as a function of r.

10. The height h of a tin can in the shape of a cylinder is equal to its diameter d.

(a) Express the volume V of the tin can as a function of h.
(b) Express the surface area S of the tin can as a function of d.

11. The point $P(x, y)$ lies on the graph of the circle $x^2 + y^2 = 1$ as shown in the figure

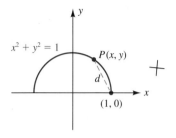

(a) Express the distance d from point P to the point $(1, 0)$ as a function of the x-coordinate of P.
(b) State the domain of the function in part (a).

12. The point $P(x, y)$ lies on the graph of the parabola $2y = x^2$ shown in the figure.

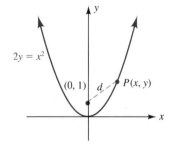

(a) Express the distance d from the point P to the point $(0, 1)$ as a function of the y-coordinate of P.
(b) State the domain of the function in part (a).

13. The point $Q(x, y)$ lies on the graph of $y = \sqrt{x}$ shown in the figure.

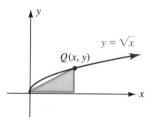

(a) Express the area A of the shaded right triangle as a function of the x-coordinate of Q.
(b) Express the perimeter P of the shaded right triangle as a function of the x-coordinate of Q.

14. The point $Q(x, y)$ lies on the graph of the semicircle $y = \sqrt{25 - x^2}$ shown in the figure.

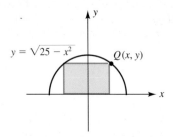

(a) Express the area A of the shaded rectangle as a function of the x-coordinate of Q.

(b) Express the perimeter P of the shaded rectangle as a function of the x-coordinate of Q.

15. The volume of a rectangular box with a square base is 64 cubic inches. Express its surface area S as a function of the width x of its base.

16. The volume of a rectangular box with a square base and an open top is 100 cubic meters. The material used to construct the base costs $4 per square meter, and the material for the sides cost $2.50 per square meter. Express the cost C to construct the rectangular box as a function of the width x of its base.

17. The ends of a water trough are isosceles triangles, as shown in the figure.

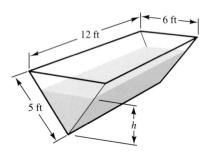

(a) Express the volume V of water in the trough as a function of the height h of the water in the trough.

(b) Give the domain of the function in part (a), then sketch its graph.

18. Suppose the trough described in Exercise 17 is initially empty, and begins to fill with water at the rate of 1 cubic foot per minute. Express the height h of the water in the trough as a function of the time t in minutes that the water flows into the trough.

19. A baseball diamond is a square with 90-foot base paths, as shown in the figure. Suppose a runner on first base runs toward second base at the rate of 30 feet per

second as soon as the pitcher throws the ball to home plate.

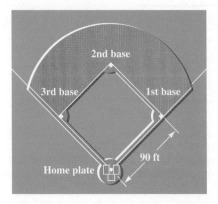

(a) Express the runner's distance d from home plate as a function of the time t in seconds after the pitcher throws the ball.

(b) Describe the domain and range of the function in part (a).

20. The fixed cost to run a wood-stove company is $25,000 per month, and the variable cost to produce each stove is $200 per unit. If the stoves sell for x dollars, then the number of stoves sold per month is estimated to be $1000 - 2x$. Express the monthly profit of the company as a function of x. (*Hint*: Recall from Section 2.2 that Profit = Total revenue − total cost.)

In Exercises 21–32, the functional relationship between the quantities is given in terms of variation.

21. Hooke's law states that the distance d a spring stretches varies directly as the force F applied to the spring, as long as the elastic limit of the spring is not exceeded. Assume that the elastic limit of a spring occurs at 2400 Newtons.

(a) Express d as a function of F if $d = 8$ cm when $F = 400$ Newtons.

(b) State the domain of the function in part (a), then sketch its graph.

22. The electrical resistance R of a wire is directly proportional to the length l of the wire.

(a) Express R as a function of l if $R = 2$ ohms when $l = 30$ meters.

(b) State the domain of the function in part (a), then sketch its graph.

23. The weight W of a steel beam varies directly as its length l.

(a) Express W as a function of l if $W = 1500$ pounds when $l = 12$ feet.

(b) State the domain of the function in part (a), then sketch its graph.

24. The weekly payroll P of a company varies directly as the number n of workers assigned to the job.

   (a) Express P as a function of n if $P = \$5200$ when $n = 13$ workers.
   (b) State the domain of the function in part (a), then sketch its graph.

25. The number N of long-distance phone calls per day between two towns, each with a population of approximately 25,000 people, varies inversely as the distance d between the towns.

   (a) Express N as a function of d if $N = 20$ calls when $d = 500$ miles.
   (b) State the domain of the function in part (a), then sketch its graph.

26. Boyle's law states that when the temperature of a confined gas remains constant, the pressure P it exerts varies inversely as the volume V it occupies.

   (a) Express P as a function of V if $P = 4$ pounds per square inch when $V = 30$ cubic inches.
   (b) State the domain of the function in part (a), then sketch its graph.

27. The wavelength w of a radio wave varies inversely as its frequency f.

   (a) Express w as a function of f if $w = 40$ meters when $f = 7.5$ megahertz.
   (b) State the domain of the function in part (a), then sketch its graph.

28. If the voltage in an electrical circuit is constant, the current I through a resistor varies inversely as its resistance R.

   (a) Express I as a function of R if $I = 2$ milliamps when $R = 100$ ohms.
   (b) State the domain of the function in part (a), then sketch its graph.

29. The gravitational force F between two objects varies inversely as the square of the distance d between them.

   (a) Express F as a function of d if $F = 100$ Newtons when $d = 40$ meters.
   (b) State the domain of the function in part (a), then sketch its graph.

30. The lift L of an airplane wing varies directly as the square of the speed v of air flowing over it.

   (a) Express L as a function of v if $L = 225$ pounds per square foot when $v = 150$ miles per hour.
   (b) State the domain of the function in part (a), then sketch its graph.

31. The centrifugal force acting on an object traveling in a circular path varies directly as the square of its velocity and inversely as the radius of the circle. Describe what happens to the centrifugal force acting on the object if the velocity and radius are both tripled.

32. The power produced by an electric generator varies jointly as the load resistance and the square of the current. Describe what happens to the power produced by the generator if the load resistance is doubled and the current is halved.

 Critical Thinking

33. A church window has the shape of a rectangle surmounted by a semicircle, as shown in the figure. Express the area A of the window as a function of the radius r of the semicircle if the perimeter of the window is 30 feet.

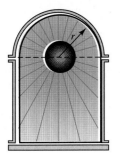

34. A football stadium has the shape of a rectangle with semicircular ends, as shown in the figure. Express the area A of the stadium as a function of the radius r of the semicircle if the perimeter of the stadium is 1 mile.

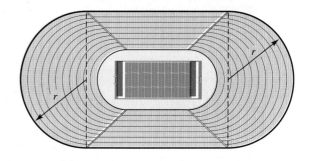

35. A baseball player hits the ball from home plate to the outfield and tries for a double running directly on the base paths. (See the figure in Exercise 19.) Suppose he can run at the rate of 30 feet per second. Express the straight-line distance d from home plate to the runner as a function of the time t in seconds after he hits the ball. (*Hint*: Use a piecewise-defined function.)

36. An author is paid a royalty of $3.00 per book for the first 1000 books sold. The royalty increases to $3.05 per book for each book sold in excess of 1000. Express the author's royalty R as a function of the number n of books sold. (*Hint*: Use a piecewise-defined function.)

37. The time required for an elevator to lift its passengers varies jointly as the weight of the passengers and the distance they are lifted, and inversely as the horsepower of the motor that is used. Suppose it takes 10 seconds for an elevator to lift 800 pounds to a height of 40 feet with a 20 horsepower motor. Determine the time it takes this elevator to lift 1000 pounds to a height of 80 feet with the same motor.

38. The safe-load capacity of a wooden rectangular beam supported at both ends varies jointly as its width and the square of its depth, and inversely as the distance between the supports. Suppose the safe-load capacity of a beam 4 inches wide and 12 inches deep is 3600 pounds when the distance between the supports is 16 feet. Determine the safe-load capacity of a similar beam that is 8 inches wide and 8 inches deep if the distance between the supports remains 16 feet.

 Calculator Activities

39. The length of the hypotenuse of a right triangle is 16.2 inches.

   (a) Express the area A of the triangle as a function of the length x of one of the legs.
   (b) Find the area when $x = 12.4$ inches.

40. Two joggers start from the same place at the same time. One runs due east at 8.2 miles per hour (mph) and the other runs due north at 6.3 mph.

   (a) Determine the distance d between the joggers as a function of the time t in hours that they have been running.
   (b) Find the distance between the joggers after 1 hour 24 minutes.

41. One thousand feet of fencing is to be used to enclose a rectangular pasture along a river, as shown in the figure. No fencing is needed along the river.

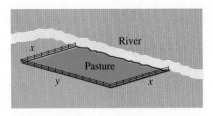

   (a) Express the area A of the pasture as a function of the length x of the pasture.
   (b) State the domain of the function defined in part (a).
   (c) Use a graphing calculator to generate the graph of the function defined in part (a). Trace to the peak of this curve, then state the dimensions of the rectangular pasture whose area is the *largest* possible.

42. Three adjacent rectangular corrals are to be built with 120 feet of fencing, as shown in the figure.

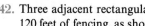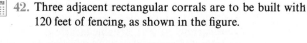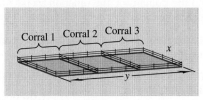

   (a) Express the total enclosed area A as a function of the common length x.
   (b) State the domain of the function defined in part (a).
   (c) Use a graphing calculator to generate the graph of the function defined in part (a). Trace to the peak of this curve, then state the overall dimensions of the corrals so that the total enclosed area is the *largest* possible.

43. The period T of a pendulum is directly proportional to the square root of the length l of the pendulum.

   (a) Express T as a function of l if $T = 2.1$ seconds when $l = 1.095$ meters.
   (b) Find the period when the length of the pendulum is 2.405 meters.

44. The horsepower H required to propel a motorboat through the water varies directly as the cube of the speed s of the boat.

   (a) Express H as a function of s if $H = 48$ horsepower when $s = 11.2$ knots.
   (b) Find the horsepower when the speed is 15.6 knots.

Chapter 3 Review

Questions for Group Discussion

1. What are *Cartesian coordinates*?

2. Explain in words how to find the *midpoint* of a line segment.

3. What is the difference between the *independent variable* and the *dependent variable* in an equation?

4. Describe the *point-plotting method* of sketching the graph of an equation.

5. List some aids that can be used to help sketch the graph of an equation.

6. What is meant by the *x-intercept* and *y-intercept* of a graph? How can the intercepts be determined from a given equation?

7. What is the advantage of writing the equation of a circle in *standard form*?

8. Is the graph of $x^2 + y^2 + Dx + Ey + F = 0$ always a circle? Explain.

9. How can the *vertical line test* be used to determine if an equation defines y as a function of x?

10. What is a *function*? What is meant by its *domain* and *range*?

11. How can the *zeros* of a function be determined?

12. What is the difference between an *even function* and an *odd function*? Give an example of each.

13. What can we learn about a function from its *graph*?

14. From the graph of a function, how is it possible to determine if the function is *one-to-one*?

15. Does every function have an *inverse function* associated with it? Explain.

16. How does the graph of a function relate to the graph of its inverse?

17. Explain the algebraic procedure for finding the inverse of a one-to-one function.

18. Can the graph of a function be symmetric with respect to the *x*-axis? Explain.

19. How are the graphs of the following functions related to the graph of $y = x^2$?
    (a) $y = x^2 + 3$ (b) $y = (x - 2)^2$ (c) $y = -x^2$ (d) $y = (x + 2)^2 - 1$

20. What is the difference between $f \cdot g$ and $f \circ g$? Illustrate with examples.

21. How is *composition* used to show that the functions f and g are inverse functions?

22. If f is an even function, what can you conclude about f^{-1}? Explain.

23. Can a function be its own inverse? If so, give an example.

24. Explain how the *distance formula* can be used to determine if three points are collinear.

25. Suppose *y varies inversely as x* and *x varies inversely as t*. What can you conclude about y and t?

26. Give an example of how a *composite function* might come up in the real world.

Review Exercises

In Exercises 1–6, find

(a) *the length of the line segment joining the points A and B.*
(b) *the coordinates of the midpoint M of the line segment joining the points A and B.*
(c) *the standard form of the equation of a circle whose diameter is the line segment joining the points A and B.*

1. $A(0, 3)$; $B(4, 6)$

2. $A(2, 0)$; $B(7, 12)$

3. $A(-2, 1)$; $B(2, 3)$

4. $A(4, -1)$; $B(1, -4)$

5. $A(-3, -5)$; $B(6, -1)$

6. $A(7, -3)$; $B(-3, -5)$

7. Show that the quadrilateral that joins the points $A(2, 1)$, $B(4, 3)$, $C(2, 5)$, and $D(0, 3)$ is a square. Then find its perimeter and area.

8. Show that the triangle that joins the points $A(-1, 8)$, $B(5, -2)$, and $C(15, 4)$ is an isosceles triangle. Then find its perimeter and area.

For Exercises 9–12, refer to the triangle with vertices A, O, and B as shown:

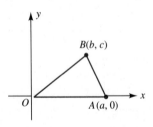

9. Find the length of the line segment joining the midpoints of the sides $\overline{OB}$ and $\overline{AB}$. Compare this length to OA, the length of the third side of the triangle.

10. Find the length of the line segment joining the midpoints of the sides $\overline{OA}$ and $\overline{AB}$. Compare this length to OB, the length of the third side of the triangle.

11. Find the length of the line segment joining the midpoints of the sides $\overline{OA}$ and $\overline{OB}$. Compare this length to AB, the length of the third side of the triangle.

12. What conclusion can be made about the length of a line segment joining the midpoints of any two sides of a triangle with respect to the length of the third side of the triangle.

For the equations in Exercises 13–30,

(a) *determine any x-intercept and y-intercept.*
(b) *check for symmetry with respect to the x-axis, y-axis, and origin.*
(c) *sketch the graph.*
(d) *determine if the equation defines y as a function of x.*

13. $y = \frac{1}{2}x^3$

14. $4x = y^2$

15. $3x - |y| = 0$

16. $y - 2\sqrt{x} = 0$

17. $x^2 y = 1$

18. $xy = 2$

19. $2x + 3y = 6$

20. $3x - y = 9$

21. $x^2 - 3y = 9$

22. $x + |y| = 3$

23. $|x| + 2|y| = 2$

24. $|x + y| = 1$

25. $x^2 + y^2 - 16 = 0$

26. $x^2 + y^2 = 1$

27. $x^2 + y^2 - 6x = 7$

28. $x^2 + y^2 - 6y - 16 = 0$

29. $x^2 + y^2 - 10x - 8y + 16 = 0$

30. $4x^2 + 4y^2 - 8x + 12y = 3$

Given the functions f, g, and h, defined by

$$f(x) = 3x + 4$$
$$g(x) = 2x^2 - x$$

and

$$h(x) = \frac{x + 1}{x - 3}$$

compute the functional values in Exercises 31–50.

31. $f(-1)$

32. $g(3)$

33. $h(4)$

34. $f(2x)$

35. $g(x - 3)$

36. $h\left(\dfrac{a}{2}\right)$

37. $f(\frac{1}{3}) - f(2)$

38. $(g \cdot f)(x)$

39. $\dfrac{f(7)}{h(4)}$

40. $(f \circ h)(2)$

41. $(g \circ f)(x^2)$

42. $(f \circ f)(-5)$

43. $h^{-1}(x)$

44. $f(x) + f^{-1}(x)$

45. $f^{-1}(x + 2)$

46. $h^{-1}(-x)$

47. $(h \circ h^{-1})(2)$ **48.** $(g \circ f^{-1})(x)$

49. $\dfrac{g(x + \Delta x) - g(x)}{\Delta x}$ **50.** $\dfrac{f^{-1}(a + h) - f^{-1}(a)}{h}$

In Exercises 51–70,

(a) *find the domain of the function.*
(b) *find any zeros of the function.*
(c) *determine if the function is even, odd, or neither.*
(d) *sketch the graph of the function (if it is not given).*
(e) *determine the intervals where the function is increasing,
    decreasing, or constant.*
(f) *determine if the function is one-to-one.*
(g) *use the graph to find the range of the function.*

51. $f(x) = x + 3$ **52.** $g(x) = -2x$

53. $h(x) = \sqrt[3]{2x}$ **54.** $F(x) = \sqrt[3]{x - 1}$

55. $G(x) = x^2 - 9$ **56.** $H(x) = |x| + 5$

57. $g(x) = (x + 3)^2$ **58.** $F(x) = \sqrt{3 - x}$

59. $G(x) = x^3 + 8$ **60.** $h(x) = \dfrac{1}{x - 4}$

61. $F(x) = |x - 3| - 1$ **62.** $H(x) = 8 - (x + 1)^3$

63. $H(x) = 6x - x^2$

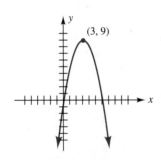

64. $f(x) = |x + 1| + |x - 1|$

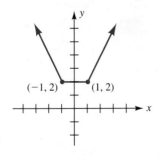

65. $f(x) = \sqrt{100 - x^2}$

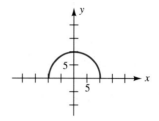

66. $g(x) = \sqrt{x^2 - 49}$

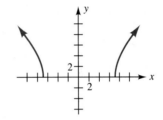

67. $g(x) = 2x^2 - x^4$

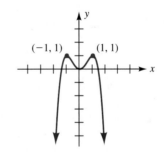

68. $H(x) = 2x^3 - 6x$

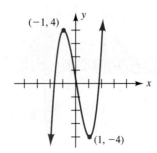

69. $F(x) = \begin{cases} x^2 + 1 & \text{if } x \le 0 \\ 1 & \text{if } 0 < x \le 3 \end{cases}$

70. $h(x) = \begin{cases} \sqrt{x - 1} & \text{if } 1 \le x \le 5 \\ x - 3 & \text{if } x > 5 \end{cases}$

In Exercises 71–76, use the graph of $y = x^{2/3}$ in the figure and the techniques of shifting, reflecting, shrinking, and stretching, to sketch the graph of the function defined by each equation.

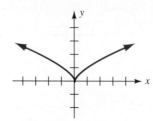

71. $y = x^{2/3} + 2$

72. $y = (x - 3)^{2/3}$

73. $y = (x + 1)^{2/3} - 4$

74. $y = 3 - x^{2/3}$

75. $y = -3x^{2/3}$

76. $y = \dfrac{x^{2/3}}{3}$

In Exercises 77–84, each function f is one-to-one.

(a) Find the inverse function f^{-1}.
(b) Sketch the graphs of f and f^{-1} on the same coordinate plane.

77. $f(x) = x^2, \quad x \leq 0$

78. $f(x) = 7x - 8$

79. $f(x) = 3 - x^3$

80. $f(x) = \dfrac{1}{x} - 3$

81. $f(x) = \sqrt{x - 2}$

82. $f(x) = \sqrt{x} + 3$

83. $f(x) = \begin{cases} x^3 & \text{if } x < 1 \\ \dfrac{x + 1}{2} & \text{if } x \geq 1 \end{cases}$

84. $f(x) = \begin{cases} x^{2/3} & \text{if } x \geq 0 \\ -(x^2 + 2) & \text{if } x < 0 \end{cases}$

In Exercises 85–92, refer to the graph of the function f that is shown in the figure.

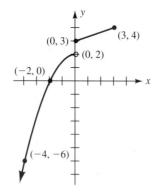

85. Specify the domain of f.

86. Specify the range of f.

87. Find $f(0)$.　　　　**88.** Find $f(-2)$.

89. Find $f(f(0))$.　　　**90.** Find $f(f(-2))$.

91. Does f possess an inverse? If so, determine each value:
(a) $f^{-1}(0)$　　(b) $f^{-1}(f^{-1}(4))$　　(c) $f(f^{-1}(-6))$

92. Is f continuous on the interval $(-2, 3]$? Explain.

93. If the function f is one-to-one and defined by

$$f(x) = \frac{x + a}{x + b},$$

where a and b are constants and $a \neq b$, find the inverse function f^{-1}.

94. A function f is its own inverse if $f(f(x)) = x$. Show that the function f defined by

$$f(x) = \frac{x + k}{x - 1},$$

where k is a constant, is its own inverse for any real number k if $k \neq -1$.

95. The cost C in dollars for a daily truck rental is a function of the miles x driven and is given by

$$C(x) = 32 + 0.25x.$$

(a) Evaluate $C(125.7)$, rounding to the nearest cent.
(b) Determine the miles driven when $C(x) = \$138.40$.

96. The volume of a rectangular box with a square base and open top is 9 cubic feet. Express the total surface area S as a function of the width x of its base.

97. The monthly charge for water in a small town is $0.015 per gallon for the first 1000 gallons that are used and $0.02 per gallon for each gallon in excess of 1000 gallons used. Express the charge C for the water as a function of the number n of gallons used.

98. The fixed cost to run a company that manufactures picnic tables is \$10,000 per month, and the variable cost to produce each table is \$100 per unit. If each table sells for x dollars, then the number of tables sold per month is estimated to be $800 - 2x$. Express the monthly profit P of the company as a function of x. (*Hint:* Recall from Section 2.2 that Profit = Total revenue − total cost.)

99. The point $P(x, y)$ lies on the right portion of the parabola $y = 3 - x^2$ with $y \geq 0$ as shown in the figure.

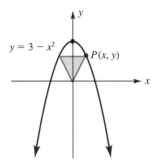

(a) Express the area A of the shaded isosceles triangle as a function of the x-coordinate of P.

(b) State the domain of the function defined in part (a).

100. A street light 30 feet above level ground casts the shadow of a man 6 foot tall.

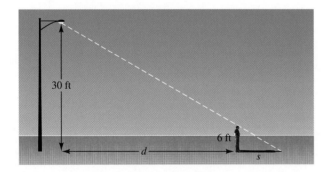

(a) Express the shadow length s as a function of the man's distance d from the base of the light pole.

(b) If the man stands 8 feet from the base of the light pole and then walks away from the pole at the rate of 4 feet per second, express the shadow length s (in feet) as a function of the time t (in seconds).

101. The excise tax T on an automobile varies directly as its actual value V.

(a) Express T as a function of V if $T = \$200$ when $V = \$5000$.

(b) State the domain of the function defined in part (a), then sketch its graph.

102. The intensity I of a light source varies inversely as the square of the distance d from the light source.

(a) Express I as a function of d if $I = 4$ candlepower when $d = 5$ feet.

(b) State the domain of the function defined in part (a), then sketch its graph.

103. A large field is capable of supporting life for a maximum of 2000 field mice. Suppose the rate of growth G of the field mouse population varies jointly as the number n of field mice present and the difference between the maximum number supportable by the field and the number present.

(a) Express G as a function of n if $G = 50$ field mice per week when $n = 1200$ mice.

(b) Find the rate of growth when 1500 field mice are living in the field.

104. The time t required for the excavation of a sewer line varies jointly as its length l, width w, and square of its depth d, and inversely as the number n of backhoes used. Describe what happens to the time for excavation if the depth is halved and the number of backhoes is doubled.

Linear and Quadratic Functions

A four-year-old car has a value of $3000. When one year
old, its value was $9000.
(a) Assuming the car's depreciation is linear, express the
    value V of the car as a function of its age x (in years).
(b) What is the age of the car when its value is fully
    depreciated?

(For the solution, see Example 6 in Section 4.2.)

4.1   **Linear Functions**

◆ **Introductory Comments**

Recall from Section 3.4 that the graph of the identity function $f(x) = x$ is a straight line that passes through the origin and splits quadrants I and III in half. [See Figure 3.31(b) in Section 3.4.] By applying the vertical shift rule and the vertical stretch and shrink rule (Section 3.4), we can show that the graph of

$$F(x) = ax + b,$$

where a and b are real numbers and $a \neq 0$, is also a straight line but with a y-intercept of b. By the x-axis reflection rule (Section 3.4), if $a > 0$ the line slants up as we move from left to right along the x-axis [see Figure 4.1(a)], and if $a < 0$ the line slants down as we move from left to right along the x-axis [see Figure 4.1(b)].

FIGURE 4.1
Graph of $F(x) = ax + b$

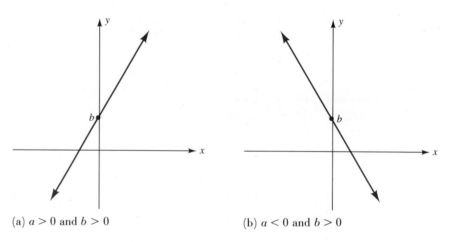

(a) $a > 0$ and $b > 0$ (b) $a < 0$ and $b > 0$

We refer to the function F defined by $F(x) = ax + b$ as a *linear function*, since its graph is always a straight line.

◆ **Linear Function**

> If a and b are real numbers with $a \neq 0$, then the function F defined by
>
> $$F(x) = ax + b$$
>
> is a **linear function** and its graph is a **straight line**.

In this section, we study linear functions and related equations whose graphs are straight lines. We begin our discussion by defining the *slope* of a line.

◆ The Slope of a Line

The *slope* of a nonvertical line is a measure of the line's steepness and is defined as the ratio of the vertical rise to the horizontal run between any two distinct points on the line.

$$\text{Slope} = \frac{\text{Vertical rise}}{\text{Horizontal run}}$$

Consider the line through the points $P(x_1, y_1)$ and $Q(x_2, y_2)$ shown in Figure 4.2. The vertical rise between the points P and Q is $y_2 - y_1$ and the horizontal run between these points is $x_2 - x_1$. Thus, the ratio of $(y_2 - y_1)$ to $(x_2 - x_1)$ defines the slope of this line. It is common practice to designate the slope of a line with the lowercase letter m.

FIGURE 4.2

The slope of the line through the points P and Q is the ratio of the vertical rise $(y_2 - y_1)$ to the horizontal run $(x_2 - x_1)$.

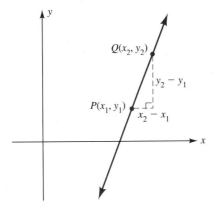

Slope Formula

> The **slope** m of a nonvertical line through the distinct points $P(x_1, y_1)$ and $Q(x_2, y_2)$, where $x_1 \neq x_2$, is
>
> $$m = \frac{y_2 - y_1}{x_2 - x_1}.$$
>
> If $x_1 = x_2$, the line is vertical and its slope is *undefined*.

Regardless of which two distinct points we choose on the line, the slope of the line is always the same. To show this, consider the straight line that passes through the points $P(x_1, y_1)$ and $Q(x_2, y_2)$ in Figure 4.2 and also through two other distinct points $R(x_3, y_3)$ and $S(x_4, y_4)$, as shown in Figure 4.3. Notice in Figure 4.3 that the triangles PAQ and RBS are similar, since their corresponding angles are equal. Thus, the corresponding sides of these triangles are proportional, and we have

$$\frac{y_2 - y_1}{x_2 - x_1} = \frac{y_4 - y_3}{x_4 - x_3}.$$

FIGURE 4.3
Since the triangles *PAQ* and *RBS* are similar, the slope of the line is the same value regardless of which two distinct points we choose on the line.

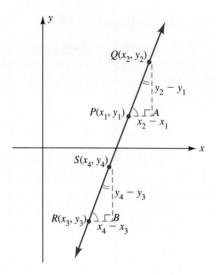

Since these two ratios are equal, and we have already defined the slope of this line as $m = \dfrac{y_2 - y_1}{x_2 - x_1}$, we know that $\dfrac{y_4 - y_3}{x_4 - x_3}$ also defines the slope of this line.

E X A M P L E   1   Determine the slope of the line that passes through the given points.

(a) $(-1, 2)$ and $(3, 4)$ (b) $(2, -3)$ and $(-2, 4)$

(c) $(-2, 3)$ and $(3, 3)$ (d) $(-1, 1)$ and $(-1, -4)$

S O L U T I O N

(a) Let $(x_1, y_1) = (-1, 2)$ and $(x_2, y_2) = (3, 4)$. Then

$$m = \frac{y_2 - y_1}{x_2 - x_1} = \frac{4 - 2}{3 - (-1)} = \frac{2}{4} = \frac{1}{2}   \text{(See Figure 4.4.)}$$

(b) Let $(x_1, y_1) = (2, -3)$ and $(x_2, y_2) = (-2, 4)$, then

$$m = \frac{y_2 - y_1}{x_2 - x_1} = \frac{4 - (-3)}{-2 - 2} = -\frac{7}{4}   \text{(See Figure 4.5.)}$$

(c) Let $(x_1, y_1) = (-2, 3)$ and $(x_2, y_2) = (3, 3)$, then

$$m = \frac{y_2 - y_1}{x_2 - x_1} = \frac{3 - 3}{3 - (-2)} = \frac{0}{5} = 0   \text{(See Figure 4.6.)}$$

(d) Since $x_1 = x_2 = -1$, the slope is undefined (see Figure 4.7).   ◆

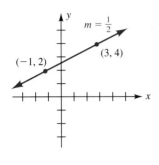

FIGURE 4.4
A line with a *positive* slope.

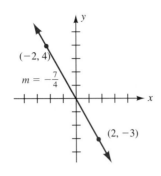

FIGURE 4.5
A line with a *negative* slope.

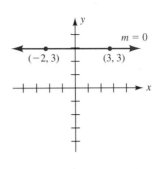

FIGURE 4.6
A line with a *zero* slope.

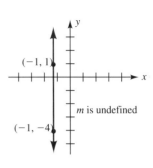

FIGURE 4.7
A line with an *undefined* slope.

PROBLEM 1   When determining the slope, does it matter which point is labeled (x_1, y_1) or (x_2, y_2)? Interchange the labeling of the points (x_1, y_1) and (x_2, y_2) in parts (a)–(c) of Example 1, and then determine the slope. What do you conclude?   ◆

We can use Example 1 to make some observations about the slope of a line:

1. If the line slants *upward* (increases) as we move from left to right along the x-axis, as in Figure 4.4, then the slope of the line is *positive*.

2. If the line slants *downward* (decreases) as we move from left to right along the x-axis, as in Figure 4.5, then the slope of the line is *negative*.

3. If the line is *horizontal* (parallel to the x-axis), as in Figure 4.6, then $y_1 = y_2$, and therefore the slope of the line is *zero*.

4. If the line is *vertical* (perpendicular to the x-axis), as in Figure 4.7, then $x_1 = x_2$, and therefore the slope of the line is *undefined*.

EXAMPLE 2   Sketch the line that passes through the point $(2, -1)$ and has the given slope.

(a) $m = 3$ **(b)** $m = -\frac{2}{3}$

SOLUTION

(a) A slope of $3 = \frac{3}{1}$ represents a vertical rise of 3 to a horizontal run of 1 (or a vertical rise of 6 to a horizontal run of 2, and so on). Thus, starting at the point $(2, -1)$, we can go *up 3 units* and run to the *right 1 unit* to locate another point on the line. Connecting these points with a straightedge gives us a sketch of this straight line, as shown in Figure 4.8.

(b) A slope of $-\frac{2}{3}$ represents a vertical rise of -2 to a horizontal run of 3 (or a vertical rise of -4 to a horizontal run of 6, and so on). Thus, starting at the point $(2, -1)$, we can go *down 2 units* and run to the *right 3 units* to locate another point on the line. Connecting these points with a straightedge gives us a sketch of this straight line, as shown in Figure 4.9.   ◆

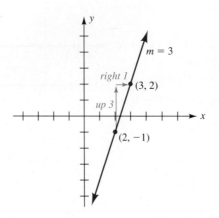

FIGURE 4.8
A line passing through the point $(2, -1)$
with a slope of 3.

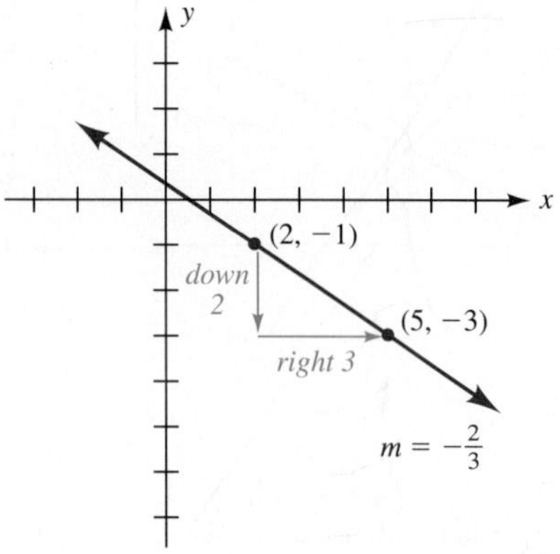

FIGURE 4.9
A line passing through the point $(2, -1)$
with a slope of $-\frac{2}{3}$.

PROBLEM 2   Sketch the line that passes through the point $(-3, 0)$ and has a slope of $\frac{5}{2}$.   ◆

◆ **Sketching the Graph of a Linear Function**

The slope m of the line with the equation $y = ax + b$ is the number a. To show this, we let $P(x_1, y_1)$ and $Q(x_2, y_2)$ be any two distinct points on this line. Since these two points must satisfy the equation $y = ax + b$, we have

$$y_1 = ax_1 + b \qquad \text{and} \qquad y_2 = ax_2 + b.$$

Thus,

$$m = \frac{y_2 - y_1}{x_2 - x_1} = \frac{(ax_2 + b) - (ax_1 + b)}{x_2 - x_1}$$
$$= \frac{a(x_2 - x_1)}{x_2 - x_1}$$
$$= a$$

When we replace a with m in the equation $y = ax + b$, we obtain

$$\boxed{y = mx + b}$$

which is the **slope-intercept form** for the equation of a straight line. As illustrated in the next example, we can quickly sketch the graph of a linear function by simply noting its slope and y-intercept.

EXAMPLE 3   Identify the slope and y-intercept and then sketch the graph of the linear function defined as follows.

(a) $f(x) = \frac{4}{3}x - 2$ (b) $y = 3 - 5x$

SOLUTION

(a) The linear function f defined by

$$f(x) = \tfrac{4}{3}x + (-2),$$

has slope $\tfrac{4}{3}$ and y-intercept -2. To graph this linear function, we begin by plotting the point $(0, -2)$. Next, remember that the slope is the ratio of the rise to the run. Thus, from the point $(0, -2)$, we go *up 4 units* and run to the *right 3 units* to locate another point on the line. Connect these two points to form the graph of $f(x) = \tfrac{4}{3}x - 2$, as shown in Figure 4.10.

(b) For the linear function defined by the equation

$$y = 3 - 5x = -5x + 3,$$

the slope is -5, or $-\tfrac{5}{1}$, and the y-intercept is 3. Since the slope is negative, we start at the point $(0, 3)$ and go *down 5 units*, then run *right 1 unit* to locate another point on the line. Connecting these two points, we have the graph of $y = 3 - 5x$, as shown in Figure 4.11.   ◆

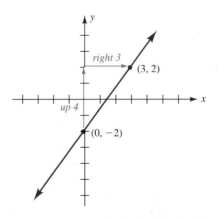

FIGURE 4.10
Graph of $f(x) = \tfrac{4}{3}x - 2$ has slope $\tfrac{4}{3}$ and
y-intercept -2.

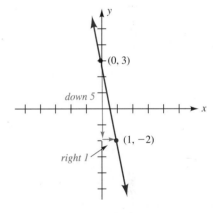

FIGURE 4.11
Graph of $y = 3 - 5x$ has slope -5 and
y-intercept 3.

PROBLEM 3   Repeat Example 3 for $g(x) = 3x - \tfrac{1}{2}$.   ◆

◆ **General Form for the Equation of a Straight Line**

Consider the equation $Ax + By + C = 0$, where A, B, and C are constants with $B \neq 0$. Solving this equation for y, we obtain

$$y = \left(-\frac{A}{B}\right)x + \left(-\frac{C}{B}\right),$$

with labels: slope over the $\left(-\frac{A}{B}\right)$ term and y-intercept under the $\left(-\frac{C}{B}\right)$ term.

which we know is the *slope-intercept form* for the equation of a straight line. Thus, we also know that the graph of the equation $Ax + By + C = 0$ is a straight line with slope $-A/B$ and y-intercept $-C/B$, provided $B \neq 0$. We refer to the equation

$$Ax + By + C = 0$$

as the **general form** for the equation of a straight line.

Suppose $A = 0$ in the equation $Ax + By + C = 0$ with $B \neq 0$. Solving for y, we obtain

$$y = \overset{\text{slope}}{0}x + \left(-\frac{C}{B}\right).$$

Since the slope of the line is zero, we conclude that $0x + By + C = 0$, or

$$By + C = 0$$

is the equation of a *horizontal line* with y-intercept $-C/B$. Actually, the equation $By + C = 0$ defines a *constant function*.

If $B = 0$ in the equation $Ax + By + C = 0$ with $A \neq 0$, then we cannot write the equation in the slope-intercept form $y = mx + b$, since division by zero is not defined. However, regardless of what value we choose for y in the equation $Ax + 0y + C = 0$, the quantity $0y$ is always zero. This means that ordered pairs that satisfy $Ax + 0y + C = 0$, or $Ax + C = 0$, can have any y-value as long as the x-value is $-C/A$. Thus, we conclude that

$$Ax + C = 0$$

is the equation of a *vertical line* with an x-intercept $-C/A$. Although the graph of $Ax + C = 0$ is a straight line, the equation $Ax + C = 0$ does *not* define a linear function, because a vertical line has an *undefined* slope.

EXAMPLE 4   Sketch the graph of each equation.

(a) $2x - 3y = 12$

(b) $2y - 7 = 0$

(c) $2x + 6 = 0$

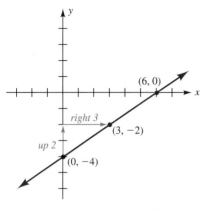

FIGURE 4.12
Graph of $2x - 3y = 12$.

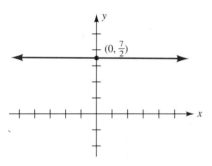

FIGURE 4.13
Graph of $2y - 7 = 0$.

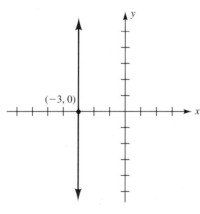

FIGURE 4.14
Graph of $2x + 6 = 0$.

SOLUTION

(a) Solving for y, we obtain

$$2x - 3y = 12$$
$$-3y = -2x + 12$$
$$y = \tfrac{2}{3}x + (-4)$$

This last equation is the slope-intercept form for the equation of a line. The graph of this equation is a straight line with slope $\tfrac{2}{3}$ and y-intercept -4, as shown in Figure 4.12.

(b) The equation $2y - 7 = 0$ can be written in the slope-intercept form as

Since the slope is zero, the graph of this equation is a horizontal line with a y-intercept of $\tfrac{7}{2}$, as shown in Figure 4.13.

(c) Although the equation $2x + 6 = 0$ cannot be written in slope-intercept form, we can write it in general form as

$$2x + 0y + 6 = 0.$$

Ordered pairs that satisfy this equation can have any y-value as long as $x = -3$. Thus, the graph of this equation is a vertical line with x-intercept -3, as shown in Figure 4.14.   ◆

An alternate method of graphing the equation $Ax + By + C = 0$, where A, B, and C are nonzero numbers, is to find the x-intercept and y-intercept. Connecting these intercepts with a straightedge also gives us the graph of the straight line. For the equation $2x - 3y = 12$ in Example 4(a), we have

x-intercept of $2x - 3y = 12$	y-intercept of $2x - 3y = 12$
Let $y = 0$: $2x - 3(0) = 12$	Let $x = 0$: $2(0) - 3y = 12$
$2x = 12$	$-3y = 12$
$x = 6$	$y = -4$

Connecting the points $(6, 0)$ and $(0, -4)$ also gives us the graph of the straight line as illustrated in Figure 4.12.

PROBLEM 4 Find the *x*-intercept and *y*-intercept for the equation $3x + 8y = 12$. Then sketch its graph. ◆

◆ **Application: Tax Rate Schedule**

The amount of federal income tax you pay each year is a function of your taxable income for that year. This amount is usually given by the Internal Revenue Service (IRS) in terms of a *tax rate schedule*. Each line in the tax rate schedule can be expressed as a different linear function. If a function is defined by different linear functions on distinct subsets of its domain, it is called a **piecewise linear function**.

EXAMPLE 5 A recent tax rate schedule for a single person with a taxable income up to $93,130 is given by the IRS as follows.

Taxable Income	Tax
$0 to $18,550	15% of the amount over $0
$18,550 to $44,900	$2782.50 + 28% of the amount over $18,550
$44,900 to $93,130	$10,160.50 + 33% of the amount over $44,900

Find the piecewise linear function defined by this tax rate schedule and sketch its graph.

SOLUTION Using this tax rate schedule, we can write three equations:

1. If the taxable income x is $0 to $18,550, then the tax T in dollars is

$$T(x) = 0.15x$$

2. If the taxable income x is more than $18,550 but not more than $44,900, the tax T in dollars is

$$T(x) = 2782.50 + 0.28(x - 18,550)$$
$$= 0.28x - 2411.50$$

3. If the taxable income x is more than $44,900 but not more than $93,130, the tax T in dollars is

$$T(x) = 10,160.50 + 0.33(x - 44,900)$$
$$= 0.33x - 4656.50$$

Thus, the piecewise linear function defined by this tax rate schedule has three parts:

$$T(x) = \begin{cases} 0.15x & \text{if } 0 \le x \le 18,550 \\ 0.28x - 2411.50 & \text{if } 18,550 < x \le 44,900 \\ 0.33x - 4656.50 & \text{if } 44,900 < x \le 93,130 \end{cases}$$

The graph of this piecewise linear function is shown in Figure 4.15.

FIGURE 4.15

Graph of the piecewise linear function defined by the tax rate schedule in Example 5.

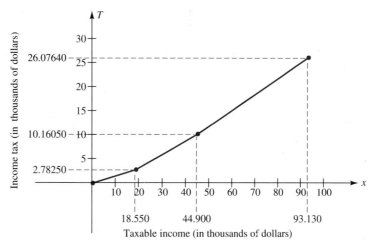

PROBLEM 5   Use the graph in Figure 4.15 to estimate the taxable income of a single person whose income tax is $15,000. Then use the piecewise linear function defined in Example 5 to find the taxable income to the nearest cent.   ◆

Exercises 4.1

Basic Skills

In Exercises 1–12, find the slope (if it exists) of the line that passes through the given points.

1. $(2, 0)$ and $(3, 2)$
2. $(4, 1)$ and $(-1, -1)$
3. $(-3, 4)$ and $(6, -2)$
4. $(3, -2)$ and $(4, -3)$
5. $(-4, 3)$ and $(-4, 1)$
6. $(-2, 5)$ and $(4, 5)$
7. $\left(\frac{3}{2}, -1\right)$ and $\left(-\frac{1}{2}, 7\right)$
8. $\left(\frac{3}{4}, \frac{2}{3}\right)$ and $\left(-1, \frac{1}{6}\right)$
9. $(4, 3)$ and $(0, b)$
10. $(a, 0)$ and $(0, b)$
11. (a, b) and (b, a)
12. $\left(a, \frac{a}{b}\right)$ and $\left(b, \frac{b}{a}\right)$

In Exercises 13–18, sketch the line that passes through the given point and has the given slope.

13. Through $(3, 1)$ with slope 3
14. Through $(-3, -1)$ with slope $\frac{4}{3}$
15. Through $\left(0, -\frac{2}{3}\right)$ with slope $-\frac{1}{3}$
16. Through $\left(\frac{1}{2}, \frac{3}{2}\right)$ with slope -2
17. Through $(-1, 2)$ with slope 0
18. Through $(2, 3)$ with undefined slope

In Exercises 19–30, identify the slope and y-intercept, and then sketch the graph of the linear function.

19. $y = 3x$
20. $y = -2x$
21. $f(x) = -\frac{1}{3}x$
22. $F(x) = \frac{3x}{4}$
23. $y = 2x + 3$
24. $y = 3x - 4$
25. $g(x) = -4x + 1$
26. $G(x) = 5 - 2x$
27. $y = \frac{x}{2} - 1$
28. $y = \frac{2}{3}x - 2$
29. $h(x) = \frac{1 - 3x}{4}$
30. $H(x) = \frac{-x + 3}{2}$

In Exercises 31–42, sketch the graph of the equation, state the slope, and label the x-intercept and y-intercept (if they exist).

31. $x + y = 2$
32. $x - y = -3$
33. $2x + 5y = 10$
34. $3x + y = -6$
35. $21x - 28y = -56$
36. $9x - 18y = 60$
37. $x = 2y$
38. $4x - 3y = 0$

39. $x = -2$

40. $y = 4$

41. $\frac{4}{3}y - 3 = 0$

42. $2.5 - 0.2x = 0$

In Exercises 43–48, sketch the graph of each piecewise linear function.

43. $f(x) = \begin{cases} x & \text{if } x \le 2 \\ 2x - 2 & \text{if } x > 2 \end{cases}$

44. $F(x) = \begin{cases} 2x & \text{if } x \le 1 \\ 3 - x & \text{if } x > 1 \end{cases}$

45. $h(x) = \begin{cases} 2x + 5 & \text{if } x \le -2 \\ x + 3 & \text{if } x > -2 \end{cases}$

46. $H(x) = \begin{cases} -(3x + 4) & \text{if } x \le -1 \\ \dfrac{x - 2}{3} & \text{if } x > -1 \end{cases}$

47. $g(x) = \begin{cases} 3x - 1 & \text{if } x \le 0 \\ x - 1 & \text{if } 0 < x \le 2 \\ 6 - x & \text{if } x > 2 \end{cases}$

48. $G(x) = \begin{cases} x + 3 & \text{if } x \le 2 \\ 7 - 2x & \text{if } 2 < x \le 4 \\ x - 5 & \text{if } x > 4 \end{cases}$

49. The linear equation $F = \frac{9}{5}C + 32$ expresses the Fahrenheit temperature F as a function of the Celsius temperature C.

(a) Identify the slope and F-intercept.
(b) Sketch the graph of the linear equation.

50. The linear equation $C = \frac{5}{9}(F - 32)$ expresses the Celsius temperature C as a function of the Fahrenheit temperature F.

(a) Identify the slope and C-intercept.
(b) Sketch the graph of the linear equation.

51. The linear equation $v = 44 - 3t$ expresses the velocity v (in feet per second) of an object as a function of the time t (in seconds) it travels at a constant deceleration of 3 ft/s². (*Note:* ft/s² is a more compact way of writing feet per second per second, $\dfrac{\text{ft/s}}{\text{s}}$.)

(a) Identify the slope and v-intercept.
(b) Sketch the graph of the linear equation.
(c) What is the significance of the t-intercept?

52. The linear equation $A = 500 + 3.75t$ expresses the amount A (in dollars) accumulated in a certain savings account after t months ($0 \le t \le 12$) if the money earns 9% simple interest per year.

(a) Identify the slope and A-intercept.
(b) Sketch the graph of the linear equation for $0 \le t \le 12$.
(c) What is the significance of the slope and A-intercept?

 Critical Thinking

53. The center of a circle of radius 5 has coordinates $(-1, 2)$. What are the coordinates of the endpoints of a diameter whose slope is $-\frac{3}{4}$.

54. The midpoint of a line segment 20 units long is $(2, -1)$. If the line segment has slope $\frac{4}{3}$, what are the coordinates of its endpoints?

55. For what value of a will the line through the points $(3, -2)$ and $(a, 2)$ have the given slope?

(a) $m = -2$ (b) $m = \frac{2}{3}$ (c) $m = -\frac{1}{2}$

56. For what value of a will the points $(-1, 3)$, $(2, a)$, and $(3, -3)$ be collinear (lie on the same line)?

57. Find the inverse function f^{-1} for the linear function $f(x) = mx + b$ with $m \ne 0$.

58. Suppose f and g are linear functions defined by $f(x) = m_1x + b_1$ and $g(x) = m_2x + b_2$, where $m_1 \ne 0$ and $m_2 \ne 0$.

(a) Find $f \circ g$.   (b) Find $g \circ f$.
(c) Is $f \circ g$ a linear function? If so, give its slope and y-intercept.
(d) Is $g \circ f$ a linear function? If so, give its slope and y-intercept.

 Calculator Activities

In Exercises 59–62, approximate to three significant digits the slope of the line that passes through the given points.

59. $(4.4, -1.9)$ and $(-6.8, -4.7)$

60. $(-0.41, -0.92)$ and $(0.87, 0.75)$

61. $(-0.987, 4.873)$ and $(4.256, -0.655)$

62. $(12.65, -24.76)$ and $(42.55, -116.78)$

In Exercises 63–66, approximate to three significant digits the slope of the given line.

63. $6.45x - 8.76y = 34.5$

64. $345.6x - 124.8y + 187.4 = 0$

65. $7.89y - 1.23x = 14.8$

66. $18.4x = 67.3 - 18.7y$

To display the graph of the line $Ax + By = C$ on a graphing calculator, we must enter the equation in its slope-intercept form, $y = mx + b$. In Exercises 67–70, use a graphing calculator to generate the graph of each equation in the viewing rectangle. Then estimate the x-intercept and y-intercept by tracing to these points. Check each answer by finding the exact intercepts.

67. $7x + 4y = 24$

68. $16y - 18x = -33$

69. $14x - 18y + 57 = 0$

70. $124x = 234y - 167$

71. The cost C of electricity is a function of the amount x of electricity consumed (in kilowatt hours). Suppose the cost of electricity is given by the following rate schedule.

Usage	Cost
First 30 kilowatt hours or less	$3.00
Next 70 kilowatt hours	$0.06 per kilowatt hour
Next 200 kilowatt hours	$0.04 per kilowatt hour
All kilowatt hours over 300	$0.01 per kilowatt hour

(a) Find the piecewise linear function that describes this rate schedule.
(b) Sketch the graph of the piecewise linear function in part (a).
(c) Use the rate table to find the cost of using 240 kilo-

watt hours of electricity. Then use the piecewise linear function defined in part (a) to find the cost.
(d) Use the graph sketched in part (b) to estimate the number of kilowatt hours consumed if the cost is $14.08. Then use the piecewise linear function from part (a) to find the exact number of kilowatt hours consumed.

72. The amount of federal income tax T you pay is a function of your taxable income x. A recent tax rate schedule for a single person is given:

Taxable Income	Income Tax
$0 to $1800	11% of the amount over $0
$1800 to $16,800	$198 + 15% of the amount over $1800
$16,800 to $27,000	$2448 + 28% of the amount over $16,800
$27,000 to $54,000	$5304 + 35% of the amount over $27,000
$54,000 and up	$14,754 + 38.5% of the amount over $54,000

(a) Find the piecewise linear function that describes this rate schedule.
(b) Sketch the graph of the piecewise linear function defined in part (a).
(c) Use the tax rate table to find the tax a single person must pay on a taxable income of $35,000. Then use the piecewise linear function from part (a) to find the tax.
(d) Use the graph sketched in part (b) to estimate the taxable income of a single person whose income tax is $7,000. Then use the piecewise linear function defined in part (a) to find this taxable income to the nearest cent.

4.2   Determining the Equation of a Line

In Section 4.1 we were given the equation of a line and then asked to find some information about the line, such as its slope or x- and y-intercepts. In this section, we reverse the procedure—from some given information about a line, we determine its equation. We begin by developing the *point-slope form* for the equation of a straight line. This form is useful for determining the equation of a line when we are given the slope of the line and the coordinates of a fixed point on the line.

◆ Point-Slope Form for the Equation of a Line

Let m be the given slope of a line and let $P(x_1, y_1)$ be a given point on the line. Also, let $Q(x, y)$ be *any* point other than P on the given line, as shown in Figure 4.16. By the slope formula (Section 4.1), we have

$$m = \frac{y - y_1}{x - x_1}$$

$$y - y_1 = m(x - x_1) \qquad \text{Multiply both sides by } (x - x_1)$$

This last equation is called the **point-slope form** for the equation of a straight line.

FIGURE 4.16

The slope m of the line through a fixed point $P(x_1, y_1)$ and an arbitrary point $Q(x, y)$ on the line is 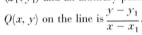.

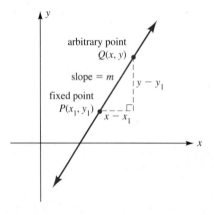

◆ **Point-Slope Form**

> The equation of a line having a given slope m and passing through the fixed point $P(x_1, y_1)$ and an arbitrary point $Q(x, y)$ is
>
> $$y - y_1 = m(x - x_1).$$

E X A M P L E 1   Find the equation of a line that passes through the point $(2, -1)$ and has a slope of 3. Write the equation in slope-intercept form (Section 4.1).

S O L U T I O N   The line that passes through the point $(2, -1)$ and has a slope of 3 is shown in Figure 4.17. Thus, we replace x_1 with 2, y_1 with -1, and m with 3 in the point-slope form $y - y_1 = m(x - x_1)$:

$$y - (-1) = 3(x - 2) \qquad \text{Substitute}$$

$$y + 1 = 3x - 6 \qquad \text{Multiply}$$

$$y = 3x - 7 \qquad \text{Write in slope-intercept form} \qquad ◆$$

FIGURE 4.17

Sketch of a line that passes through the
point $(2, -1)$ and has slope 3.

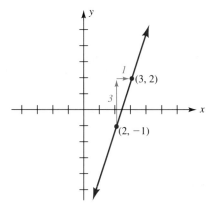

PROBLEM 1   Repeat Example 1 for a line with slope $\frac{1}{2}$ and x-intercept -3.   ◆

As illustrated in the next example, if we are given two points on a line, then
we can determine the equation of the line by using the slope formula (Section 4.1) and the point-slope form for the equation of a line.

EXAMPLE 2   Find the equation of the line that passes through the points $(-2, -3)$ and
$(2, 1)$. Write the equation in slope-intercept form.

SOLUTION   The line that passes through the points $(-2, -3)$ and $(2, 1)$ is
shown in Figure 4.18. The slope of the line is

$$m = \frac{1 - (-3)}{2 - (-2)}$$

$$= \frac{4}{4} = 1.$$

Now using $(x_1, y_1) = (-2, -3)$ as the fixed point and $m = 1$ as the slope,
we have

$$y - y_1 = m(x - x_1)$$

$$y - (-3) = 1[x - (-2)]   \qquad \textbf{Substitute}$$

$$y + 3 = x + 2   \qquad \textbf{Simplify}$$

$$y = x - 1   \qquad \textbf{Write in slope-intercept form}   \qquad ◆$$

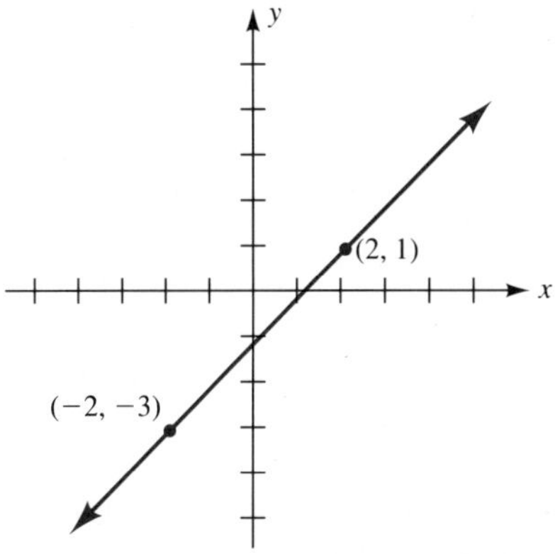

FIGURE 4.18

Sketch of a line that passes through the
points $(-2, -3)$ and $(2, 1)$.

PROBLEM 2   Referring to Example 2, use $(x_1, y_1) = (2, 1)$ as the fixed point and $m = 1$ as
the slope to determine the equation of the line. Your answer should agree
with the result in the example.   ◆

◆ Intercept Form for the Equation of a Line

Consider a line in which the x-intercept a and the y-intercept b are given, with $a \neq 0$ and $b \neq 0$, as shown in Figure 4.19. The slope of the line is

$$m = \frac{b-0}{0-a} = -\frac{b}{a}.$$

Using the point-slope form with a fixed point $(0, b)$ and slope $-b/a$, we can find the equation of the line:

$$y - b = -\frac{b}{a}(x - 0)$$

$$\frac{y}{b} - 1 = -\frac{x}{a} \qquad \text{Divide both sides by } b \text{ with } b \neq 0$$

$$\frac{x}{a} + \frac{y}{b} = 1 \qquad \text{Add } 1 + \tfrac{x}{a} \text{ to both sides}$$

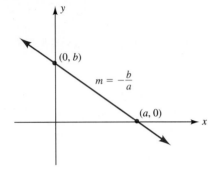

FIGURE 4.19

A line with x-intercept a and y-intercept b has slope $-b/a$.

This last equation is called the **intercept form** for the equation of a straight line. It is useful for determining the equation of a straight line when we are given the x- and y-intercepts.

Intercept Form

> The equation of a line whose x- and y-intercepts are a and b, respectively, is
>
> $$\frac{x}{a} + \frac{y}{b} = 1, \qquad \text{where } a \neq 0, b \neq 0.$$

E X A M P L E 3   Find the equation of a line with x-intercept -3 and y-intercept 2. Write the equation in general form (Section 4.1) with integer coefficients.

S O L U T I O N   The line with x-intercept -3 and y-intercept 2 is shown in Figure 4.20. Replacing a with -3 and b with 2 in the intercept form, we obtain

$$\frac{x}{a} + \frac{y}{b} = 1$$

$$\frac{x}{-3} + \frac{y}{2} = 1 \qquad \text{Substitute}$$

$$-2x + 3y = 6 \qquad \text{Multiply both sides by 6}$$

$$2x - 3y + 6 = 0 \qquad \text{Write in general form} \qquad ◆$$

FIGURE 4.20

Sketch of a line with x-intercept -3 and y-intercept 2.

PROBLEM 3   Repeat Example 3 if the x-intercept is $\frac{1}{2}$ and the y-intercept is 3.   ◆

◆ **Parallel Lines**

Suppose two distinct, nonvertical lines $y = m_1x + b_1$ and $y = m_2x + b_2$ are parallel, as shown in Figure 4.21.

FIGURE 4.21

If the lines $y = m_1x + b_1$ and $y = m_2x + b_2$ are *parallel,* then the triangles *ABC* and *DEF* are similar and their corresponding sides are proportional.

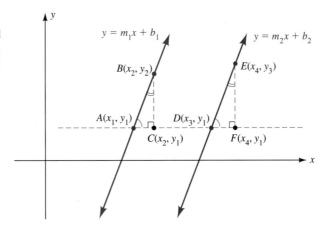

The triangles *ABC* and *DEF* are similar, since their corresponding angles are equal. Thus, their corresponding sides are proportional, and we have

$$\frac{y_2 - y_1}{x_2 - x_1} = \frac{y_3 - y_1}{x_4 - x_3}.$$

However, $\dfrac{y_2 - y_1}{x_2 - x_1} = m_1$ and $\dfrac{y_3 - y_1}{x_4 - x_3} = m_2.$

Thus, if two nonvertical lines are parallel, they have the same slope. The converse of this statement is also true: If two distinct nonvertical lines have the same slope, then the lines are parallel.

◆ **Parallel Lines**

> Two distinct nonvertical lines with slopes m_1 and m_2 are **parallel** if and only if
>
> $$m_1 = m_2.$$

EXAMPLE 4   Find the equation of the line that passes through the point $(1, 3)$ and is parallel to the line that passes through the points $(-2, -4)$ and $(2, -1)$. Write the equation in slope-intercept form.

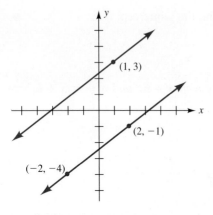

FIGURE 4.22

A sketch of the line that passes through the point (1, 3) and is parallel to the line that passes through the points $(-2, -4)$, and $(2, -1)$.

SOLUTION   The given information is sketched in Figure 4.22. The slope of the line that passes through the points $(-2, -4)$ and $(2, -1)$ is

$$m = \frac{-1-(-4)}{2-(-2)} = \frac{3}{4}.$$

Since the slopes of two parallel lines are the same, we know that the slope of the line that passes through the point (1, 3) is also $\frac{3}{4}$. Now, using the point-slope form for the equation of a line with fixed point $(x_1, y_1) = (1, 3)$ and slope $m = \frac{3}{4}$, we can find the equation of the desired line as follows:

$$y - y_1 = m(x - x_1)$$

$$y - 3 = \tfrac{3}{4}(x - 1) \qquad \textbf{Substitute}$$

$$y - 3 = \tfrac{3}{4}x - \tfrac{3}{4} \qquad \textbf{Multiply}$$

$$y = \tfrac{3}{4}x + \tfrac{9}{4} \qquad \textbf{Write in slope-intercept form}$$ ◆

PROBLEM 4   Repeat Example 4 for a line that has x-intercept -3 and is parallel to a line whose slope is $\frac{2}{3}$.   ◆

◆   **Perpendicular Lines**

Suppose two nonvertical lines $y = m_1x + b_1$ and $y = m_2x + b_2$ are perpendicular, as shown in Figure 4.23.

FIGURE 4.23

If the lines $y = m_1x + b_1$ and $y = m_2x + b_2$ are *perpendicular*, then the triangles BAC and DBE are similar and their corresponding sides are proportional.

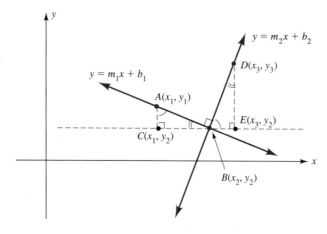

Angle BAC is complementary to angle ABC, and angle ABC is complementary to angle DBE. Thus, angles BAC and DBE are equal and triangles BAC and DBE are similar. Using the fact that the corresponding sides of these triangles are proportional, we have

$$\frac{y_3 - y_2}{x_3 - x_2} = \frac{x_2 - x_1}{y_1 - y_2}$$

However,

$$\frac{y_3 - y_2}{x_3 - x_2} = m_2,$$

and

$$\frac{x_2 - x_1}{y_1 - y_2} = -\frac{x_1 - x_2}{y_1 - y_2} = -\frac{1}{\dfrac{y_1 - y_2}{x_1 - x_2}} = -\frac{1}{m_1}.$$

Thus, if two nonvertical lines are perpendicular, the slope of one line is the negative reciprocal of the slope of the other. The converse of this statement is also true: If the slope of one line is the negative reciprocal of the slope of another, then the lines are perpendicular.

Perpendicular Lines

> Two nonvertical lines with slopes m_1 and m_2 are **perpendicular** if and only if
>
> $$m_2 = -\frac{1}{m_1}.$$

E X A M P L E   5   Find the equation of a line that passes through the point (2, 1) and is perpendicular to the line $3x - 5y = 10$. Write the equation in slope-intercept form.

S O L U T I O N   We can determine the slope of the line $3x - 5y = 10$ by writing it in slope-intercept form:

$$3x - 5y = 10$$
$$-5y = -3x + 10$$
$$y = \tfrac{3}{5}x - 2$$
$$\uparrow$$
$$\boxed{\text{slope}}$$

The graph of the line that passes through the point (2, 1) and is perpendicular to $3x - 5y = 10$ is shown in Figure 4.24. Since the slope of the line $3x - 5y = 10$ is $\tfrac{3}{5}$, the slope m of the perpendicular line is

$$m = -\frac{1}{3/5} = -\frac{5}{3}.$$
$$\boxed{\text{Negative reciprocal of } \tfrac{3}{5}}$$

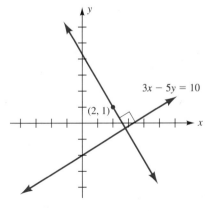

FIGURE 4.24

Graph of the line that passes through the point (2, 1) and is perpendicular to $3x - 5y = 10$.

Now, using the point-slope form for the equation of a line with fixed point $(x_1, y_1) = (2, 1)$ and slope $m = -\frac{5}{3}$, we can find the desired equation:

$$y - y_1 = m(x - x_1)$$

$$y - 1 = -\tfrac{5}{3}(x - 2) \qquad \text{Substitute}$$

$$y - 1 = -\tfrac{5}{3}x + \tfrac{10}{3} \qquad \text{Multiply}$$

$$y = -\tfrac{5}{3}x + \tfrac{13}{3} \qquad \text{Write in slope-intercept form} \qquad \blacklozenge$$

PROBLEM 5   Repeat Example 5 for a line that passes through the point $(3, -4)$ and is perpendicular to a line that passes through the points $(3, 0)$ and $(2, -3)$. $\blacklozenge$

◆ **Application: Linear Depreciation**

One type of applied problem that occurs frequently in business is *linear depreciation*. In linear depreciation, the value of an asset decreases linearly over time.

EXAMPLE 6   A four-year-old car has a value of $3000. When one year old, its value was $9000.

(a)   Assuming the car's depreciation is linear, express the value V of the car as a function of its age x (in years).

(b)   What is the age of the car when its value is fully depreciated?

SOLUTION

(a)   A graph of the given information is shown in Figure 4.25. The slope of the line segment is

$$m = \frac{\$9000 - \$3000}{1 \text{ yr} - 4 \text{ yr}} = -\$2000 \text{ per year.}$$

Note that the slope tells us the amount of *depreciation per year*. Now, to express the value V of the car as a function of its age x, we use the point-slope form for the equation of a line with fixed point $(1, 9000)$ and slope -2000. Thus,

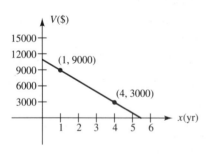

$$V - V_1 = m(x - x_1)$$

$$V - 9000 = -2000(x - 1) \qquad \text{Substitute}$$

$$V - 9000 = -2000x + 2000 \qquad \text{Multiply}$$

$$V = 11{,}000 - 2000x \qquad \text{Write in slope-intercept form}$$

FIGURE 4.25

Graph showing the linear depreciation of the car.

(b)   The car's value is fully depreciated when $V = 0$. Assuming the car continues to depreciate according to the equation $V = 11{,}000 - 2000x$,

we have

$$0 = 11{,}000 - 2000x$$

$$2000x = 11{,}000$$

$$x = 5\tfrac{1}{2}$$

Thus, the car's value is fully depreciated when it is $5\tfrac{1}{2}$ years old. ◆

PROBLEM 6   Referring to Example 6, what is the significance of the V-intercept in the equation $V = 11{,}000 - 2000x$? ◆

Exercises 4.2

 Basic Skills

In Exercises 1–38, find the equation of a line that satisfies the indicated conditions. Write the equation in each of the following forms:

(a) general form with integer coefficients
(b) slope-intercept form

1. y-intercept 3 and slope $\frac{3}{4}$

2. y-intercept -2 and slope $-\frac{1}{2}$

3. Through $(-1, 2)$ with slope 3

4. Through $(3, 4)$ with slope -2

5. x-intercept 4 and slope $-\frac{7}{3}$

6. x-intercept -3 and slope $\frac{2}{5}$

7. Through the origin with slope -3

8. Through the origin and $(6, 5)$

9. Through $(4, -1)$ and $(2, 5)$

10. Through $(-1, -2)$ and $(4, -3)$

11. Through $(-2, \frac{4}{3})$ and $(3, -1)$

12. Through $(\frac{6}{5}, -2)$ and $(-\frac{3}{2}, 1)$

13. Through $(-3, 4)$ with y-intercept $\frac{1}{6}$

14. Through $(5, 3)$ with x-intercept -3

15. x-intercept 3 and y-intercept -2

16. x-intercept -4 and y-intercept 6

17. x-intercept $\frac{1}{4}$ and y-intercept 6

18. x-intercept -3 and y-intercept $-\frac{5}{3}$

19. y-intercept $\frac{3}{2}$ and slope 0

20. Through $(4, -3)$ and $(-1, -3)$

21. Through $(3, -2)$ and $(3, 4)$

22. x-intercept $-\frac{2}{3}$ with undefined slope

23. Through $(5, -1)$ with same y-intercept as $3x - 2y = 4$

24. Through $(\frac{1}{2}, -1)$ with same x-intercept as $4x - y + 4 = 0$

25. Through $(2, 3)$ with x-intercept a and y-intercept a $(a \neq 0)$

26. Through $(\frac{1}{3}, -5)$ with x-intercept a and y-intercept $-a$ $(a \neq 0)$

27. x-intercept -2 and parallel to a line with slope $-\frac{1}{2}$

28. y-intercept $\frac{4}{3}$ and parallel to a line with slope -4

29. Through $(-3, -1)$ and parallel to the line that passes through $(1, 0)$ and $(2, 3)$

30. Through $(4, 5)$ and parallel to the line with x-intercept 3 and y-intercept -2

31. Through $(2, -3)$ and parallel to the line $2x - 3y + 6 = 0$

32. Through the origin and parallel to the line $x - 2y - 3 = 0$

33. y-intercept 3 and perpendicular to a line with slope -4

34. x-intercept $\frac{5}{2}$ and perpendicular to a line with slope $\frac{8}{5}$

35. Through $(-1, 0)$ and perpendicular to the line that passes through $(2, 0)$ and $(-1, 4)$

36. Through $(5, -2)$ and perpendicular to the y-axis

37. Through the origin and perpendicular to the line $2x + 4y = 5$

38. Through $(2, -7)$ and perpendicular to the line
$3x - 2y - 9 = 0$

39. At $60\,°$F a cricket chirps 50 times per minute and at $80\,°$F it chirps 100 times per minute.

    (a) Assuming that the rate of chirping varies linearly with the temperature, express the cricket's chirping rate r as a function of the temperature T.

    (b) Determine the cricket's chirping rate when $T = 40\,°$F.

40. A salesman's weekly pay is $450 when his sales are $3000, and his weekly pay is $500 when his sales are $4000.

    (a) Assuming that his pay varies linearly with the amount of his sales, express his pay P as a function of the amount of his sales S.

    (b) Determine his base pay and his rate of commission.

41. A new dump truck cost $60,000 and depreciates linearly to a value of $6000 after 6 years.

    (a) Express the value V of the truck as a function of its age x (in years).

    (b) What is the value of the truck after 4 years?

42. A factory owner buys a new machine for $25,000. The value of the machine depreciates linearly, and the machine has no resale value after 10 years.

    (a) Express the value V of the machine as a function of its age x (in years).

    (b) What is the resale value of the machine after 6 years?

43. A house cost $50,000 to build eight years ago. Its value has appreciated linearly and today the house is valued at $170,000.

    (a) Express the value V of the house as a function of its age x (in years).

    (b) What is the projected value of the house four years from now?

44. A diamond ring was purchased for $600. Fifteen years later the ring was appraised for $2400. Assume the ring appreciated linearly.

    (a) Express the value V of the ring as a function of its age x (in years).

    (b) Find the value of the ring 20 years after it was purchased.

Critical Thinking

45. The *perpendicular bisector* of a line segment is a line that is perpendicular to the segment at its midpoint. Find the equation of the perpendicular bisector of the line segment that joins the given points.

    (a) $(4, 0)$ and $(0, -12)$ (b) $(3, -2)$ and $(-1, 1)$

46. Use slopes to show that the triangle with vertices $A(-2, 1)$, $B(4, -8)$ and $C(7, -6)$ is a right triangle. Then find the area of this right triangle.

In Exercises 47–50, find the linear function that satisfies the indicated conditions.

47. $f(2) = 7$ and $f(-1) = 3$

48. $h(3) = 1$ and $h(0) = 2$

49. $g(2) = -1$ and the graph of g has slope 3

50. $F(-1) = -3$ and the graph of F has slope -2

51. Find $H(3)$ given that H is a linear function with $H(5) = 4$ and $H(0) = 0$.

52. Find $G(-4)$ given that G is a linear function with $G(-7) = 3$ and $G(5) = 1$.

53. Find the value of k such that the line $kx + 2y + 4 = 0$ satisfies the given conditions.

    (a) passes through $(3, 1)$

    (b) is parallel to the x-axis

    (c) is parallel to $y = 3x - 4$

    (d) is perpendicular to $2x - 4y = 5$

54. Find the value of k such that the line $3x + ky = 6$ satisfies the given conditions.

    (a) passes through $(-2, 4)$

    (b) is perpendicular to the x-axis

    (c) is parallel to $y = 5 - 2x$

    (d) is perpendicular to $x - y = 0$

$\boxed{\frac{\Delta y}{\Delta x}}$ **55.** A *tangent line* to a curve is a line that just touches the curve at a single point and is closer to the curve in the vicinity of the point than any other line drawn through the point. Using calculus, it can be shown that the *slope of the tangent line* (denoted m_{tan}) to the curve $y = 1 - 3x^2$ at any point (x, y) on the curve is given by $m_{\text{tan}} = -6x$. Find the equation of the tangent line that touches the point $(1, -2)$, as shown in the sketch.

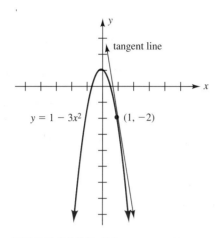

EXERCISE 55

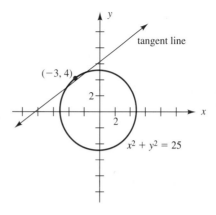

EXERCISE 56

 56. Using calculus, it can be shown that the slope of the tangent line (see Exercise 55) to the circle $x^2 + y^2 = 25$ at any point (x, y) on the circle is given by $m_{tan} = -x/y$. Find the equation of the tangent line that touches the point $(-3, 4)$, as shown in the sketch.

57. Find the equation of the line that passes through the point $P(x_1, y_1)$ and is parallel to the line $Ax + By + C = 0$. Write the equation in slope-intercept form.

58. Find the equation of the line that passes through the origin and is perpendicular to the line $Ax + By + C = 0$. Write the equation in general form.

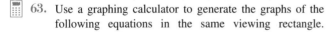

 Calculator Activities

In Exercises 59–62, find the equation of a line that satisfies the indicated conditions. Record your answer in slope-intercept form, rounding the slope and y-intercept to three significant digits.

59. Passes through (2.62, 1.98) and (4.44, −1.84)

60. Passes through (−42.6, −38.3) with same y-intercept as $2.7x − 3.2y = 66.7$

61. Passes through (6.2, −1.9) and is parallel to the line that passes through (−4.3, −3.2) and (−2.6, 4.9)

62. Passes through (0.876, 0.342) and is perpendicular to $3.4x + 5.7y = 10.4$.

63. Use a graphing calculator to generate the graphs of the following equations in the same viewing rectangle.

$$2.795x − 1.625y = 8.325$$

$$6.622x − 3.850y = −5.765$$

Use the picture to determine if the lines appear to be parallel to each other. Verify your answer.

64. Use a graphing calculator to generate the graphs of the following equations in the same viewing rectangle:

$$2.740x + 0.8768y = 10.65$$

$$0.6672x − 2.085y = 15.76$$

Use the picture to determine if the lines appear to be perpendicular to each other. Verify your answer.

65. Trolley tracks are laid with expansion gaps between the steel rails so that the rails can expand without distortion. As the temperature increases, the width of the gap between the rails decreases linearly. Suppose the gap is 2.5 mm wide when the temperature is 51 °F and 1.2 mm wide when the temperature is 83 °F.

(a) Express the width w of the expansion gap as a function of the temperature T.
(b) What is the approximate width of the expansion gap when the temperature is 35 °F?
(c) At what temperature will the rails just touch?

66. The fuel tank of an automobile is filled with 22.3 gallons of gasoline. When the auto is traveling at a constant rate of speed, the amount of fuel remaining in the tank decreases linearly. After 1 hour 12 minutes of travel, the tank contains 18.1 gallons of fuel.

(a) Express the number N of gallons of gasoline in the tank as a function of the amount of time t driven.
(b) Approximately how many gallons of gasoline remain after traveling for 2 hours 21 minutes?
(c) When does the automobile run out of gasoline?

4.3 Quadratic Functions

◆ **Introductory Comments**

Recall from Section 3.4 that the graph of the squaring function $f(x) = x^2$ is a cup-shaped curve [see Figure 3.31(d) in Section 3.4]. Its graph is symmetric with respect to the vertical line $x = 0$ (the y-axis). We refer to the vertical line $x = 0$ as the **axis of symmetry** of this cup-shaped curve and the point where the axis of symmetry intersects the curve as its **vertex**. The vertex for the graph of $f(x) = x^2$ is the origin $(0, 0)$.

Using the vertical and horizontal shift rules and the vertical stretch and shrink rule (Section 3.4), we can show that the graph of

$$F(x) = a(x - h)^2 + k,$$

where a, h, and k are real numbers with $a \neq 0$, is also a cup-shaped curve. However, the point (h, k) is now the vertex and the line $x = h$ is the axis of symmetry. By the x-axis reflection rule (Section 3.4), this cup-shaped curve opens *upward* if $a > 0$ [see Figure 4.26(a)] or downward if $a < 0$ [see Figure 4.26(b)].

FIGURE 4.26
Graph of $F(x) = a(x - h)^2 + k$

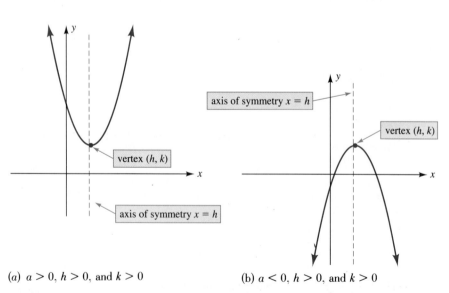

(a) $a > 0$, $h > 0$, and $k > 0$ (b) $a < 0$, $h > 0$, and $k > 0$

We refer to the function F defined by $F(x) = a(x - h)^2 + k$ as a **quadratic function** in standard form, and its graph as a **parabola** with vertex (h, k) and axis of symmetry $x = h$.

Quadratic Function in Standard Form

If a, h, and k are real numbers with $a \neq 0$, then the function F defined by

$$F(x) = a(x - h)^2 + k$$

is a **quadratic function** in standard form. The graph of this function is a **parabola** with vertex (h, k) and axis of symmetry $x = h$.

In this section, we study quadratic functions and their applications.

◆ **Working with Quadratic Functions in Standard Form**

To sketch the graph of the quadratic function $F(x) = a(x - h)^2 + k$, we simply locate the vertex (h, k) and determine any x- and y-intercepts.

EXAMPLE 1   Determine the vertex and any x- and y-intercepts for the graph of the quadratic function $f(x) = -2(x + 1)^2 + 8$. Then sketch the graph of f.

SOLUTION   We begin by writing this quadratic function in standard form:

$$f(x) = -2[x - (-1)]^2 + 8$$

with $a = -2$, $h = -1$, $k = 8$.

Since $a = -2$ $(a < 0)$, we know the graph is a parabola that opens *downward*. Since $h = -1$ and $k = 8$, its vertex is $(-1, 8)$.

To find the y-intercept, we let $x = 0$ and determine $f(0)$ as follows:

$$f(0) = -2(0 + 1)^2 + 8 = -2 + 8 = 6 \quad \text{(y-intercept)}$$

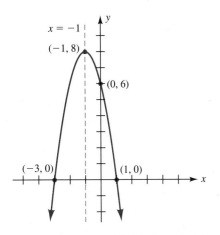

$x = -1$

$(-1, 8)$

$(0, 6)$

$(-3, 0)$ $(1, 0)$

FIGURE 4.27
The graph of $f(x) = -2(x + 1)^2 + 8$ is a parabola that opens *downward* with vertex $(-1, 8)$. The graph is symmetric with respect to the vertical line $x = -1$.

To find the x-intercepts, we let $f(x) = 0$ and determine x:

$$0 = -2(x + 1)^2 + 8$$
$$(x + 1)^2 = 4$$
$$x + 1 = \pm 2$$
$$x = 2 - 1 \quad \text{or} \quad x = -2 - 1$$
$$x = 1 \qquad\qquad x = -3 \quad \text{(x-intercepts)}$$

The graph of $f(x) = -2(x + 1)^2 + 8$ is the parabola shown in Figure 4.27.   ◆

PROBLEM 1   Repeat Example 1 for $g(x) = (x - 3)^2 - 1$.   ◆

As illustrated in the next example, we can determine the equation of a quadratic function if we know the vertex of the parabola and the coordinates of just one other point on the parabola.

EXAMPLE 2   Determine the equation of the quadratic function whose graph passes through the point (4, 1) and has its vertex at (2, 3).

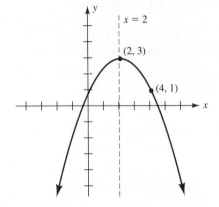

SOLUTION   The graph of the quadratic function is shown in Figure 4.28. The left-hand side of the curve is drawn by using the fact that the parabola is symmetric with respect to the line $x = 2$. Since the vertex is (2, 3), the quadratic function must have the form

$$y = a(x - 2)^2 + 3.$$

To determine a, we replace x with 4 and y with 1 as follows:

$$y = a(x - 2)^2 + 3$$
$$1 = a(4 - 2)^2 + 3$$
$$1 = 4a + 3$$
$$-2 = 4a$$
$$a = -\tfrac{1}{2}.$$

FIGURE 4.28
Graph of the quadratic function that passes through the point (4, 1) with vertex (2, 3).

Thus, the required quadratic function is $y = -\tfrac{1}{2}(x - 2)^2 + 3$.   ◆

PROBLEM 2   Repeat Example 2 if the graph of the quadratic function passes through the point (2, 3) and has its vertex at (4, 1).   ◆

◆ **Quadratic Functions in General Form**

If we square the expression $(x - h)$ in $F(x) = a(x - h)^2 + k$, we obtain

$$F(x) = a(x^2 - 2hx + h^2) + k$$
$$= ax^2 + (-2ah)x + (ah^2 + k)$$
$$= ax^2 + \quad bx \quad + \quad c$$

Replace the constant $-2ah$ with b and the constant $ah^2 + k$ with c.

where a, b, and c are constants with $a \neq 0$. We refer to a function of this form as a **quadratic function** in general form.

Quadratic Function in
General Form

If a, b, and c are real numbers with $a \neq 0$, then the function F defined by

$$F(x) = ax^2 + bx + c$$

is a **quadratic function** in general form.

If a quadratic function is written in the general form $F(x) = ax^2 + bx + c$, we know that its graph is a parabola and we can find its vertex by writing the function in the standard form $F(x) = a(x - h)^2 + k$ and noting the values of h and k. To accomplish this task, we use the process of completing the square (Section 2.3) and proceed as follows:

$F(x) = ax^2 + bx + c$

$$= a\left(x^2 + \frac{b}{a}x \quad\quad\right) + c \qquad \textbf{Factor out } a$$

$$= a\left[\left(x^2 + \frac{b}{a}x + \frac{b^2}{4a^2}\right) - \frac{b^2}{4a^2}\right] + c \qquad \textbf{Complete the square by adding}$$
$$\textbf{then subtracting } \frac{b^2}{4a^2} \textbf{ as}$$
$$\textbf{shown}$$

$$\left(\text{half of } \frac{b}{a}\right)^2$$

$$= a\left(x^2 + \frac{b}{a}x + \frac{b^2}{4a^2}\right) - \frac{b^2}{4a} + c \qquad \textbf{Multiply through by } a \textbf{ to}$$
$$\textbf{eliminate the bracket}$$

$$= a\left(x + \frac{b}{2a}\right)^2 + \frac{4ac - b^2}{4a} \qquad \textbf{Factor the perfect square}$$
$$\textbf{trinomial and add fractions}$$

$$= a\left[x - \left(-\frac{b}{2a}\right)\right]^2 + \frac{4ac - b^2}{4a} \qquad \textbf{Write in standard form}$$

x-coordinate y-coordinate
of the vertex   of the vertex

Note: It is only necessary to remember the fact that the x-coordinate is $-\dfrac{b}{2a}$. The y-coordinate for the vertex of the parabola can be determined by evaluating $F\left(-\dfrac{b}{2a}\right)$.

Vertex Formula

The graph of the quadratic function $F(x) = ax^2 + bx + c$ is a parabola with axis of symmetry $x = -\dfrac{b}{2a}$ and vertex

$$\left(-\frac{b}{2a}, F\left(-\frac{b}{2a}\right)\right).$$

EXAMPLE 3   Determine the vertex and any x- and y-intercepts for the graph of each quadratic function. Then sketch the graph of the function.

(a) $f(x) = x^2 - 3x - 4$ **(b)** $y = -2x^2 + 4x - 5$

SOLUTION

(a) For $f(x) = x^2 - 3x - 4$, we have $a = 1$, $b = -3$, and $c = -4$. Thus, the x-coordinate of the vertex of the parabola is

$$x = -\frac{b}{2a} = -\frac{-3}{2(1)} = \frac{3}{2}$$

and the y-coordinate is

$$y = f\left(-\frac{b}{2a}\right) = f\left(\frac{3}{2}\right) = \left(\frac{3}{2}\right)^2 - 3\left(\frac{3}{2}\right) - 4 = -\frac{25}{4}.$$

Hence, the vertex of the parabola is $(\frac{3}{2}, -\frac{25}{4})$. Since $a = 1$ ($a > 0$), we know the parabola opens upward from the point $(\frac{3}{2}, -\frac{25}{4})$. We find the y-intercept by letting $x = 0$ and evaluating $f(0)$:

$$f(0) = (0)^2 - 3(0) - 4 = -4$$

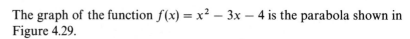

We find the x-intercepts by letting $f(x) = 0$ and solving for x:

$$0 = x^2 - 3x - 4$$
$$0 = (x - 4)(x + 1)$$
$$x - 4 = 0 \quad \text{or} \quad x + 1 = 0$$
$$x = 4 \qquad\qquad x = -1$$

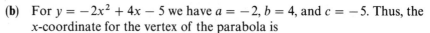

The graph of the function $f(x) = x^2 - 3x - 4$ is the parabola shown in Figure 4.29.

(b) For $y = -2x^2 + 4x - 5$ we have $a = -2$, $b = 4$, and $c = -5$. Thus, the x-coordinate for the vertex of the parabola is

$$x = -\frac{b}{2a} = -\frac{4}{2(-2)} = 1.$$

To find the y-coordinate of the vertex, we replace x with 1 in the equation $y = -2x^2 + 4x - 5$ as follows:

$$y = -2(1)^2 + 4(1) - 5 = -3$$

Hence, the vertex has coordinates $(1, -3)$ and, since $a = -2$ ($a < 0$), the parabola opens downward from the point $(1, -3)$. Therefore, there is no x-intercept (see Figure 4.30). We find the y-intercept by letting $x = 0$ and

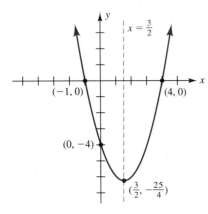

FIGURE 4.29

The graph of $f(x) = x^2 - 3x - 4$ is a parabola that opens upward with vertex $(\frac{3}{2}, -\frac{25}{4})$.

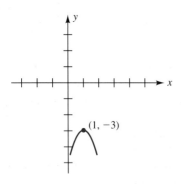

FIGURE 4.30

A parabola that opens *downward* with vertex $(1, -3)$ cannot have an x-intercept.

solving for y as follows:

$$y = -2(0)^2 + 4(0) - 5 = -5$$

y-intercept

The graph of $y = -2x^2 + 4x - 5$ is shown in Figure 4.31. We draw the left-hand side of the parabola by using the vertex and the y-intercept. We then use the fact that the parabola is symmetric with respect to the line $x = 1$ to draw the right-hand side of the curve.   ◆

FIGURE 4.31

The graph of $y = -2x^2 + 4x - 5$ is a parabola that opens downward with vertex $(1, -3)$.

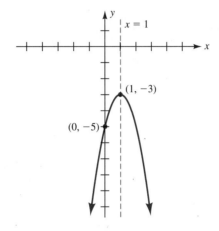

PROBLEM 3   Referring to Example 3(b), try to find any x-intercept by letting $y = 0$ and solving the equation $0 = -2x^2 + 4x - 5$ by using the quadratic formula. What dilemma do you confront?   ◆

◆ Maximum and Minimum Values

If $a > 0$, then the vertex (h, k) is the *lowest point* on the graph of $F(x) = a(x - h)^2 + k$. Hence, we say that the function F has a **minimum value** of k when $x = h$. For example, Figure 4.29 indicates that the lowest point on the graph of $f(x) = x^2 - 3x - 4$ is its vertex $(\frac{3}{2}, -\frac{25}{4})$. We refer to $-\frac{25}{4}$ as the *minimum value* for this quadratic function and say that it occurs when $x = \frac{3}{2}$.

If $a < 0$, then the vertex (h, k) is the *highest point* on the graph of $F(x) = a(x - h)^2 + k$. Hence, we say that the function F has a **maximum value** of k when $x = h$. Figure 4.31 indicates that the highest point on the graph of $y = -2x^2 + 4x - 5$ is its vertex $(1, -3)$. We refer to -3 as the *maximum value* for this quadratic function and say that it occurs when $x = 1$.

In general, calculus is needed to find the *maximum and minimum values* of a function. However, if the function is quadratic, we may use the vertex formula to find the maximum or minimum value of the function.

EXAMPLE 4   Determine if the function $g(x) = 3x^2 + 6x - 8$ has a maximum or minimum value. Then find this value.

SOLUTION   Since the function g is quadratic, we can use the vertex formula to determine the maximum or minimum value. Since $a = 3$ ($a > 0$), we know the parabola opens upward. Thus, g has a *minimum value* at

$$x = -\frac{b}{2a} = -\frac{6}{2(3)} = -1.$$

To find the minimum value, we replace x with -1 and evaluate $g(-1)$ as follows:

$$g(-1) = 3(-1)^2 + 6(-1) - 8 = -11.$$

Thus, the minimum value for the quadratic function $g(x) = 3x^2 + 6x - 8$ is -11 and it occurs when $x = -1$.   ◆

PROBLEM 4   Repeat Example 4 for $y = 3 - 3x^2$.   ◆

◆ **Application: Maximizing the Area of a Garden**

The idea of maximum and minimum values occurs in many applied problems. We illustrate one such application in our next example.

EXAMPLE 5   One hundred feet of fencing is to be used to enclose a rectangular garden that abuts a barn. No fencing is needed along the barn. What is the *largest* possible area that can be enclosed?

SOLUTION   In this problem we want the *area* to be the *largest* possible. Thus, we express the area as a quadratic function and then find its *maximum value* by using the vertex formula. We begin by letting

w = the width of the rectangle   and l = the length of the rectangle,

as shown in Figure 4.32. Then the area A of the rectangle is

$$A = lw.$$

Since 100 ft of fencing is available to enclose the rectangular garden, we have

$$100 = l + 2w   \text{or}   l = 100 - 2w.$$

Substituting $100 - 2w$ for l in the area formula $A = lw$, we obtain

$$A = (100 - 2w)w$$

$$A = 100w - 2w^2$$

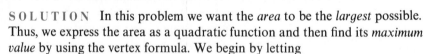

A quadratic function with $a = -2$, $b = 100$, and $c = 0$

FIGURE 4.32
A rectangular garden of length l and width w with no fencing along the barn.

Since $a = -2$ $(a < 0)$, we know that the graph of $A = 100w - 2w^2$ is a parabola that opens downward. Thus, this quadratic function has a maximum value that occurs at

$$w = -\frac{b}{2a} = -\frac{100}{2(-2)} = 25.$$

Replacing w with 25 in $A = 100w - 2w^2$, we obtain

$$A = 100(25) - 2(25)^2 = 1250.$$

Maximum value

Thus, the largest possible area that can be enclosed is 1250 square feet. This occurs when the width w is 25 feet and the length l is 50 feet.   ◆

PROBLEM 5   Suppose 100 feet of fencing is to be used to enclose a rectangular garden on all four sides. What is the *largest* possible area that can be enclosed?   ◆

Exercises 4.3

    Basic Skills

In Exercises 1–24, determine the vertex and any x- and y-intercepts for the graph of each quadratic function. Then sketch the graph of the function.

1. $y = (x - 1)^2 + 3$
2. $y = (x + 2)^2 - 1$
3. $f(x) = -(x + 3)^2 + 4$
4. $F(x) = -2(x - 1)^2 - 2$
5. $y = 3 - 3(x + 1)^2$
6. $y = \frac{1}{4}(x - 4)^2$
7. $G(x) = \frac{1}{2}(x - 4)^2 - 1$
8. $y = 2(x - \frac{1}{2})^2 - \frac{3}{2}$
9. $y = x^2 - 1$
10. $F(x) = 2 - \frac{1}{2}x^2$
11. $f(x) = 4x - x^2$
12. $g(x) = x^2 - 6x$
13. $y = x^2 + 2x - 3$
14. $G(x) = x^2 + 8x + 12$
15. $f(x) = 5 + 4x - x^2$
16. $y = -x^2 - 6x + 7$
17. $y = x^2 + x - 6$
18. $f(x) = 2x^2 - 3x - 5$
19. $h(x) = -2x^2 + 8x - 3$
20. $y = 3x^2 - 6x + 1$
21. $y = 3x^2 - 2x + 4$
22. $y = 3 + 5x - 5x^2$
23. $y = 1 + 4x + 4x^2$
24. $H(x) = 4x^2 - 12x + 9$

In Exercises 25–32, determine the equation of the quadratic function whose graph satisfies the indicated conditions.

25. Vertex at $(0, 0)$ and passing through $(2, 1)$
26. Vertex at the origin and passing through $(-1, -3)$

27. Vertex $(0, 2)$ and x-intercepts ± 2
28. Vertex $(0, -4)$ and passing through $(1, 4)$
29. Vertex $(3, 2)$ and passing through the origin
30. Vertex $(2, -1)$ and passing through $(-1, -2)$
31. Vertex $(-1, \frac{4}{3})$ and y-intercept 2
32. Vertex $(-\frac{3}{4}, -\frac{49}{8})$ and passing through $(-1, 0)$

In Exercises 33–38, determine whether the quadratic function has a maximum or minimum value. Then find this value.

33. $y = 8x - x^2$
34. $f(x) = x^2 + 5x$
35. $G(x) = 4x^2 + 6x + 3$
36. $y = -2x^2 - 4x + 8$
37. $h(x) = -3(x - 7)^2 + 9$
38. $F(x) = 5(x + 4)^2 - 11$

39. If a ball is thrown vertically upward with an initial velocity of 64 ft/s from a height of 80 ft above the ground, its height h (in feet) above the ground after t seconds is $h(t) = -16t^2 + 64t + 80$.

    (a) When does the ball reach its maximum height?
    (b) What is its maximum height?
    (c) How long is the ball in flight?

40. When a golf ball is driven from a tee with an initial velocity of 160 ft/s at an angle of 45° with respect to the horizontal, the height h (in feet) of the ball above the

ground is a function of the distance x (in feet) from the tee and is given by $h(x) = x - \dfrac{x^2}{800}$.

(a) What is the horizontal distance from the tee when the golf ball reaches its maximum height?
(b) What is its maximum height?
(c) How far away from the tee does the ball strike the ground?

41. The daily manufacturing cost C (in dollars) for a ski company is given by $C(x) = x^2 - 100x + 4800$, where x is the number of pairs of skis produced per day.

(a) How many pairs of skis should be produced to minimize the daily manufacturing cost?
(b) What is the minimum daily manufacturing cost?

42. The daily profit P (in dollars) for a company that makes tennis rackets is given by $P(x) = -x^2 + 240x - 5400$, where x is the number of tennis rackets produced per day.

(a) How many tennis rackets should be produced to maximize the daily profit?
(b) What is the maximum daily profit?

43. Find two numbers whose sum is 124 such that their product is the largest possible.

44. Find two numbers whose sum is -30 such that the sum of their squares is a minimum.

45. Of all rectangles with a perimeter of 44 inches, what are the dimensions of the one with maximum area?

46. A long piece of sheet metal 12 inches wide is to be made into a rain gutter by bending up the two long edges to make straight sides. The gutter will have a rectangular cross-sectional area, as shown in the figure. What depth will give the greatest carrying capacity?

47. Two hundred feet of fencing is to be used to enclose two identical rectangular animal pens that abut a barn, as shown in the figure. No fencing is needed along the barn. What dimensions for the total enclosure make the area of the pens the largest possible?

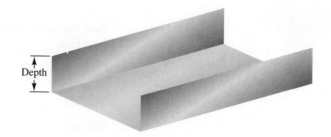

Depth

EXERCISE 46

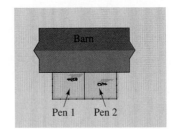

Barn

Pen 1   Pen 2

EXERCISE 47

48. A football stadium is to be built in the shape of a rectangle with semicircular ends, as shown in the figure. If the perimeter of the stadium is 1 mile, find the dimensions of x and r so that the area of the rectangular part of the stadium is the largest possible.

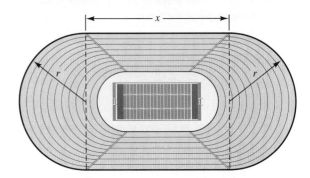

x

r r

 Critical Thinking

49. If a parabola has two distinct x-intercepts a and b, what is the x-coordinate of its vertex?

50. Find all values of k such that the vertex of the graph of $y = x^2 + kx + 9$ lies on the x-axis.

51. Determine the quadratic function f with zeros -1 and 3 and range $[-2, \infty)$.

52. Find the sum of the zeros of the quadratic function $f(x) = ax^2 + bx + c$.

53. If the graph of the following quadratic function $f(x) = ax^2 + bx + c$ passes through the origin, what is the value of c?

54. Given the linear function $f(x) = mx + b$ and the quadratic function $g(x) = a(x - h)^2 + k$, where $m \neq 0$ and

$a \neq 0$,

(a) find $f \circ g$.   (b) find $g \circ f$.
(c) classify the functions $f \circ g$ and $g \circ f$ as linear, quadratic, or neither.

55. A bus service between Plymouth and Boston charges $8 per person and carries 400 passengers per day. Research shows that for each $1 increase in fare the company loses 25 passengers per day.

(a) What daily rate gives the company the maximum revenue?

(b) What is the maximum revenue?

56. A math tutor charges $30 per hour for her services and averages 21 students per week in one-hour sessions. She has found that for each $2 decrease in her hourly rate, she acquires 3 more students per week to tutor.

(a) What hourly rate should she charge for maximum revenue?

(b) What is her maximum revenue?

Calculator Activities

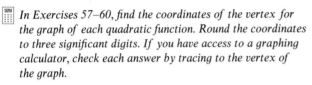
In Exercises 57–60, find the coordinates of the vertex for the graph of each quadratic function. Round the coordinates to three significant digits. If you have access to a graphing calculator, check each answer by tracing to the vertex of the graph.

57. $f(x) = 12.5x^2 - 18.6x + 10.7$

58. $f(x) = 1.24x^2 + 4.56x - 9.55$

59. $f(x) = -0.024x^2 + 0.126x + 1.344$

60. $f(x) = 107 - 425x - 121x^2$

In Exercises 61 and 62, determine a quadratic function in standard form whose graph satisfies the indicated conditions.

61. Vertex at $(2.67, -1.98)$ and y-intercept 1.05

62. Maximum value of 9.06 with x-intercepts 4.67 and -1.51

 63. Use a graphing calculator to generate the five graphs of $y = x^2 + 4x + c$ for $c = -2, -1, 0, 1$, and 2 in the same viewing rectangle. What effect does the constant c have on the axis of symmetry of these parabolas?

64. Use a graphing calculator to generate the four graphs of $y = a(x - 2)^2 + 1$ for $a = -2, -1, 1$, and 2 in the

same viewing rectangle. What effect does the constant a have on the vertex of these parabolas?

65. For the electrical circuit shown in the sketch, the power P (in watts) delivered to the load resistance R_L (in ohms) is given by

$$P = 21.6i - 0.34i^2,$$

where i is the current (in amperes) flowing through the circuit. What is the maximum power that can be delivered to the load resistance?

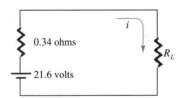

66. A wire 21.6 cm long is cut into two pieces. One of the pieces is bent into a square and the other piece is bent into a circle. Where should the wire be cut if the total area of the square and circle is to be a minimum?

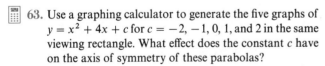

4.4   General Quadratic Equations in Two Unknowns: Conic Sections

◆ Introductory Comments

Early Greek mathematicians noted that when a double cone is sliced with a plane, a special family of curves is formed. As illustrated in Figure 4.33, this family of curves has four main members: **circle, ellipse, parabola**, and **hyperbola**. Collectively, this family of curves is called the **conic sections**.

FIGURE 4.33
The conic sections.

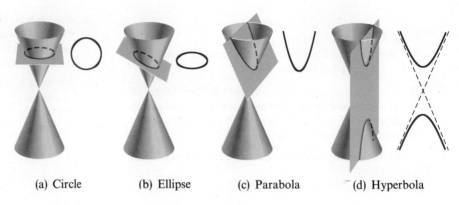

(a) Circle   (b) Ellipse   (c) Parabola   (d) Hyperbola

When the plane intersects the double cone, it is also possible to obtain a single point, one line, or a pair of intersecting lines. Do you see how? These special cases are called the **degenerate conic sections**.

An equation of the form

$$Ax^2 + Bxy + Cy^2 + Dx + Ey + F = 0$$

where A, B, C, D, E, and F are real numbers, is called a **general quadratic equation in two unknowns**. The graph of this type of equation (if a graph exists) is either a conic section or a degenerate conic section. In this section, we consider general quadratic equations in two unknowns in which $B = 0$. By choosing $B = 0$, we keep the axes of symmetry of the curves vertical or horizontal. As we will see, the values of A, C, D, E, and F determine which conic section we obtain.

◆ **Equations of Parabolas**

Recall from Section 4.3 that the graph of the equation $y = ax^2 + bx + c$ is a parabola with *vertical* axis of symmetry $x = -\dfrac{b}{2a}$. The parabola opens upward if $a > 0$ and downward if $a < 0$. If we interchange x and y in the equation $y = ax^2 + bx + c$, we obtain

$$x = ay^2 + by + c.$$

Interchanging x and y reflects a graph about the line $y = x$. Thus, the graph of $x = ay^2 + by + c$ is still a parabola, but with *horizontal axis* of symmetry $y = -\dfrac{b}{2a}$ as shown in Figure 4.34. The parabola opens to the *right* if $a > 0$ or to the *left* if $a < 0$.

The equation $y = ax^2 + bx + c$ may appear in the form

$$Ax^2 + Dx + Ey + F = 0,$$

where A, D, E, and F are real numbers with $A \neq 0$ and $E \neq 0$, and the equation

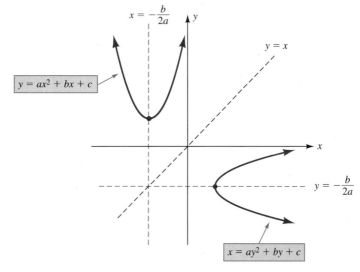

$x = ay^2 + by + c$ may appear in the form

$$Cy^2 + Dx + Ey + F = 0,$$

where C, D, E, and F are real numbers with $C \neq 0$ and $D \neq 0$. We refer to

$$Ax^2 + Dx + Ey + F = 0$$

and

$$Cy^2 + Dx + Ey + F = 0$$

as the **equations of a parabola in general form** with vertical and horizontal axes of symmetry, respectively. These general quadratic equations in two unknowns are characterized by the presence of either an x^2 or y^2 term, but never both.

EXAMPLE 1   Sketch the graph of the equation $y^2 - 2x - 4y + 10 = 0$.

SOLUTION   We can write this equation in the form $x = ay^2 + by + c$ by solving for x:

$$y^2 - 2x - 4y + 10 = 0$$

$$2x = y^2 - 4y + 10 \qquad \text{Add } 2x \text{ to both sides}$$

$$x = \tfrac{1}{2}y^2 - 2y + 5 \qquad \text{Divide both sides by 2}$$

Hence, we know the graph of this equation is a parabola with horizontal axis of symmetry

$$y = -\frac{b}{2a} = -\frac{-2}{2(\frac{1}{2})} = 2.$$

To determine the x-coordinate of the vertex, we substitute $y = 2$ in the equation $x = \frac{1}{2}y^2 - 2y + 5$:

$$x = \frac{1}{2}(2)^2 - 2(2) + 5 = 3.$$

Thus the vertex has coordinates (3, 2) and, since $a = \frac{1}{2}$ ($a > 0$), the parabola opens to the right from the point (3, 2). Hence, the graph has no y-intercept. We find the x-intercept by letting $y = 0$ and solving for x as follows:

$$x = \frac{1}{2}(0)^2 - 2(0) + 5 = 5$$

x-intercept

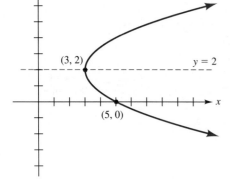

FIGURE 4.35
The graph of $y^2 - 2x - 4y + 10 = 0$ is a parabola that opens to the right with vertex (3, 2).

The graph of the equation $y^2 - 2x - 4y + 10 = 0$ is shown in Figure 4.35. We draw the bottom half of the parabola by connecting the vertex (3, 2) and x-intercept (5, 0). We then use the fact that the parabola is symmetric with respect to the line $y = 2$ to draw the upper part.   ◆

PROBLEM 1   Repeat Example 1 for $x^2 - 2y - 4x + 10 = 0$.   ◆

◆ **Equations of Circles and Ellipses**

Consider the equation

$$\frac{x^2}{a^2} + \frac{y^2}{b^2} = 1,$$

where $a > 0$ and $b > 0$. This is a general quadratic equation in two unknowns with $A = 1/a^2$, $C = 1/b^2$, $F = -1$, and $B = D = E = 0$. If $a = b$, then we have

$$\frac{x^2}{a^2} + \frac{y^2}{a^2} = 1 \quad \text{or} \quad x^2 + y^2 = a^2.$$

Recall from Section 3.2 that the graph of $x^2 + y^2 = a^2$ is a circle with center at the origin and radius a. If a and b have different values, then the circle is flattened or stretched to form an egg-shaped curve called an **ellipse**. The center of the ellipse is at the origin with x-intercepts $\pm a$ and y-intercepts $\pm b$. Figure 4.36 shows the three possibilities for the graph of $(x^2/a^2) + (y^2/b^2) = 1$.

In each of the graphs sketched in Figure 4.36, the horizontal and vertical line segments $\overline{AB}$ and $\overline{CD}$ are the **axes** of the ellipse and their lengths are $2a$ and $2b$, respectively. We refer to the longer line segment as the **major axis** and the shorter line segment as the **minor axis**. If $a > b$ as in Figure 4.36(b), then

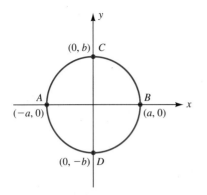

(a) If $a = b$, the graph is a circle with center at the origin.

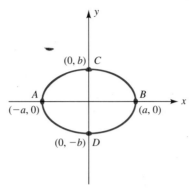

(b) If $a > b$, the graph is an ellipse elongated horizontally with center at the origin.

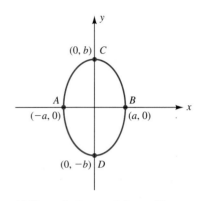

(c) If $a < b$, the graph is an ellipse elongated vertically with center at the origin.

FIGURE 4.36   The three possibilities for the graph of $\dfrac{x^2}{a^2} + \dfrac{y^2}{b^2} = 1$

$\overline{AB}$ is the major axis and $\overline{CD}$ is the minor axis. However, if $a < b$ as in Figure 4.36(c), then $\overline{CD}$ is the major axis and $\overline{AB}$ is the minor axis. We refer to the endpoints of the major axis as the **vertices** of the ellipse and the midpoint of the major axis, and minor axis, as its **center**.

If we replace x with $(x - h)$ and y with $(y - k)$ in the equation $(x^2/a^2) + (y^2/b^2) = 1$, we obtain

$$\frac{(x - h)^2}{a^2} + \frac{(y - k)^2}{b^2} = 1.$$

If a and b are positive real numbers such that $a = b$, we have

$$\frac{(x - h)^2}{a^2} + \frac{(y - k)^2}{a^2} = 1 \quad \text{or} \quad (x - h)^2 + (y - k)^2 = a^2,$$

which is the equation of a circle in standard form with center (h, k) and radius a (see Section 3.2). If a and b have different positive values, we conclude $(x - h)^2/a^2 + (y - k)^2/b^2 = 1$ is the **equation of an ellipse in standard form** with center (h, k), horizontal axis of length $2a$, and vertical axis of length $2b$.

 Equation of an Ellipse in Standard Form

The **equation of an ellipse in standard form** with center (h, k), horizontal axis of length $2a$, and vertical axis of length $2b$ is

$$\frac{(x - h)^2}{a^2} + \frac{(y - k)^2}{b^2} = 1.$$

EXAMPLE 2   Determine the equation of an ellipse in standard form if its vertices are (3, 3) and (−5, 3) and its minor axis has length 4.

SOLUTION   The vertices of an ellipse are the endpoints of its major axis. Since the y-coordinates of the vertices are equal, we know the major axis is horizontal. Thus, the minor axis is vertical. Now, if the minor axis is vertical and has a length of 4, we have

$$2b = 4 \quad \text{or} \quad b = 2.$$

The length of the major axis is the distance between the x-coordinates of the vertices or $|3 - (-5)| = 8$. Thus,

$$2a = 8 \quad \text{or} \quad a = 4.$$

The midpoint of the major axis is the center of the ellipse. Using the midpoint formula (Section 3.1), we find the coordinates of the center:

$$(h, k) = \left(\frac{3 + (-5)}{2}, \frac{3 + 3}{2}\right) = (-1, 3)$$

Thus, the equation of the ellipse in standard form is

$$\frac{(x - h)^2}{a^2} + \frac{(y - k)^2}{b^2} = 1$$

$$\frac{[x - (-1)]^2}{4^2} + \frac{(y - 3)^2}{2^2} = 1 \qquad \textbf{Substitute}$$

$$\frac{(x + 1)^2}{16} + \frac{(y - 3)^2}{4} = 1 \qquad \textbf{Simplify}$$

The graph of this equation is shown in Figure 4.37.   ◆

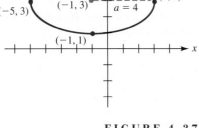

FIGURE 4.37
The equation of an ellipse with vertices (−5, 3) and (3, 3) and minor axis of length 4 is $\dfrac{(x + 1)^2}{16} + \dfrac{(y - 3)^2}{4} = 1$

PROBLEM 2   Determine the y-intercepts for the ellipse in Example 2.   ◆

If we take the equation of an ellipse in standard form and multiply both sides of the equation by $a^2 b^2$, we obtain an equation of the form

$$A(x - h)^2 + C(y - k)^2 = N,$$

where $A = b^2$, $C = a^2$, and $N = a^2 b^2$. If we expand $(x - h)^2$ and $(y - k)^2$, we obtain

$$A(x^2 - 2hx + h^2) + C(y^2 - 2ky + k^2) = N$$

$$Ax^2 + Cy^2 - 2Ahx - 2Cky + (Ah^2 + Ck^2 - N) = 0.$$

Now, replacing the constants $-2Ah$, $-2Ck$, and $(Ah^2 + Ck^2 - N)$ with D, E, and *F*, respectively, gives us

$$Ax^2 + Cy^2 + Dx + Ey + F = 0$$

which is the **equation of an ellipse in general form** with vertical and horizontal axes of symmetry. This general quadratic equation in two unknowns is characterized by the presence of x^2 and y^2 terms that have *different coefficients* but *like signs*. The difference between the general form of the equation of an ellipse and that of a circle is that the coefficients of the x^2 and y^2 terms in an ellipse are different, whereas those in a circle are the same.

EXAMPLE 3   Sketch the graph of the equation.

(a) $4x^2 + y^2 = 9$ (b) $x^2 + 9y^2 + 8x - 36y + 43 = 0$

SOLUTION

(a)   We begin by writing the equation in standard form:

$$4x^2 + y^2 = 9$$

$$\frac{4x^2}{9} + \frac{y^2}{9} = 1 \qquad \text{Divide both sides by 9}$$

$$\frac{x^2}{\frac{9}{4}} + \frac{y^2}{9} = 1 \qquad \text{Invert } \tfrac{4}{9} \text{ and divide}$$

$$\frac{(x-0)^2}{(\frac{3}{2})^2} + \frac{(y-0)^2}{3^2} = 1 \qquad \text{Write in standard form}$$

This last equation tells us that the graph of $4x^2 + y^2 = 9$ is an ellipse with $a = \tfrac{3}{2}$, $b = 3$, and center at the origin. Since $b > a$, the ellipse is elongated vertically with major axis of length $2b = 6$ and minor axis of length $2a = 3$. The graph of the equation is shown in Figure 4.38.

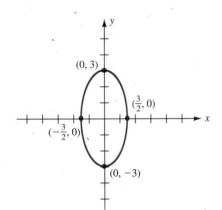

FIGURE 4.38
The graph of $4x^2 + y^2 = 9$ is an ellipse with center $(0, 0)$ and vertices $(0, 3)$ and $(0, -3)$.

(b)   We begin by writing the equation in standard form. To do this, we use the process of completing the square (Section 2.3), as follows:

$$x^2 + 9y^2 + 8x - 36y + 43 = 0$$

$$(x^2 + 8x \quad) + (9y^2 - 36y \quad) = -43 \qquad \text{Regroup}$$

$$(x^2 + 8x \quad) + 9(y^2 - 4y \quad) = -43 \qquad \text{Factor out 9}$$

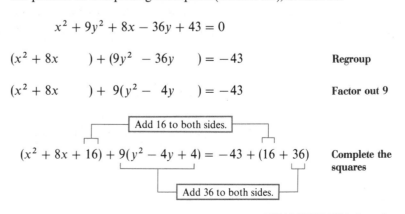

$$(x^2 + 8x + 16) + 9(y^2 - 4y + 4) = -43 + (16 + 36) \qquad \begin{array}{l}\text{Complete the}\\\text{squares}\end{array}$$

EXAMPLE 3(b) (*continued*)

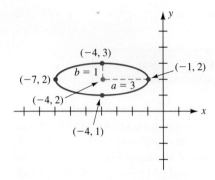

FIGURE 4.39

The graph of
$x^2 + 9y^2 + 8x - 36y + 43 = 0$ is an
ellipse with center $(-4, 2)$ and vertices
$(-7, 2)$ and $(-1, 2)$.

$$(x + 4)^2 + 9(y - 2)^2 = 9 \qquad \text{Factor}$$

$$\frac{(x + 4)^2}{9} + \frac{(y - 2)^2}{1} = 1 \qquad \begin{array}{l}\text{Divide both sides}\\ \text{by 9}\end{array}$$

$$\frac{[x-(-4)]^2}{3^2} + \frac{(y - 2)^2}{1^2} = 1 \qquad \begin{array}{l}\text{Write in standard}\\ \text{form}\end{array}$$

This last equation tells us that the graph of $x^2 + 9y^2 + 8x - 36y + 43 = 0$ is an ellipse with $a = 3$, $b = 1$, and center $(h, k) = (-4, 2)$. Since $a > b$, the ellipse is elongated horizontally with major axis of length $2a = 6$ and minor axis of length $2b = 2$. The graph of the equation is shown in Figure 4.39. ◆

An ellipse is not the graph of every equation of the form $Ax^2 + Cy^2 + Dx + Ey + F = 0$, where the x^2 and y^2 terms have different coefficients but like signs. For example, if this equation is written in the form

$$A(x - h)^2 + C(y - k)^2 = N \qquad \text{and} \qquad N = 0,$$

the graph of the equation is the single point (h, k). We refer to this single point as a *degenerate ellipse*. Also, if

$$A(x - h)^2 + C(y - k)^2 = N \qquad \text{and} \qquad N < 0,$$

this equation has no graph, since no point (x, y) with real coordinates satisfies this equation.

PROBLEM 3   Is $3x^2 + 2y^2 - 6x + 8y + 15 = 0$ the equation of an ellipse? Explain. ◆

◆ Equations of Hyperbolas

Next, we consider the equation

$$\frac{x^2}{a^2} - \frac{y^2}{b^2} = 1$$

with $a > 0$ and $b > 0$. This is a general quadratic equation in two unknowns with $A = 1/a^2$, $C = -1/b^2$, $F = -1$, and $B = D = E = 0$. According to the symmetry tests in Section 3.2, we know that the graph of this equation is symmetric to both coordinate axes and to the origin, since replacing x with $-x$ and y with $-y$ does not change the equation. Letting $y = 0$, we find that the x-intercepts are $\pm a$. However, letting $x = 0$, we find that the graph has no y-intercept, since the equation

$$-\frac{y^2}{b^2} = 1 \qquad \text{or} \qquad y^2 = -b^2$$

has no real solution. If we solve the equation $(x^2/a^2) - (y^2/b^2) = 1$ for y, we obtain

$$y = \pm \frac{b}{a}\sqrt{x^2 - a^2}.$$

Now for y to be a real number, $x^2 - a^2$ must be either positive or zero; that is,

$$x^2 - a^2 \geq 0$$

$$x \geq a \qquad \text{or} \qquad x \leq -a$$

This means that the graph exists only to the right of the x-intercept a and to the left of the x-intercept $-a$. Intuitively, we can see that for very large values of x (either negative or positive),

$$y = \pm \frac{b}{a}\sqrt{x^2 - a^2} \approx \pm \frac{b}{a}\sqrt{x^2} = \pm \frac{b}{a}x,$$

since x^2 is so much larger than a^2. That is, the graph of $(x^2/a^2) - (y^2/b^2) = 1$ approaches the lines $y = \pm(b/a)x$ as $|x|$ increases without bound. We refer to the lines

$$y = \pm \frac{b}{a}x$$

as the **asymptotes** of the curve.

The diagonals of a rectangle with vertices $(a, b), (-a, b), (-a, -b)$, and $(a, -b)$ have slopes $\pm b/a$. Extending these diagonals gives us the asymptotes of the curve, as shown in Figure 4.40. Using the x-intercepts and the asymptotes, we can sketch the graph of $(x^2/a^2) - (y^2/b^2) = 1$. The graph is called a **hyperbola** and is shown in Figure 4.41.

In Figure 4.41, the points $(-a, 0)$ and $(a, 0)$ are the **vertices** of the hyperbola and the horizontal and vertical line segments $\overline{AB}$ and $\overline{CD}$ are the **axes** of the hyperbola. Note that the lengths of $\overline{AB}$ and $\overline{CD}$ are $2a$ and $2b$, respectively. The line segment that connects the vertices of the hyperbola is called the **transverse axis** and the other axis of the hyperbola is the **conjugate axis**. In this figure the *horizontal* line segment $\overline{AB}$ is the transverse axis and the *vertical* line segment $\overline{CD}$ is the conjugate axis. The midpoint of both the transverse axis and the conjugate axis is the **center** of the hyperbola. The hyperbola in Figure 4.41 has its center at the origin.

The graph of the equation

$$\frac{y^2}{b^2} - \frac{x^2}{a^2} = 1$$

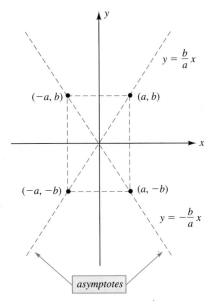

FIGURE 4.40
The asymptotes $y = \pm(b/a)x$ can be constructed from the diagonals of the rectangle with vertices $(a, b), (-a, b), (-a, -b)$, and $(a, -b)$.

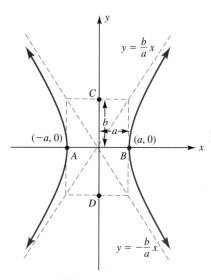

FIGURE 4.41
The graph of $(x^2/a^2) - (y^2/b^2) = 1$ is called a hyperbola. The x-intercepts are $\pm a$ and the asymptotes are $y = \pm(b/a)x$.

with $a > 0$ and $b > 0$ is also a hyperbola. However, the y-intercepts are now $\pm b$ and the graph has no x-intercept. Thus, the vertices of this hyperbola are $(0, b)$ and $(0, -b)$. The transverse axis is *vertical* with length $2b$ and the conjugate axis is *horizontal* with length $2a$. The asymptotes are still the two lines $y = \pm(b/a)x$ and the center of the hyperbola remains at the origin. The graph of $(y^2/b^2) - (x^2/a^2) = 1$ is shown in Figure 4.42.

FIGURE 4.42
The graph of $(y^2/b^2) - (x^2/a^2) = 1$ is a hyperbola with y-intercepts $\pm b$ and asymptotes $y = \pm(b/a)x$.

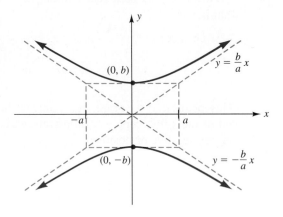

The graphs of the equations

$$\frac{(x - h)^2}{a^2} - \frac{(y - k)^2}{b^2} = 1 \qquad \text{and} \qquad \frac{(y - k)^2}{b^2} - \frac{(x - h)^2}{a^2} = 1$$

are the same as the graphs of $(x^2/a^2) - (y^2/b^2) = 1$ and $(y^2/b^2) - (x^2/a^2) = 1$, respectively, except the center of the hyperbola is now (h, k). We refer to

$$\frac{(x - h)^2}{a^2} - \frac{(y - k)^2}{b^2} = 1 \qquad \text{and} \qquad \frac{(y - k)^2}{b^2} - \frac{(x - h)^2}{a^2} = 1$$

with $a > 0$ and $b > 0$ as the *equations of a hyperbola in standard form*.

◆ **Equations of a Hyperbola in Standard Form**

The **equation of a hyperbola in standard form** with center (h, k), *horizontal* transverse axis of length $2a$, and vertical conjugate axis of length $2b$ is

$$\frac{(x - h)^2}{a^2} - \frac{(y - k)^2}{b^2} = 1.$$

The **equation of a hyperbola in standard form** with center (h, k), *vertical* transverse axis of length $2b$, and horizontal conjugate axis of length $2a$ is

$$\frac{(y - k)^2}{b^2} - \frac{(x - h)^2}{a^2} = 1.$$

EXAMPLE 4   Determine the equation of a hyperbola in standard form if its asymptotes are $y = \frac{1}{2}x + 1$ and $y = -\frac{1}{2}x + 1$ and its horizontal transverse axis has length 6.

SOLUTION   The intersection point of the asymptotes is the center of the hyperbola. Since the lines $y = \frac{1}{2}x + 1$ and $y = -\frac{1}{2}x + 1$ both have a y-intercept of 1, we conclude that the center of the hyperbola is

$$(h, k) = (0, 1).$$

Since the horizontal transverse axis has length 6, we have

$$2a = 6 \qquad \text{or} \qquad a = 3.$$

We can determine b by using the fact that the slopes of the asymptotes are $\pm b/a = \pm\frac{1}{2}$. Since $a = 3$, we have

$$\pm\frac{b}{3} = \pm\frac{1}{2} \qquad \text{or} \qquad b = \frac{3}{2}.$$

Now the equation of the hyperbola is

$$\frac{(x - h)^2}{a^2} - \frac{(y - k)^2}{b^2} = 1$$

$$\frac{(x - 0)^2}{3^2} - \frac{(y - 1)^2}{(3/2)^2} = 1 \qquad \textbf{Substitute}$$

$$\frac{x^2}{9} - \frac{4(y - 1)^2}{9} = 1 \qquad \textbf{Simplify}$$

The graph of this equation is shown in Figure 4.43.   ◆

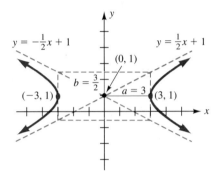

FIGURE 4.43

The equation of the hyperbola with asymptotes $y = \pm\frac{1}{2}x + 1$ and horizontal transverse axis of length 6 is $\frac{x^2}{9} - \frac{4(y - 1)^2}{9} = 1.$

PROBLEM 4   Rework Example 4 if the transverse axis is vertical with length 6.   ◆

If we take the equations of a hyperbola in standard form and multiply both sides of these equations by $a^2 b^2$, we obtain equations of the forms

$$A(x - h)^2 - C(y - k)^2 = N \qquad \text{and} \qquad C(y - k)^2 - A(x - h)^2 = N,$$

where $A = b^2$, $C = a^2$, and $N = a^2 b^2$.
    If we expand $(x - h)^2$ and $(y - k)^2$ in the equation $A(x - h)^2 - C(y - k)^2 = N$, we obtain

$$A(x^2 - 2hx + h^2) - C(y^2 - 2ky + k^2) = N$$

$$Ax^2 - Cy^2 - 2Ahx + 2Cky + (Ah^2 - Ck^2 - N) = 0.$$

Now, replacing the constants $-2Ah$, $2Ck$, and $(Ah^2 - Ck^2 - N)$ with D, E, and

F, respectively, gives us

$$Ax^2 - Cy^2 + Dx + Ey + F = 0.$$

If we expand $(x - h)^2$ and $(y - k)^2$ in the equation $C(y - k)^2 - A(x - h)^2 = N$, we obtain

$$C(y^2 - 2ky + k^2) - A(x^2 - 2hx + h^2) = N$$
$$Cy^2 - Ax^2 + 2Ahx - 2Cky + (Ck^2 - Ah^2 - N) = 0.$$

Now, replacing the constants $2Ah$, $-2Ck$, and $(Ck^2 - Ah^2 - N)$ with D, E, and F, respectively, gives us

$$Cy^2 - Ax^2 + Dx + Ey + F = 0.$$

We refer to

$$\boxed{Ax^2 - Cy^2 + Dx + Ey + F = 0}$$

and

$$\boxed{Cy^2 - Ax^2 + Dx + Ey + F = 0}$$

as the **equations of a hyperbola in general form** with vertical and horizontal axes of symmetry. These general quadratic equations in two unknowns are characterized by the presence of x^2 and y^2 terms that have *different* signs.

EXAMPLE 5   Sketch the graph of the equation.

(a) $4x^2 - 5y^2 = 20$ (b) $4x^2 - 9y^2 + 8x + 36y + 4 = 0$

SOLUTION

(a)   We begin by writing the equation in standard form:

$$4x^2 - 5y^2 = 20$$

$$\frac{x^2}{5} - \frac{y^2}{4} = 1 \qquad \text{Divide both sides by 20}$$

$$\frac{(x - 0)^2}{(\sqrt{5})^2} - \frac{(y - 0)^2}{2^2} = 1 \qquad \text{Write in standard form}$$

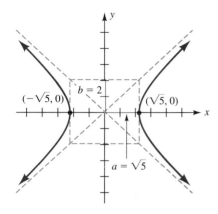

FIGURE 4.44
The graph of $4x^2 - 5y^2 = 20$ is a hyperbola with center at the origin and vertices $(\pm\sqrt{5}, 0)$

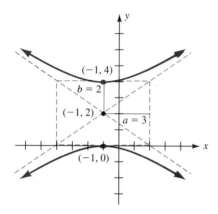

FIGURE 4.45
The graph of $4x^2 - 9y^2 + 8x + 36y + 4 = 0$ is a hyperbola with center $(-1, 2)$ and vertices $(-1, 4)$ and $(-1, 0)$.

This last equation tells us that the graph of $4x^2 - 5y^2 = 20$ is a hyperbola with $a = \sqrt{5}$, $b = 2$, center at the origin, and a horizontal transverse axis of length $2a = 2\sqrt{5}$. As an aid in graphing the hyperbola, we may draw a rectangle using the values of a and b, and then extend the diagonals of the rectangle to form the asymptotes. The graph of the hyperbola is shown in Figure 4.44.

(b) We begin by writing the equation in standard form. To do this, we use the process of completing the square (Section 2.3) as follows:

$$4x^2 - 9y^2 + 8x + 36y + 4 = 0$$

$$(4x^2 + 8x \quad) + (-9y^2 + 36y \quad) = -4 \qquad \text{Regroup}$$

$$4(x^2 + 2x \quad) - 9(y^2 - 4y \quad) = -4 \qquad \begin{array}{l}\text{Factor out 4}\\\text{and } -9\end{array}$$

Add 4 to both sides.

$$4(x^2 + 2x + 1) - 9(y^2 - 4y + 4) = -4 + (4 - 36) \qquad \begin{array}{l}\text{Complete the}\\\text{squares}\end{array}$$

Add -36 to both sides.

$$4(x + 1)^2 - 9(y - 2)^2 = -36 \qquad \text{Factor}$$

$$\frac{(y - 2)^2}{4} - \frac{(x + 1)^2}{9} = 1 \qquad \begin{array}{l}\text{Divide both}\\\text{sides by } -36\end{array}$$

$$\frac{(y - 2)^2}{2^2} - \frac{(x + 1)^2}{3^2} = 1 \qquad \begin{array}{l}\text{Write in standard}\\\text{form}\end{array}$$

The last equation tells us that the graph of $4x^2 - 9y^2 + 8x + 36y + 4 = 0$ is a hyperbola with $a = 3$, $b = 2$, center $(h, k) = (-1, 2)$, and a vertical transverse axis of length $2b = 4$. The graph of this hyperbola is shown in Figure 4.45. ◆

A hyperbola is not the graph of every equation of the form $Ax^2 - Cy^2 + Dx + Ey + F = 0$ or $Cy^2 - Ax^2 + Dx + Ey + F = 0$, where the x^2 and y^2 terms have different signs. For example, when $Ax^2 - Cy^2 + Dx + Ey + F = 0$ is written in the form $A(x - h)^2 - C(y - k)^2 = N$ with $N = 0$, the graph of the equation is a pair of intersecting lines. We refer to those two lines as a *degenerate hyperbola*.

PROBLEM 5 Is $x^2 - 4y^2 - 6x + 8y + 5 = 0$ the equation of a hyperbola? Explain. ◆

◆ **Concluding Comments**

In this section we have examined the graphs of general quadratic equations in two unknowns of the form

$$Ax^2 + Cy^2 + Dx + Ey + F = 0,$$

where A, C, D, E, and F are real numbers. We conclude with a summary of the graph of an equation in this form.

  Graph of a General Quadratic Equation in Two Unknowns

If the graph of a general quadratic equation in two unknowns

$$Ax^2 + Cy^2 + Dx + Ey + F = 0$$

exists and is not degenerate, then the graph is

1. a *parabola* if either $A = 0$ or $C = 0$, but not both.

2. a *circle* if $A = C \neq 0$.

3. an *ellipse* if $A \neq C$ and $AC > 0$.

4. a *hyperbola* if $AC < 0$.

Exercises 4.4

  Basic Skills

In Exercises 1–12, sketch the graph of each parabola. Label the vertex and any x- and y-intercepts.

1. $x = y^2$
2. $x = -3y^2$
3. $x = 4y - y^2$
4. $4x = y^2 - 4$
5. $x = y^2 - 3y - 4$
6. $x = 6 - y - \frac{1}{3}y^2$
7. $y^2 - 2x - 4y + 2 = 0$
8. $y^2 + 3x + 6y + 3 = 0$
9. $x^2 + 8x - 2y + 16 = 0$
10. $x^2 - 3y - 6x + 15 = 0$
11. $2y^2 + 3x + 16y + 26 = 0$
12. $3y^2 + 4x - 6y + 3 = 0$

In Exercises 13–28, sketch the graph of each ellipse. Label the center of the ellipse, if it is not at the origin, and label the endpoints of the major and minor axes.

13. $\dfrac{x^2}{16} + \dfrac{y^2}{4} = 1$
14. $\dfrac{x^2}{4} + \dfrac{y^2}{9} = 1$
15. $25x^2 + 4y^2 = 100$
16. $16x^2 + 25y^2 = 400$

17. $x^2 + 16y^2 = 9$
18. $4x^2 + 9y^2 = 16$
19. $3x^2 + 4y^2 - 9 = 0$
20. $x^2 + 5y^2 - 10 = 0$
21. $\dfrac{(x - 2)^2}{16} + \dfrac{(y + 1)^2}{4} = 1$
22. $\dfrac{x^2}{9} + \dfrac{(y - 3)^2}{25} = 1$
23. $9x^2 + 4(y + 2)^2 = 36$
24. $16(x - 4)^2 + (y - 5)^2 = 16$
25. $25x^2 + 4y^2 - 50x + 24y + 45 = 0$
26. $9x^2 + 25y^2 - 36x + 50y = 39$
27. $4x^2 + y^2 + 40x - 4y + 103 = 0$
28. $2x^2 + y^2 + 8x + 8y + 23 = 0$

In Exercises 29–44, sketch the graph of each hyperbola. Label the center of the hyperbola, if it is not at the origin, and label the vertices of the transverse axis.

29. $\dfrac{x^2}{4} - \dfrac{y^2}{16} = 1$
30. $\dfrac{y^2}{4} - \dfrac{x^2}{4} = 1$

31. $9y^2 - 4x^2 = 36$ **32.** $9x^2 - 16y^2 = 144$

33. $16x^2 - y^2 = 9$ **34.** $4y^2 - 16x^2 = 25$

35. $9x^2 - 25y^2 = -4$ **36.** $4y^2 - x^2 + 12 = 0$

37. $\dfrac{x^2}{9} - \dfrac{(y+2)^2}{4} = 1$ **38.** $\dfrac{(y-1)^2}{16} - (x+3)^2 = 1$

39. $(y-3)^2 - (x-1)^2 = 36$

40. $9(x+1)^2 - 25(y-4)^2 = 225$

41. $36x^2 - y^2 - 144x - 6y + 126 = 0$

42. $16y^2 - 9x^2 + 90x = 261$

43. $9y^2 - 3x^2 - 24x + 18y = 120$

44. $2x^2 - 4y^2 + 12x + 16y + 1 = 0$

In Exercises 45–54, determine the equation of the conic section with the given characteristics.

45. Parabola with vertex at the origin, passing through the point (2, 4), and symmetric with respect to the *x*-axis

46. Parabola with vertex at (2, 1), passing through the origin, and symmetric with respect to the line $x = 2$

47. Ellipse with center at the origin, minor axis of length 2, and horizontal major axis of length 5

48. Ellipse with vertices (0, ±3) and minor axis of length 2

49. Ellipse with vertices (2, −6) and (2, 2) and minor axis of length 3

50. Ellipse with vertices (−1, 2) and (5, 2) and endpoints of minor axis (2, 4) and (2, 0)

51. Hyperbola with vertices (0, ±4) and asymptotes $4y - x = 0$ and $4y + x = 0$

52. Hyperbola with vertices (2, −2) and (−2, −2) and endpoints of conjugate axis (0, 0) and (0, −4)

53. Hyperbola with asymptotes $y = \pm x$ and horizontal transverse axis of length 4

54. Hyperbola with asymptotes $y = -x$ and $y = x + 2$ and vertical transverse axis of length 4

Critical Thinking

In Exercises 55–66, without graphing each equation, determine if its graph is a parabola, circle, ellipse, hyperbola, or degenerate conic section.

55. $x^2 + 4y^2 - 8y = 0$ **56.** $x^2 + y^2 + 16x = 0$

57. $y^2 - 2x - 4y + 10 = 0$

58. $y^2 - x^2 + 6x - 2y + 1 = 0$

59. $9x^2 - y^2 + 18x - 27 = 0$

60. $x^2 - 6x - 3y + 6 = 0$

61. $16x^2 + 9y^2 - 32x - 36y + 43 = 0$

62. $9x^2 + y^2 - 18x - 10y + 30 = 0$

63. $x^2 + y^2 - 6x - 8y + 25 = 0$

64. $9x^2 + 4y^2 + 54x - 16y + 97 = 0$

65. $4x^2 - y^2 - 16x - 4y + 12 = 0$

66. $y^2 - 9x^2 + 36x - 10y - 11 = 0$

If a and b are positive real numbers, then solving $(x^2/a^2) + (y^2/b^2) = 1$ for y yields $y = \pm(b/a)\sqrt{a^2 - x^2}$. Taken separately, the equations

$$y = (b/a)\sqrt{a^2 - x^2} \quad \text{and} \quad y = -(b/a)\sqrt{a^2 - x^2}$$

define y as a function of x, and the graph of each function is a semiellipse (half an ellipse) whenever $a \neq b$.

In Exercises 67–70, state the domain and range of each function and sketch its graph.

67. $y = \frac{2}{3}\sqrt{9 - x^2}$ **68.** $y = \frac{1}{2}\sqrt{16 - x^2}$

69. $y = -3\sqrt{4 - x^2}$ **70.** $y = -\sqrt{25 - 9x^2}$

If a and b are positive real numbers, then solving $(x^2/a^2) - (y^2/b^2) = 1$ and $(y^2/b^2) - (x^2/a^2) = 1$ for y yields $y = \pm(b/a)\sqrt{x^2 - a^2}$ and $y = \pm(b/a)\sqrt{x^2 + a^2}$, respectively. Taken separately, the equations

$$y = (b/a)\sqrt{x^2 - a^2} \qquad y = -(b/a)\sqrt{x^2 - a^2}$$

$$y = (b/a)\sqrt{x^2 + a^2} \qquad y = -(b/a)\sqrt{x^2 + a^2}$$

define y as a function of x, and the graph of each function is half a hyperbola.

In Exercises 71–74, state the domain and range of each function and sketch its graph.

71. $y = \frac{1}{2}\sqrt{x^2 - 4}$ **72.** $y = -3\sqrt{x^2 - 25}$

73. $y = -\sqrt{x^2 + 9}$ **74.** $y = \frac{3}{4}\sqrt{4x^2 + 1}$

75. A road passes through a tunnel whose cross section is a semiellipse 12 feet high at the center and 30 feet wide, as shown in the sketch. What is the tallest tractor-trailer rig that can fit through the tunnel if the trailer is 10 feet wide?

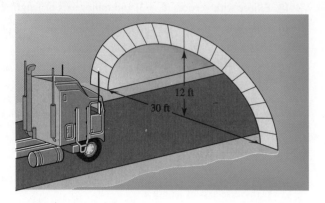

76. The hyperbolic path of a UFO that starts toward earth along the line $y = x/4$ and comes within 6000 miles of the earth's surface is shown in the figure. Assuming the radius of the earth is 4000 miles, determine the equation of the path of the UFO.

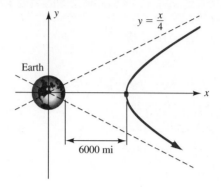

 Calculator Activities

77. Find the coordinates of the vertex of the parabola with the given equation, and round each coordinate to three significant digits.

(a) $23.4x^2 - 45.8x - 12.6y = 0$
(b) $y^2 - 2.34x - 5.74y = -6.87$

78. Find the coordinates of the vertices of the ellipse with the given equation, and round each coordinate to three significant digits.

(a) $23.4x^2 + 19.6y^2 = 87.3$
(b) $3.67x^2 + 8.67y^2 = 9.87x$

79. Find the coordinates of the vertices of the hyperbola with the given equation, and round each coordinate to three significant digits.

(a) $356x^2 - 129y^2 = 926$
(b) $1.25x^2 - 2.70y^2 + 8.55y = 2.35$

In Exercises 80–82, determine the equation of the conic section with the given characteristics.

80. Parabola with vertex $(-2.43, 1.67)$, horizontal axis of symmetry, and passing through the origin

81. Ellipse with vertices $(3.54, -5.67)$ and $(3.54, 1.53)$ and minor axis of length 2.40

82. Hyperbola with asymptotes $y = \pm 0.34x$ and horizontal transverse axis of length 6.24

83. A stream 2.3 meters deep passes through an elliptical culvert as shown in the sketch. What is the width w of the stream?

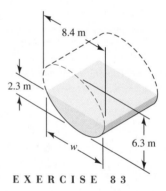

E X E R C I S E   8 3

84. In an engineer's design for the horn on a megaphone a hyperbola is rotated about the x-axis (see figure). If the distance from A to B is 10.26 cm, determine the radius of each end of the megaphone.

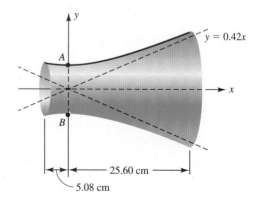

4.5 **Finding the Intersection Points of Two Curves**

Calculus can be used to find the area of a region bounded by two or more curves. However, in order to find the area, we must first find the intersection points of the curves. In this section, we discuss methods for finding these inter-section points. We begin by discussing the method for finding the coordinates of the point where two lines intersect.

◆ **Intersection Point of Two Lines**

Figure 4.46 illustrates the three possibilities for the intersection of two lines graphed on the same coordinate plane.

FIGURE 4.46

The three possibilities for the intersection of two lines graphed on the same coordinate axes.

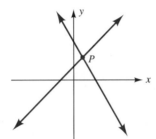 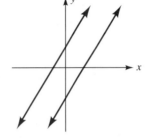 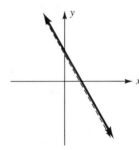

(a) The lines intersect at one point.

(b) The lines are parallel and do not intersect.

(c) The lines coincide and intersect at an infinite number of points.

If two lines are graphed on the same set of coordinate axes and intersect at point P, then the coordinates of the point P must satisfy the equation of each line. One method of finding the coordinates of the intersection point is to solve one of the equations for one of its variables and then *substitute* this quantity into the other equation to obtain one equation in one unknown. The procedure is called *solving the equations simultaneously* by using **substitution**.

EXAMPLE 1   Find the coordinates of the intersection point for the lines with the given equations.

(a) $5x + y = 3$ and $3x - 2y = 8$ **(b)** $y = 2x + 4$ and $y = 2x - 1$

SOLUTION

(a) First we graph the lines on the same set of coordinate axes by using the methods discussed in Section 4.1. According to Figure 4.47, the x-coordinate of the intersection point P appears to be about 1, while the y-coordinate appears to be between -2 and -3. One method of finding the exact coordinates of the intersection point is to solve the equations of the lines simultaneously by using substitution. Solving the equation $5x + y = 3$ for y, we obtain

$$y = 3 - 5x.$$

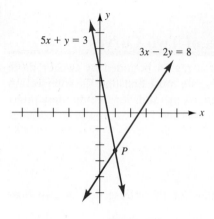

FIGURE 4.47

The lines with equations $5x + y = 3$ and $3x - 2y = 8$ intersect at point P

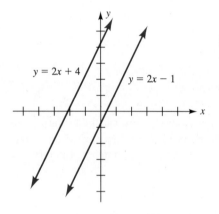

FIGURE 4.48

The lines $y = 2x + 4$ and $y = 2x - 1$ are *parallel*. Hence there is *no intersection point*.

Now, we substitute $3 - 5x$ for y in the equation $3x - 2y = 8$ and solve for x as follows:

$$3x - 2y = 8$$

$$3x - 2(3 - 5x) = 8 \qquad \text{Substitute } 3 - 5x \text{ for } y$$

$$3x - 6 + 10x = 8 \qquad \text{Solve for } x$$

$$13x = 14$$

$$x = \tfrac{14}{13}$$

To find the y-coordinate of the intersection point we substitute $\tfrac{14}{13}$ for x in the equation $y = 3 - 5x$:

$$y = 3 - 5(\tfrac{14}{13}) = 3 - \tfrac{70}{13} = -\tfrac{31}{13}$$

Thus, the lines intersect at the point $P(\tfrac{14}{13}, -\tfrac{31}{13})$.

(b)   The lines $y = 2x + 4$ and $y = 2x - 1$ have the same slope, 2, but different y-intercepts (4 and -1). Thus, we conclude the lines are *parallel* and hence have *no intersection point* (see Figure 4.48). If we solve the equations simultaneously by substituting $2x + 4$ for y in the equation $y = 2x - 1$, we find that

$$y = 2x - 1$$

$$2x + 4 = 2x - 1 \qquad \text{Substitute } 2x + 4 \text{ for } y$$

$$4 = -1 \qquad \text{Subtract } 2x \text{ on both sides}$$

| A false statement |

When the variables drop out and the resulting statement is false, we conclude that the lines are parallel and have no intersection point. ◆

PROBLEM 1   Check to see if the point $P(\tfrac{14}{13}, -\tfrac{31}{13})$ satisfies the equations of both lines in Example 1(a). ◆

We now discuss an alternative method for solving two equations simultaneously. Consider the two equations $A = B$ and $C = D$. The equation $C = D$ states that C is the same as D. Thus, if we add C to the left-hand side and D to the right-hand side of the equation $A = B$, we are actually adding the *same* quantity to both sides of the equation $A = B$. Therefore, we can state that

if $\qquad A = B$

and $\qquad \dfrac{C = D}{}$

then $\quad \overline{A + C = B + D} \qquad \text{Add the left-hand and right-hand sides}$

This idea gives us an alternative method for finding the intersection point of two lines. We solve the equations simultaneously by *adding* the left-hand and right-hand sides of the equations in order to eliminate one of the variables. To accomplish this task, it may be necessary to first multiply one of the equations by a constant so that the coefficient of a variable (either x or y) in one equation is the negative of the same variable's coefficient in the other equation. In fact, we must sometimes multiply both equations (by different constants) to obtain coefficients of a variable that can be eliminated by adding the equations. We refer to the method as the **addition method** and illustrate the procedure in the next example.

EXAMPLE 2 Find the coordinates of the intersection point for the lines with the given equations.

(a) $3x - 3y = 10$ and $5x + 6y = 2$

(b) $-6x + 2y = 1$ and $9x - 3y = -\frac{3}{2}$

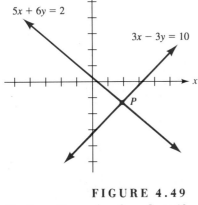

FIGURE 4.49
The lines with equations $3x - 3y = 10$ and $5x + 6y = 2$ intersect at point P.

SOLUTION

(a) The lines are sketched in Figure 4.49 and intersect at point P. The x-coordinate of P appears to be about 2, and the y-coordinate of P appears to be between -1 and -2. One method of finding the exact coordinates of the intersection point is to use the addition method and proceed as follows:

$$3x - 3y = 10 \xrightarrow{\substack{\text{Multiply both} \\ \text{sides by 2}}} 6x - 6y = 20$$
$$5x + 6y = 2 \xrightarrow{\text{Leave as is}} \underline{5x + 6y = 2}$$
$$11x + 0y = 22 \qquad \textbf{Add columns}$$
$$x = 2 \qquad \textbf{Solve for } x$$

To find the y-coordinate of the intersection point, we substitute 2 for x in either of the original equations and solve for y. Replacing x with 2 in the equation $3x - 3y = 10$, we find $y = -\frac{4}{3}$. Thus, the lines intersect at the point $P(2, -\frac{4}{3})$.

(b) Writing both equations in slope-intercept form, $y = mx + b$, we find

$$-6x + 2y = 1 \qquad\qquad 9x - 3y = -\tfrac{3}{2}$$
$$2y = 6x + 1 \qquad\qquad -3y = -9x - \tfrac{3}{2}$$
$$y = 3x + \tfrac{1}{2}. \qquad\qquad y = 3x + \tfrac{1}{2}.$$

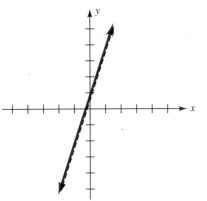

FIGURE 4.50
The lines $-6x + 2y = 1$ and $9x - 3y = -\frac{3}{2}$ *coincide*. Hence, the lines have an *infinite number of intersection points*.

Since these equations have the same slope and y-intercept, we conclude that the lines *coincide* (see Figure 4.50). Thus, every point on each line satisfies the equation of the other line, and we conclude that there is an *infinite number of intersection points*.

If we solve the equations simultaneously by using the addition method, we find that

$$-6x + 2y = 1 \xrightarrow{\text{Multiply both sides by 3}} -18x + 6y = 3$$
$$9x - 3y = -\tfrac{3}{2} \xrightarrow{\text{Multiply both sides by 2}} \underline{18x - 6y = -3}$$
$$0 = 0$$

A true statement

When the variables drop out and the resulting statement is true, we conclude that the lines coincide and have an infinite number of intersection points. ◆

PROBLEM 2   Find the coordinates of the intersection point for the lines with the equations $y = 2x - 3$ and $y = 4x + 5$ by using each method.

(a) substitution   **(b)** the addition method   ◆

◆ **Intersection Points of a Line and a Parabola**

Figure 4.51 illustrates the three possibilities for the intersection of a line and a parabola sketched on the same coordinate plane.

FIGURE 4.51

The three possibilities for the intersection of a line and a parabola graphed on the same coordinate axes.

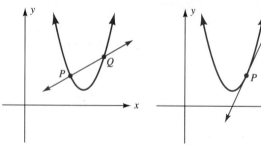

(a) The line and the parabola intersect at two points.

(b) The line and the parabola intersect at one point.

(c) The line and the parabola do not intersect.

Note: Although the notion of *tangency* is dealt with in calculus, we mention the idea at this time: If a line is not parallel to the axis of a parabola and touches a parabola at only one point, as shown in Figure 4.51(b), then the line is said to be ***tangent*** to the parabola at that point.

We may find the coordinates of the intersection points of a line and a parabola by solving the equations of the curves simultaneously, by using substitution. The coordinates of the intersection points must satisfy *both* the equation of the line and the equation of the parabola.

EXAMPLE 3   Find the coordinates of the intersection points for the line and the parabola with the given equations.

(a) $3x - 2y = 2$ and $y = x^2 - 2x - 3$

(b) $y = x + 1$ and $x = y^2 - 2y + 3$

SOLUTION

(a) We use the methods discussed in Sections 4.1 and 4.3 to sketch the graphs of the line and the parabola, as shown in Figure 4.52. To find the coordinates of the intersection points P and Q, we solve the equations simultaneously by using substitution:

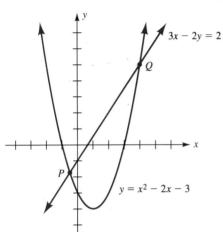

FIGURE 4.52

The line $3x - 2y = 2$ and the parabola $y = x^2 - 2x - 3$ intersect at points P and Q.

$$3x - 2y = 2$$

$$3x - 2(x^2 - 2x - 3) = 2 \qquad \text{Substitute } x^2 - 2x - 3 \text{ for } y$$

$$3x - 2x^2 + 4x + 6 = 2 \qquad \text{Simplify}$$

$$2x^2 - 7x - 4 = 0 \qquad \begin{array}{l}\text{Write the quadratic equation}\\\text{in standard form}\end{array}$$

$$(2x + 1)(x - 4) = 0 \qquad \text{Solve the quadratic equation}$$

$$2x + 1 = 0 \quad \text{or} \quad x - 4 = 0$$

$$x = -\tfrac{1}{2} \qquad\qquad x = 4$$

$$\boxed{\begin{array}{l}\text{x-coordinates of the}\\\text{intersection points}\end{array}}$$

To find the y-coordinates of the intersection points that correspond to these x-coordinates, we substitute the x-values of $-\tfrac{1}{2}$ and 4 in either of the original equations and solve for y. Substituting $x = -\tfrac{1}{2}$ and $x = 4$ into the equation $y = x^2 - 2x - 3$, we have

$$y = (-\tfrac{1}{2})^2 - 2(-\tfrac{1}{2}) - 3 = -\tfrac{7}{4} \quad \text{and} \quad y = (4)^2 - 2(4) - 3 = 5.$$

$$\boxed{\begin{array}{l}\text{y-coordinates of the}\\\text{intersection points}\end{array}}$$

Thus, the line and the parabola intersect at the points $P(-\tfrac{1}{2}, -\tfrac{7}{4})$ and $Q(4, 5)$.

(b) The graphs of the line $y = x + 1$ and the parabola $x = y^2 - 2y + 3$ are shown in Figure 4.53 and do not appear to intersect. If we solve the equations simultaneously by substitution, we find

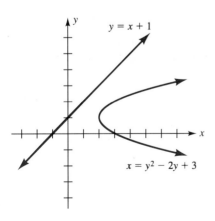

FIGURE 4.53

The line $y = x + 1$ and the parabola $x = y^2 - 2y + 3$ do *not* intersect.

$$y = x + 1$$

$$y = (y^2 - 2y + 3) + 1 \qquad \text{Substitute } y^2 - 2y + 3 \text{ for } x$$

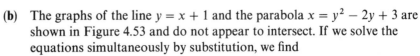
EXAMPLE 3(b) (*continued*)

$$y^2 - 3y + 4 = 0$$

Write the quadratic equation in standard form and solve by the quadratic formula

$$y = \frac{-(-3) \pm \sqrt{(-3)^2 - 4(1)(4)}}{2(1)} = \frac{3 \pm \sqrt{-7}}{2}$$

$\boxed{\text{Not a real solution}}$

If the equations of a line and a parabola have no real solution when solved simultaneously, then the line and the parabola do not intersect.   ◆

PROBLEM 3   Repeat Example 3 for the line $4x - y = 6$ and the parabola $y = x^2 - 2x + 3$.   ◆

◆   **Intersection Points of Other Curves**

As illustrated in the next example, substitution and the addition method can be used to find the intersection points of other curves as well.

EXAMPLE 4   Find the coordinates of the intersection points of the curves with the given equations.

(a) $y = \sqrt{x}$ and $x + y = 6$ (b) $x^2 + y^2 = 4$ and $x^2 - 3y = 0$

SOLUTION

(a)   The graphs of the square root function $y = \sqrt{x}$ (Section 3.4) and the line $x + y = 6$ intersect a point P, as shown in Figure 4.54. To find the coordinates of the intersection point P, we solve the equations simultaneously by substituting $\sqrt{x}$ for y in the equation $x + y = 6$ as follows:

$$x + y = 6$$

$$x + \sqrt{x} = 6 \qquad \text{Substitute } \sqrt{x} \text{ for } y$$

$$\sqrt{x} = 6 - x \qquad \text{Isolate the radical}$$

$$x = 36 - 12x + x^2 \qquad \text{Square both sides}$$

$\boxed{\text{Squaring both sides may introduce extraneous roots.}}$

$$0 = 36 - 13x + x^2 \qquad \text{Subtract } x \text{ from both sides}$$

$$0 = (9 - x)(4 - x) \qquad \text{Solve the quadratic equation}$$

$$9 - x = 0 \quad \text{or} \quad 4 - x = 0$$

$$x = 9 \qquad\qquad x = 4$$

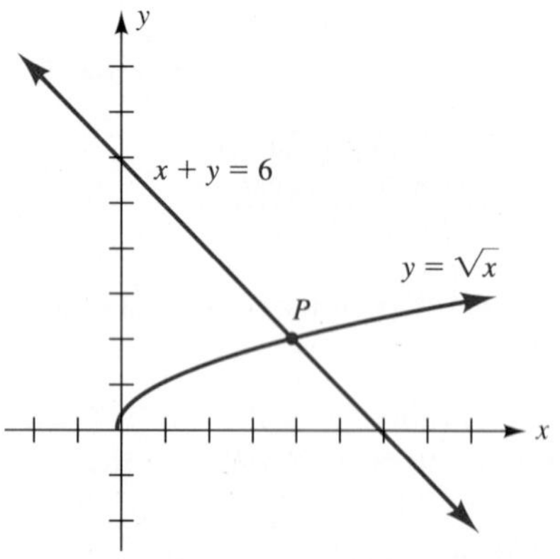

FIGURE 4.54
The curves with the equations $y = \sqrt{x}$ and $x + y = 6$ intersect at point P.

Notice there appear to be two x-coordinates, but this disagrees with Figure 4.54. Substituting $x = 9$ into the equation $y = \sqrt{x}$, we find

$$y = \sqrt{9} = 3.$$

However, the coordinates $(9, 3)$ do not satisfy the equation $x + y = 6$. Thus, $(9, 3)$ is not an intersection point of these curves. Substituting $x = 4$ into the equation $y = \sqrt{x}$, we find

$$y = \sqrt{4} = 2.$$

Since the coordinates $(4, 2)$ also satisfy the equation $x + y = 6$, we conclude that the only intersection point of these two curves is $P(4, 2)$.

(b) Recall from Section 3.2 that $x^2 + y^2 = 4$ is the equation of a circle with center at the origin and radius 2. Solving the equation $x^2 - 3y = 0$ for y, we obtain

$$y = \tfrac{1}{3}x^2,$$

which is the equation of a parabola that has its vertex at the origin and opens upward. A sketch of the circle and parabola is shown in Figure 4.55. As shown in the figure, these curves intersect at points P and Q. To find the coordinates of the intersection points, we solve the equations simultaneously by using the addition method:

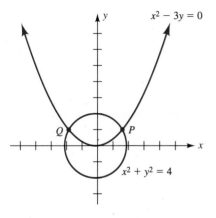

FIGURE 4.55

The curves with the equations $x^2 + y^2 = 4$ and $x^2 - 3y = 0$ intersect at points P and Q.

$$
\begin{array}{llll}
x^2 + y^2 = 4 & \xrightarrow{\text{Leave as is}} & x^2 + y^2 = 4 & \\
x^2 - 3y = 0 & \xrightarrow{\text{Multiply both sides by } -1} & -x^2 + 3y = 0 & \\
& & \overline{\; y^2 + 3y = 4} & \begin{array}{l}\text{Add and solve}\\ \text{the quadratic}\\ \text{equation for } y.\end{array}
\end{array}
$$

$$y^2 + 3y - 4 = 0$$

$$(y - 1)(y + 4) = 0$$

$$y - 1 = 0 \quad \text{or} \quad y + 4 = 0$$

$$y = 1 \qquad\qquad y = -4.$$

Substituting $y = 1$ into the equation $x^2 - 3y = 0$ gives us

$$x^2 = 3 \qquad \text{or} \qquad x = \pm\sqrt{3}.$$

Substituting $y = -4$ into the equation $x^2 - 3y = 0$ gives us

$$x^2 = -12,$$

which has no real solution. Thus, the only intersection points of the given curves are $P(\sqrt{3}, 1)$ and $Q(-\sqrt{3}, 1)$. ◆

PROBLEM 4   Repeat Example 4 for the parabolas with the equations $y = x^2$ and $y = 4x - x^2$.

◆

◆ **Approximating the Intersection Points of Two Curves**

The substitution procedure sometimes yields an equation in one unknown that may be quite challenging to solve. For problems of this nature, we can use a calculator or computer with graphing capability to find the approximate points of intersection.

EXAMPLE 5   Find the approximate coordinates of the intersection points for the curves with equations $4x^2 + y^2 = 16$ and $y = x^3$.

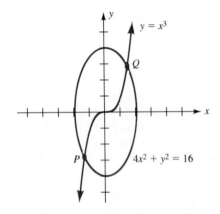

FIGURE 4.56

The curves with the equations $4x^2 + y^2 = 16$ and $y = x^3$ intersect at points P and Q.

SOLUTION   The graphs of the ellipse $4x^2 + y^2 = 16$ (see Section 4.4) and the cubing function $y = x^3$ (see Section 3.4) intersect at two points P and Q, as shown in Figure 4.56. To determine the coordinates of points P and Q, we substitute x^3 for y in the equation $4x^2 + y^2 = 16$, and obtain

$$4x^2 + (x^3)^2 = 16 \qquad \text{or} \qquad x^6 + 4x^2 - 16 = 0.$$

Note that the equation $x^6 + 4x^2 - 16 = 0$ is not of quadratic type, and hence, the chance of solving this equation by any of the algebraic methods discussed in Chapter 2 seems bleak. However, we can approximate the coordinates of P and Q by using a graphing calculator.

We begin by choosing a viewing rectangle by pressing the [RANGE] key. Figure 4.56 suggests a viewing rectangle:

$$[-2, 2] \qquad \text{by} \qquad [-4, 4]$$

x-min   x-max   y-min   y-max

To display the graph of a conic section, such as an ellipse, on a graphing calculator, we must solve the equation of the ellipse for y and then graph the two functions that result. Solving the equation $4x^2 + y^2 = 16$ for y, we obtain

$$y = \pm 2\sqrt{4 - x^2}.$$

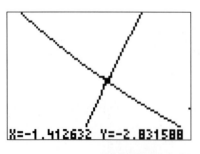

X=-1.412632 Y=-2.831588

FIGURE 4.57

Picture in viewing rectangle after *zooming in* by a factor of 10.

Now, by entering the ellipse equations $y = 2\sqrt{4 - x^2}$ and $y = -2\sqrt{4 - x^2}$ and the cubing function $y = x^3$, we obtain their graphs in the viewing rectangle.

Next, we activate the [TRACE] key and move the blinking cursor to the first intersection point of the two curves, point P. To the nearest tenth, we find that the coordinates of P are $(-1.4, -2.8)$. To obtain more precise values, we may zoom in on point P by using the Zoom feature on the graphing calculator. Figure 4.57 illustrates the viewing rectangle after we zoom in on point P by a factor of 10. After repeating this process of tracing and zooming in, we find that the coordinates of point P are $(-1.414, -2.828)$ to the nearest thousandth. By symmetry, we know that the coordinates of point Q are the same as the coordinates of point P, except with opposite signs. In summary, the approximate intersection points are, to the nearest thousandth,

$$P(-1.414, -2.828) \qquad \text{and} \qquad Q(1.414, 2.828).$$ ◆

PROBLEM 5   Referring to Example 5, when we substitute x^3 for y in the equation $4x^2 + y^2 = 16$, we obtain $x^6 + 4x^2 - 16 = 0$. With some *creative* factoring, we can find the *exact* solutions of this equation. First, we rewrite this equation as

$$(x^6 - 8) + 4x^2 - 8 = 0.$$

Now, factor $x^6 - 8$ as the difference of cubes (Section 1.6) and then look for a common binomial factor between this expression and $4x^2 - 8$. What are the exact coordinates of the intersection points P and Q? ◆

Exercises 4.5

Basic Skills

In Exercises 1–12, determine the coordinates of any intersection point for the lines with the given equations.

1. $3x + 5y = 8$ and $x = 5y$

2. $y = x + 6$ and $3x + 5y = 4$

3. $y = 8x + 3$ and $2x - 3y = 2$

4. $y = 3x - 5$ and $y = 5 - 2x$

5. $y = 4 - 3x$ and $y = 3 - 5x$

6. $y + 2x = 0$ and $3x - 2y = 14$

7. $7x - y = 2$ and $5x - y = 8$

8. $2x + 3y = 7$ and $3x - 2y = 4$

9. $3x - 3y = 10$ and $x - y = 1$

10. $4x - 3y = 8$ and $4x - 2y = 7$

11. $5x - 2y = 10$ and $3x + 6y = 12$

12. $4y = x + 6$ and $3x - 12y = -18$

In Exercises 13–28, determine the coordinates of any intersection point for the line and parabola with the given equations.

13. $y = 2x$ and $y = 3 - x^2$

14. $x + y = 6$ and $y = x^2$

15. $2x - y = 3$ and $x = 2y^2$

16. $x - y = 1$ and $x = 3 - y^2$

17. $y = x + 3$ and $y = 2x^2 - 3$

18. $4x - 3y = 4$ and $y^2 = 4x$

19. $3x - 2y = 5$ and $y = x^2 - 4x$

20. $x + y = -6$ and $y = x^2 - 2x - 8$

21. $y = \frac{5}{2}x + 2$ and $y = 3 - x - 2x^2$

22. $y = \frac{1}{4}x + 2$ and $y = \frac{1}{4}x^2 - x + 3$

23. $x - 2y = -1$ and $x = y^2 - 6y + 15$

24. $6x - 2y = 31$ and $x^2 - 2y - 6x + 5 = 0$

25. $2y = x$ and $x^2 - y + 6x + 5 = 0$

26. $y = 3x$ and $x = 1 - y^2$

27. $y = \frac{1}{3}x + 2$ and $y^2 = 2x$

28. $5x + 2y = 10$ and $y = 4 - 2x^2$

In Exercises 29–58, determine the coordinates of the intersection points for the curves with the given equations.

29. $y = \frac{1}{4}x^2$ and $x = \frac{1}{4}y^2$

30. $y = x^2 - 4x$ and $y = 6x - x^2$

31. $y = x^2 - 4x + 3$ and $y = 3 + 2x - x^2$

32. $y = x^2$ and $x^2 - 4x + 3y - 8 = 0$

33. $x^2 + y^2 = 36$ and $y^2 - 5x = 0$

34. $x^2 + y^2 = 1$ and $y = 2x^2 - 1$

35. $x^2 + y^2 = 25$ and $y = \frac{1}{3}x + 3$

36. $x^2 + y^2 = 4$ and $y = 2/x$

37. $y = \dfrac{1}{x}$ and $2x - y = 1$

38. $y = \sqrt{x}$ and $y = x - 2$

39. $x - 2y = 4$ and $y = \sqrt{x - 4}$

40. $y = \sqrt{x + 4}$ and $x = -y^2$

41. $y = \frac{1}{3}x + 1$ and $y = \sqrt{x}$

42. $x + y = 3$ and $xy = 3$

43. $9x^2 + y^2 = 36$ and $x^2 + y^2 = 28$

44. $x^2 + 2y^2 = 18$ and $2x^2 + y^2 = 15$

45. $y = x + 1$ and $x^2 + 4y^2 = 4$

46. $y = x^2 - 1$ and $4x^2 + y^2 = 16$

47. $x^2 + 2y^2 = 8$ and $y = \sqrt{x}$

48. $9x^2 + 4y^2 = 25$ and $y = 2/x$

49. $9y^2 - 4x^2 = 36$ and $2x^2 + y^2 = 4$

50. $4x^2 - y^2 = 16$ and $x^2 + y^2 = 9$

51. $y = x - 2$ and $x^2 - y^2 = 1$

52. $y = x^2 - 1$ and $x^2 - 4y^2 = 1$

53. $x^2 - 5y^2 = 36$ and $x = y^2$

54. $10y^2 - x^2 = 16$ and $y = \sqrt{x}$

55. $4x^2 + y^2 - 8x = 0$ and $y = 6x - 4$

56. $x^2 + 4y^2 - 6x + 5 = 0$ and $y = 2x - 5$

57. $x + y = -4$ and $x^2 - 4y^2 - 6x = 0$

58. $x^2 - y^2 - 4x + 6y + 4 = 0$ and $x + 2y = 2$

 Critical Thinking

59. Find the coordinates of the vertices of a triangle whose sides lie on the lines $x + y = 4$, $3x - y = 4$, and $x - y = 6$.

60. Show that the triangle in Exercise 59 is a right triangle, then find its area.

61. Two sides of a parallelogram lie on the lines $2x - y = -3$ and $x + 2y = 11$ and one of its vertices is $(7, -3)$. Find the coordinates of the other three vertices of the parallelogram.

62. Show that the parallelogram in Exercise 61 is a rectangle, then find its area.

63. Find the length of the line segment that joins the intersection points of the line $y = 2x + 3$ and the parabola opening upward with vertex $(-2, -4)$ and y-intercept 0.

64. Find the coordinates of the midpoint of the line segment that joins the intersection points of the parabola $x = 3 - y^2$ and the line having slope 1 and y-intercept -1.

In Exercises 65–70, find the coordinates of the intersection points P, Q, R and, when applicable, S.

66.

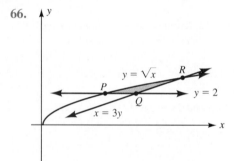

67.

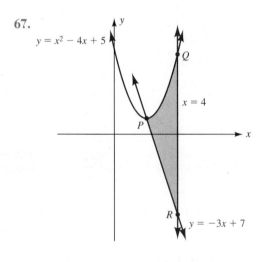

65.

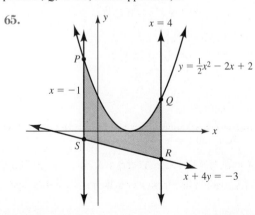

68.

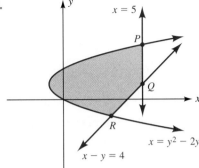

69.

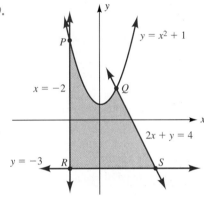

70.

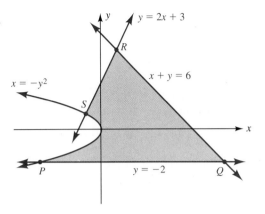

Calculator Activities

In Exercises 71-76, approximate (to the nearest hundredth) the coordinates of the intersection points of the given curves. If you have access to a graphing calculator, check each answer by tracing to the points of intersection.

71. $y = 9.24x$ and $y = 18.21x - 32.40$

72. $2.65x + 5.95y = 12.63$ and $y = 4.02x - 1.63$

73. $y = 6.09 - 2.72x$ and $y = 3.60x^2 - 2.13x$

74. $27.3x - 42.3y = 75.6$ and $y^2 = 0.325x$

75. $x^2 + y^2 = 295.6$ and $y = 9.8x - 17.2$

76. $2.8x^2 + 2.9y^2 = 14.6$ and $y = \sqrt{1.4x}$

In Exercises 77–84, use a graphing calculator to approximate

(to the nearest thousandth) the coordinates of the inter-section points of the given curves.

77. $y = 3x - 1$ and $y = x^3$

78. $y = 2x^2 - 3x - 5$ and $y = x^3$

79. $x^2 + y^2 = 16$ and $y = x^2 - 2x$

80. $x^2 - y^2 = 4$ and $y = 4x - x^2$

81. $x^2 + 4y^2 - 4x = 0$ and $y = 1/x$

82. $y^2 + 16x = 0$ and $y^2 - 9x^2 - 18x = 18$

83. $y = 4 - 3x - x^2$ and $y = 2x^4$

84. $y = \sqrt[3]{x}$ and $y = 1 - x^2$

Chapter   4   Review

Questions for Group Discussion

1. What is meant by the *slope* of a straight line? Does every line have a slope? Explain.

2. How can you find the slope of a straight line from its equation? from its graph?

3. Describe the graph of a *linear function*. Explain the procedure for graphing a linear function.

4. Explain the procedure for finding the equation of a straight line when two points on the line are given.

5. Draw two lines on the same set of coordinate axes that have the given conditions.
   (a) negative reciprocal slopes and the same *y*-intercept
   (b) the same slope and different *y*-intercepts

6. Describe the graph of a *quadratic function*. Explain the procedure for graphing a quadratic function.

7. How can we determine whether a quadratic function has a *maximum* or *minimum value*? Explain how the maximum or minimum value can be found.

8. How can you tell from the equation of a *parabola* whether the curve opens upward, downward, to the left, or to the right?

9. Illustrate with sketches some possibilities of intersection for two lines graphed on the same set of coordinate axes.

10. Describe methods that can be used to determine the coordinates of the intersection points of two curves.

11. What can we conclude if the equations of two curves have no real solution when solved simultaneously?

12. Name the four main *conic sections*. Is the graph of every equation of the form $Ax^2 + Cy^2 + Dx + Ey + F = 0$ a conic section? Explain.

13. Show how the *degenerate conic sections* can be formed by slicing a double cone with a plane.

14. What is meant by the *major axis* of an ellipse? How can you determine from the equation of an ellipse in standard form whether the major axis is horizontal or vertical?

15. What is meant by the *transverse axis* of a hyperbola? How can you determine from the equation of a hyperbola in standard form whether the transverse axis is horizontal or vertical?

16. How can *asymptotes* be used as an aid in graphing a hyperbola?

Review Exercises

In Exercises 1–24, sketch the graph of each equation. If the graph is a line, give its slope and label any x-intercept and y-intercept. If the graph is a parabola, label its vertex and any x- and y-intercepts. If the graph is an ellipse or hyperbola, label its center and vertices.

1. $y = 2x - 3$

2. $y = 1 - 4x$

3. $y = -\frac{3}{4}x - 1$

4. $y = \frac{2}{3}x + 2$

5. $y = (x - 2)^2 - 3$

6. $y = -2(x + 1)^2 + 2$

7. $y = 15 + 2x - x^2$

8. $y = 2x^2 + x - 10$

9. $x = 2y^2 + 4y - 3$

10. $x = 6y - y^2$

11. $6x - 2y = 8$

12. $3x - 4y + 18 = 0$

13. $3y - 18 = 0$

14. $2x + 5 = 0$

15. $x^2 + 2x - 2y + 5 = 0$

16. $2y^2 - 8y - 3x + 5 = 0$

17. $x^2 + 9y^2 = 16$

18. $4x^2 + 3y^2 = 9$

19. $x^2 - y^2 = 25$

20. $4y^2 - x^2 = 49$

21. $4x^2 + y^2 - 8x + 4y - 8 = 0$

22. $x^2 + 16y^2 - 32y - 9 = 0$

23. $4y^2 - 9x^2 - 8y + 5 = 0$

24. $16x^2 - 9y^2 - 32x - 36y - 24 = 0$

In Exercises 25–40, determine the equation from the given information.

25. Line with slope $\frac{3}{4}$ and y-intercept -2

26. Line with x-intercept 4 and y-intercept -2

27. Line through $(-2, 4)$ with slope 3

28. Line through the origin and parallel to $x - 3y = 4$

29. Line through $(-2, 1)$ and perpendicular to the line through $(1, 1)$ and $(4, 3)$

30. Line through $(-3, 2)$ and perpendicular to the x-axis

31. Parabola with vertex $(0, -4)$ and x-intercepts ± 2

32. Parabola with vertex $(2, 1)$ and y-intercepts -1 and 3

33. Parabola passing through the origin, with vertex $(-2, -3)$ and symmetric with respect to $y = -3$

34. Parabola passing through $(-1, 0)$, with vertex $(2, 3)$ and symmetric with respect to $x = 2$.

35. Ellipse with vertices $(0, 0)$ and $(6, 0)$ and minor axis of length 4

36. Ellipse with center $(2, 5)$, minor axis of length 2, and horizontal major axis of length 8

37. Hyperbola with asymptotes $y = \pm \frac{2}{3}x$ and vertical transverse axis of length 6

38. Hyperbola with center $(2, -1)$, horizontal transverse axis of length 10, and conjugate axis of length 4

39. Line through the vertices of $y = x^2 - 3x + 4$ and $y = -2x^2 + 5x$

40. Line through the centers of $x^2 + y^2 + 4x - 6y = 0$ and $4x^2 - y^2 + 8x - 8y - 28 = 0$

In Exercises 41–50, determine the intersection points of the curves with the given equations.

41. $y = 4x - 5$ and $2x - 3y = 10$

42. $3x - 4y = 6$ and $6x - 3y = 7$

43. $y = x^2 - 4x + 6$ and $y = x + 2$

44. $x = 4y^2 - 8y$ and $x - 8y = -12$

45. $x^2 + y^2 = 36$ and $y = x^2 - 6$

46. $y = 2 - 2x$ and $4x^2 + 9y^2 = 36$

47. $y^2 - x^2 = 1$ and $x^2 + 4y^2 = 9$

48. $y = \sqrt{x + 1}$ and $x^2 + 4y^2 = 9$

49. $x^2 - 4y^2 - 2x - 15 = 0$ and $x - 4y = 0$

50. $y = x^2 - 2x - 3$ and $4x^2 - 12x + 3y = 0$

In Exercises 51 and 52, use a calculator or computer with graphing capability to approximate (to the nearest thousandth) the coordinates of the intersection points of the given curves.

51. $3x + 2y = 6$ and $y = x^3$

52. $4x^2 + y^2 = 16$ and $y = x^2 - 4x$

53. Given the two equations $A_1x + B_1y = C_1$ and $A_2x + B_2 y = C_2$, solve the equations simultaneously (a) for x and (b) for y.

54. Given the two equations $A_1x^2 + B_1y^2 = C_1$ and $A_2x^2 + B_2y^2 = C_2$, solve the equations simultaneously (a) for x and (b) for y.

55. The graph of the line $Ax + By + C = 0$ with $A \neq 0$ and $B \neq 0$ is sketched in the figure.

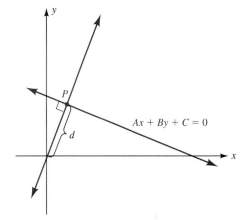

(a) Determine the equation of the line that passes through the origin and is perpendicular to the line $Ax + By + C = 0$.

(b) Determine the coordinates of the intersection point P.

(c) Show that the distance d from the line $Ax + By + C = 0$ to the origin is given by

$$d = \frac{|C|}{\sqrt{A^2 + B^2}}.$$

56. Use the formula from Exercise 55(c) to find

(a) the distance from the origin to the line $3x + 4y - 12 = 0$.

(b) the distance between the parallel lines $3x + y = 4$ and $3x + y = 1$.

57. Two sides of a parallelogram lie along the lines $y = x + 4$ and $x + 2y = 2$. One diagonal of the parallelogram is on the line $5x - 2y = -4$. What are the coordinates of the vertices of the parallelogram?

58. Find the area of the triangle bounded by the lines $2x - y = 6$, $x + 2y = 18$, and $3x + y = 4$.

59. Determine whether the quadratic function $f(x) = 3x^2 - 4x + 5$ has a maximum or minimum value. Then find this value.

60. Determine whether the quadratic function $f(x) = 5x - 2x^2$ has a maximum or minimum value. Then find this value.

61. Determine the value of a such that the quadratic function $f(x) = ax^2 - 4x + 3$ has a maximum value of 5.

62. Let $f(x) = 2x + 3$ and $g(x) = x^2 - 3x + 4$.
    (a) Determine the minimum value of $f \circ g$.
    (b) Determine the minimum value of $g \circ f$.

63. The value of a personal computer depreciates linearly from $2400 at time of purchase to $400 after 5 years.
    (a) Express the value V of the computer as a function of its age x in years.
    (b) What is the age of the computer when its value is fully depreciated?

64. Each week a salesperson gets paid a base salary plus commission, which is a fixed percentage of her sales. When her sales are $1000 her total pay is $280, and when her sales are $3000 her total pay is $440.
    (a) Express her total pay P as a function of her sales x.
    (b) What is her base salary?
    (c) What is the significance of the slope of the graph of the function defined in part (a)?

65. In a certain city, the yearly cost C for water that is supplied to each household is a function of the number x of gallons that are used. Suppose the cost of water is given by the following rate schedule.

Usage	Cost
First 5000 gallons or less	$120
Next 10,000 gallons	$0.018 per gallon
Next 15,000 gallons	$0.011 per gallon
All gallons over 30,000	$0.003 per gallon

Find the piecewise linear function that describes this rate schedule, then sketch its graph.

66. Five hundred feet of fencing is to be used to build three adjacent rectangular corrals (see figure). Find the overall dimensions of the corrals if the total enclosed area is to be the largest possible.

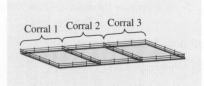

EXERCISE 66

67. A triangular piece of land has 400 feet of frontage on a tar road and 250 feet of frontage on a gravel road that is perpendicular to the tar road. What are the dimensions of the largest rectangular building lot that can be inscribed in this triangular piece of land if the sides of the rectangular lot lie along the tar and gravel roads?

68. When the owner of an apartment complex charges $600 per month for rent, all 60 apartments are occupied. Research shows that for each $50 increase in the monthly rent, three apartments become vacant. What monthly charge for rent gives the owner maximum revenue?

69. The cable that supports the weight of a suspension bridge approximates a parabola with vertex halfway between the supports, as shown in the sketch. Determine the length x of the vertical cable that is 30 meters from the end of a support.

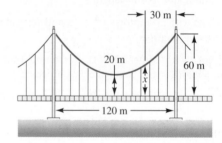

70. A square is inscribed in the ellipse $9x^2 + 16y^2 = 144$ that is sketched in the figure. What is the length of the side of the square?

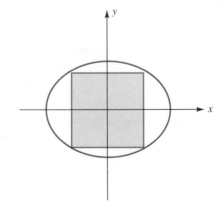

CHAPTER 5

Polynomial and Rational Functions

A silo has the shape of a right circular cylinder surmounted by a hemisphere and holds 1000π cubic feet of grain. If the height of the silo is 60 feet, what is the radius of its base?

(For the solution, see Example 5 in Section 5.4.)

5.1   Polynomial Division

◆ Introductory Comments

Any function P defined by

$$P(x) = a_n x^n + a_{n-1} x^{n-1} + a_{n-2} x^{n-2} + \cdots + a_2 x^2 + a_1 x + a_0$$

where $a_0, a_1, a_2, \ldots, a_n$ are real numbers with $a_n \neq 0$ and n is a nonnegative integer, is called a **polynomial function of degree** n with real coefficients.

A polynomial function of degree 0 is a constant function $f(x) = a_0$; of degree 1 is a linear function $g(x) = a_1 x + a_0$; and of degree 2 is a quadratic function $h(x) = a_2 x^2 + a_1 x + a_0$. Since we have discussed constant functions, linear functions, and quadratic functions in previous chapters, we devote this chapter to polynomial functions of degree 3 or greater, such as the following:

1. Cubic function: $F(x) = a_3 x^3 + a_2 x^2 + a_1 x + a_0$
2. Quartic function: $G(x) = a_4 x^4 + a_3 x^3 + a_2 x^2 + a_1 x + a_0$
3. Quintic function: $H(x) = a_5 x^5 + a_4 x^4 + a_3 x^3 + a_2 x^2 + a_1 x + a_0$

In order to work with polynomial functions of degree 3 or greater, we must be able to divide one polynomial by another. Thus, we begin with a discussion of *polynomial division.*

◆ Polynomial Long Division

To divide 349 by 15, we can use the long division scheme from arithmetic:

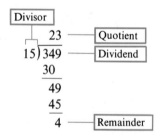

Since 23 is the quotient and 4 is the remainder, we write

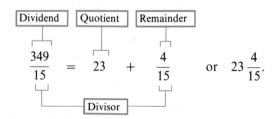

In a similar manner, we can divide one polynomial $P(x)$ by another polynomial $D(x)$, where $D(x) \neq 0$. This long division scheme, which is called the **division algorithm for polynomials**, yields a quotient polynomial $Q(x)$ and a remainder polynomial $R(x)$ of degree less than that of the divisor $D(x)$ such that

$$\frac{P(x)}{D(x)} = Q(x) + \frac{R(x)}{D(x)}$$

EXAMPLE 1   Divide $P(x) = 2x^3 - 5x^2 + 3x - 4$ by $D(x) = x - 2$ using polynomial long division.

SOLUTION   We begin by dividing the first term of the divisor, x, into the first term of the dividend, $2x^3$, to obtain the first term of the quotient:

$$
\begin{array}{r}
\boxed{\text{Divisor } D(x)} \qquad \boxed{\text{First term of quotient } Q(x)} \\[4pt]
2x^2 \phantom{{}-5x^2+3x-4} \\
x - 2 \overline{\smash{)}\, 2x^3 - 5x^2 + 3x - 4} \quad \boxed{\text{Dividend } P(x)}
\end{array}
$$

Next, we multiply the first term of the quotient by the divisor and *subtract* this product from the dividend to obtain the first remainder:

$$
\begin{array}{r}
\boxed{\text{Divisor } D(x)} \qquad \boxed{\text{First term of quotient } Q(x)} \\[4pt]
2x^2 \phantom{{}-5x^2+3x-4} \\
x - 2 \overline{\smash{)}\, 2x^3 - 5x^2 + 3x - 4} \quad \boxed{\text{Dividend } P(x)} \\
\underline{2x^3 - 4x^2} \phantom{{}+3x-4} \\
-x^2 + 3x - 4 \quad \boxed{\text{First remainder}}
\end{array}
$$

Now, using the first remainder as the *new* dividend, we continue this process until a remainder occurs that is either zero or a polynomial of degree lower than that of the divisor.

$$
\begin{array}{r}
\boxed{\text{Divisor } D(x)} \\[4pt]
2x^2 - x + 1 \quad \boxed{\text{Quotient } Q(x)} \\
x - 2 \overline{\smash{)}\, 2x^3 - 5x^2 + 3x - 4} \quad \boxed{\text{Dividend } P(x)} \\
\underline{2x^3 - 4x^2} \phantom{{}+3x-4} \\
-x^2 + 3x - 4 \\
\underline{-x^2 + 2x} \phantom{{}-4} \\
x - 4 \\
\underline{x - 2} \\
-2 \quad \boxed{\text{Remainder } R(x)}
\end{array}
$$

Thus, we write

$$\frac{2x^3 - 5x^2 + 3x - 4}{x - 2} = 2x^2 - x + 1 + \frac{-2}{x - 2}.$$

◆

To check polynomial long division, we use the same method of checking as for arithmetic division, and show that

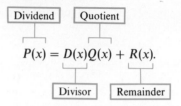

$$P(x) = D(x)Q(x) + R(x).$$

PROBLEM 1   Check the long division in Example 1.   ◆

Before beginning the long division process, we usually write the dividend and divisor in descending powers of x, using 0 as a coefficient of any term that is missing from either the dividend or divisor. This helps us to align like terms in the same column when we carry out the division.

EXAMPLE 2   Divide $P(x) = 3x^5 + 4x^4 - 2x^2 + 3x - 2$ by $D(x) = x^2 - 1$ using long division.

SOLUTION   First arranging the dividend and divisor in descending powers of x, and using 0 as a coefficient of any missing term, we then carry out the long division as follows:

$$
\begin{array}{r}
3x^3 + 4x^2 + 3x \ + 2 \qquad\qquad Q(x) \\
x^2 + 0x - 1 \overline{)\,3x^5 + 4x^4 + 0x^3 - 2x^2 + 3x - 2} \qquad P(x) \\
\underline{3x^5 + 0x^4 - 3x^3} \qquad\qquad\qquad\qquad\qquad \\
4x^4 + 3x^3 - 2x^2 + 3x - 2 \\
\underline{4x^4 + 0x^3 - 4x^2} \qquad\qquad\qquad \\
3x^3 + 2x^2 + 3x - 2 \\
\underline{3x^3 + 0x^2 - 3x} \qquad\qquad \\
2x^2 + 0x - 2 \\
\underline{2x^2 + 0x - 2} \\
0 \qquad R(x)
\end{array}
$$

Since the remainder is 0, we write the answer as the quotient polynomial $Q(x)$:

$$\frac{3x^5 + 4x^4 - 2x^2 + 3x - 2}{x^2 - 1} = 3x^3 + 4x^2 + 3x + 2. \qquad\qquad ◆$$

PROBLEM 2   Divide $P(x) = 6x^3 - 3x^2$ by $D(x) = 3x^2 - 2$ using long division.   ◆

◆ **Synthetic Division**

As we illustrated in Example 1, when a polynomial $P(x)$ of degree $n \geq 1$ is divided by a first-degree polynomial $D(x) = x - r$, the remainder $R(x)$ is a constant k

and the degree of the quotient polynomial $Q(x)$ is one less than the degree of $P(x)$. That is,

$$\frac{P(x)}{x-r} = Q(x) + \frac{k}{x-r}.$$

A shortcut called *synthetic division* may be used to divide a polynomial of degree $n \geq 1$ by a first-degree polynomial $x - r$. In **synthetic division**, we use only the essential data from the long division process by eliminating any power of x or coefficient that is a duplicate of that directly above it (in the same column). Also, we replace the divisor $x - r$ by the *synthetic divisor r* so that we may *add* rather than subtract columns. In the next example, we rework Example 1 using synthetic division.

E X A M P L E   3   Divide $P(x) = 2x^3 - 5x^2 + 3x - 4$ by $D(x) = x - 2$ using synthetic division.

S O L U T I O N   We begin by replacing the divisor $x - 2$ with the synthetic divisor 2 and writing the coefficients of the dividend:

Synthetic divisor ——— $2\overline{)2 \quad -5 \quad 3 \quad -4}$ ——— Coefficients of $P(x)$

When a polynomial is written in descending powers of x, the coefficient of the term with the highest degree is the *leading coefficient*. We bring down the leading coefficient two rows, multiply the synthetic divisor by the leading coefficient, and record this product in the middle row under the second coefficient of $P(x)$. Then we add to obtain the sum of the second column.

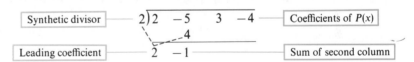

Now, we multiply the synthetic divisor by the sum of the second column; record this product in the middle row under the third coefficient of $P(x)$; then add to obtain the sum of the third column. We continue in this manner until all columns have a sum.

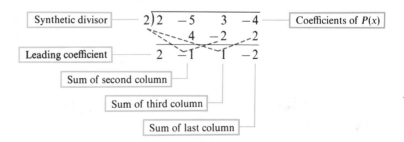

The sum of the last column is the constant remainder k. The preceding sums, to the left of the last sum, are the coefficients of the successive terms in the quotient polynomial $Q(x)$ whose degree is one less than that of $P(x)$. Thus, from the synthetic division

we can state

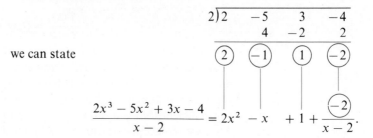

$$\frac{2x^3 - 5x^2 + 3x - 4}{x - 2} = 2x^2 - x + 1 + \frac{-2}{x - 2}.$$

This result agrees with the quotient and remainder polynomials we obtained in Example 1.   ◆

P R O B L E M 3   Divide $P(x) = x^3 - 5x^2 + 7x - 3$ by $D(x) = x - 1$.   ◆

The shortcut of synthetic division works only when the divisor $D(x)$ is a first-degree polynomial $x - r$. For problems such as Example 2, where the divisor $D(x)$ is quadratic, synthetic division does not apply.

In synthetic division, the coefficients of the dividend $P(x)$ must be recorded in the order that they appear when $P(x)$ is written in descending powers of x, and if a power of x is missing, we must record 0 as its coefficient.

E X A M P L E 4   Divide $P(x) = 2x^4 - 15x^2 + 9x + 1$ by $D(x) = x + 3$ using synthetic division.

S O L U T I O N   Since the divisor $D(x) = x + 3 = x - (-3)$, we use -3 as the synthetic divisor. Now, after inserting a 0 for the coefficient of the missing x^3 term in the dividend, we carry out the synthetic division as follows:

```
  Synthetic divisor

    -3)2    0   -15    9    1 ——— Coefficients of P(x)
          -6    18   -9    0
        2  -6     3    0   (1) ——— Constant remainder k

          Coefficients of Q(x) whose
          degree is 1 less than P(x)
```

Thus,

$$\frac{2x^4 - 15x^2 + 9x + 1}{x + 3} = 2x^3 - 6x^2 + 3x + \frac{1}{x + 3}.$$ ◆

We can check the synthetic division process by showing that

$$P(x) = (x - r)Q(x) + k.$$

PROBLEM 4   Check the synthetic division in Example 4.   ◆

◆ **The Remainder Theorem**

If $Q(x)$ is the quotient polynomial and k the constant remainder obtained when a polynomial $P(x)$ is divided by $x - r$, then we have

$$\frac{P(x)}{x - r} = Q(x) + \frac{k}{x - r}.$$

Multiplying both sides of this equation by $x - r$, we obtain

$$P(x) = (x - r)Q(x) + k.$$

Now, replacing x with r, we find

$$P(r) = (r - r)Q(r) + k$$
$$= 0 \cdot Q(r) + k$$
$$= k.$$

In other words, $P(r)$ is the same number as the remainder k obtained by dividing $P(x)$ by $x - r$. We refer to this fact about polynomial division as the **remainder theorem**.

◆ Remainder Theorem

> When a polynomial $P(x)$ is divided by $x - r$, the remainder is $P(r)$.

EXAMPLE 5   Use the remainder theorem to determine the remainder when

$$P(x) = x^4 - 3x^3 - 4x^2 + 13x - 8 \qquad \text{is divided by} \qquad x - 3.$$

SOLUTION   The remainder theorem tells us that when $P(x)$ is divided by $x - 3$ the remainder is $P(3)$. Evaluating $P(3)$, we find

$$P(3) = (3)^4 - 3(3)^3 - 4(3)^2 + 13(3) - 8$$
$$= 81 - 81 - 36 + 39 - 8$$
$$= -5$$

Thus, when $P(x)$ is divided by $x - 3$, the remainder is -5.   ◆

PROBLEM 5   Verify the remainder in Example 5 by dividing $P(x)$ by $x - 3$ using synthetic division.   ◆

As illustrated in the following example, we can evaluate $P(r)$ by using the remainder theorem and synthetically dividing $P(x)$ by $x - r$. This approach is often easier than finding $P(r)$ by direct substitution.

EXAMPLE 6   Given that $P(x) = 4x^3 - 6x^2 + 4x - 3$, find $P(-\frac{1}{2})$ by using synthetic division and the remainder theorem.

SOLUTION   According to the remainder theorem, $P(-\frac{1}{2})$ is the remainder obtained when $P(x)$ is divided by $x - (-\frac{1}{2}) = x + \frac{1}{2}$. Using synthetic division, we find the remainder:

$$
\begin{array}{r|rrrr}
-\frac{1}{2} & 4 & -6 & 4 & -3 \\
 & & -2 & 4 & -4 \\
\hline
 & 4 & -8 & 8 & \boxed{-7}
\end{array}
$$

Remainder

Thus, $P(-\frac{1}{2}) = -7$.   ◆

PROBLEM 6   Referring to Example 6, verify that $P(-\frac{1}{2}) = -7$ by using direct substitution.   ◆

If $P(x)$ is a polynomial and r is a constant such that $P(r) = 0$, we say that r is a **zero** of the polynomial function P or a **root** of the polynomial equation $P(x) = 0$. Also, recall from Section 3.3, any *real zero* of a function P represents an x-intercept of the graph of P.

EXAMPLE 7   Given the function P defined by $P(x) = 3x^3 - 8x^2 + 10x - 4$, use synthetic division to determine if each expression is a zero of P:

(a) $\frac{2}{3}$ **(b)** $1 + i$

SOLUTION

(a)   If $\frac{2}{3}$ is a zero of the function P, then $P(\frac{2}{3}) = 0$. Using synthetic division, we find

$$
\begin{array}{r|rrrr}
\frac{2}{3} & 3 & -8 & 10 & -4 \\
 & & 2 & -4 & 4 \\
\hline
 & 3 & -6 & 6 & \boxed{0}
\end{array}
$$

Remainder

Since the remainder is 0, we know $P(\frac{2}{3}) = 0$, and so $\frac{2}{3}$ is a zero of the polynomial function P defined by $P(x) = 3x^3 - 8x^2 + 10x - 4$.

(b)   If $1 + i$ is a zero of the function P, then $P(1 + i) = 0$. Using synthetic division and recalling from Section 1.5 that $i^2 = -1$, we find

$$
\begin{array}{r|rrrr}
1 + i & 3 & -8 & 10 & -4 \\
 & & 3 + 3i & -8 - 2i & 4 \\
\hline
 & 3 & -5 + 3i & 2 - 2i & \boxed{0}
\end{array}
$$

Remainder

Since the remainder is 0, we know $P(1 + i) = 0$, and so $1 + i$ is a zero of the polynomial function P defined by $P(x) = 3x^3 - 8x^2 + 10x - 4$.   ◆

PROBLEM 7   Use synthetic division to determine if $1 - i$ is a zero of the polynomial function P defined in Example 7.   ◆

Exercises 5.1

 Basic Skills

In Exercises 1–16, divide P(x) by D(x) using long division.

1. $P(x) = 3x^3 - 2x^2 + x - 3$
   $D(x) = x - 1$

2. $P(x) = 2x^4 + 3x^3 - 4x^2 - 2x + 1$
   $D(x) = x + 2$

3. $P(x) = 3x^4 - 10x^3 + 16x^2 - 14x + 3$
   $D(x) = x^2 - 2x + 3$

4. $P(x) = 2x^3 - 7x^2 + 11x - 4$
   $D(x) = x^2 - 3x + 4$

5. $P(x) = 4x^3 + 3x - 3$
   $D(x) = 2x - 1$

6. $P(x) = 9x^4 + 6x^3 + 3x - 2$
   $D(x) = 3x + 2$

7. $P(x) = 2x^5 + 2x^3 - 2x^2 + 5$
   $D(x) = 2x^3 + 1$

8. $P(x) = x^6 - 2x^2 - 3x + 4$
   $D(x) = x^3 - 3x$

9. $P(x) = x^3$
   $D(x) = x + 1$

10. $P(x) = x^4$
    $D(x) = x^2 + 1$

11. $P(x) = 6x^5 - x^3 + 2x^2 - x - 1$
    $D(x) = 1 - 2x^2$

12. $P(x) = 3x^4 + 4x^3 - 15x^2 - 4x - 12$
    $D(x) = 6 - x - x^2$

13. $P(x) = 3 - 6x$ 14. $P(x) = 8 - 2x - x^2$
    $D(x) = 2x + 1$ $D(x) = x^2 + 1$

15. $P(x) = 8x^3$ 16. $P(x) = (x + 1)^4$
    $D(x) = (2x - 1)^2$ $D(x) = x^2 + 1$

In Exercises 17–30, divide P(x) by D(x) using synthetic division. In each case, verify the remainder by using the remainder theorem.

17. $P(x) = 2x^2 - 3x + 4$
    $D(x) = x - 1$

18. $P(x) = 3x^2 - 6x - 5$
    $D(x) = x + 3$

19. $P(x) = x^3 + 5x^2 - 2x + 4$
    $D(x) = x + 2$

20. $P(x) = x^4 - 6x^3 + 4x^2 + 15x + 2$
    $D(x) = x - 4$

21. $P(x) = 2x^4 - 7x^3 - 3x^2 - 16$
    $D(x) = x - 4$

22. $P(x) = 3x^3 - 70x + 25$
    $D(x) = x + 5$

23. $P(x) = 4x^4 + 25x^3 - 32x$
    $D(x) = x + 6$

24. $P(x) = x^5 - 6x^3 + 7x^2 - 3$
    $D(x) = x - 2$

25. $P(x) = 3 - x^5$
    $D(x) = x - 2$

26. $P(x) = 7 + x - 2x^3$
    $D(x) = x + 1$

27. $P(x) = 3x^3 - 8x^2 + 19x - 10$
    $D(x) = x - \frac{2}{3}$

28. $P(x) = 4x^3 - 8x^2 + x + 3$
    $D(x) = x + \frac{1}{2}$

29. $P(x) = 2x^4 + x^3 - 19x + 8$
    $D(x) = x + \frac{5}{2}$

30. $P(x) = 4x^4 + 3x^2 - 27$
    $D(x) = x - \frac{3}{2}$

In Exercises 31–44, use synthetic division and the remainder theorem to find the indicated functional value.

31. $P(x) = 2x^3 - 3x^2 + 4x + 5; \quad P(1)$

32. $Q(x) = 3x^3 - 4x^2 + x - 9; \quad Q(-1)$

33. $F(x) = x^4 + 2x^3 - 5x - 3; \quad F(-2)$

34. $g(x) = x^3 - 2x^2 - 5x + 6; \quad g(3)$

35. $P(x) = 2x^3 - 7x^2 + 2x + 1; \quad P(4)$

36. $Q(x) = -x^3 + 4x^2 + 12x + 2; \quad Q(6)$

37. $Q(x) = 4x^3 + 2x^2 - 1; \quad Q(\frac{1}{2})$

38. $P(x) = 3x^4 - x^3 + x - 1; P(-\frac{2}{3})$

39. $f(x) = 3x^4 - 5x^3 + 7x^2 + 9x + 2; \quad f(-\frac{1}{3})$

40. $H(x) = 5x^4 - 7x^2 + 12x - 4; \quad H(\frac{2}{5})$

41. $h(x) = 2x^3 - 2x^2 + 4x + 7; \quad h(1 - i)$

42. $G(x) = 2x^4 + x^3 - 8x^2 - 1; \quad G(-2i)$

43. $P(x) = x^4 - x^3 + 3x^2 + 2x + 2; \quad P(\sqrt{2})$

44. $P(x) = 3x^3 - 21x - 4; \quad P(2 + \sqrt{3})$

In Exercises 45–52, use synthetic division to determine if the given value of x is a zero of the polynomial function P.

45. $P(x) = 2x^3 + x^2 - 14x - 3; x = -3$

46. $P(x) = 4x^3 + 2x^2 - 11x - 18; x = 2$

47. $P(x) = -2x^4 + 3x^3 + 8x^2 - 18; x = \frac{3}{2}$

48. $P(x) = -6x^4 + x^3 + 2; x = \frac{2}{3}$

49. $P(x) = x^4 - 3x^3 + 13x^2 - 27x + 36; x = 3i$

50. $P(x) = 4x^3 - 17x^2 + 24x - 5; x = 2 - i$

51. $P(x) = 3x^3 - 10x^2 + 5x + 4; x = 1 - \sqrt{2}$

52. $P(x) = x^4 - 2x^3 + x^2 + 6x - 12; x = \sqrt{3}$

In Exercises 53–60, use synthetic division to determine if the given value of x is a root of the polynomial equation.

53. $3x^4 - 5x^3 - 12x^2 + 19x + 2 = 0; x = -2$

54. $-6x^3 + 2x^2 + 5x + 21 = 0; x = 3$

55. $6x^3 + 2x^2 + 3x + 1 = 0; x = -\frac{1}{3}$

56. $25x^5 + x^3 - 2x^2 + 5x + 2 = 0; x = \frac{2}{5}$

57. $2x^3 - 3x^2 + 8x - 12 = 0; x = -2i$

58. $x^4 + 2x^3 + x^2 - 2x - 2 = 0; x = -1 + i$

59. $3x^4 - 2x^3 - 8x^2 + 4x + 4 = 0; x = \sqrt{2}$

60. $x^4 - 2x^3 - 6x^2 + 8x + 8 = 0; x = 1 + \sqrt{3}$

 ## Critical Thinking

61. When the polynomial $P(x)$ is divided by $x + 6$, the quotient is $x^3 + x^2 + x + 5$ and the remainder is -12. Find $P(x)$.

62. When $P(x) = 2x^3 - 5x^2 + 8x - 11$ is divided by the polynomial $D(x)$, the quotient is $x^2 - 2x + 3$ and the remainder is -8. Find $D(x)$.

Synthetic division may also be used to divide a polynomial $P(x)$ of degree $n \geq 1$ by a first-degree polynomial $D(x) = ax + b$. Since

$$\frac{P(x)}{ax + b} = \frac{P(x)}{a\left(x + \frac{b}{a}\right)} = \frac{\frac{1}{a}P(x)}{x + \frac{b}{a}},$$

we can obtain the quotient and remainder by using a synthetic divisor $-b/a$ with dividend $(1/a)P(x)$.

In Exercises 63–68, divide $P(x)$ by $D(x)$ using synthetic division.

63. $P(x) = 6x^3 + 7x^2 - 18x + 16$
    $D(x) = 3x + 8$

64. $P(x) = 8x^4 + 2x^3 - 3x^2 + 2x - 1$
    $D(x) = 2x - 1$

65. $P(x) = 32x^4 - 4x^3 + 7x^2 - 10x + 2$
    $D(x) = 4x - 1$

66. $P(x) = 3x^3 + x^2 + 2x + 8$
    $D(x) = 3x + 4$

67. $P(x) = x^4 - 2x^3 + 2x^2 - 7x + 6$
    $D(x) = 2 - x$

68. $P(x) = 24 - 20x - 12x^2 + 28x^3 - 15x^4$
    $D(x) = 6 - 5x$

In Exercises 69 and 70, use synthetic division to determine a value of k such that -2 is a zero of the given polynomial function.

69. $P(x) = x^4 + 2x^3 + 6x^2 + kx + 8$

70. $P(x) = kx^3 + 3x^2 + 6x - 16$

In Exercises 71 and 72, use synthetic division to determine a value of k such that 4 is a root of the given polynomial equation.

71. $kx^3 - 10x^2 + kx + 24 = 0$

72. $2x^3 + k^2x^2 - 2x + 88k = 0$

 ## Calculator Activities

In Exercises 73–76, divide $P(x)$ by $D(x)$ using synthetic division.

73. $P(x) = 14.5x^3 + 266.4x^2 - 828.1$
    $D(x) = x + 18.2$

74. $P(x) = 2.5x^3 - 6.0x^2 - 22.3x + 41.4$
    $D(x) = x - 3.6$

75. $P(x) = 275x^4 - 126x^2 - 627x - 1350$
    $D(x) = x - 1.8$

76. $P(x) = 125x^4 + 2680x^3 - 1625x^2 - 1824x + 1012$
$D(x) = x + 22$

In Exercises 77 and 78, use synthetic division and the remainder theorem to find the indicated functional value. Check each answer by using direct substitution.

77. $P(x) = 2.6x^3 - 9.8x^2 + 3.7x + 4.2; \quad P(1.7)$

78. $P(x) = 0.035x^4 + 1.025x^3 - 2134; \quad P(-32.2)$

79. When a Patriot missile is fired from its launchpad, the height h (in meters) of the missile above the ground is a function of the horizontal distance x (in kilometers) from the launchpad and is given by

$$h(x) = 244x + 35x^2 - 24x^3, \quad 0 \le x \le 4.$$

Use synthetic division and the remainder theorem to find $h(2.45)$. Round the answer to three significant digits.

80. The current i (in milliamperes) flowing through an electrical circuit is a function of time t (in seconds) and is given by

$$i(t) = 545 + 8.2t^2 - 1.2t^4, \quad 0 \le t \le 5.$$

Use synthetic division and the remainder theorem to find $i(3.8)$. Round the answer to three significant digits.

5.2 ◆ Factors and Zeros of Polynomial Functions

Finding the zeros of a polynomial function of degree 2 is simple: we apply the quadratic formula. Finding the zeros of a polynomial function of degree 3 or greater is usually much more difficult, but mathematicians have developed formulas for this process. Unlike the quadratic formula, however, formulas for finding the zeros of a polynomial function with degree 3 or 4 are extremely complicated. Mathematicians have also shown that no such general formulas exist for finding the zeros of polynomial functions with degree 5 or greater.

In this section, we discuss theorems that we can use to determine the number and nature of the zeros of a polynomial function. In sections 5.3 and 5.4 we discuss methods for determining rational and irrational zeros. We begin by proving the *factor theorem*, which shows a relationship between the zeros and factors of a polynomial function.

◆ The Factor Theorem

If $Q(x)$ is the quotient polynomial and k the constant remainder obtained when a polynomial $P(x)$ is divided by $x - r$, then we have

$$\frac{P(x)}{x - r} = Q(x) + \frac{k}{x - r}.$$

Multiplying both sides of this equation by $x - r$, we obtain

$$P(x) = (x - r)Q(x) + k.$$

Now by the remainder theorem, we know that $P(r) = k$. So we can also write

$$P(x) = (x - r)Q(x) + P(r).$$

However, if the remainder $P(r) = 0$, then

$$P(x) = (x - r)Q(x),$$

and we conclude that $(x - r)$ is a factor of $P(x)$. Conversely, if $x - r$ is a factor of $P(x)$, then

$$P(x) = (x - r)Q(x),$$

and it follows that

$$
\begin{aligned}
P(r) &= (r - r)Q(r) \\
&= 0 \cdot Q(r) \\
&= 0
\end{aligned}
$$

We refer to this consequence of the remainder theorem as the **factor theorem**.

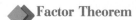
Factor Theorem

> A first-degree polynomial $x - r$ is a factor of a polynomial $P(x)$ if and only if $P(r) = 0$, or equivalently, if and only if r is a zero of P.

It is important to recognize that the concepts of factors, roots, and zeros are closely related. In general if P is a polynomial function and r is a constant, then the following statements are equivalent:

1. $x - r$ is a factor of the polynomial $P(x)$.
2. r is a zero of the function P.
3. r is a root of the equation $P(x) = 0$.

EXAMPLE 1   Given that $P(x) = 2x^3 - 3x^2 + 8x - 12$, determine if each expression is a factor of $P(x)$:

(a) $x - 2$ **(b)** $x + 2i$

SOLUTION

(a) According to the factor theorem, $x - 2$ is a factor of $P(x)$ if $P(2) = 0$. To determine $P(2)$, we can use either direct substitution or synthetic division. Choosing synthetic division, we find

$$
\begin{array}{r|rrrr}
2) & 2 & -3 & 8 & -12 \\
   &   & 4 & 2 & 20 \\
\hline
   & 2 & 1 & 10 & \circled{8}
\end{array}
$$
$\circled{8} \longrightarrow \boxed{P(2)}$

Since $P(2) = 8 \neq 0$, we conclude that $x - 2$ is not a factor of $P(x)$.

(b) According to the factor theorem, $x + 2i$ is a factor of $P(x)$ if $P(-2i) = 0$. To determine $P(-2i)$, we can use either direct substitution or synthetic division. Choosing synthetic division and recalling from Section 1.5 that

$i^2 = -1$, we find

$$
\begin{array}{r|rrrr}
-2i & 2 & -3 & 8 & -12 \\
 & & -4i & 6i-8 & 12 \\
\hline
 & 2 & -3-4i & 6i & \boxed{0} \quad\boxed{P(-2i)}
\end{array}
$$

Since $P(-2i) = 0$, we conclude that $x + 2i$ is a factor of $P(x)$. ◆

PROBLEM 1   Given the polynomial $P(x)$ in Example 1, determine if $x - 2i$ is a factor of $P(x)$.
◆

◆ **Fundamental Theorem of Algebra**

A linear function $f(x) = ax + b$, with $a \neq 0$, has just one zero, $-b/a$. A quadratic function $g(x) = ax^2 + bx + c$, with $a \neq 0$, has two zeros,

$$
\frac{-b + \sqrt{b^2 - 4ac}}{2a} \quad \text{and} \quad \frac{-b - \sqrt{b^2 - 4ac}}{2a}.
$$

How many zeros for a cubic function? a quartic function? a quintic function? The **fundamental theorem of algebra** begins to answer these questions.

Fundamental Theorem of Algebra

Every polynomial function P of degree $n \geq 1$ has at least one zero in the set of complex numbers.

The proof of this theorem was provided by Carl Friedrich Gauss in 1799, but is beyond the scope of this book. Thus, we shall accept its validity and use its results.

Consider the polynomial function P of degree $n \geq 1$ defined by

$$
P(x) = a_n x^n + a_{n-1} x^{n-1} + \cdots + a_1 x + a_0,
$$

where $a_n \neq 0$. According to the fundamental theorem of algebra, the function P has at least one zero. If this zero is r_1, then, by the factor theorem, $x - r_1$ is a factor of the polynomial $P(x)$. Thus, we can write

$$
P(x) = (x - r_1)Q_1(x),
$$

where the polynomial $Q_1(x)$ is of degree $n - 1$. Now if $n - 1 \geq 1$, then by the fundamental theorem of algebra, the function Q_1 has at least one zero. If this zero is r_2, then, by the factor theorem, $x - r_2$ is a factor of the polynomial $Q_1(x)$ and we have

$$
Q_1(x) = (x - r_2)Q_2(x),
$$

where the polynomial $Q_2(x)$ is of degree $n - 2$. Thus, using substitution,

we obtain

$$P(x) = (x - r_1)(x - r_2)Q_2(x).$$

If $n - 2 \geq 1$, we continue this procedure until we obtain a quotient polynomial $Q_n(x)$ of degree 0, that is, until $Q_n(x)$ is the constant a_n. We can then write

$$P(x) = \underbrace{(x - r_1)(x - r_2) \cdots (x - r_n)}_{n \text{ linear factors}} a_n,$$

and conclude that P has exactly n linear factors. Hence, P has exactly n zeros, namely $r_1, r_2, \ldots, r_n$.

Extension of the Fundamental Theorem

> Every polynomial function P of degree $n \geq 1$ can be expressed as the product of n linear factors and, consequently, P has exactly n zeros in the set of complex numbers.

Thus, a cubic function can be expressed as the product of three linear factors, and it has three zeros; a quartic function can be expressed as the product of four linear factors, and it has four zeros; a quintic function can be expressed as the product of five linear factors, and it has five zeros; and so on.

The n zeros of a polynomial function of degree $n \geq 1$ are not necessarily distinct. For example, consider the polynomial (quadratic) function P defined by $P(x) = x^2 - 6x + 9$. We can write $P(x)$ as the product of two linear factors:

$$P(x) = x^2 - 6x + 9 = \underbrace{(x - 3)(x - 3)}_{\text{Two linear factors}}$$

Since the linear factor $(x - 3)$ appears twice, we say that 3 is a zero of *multiplicity two*. In general, if a linear factor $x - r$ appears k times, then *r is a zero of multiplicity k*. In the next example, we use the factoring techniques from Section 1.6 and the factor theorem to find all the zeros of a polynomial function.

EXAMPLE 2   If $P(x) = 2x^5 - 4x^4 + 2x^3$, find all the zeros of the function P, and give the multiplicity of each zero.

SOLUTION   This fifth-degree polynomial must have five linear factors and, consequently, five zeros in the set of complex numbers. Using the factoring techniques from Section 1.6, we have

$$P(x) = 2x^5 - 4x^4 + 2x^3$$

$$= 2x^3(x^2 - 2x + 1) \qquad \text{Factor out } 2x^3$$

$$= 2x^3(x - 1)^2 \qquad \textbf{Factor the trinomial}$$

$$= 2\underbrace{(x - 0)(x - 0)(x - 0)(x - 1)(x - 1)}_{\text{Five linear factors}}$$

By the factor theorem, the zeros are 0 of multiplicity three and 1 of multiplicity two.   ◆

If we know the *n* zeros of a polynomial function *P* of degree *n*, then we can write the general form of the polynomial function. For example, if the zeros of a polynomial function of degree five are 0 of multiplicity three and 1 of multiplicity two, then the polynomial function can be expressed as

$$P(x) = a(x - 0)(x - 0)(x - 0)(x - 1)(x - 1),$$
$$= a(x^3)(x - 1)^2,$$
$$= a(x^5 - 2x^4 + x^3),$$

where *a* is any nonzero number.

PROBLEM 2   Write the general form of a polynomial function *P* of degree four if its zeros are 1 of multiplicity two and -1 of multiplicity two.   ◆

Unfortunately, the zeros of a polynomial function are not always as easily attainable as those found in Example 2, because not every polynomial can be factored by the methods discussed in Section 1.6. However, as illustrated in the next two examples, if enough linear factors of a polynomial function are known, such that the remaining factor is quadratic, then we can find all zeros of the function.

EXAMPLE 3   Find all zeros of $P(x) = 2x^3 + 9x^2 + 7x - 6$, given that $x + 2$ is a factor.

SOLUTION   This third-degree polynomial function must have three zeros in the set of complex numbers. Since $x + 2$ is a factor of the polynomial $P(x)$, we can write

$$P(x) = (x + 2)Q(x).$$

Now, we can find $Q(x)$ by dividing $P(x)$ by $x + 2$, using synthetic division:

$$
\begin{array}{r|rrrr}
-2) & 2 & 9 & 7 & -6 \\
    &   & -4 & -10 & 6 \\
\hline
    & 2 & 5 & -3 & \boxed{0} \; \longrightarrow \text{Remainder}
\end{array}
$$

Coefficients of $Q(x)$ whose degree is 1 less than $P(x)$

Thus, $Q(x) = 2x^2 + 5x - 3$. Since Q is a quadratic function, we can find the zeros of Q by the quadratic formula or by simply factoring, as follows:

$$
\begin{aligned}
P(x) = (x + 2)Q(x) &= (x + 2)(2x^2 + 5x - 3) \\
&= (x + 2)(2x - 1)(x + 3) \\
&= 2(x + 2)(x - \tfrac{1}{2})(x + 3)
\end{aligned}
$$

Therefore, by the factor theorem, the three zeros of P are $-2, \tfrac{1}{2}$, and -3.   ◆

PROBLEM 3   Find all zeros of $P(x) = x^3 - x^2 - 10x - 8$, given that 4 is a zero.   ◆

EXAMPLE 4   Write $P(x) = x^4 - 8x^3 + 26x^2 - 48x + 45$ as the product of linear factors, given that 3 is a zero of multiplicity two.

SOLUTION   This fourth-degree polynomial function can be written as the product of four linear factors. Since 3 is a zero of multiplicity two, $(x - 3)^2$ is a factor of the polynomial $P(x)$. So we can write

$$
P(x) = (x - 3)^2 Q(x).
$$

Now we can find $Q(x)$ either by dividing $P(x)$ by $(x - 3)^2 = x^2 - 6x + 9$, using long division, or by dividing $P(x)$ by $x - 3$, using synthetic division and then dividing the result of this division by $x - 3$, using synthetic division again. Choosing synthetic division, we have

$$
\begin{array}{r|rrrrr}
3) & 1 & -8 & 26 & -48 & 45 \\
   &   & 3 & -15 & 33 & -45 \\ \hline
3) & 1 & -5 & 11 & -15 & 0 \\
   &   & 3 & -6 & 15 \\ \hline
   & 1 & -2 & 5 & 0
\end{array}
$$

Coefficients of $Q(x)$ whose degree is 2

Thus, $Q(x) = x^2 - 2x + 5$, and so we have

$$
P(x) = (x - 3)^2 Q(x) = (x - 3)^2 (x^2 - 2x + 5).
$$

We can find the zeros of the quadratic function $Q(x) = x^2 - 2x + 5$ from the quadratic formula:

$$
\begin{aligned}
x &= \frac{-(-2) \pm \sqrt{(-2)^2 - 4(1)(5)}}{2(1)} = \frac{2 \pm \sqrt{-16}}{2} \\
&= \frac{2 \pm 4i}{2} = 1 \pm 2i.
\end{aligned}
$$

Thus, the four zeros of the function P are 3 of multiplicity two, $1 + 2i$, and $1 - 2i$. By the factor theorem, we can write $P(x)$ as the product of four linear factors:

$$P(x) = \underbrace{(x - 3)(x - 3)[x - (1 + 2i)][x - (1 - 2i)]}_{\text{Four linear factors}}$$

◆

PROBLEM 4   To check Example 4, find the product $(x - 3)^2[x - (1 + 2i)][x - (1 - 2i)]$
You should obtain the original form of the polynomial.   ◆

◆ **Conjugate Pairs**

In Example 4 we found that two zeros of the polynomial function

$$P(x) = x^4 - 8x^3 + 26x^2 - 48x + 45$$

are the imaginary numbers $1 + 2i$ and $1 - 2i$. Together, these imaginary numbers are called a *conjugate pair*. If a polynomial function with real coefficients has imaginary zeros, these zeros always occur as conjugate pairs. We refer to this fact as the **conjugate pair theorem**.

◆ Conjugate Pair Theorem

> If P is a polynomial function of degree $n \geq 1$ with real coefficients, and if the imaginary number $a + bi$ is a zero of P, then its conjugate $a - bi$ is also a zero of P.

We can state two consequences of the conjugate pair theorem:

A polynomial function with real coefficients of *odd* degree must have at least one real zero. In fact, such a polynomial function must have an odd number of real zeros (see Example 3).

A polynomial function with real coefficients of *even* degree must have either an even number of real zeros or no real zero at all (see Examples 4 and 5).

EXAMPLE 5   Find all zeros of $P(x) = x^4 - 6x^3 + 15x^2 - 18x + 10$, given that $2 - i$ is a zero of P.

SOLUTION   This fourth-degree polynomial function must have four zeros in the set of complex numbers. By the conjugate pair theorem, since $2 - i$ is a zero of P and P has real coefficients, we know that $2 + i$ is also a zero of P. Thus, $x - (2 - i)$ and $x - (2 + i)$ are factors of the polynomial $P(x)$. Hence, we can write

$$P(x) = [x - (2 - i)][x - (2 + i)]Q(x).$$

Now we can find $Q(x)$ by dividing $P(x)$ by $x - (2 - i)$, using synthetic division, and then dividing the result of this division by $x - (2 + i)$, using synthetic division again:

$$
\begin{array}{r|rrrr}
2-i & 1 & -6 & 15 & -18 & 10 \\
    &   & 2-i & -9+2i & 14-2i & -10 \\
\hline
2+i & 1 & -4-i & 6+2i & -4-2i & 0 \\
    &   & 2+i & -4-2i & 4+2i & \\
\hline
    & 1 & -2 & 2 & 0 &
\end{array}
$$

Coefficients of $Q(x)$ whose degree is 2

Thus, $Q(x) = x^2 - 2x + 2$. We can determine the zeros of this quadratic function from the quadratic formula:

$$
x = \frac{-(-2) \pm \sqrt{(-2)^2 - 4(1)(2)}}{2(1)} = \frac{2 \pm \sqrt{-4}}{2} = 1 \pm i.
$$

Thus, the four zeros of P are the conjugate pairs $2 \pm i$ and $1 \pm i$.   ◆

PROBLEM 5   Express the polynomial function in Example 5 as the product of four linear factors.   ◆

Exercises 5.2

◆ Basic Skills

In Exercises 1–12, use the factor theorem to determine if the first polynomial is a factor of the second polynomial.

1. $x - 1$; $x^3 - 4x^2 + x - 4$

2. $x + 2$; $x^3 - 4x^2 + x - 4$

3. $x + 3$; $x^4 - 7x^3 + 10x^2 + 3x - 15$

4. $x - 5$; $x^4 - 7x^3 + 10x^2 + 3x - 15$

5. $x - 4$; $x^4 - 3x^3 - 12x - 16$

6. $x + 1$; $x^4 - 3x^3 - 12x - 16$

7. $x + \frac{2}{3}$; $6x^3 + x^2 + 7x + 6$

8. $x - \frac{1}{3}$; $6x^3 + x^2 + 7x + 6$

9. $x - i$; $x^4 - 2x^2 + 1$

10. $x + i$; $x^4 - 2x^2 + 1$

11. $x - (1 + i)$; $x^3 - 2x + 4$

12. $x - (1 - i)$; $x^3 - 2x + 4$

In Exercises 13–24, (a) give the degree of each polynomial function, and (b) find all the zeros for each polynomial function, stating the multiplicity of any zero that is not distinct.

13. $P(x) = x^2(x - 2)^3$

14. $P(x) = 2x(x + 3)^4$

15. $f(x) = 3x^3(x + 1)^2(x - 4)^3$

16. $G(x) = 4x^4(x - \frac{1}{2})^2(x + 6)^2$

17. $P(x) = 3x^4 - 12x^3 + 12x^2$

18. $Q(x) = 2x^5 - 12x^4 + 18x^3$

19. $f(x) = x^4 - 2x^2 + 1$

20. $h(x) = 16x^2 - 8x^4 + x^6$

21. $D(x) = x^3 + 2x^2 - 4x - 8$

22. $P(x) = 8x^3 + 4x^2 - 2x - 1$

23. $F(x) = x^4 - 16$

24. $H(x) = 3x^6 + 6x^5 + 6x^4$

In Exercises 25–30, determine the general form of a polynomial function P whose degree and zeros are given.

25. Degree 3; zeros are 1, −1, 2.

26. Degree 3; zeros are 2 of multiplicity three.

27. Degree 4; zeros are ± 1, $\pm i$.

28. Degree 4; zeros are $1 \pm i$ of multiplicity two.

29. Degree 5; zeros are 0 of multiplicity three and $\pm \sqrt{2}$.

30. Degree 6; zeros are 0 of multiplicity two, 1 of multiplicity two, and $1 \pm \sqrt{3}$.

Given the polynomial function P defined in Exercises 31–42, (a) write P(x) as the product of linear factors and (b) find all the zeros of P.

31. $P(x) = x^3 + x^2 - 5x + 3$;   1 is a zero

32. $P(x) = 2x^3 + 9x^2 + x - 12$;   −4 is a zero

33. $P(x) = 6x^3 + x^2 - 9x - 4$; $(x + \frac{1}{2})$ is a factor

34. $P(x) = 18x^3 - 33x^2 + 20x - 4$; $\frac{2}{3}$ is a zero

35. $P(x) = x^4 - 4x^3 + 5x^2 - 4x + 4$;   2 is a zero of multiplicity two

36. $P(x) = x^5 - 5x^4 + 10x^3 - 10x^2 + 5x - 1$; $(x - 1)^3$ is a factor

37. $P(x) = 2x^4 - 17x^3 + 53x^2 - 72x + 36$; $\frac{3}{2}$ is a zero and $(x - 3)$ is a factor

38. $P(x) = 8x^4 - 14x^3 - 71x^2 - 10x + 24$; $-\frac{3}{4}$ and $\frac{1}{2}$ are zeros

39. $P(x) = x^4 - 3x^3 - 12x - 16$; $(x + 2i)$ is s a factor

40. $P(x) = 2x^4 - 3x^3 - 3x - 2$; i is a zero

41. $P(x) = x^4 - 4x^3 - 7x^2 + 50x - 50$; $3 - i$ is a zero

42. $P(x) = x^4 - 6x^3 + 18x^2 - 30x + 25$; $[x - (1 + 2i)]$ is a factor

In Exercises 43–48, find all the roots of the polynomial equation, given the information provided about its roots.

43. $x^3 - 5x^2 - 8x + 12 = 0$;   1 is a root

44. $x^3 + 8x^2 + x - 22 = 0$;   −2 is a root

45. $x^4 - 2x^3 - 10x^2 + 16x + 40 = 0$;   −2 is a double root

46. $9x^4 + 18x^3 - 115x^2 + 96x - 20 = 0$; $\frac{1}{3}$ and $\frac{2}{3}$ are roots

47. $3x^4 - 13x^3 + 15x^2 + 11x - 20 = 0$; $2 + i$ is a root

48. $x^6 - x^5 - 2x^3 - 3x^2 - x - 2 = 0$; i is a double root

---

Critical Thinking

49. Use the sum of cubes formula (Section 1.6) to factor $x^3 + 27$ into the product of linear and quadratic expressions. Then find all the roots of the equation $x^3 + 27 = 0$.

50. Use the difference of squares, sum of cubes, and difference of cubes formulas (Section 1.6) to factor $x^6 - 64$ into the product of linear and quadratic expressions. Then find all the roots of the equation $x^6 - 64 = 0$.

51. Is $x + a$, $a > 0$, a factor of $x^n - a^n$ when n is even? when n is odd? Explain.

52. Is $x + a$, $a > 0$, a factor of $x^n + a^n$ when n is even? when n is odd? Explain.

53. Determine the value of k if $x - 2$ is a factor of $x^4 + 3x^3 + kx^2 - 5x - 2$.

54. Determine the value of k if $x + 3$ is a factor of $2x^4 + 4x^3 - 5x^2 + 2k + k^2$.

In Exercises 55–60, determine the possibilities for the number of times that the graph of P crosses the x-axis, that is, goes from below the x-axis to above the x-axis or vice versa. (Hint: Refer to the two statements in the text that follow the conjugate pair theorem.)

55. $P(x) = x^3 - 2x^2 + 4x + 7$

56. $P(x) = 4x^4 - 2x^2 + x + 1$

57. $P(x) = -2x^6 + 5x^4 - 2x^3 - 8$

58. $P(x) = x^5 + 1$

59. $P(x) = 6x^9 - 3x^7 + 2$

60. $P(x) = 2x^{10} - 7x^2 + 3x - 1$

61. The graph of a cubic function is shown in the sketch. Determine the equation that defines this function.

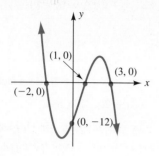

62. The graph of a quartic function is sketched in the figure. Determine the equation that defines this function.

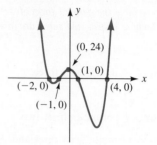

Calculator Activities

In Exercises 63–66, use the information given to find all the zeros of P. Round each zero to three significant digits.

63. $P(x) = 5.0x^3 + 3.8x^2 - 16.2x + 3.6$;
    $(x - 0.240)$ is a factor

64. $P(x) = -8.0x^3 - 9.6x^2 + 138.7x + 280.0$;
    -3.20 is a zero

65. $P(x) = 6.8x^4 - 32.3x^3 - 187.2x^2 + 49.4x + 270.4$;
    -3.25 and 8.00 are zeros

66. $P(x) = 4.4x^4 + 196.4x^3 + 1424.7x^2 - 18635.4x - 5462.1$;
    -25.5 is a zero of multiplicity two

In Exercises 67 and 68, use the given information to find all roots of the polynomial equation. Round each root to three significant digits.

67. $2.5x^3 - 5.4x^2 - 10.44x - 2.16 = 0$; -0.240 is a root

68. $3.2x^4 + 13.1x^3 - 556.75x^2 - 1466.3x + 1610 = 0$;
    $(x - 12.5)$ and $(x + 14.0)$ are factors

69. The velocity v (in feet per second, or ft/s) of the current at the mouth of a tidal river is a function of the time t

(in hours) after high tide and is given by

$$v(t) = -0.240t^3 + 0.064t^2 + 8.527t - 2.982,$$

$$0 \le t \le 6.$$

Twenty-one minutes after high tide the velocity of the current is 0 ft/s. Approximate to the nearest minute any other time t for which the velocity is zero.

70. The deflection d (in millimeters, mm) of a 34.0-meter steel beam is a function of the distance x (in meters, m) from the left end of the beam and is given by

$$d(x) = 0.002x^4 - 0.129x^3 + 2.480x^2 - 13.804x,$$

$$0 \le x \le 34.$$

The deflection at both ends of the beam is 0 mm. Approximate to three significant digits any other distances x for which the deflection is zero.

5.3 Rational Zeros of Polynomial Functions

From Section 5.2, if P is a polynomial function with real coefficients of degree $n \ge 3$, then we know the following facts about its zeros:

1. P has n zeros.

2. Complex zeros of P appear in conjugate pairs.

3. Once $n - 2$ of the zeros of P are known, the quadratic formula may be used to find the remaining two zeros.

In this section, we extend these ideas to show that if P has *integer* coefficients, then any *rational zero* of P must be a factor of the constant term divided by a factor of the coefficient of the highest power term in the polynomial.

◆ Rational Zero Theorem

Consider the polynomial function P of degree n:

$$P(x) = a_n x^n + a_{n-1} x^{n-1} + \cdots + a_1 x + a_0.$$

Suppose that its coefficients $a_n, a_{n-1}, \ldots, a_1, a_0$ are *integers* and that $a_n \neq 0$, $a_0 \neq 0$. Furthermore, suppose that b and c are *integers* such that the rational number b/c is a zero of P. Then we can write the polynomial equation

$$P\left(\frac{b}{c}\right) = a_n \left(\frac{b}{c}\right)^n + a_{n-1}\left(\frac{b}{c}\right)^{n-1} + \cdots + a_1\left(\frac{b}{c}\right) + a_0 = 0.$$

If we multiply both sides of this equation by c^n, we obtain

$$a_n b^n + a_{n-1} b^{n-1} c + \cdots + a_1 b c^{n-1} + a_0 c^n = 0.$$

Now solving for $a_0 c^n$, we have

$$
\begin{aligned}
a_0 c^n &= -a_n b^n - a_{n-1} b^{n-1} c - \cdots - a_1 b c^{n-1} \\
&= b(\underbrace{-a_n b^{n-1} - a_{n-1} b^{n-2} c - \cdots - a_1 c^{n-1}}_{\boxed{\text{An integer}}})
\end{aligned}
$$

Hence, b is a factor of $a_0 c^n$. However, if b/c is given in lowest terms, b and c are relatively prime and so b and c^n must also be relatively prime. Therefore, we can conclude that b is a factor of a_0. In a similar manner, by solving the equation

$$a_n b^n + a_{n-1} b^{n-1} c + \cdots + a_1 b c^{n-1} + a_0 c^n = 0$$

for $a_n b^n$, we find that c is a factor of a_n. In summary, if b/c is a zero of P and b/c is in lowest terms, then b is a factor of a_0 and c is a factor of a_n. We refer to this observation about rational zeros as the **rational zero theorem**.

◆ Rational Zero Theorem

Let

$$P(x) = a_n x^n + a_{n-1} x^{n-1} + \cdots + a_1 x + a_0,$$

be a polynomial function with integer coefficients and with $a_n \neq 0$, $a_0 \neq 0$. If the rational number b/c, in lowest terms, is a zero of P, then b must be a factor of the constant term a_0, and c must be a factor of the coefficient of the highest degree term a_n.

The rational zero theorem does not say that a polynomial function with integer coefficients has rational zeros. It simply enables us to develop a list of all *possible* rational zeros of a polynomial function P with integer coefficients. By using direct substitution or synthetic division, we can then determine whether any of the rational numbers in this list are *actual* zeros of P.

EXAMPLE 1   Given $P(x) = 2x^3 - 11x^2 + 13x - 4$, list all possible rational zeros of P.

SOLUTION   For $P(x) = 2x^3 - 11x^2 + 13x - 4$, the constant term is -4 and the coefficient of the highest degree term is 2. Now the factors of -4 are ± 1, ± 2, and ± 4, and the factors of 2 are ± 1 and ± 2. Thus, according to the rational zero theorem, the possible rational zeros of P are

$$\frac{\pm 1}{\pm 1}, \quad \frac{\pm 2}{\pm 1}, \quad \frac{\pm 4}{\pm 1}, \quad \text{and} \quad \frac{\pm 1}{\pm 2}, \quad \frac{\pm 2}{\pm 2}, \quad \frac{\pm 4}{\pm 2},$$

which simplify to ± 1, ± 2, ± 4, and $\pm \frac{1}{2}$.   ◆

PROBLEM 1   Repeat Example 1 for $P(x) = 3x^3 + 22x^2 + 25x + 6$.   ◆

◆ Descartes' Rule of Signs

To determine if any of the eight possible rational zeros found in Example 1 are actual zeros of $P(x) = 2x^3 - 11x^2 + 13x - 4$, we can use direct substitution or synthetic division. However, checking each of these eight possible rational zeros can be quite tedious. Fortunately, **Descartes' rule of signs** often enables us to narrow the possibilities. As you read the rule, notice that it pertains to *real* zeros, not simply *rational* zeros. Also, by a **variation in sign**, we mean that two consecutive (nonzero) coefficients have opposite signs.

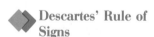
Descartes' Rule of Signs

If $P(x)$ is a polynomial with *real coefficients* and is written in descending powers of x, then

1. the number of *positive real zeros* of P either is equal to the number of variations in sign of $P(x)$ or is less than this number by an even number.

2. The number of *negative real zeros* of P either is equal to the number of variations in sign of $P(-x)$ or is less than this number by an even number.

EXAMPLE 2   Find all the rational zeros of the function P defined in Example 1.

SOLUTION   In Example 1 we found that the possible rational zeros of P are ± 1, ± 2, ± 4, and $\pm \frac{1}{2}$. From Descartes' rule of signs, we can determine that

$$P(x) = +2x^3 - 11x^2 + 13x - 4$$

$$\underbrace{\qquad}_{1} \quad \underbrace{\qquad}_{2} \quad \underbrace{\qquad}_{3}$$

has three variations in sign so the function has either 1 or 3 positive real zeros. Also, since

$$P(-x) = -2x^3 - 11x^2 - 13x - 4$$

has no variation in sign, Descartes' rule of signs tells us that the function P has no negative real zero. Hence, we discard the possible negative rational zeros $(-1, -2, -4,$ and $-\frac{1}{2})$ and check only the possible positive rational zeros $(1, 2, 4,$ and $\frac{1}{2})$.

Testing 1 using synthetic division, we find

$$
\begin{array}{r|rrrr}
1) & 2 & -11 & 13 & -4 \\
   &   & 2 & -9 & 4 \\
\hline
   & 2 & -9 & 4 & \boxed{0} \text{ —— Remainder} \\
\end{array}
$$

Coefficients of $Q(x)$

Since the remainder is 0 when we divide by the synthetic divisor 1, we know that 1 is a rational zero of P, and hence $x - 1$ is a factor of $P(x)$. Now we can write P as the product of linear factors:

$$
\begin{aligned}
P(x) = (x - 1)Q(x) &= (x - 1)(2x^2 - 9x + 4) \\
&= (x - 1)(x - 4)(2x - 1) \\
&= 2(x - 1)(x - 4)(x - \tfrac{1}{2}).
\end{aligned}
$$

Hence, by the factor theorem, the rational zeros of P are 1, 4, and $\frac{1}{2}$. Note that 4 and $\frac{1}{2}$ (the zeros of $Q(x) = 2x^2 - 9x + 4$) were also in our list of possible positive rational zeros of P. ◆

PROBLEM 2   Find all the rational zeros of the function P defined in Problem 1. ◆

◆ **Upper and Lower Bound Rule**

An **upper bound** of a polynomial function P is any real number that is greater than or equal to the largest real zero of P. Similarly, a **lower bound** of a polynomial function P is any real number that is less than or equal to the smallest real zero of P. We can use information about bounds to confirm or discard possible rational zeros of a polynomial function, and we state this information as the **upper and lower bound rule**. As you read the rule, notice that it pertains to *real* zeros, not simply rational zeros.

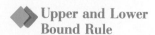

**Upper and Lower
Bound Rule**

If $P(x)$ is a polynomial with real coefficients and is written in descending powers of x with a positive leading coefficient, then

1. if $P(x)$ is divided synthetically by $x - r$, where $r > 0$, and all numbers in the final row of the synthetic division are nonnegative, then r is an *upper bound* for the real zeros of P.

2. if $P(x)$ is divided synthetically by $x - r$, where $r < 0$, and the numbers in the final row of the synthetic division alternate sign (are alternately nonpositive and nonnegative), then r is a *lower bound* for the real zeros of P.

Note: The number 0 in the final row of the synthetic division may be considered either positive or negative, as needed.

The *converse* of the upper bound rule states that if r is an upper bound for the real zeros of a polynomial function P, then all numbers in the final row of the corresponding synthetic division are positive. *The converse of the upper bound rule is not necessarily true.* For example, 5 is an upper bound for the function P defined in Example 2, since 5 is greater than or equal to 4 (the largest real zero of P). However,

$$
\begin{array}{r|rrrr}
5 & 2 & -11 & 13 & -4 \\
  &   & 10 & -5 & 40 \\
\hline
  & 2 & -1 & 8 & 36
\end{array}
$$

All numbers in the final
row are not positive

Also, by using other examples, we can show that *the converse of the lower bound rule is not necessarily true.*

EXAMPLE 3   Given the polynomial function P defined by $P(x) = 4x^3 - 18x^2 - 10x - 9$,

(a)   show that 6 is an upper bound for the zeros of P, and

(b)   show that -1 is a lower bound for the zeros of P.

SOLUTION

(a)   Using synthetic division, we find

$$
\begin{array}{r|rrrr}
6 & 4 & -18 & -10 & -9 \\
  &   & 24 & 36 & 156 \\
\hline
  & 4 & 6   & 26 & 147
\end{array}
$$

Since the synthetic divisor $6 > 0$ and all numbers in the final row of this synthetic division are nonnegative, we conclude that 6 is an upper bound for the zeros of P.

(b) Using synthetic division, we have

$$
\begin{array}{r|rrrr}
-1 & 4 & -18 & -10 & -9 \\
   &   & -4 & 22 & -12 \\
\hline
   & 4 & -22 & 12 & -21
\end{array}
$$

Since the synthetic divisor $-1 < 0$ and the numbers in the last row of this synthetic division alternate sign, we conclude that -1 is a lower bound for the zeros of P. ♦

If an upper bound and a lower bound for the zeros of a polynomial function are known, we can conclude that all real zeros of the function lie between these two bounds. Thus, referring to Example 3, we can conclude that all real zeros of P must occur in the interval $[-1, 6]$. Although this interval contains all the real zeros of the polynomial function $P(x) = 4x^3 - 18x^2 - 10x - 9$, it is *not* necessarily the *smallest* interval to contain the zeros.

P R O B L E M   3   Show that all real zeros of $P(x) = 2x^4 - 5x^3 - 2x^2 + 5x - 24$ must lie in the interval $[-2, 3]$. ♦

♦ **General Procedure for Finding Rational Zeros**

Next, we outline a general four-step procedure for finding the rational zeros of a polynomial function P with integer coefficients and of degree $n \geq 3$.

General Procedure for Finding Rational Zeros

The rational zeros of a polynomial function P with integer coefficients and of degree $n \geq 3$ can be found as follows:

Step 1 Use the rational zero theorem to list all possible rational zeros.

Step 2 Use Descartes' rule of signs to determine the number of positive and negative real zeros.

Step 3 Use synthetic division along with the upper and lower bound rule to find a rational zero r from the possible zeros listed in step 1.

Step 4 Use the rational zero from step 3 to write

$$P(x) = (x - r)Q(x)$$

and find the rational zeros of Q, repeating the previous steps if necessary, until all rational zeros are found.

EXAMPLE 4   Find all the rational zeros of $P(x) = 2x^3 - 11x^2 + 12x + 9$.

SOLUTION   We follow the four-step procedure for finding rational zeros:

Step 1   For $P(x) = 2x^3 - 11x^2 + 12x + 9$, the constant term is 9 and the coefficient of the highest power term is 2. The factors of 9 are ± 1, ± 3, and ± 9, and the factors of 2 are ± 1, ± 2. According to the rational zero theorem, the possible rational zeros of P are

$$\frac{\pm 1}{\pm 1}, \quad \frac{\pm 3}{\pm 1}, \quad \frac{\pm 9}{\pm 1} \quad \text{and} \quad \frac{\pm 1}{\pm 2}, \quad \frac{\pm 3}{\pm 2}, \quad \frac{\pm 9}{\pm 2},$$

which simplify to ± 1, ± 3, ± 9, $\pm \frac{1}{2}$, $\pm \frac{3}{2}$, and $\pm \frac{9}{2}$.

Step 2   The function P defined by

$$P(x) = \underbrace{+2x^3 - 11x^2}_{1} \underbrace{+ 12x}_{2} + 9$$

has two variations in sign, and

$$P(-x) = -2x^3 - 11x^2 \underbrace{- 12x + 9}_{1}$$

has one variation in sign. Thus, according to Descartes' rule of signs, the function P has either zero or two positive real zeros and exactly one negative real zero.

Step 3   Since we are looking for exactly one negative real zero, we begin by testing -1:

$$
\begin{array}{r|rrrr}
-1) & 2 & -11 & 12 & 9 \\
    &   & -2 & 13 & -25 \\
\hline
    & 2 & -13 & 25 & \boxed{-16} \\
\end{array}
\;\longleftarrow\; \fbox{Remainder}
$$

Since the remainder is -16, we know that -1 is not a zero of P. However, notice the numbers in the final row of the synthetic division alternate sign. Thus, according to the upper and lower bound rule, -1 is a lower bound for the zeros of P. We can now discard -3, -9, $-\frac{3}{2}$, and $-\frac{9}{2}$ as possibilities. The only other possible negative rational zero to test is $-\frac{1}{2}$. If $-\frac{1}{2}$ is not a zero, we will conclude that the negative real zero of P is irrational. Testing $-\frac{1}{2}$, we find

$$
\begin{array}{r|rrrr}
-\frac{1}{2}) & 2 & -11 & 12 & 9 \\
              &   & -1 & 6 & -9 \\
\hline
              & 2 & -12 & 18 & \boxed{0} \\
\end{array}
\;\longleftarrow\; \fbox{Remainder}
$$

$$\underbrace{}_{\fbox{Coefficients of $Q(x)$}}$$

Since the remainder is 0, we know that $-\frac{1}{2}$ is our negative real zero.

Step 4   We can now write $P(x)$ as the product of linear factors:

$$P(x) = (x + \tfrac{1}{2})Q(x)$$
$$= (x + \tfrac{1}{2})(2x^2 - 12x + 18)$$
$$= 2(x + \tfrac{1}{2})(x^2 - 6x + 9)$$
$$= 2(x + \tfrac{1}{2})(x - 3)(x - 3)$$

Hence, the rational zeros of P are 3 of multiplicity two and $-\tfrac{1}{2}$.   ◆

PROBLEM 4   Find all the rational zeros of $P(x) = x^3 - 19x + 30$.   ◆

EXAMPLE 5   Factor completely $P(x) = 2x^5 + x^4 + x^3 + 13x^2 + 5x - 6$ over the set of integers.

SOLUTION

Step 1   For $P(x) = 2x^5 + x^4 + x^3 + 13x^2 + 5x - 6$, the constant term is -6 and the coefficient of the highest-power term is 2. The factors of -6 are ± 1, ± 2, ± 3, and ± 6, and the factors of 2 are ± 1 and ± 2. According to the rational zero theorem, the possible rational zeros of P are

$$\frac{\pm 1}{\pm 1}, \quad \frac{\pm 2}{\pm 1}, \quad \frac{\pm 3}{\pm 1}, \quad \frac{\pm 6}{\pm 1} \quad \text{and} \quad \frac{\pm 1}{\pm 2}, \quad \frac{\pm 2}{\pm 2}, \quad \frac{\pm 3}{\pm 2}, \quad \frac{\pm 6}{\pm 2},$$

which simplify to ± 1, ± 2, ± 3, ± 6, $\pm \tfrac{1}{2}$, and $\pm \tfrac{3}{2}$.

Step 2   We find that

$$P(x) = +2x^5 + x^4 + x^3 + 13x^2 + 5x - 6$$

has one variation in sign, and

$$P(-x) = -2x^5 + x^4 - x^3 + 13x^2 - 5x - 6,$$

has four variations in sign. Thus, according to Descartes' rule of signs, the function P has exactly one positive real zero and either zero, two, or four negative real zeros.

Step 3   Testing 1 as a possible positive rational zero, we find

1)	2	1	1	13	5	−6
		2	3	4	17	22
	2	3	4	17	22	⑯ — Remainder

Since the remainder is 16, we know that 1 is not a zero of P. However, notice that all the numbers in the bottom row of the synthetic division are non-negative. Thus, according to the upper and lower bound rule, 1 is an upper bound for the zeros of P. Thus, we discard 2, 3, 6, and $\frac{3}{2}$ as possibilities. The only other possible positive rational zero is $\frac{1}{2}$. Testing $\frac{1}{2}$, we find

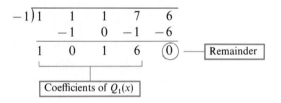

$$\frac{1}{2}\overline{)2 \quad 1 \quad 1 \quad 13 \quad 5 \quad -6}$$
$$\phantom{\frac{1}{2})}\; 1 \quad 1 \quad 1 \quad 7 \quad 6$$
$$\overline{\phantom{\frac{1}{2})}2 \quad 2 \quad 2 \quad 14 \quad 12 \quad \textcircled{0}} \;\text{—}\boxed{\text{Remainder}}$$

Coefficients of $Q(x)$

Since the remainder is 0, we conclude that $\frac{1}{2}$ is our positive real zero.

Step 4   We now write

$$P(x) = (x - \tfrac{1}{2})(2x^4 + 2x^3 + 2x^2 + 14x + 12)$$
$$= (x - \tfrac{1}{2}) \cdot 2(x^4 + x^3 + x^2 + 7x + 6)$$
$$= (2x - 1)(x^4 + x^3 + x^2 + 7x + 6)$$

and proceed to find the zeros of $Q(x) = x^4 + x^3 + x^2 + 7x + 6$. Using the rational zero theorem and the fact that we have already found the only positive real zero of P, we can say that the possible rational zeros of Q are -1, -2, -3, and -6. Testing -1 as a possible negative rational zero for Q, we find

$$-1\overline{)1 \quad 1 \quad 1 \quad 7 \quad 6}$$
$$\; -1 \quad 0 \quad -1 \quad -6$$
$$\overline{1 \quad 0 \quad 1 \quad 6 \quad \textcircled{0}} \;\text{—}\boxed{\text{Remainder}}$$

Coefficients of $Q_1(x)$

Since the remainder is 0, we know that -1 is a negative rational zero of Q and, therefore, a negative rational zero of P. We now write

$$P(x) = (2x - 1)(x + 1)(x^3 + x + 6),$$

and proceed to find the zeros of $Q_1(x) = x^3 + x + 6$. Using the rational zero theorem and the fact that we have already found the only positive real zero of P, we can say that the possible rational zeros of Q_1 are -1, -2, -3, and -6. Also, by Descartes' rule of signs, we know that Q_1 has exactly one negative real zero. Therefore, P has exactly two negative real zeros. Checking each of our rational possibilities, we find that neither -1, -2, -3, nor -6 are actual rational zeros of Q_1. Hence, the other negative real zero must be irrational. Thus, factored completely over the set of integers,

$$P(x) = (2x - 1)(x + 1)(x^3 + x + 6).$$

In Section 5.4, we will discuss methods for approximating irrational zeros.   ◆

PROBLEM 5   Find all rational zeros of $P(x) = 3x^4 + x^3 + 10x^2 - 26x + 12$.   ◆

◆ Application: Ash Pan for a Coal Stove

For some applied problems, it is necessary to solve a polynomial equation of the form

$$\underbrace{a_n x^n + a_{n-1} x^{n-1} + \cdots + a_1 x + a_0}_{\boxed{P(x)}} = 0,$$

where $n \geq 3$. If the polynomial $P(x)$ can be written as the product of linear and quadratic factors, then we can solve the equation by applying the methods discussed in this section.

EXAMPLE 6   An ash pan for a small coal stove is made from a piece of sheet metal, 8 inches by 15 inches, by cutting out a square from each corner and turning up the resulting sides. What are the *two* possible dimensions of the square cutout if the volume of the pan that is formed is 90 cubic inches?

SOLUTION   Recall from elementary geometry that the volume V of a rectangular box is the product of its length, width, and height. If we let

$$x = \text{the length of the side of the square cutout,}$$

then the dimensions of the ash pan are

$$\text{height} = x, \quad \text{length} = 15 - 2x, \quad \text{and} \quad \text{width} = 8 - 2x,$$

as shown in Figure 5.1. Thus,

$$V = (15 - 2x)(8 - 2x)x = 4x^3 - 46x^2 + 120x.$$

FIGURE 5.1
The dimensions of the ash pan

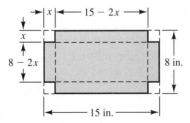 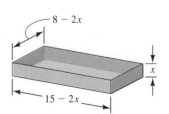

Now to find the values of x when $V = 90$ cubic inches, we must solve the equation

$$90 = 4x^3 - 46x^2 + 120x$$

$$4x^3 - 46x^2 + 120x - 90 = 0$$

$$\underbrace{2x^3 - 23x^2 + 60x - 45}_{\boxed{P(x)}} = 0$$

By the rational zero theorem, the possible rational zeros of P are

$$\pm 1, \quad \pm 3, \quad \pm 5, \quad \pm 9, \quad \pm 15, \quad \pm 45, \quad \pm\tfrac{1}{2}, \quad \pm\tfrac{3}{2}, \quad \pm\tfrac{5}{2}, \quad \pm\tfrac{9}{2}, \quad \pm\tfrac{15}{2}, \quad \text{and} \quad \pm\tfrac{45}{2}.$$

However, the length of the side of the cutout cannot be a negative number, nor can it be greater than 4 inches (since the piece of metal is 8 inches wide). Since $0 \le x \le 4$, we check only the possible rational zeros $1, 3, \tfrac{1}{2}, \tfrac{3}{2}$, and $\tfrac{5}{2}$. Testing $\tfrac{3}{2}$, we find

$$
\begin{array}{r|rrrr}
\tfrac{3}{2} & 2 & -23 & 60 & -45 \\
  &   & 3 & -30 & 45 \\
\hline
  & 2 & -20 & 30 & \enspace\textcircled{0}
\end{array}
$$

Coefficients of $Q(x)$ ← (bracket under $2\;-20\;30$)

Remainder ← (box pointing to $\textcircled{0}$)

Since the remainder is 0, we know that $x = \tfrac{3}{2}$ is a root of the equation $P(x) = 0$. We can now write

$$
\begin{aligned}
P(x) = (x - \tfrac{3}{2})\, Q(x) &= (x - \tfrac{3}{2})(2x^2 - 20x + 30) \\
&= (x - \tfrac{3}{2})[2(x^2 - 10x + 15)] \\
&= (2x - 3)(x^2 - 10x + 15)
\end{aligned}
$$

We can find the zeros of the quadratic function $Q_1(x) = x^2 - 10x + 15$ from the quadratic formula, as follows:

$$x = \frac{-(-10) \pm \sqrt{(-10)^2 - 4(1)(15)}}{2(1)} = 5 \pm \sqrt{10}.$$

Since $0 \le x \le 4$, we discard $5 + \sqrt{10} \approx 8.16$. In summary, the volume of the pan is 90 cubic inches when $x = \tfrac{3}{2}$ inches or $x = 5 - \sqrt{10} \approx 1.84$ inches.   ◆

PROBLEM 6   What is the length and width of the ash pan in Example 6 when the volume of the pan is 90 cubic inches?   ◆

Exercises 5.3

Basic Skills

In Exercises 1–10,

(a) use the rational zero theorem to list all possible rational zeros of P.

(b) use Descartes' rule of signs to determine the possibilities for the number of positive and negative real zeros of P.

1. $P(x) = x^3 + 5x^2 + 7x + 6$

2. $P(x) = x^4 + 7x^3 - 3x^2 + 2x + 10$

3. $P(x) = 2x^4 - 3x^3 + 2x^2 - 2x + 5$

4. $P(x) = 3x^3 - 5x^2 + 2x - 4$

5. $P(x) = 4x^3 - 3x^2 + 5x + 8$

6. $P(x) = 6x^5 - 2x^4 + 3x^2 - 9$

7. $P(x) = 3x^6 + x^2 + x - 12$

8. $P(x) = 8x^4 - 2x + 15$

9. $P(x) = 12x^5 + x^3 + x^2 + x - 3$

10. $P(x) = 2x^7 - 2x^3 - 2x^2 + x - 18$

In Exercises 11–18, use the upper and lower bound rule to show that all real zeros of P are in the given interval [a, b].

11. $P(x) = x^3 - 2x^2 + 5x + 4; \quad [-1, 2]$

12. $P(x) = x^4 - 2x^2 - 7x - 2; \quad [-2, 3]$

13. $P(x) = 3x^4 - 8x^3 - 15x^2 - 12; \quad [-2, 4]$

14. $P(x) = 2x^3 - 6x^2 - 21x - 5; \quad [-3, 6]$

15. $P(x) = 8x^5 - 60x - 5; \quad [-2, 2]$

16. $P(x) = 4x^4 + 20x^3 - 17x^2 + 41x - 6; \quad [-6, 1]$

17. $P(x) = 2x^4 - 3x^3 + 6x^2 + x - 15; \quad [-2, \frac{3}{2}]$

18. $P(x) = 3x^4 + x^3 + 7x^2 - 6x + 2; \quad [-\frac{1}{3}, \frac{2}{3}]$

In Exercises 19–36, (a) find all rational zeros of each polynomial function and (b) factor the polynomial completely over the set of integers.

19. $P(x) = x^3 - 6x^2 + 11x - 6$

20. $P(x) = x^3 + 7x^2 + 16x + 12$

21. $f(x) = 2x^3 - 5x^2 - 14x + 8$

22. $F(x) = 6x^3 + 35x^2 - 8x - 12$

23. $H(x) = 4x^3 + 9x^2 + 26x + 6$

24. $G(x) = 2x^3 - 9x^2 + 14x - 10$

25. $P(x) = 12x^3 + 4x^2 - 17x + 6$

26. $Q(x) = 24x^3 - 70x^2 + 63x - 18$

27. $f(x) = 2x^4 + x^3 + 3x - 18$

28. $F(x) = x^4 - 11x^3 - 14x - 26$

29. $g(x) = 3x^4 - 14x^3 + 16x^2 - 3x + 8$

30. $h(x) = 4x^4 - 3x^3 - 12x^2 + 25x - 12$

31. $P(x) = 2x^4 + 9x^3 + 14x^2 + 9x + 2$

32. $f(x) = 4x^4 - 28x^3 + 67x^2 - 63x + 18$

33. $f(x) = x^5 + x^4 - 6x^3 + 8x - 16$

34. $P(x) = 6x^5 - 8x^4 + 5x^3 - 16x^2 - 25x + 10$

35. $F(x) = 2x^5 + 5x^4 - 4x^3 - 11x^2 + 4x + 4$

36. $D(x) = 16x^5 - 80x^4 + 127x^3 - 59x^2 - 8x + 4$

37. The total profit P a company makes is a function of the number of units x it sells and is given by $P(x) = x^3 - x^2 - 12x - 210$. At least how many units must the company sell to make a profit?

38. Under certain conditions, the velocity v (in feet per second, ft/s) of an object as a function of time t (in seconds) is given by $v(t) = 4t^3 - 11t^2 - 6t + 9$, where $0 \le t \le 4$. Determine the times t when $v(t) = 0$ ft/s.

39. The width of a rectangular toy box is twice its height and the length of the box is 1 foot more than its width, as shown in the figure. The volume of the box is 18 cubic feet. What are the dimensions of the box?

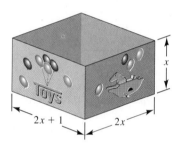

40. Recall from elementary geometry that the volume V of a cone with radius r and height h is $V = \frac{1}{3}\pi r^2 h$. If a conical sandpile has a slant height of 10 meters, as shown in the sketch, for what rational value of h is $V = 128\pi$ cubic meters?

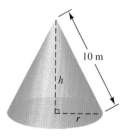

Critical Thinking

In Exercises 41–46, find all rational zeros of each polynomial function. [Hint: To extend the rational zero theorem to include a polynomial function with rational coefficients, simply multiply both sides by the LCD to eliminate all fractions. The function that is formed has the same zeros as the original function.]

41. $P(x) = \frac{1}{2}x^3 + \frac{5}{6}x^2 + \frac{8}{3}x - 1$

42. $f(x) = \frac{1}{2}x^3 + x^2 - \frac{1}{8}x - \frac{1}{4}$

43. $g(x) = \frac{3}{4}x^4 + x^3 - \frac{1}{4}x^2 + x - 1$

44. $G(x) = \frac{1}{10}x^4 - \frac{1}{2}x^2 - x - \frac{3}{5}$

45. $f(x) = 0.2x^3 - 2.3x^2 + 3.1x + 1.4$

46. $P(x) = 0.6x^4 - 1.1x^3 + 0.7x^2 - 2.2x + 2.4$

47. Find all integers k for which the polynomial function $P(x) = x^3 - 2x^2 + kx - 3$ has one rational zero and two imaginary zeros.

48. What are the possible rational zeros of

$$P(x) = x^n + a_{n-1}x^{n-1} + a_{n-2}x^{n-2} + \cdots + a_1x + a_0$$

if each coefficient is an integer and a_0 is a prime number?

49. When the length of one side of a cube is reduced by 2 inches, the volume of the remaining rectangular solid is 75 cubic inches. What is the length of the side of the original cube?

50. A rectangular box has length 2 inches, width 2 inches, and height 3 inches. Each of these dimensions is increased uniformly until a new box is formed whose volume is triple that of the original box. What are the dimensions of the new box?

Calculator Activities

 In Exercises 51–54,

(a) *use a graphing calculator to generate the graphs of the two equations in the same viewing rectangle. Then* approximate *the coordinates of the points where the graphs intersect by tracing to the intersection points.*

(b) *determine the* exact *coordinates of the intersection points by solving the two equations simultaneously and applying the rational zero theorem.*

51. $y = x^3$; $21x - 4y = 10$

52. $y = x^4$; $20x + 27y = 7$

53. $y = \dfrac{6}{x}$; $y = x^2 + 2x - 5$

54. $y = \dfrac{3}{x}$; $x^2 + y^2 = 10$

55. A piece of cardboard, 4 cm by 5 cm, is made into an open rectangular box by cutting out a square from each corner and turning up the resulting sides, as shown in the figure. If the volume of the box that is formed is 3 cubic centimeters, what are the *two* possibilities for the length of the side of the square cutout?

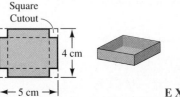

E X E R C I S E   5 5

56. Recall from elementary geometry that the volume V of a cylinder with radius r and height h is $V = \pi r^2 h$. Suppose the slant distance from the top of an oil drum to the bottom is 12 feet, as sketched in the figure. Find *two* values of h if the volume of the drum is 160π cubic feet.

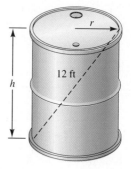

Approximating Irrational Zeros of Polynomial Functions

5.4

The graph of a polynomial function with real coefficients is continuous with smooth, rounded turns; that is, its graph contains no gap, break, jump or sharp turn. The graph of a typical polynomial function is shown in Figure 5.2. Throughout this section we will refer to the intervals and pionts labeled on this graph to discuss its zeros.

FIGURE 5.2
The graph of a typical polynomial function P.

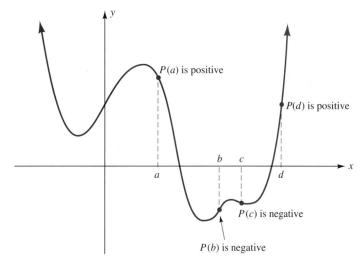

In this section, we use the notion of continuity to help develop a method for approximating the irrational zeros of a polynomial function with real coefficients.

◆ Opposite Sign Property

Notice in Figure 5.2 that if a polynomial function P changes sign in an interval $[a, b]$, then the graph of P crosses the x-axis between a and b. Hence, the polynomial function P has a real zero between a and b. We refer to this observation about a zero of a polynomial function as the **opposite sign property**.

 **Opposite Sign Property**

> If $P(x)$ is a polynomial with real coefficients and if, for the real numbers a and b, $P(a)$ and $P(b)$ have opposite signs, then there exists at least one real zero between a and b.

The converse of the opposite sign property states that if there is at least one real zero of P between the real numbers a and b, then $P(a)$ and $P(b)$ have opposite signs. *The converse of the opposite sign property is not necessarily true.* For example, referring to Figure 5.2, there are two real zeros between a and d since the graph of P crosses the x-axis twice between these two values. However, $P(a)$ and $P(d)$ do not have opposite signs. Both $P(a)$ and $P(d)$ are positive.

EXAMPLE 1   Show that $P(x) = x^3 - 2x^2 - 1$ has at least one real zero in the interval $[2, 3]$.

SOLUTION   Using direct substitution or synthetic division, we evaluate $P(2)$ and $P(3)$. Choosing synthetic division, we find

$$
\begin{array}{r|rrrr}
2 & 1 & -2 & 0 & -1 \\
  &   & 2 & 0 & 0 \\
\hline
  & 1 & 0 & 0 & \boxed{-1} \leftarrow \fbox{$P(2)$}
\end{array}
\qquad
\begin{array}{r|rrrr}
3 & 1 & -2 & 0 & -1 \\
  &   & 3 & 3 & 9 \\
\hline
  & 1 & 1 & 3 & \boxed{8} \leftarrow \fbox{$P(3)$}
\end{array}
$$

Opposite signs

Since $P(2)$ and $P(3)$ have opposite signs, we conclude, by the opposite sign property, that the function P has at least one real zero in the interval $[2, 3]$.

◆

PROBLEM 1   Show that $P(x) = x^3 + 2x - 5$ has at least one real zero in the interval $[1, 2]$.

◆

◆ **Method of Successive Approximations**

By the rational zero theorem (Section 5.3), the possible rational zeros for the polynomial function in Example 1, $P(x) = x^3 - 2x^2 - 1$, are ± 1. However, using synthetic division, we find

$$
\begin{array}{r}
1)\overline{1 \quad -2 \quad 0 \quad -1} \\
\quad 1 \quad -1 \quad -1 \\
\hline
1 \quad -1 \quad -1 \quad -2
\end{array}
\qquad
\begin{array}{r}
-1)\overline{1 \quad -2 \quad 0 \quad -1} \\
\quad -1 \quad 3 \quad -3 \\
\hline
1 \quad -3 \quad 3 \quad -4
\end{array}
$$

Since neither remainder is zero, we conclude that ± 1 are not rational zeros of P. Hence P has no rational zero. However, we know that

$$P(x) = +x^3 - 2x^2 - 1$$
$$\underset{1}{\overleftrightarrow{}}$$

has one variation in sign, and

$$P(-x) = -x^3 - 2x^2 - 1,$$

has no variation in sign. Thus, according to Descartes' rule of signs (Section 5.3), the function P has one positive real zero and no negative real zero. Since the positive real zero is not rational, it must be irrational. From Example 1, we know that this irrational zero must be in the interval $[2, 3]$. We can approximate this irrational zero by using the **method of successive approximations**.

Method of Successive Approximations

If $P(x)$ has one real zero in the interval $[a, b]$, where a and b are consecutive integers, then this zero can be approximated as follows:

Step 1   Divide the interval $[a, b]$ into ten parts to form ten equal subintervals.

Step 2   Determine the subinterval in step 1 where $P(x)$ changes sign.

Step 3   Divide the interval found in step 2 into ten equal parts to form ten subintervals.

Step 4   Determine the subinterval in step 3 where $P(x)$ changes sign.

Step 5   Continue the process until the desired accuracy is achieved.

Note: The method of successive approximations involves working with decimal numbers. To help make the computations, use a calculator.

EXAMPLE 2   Use the method of successive approximations to determine to the nearest hundredth the irrational zero for the polynomial function defined in Example 1.

SOLUTION   From Example 1 and our preceding discussion, we know that $P(x) = x^3 - 2x^2 - 1$ has one irrational zero in the interval $[2, 3]$.

Step 1   Divide the interval $[2, 3]$ into ten equal subintervals:

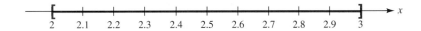

Step 2   Determine the subinterval where $P(x)$ changes sign. Using either direct substitution or synthetic division, and making the computations with a calculator, we find

$$P(2.2) = -0.032 \qquad \text{and} \qquad P(2.3) = +0.587.$$

Opposite signs

Thus, by the opposite sign property, the irrational zero is in the interval $[2.2, 2.3]$.

Step 3   Divide the interval $[2.2, 2.3]$ into ten equal subintervals:

Step 4   Again, we determine the subinterval where $P(x)$ changes sign. Using either direct substitution or synthetic division and a calculator, we find

$$P(2.20) = -0.032 \qquad \text{and} \qquad P(2.21) = +0.025661$$

Opposite signs

Thus, by the opposite sign property, the irrational zero is in the interval $[2.20, 2.21]$.

Step 5   Continuing the process, we divide the interval $[2.20, 2.21]$ into ten equal subintervals, and then determine in which one of these subintervals $P(x)$ changes sign. You may verify that

$$P(2.205) \approx -0.0032849 \quad \text{and} \quad P(2.206) \approx 0.0024858.$$

Thus, by the opposite sign property, the irrational zero is in the interval $[2.205, 2.206]$. Now rounding back to two decimal places, we conclude that the irrational zero is 2.21 to the nearest hundredth.   ◆

PROBLEM 2   For the polynomial function defined in Problem 1, use the method of successive approximations to determine to the nearest hundredth the irrational zero in the interval $[1, 2]$.   ◆

◆ Using a Graphing Calculator to Approximate Zeros

The irrational zeros of a polynomial function may also be approximated by using a calculator or computer with graphing capabilities.

EXAMPLE 3   Given $P(x) = 7x^3 - 4x + 1$,

(a)   show that P has two irrational zeros in the interval $[0, 1]$ and

(b)   use a graphing calculator to find the approximate values of these two zeros to the nearest thousandth.

SOLUTION

(a)   According to Descartes' rule of signs (Section 5.3), P has either zero or two positive real zeros and exactly one negative real zero. Dividing the interval $[0, 1]$ into two equal subintervals, we see that

$$P(0) = 1, \qquad P(\tfrac{1}{2}) = -\tfrac{1}{8}, \qquad \text{and} \qquad P(1) = 4.$$

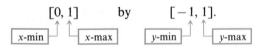

Thus, by the opposite sign property, we conclude that P does have two real zeros in the interval $[0, 1]$. By the rational zero theorem (Section 5.3) the only possible rational zeros of P in this interval are 1 and $\tfrac{1}{7}$. However, since $P(1) = 4$ and $P(\tfrac{1}{7}) = \tfrac{22}{49}$, we conclude that there is no rational zero in the interval $[0, 1]$. Hence, the zeros in the interval $[0, 1]$ must be irrational.

(b)   To display the graph of $P(x) = y = 7x^3 - 4x + 1$ on a calculator, we begin by choosing a viewing rectangle by pressing the RANGE key. Since these two zeros are in the interval $[0, 1]$, a reasonable choice for the viewing rectangle is

$$[0, 1] \qquad \text{by} \qquad [-1, 1].$$

Now, by entering the equation $y = 7x^3 - 4x + 1$, we obtain the desired portion of the graph in the viewing rectangle (see Figure 5.3). Next, activate the TRACE key and move the blinking cursor to the first x-intercept. To the nearest tenth, we find that $x = 0.3$. Tracing to the second x-intercept, we find that $x = 0.6$ (to the nearest tenth).

   To obtain more precise values, we can *zoom in* on these x-intercepts by activating the Zoom feature on the calculator. Repeating the process of tracing and zooming in, we find that the approximate irrational zeros of

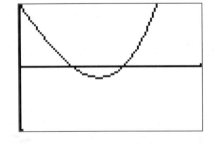

FIGURE 5.3

The graph of
$y = 7x^3 - 4x + 1$,
$0 \le x \le 1$, shown in the
viewing rectangle $[0, 1]$ by
$[-1, 1]$.

P in the interval $[0, 1]$ are

$$x = 0.295 \qquad \text{and} \qquad x = 0.564,$$

to the nearest thousandth.   ◆

PROBLEM 3   Referring to Example 3, show that the negative real zero of P is in the interval $[-1, 0]$. Then use a calculator or computer with graphing capabilities to find the approximate value of this zero, to the nearest hundredth.   ◆

◆   **General Procedure for Finding Real Zeros**

We now outline a general four-step procedure for finding all *real* zeros of a polynomial function P of degree $n \geq 3$.

General Procedure for Finding Real Zeros

If P is a polynomial function with integer coefficients and of degree $n \geq 3$, then the rational zeros of P can be found exactly and the irrational zeros of P can be found approximately as follows:

Step 1   Find all the rational zeros $r_1, r_2, \ldots, r_k$ by using the procedure outlined in Section 5.3.

Step 2   Use the rational zeros from step 1 to write

$$P(x) = (x - r_1)(x - r_2) \cdots (x - r_k)Q(x)$$

then use Descartes' rule of signs to determine if $Q(x)$ has any positive or negative irrational zeros.

Step 3   Use synthetic division along with the opposite sign property to determine an interval on which each irrational zero is located.

Step 4   Use the method of successive approximations or a calculator with graphing capabilities to approximate the irrational zeros to the desired degree of accuracy.

EXAMPLE 4   Find all real zeros for $P(x) = 4x^4 - 2x^3 + 2x^2 + 7x - 4$. Record irrational zeros to the nearest hundredth.

SOLUTION

Step 1   By the rational zero theorem (Section 5.3), the possible rational zeros of P are ± 1, ± 2, ± 4, $\pm \frac{1}{2}$, and $\pm \frac{1}{4}$. Testing each possibility, we find that $\frac{1}{2}$ is the only rational zero for P:

$$
\begin{array}{r|rrrrr}
\frac{1}{2}) & 4 & -2 & 2 & 7 & -4 \\
 &   & 2 & 0 & 1 & 4 \\
\hline
 & 4 & 0 & 2 & 8 & 0
\end{array}
$$

Step 2   So we can write

$$P(x) = (x - \tfrac{1}{2})(4x^3 + 2x + 8)$$
$$= 2(x - \tfrac{1}{2})(2x^3 + x + 4)$$

For

$$Q(x) = +2x^3 + x + 4,$$

there are no sign changes, and for

$$Q(-x) = -2x^3 - x + 4,$$

there is one sign change. Thus by Descartes' rule of signs (Section 5.3), Q has no positive real zero and exactly one negative real zero. Since this zero is not rational, it must be irrational.

Step 3   We now search for two consecutive integers a and b such that $Q(a)$ and $Q(b)$ have opposite signs. Notice that

```
 -2)2    0    1    4              -1)2    0    1    4
      -4    8  -18                    -2    2   -3
 ─────────────────              ──────────────────
    2   -4    9  (-14)← Q(-2)       2   -2    3   (1)← Q(-1)
```

Opposite signs

Since $Q(-2) = -14$ and $Q(-1) = 1$, the opposite sign property tells us that the negative irrational zero is in the interval $[-2, -1]$,

Step 4   Using the method of successive approximations on the interval $[-2, -1]$, we find that

$$Q(-1.1) = 0.238 \qquad \text{and} \qquad Q(-1.2) = -0.656,$$
$$Q(-1.12) = 0.070144 \qquad \text{and} \qquad Q(-1.13) = -0.015794,$$
$$Q(-1.128) \approx 0.0015017 \qquad \text{and} \qquad Q(-1.129) \approx -0.00713938.$$

Thus, we know that the irrational zero is in the interval $[-1.129, -1.128]$. In conclusion, the polynomial function P has two real zeros:

$$\frac{1}{2} \qquad \text{and} \qquad -1.13 \text{ (to the nearest hundredth)} \qquad \blacklozenge$$

PROBLEM 4   Find all real zeros for $P(x) = x^4 + 2x^3 + x^2 - 2x - 8$. Record irrational zeros to the nearest hundredth. $\qquad \blacklozenge$

◆ **Application: Silo for Storing Grain**

Many applied problems involve the solution of a polynomial equation of the form

$$a_n x^n + a_{n-1} x^{n-1} + \cdots + a_1 x + a_0 = 0,$$

where $n \geq 3$. Such equations often have solutions that are irrational numbers. We illustrate one such application in the next example.

EXAMPLE 5   A silo has the shape of a right circular cylinder surmounted by a hemisphere and holds 1000π cubic feet of grain. If the height of the silo is 60 feet, what is the radius of its base?

SOLUTION   Let

$$x = \text{the radius of the cylindrical part of the silo.}$$

Since the radius of the hemisphere is the same as the radius of the cylinder, the height of the cylindrical part of the silo is $60 - x$, as shown in Figure 5.4. Now the volume V of the silo is the sum of the volumes of the cylinder and hemisphere:

$$V = \pi r^2 h + \tfrac{2}{3}\pi r^3$$

$$1000\pi = \pi x^2(60 - x) + \tfrac{2}{3}\pi x^3$$

$$3000 = 3x^2(60 - x) + 2x^3 \qquad \text{Multiply both sides by } 3/\pi$$

$$\underbrace{x^3 - 180x^2 + 3000}_{P(x)} = 0 \qquad \text{Write in the form } P(x) = 0$$

The function P defined by $P(x) = x^3 - 180x^2 + 3000$ has two variations in sign and $P(-x) = -x^3 - 180x^2 + 3000$ has one variation in sign. Thus, according to Descartes' rule of signs (Section 5.3), the function P has either zero or two positive real zeros and exactly one negative real zero. Since the radius must be positive, we disregard the negative possibility and search for a positive interval $[a, b]$ where $P(a)$ and $P(b)$ have opposite signs. Observe that

$$
\begin{array}{r|rrrr}
4) & 1 & -180 & 0 & 3000 \\
   &   & 4 & -704 & -2816 \\
\hline
   & 1 & -176 & -704 & \boxed{184} \leftarrow \fbox{$P(4)$}
\end{array}
\qquad
\begin{array}{r|rrrr}
5) & 1 & -180 & 0 & 3000 \\
   &   & 5 & -875 & -4375 \\
\hline
   &   & -175 & -875 & \boxed{-1375} \leftarrow \fbox{$P(5)$}
\end{array}
$$

Opposite signs

Thus, by the opposite sign property, a zero occurs in the interval $[4, 5]$. Using the method of successive approximations on the interval $[4, 5]$, we find

$$P(4.1) = 43.121 \qquad \text{and} \qquad P(4.2) = -101.112$$

$$P(4.13) = 0.202997 \qquad \text{and} \qquad P(4.14) = -14.170056.$$

Since $P(4.13) = 0.202997$ is so much closer to zero than is $P(4.14) = -14.170056$, it is reasonable to conclude that a root of the equation $P(x) = 0$ is 4.13 (to the nearest hundredth). Thus, the radius of the base of the silo is approximately 4.13 feet. ◆

FIGURE 5.4
A silo with height 60 ft and radius x

PROBLEM 5   According to Descartes' rule of signs, the function P in Example 5 has another positive real zero. Show that this positive real zero is in the interval [179, 180]. Why isn't this zero considered a solution to the given application problem?   ◆

Exercises 5.4

 Basic Skills

In Exercises 1–10,

(a) *show that the given polynomial function has exactly one irrational zero in the indicated interval.*

(b) *use the method of successive approximations or a calculator with graphing capabilities to approximate this irrational zero to the nearest hundredth.*

1. $P(x) = x^3 + 2x - 1$; $[0, 1]$

2. $Q(x) = x^3 + 3x - 8$; $[1, 2]$

3. $f(x) = 2x^3 - 4x^2 - 3$; $[2, 3]$

4. $R(x) = 3x^3 + x + 6$; $[-2, -1]$

5. $H(x) = 4x^3 + 6x^2 - 5x + 11$; $[-3, -2]$

6. $f(x) = x^3 + 7x^2 - 6x + 8$; $[-8, -7]$

7. $g(x) = x^4 - 3x^3 + 2x^2 - 3$; $[-1, 0]$

8. $P(x) = 5x^4 - 2x - 1$; $[0, 1]$

9. $F(x) = 2x^5 - 3x^2 - 4$; $[1, 2]$

10. $h(x) = x^6 + 2x^2 + 5x - 1$; $[-2, -1]$

Each of the polynomial functions in Exercises 11–14 has more than one real zero in the indicated interval. Find these zeros (nearest hundredth).

11. $P(x) = 8x^3 - 8x + 1$; $[0, 1]$

12. $P(x) = 4x^3 - 12x^2 + 8x + 1$; $[1, 2]$

13. $P(x) = 25x^3 - 35x^2 + 13x - 1$; $[0, 1]$

14. $P(x) = 10x^3 + 45x^2 + 67x + 33$; $[-2, -1]$

In Exercises 15–34, find all real zeros for the given polynomial function. Record irrational zeros to the nearest hundredth.

15. $P(x) = x^3 + x - 4$ 16. $P(x) = 2x^3 + x + 9$

17. $F(x) = 2x^3 + x^2 + 5$

18. $f(x) = x^3 - 6x^2 - 5$

19. $G(x) = x^3 - 3x^2 - 6x + 6$

20. $h(x) = 2x^3 + 4x^2 - 4x - 5$

21. $P(x) = 3x^3 - 2x^2 - 8x - 2$

22. $Q(x) = x^3 + 5x^2 + 2x - 3$

23. $f(x) = x^4 - 4x - 3$

24. $P(x) = x^4 + 3x^3 + 2$

25. $h(x) = 2x^4 + 2x^3 + x^2 - 3x - 4$

26. $G(x) = 4x^4 - 4x^2 + x - 1$

27. $P(x) = 2x^4 + 3x^3 + 10x^2 + 17x + 3$

28. $P(x) = 2x^4 - 7x^3 + 3x^2 - 4x + 2$

29. $f(x) = 3x^5 - 4x^4 - 5x^3 - 2x^2 + 4x + 8$

30. $F(x) = 2x^5 - 7x^4 - 2x^3 - 5x^2 - 11x - 4$

31. $P(x) = 2x^5 + 6x^4 - 10x^3 - 25x^2 + 8x + 4$

32. $P(x) = 3x^5 + 11x^4 + 3x^3 - 22x^2 - 12x + 8$

33. $h(x) = 3x^6 - 3x^5 + 2x^3 - 2x^2 + 6x - 6$

34. $G(x) = 4x^6 - 12x^5 - x^4 + 3x^3 - 8x^2 + 2$

35. As a result of poaching, the number of elephants in a certain herd is declining rapidly. It is estimated that the population P of the herd is a function of time t (in years) and is given by

$$P(t) = -t^3 - t^2 + t + 150.$$

In approximately how many years will the herd be extinct?

36. The charge q (in coulombs) on a certain electrical capacitor is a function of the amount of time t (in seconds) that the capacitor has been charging and is given by

$$q(t) = 3t^3 + 4t.$$

After how many seconds is the charge on the capacitor 10 coulombs?

37. The rectangular jewelry box shown in the figure has a square base, and its height is 2 inches less than the length of its base. Find the dimensions of the box if the volume of the box is 48 cubic inches.

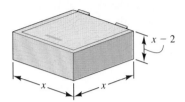

38. The allergy capsule shown in the figure has the shape of a cylinder surmounted by a hemisphere on each end. The total length of the capsule is 10 millimeters and it holds 25π cubic millimeters of medicine. Find the diameter d of the capsule.

Critical Thinking

In Exercises 39–44,

(a) sketch the graph of the two equations on the same coordinate plane.

(b) by solving the equations simultaneously, determine to the nearest hundredth the coordinates of the point(s) where their graphs intersect.

39. $y = x^3$ and $y = x + 1$

40. $y = x^3$ and $y = x^2 - 2x - 8$

41. $y = x^4$ and $x^2 + y = 1$

42. $y = x^4$ and $x + y = 1$

43. $y = \dfrac{1}{x}$ and $y = x^2 - 4$

44. $y = \dfrac{1}{x^2}$ and $x = 4 - y^2$

45. Increasing the length of each edge of a cube by 1 cm doubles the volume of the cube. Find the length of an edge of the original cube.

46. The hypotenuse of a right triangle is 2 cm longer than one of the legs of the triangle. The area of the triangle is 8 square centimeters. Find the lengths of the sides of the triangle.

Calculator Activities

A cubic formula may be used to solve cubic equations of the form $x^3 + ax + b = 0$, where a and b are real numbers. The formula states that the three roots of this equation are

$$A + B, \; -\frac{A + B}{2} + \frac{A - B}{2}\sqrt{-3},$$

and

$$-\frac{A + B}{2} - \frac{A - B}{2}\sqrt{-3},$$

where

$$A = \sqrt[3]{-\frac{b}{2} + \sqrt{\frac{b^2}{4} + \frac{a^3}{27}}}$$

and

$$B = \sqrt[3]{-\frac{b}{2} - \sqrt{\frac{b^2}{4} + \frac{a^3}{27}}}.$$

In Exercises 47 and 48, use the cubic formula and a calculator to approximate the three roots of the given equation.

47. $x^3 + x - 4 = 0$ 48. $2x^3 + x + 9 = 0$

The cubic equation $y^3 + py^2 + qy + r = 0$ can be written in the form $x^3 + ax + b = 0$ by replacing y with $x - \dfrac{p}{3}$. In Exercises 49 and 50,

(a) use this substitution to write the given equation in the form $x^3 + ax + b = 0$.

(b) use the cubic formula from Exercises 47–48 and a calculator to approximate the three roots of the given equation.

49. $y^3 - 3y^2 - 9y + 27 = 0$ 50. $y^3 - 6y^2 - 5 = 0$

5.5 ◆ Graphs of Polynomial Functions

In this section, we discuss a method for sketching the graph of a polynomial function P of degree $n \geq 3$. If we can find the real zeros of P, then we have the x-intercepts of the graph of P. We can usually draw a fairly accurate sketch of the graph of P by following these three steps:

1. Use the x- and y-intercepts.

2. Investigate the left and right behavior of the graph.

3. Determine the possible shapes for the middle of the graph.

◆ **Left and Right Behavior**

Consider sketching the graph of the polynomial function

$$P(x) = a_n x^n + a_{n-1} x^{n-1} + \cdots + a_1 x + a_0,$$

where the coefficients $a_n, a_{n-1}, \ldots, a_1, a_0$ are real numbers with $a_n \neq 0$. As $|x|$ gets larger and larger, the highest-degree term of the polynomial function, $a_n x^n$, becomes dominant in relation to the other terms of the polynomial. Because of this, the graph of a polynomial function P eventually rises or falls depending on

(i) the sign of its leading coefficient a_n, and

(ii) whether n is even or odd.

Figure 5.5 illustrates four possibilities for the left and right behavior of the graph of a polynomial function P as x decreases without bound ($x \to -\infty$) and as x increases without bound ($x \to \infty$). We will refer to these graphs in the following discussion.

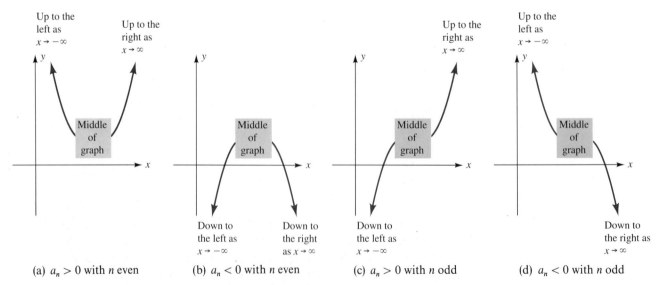

(a) $a_n > 0$ with n even   (b) $a_n < 0$ with n even   (c) $a_n > 0$ with n odd   (d) $a_n < 0$ with n odd

FIGURE 5.5   Left and right behavior of the graph of $P(x) = a_n x^n + a_{n-1} x^{n-1} + \cdots + a_1 x + a_0$.

EXAMPLE 1   Determine the left and right behavior of the graph of each function:

(a) $P(x) = 3x^4 + 3x^2 - 6x - 1$

(b) $P(x) = -x^5 - 6x^3 + 6x - 3$

SOLUTION

(a)   For $P(x) = 3x^4 + 3x^2 - 6x - 1$, we have $a_n = 3$ and $n = 4$. Since $a_n > 0$ and n is even, we know the graph of P goes up to the left as $x \to -\infty$ and up to the right as $x \to \infty$ [see Figure 5.5(a)].

(b)   For $P(x) = -x^5 - 6x^3 + 6x - 3$, we have $a_n = -1$ and $n = 5$. Since $a_n < 0$ and n is odd, we know that the graph of P goes up to the left as $x \to -\infty$ and down to the right as $x \to \infty$ [see Figure 5.5(d)].   ◆

PROBLEM 1   Repeat Example 1 for each function:

(a) $P(x) = -2x^6 - x^3 + 3x - 2$

(b) $P(x) = 4x^3 - x^2 + x - 1$ ◆

◆ **Relative Maxima and Minima**

In Section 5.4 we illustrated the graph of a polynomial function

$$P(x) = a_n x^n + a_{n-1} x^{n-1} + \cdots + a_1 x + a_0,$$

with real coefficients $a_n, a_{n-1}, \ldots, a_1, a_0$ ($a_n \neq 0$) (see Figure 5.2). The graph is *continuous* with smooth, rounded turns—it contains no gap, break, jump, or sharp turn. The sketches in Figure 5.6 illustrate the differences between the graphs of polynomial and nonpolynomial functions.

FIGURE 5.6

Graphs of polynomial and
nonpolynomial functions

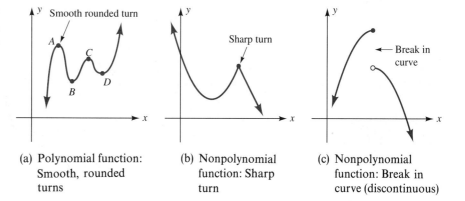

(a) Polynomial function:
   Smooth, rounded
   turns

(b) Nonpolynomial
   function: Sharp
   turn

(c) Nonpolynomial
   function: Break in
   curve (discontinuous)

On the graph of a polynomial function, points where the function changes from increasing to decreasing or from decreasing to increasing are called **relative extrema**. The graph of the polynomial function in Figure 5.6(a) has four relative extrema, at points A, B, C, and D. The points A and C are called **relative maxima** (peaks), and the points B and D are **relative minima** (valleys) of the function. A single point of change is called a *relative extremum*. Using calculus, it can be shown that if P is a polynomial function of degree n, then the number of relative extrema either is equal to $n - 1$ or is less than this number by an even number. This fact enables us to determine the possible shapes for the middle portion of the graphs in Figure 5.5.

◆◆ Relative Extrema
_____

> If P is a polynomial function of degree n with real coefficients, then the number of relative extrema either is equal to $n - 1$ or is less than this number by an even number.

E X A M P L E 2   Given that P is a polynomial function of degree 3 with a positive leading coefficient, sketch two possible shapes for the graph of P.

S O L U T I O N   Since the degree of P is odd and the leading coefficient is positive, we know the left and right behavior of tne graph is like that shown in Figure 5.5(c). Since the degree of P is 3, the graph has either $3 - 1 = 2$ relative extrema or no relative extremum for the middle portion of this graph. Sketches of two possible shapes for the graph of P are shown in Figure 5.7.

FIGURE 5.7
Sketches of two possible shapes for the graph of a polynomial function of degree 3 with a positive leading coefficient.

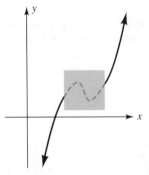

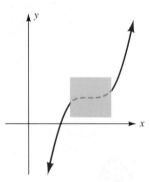

(a) Down to the left and up to the right with two relative extrema

(b) Down to the left and up to the right with no relative extremum   ◆

P R O B L E M 2   Try to sketch a polynomial function with the characteristics described in Example 2 such that it has one relative extremum. What can you conclude?   ◆

◆ **General Procedure for Graphing a Polynomial Function**

We now present a general five-step procedure for sketching the graph of a polynomial function of degree $n \geq 3$.

General Procedure for Graphing a Polynomial Function

To sketch the graph of a polynomial function P with real coefficients and of degree $n \geq 3$, proceed as follows:

Step 1 Determine the left and right behavior of the graph. Then determine the possible shapes for the middle of the graph by considering relative extrema.

Step 2 Determine the y-intercept by evaluating $P(0)$.

Step 3 Determine the real zeros of the function P. These real zeros are the x-intercepts for the graph of P.

Step 4 Determine the intervals into which the x-intercepts divide the x-axis. Then select a few values of x from each of these intervals and determine their corresponding outputs, $P(x)$.

Step 5 Plot the points gathered in steps 2–4, and connect the points to form a smooth curve.

EXAMPLE 3 Sketch the graph of $P(x) = 9 + 8x^2 - x^4$.

SOLUTION

Step 1 The highest-degree term of this polynomial function is $-x^4$. Since $a_n = -1 < 0$, and $n = 4$ (even), we know that the left and right behavior of this graph is like that shown in Figure 5.5(b), that is, down to the left and down to the right. The graph of this function has either $4 - 1 = 3$ relative extrema or one relative extremum. Two possible shapes of the graph of this function are shown in Figure 5.8.

FIGURE 5.8
Possible shapes of the graph of
$P(x) = 9 + 8x^2 - x^4$

Down to the left and
down to the right
with three relative
extrema

Down to the left and
down to the right
with one relative
extremum

Step 2 Since $P(0) = 9$, the y-intercept is 9.

Step 3 The real zeros of the function can be obtained by using the factoring techniques discussed in Section 1.6. Letting $P(x) = 0$, we find

$$0 = 9 + 8x^2 - x^4$$

$$0 = (9 - x^2)(1 + x^2)$$

$$0 = (3 + x)(3 - x)(1 + x^2)$$

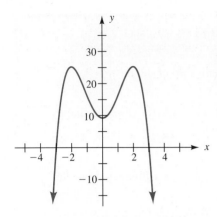

FIGURE 5.9
The graph of $P(x) = 9 + 8x^2 - x^4$ has two relative maxima and one relative minimum.

Thus, the real zeros are ± 3. Hence, the x-intercepts are ± 3.

Step 4 The two x-intercepts ± 3 divide the x-axis into three intervals. Selecting some arbitrary x values from each of these intervals, we determine their corresponding outputs $P(x)$ by using either direct substitution or synthetic division. Notice that $P(x) = P(-x)$. Thus, P is an even function. Hence, the graph must be symmetric with respect to the y-axis.

x	± 4	± 2	± 1
$P(x)$	-119	25	16

Step 5 The graph of the function P is shown in Figure 5.9. Note we have chosen convenient scales for the x- and y-axes.   ◆

PROBLEM 3   Sketch the graph of $P(x) = x^3 + 2x^2 + 5x$.   ◆

If r is a zero of multiplicity k, with $k \geq 2$, for a polynomial function P, then the graph of P is *tangent* to the x-axis at $x = r$. If k is even, the tangency looks similar to either of the sketches shown in Figure 5.10. If k is odd, the tangency looks similar to either of the sketches shown in Figure 5.11.

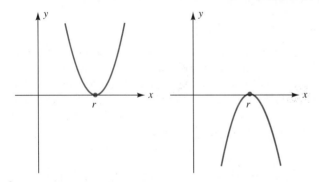

FIGURE 5.10   Appearance of tangency when r is a zero of multiplicity k and k is even

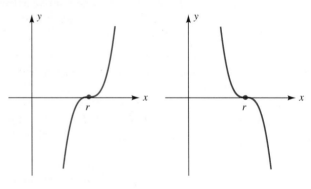

FIGURE 5.11   Appearance of tangency when r is a zero of multiplicity k and k is odd

EXAMPLE 4   Sketch the graph of $P(x) = 2x^3 - 11x^2 + 12x + 9$.

SOLUTION

Step 1 Observe that P is a polynomial function of degree 3 with a *positive* leading coefficient. Two possible shapes for the graph of this function are shown in Figure 5.7.

Step 2 Since $P(0) = 9$, the y-intercept for the graph of P is 9.

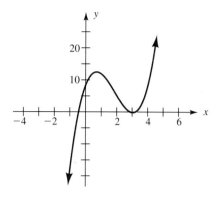

FIGURE 5.12
The graph of
$P(x) = 2x^3 - 11x^2 + 12x + 9$ is
tangent to the x-axis at $x = 3$.

Step 3   The real zeros of P are 3 of multiplicity two and $-\frac{1}{2}$ (see Example 4 in Section 5.3 for the method of determining these zeros). Thus, the x-intercepts for the graph of P are 3 and $-\frac{1}{2}$. Since 3 is a zero of *even* multiplicity, the graph of P is tangent to the x-axis at $x = 3$, and the appearance of the graph around the point of tangency is similar to one of those shown in Figure 5.10.

Step 4   The x-intercepts divide the x-axis into three intervals. Selecting some arbitrary x values from each of these intervals, we determine their corresponding outputs $P(x)$ by using either direct substitution or synthetic division.

x	-2	-1	1	2	4	5
$P(x)$	-75	-16	12	5	9	44

Step 5   The graph of the function P is shown in Figure 5.12. Note we have chosen convenient scales for the x- and y-axes.   ◆

Note: The relative maximum for the polynomial function in Example 4 occurs when $x \approx 1$. The exact location of the relative maximum can be determined by using calculus. In this function, the relative maximum occurs when $x = \frac{2}{3}$.

PROBLEM 4   Referring to Example 4, evaluate $P(\frac{2}{3})$ and compare your result to $P(1)$. Which is larger?   ◆

◆ **Application: Bending Moment of a Beam**

When a load is applied to a steel or wood beam, the force produced by the load creates a *bending moment*. Due to this bending moment, the beam will deform slightly, resulting in a curvature of the beam. Structural engineers must investigate the bending moment at various points along a beam in order to design and select a beam of proper size to safely carry a particular load. Bending moments are described frequently by polynomial functions of degree $n \geq 3$.

EXAMPLE 5   For a simply supported beam of length L that carries a uniformly increasing load of weight W, as shown in Figure 5.13, the bending moment M at any distance x from the left end of the beam is given by

$$M(x) = \frac{Wx}{3L^2}(L^2 - x^2).$$

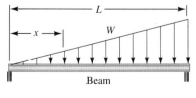

FIGURE 5.13
Simple beam with uniformly
increasing load.

(a)   Determine the bending moment function when $W = 300$ pounds and $L = 10$ feet. Then sketch the graph of this function.

(b)   Use a calculator or computer with graphing capabilities to find the approximate distance x from the left end of the beam where the bending moment is greatest.

SOLUTION

(a) Replacing W with 300 and L with 10 gives us

$$M(x) = x(100 - x^2) = 100x - x^3.$$

Observe that M is a polynomial function of degree 3. Thus, the graph of M has either zero or two relative extrema. Since the coefficient of x^3 is negative, the graph of M goes up to the left as $x \to -\infty$ and down to the right as $x \to \infty$.

The x-intercepts are found by solving the equation $M(x) = 0$, as follows:

$$x(100 - x^2) = 0$$

$$x(10 + x)(10 - x) = 0$$

$$x = 0, \pm 10$$

x-intercepts

These three x-intercepts divide the x-axis into four intervals. After selecting some arbitrary x-values from these intervals, we determine their corresponding outputs $M(x)$:

x	-12	-5	5	12
$M(x)$	528	-375	375	-528

Note that $M(-x) = -M(x)$ for all x. Hence, M is an odd function, and its graph is symmetric with respect to the origin. The graph of M is shown in Figure 5.14. Since the length of the beam is 10 ft, we must have $0 \le x \le 10$. Note we have darkened this portion of the graph.

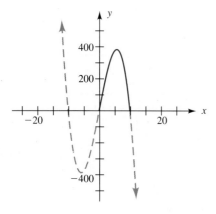

FIGURE 5.14

Graph of $M(x) = 100x - x^3$.

$0 \le x \le 10$

(b) To display the graph of $M(x) = y = 100x - x^3$ on a calculator, we begin by choosing a viewing rectangle by pressing the [RANGE] key. We use our work in part (a) to make a reasonable choice:

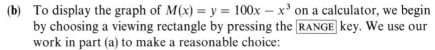

[0, 10]   by   [0, 500]

x-min   x-max   y-min   y-max

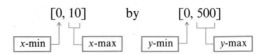

Now, by entering the equation $y = 100x - x^3$, we obtain the desired portion of the graph in the viewing rectangle (see Figure 5.15).

Next, we activate the [TRACE] key and move the blinking cursor to the relative maximum point. Reading the coordinates of this point, we find (5.8, 384.9) to the nearest tenth. To obtain more precise values, we may zoom in on this maximum point by activating the Zoom feature and tracing to the maximum point once again. By repeating this process of tracing and zooming in, we find that the coordinates of the maximum point are

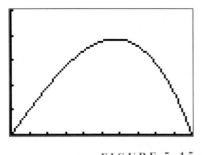

FIGURE 5.15

The graph of $y = 100x - x^3$.

$0 \le x \le 10$. in the viewing rectangle

[0, 10] by [0, 500].

(5.774, 384.900) to the nearest thousandth. In summary, the maximum bending moment along the beam is approximately 384.900 pound-feet (lb-ft), and this occurs when $x \approx 5.774$ feet.   ◆

Note: Using calculus, it can be shown that the *exact* maximum bending moment is $\dfrac{2000\sqrt{3}}{9}$ lb-ft, and this occurs when $x = \dfrac{10\sqrt{3}}{3}$ feet.

PROBLEM 5   Referring to the beam in Example 5, at what distances x from the left end of the beam is the bending moment 375 lb-ft?   ◆

Exercises 5.5

Basic Skills

For each polynomial function in Exercises 1–8, determine the following features of its graph: (a) the left and right behavior and (b) the maximum number of relative extrema.

1. $P(x) = 2x^3 - 7x^2 + 3x - 2$

2. $P(x) = 7 - 5x - 4x^3$

3. $F(x) = 6 - 2x + 8x^2 - 5x^4$

4. $f(x) = 3x^8 - 2x^7 + x^6 - 3x + 1$

5. $P(x) = -x^7 - 2x^5 + 4x - 3$

6. $P(x) = 2x^5 - x^4 + 3x - 9$

7. $g(x) = 3x^6 - 2x^4 + 3x^2 + 1$

8. $G(x) = -5x^6 + 3x^5 + x^4 - 1$

In Exercises 9–16, sketch the possible shapes of the graph of a polynomial function with the given characteristics.

9. Degree 3 with a negative leading coefficient

10. Degree 2 with a positive leading coefficient

11. Degree 4 with a positive leading coefficient

12. Degree 4 with a negative leading coefficient

13. Degree 5 with a positive leading coefficient

14. Degree 7 with a negative leading coefficient

15. Degree 6 with a negative leading coefficient

16. Degree 6 with a positive leading coefficient

In Exercises 17–36, sketch the graph of each polynomial function. Label the x- and y-intercepts.

17. $P(x) = x^3 + x^2 - 2x$

18. $P(x) = 12 + 4x - 3x^2 - x^3$

19. $G(x) = x^4 - 10x^2 + 9$

20. $P(x) = 3 - 2x^2 - x^4$

21. $h(x) = 3x^3 - x^4$

22. $f(x) = 4x^3 - 4x^2 + x$

23. $P(x) = x^5 - x^4$

24. $F(x) = 9x^3 - 4x^5$

25. $P(x) = x^3 + x - 10$

26. $P(x) = x^3 - 3x^2 - 6x + 8$

27. $F(x) = 18 + 9x + 7x^2 - x^3 - x^4$

28. $H(x) = -x^4 + 2x^3 + 7x^2 - 8x - 12$

29. $f(x) = x^4 + 2x^3 + 1$ 30. $P(x) = x^3 + x - 5$

31. $f(x) = 4x^4 - 28x^3 + 61x^2 - 42x + 9$

32. $G(x) = 2x^4 - x^3 - 19x^2 + 2x + 30$

33. $P(x) = x^5 - 3x^4 - 6x^3 + 26x^2 - 27x + 9$

34. $P(x) = x^4 - 8x^3 + 24x^2 - 32x + 16$

35. $P(x) = 3x^6 - 3x^5 + 2x^3 - 2x^2 + 6x - 6$

36. $g(x) = 3x^5 + 11x^4 + 3x^3 - 22x^2 - 12x + 8$

In Exercises 37 and 38, sketch the graph of the function described.

37. The response time T (in microseconds) to a certain stimulus is a function of one's age x (in years) and is given by

$$T(x) = 20 + 79x + 76x^2 - 4x^3, \qquad 10 \le x \le 18.$$

38. The total profit P (in thousands of dollars) that a manufacturer makes is a function of the number of units x that are produced and is given by

$$P(x) = x^3 - 4x - 15; \qquad 0 \le x \le 7.$$

Critical Thinking

39. Sketch the graphs of the two given equations on the same coordinate plane. Then, by solving the equations simultaneously, determine the coordinates of the point(s) where the graphs intersect.

(a) $y = x^3 - x$ and $y = 3x + 6$
(b) $y = x^4 + 3x^3 - 4x$ and $x + y = 1$

40. Determine the maximum number of times the graphs of two polynomial functions P and Q may intersect if the degree of P is m and the degree of Q is n, where $m > n$. Explain.

41. The graph of a cubic function is shown in the figure. Determine the equation that defines this function.

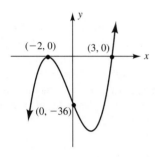

42. The graph of a quartic function is shown in the sketch. Determine the equation that defines this function.

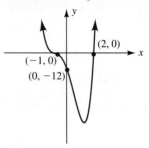

In Exercises 43–46,

(a) *sketch the graph of a polynomial function P with the given characteristics.*
(b) *determine the equation that defines the function P.*

43. Degree 2; $P(-3) = P(2) = 0$; $P(1) = 8$

44. Degree 3;   an odd function; $P(-2) = 0$; $P(4) = -12$

45. Degree 4;   an even function; $P(0) = -27$;   graph is tangent to the x-axis at $(3, 0)$

46. Degree 5;   graph is tangent to the x-axis at $(-3, 0)$;   relative maximum at $(0, 72)$;   relative minimum at $(2, 0)$;   no other relative extrema

Calculator Activities

47. A piece of cardboard, 8 cm by 16 cm, is cut and folded as shown in the sketch to form a rectangular box with a closed top and flap.

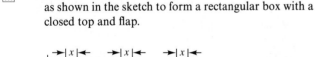

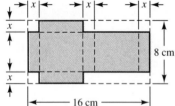

(a) Express the volume V of this box as a function of x.
(b) Sketch the graph of the function in part (a).
(c) Use a calculator or computer with graphing capabilities to find the approximate value of x when the volume of the box is greatest.

48. A beam of length L is fixed at one end and simply supported at the other. It carries a uniformly increasing load of weight W, as shown in the figure. The bending moment M at any distance x from the left end of this beam is given by

$$M(x) = \frac{Wx}{15L^2}(3L^2 - 5x^2).$$

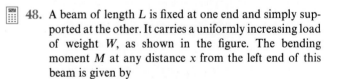
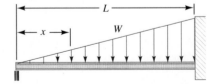

(a) Determine the bending moment function when $W = 1500$ pounds and $L = 10$ feet. Then sketch the graph of this function.

(b) Use a calculator or computer with graphing capabilities to find the approximate distance x from the left end of the beam where the bending moment is greatest.

(c) At what distances x from the left end of the beam is the bending moment 880 pound-feet?

5.6 Rational Functions

Any function f of the form

$$f(x) = \frac{P(x)}{D(x)},$$

where $P(x)$ and $D(x)$ are polynomials and $D(x) \neq 0$, is called a **rational function**. Since division by zero is undefined, the domain of a rational function f is the set of all real numbers other than those numbers that make its denominator $D(x) = 0$. Although the domain of a rational function f does not include the zeros of D, these zeros of D play an important part in sketching the graph of the function. In this section, we discuss a procedure for graphing rational functions.

◆ Vertical Asymptotes

The reciprocal function $f(x) = 1/x$ (see Section 3.4) is one of the simplest rational functions. Its domain is the set of all real numbers except 0. Using interval notation, we write the domain as $(-\infty, 0) \cup (0, \infty)$. Figure 5.16 shows the graph of the reciprocal function. We say that the y-axis (the line $x = 0$) is a *vertical asymptote* of the graph of f, since the graph approaches the y-axis as x gets closer and closer to zero. As x approaches zero from the left $(x \to 0^-)$, the value of $f(x)$ decreases without bound $[f(x) \to -\infty]$, and as x approaches zero from the right $(x \to 0^+)$, the value of $f(x)$ increases without bound $[f(x) \to \infty]$.

FIGURE 5.16
Graph of $f(x) = \frac{1}{x}$

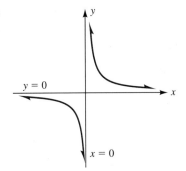

We define a vertical assymptote as follows:

Vertical Asymptotes

The line $x = a$ is a **vertical asymptote** of the graph of a rational function f if at least one of the following statements is true.

(a)   As $x \to a^-$, $f(x) \to \infty$.

(b)   As $x \to a^+$, $f(x) \to \infty$.

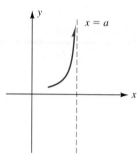

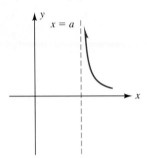

(c)   As $x \to a^-$, $f(x) \to -\infty$.

(d)   As $x \to a^+$, $f(x) \to -\infty$.

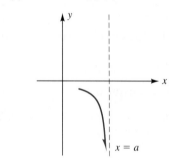

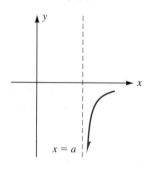

To determine the vertical asymptote(s) of a rational function $f(x) = P(x)/D(x)$, where $P(x)/D(x)$ is reduced to lowest terms, we simply find the real zeros of D. If a is a real zero of D, then the vertical line $x = a$ is a vertical asymptote of the graph of f.

E X A M P L E 1   Given $f(x) = \dfrac{x + 3}{x^2 - 2x - 8}$,

(a)   find the vertical asymptotes of the graph of f, and

(b)   determine the appearance of the graph of f as x approaches each vertical asymptote from the left and from the right.

S O L U T I O N

(a)   Factoring the denominator, we have

$$f(x) = \frac{x + 3}{x^2 - 2x - 8} = \frac{x + 3}{(x + 2)(x - 4)}.$$

Thus, the zeros of $D(x) = x^2 - 2x - 8$ are -2 and 4. Since $P(x)/D(x)$ is in reduced form, we conclude that $x = -2$ and $x = 4$ are vertical asymptotes of the graph of f.

(b)   To determine whether $f(x) \to \infty$ or $f(x) \to -\infty$ as x approaches each vertical asymptote, we check the algebraic sign of $f(x)$ near each asymptote, as follows:

$$\text{As } x \to -2^-, \qquad f(x) = \frac{x + 3}{(x + 2)(x - 4)} = \frac{(+)}{(-)(-)} = +.$$

$$\text{As } x \to -2^+, \qquad f(x) = \frac{x + 3}{(x + 2)(x - 4)} = \frac{(+)}{(+)(-)} = -.$$

$$\text{As } x \to 4^-, \qquad f(x) = \frac{x + 3}{(x + 2)(x - 4)} = \frac{(+)}{(+)(-)} = -.$$

$$\text{As } x \to 4^+, \qquad f(x) = \frac{x + 3}{(x + 2)(x - 4)} = \frac{(+)}{(+)(+)} = +.$$

Thus, we conclude $f(x) \to \infty$ as $x \to -2^-$, $f(x) \to -\infty$ as $x \to -2^+$, $f(x) \to -\infty$ as $x \to 4^-$, and $f(x) \to \infty$ as $x \to 4^+$ (see Figure 5.17).

FIGURE 5.17

Appearance of the graph of
$$f(x) = \frac{x + 3}{x^2 - 2x - 8} \text{ as } x$$
approaches each vertical asymptote.

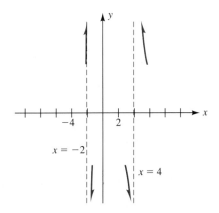

PROBLEM 1   State the domain of the rational function defined in Example 1.

◆ **Horizontal Asymptotes**

Returning to the reciprocal function, $f(x) = 1/x$, sketched in Figure 5.16, we say that the x-axis (the line $y = 0$) is a *horizontal asymptote* of the graph of f, since the graph approaches the x-axis as $|x|$ gets larger and larger. As x increases without bound ($x \to \infty$), $f(x)$ approaches zero through values *greater than* zero $[f(x) \to 0^+]$ and as x decreases without bound ($x \to -\infty$), $f(x)$ approaches zero through values less than zero $[f(x) \to 0^-]$.

We define a horizontal asymptote as follows:

◆ Horizontal Asymptotes

The line $y = b$ is a **horizontal asymptote** of the graph of a rational function f if at least one of the following statements is true.

(a)   As $x \to \infty$, $f(x) \to b^+$.

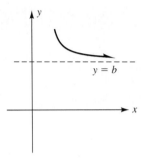

(b)   As $x \to \infty$, $f(x) \to b^-$.

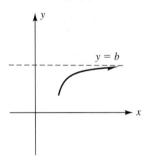

(c)   As $x \to -\infty$, $f(x) \to b^+$.

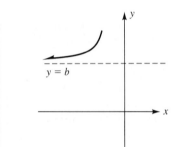

(d)   As $x \to -\infty$, $f(x) \to b^-$.

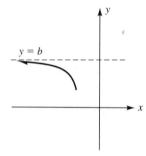

To determine if a rational function has a horizontal asymptote, we begin by dividing each term of the numerator and denominator by the highest power of x that appears in the rational expression. In doing so, we develop fractions of the form k/x^n, where k is a constant and n is a positive integer. Now as $|x|$ becomes very large, expressions of the form k/x^n become very small. We indicate this by writing

$$\frac{k}{x^n} \to 0 \quad \text{as} \quad |x| \to \infty.$$

Although a rational function may have many vertical asymptotes, it can have at most one horizontal asymptote. By using the method described above, it can be shown that the rational function $f(x) = P(x)/D(x)$, where $P(x)/D(x)$ is reduced to lowest terms, follows these rules for the existence of a horizontal asymptote:

1. one horizontal asymptote if the degree of $P(x)$ is less than or equal to the degree of $D(x)$.

2. no horizontal asymptote if the degree of $P(x)$ is greater than the degree of $D(x)$.

EXAMPLE 2   Given the rational function $f(x) = \dfrac{x + 3}{x^2 - 2x - 8}$,

(a) find the horizontal asymptote for the graph of f.

(b) determine the appearance of the graph of f as $|x| \to \infty$.

SOLUTION

(a) The highest power of x that appears in the rational expression is x^2. Dividing each term of the numerator and denominator by x^2, we obtain

$$f(x) = \frac{x + 3}{x^2 - 2x - 8} = \frac{\dfrac{1}{x} + \dfrac{3}{x^2}}{1 - \dfrac{2}{x} - \dfrac{8}{x^2}}.$$

Now, as $|x| \to \infty$, each of the fractions $\dfrac{1}{x}, \dfrac{3}{x^2}, \dfrac{2}{x}$, and $\dfrac{8}{x^2}$ approaches zero. Thus,

$$\text{as } |x| \to \infty, \qquad f(x) \to \frac{0 + 0}{1 - 0 - 0} = 0.$$

Hence, the line $y = 0$ (the x-axis) is a horizontal asymptote of the graph of f.

(b) To determine if $f(x)$ approaches the x-axis through values greater than zero $[f(x) \to 0^+]$ or through values less than zero $[f(x) \to 0^-]$ as $|x| \to \infty$, we select any large value of $|x|$ and determine if the approach is from above or from below the x-axis. Choosing $|x| = 100$, we find

$$f(100) = \frac{(100) + 3}{(100)^2 - 2(100) - 8} = +\frac{103}{9792},$$

> Approaching 0 through values greater than 0

$$f(-100) = \frac{(-100) + 3}{(-100)^2 - 2(-100) - 8} = -\frac{97}{10192}.$$

> Approaching 0 through values less than 0

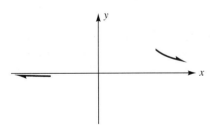

FIGURE 5.18

Appearance of the graph of
$f(x) = \dfrac{x + 3}{x^2 - 2x - 8}$ as
$|x| \to \infty$.

Thus, we conclude that $f(x) \to 0^+$ as $x \to \infty$ and $f(x) \to 0^-$ as $x \to -\infty$. The appearance of the graph of f as $|x| \to \infty$ is shown in Figure 5.18. ◆

PROBLEM 2   Find the vertical asymptote(s) and horizontal asymptote for the graph of
$f(x) = \dfrac{2x - 4}{x^2 - 2x + 1}.$ ◆

◆ General Procedure for Graphing a Rational Function

Vertical and horizontal asymptotes are valuable aids for graphing a rational function. By combining Figures 5.17 and 5.18, as shown in Figure 5.19, we begin to see the shape of the graph of $f(x) = \dfrac{x + 3}{x^2 - 2x - 8}$.

FIGURE 5.19

The graph of $f(x) = \dfrac{x + 3}{x^2 - 2x - 8}$ has two vertical asymptotes, $x = -2$ and $x = 4$, and a horizontal asymptote, $y = 0$.

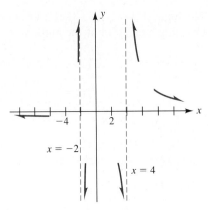

To complete the graph of f, we determine the x- and y-intercepts. The x-intercepts of a rational function $f(x) = P(x)/D(x)$, where $P(x)/D(x)$ is reduced to lowest terms, are the real zeros of the numerator $P(x)$. The y-intercept of a rational function is found by evaluating $f(0)$. For the rational function $f(x) = \dfrac{x + 3}{x^2 - 2x - 8}$, the x-intercept is -3, and the y-intercept is

$$f(0) = \frac{(0) + 3}{(0)^2 - 2(0) - 8} = -\frac{3}{8}.$$

Plotting the intercepts, we sketch the graph of this function as shown in Figure 5.20.

FIGURE 5.20

Graph of $f(x) = \dfrac{x + 3}{x^2 - 2x - 8}$

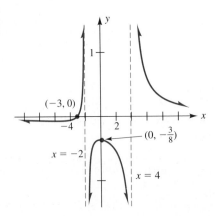

We now present a general four-step procedure for graphing a rational function in which the degree of the numerator is less than or equal to the degree of the denominator.

General Procedure for Graphing a Rational Function

To graph a rational function $$f(x) = \frac{P(x)}{D(x)},$$

where the degree of $P(x)$ is less than or equal to the degree of $D(x)$ and $P(x)/D(x)$ is reduced to lowest terms, proceed as follows:

Step 1   Find the vertical asymptote(s) of the graph of f and determine the appearance of the graph of f as x approaches each vertical asymptote from the left and from the right.

Step 2   Find the horizontal asymptote of the graph of f and determine the appearance of the graph of f as $|x| \to \infty$.

Step 3   **(a)**   Find the x-intercepts for the graph of f.
        (b)   Find the y-intercept for the graph of f.
        (c)   Determine whether the graph crosses its horizontal asymptote.

Step 4   Plot the points from step 3 and, if necessary, plot a few additional points. Then sketch the graph.

EXAMPLE 3   Sketch the graph of $f(x) = \dfrac{2x^2 + x - 1}{x^2 + 1}$.

SOLUTION

Step 1   *Vertical asymptotes:* Since $D(x) = x^2 + 1$ has no real zero, the graph of f has no vertical asymptote.

Step 2   *Horizontal asymptote:* Dividing each term of the numerator and denominator by x^2 (the highest power of x), we obtain

$$f(x) = \frac{2x^2 + x - 1}{x^2 + 1} = \frac{2 + \dfrac{1}{x} - \dfrac{1}{x^2}}{1 + \dfrac{1}{x^2}}.$$

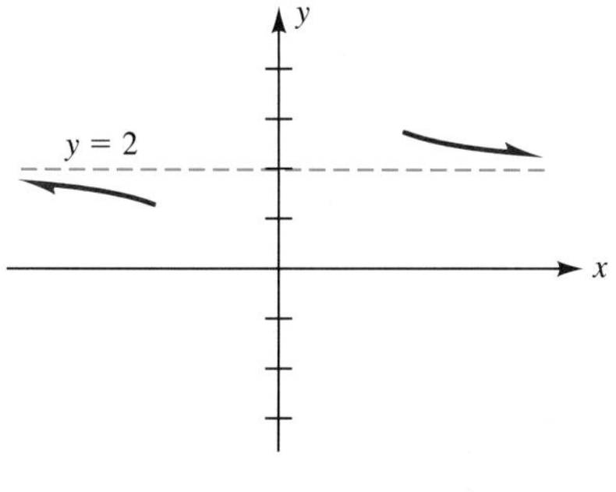

Now as $|x| \to \infty$, $f(x) \to 2$. Thus, the horizontal asymptote of the graph of f is $y = 2$. Selecting any large value for $|x|$, we see that $f(x) \to 2^+$ as $x \to \infty$ and $f(x) \to 2^-$ as $x \to -\infty$. The appearance of the graph of f as $|x| \to \infty$ is shown in Figure 5.21.

Step 3   *Additional points:* The x-intercepts are the real zeros of the numerator. Since

$$P(x) = 2x^2 + x - 1 = (2x - 1)(x + 1),$$

FIGURE 5.21
Appearance of the graph of
$f(x) = \dfrac{2x^2 + x - 1}{x^2 + 1}$ as $|x| \to \infty$

$y = 2$

the x-intercepts are $\frac{1}{2}$ and -1. To find the y-intercept, we evaluate $f(0)$. The y-intercept is

$$f(0) = \frac{2(0)^2 + (0) - 1}{(0)^2 + 1} = -1.$$

To determine whether the graph crosses its horizontal asymptote $y = 2$, we solve the equation $f(x) = 2$, as follows:

$$2 = \frac{2x^2 + x - 1}{x^2 + 1}$$

$$2(x^2 + 1) = 2x^2 + x - 1 \qquad \text{Multiply both sides by } x^2 + 1$$

$$x = 3 \qquad \text{Solve for } x$$

Hence the graph crosses its horizontal asymptote at $(3, 2)$.

Step 4 *Graph:* Plotting the points from step 3, we draw the graph of the function as shown in Figure 5.22.   ◆

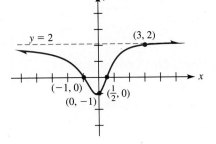

FIGURE 5.22

Graph of $f(x) = \dfrac{2x^2 + x - 1}{x^2 + 1}$

PROBLEM 3   Sketch the graph of $f(x) = \dfrac{2x - 4}{x^2 - 2x + 1}$.   ◆

◆ **Oblique Asymptotes**

If the degree of the numerator $P(x)$ of a rational function is *one more than* the degree of the denominator $D(x)$ and $P(x)/D(x)$ is reduced to lowest terms, then the graph of the rational function $f(x) = P(x)/D(x)$ has a slanting line (one that is neither vertical nor horizontal) as an asymptote. An asymptote that is neither a vertical line nor a horizontal line is called an **oblique asymptote**. To determine the oblique asymptote of the rational function $f(x) = P(x)/D(x)$, where the degree of $P(x)$ is one more than the degree of $D(x)$, we use polynomial long division to write the rational function in the form

$$f(x) = (mx + b) + \frac{R(x)}{D(x)},$$

where the degree of $R(x)$ is less than the degree of $D(x)$. Now,

$$\text{as } |x| \to \infty, \qquad \frac{R(x)}{D(x)} \to 0.$$

Hence, $f(x) \to mx + b$, which we recognize as the equation of a straight line with slope m and y-intercept b.

EXAMPLE 4   Given the rational function $f(x) = \dfrac{x^2}{x + 1}$,

(a) determine the oblique asymptote of the function, and **(b)** sketch its graph.

SOLUTION

(a) Using polynomial long division, we divide x^2 by $x + 1$, as follows:

$$
\begin{array}{r}
x - 1 \\
x + 1\overline{)\,x^2 + 0x + 0} \\
\underline{x^2 +\ \ x} \\
-x + 0 \\
\underline{-x - 1} \\
1
\end{array}
$$

So we can write

$$f(x) = \frac{x^2}{x + 1} = (x - 1) + \frac{1}{x + 1}.$$

Now, as $|x| \to \infty$, $\dfrac{1}{x + 1} \to 0$. Hence, $f(x) \to x - 1$. Thus, the line $y = x - 1$ is an oblique asymptote of the graph of f.

(b) To sketch the graph of this rational function, we use our general procedure (see page 347), along with the fact that $y = x - 1$ is an oblique asymptote.

Step 1   *Vertical asymptote:* $x = -1$

$$\text{As } x \to -1^-, \quad f(x) = \frac{x^2}{x + 1} = \frac{(+)}{(-)} = -.$$

$$\text{As } x \to -1^+, \quad f(x) = \frac{x^2}{x + 1} = \frac{(+)}{(+)} = +.$$

Thus, $f(x) \to -\infty$ as $x \to -1^-$ and $f(x) \to \infty$ as $x \to -1^+$.

Step 2   *Horizontal asymptote:* None, because the degree of the numerator is greater than the degree of the denominator.

Step 3   *Additional points:* Since $f(0) = 0$, the graph passes through the origin. To determine whether the graph crosses its oblique asymptote, we solve the equations $y = \dfrac{x^2}{x + 1}$ and $y = x - 1$ simultaneously:

$$\frac{x^2}{x + 1} = x - 1$$

$$x^2 = (x + 1)(x - 1)$$

$$x^2 = x^2 - 1$$

$$0 = -1, \quad \text{which is impossible.}$$

Thus, we conclude that the graph does not cross its oblique asymptote.

Step 4   *Graph:* The graph of this function in shown in Figure 5.23.   ◆

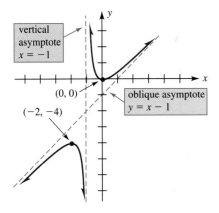

FIGURE 5.23

Graph of $f(x) = \dfrac{x^2}{x + 1}$

PROBLEM 4   Determine the oblique asymptote for the graph of $f(x) = \dfrac{2x^3 + 3}{x^2}$.   ◆

◆ **Rational Functions with Common Factors**

To sketch the graph of a rational function $f(x) = P(x)/D(x)$, where $P(x)$ and $D(x)$ have common factors, we begin by reducing $P(x)/D(x)$ to lowest terms. The graph of f is the same as the graph of the new function that is formed by the reducing process, except it may have "holes" to indicate that f is not a continuous function.

EXAMPLE 5   Sketch the graph of $f(x) = \dfrac{x^2 + x - 2}{x - 1}$.

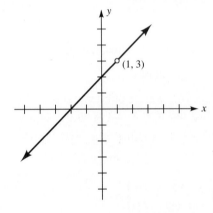

SOLUTION   Factoring the numerator, we obtain

$$f(x) = \frac{(x + 2)(x - 1)}{x - 1},$$

and reducing to lowest terms gives us the new function

$$F(x) = x + 2.$$

Note that the functions f and F are *not* the same functions. The domain of f is the set of all real numbers except 1, whereas the domain of F is the set of all real numbers. The graph of F is a straight line with slope 1 and y-intercept 2. The graph of f is the same as the graph of F, except for a hole at $(1, 3)$, which indicates that f is undefined at this point. The graph of f is shown in Figure 5.24.   ◆

FIGURE 5.24

The graph of $f(x) = \dfrac{x^2 + x - 2}{x - 1}$ is discontinuous at $(1, 3)$.

PROBLEM 5   Sketch the graph of $f(x) = \dfrac{x^2 + 2x}{x + 2}$.   ◆

◆ **Application: Average Cost Function**

The *total cost* to run a business is the sum of the *fixed cost* (costs that do not depend on the level of production, such as mortgage payment, insurance, utilities, and so on) and *variable cost* (costs that vary with the level of production, such as maintenance of equipment, cost of materials, labor, and so on). If $T(x)$ is the total cost of producing x units of a product, then the **average cost function** C is defined by

$$C(x) = \frac{T(x)}{x}$$

EXAMPLE 6

A certain manufacturer produces a kit for building a log home. The fixed cost for the company is $141,120 per year and the variable cost (in dollars) of producing x log home kits per year is $12,000x + 20x^2$.

(a) Define a function that represents the *average cost per kit*, and sketch the graph of this function.

(b) Use a computer or calculator with graphing capabilities to find the approximate number of kits that should be produced per year in order to minimize the average cost per kit.

SOLUTION

(a) If $T(x)$ is the total cost of producing x log-home kits per year, then

$$T(x) = \underbrace{141,120}_{\text{Fixed cost}} + \underbrace{(12,000x + 20x^2)}_{\text{Variable cost}}, \quad x \geq 0.$$

Hence, the average cost per kit is

$$C(x) = \frac{141,120 + 12,000x + 20x^2}{x}, \qquad x > 0.$$

To sketch the graph of this rational function for values of x greater than zero, we proceed in the usual manner.

Vertical asymptote: $x = 0$

$$\text{As } x \to 0^+, \qquad C(x) = \frac{141,120 + 12,000x + 20x^2}{x} = \frac{(+)}{(+)} = +.$$

Thus $C(x) \to \infty$ as $x \to 0^+$.

Horizontal asymptote: None; the degree of the numerator is greater than the degree of the denominator.

Oblique asymptote: Dividing numerator and denominator by x, we obtain

$$C(x) = \frac{141,120}{x} + 12,000 + 20x.$$

Now,

$$\text{as } |x| \to \infty, \qquad \frac{141,120}{x} \to 0.$$

Hence, $C(x) \to 12,000 + 20x$. Thus, the line $y = 12,000 + 20x$ is an oblique asymptote of the graph of C. Solving the equations

$$y = \frac{141,120 + 12,000x + 20x^2}{x} \qquad \text{and} \qquad y = 12,000 + 20x$$

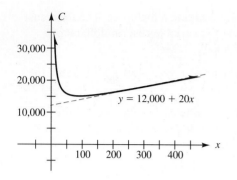

FIGURE 5.25
Graph of
$$C(x) = \frac{141{,}120 + 12{,}000x + 20x^2}{x}$$
for $x > 0$

simultaneously leads to a contradiction. Thus, we conclude the graph does not cross its oblique asymptote.

Since $C(x) = 0$ has no solution in the interval $(0, \infty)$, and $C(0)$ is undefined, we conclude the graph of C has neither an x-intercept nor a y-intercept. Selecting a few arbitrary values for x, we find their corresponding outputs $C(x)$ as shown:

x	10	50	300
$C(x)$	26,312	15,822.40	18,470.40

Using all the information we have gathered enables us to sketch the graph of the average cost function C. The graph is shown in Figure 5.25.

(b) To display the graph of

$$C(x) = y = \frac{141{,}120 + 12{,}000x + 20x^2}{x}$$

on a calculator, we first choose a viewing rectangle by pressing the RANGE key. Referring to our work in part (a), we select a reasonable rectangle:

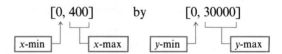

By entering the equation

$$y = \frac{141{,}120 + 12{,}000x + 20x^2}{x}$$

we obtain this portion of the graph in the viewing rectangle (see Figure 5.25 for a comparison).

Next, we activate the TRACE key and move the blinking cursor to the relative minimum point. The coordinates of this point are (84, 15360) to the nearest whole number. In summary, the approximate minimum cost per kit is $15,360, and this minimum occurs when approximately 84 kits are sold per year.   ◆

Note: Using calculus, it can be shown that the *exact* minimum average cost per kit is $15,360 and this occurs when 84 kits are produced per year. By repeatedly using the zoom in feature on a graphing calculator, you can magnify the graph around the minimum point to obtain these values.

PROBLEM 6   Referring to Example 6, how many kits are produced if the average cost per kit is $16,640?   ◆

Exercises 5.6

Basic Skills

In Exercises 1–10,

(a) *find the vertical asymptotes of the graph of f, if any exist.*

(b) *if the graph of f has vertical asymptotes, determine the appearance of the graph as x approaches each vertical asymptote from the left and from the right.*

1. $f(x) = \dfrac{x}{x + 2}$

2. $f(x) = \dfrac{x + 1}{2x - 3}$

3. $f(x) = \dfrac{1}{x^2 + 4}$

4. $f(x) = \dfrac{x + 3}{x^2 - 4}$

5. $f(x) = \dfrac{x^3 - 1}{2x^2 + 5x - 3}$

6. $f(x) = \dfrac{-1}{x^2 + x + 4}$

7. $f(x) = \dfrac{x + 4}{x^3 - 2x - 4}$

8. $f(x) = \dfrac{x^2 + 2}{x^3 + x^2 - 2}$

9. $f(x) = \dfrac{2 - 3x}{x^3 - 5x^2 + 2x + 8}$

10. $f(x) = \dfrac{2x + 1}{2x^4 - x^3 - 18x^2 + 9x}$

In Exercises 11–20,

(a) *find the horizontal asymptote of the graph of f, if one exists.*

(b) *if the graph of f has a horizontal asymptote, determine the appearance of the graph as $|x| \to \infty$.*

(c) *if the graph of f has a horizontal asymptote, determine whether the graph crosses the horizontal asymptote and, if so, where.*

11. $f(x) = \dfrac{x + 2}{x - 3}$

12. $f(x) = \dfrac{x^2 - 4}{2x + 1}$

13. $f(x) = \dfrac{x^2 - 1}{2x^2 + 7x - 15}$

14. $f(x) = \dfrac{2x^2 - 3x + 4}{x^3 + x - 1}$

15. $f(x) = \dfrac{x^3 + 3}{1 - 3x + 2x^2}$

16. $f(x) = \dfrac{x^3 + 2x - 1}{x^3 + 8}$

17. $f(x) = \dfrac{3x^2 - 2x + 1}{4x - x^2}$

18. $f(x) = \dfrac{1 - 4x^2}{x^2}$

19. $f(x) = \dfrac{3x - 1}{x^3 - 27}$

20. $f(x) = \dfrac{2x^3 - 3x + 4}{1 + x - 3x^3}$

In Exercises 21–36, sketch the graph of each rational function. Label the x- and y-intercepts and the vertical and horizontal asymptotes, if they exist.

21. $f(x) = \dfrac{3x}{x - 3}$

22. $f(x) = \dfrac{2x + 1}{x + 2}$

23. $F(x) = \dfrac{x + 6}{x^2 - 9}$

24. $F(x) = \dfrac{2}{4 - x^2}$

25. $g(x) = \dfrac{-1}{x^2 + 1}$

26. $g(x) = \dfrac{3x}{x^2 + 9}$

27. $G(x) = \dfrac{x}{x^2 - 4x - 12}$

28. $G(x) = \dfrac{10 - x}{x^2 + 3x - 10}$

29. $h(x) = \dfrac{x - 2}{2x^2 - x - 3}$

30. $h(x) = \dfrac{2x - 1}{x^2 + 2x + 1}$

31. $H(x) = \dfrac{2x^2 + 2x + 2}{x^2 - 4x}$

32. $H(x) = \dfrac{x^2 + 6x + 9}{2x^2 - 6x}$

33. $f(x) = \dfrac{6}{2x^3 + 9x^2 + 3x - 4}$

34. $F(x) = \dfrac{x^2}{x^3 + 3x^2 - x - 3}$

35. $G(x) = \dfrac{x + 2}{x^4 - 2x^3 + 2x - 1}$

36. $g(x) = \dfrac{-x^4}{x^4 + 2x^3 - 7x^2 + 2x - 8}$

In Exercises 37–44,

(a) *find the oblique asymptote of the graph of f.*

(b) *determine whether the graph of f crosses the oblique asymptote and, if so, where.*

(c) *sketch the graph of f and label the x- and y-intercepts and any asymptote.*

37. $f(x) = \dfrac{x^2 - 4}{x}$

38. $f(x) = \dfrac{x^3 + 1}{x^2}$

39. $f(x) = \dfrac{x^2}{x - 3}$

40. $f(x) = \dfrac{x^2 + 1}{x + 1}$

41. $f(x) = \dfrac{x^2 - 3x - 4}{x + 3}$

42. $f(x) = \dfrac{x^2 - x}{x - 2}$

43. $f(x) = \dfrac{x^3 + 8}{x^2 - 3x - 4}$

44. $f(x) = \dfrac{x^3 - 2x - 4}{x^2 - 1}$

For each of the rational functions in Exercises 45–50, P(x)/D(x) is not reduced to lowest terms. Sketch the graph of each function.

45. $F(x) = \dfrac{x^2 + 3x - 10}{x - 2}$

46. $f(x) = \dfrac{x^3 + 2x^2}{x}$

47. $g(x) = \dfrac{x^2 - x - 6}{x^2 + 4x + 4}$

48. $R(x) = \dfrac{2x^2 - 5x - 3}{x^2 - 4x + 3}$

49. $f(x) = \dfrac{x^4 - 2x^3 - x + 2}{x^3 - 2x^2}$

50. $h(x) = \dfrac{2x^4 + x^3 - 3x^2 + 5x - 5}{x^3 + x^2 + x - 3}$

Critical Thinking

51. Can the graph of a rational function have two or more horizontal asymptotes? Explain.

52. Can the graph of a rational function have both a horizontal and oblique asymptote? Explain.

53. Give an example of a rational function whose graph crosses its horizontal asymptote more than once.

54. Give an example of a rational function whose graph crosses its oblique asymptote more than once.

55. Sketch the graph of each equation:

(a) $x + x^2 y - y = 0$ (b) $2x^2 - xy + 3y - 1 = 0$

56. Sketch the graphs of the two given equations on the same coordinate plane. Then, by solving the equations simultaneously, determine the coordinates of the points where the graphs intersect.

(a) $y = \dfrac{1}{x^2 - 4}$ and $3x - 5y = 8$

(b) $y = \dfrac{x^3 - 1}{x^2}$ and $y = -2x - 4$

57. It is estimated that the cost C (in millions of dollars) to remove x percent of the pollution from Boston Harbor is given by $C(x) = \dfrac{500x}{100 - x}$.

(a) As $x \to 100^-$, what can be said about the cost of cleaning up the harbor?

(b) Sketch the graph of this function ($0 \le x < 100$).

58. When a 100-ohm resistor is connected in parallel with a variable resistor, x, the total resistance R is given by $R(x) = \dfrac{100x}{x + 100}$.

(a) Sketch the graph of this function for $x \ge 0$.

(b) What is the significance of the horizontal asymptote?

Calculator Activities

 59. The margins at the top and bottom of a poster are 2 inches each, and the margins at each side are 4 inches, as shown in the figure. The poster contains 50 square inches of printed matter.

(a) Show that the area A of the entire poster is given by
$$A = \frac{4x^2 + 18x}{x - 8}.$$

(b) Sketch the graph of the rational function defined in part (a).

(c) Use a calculator or computer with graphing capabilities to find the overall dimensions of the poster whose area is the smallest possible.

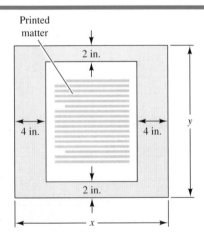

EXERCISE 59

60. A rectangular box with an open top has a square base, as sketched in the figure. The volume of the box is 64 cubic inches.

(a) Show that the surface area S of the box is given by
$$S = \frac{x^3 + 256}{x}.$$

(b) Sketch the graph of the rational function defined in part (a).

(c) Use a calculator or computer with graphing capabilities to find the dimensions of the box whose surface area is the smallest possible.

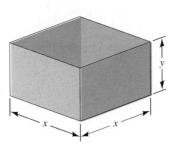

61. A certain manufacturer produces wood-burning stoves. The fixed cost for the company is $81,000 per year. The variable cost (in dollars) of producing x stoves per year is $400x + 0.1x^2$.

(a) Define a function that represents the average cost per stove, and sketch the graph of this function.

(b) Use a computer or calculator with graphing capabilities to find the approximate number of stoves that should be produced per year in order to minimize the average cost per stove.

62. A trucking company determines that the cost of operating one of its trucks is $(7 + 0.008x)$ dollars per mile when the truck is driven at x miles per hour ($x \leq 75$). In addition, the company pays the truck driver $20 per hour.

(a) Express the total cost C per mile as a function of x, and sketch the graph of this function.

(b) Use a computer or calculator with graphing capabilities to find the speed that the truck should be driven in order to minimize the total cost per mile.

5.7 Partial Fraction Decomposition of a Rational Function

◆ Introductory Comments

Recall from Section 1.7 that to find the sum of two algebraic fractions such as $\dfrac{2}{x + 1}$ and $\dfrac{4}{x - 3}$, we proceed as follows:

$$\frac{2}{x + 1} + \frac{4}{x - 3} = \frac{2(x - 3) + 4(x + 1)}{(x + 1)(x - 3)} = \frac{6x - 2}{(x + 1)(x - 3)}$$

In this section, we reverse the process; that is, we take a given rational expression such as $\dfrac{6x - 2}{(x + 1)(x - 3)}$ and express it as the sum of the simpler rational expressions $\dfrac{2}{x + 1}$ and $\dfrac{4}{x - 3}$. Each of these simpler rational expressions is called a *partial fraction*, and the *sum* of the simpler rational expressions is called the *partial fraction decomposition* of the original rational expression.

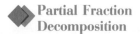

Partial Fraction Decomposition

If $P(x)/D(x)$ is a rational expression that is reduced to lowest terms, and if the degree of $P(x)$ is less than the degree of $D(x)$, then $P(x)/D(x)$ can be written as a **partial fraction decomposition** of the form

$$\frac{P(x)}{D(x)} = F_1(x) + F_2(x) + F_3(x) + \cdots + F_n(x),$$

where each **partial fraction** $F_i(x)$, $i = 1, 2, 3, \ldots, n$, has the form

$$\underbrace{\frac{A}{(a + bx)^n}}_{\text{Linear factor}} \quad \text{or} \quad \underbrace{\frac{Bx + C}{(px^2 + qx + r)^m}}_{\text{Irreducible quadratic factor}}$$

for some real numbers A, B, and C and nonnegative integers m and n.

In this section, we refer to an **irreducible quadratic factor** as one that is *prime* over the reals (see Section 1.6). Two examples of irreducible quadratic factors are

$$x^2 + 1 \quad \text{and} \quad x^2 + x + 1.$$

Whether the partial fraction decomposition of $P(x)/D(x)$ contains partial fractions of the form

$$\underbrace{\frac{A}{(a + bx)^n}}_{\text{Linear factor}} \quad \text{or} \quad \underbrace{\frac{Bx + C}{(px^2 + qx + r)^m}}_{\text{Irreducible quadratic factor}}$$

(or fractions of both forms) depends on the type of factors that appear in the denominator $D(x)$. We consider four cases:

Case 1: Distinct linear factors in the denominator.

Case 2: Repeated linear factors in the denominator.

Case 3: Distinct irreducible quadratic factors in the denominator.

Case 4: Repeated irreducible quadratic factors in the denominator.

In calculus, the task of partial fraction decomposition of a rational function is extremely important.

◆ **Case 1:   Distinct Linear Factors**

To each distinct linear factor, $ax + b$, occurring in the denominator of a rational expression $P(x)/D(x)$, we assign a single partial fraction of the form

$$\boxed{\dfrac{A}{ax + b}}$$

where A is a constant to be determined.

E X A M P L E   1   Find the partial fraction decomposition of the rational function

$$f(x) = \frac{7x - 4}{x^3 + x^2 - 2x}.$$

S O L U T I O N   We begin by factoring the denominator completely over the reals:

$$x^3 + x^2 - 2x = x(x^2 + x - 2) = x(x - 1)(x + 2).$$

Thus, we have

$$\frac{7x - 4}{x^3 + x^2 - 2x} = \frac{7x - 4}{\underbrace{x(x - 1)(x + 2)}}.$$

Distinct linear factors

Since the denominator $D(x)$ contains distinct linear factors, the partial fraction decomposition has the form

$$\frac{7x - 4}{x(x - 1)(x + 2)} = \frac{A}{x} + \frac{B}{x - 1} + \frac{C}{x + 2}.$$

Multiplying both sides of this equation by the least common denominator, $x(x - 1)(x + 2)$, we obtain the *decomposition equation*

$$7x - 4 = A(x - 1)(x + 2) + Bx(x + 2) + Cx(x - 1).$$

One method of determining the values of A, B, and C is to substitute $x = 0$, $x = 1$, and $x = -2$ [the zeros of the denominator $D(x) = x(x - 1)(x + 2)$] into the decomposition equation. By letting $x = 0$ we eliminate the B and C terms in the decomposition equation, and we obtain

$$-4 = -2A \qquad \text{or} \qquad A = 2.$$

Letting $x = 1$ eliminates the A and C terms in the decomposition equation and gives us

$$3 = 3B \qquad \text{or} \qquad B = 1.$$

Finally, letting $x = -2$ eliminates the A and B terms in the decomposition equation, and gives us

$$-18 = 6C \qquad \text{or} \qquad C = -3.$$

Hence, the partial fraction decomposition of this rational function is

$$f(x) = \frac{7x - 4}{x^3 + x^2 - 2x} = \frac{2}{x} + \frac{1}{x - 1} + \frac{-3}{x + 2}.$$ ◆

To check any partial fraction decomposition, we find the sum of the partial fraction. The sum should be equivalent to the original rational expression.

PROBLEM 1   Check the result of Example 1.   ◆

◆   **Case 2:   Repeated Linear Factors**

To each linear factor, $ax + b$, occurring r times in the denominator of a rational expression $P(x)/D(x)$, we assign a set of r partial fractions of the form

$$\frac{A_1}{ax + b}, \quad \frac{A_2}{(ax + b)^2}, \quad \frac{A_3}{(ax + b)^3}, \quad \cdots, \quad \frac{A_r}{(ax + b)^r}$$

where $A_1, A_2, A_3, \ldots, A_r$ are constants to be determined.

EXAMPLE 2   Find the partial fraction decomposition of the rational function

$$f(x) = \frac{9}{x^3 + 6x^2 + 9x}.$$

SOLUTION   We begin by factoring the denominator completely over the reals:

$$x^3 + 6x^2 + 9x = x(x^2 + 6x + 9) = x(x + 3)^2.$$

Thus, we have

$$\frac{9}{x^3 + 6x^2 + 9x} = \frac{9}{x(x + 3)^2}.$$

Distinct linear factor   Repeated linear factor

Since the denominator $D(x)$ contains both a distinct linear factor and a repeated linear factor that occurs *twice*, the partial fraction decomposition has the form

$$\frac{9}{x(x + 3)^2} = \frac{A}{x} + \frac{B}{x + 3} + \frac{C}{(x + 3)^2}.$$

Multiplying both sides of this equation by the least common denominator, $x(x + 3)^2$, we obtain the decomposition equation

$$9 = A(x + 3)^2 + Bx(x + 3) + Cx.$$

The values of A and C may be determined by substituting $x = 0$ and $x = -3$ [the zeros of the denominator, $D(x) = x(x + 3)^2$] into the decomposition equation. Letting $x = 0$ eliminates the B and C terms in the decomposition equation and gives us

$$9 = 9A \qquad \text{or} \qquad A = 1.$$

Letting $x = -3$ eliminates the A and B terms in the decomposition equation, giving us

$$9 = -3C \qquad \text{or} \qquad C = -3.$$

One method of determining the value of B is to expand the right-hand side of the decomposition equation, and then collect like terms, as follows:

$$
\begin{aligned}
9 &= A(x + 3)^2 + Bx(x + 3) + Cx \\
&= A(x^2 + 6x + 9) + B(x^2 + 3x) + Cx \\
&= (A + B)x^2 + (6A + 3B + C)x + 9A
\end{aligned}
$$

Now, since two polynomials are equal if and only if the coefficients of their like terms are equal, we can state that the x^2 terms are equal:

$$0 = A + B.$$

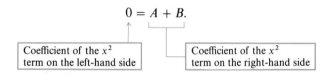

| Coefficient of the x^2 term on the left-hand side | | Coefficient of the x^2 term on the right-hand side |

However, we have already determined that $A = 1$. Substituting $A = 1$ in the equation $0 = A + B$ gives us

$$0 = (1) + B \qquad \text{or} \qquad B = -1.$$

Hence, the partial fraction decomposition of this rational function is

$$f(x) = \frac{9}{x^3 + 6x^2 + 9x} = \frac{1}{x} + \frac{-1}{x + 3} + \frac{-3}{(x + 3)^2}. \qquad \blacklozenge$$

Note: The technique of equating coefficients of like terms is especially useful when the denominator of a rational expression contains factors other than distinct linear factors.

PROBLEM 2   Referring to the expanded form of the decomposition equation in Example 2,

$$9 = (A + B)x^2 + (6A + 3B + C)x + 9A,$$

determine the value of B by equating the coefficients of the x terms on the left-hand and right-hand sides, and then substituting $A = 1$ and $C = -3$. $\qquad \blacklozenge$

◆   **Case 3:   Distinct Irreducible Quadratic Factors**

To each distinct irreducible quadratic factor, $ax^2 + bx + c$, occurring in the denominator of a rational expression $P(x)/D(x)$, we assign a single partial fraction of the form

$$\frac{Ax + B}{ax^2 + bx + c}$$

where A and B are constants to be determined.

EXAMPLE 3   Find the partial fraction decomposition of the rational function

$$f(x) = \frac{3x^2 + 5x + 18}{x^3 + 2x^2 + x + 2}.$$

SOLUTION   We begin by factoring the denominator completely over the reals:

$$x^3 + 2x^2 + x + 2 = x^2(x + 2) + 1(x + 2)$$
$$= (x + 2)(x^2 + 1)$$

Thus, we have

$$\frac{3x^2 + 5x + 18}{x^3 + 2x^2 + x + 2} = \frac{3x^2 + 5x + 18}{(x + 2)(x^2 + 1)}.$$

Distinct linear factor ⌐   Distinct irreducible quadratic factor

Since the denominator $D(x)$ contains both a distinct linear factor and a distinct irreducible quadratic factor, the partial fraction decomposition has the form

$$\frac{3x^2 + 5x + 18}{(x + 2)(x^2 + 1)} = \frac{A}{x + 2} + \frac{Bx + C}{x^2 + 1}.$$

Multiplying both sides of this equation by the least common denominator, $(x + 2)(x^2 + 1)$, we obtain the decomposition equation

$$3x^2 + 5x + 18 = A(x^2 + 1) + (Bx + C)(x + 2).$$

The value of A may be determined by substituting $x = -2$ [the real zero of the denominator, $D(x) = (x + 2)(x^2 + 1)$] into the decomposition equation. Letting $x = -2$ eliminates the B and C terms in the decomposition equation, and gives us

$$20 = 5A \qquad \text{or} \qquad A = 4.$$

To determine the values of B and C, we expand the right-hand side of the decomposition equation, and then collect like terms:

$$3x^2 + 5x + 18 = A(x^2 + 1) + (Bx + C)(x + 2)$$
$$= (Ax^2 + A) + (Bx^2 + 2Bx + Cx + 2C)$$
$$= (A + B)x^2 + (2B + C)x + (A + 2C)$$

Equating the coefficients of the x^2 terms and letting $A = 4$ gives us

$$3 = A + B$$
$$3 = (4) + B \quad \text{or} \quad B = -1.$$

Equating the coefficients of the x terms and letting $B = -1$ gives us

$$5 = 2B + C$$
$$5 = 2(-1) + C \quad \text{or} \quad C = 7.$$

Thus, the partial fraction decomposition of this rational function is

$$f(x) = \frac{3x^2 + 5x + 18}{x^3 + 2x^2 + x + 2} = \frac{4}{x + 2} + \frac{-x + 7}{x^2 + 1}. \qquad \blacklozenge$$

PROBLEM 3   Find the partial fraction decomposition of the rational function

$$f(x) = \frac{x - 8}{x^3 + 4x}. \qquad \blacklozenge$$

◆   **Case 4:   Repeated Irreducible Quadratic Factors**

To each irreducible quadratic factor, $ax^2 + bx + c$, occurring r times in the denominator of a rational expression $P(x)/D(x)$, we assign a set of r partial fractions of the form

$$\frac{A_1 x + B_1}{ax^2 + bx + c}, \quad \frac{A_2 x + B_2}{(ax^2 + bx + c)^2}, \quad \cdots, \quad \frac{A_r x + B_r}{(ax^2 + bx + c)^r}$$

where $A_1, A_2, \ldots, A_r$ and $B_1, B_2, \ldots, B_r$ are constants to be determined.

EXAMPLE 4   Find the partial fraction decomposition of the rational function

$$f(x) = \frac{2x^3 + 3}{x^4 + 6x^2 + 9}.$$

SOLUTION   We begin by factoring the denominator completely over the reals:

$$x^4 + 6x^2 + 9 = (x^2 + 3)(x^2 + 3) = (x^2 + 3)^2.$$

Thus, we have

$$\frac{2x^3 + 3}{x^4 + 6x^2 + 9} = \frac{2x^3 + 3}{(x^2 + 3)^2}.$$

Repeated irreducible quadratic factor

Since the denominator contains a repeated irreducible quadratic factor that occurs twice, the partial fraction decomposition has the form

$$\frac{2x^3 + 3}{(x^2 + 3)^2} = \frac{Ax + B}{x^2 + 3} + \frac{Cx + D}{(x^2 + 3)^2}.$$

Multiplying both sides of this equation by the least common denominator, $(x^2 + 3)^2$, we obtain the decomposition equation

$$2x^3 + 3 = (Ax + B)(x^2 + 3) + (Cx + D).$$

Expanding the right-hand side of the decomposition equation and collecting like terms, we obtain

$$2x^3 + 3 = Ax^3 + Bx^2 + (3A + C)x + (3B + D).$$

Equating the coefficients of the x^3 terms gives us $A = 2$, and equating the coefficients of the x^2 terms gives us $B = 0$. Equating the coefficients of the x terms and letting $A = 2$ gives us

$$0 = 3A + C$$
$$0 = 3(2) + C \quad \text{or} \quad C = -6.$$

Finally, equating the coefficients of the constant terms and letting $B = 0$ gives us

$$3 = 3B + D$$
$$3 = 3(0) + D \quad \text{or} \quad D = 3.$$

Thus, the partial fraction decomposition of this rational function is

$$f(x) = \frac{2x^3 + 3}{x^4 + 6x^2 + 9} = \frac{2x}{x^2 + 3} + \frac{-6x + 3}{(x^2 + 3)^2}.$$

◆

PROBLEM 4   Find the partial fraction decomposition of the rational function

$$f(x) = \frac{6x^3}{x^4 + 2x^2 + 1}.$$

◆

♦ **Improper Rational Expressions**

The procedures we have illustrated for finding the partial fraction decomposition of a rational function apply to **proper rational functions** (those in which the degree of the numerator is less than the degree of the denominator). Rational functions in which the degree of the numerator is not less than the degree of the denominator are called **improper rational functions**. To find the partial fraction decomposition of an improper rational function, we first use polynomial long division to express the function as the sum of a polynomial function and a proper rational function. The procedure is illustrated in our next example.

EXAMPLE 5   Find the partial fraction decomposition of the rational function

$$f(x) = \frac{x^5 + 10x^3 + x^2}{x^4 + 5x^2 + 4}.$$

SOLUTION   Since the degree of the numerator is greater than the degree of the denominator, this is an improper rational function. Using polynomial long division, we obtain

$$
\begin{array}{r}
x \\
x^4 + 5x^2 + 4 \overline{\smash{\big)}\ x^5 + 10x^3 + x^2 + 0x + 0} \\
\underline{x^5 + 5x^3 + 4x} \\
5x^3 + x^2 - 4x
\end{array}
$$

Thus,

$$\frac{x^5 + 10x^3 + x^2}{x^4 + 5x^2 + 4} = x + \frac{5x^3 + x^2 - 4x}{x^4 + 5x^2 + 4}$$

Polynomial   Proper rational expression

We now find the partial fraction decomposition of

$$\frac{5x^3 + x^2 - 4x}{x^4 + 5x^2 + 4} = \frac{5x^3 + x^2 - 4x}{(x^2 + 1)(x^2 + 4)}$$

Distinct irreducible quadratic factors

Since the denominator contains distinct irreducible quadratic factors, the partial fraction decomposition has the form

$$\frac{5x^3 + x^2 - 4x}{(x^2 + 1)(x^2 + 4)} = \frac{Ax + B}{x^2 + 1} + \frac{Cx + D}{x^2 + 4}.$$

Multiplying both sides of this equation by the least common denominator,

$(x^2 + 1)(x^2 + 4)$, and collecting like terms on the right-hand side, we obtain

$$5x^3 + x^2 - 4x = (Ax + B)(x^2 + 4) + (Cx + D)(x^2 + 1)$$
$$= (Ax^3 + 4Ax + Bx^2 + 4B) + (Cx^3 + Cx + Dx^2 + D)$$
$$= (A + C)x^3 + (B + D)x^2 + (4A + C)x + (4B + D).$$

Equating the coefficients of the x^3 terms and equating the coefficients of the x terms, we obtain

$$5 = A + C \qquad \text{and} \qquad -4 = 4A + C.$$

Solving this pair of equations simultaneously by the addition method (Section 4.5) gives us $A = -3$ and $C = 8$. Equating the coefficients of the x^2 terms and equating the coefficients of the constant terms, we obtain

$$1 = B + D \qquad \text{and} \qquad 0 = 4B + D.$$

Solving this pair of equations simultaneously by the addition method (Section 4.5) gives us $B = -\frac{1}{3}$ and $D = \frac{4}{3}$.

Thus, the partial fraction decomposition of this improper rational function is

$$f(x) = \frac{x^5 + 10x^3 + x^2}{x^4 + 5x^2 + 4} = x + \frac{5x^3 + x^2 - 4x}{x^4 + 5x^2 + 4}$$

$$= x + \frac{-3x - \frac{1}{3}}{x^2 + 1} + \frac{8x + \frac{4}{3}}{x^2 + 4}.$$

PROBLEM 5   Find the partial fraction decomposition of the rational function

$$f(x) = \frac{x^2 - 3x - 8}{x^2 - 2x + 1}.$$

Exercises 5.7

Basic Skills

 In Exercises 1–24, find the partial fraction decomposition of each rational function.

1. $f(x) = \dfrac{6}{x^2 - 9}$

2. $f(x) = \dfrac{-20x}{x^2 - 25}$

3. $f(x) = \dfrac{x - 2}{x^2 + 6x + 8}$

4. $f(x) = \dfrac{2x - 1}{x^2 - x - 6}$

5. $f(x) = \dfrac{x^2 - 4x - 2}{x^3 - x^2 - 2x}$

6. $f(x) = \dfrac{6}{x^3 - 4x^2 + 3x}$

7. $f(x) = \dfrac{1}{x^3 - x^2}$

8. $f(x) = \dfrac{1}{x^4 - x^3}$

9. $f(x) = \dfrac{x^2 - 2x - 8}{x^3 - 4x^2 + 4x}$

10. $f(x) = \dfrac{3x - 4}{x^3 + 2x^2 + x}$

11. $f(x) = \dfrac{3x^2 + 1}{x^3 + x^2 + x}$

12. $f(x) = \dfrac{2x - 5}{x^3 + 2x^2 + 5x}$

13. $f(x) = \dfrac{x - 2}{x^3 + 1}$

14. $f(x) = \dfrac{2x^2 + 4}{x^3 - 1}$

15. $f(x) = \dfrac{5x^2}{2x^3 + 3x^2 + 18x + 27}$

16. $f(x) = \dfrac{2x^2 + 5}{x^3 - 2x^2 + 9x - 18}$

17. $f(x) = \dfrac{7x^2 + 3}{x^4 + 3x^2 - 4}$

18. $f(x) = \dfrac{-x^3 - 9x^2 - 3x - 34}{x^4 - x^2 - 12}$

19. $f(x) = \dfrac{5x^3 + 13x + 4}{x^4 + 6x^2 + 5}$

20. $f(x) = \dfrac{8x^3 + 2x^2 + 9x + 1}{x^4 + 4x^2 + 3}$

21. $f(x) = \dfrac{x^3 - 4x}{x^4 + 2x^2 + 1}$

22. $f(x) = \dfrac{3x^2 - 5x}{x^4 + 4x^2 + 4}$

23. $f(x) = \dfrac{2x^3 - 3x^2 + 16}{x^5 + 8x^3 + 16x}$

24. $f(x) = \dfrac{x^5 - x^4 + 5x^3 + x + 1}{x^6 + 2x^4 + x^2}$

 In Exercises 25–32, find the partial fraction decomposition of each improper rational function.

25. $f(x) = \dfrac{3x^3}{x^2 + x - 2}$

26. $f(x) = \dfrac{x^2}{x^2 + 2x + 1}$

27. $f(x) = \dfrac{x^3 - 2}{x^3 - x^2}$

28. $f(x) = \dfrac{x^3 - 1}{x^3 + 2x}$

29. $f(x) = \dfrac{x^4 + 2x^2 + 8}{x^3 - 4x}$

30. $f(x) = \dfrac{x^4 + 3x^3 - 11x - 2}{x^3 + x^2 - 2x}$

31. $f(x) = \dfrac{x^6 - x^3 + 2}{x^4 + 2x^2}$

32. $f(x) = \dfrac{x^5}{x^4 + 8x^2 + 16}$

Critical Thinking

 In Exercises 33–40, use the rational zero theorem (Section 5.3) and the factor theorem (Section 5.2) to express the denominator of the given rational function as the product of linear and irreducible quadratic factors. Then find the partial fraction decomposition of the rational function.

33. $f(x) = \dfrac{x^2 + 1}{x^3 - 6x^2 + 11x - 6}$

34. $f(x) = \dfrac{6x^2 + 22x - 23}{2x^3 + x^2 - 13x + 6}$

35. $f(x) = \dfrac{5x - 1}{x^3 - x^2 + x - 6}$

36. $f(x) = \dfrac{16 - 8x}{x^3 + x^2 - x + 15}$

37. $f(x) = \dfrac{-3x^2 + 7x}{x^3 - 3x^2 + 3x - 1}$

38. $f(x) = \dfrac{x^2 - 9x + 17}{x^3 - 3x^2 + 4}$

39. $f(x) = \dfrac{x^3 - x^2 + 24x + 8}{x^4 - 9x^2 + 4x + 12}$

40. $f(x) = \dfrac{x^3 - 8x^2 - 1}{x^4 + x^3 - 5x^2 + x - 6}$

Calculator Activities

In Exercises 41–46, use a calculator to help find the partial fraction decomposition of the rational function. If you have access to a graphing calculator, show that the graph of the original function and the graph of the partial fraction decomposition of this function are identical.

41. $f(x) = \dfrac{-117x - 40{,}260}{x^2 - 244x}$

42. $f(x) = \dfrac{73x^2 - 2074x + 13{,}284}{x^3 - 36x^2 + 324x}$

43. $f(x) = \dfrac{39x^2 - 309x + 588}{x^3 - 14x^2 + 21x - 294}$

44. $f(x) = \dfrac{41x^4 + 11x^3 + 2058x^2 + 474x + 14{,}112}{x^5 + 84x^3 + 1764x}$

45. $f(x) = \dfrac{12x^3 + 35x^2 - 3427x - 9860}{x^2 - 289}$

46. $f(x) = \dfrac{42x^3 - 696x^2 + 2076x - 37{,}440}{x^3 + 52x}$

Chapter 5   Review

Questions for Group Discussion

1. Explain the method for checking *long division* of polynomials. Illustrate with an example.

2. Under what conditions is it possible to divide one polynomial by another using *synthetic division*?

3. Explain how synthetic division and the *remainder theorem* can be used to determine if r is a zero of a polynomial function P.

4. Explain how synthetic division and the *factor theorem* can be used to determine if $x - r$ is a factor of a polynomial function P.

5. Determine the possible combinations of real and imaginary zeros of a polynomial function P
   (a) if P is a *quartic function* with real coefficients.
   (b) if P is a *quintic function* with real coefficients.

6. If P is a *cubic function* with real coefficients, find (a) the maximum number of real zeros of P and (b) the minimum number of real zeros of P.

7. Explain the procedure for determining the possible *rational zeros* of a polynomial function.

8. How is the *upper and lower bound rule* used in narrowing the choices for the possible real zeros of a polynomial function?

9. What is the significance of *Descartes' rule of signs*? How are the coefficients of the polynomial $P(x)$ related to the coefficients of $P(-x)$?

10. What are *relative extrema* of a polynomial function? How are extrema used as an aid in graphing a polynomial function?

11. What can be said about the zeros of a polynomial function P of degree n if the graph of P (a) does not cross or touch the x-axis and (b) crosses the x-axis n times?

12. Under what conditions is the graph of a polynomial function *tangent* to the x-axis? Discuss the types of tangency that can occur for the graph of a polynomial function.

13. Can each of the following sketches represent the graph of a *polynomial function*? Explain.

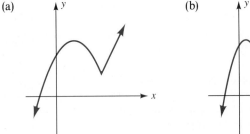

14. Does the *fundamental theorem of algebra* tell us how to find a zero of a polynomial function? If not, what does it tell us?

15. How are the *opposite sign property* and the *method of successive approximations* used to approximate an irrational zero of a polynomial function?

16. How is the domain of a *rational function* determined? Give an example of a rational function whose domain is (a) $(-\infty, a) \cup (a, \infty)$ and (b) all real numbers.

17. Is it possible for the graph of a rational function to cross its *vertical asymptote*? Explain.

18. Discuss the procedure for finding the *horizontal asymptote* of a rational function.

19. Discuss the procedure for finding the *oblique asymptote* of a rational function.

20. Can each of the following sketches represent the graph of a rational function? Explain.

(a) 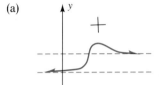   (b)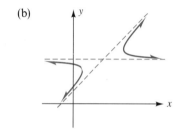

21. What is meant by the *partial fraction decomposition* of a rational function? Give the partial fraction decomposition form of a rational function whose denominator contains one distinct linear factor and one distinct quadratic factor.

22. Discuss the procedure for finding the partial fraction decomposition of an *improper rational function*.

 Review Exercises

In Exercises 1–8, divide the first polynomial by the second polynomial. Use synthetic division whenever possible.

1. $3x^3 + 2x^2 - 3x + 4$; $x^2 + 1$

2. $4x^3 + 3x^2 - x + 6$; $x - 1$

3. $2x^4 + 2x^3 + 3x - 1$; $x + 2$

4. $3x^4 + 8x^2 + 3x - 1$; $x^2 + 2x - 2$

5. $6x^4 - 2x^3 + 9x^2 - 4$; $x - \frac{1}{3}$

6. $4x^4 - x^3 + 5x^2 - 2x + 1$; $x + \frac{3}{4}$

7. $x^6 + 1$; $x^3 + x + 1$

8. $x^5 - 8x^3 - 7x + 6$; $(x + 3)^2$

In Exercises 9–12, use synthetic division and the remainder theorem to find the indicated functional value.

9. $P(x) = 2x^3 - 3x^2 + 5x - 8$; $P(2)$

10. $P(x) = 4x^3 + 7x^2 - 8x - 6$; $P(-3)$

11. $f(x) = 3x^4 - x^3 + 4x^2 - 5x - 6$; $f(-\frac{2}{3})$

12. $F(x) = 6x^4 - x^3 - 10x^2 - 5x + 3$; $F(\frac{3}{2})$

In Exercises 13–18, use the factor theorem to determine if the first polynomial is a factor of the second polynomial.

13. $x - 2$; $4x^3 - 7x^2 - 4$

14. $x + 3$; $2x^3 + 3x^2 - 7x - 9$

15. $x + \frac{4}{3}$; $3x^4 + x^3 + 2x^2 + 8x + 4$

16. $x - \frac{1}{4}$; $4x^4 + 3x^3 - x^2 - 16x + 4$

17. $x - 2i$; $x^5 + 3x^3 + 2x^2 - 4x + 8$

18. $x - (2 + i)$; $2x^4 + 7x^3 + 5x^2 - 9x - 5$

In Exercises 19–24,

(a) *use Descartes' rule of signs to determine the possibilities for the number of positive and negative real zeros of P.*

(b) *use the upper and lower bound rule to show that all real zeros of P are in the given interval $[a, b]$.*

19. $P(x) = x^3 + 2x^2 + 8x - 5$; $[-2, 1]$

20. $P(x) = 2x^3 - 4x^2 + 6x + 8$; $[-1, 2]$

21. $P(x) = 3x^4 - 4x^3 + 2x^2 + 13x + 6$; $[-2, 2]$

22. $P(x) = x^4 + 3x^3 - 18x^2 - 9x - 20$; $[-6, 4]$

23. $P(x) = 2x^5 + 6x^2 - 2x + 15$; $[-2, \frac{1}{2}]$

24. $P(x) = 3x^6 - x^2 + 11x - 12$; $[-2, 1]$

In Exercises 25–36, (a) write P(x) as the product of linear factors and (b) find all zeros of P.

25. $P(x) = x^3 - 5x^2 + 2x + 8$

26. $P(x) = 2x^3 + x^2 - 13x + 6$

27. $P(x) = 3x^3 - 10x^2 - 7x + 20$

28. $P(x) = x^3 - 10x^2 + 34x - 40$

29. $P(x) = 2x^4 + 5x^3 + 5x^2 + 20x - 12$

30. $P(x) = 2x^4 - 9x^3 + 8x^2 + 19x - 30$

31. $P(x) = 4x^5 - 4x^4 - 15x^3 + 10x^2 + 11x - 6$

32. $P(x) = 4x^5 + 16x^4 - 23x^3 - 13x^2 + 11x + 5$

33. $P(x) = x^4 - 4x^3 + 9x^2 - 4x + 8$; $x + i$ is a factor

34. $P(x) = x^4 - 6x^3 + 11x^2 - 10x + 2$;
$x - (1 - i)$ is a factor

35. $P(x) = 2x^5 - 9x^4 + 22x^3 - 41x^2 + 56x - 20$;
$2 + i$ is a zero

36. $P(x) = x^6 + 2x^5 + 5x^4 - 12x^3 - 29x^2 - 70x - 25$;
$2i - 1$ is a zero of multiplicity two

In Exercises 37–44, find all real roots of each polynomial equation. Record irrational roots to the nearest hundredth.

37. $x^3 + 2x - 8 = 0$ **38.** $2x^3 + 5x^2 + 3 = 0$

39. $2x^4 - 5x^3 + 6x^2 - 5x = 25$

40. $x^3 + 8x^2 + 2x = 3$

41. $x^4 + 6x^3 + 4 = 0$

42. $x^5 - 3x^4 - 2x^2 + x + 15 = 0$

43. $x^5 - 8x^4 + 20x^3 - 19x^2 + 12x = 12$

44. $3x^6 + 9x^5 + 10x^4 - 15x^2 - 17x = 6$

In Exercises 45–52, sketch the graph of each polynomial function. Label the x- and y-intercepts.

45. $g(x) = 2x^2 - x^4$

46. $h(x) = 4x^3 - 3x^4$

47. $f(x) = x^3 + 3x^2 - 6x - 8$

48. $F(x) = 10 - 11x - 4x^2 - 3x^3$

49. $G(x) = 4x^4 - 12x^3 - 39x^2 - 3x - 10$

50. $P(x) = x^4 - x^3 - 9x^2 + 9x$

51. $P(x) = x^5 - 6x^4 + 16x^3 - 32x^2 + 48x - 32$

52. $f(x) = x^6 - 2x^5 - 7x^4 + 16x^3 - 17x^2 + 18x - 9$

In Exercises 53–60, sketch the graph of each rational function. Label the x- and y-intercepts and any asymptotes.

53. $f(x) = \dfrac{3x - 2}{x + 1}$

54. $F(x) = \dfrac{1}{x^2 + 4}$

55. $G(x) = \dfrac{x}{x^2 - 3x - 10}$

56. $g(x) = \dfrac{x + 4}{x^3 - 4x^2 + 4x}$

57. $F(x) = \dfrac{x^2 - 1}{2x}$

58. $f(x) = \dfrac{x^2 + 4}{x - 1}$

59. $H(x) = \dfrac{x^3 + 8}{x^3 + 3x^2 - 6x - 8}$

60. $R(x) = \dfrac{x^2 - x - 12}{x^2 - 9}$

In Exercises 61–66, find the partial fraction decomposition of each rational function.

61. $f(x) = \dfrac{10x}{x^2 - 1}$

62. $f(x) = \dfrac{5x^2 - 11x - 6}{x^3 - x^2 - 4x + 4}$

63. $f(x) = \dfrac{3x^2 - 13x + 21}{x^3 - 4x^2 + x - 4}$

64. $f(x) = \dfrac{x^2 - x - 4}{x^3 - 6x^2 + 12x - 8}$

65. $f(x) = \dfrac{18}{x^5 + 6x^3 + 9x}$ **66.** $f(x) = \dfrac{x^2 + x - 6}{x^2 - x - 20}$

67. Determine a polynomial function P of degree 3 with zeros 1, -2, and 3 such that

 (a) $P(0) = 3$ (b) $P(2) = -8$

68. Determine a polynomial function P of degree 4 with zeros -1, -1, $\frac{1}{2}$, and $-\frac{3}{2}$, such that

 (a) $P(0) = -3$ (b) $P(1) = 24$

69. Sketch the graphs of the equations $y = x^3$ and $7x - 4y = -3$ on the same coordinate plane. Then determine the coordinates of the points where the graphs intersect.

70. Sketch the graphs of the functions $f(x) = 4/x$ and $g(x) = x^2 - 2x - 8$ on the same coordinate plane. Then determine to the nearest hundredth the coordinates of the points where the graphs intersect.

71. Show that the polynomial function
$P(x) = 2x^4 - 18x^2 + 5x + 3$ has no real zero greater than 3.

72. When $P(x) = 3x^4 - 8x^3 + 5x^2 - 4x + 5$ is divided synthetically by $x - 2$, the numbers in the final row of the synthetic division alternate sign as shown:

$$
\begin{array}{r|rrrrr}
2) & 3 & -8 & 5 & -4 & 5 \\
  &   & 6 & -4 & 2 & -4 \\
\hline
  & 3 & -2 & 1 & -2 & 1
\end{array}
$$

Why can't we state that 2 is a lower bound for P?

73. Given that $P(x) = x^3 + ax^2 + b$, determine the nature of the zeros of P if

(a) a and b are positive real numbers.
(b) a and b are negative real numbers.

74. Given that $P(x) = x^4 + ax + b$, determine the nature of the zeros of P if

(a) a and b are negative real numbers.
(b) a is a positive real number and b is a negative real number.

75. Determine the horizontal asymptote for

$$
f(x) = \frac{2x^2 + 5x - 7}{3x^2 + 2}.
$$

Does the graph of this function cross the horizontal asymptote? If so, find the coordinates of the point where it does cross.

76. Determine the oblique asymptote for

$$
f(x) = \frac{x^3 - 2x^2 + 3x}{x^2 - 4x + 3}.
$$

Does the graph of this function cross the oblique asymptote? If so, find the coordinates of the point where it does cross.

77. A forester estimates that the number N of 2 inch × 3 inch × 8 foot construction-grade studs that can be cut from a fir tree is given by $N(x) = 4x^3 - 8x^2 + 5x - 1$, where x is the circumference (in feet) of the base of the tree trunk, $x \geq 1$. What is the circumference of a tree that can yield 24 studs?

78. When a 20-microfarad capacitor is connected in series with a variable capacitor x, the total capacitance C is given by $C(x) = \dfrac{20x}{x + 20}$.

(a) Sketch the graph of this function ($x \geq 0$).
(b) What is the significance of the horizontal asymptote?

79. The height of a tin can is 3 centimeters more than its radius. Determine the radius of the can if its volume is 112π cubic centimeters.

80. An open rectangular box has four sides and a bottom, all of uniform thickness. The outside dimensions of the box are 3 feet by 3 feet by 2 feet, as shown in the sketch. Determine the uniform thickness (to the nearest hundredth) if the box holds 12 cubic feet of water when filled to the top.

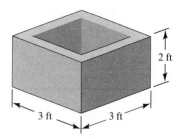

81. A piece of cardboard, 15 cm by 20 cm, is made into an open rectangular box by cutting out a square from each corner and turning up the sides, as shown in the figure.

(a) Express the volume V of this box as a function of x.
(b) Find *two* possible lengths of the side of the square cutout if the volume of the box is 250 cubic centimeters.
(c) Sketch the graph of the function found in part (a).
(d) Use a calculator or computer with graphing capabilities to find the approximate value of x when the volume of the box is greatest.

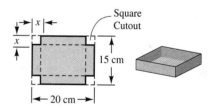

82. A beam of length L is simply supported at both ends and carries a load of weight W, that decreases uniformly from center, as shown in the figure. The bending moment M at any distance x ($x \leq L/2$) from the left end of this beam is given by

$$
M(x) = \frac{Wx}{6L^2}(3L^2 - 4x^2).
$$

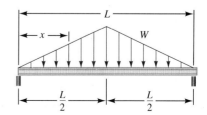

(continued)

(a) Determine the bending moment function when $W = 2400$ pounds and $L = 20$ feet. Then sketch the graph of this function.

(b) At what distance x from the left end of the beam is the bending moment 6336 pound-feet?

83. A rectangular box has a square base and open top, as shown in the sketch. The volume of the box is 100 cubic meters. The material used to construct the base costs $4 per square meter and the material for the sides cost $2.50 per square meter.

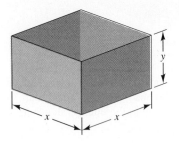

(a) Show that the cost C to construct the box is given by $C = \dfrac{4x^3 + 1000}{x}, x > 0.$

(b) Sketch the graph of the rational function defined in part (a).

(c) Use a calculator or computer with graphing capabilities to find the dimensions of the box whose cost is the smallest possible.

84. A certain manufacturer produces wood-burning stoves. The fixed cost for the company is $81,000 per year and the variable cost (in dollars) of producing x stoves per year is $400x + 0.1x^2$.

(a) Define a function that represents the average cost per stove, and sketch the graph of this function.

(b) Use a computer or calculator with graphing capabilities to find the approximate number of stoves that should be produced per year in order to minimize the average cost per stove.

CHAPTER 6

Exponential and Logarithmic Functions

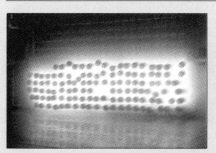

Strontium-90, a waste product from nuclear reactors, has a half-life of 28 years. It is estimated that a certain quantity of this material will be safe to handle when its mass is $\frac{1}{1000}$ of its original amount. Determine the time required to store strontium-90 until it reaches this level of safety.

(For the solution, see Example 9 in Section 6.2.)

6.1   **Properties and Graphs of Exponential Functions**
6.2   **Logarithmic Functions**
6.3   **Properties of Logarithms**
6.4   **Graphs of Logarithmic Functions**
6.5   **Logarithmic and Exponential Equations**

6.1   Properties and Graphs of Exponential Functions

Any function that can be expressed as sums, differences, products, quotients, powers, or roots of polynomials is classified as an **algebraic function**. All of the functions that we have discussed in Chapters 3, 4, and 5, are algebraic functions. Any function that goes beyond the limits of, or *transcends*, an algebraic function is a **transcendental function**. In this section we study a class of transcendental functions in which the variable appears as an exponent. We begin by reviewing the idea of rational exponents and then extending this idea to include irrational exponents.

◆ Real Exponents

In Sections 1.2 and 1.3 we evaluated expressions of the form b^x, where b is a *positive real number* and x is any *rational number*. Whether x is a positive integer, a negative integer, a common fraction, or a decimal fraction, we can evaluate b^x as a real number, as shown in the following examples:

$$4^2 = 4 \cdot 4 = 16$$

$$4^{-2} = \frac{1}{4^2} = \frac{1}{16}$$

$$4^{3/2} = (4^{1/2})^3 = (\sqrt{4})^3 = (2)^3 = 8$$

$$4^{1.4} = 4^{14/10} = 4^{7/5} \approx 6.9644 \longleftarrow$$

> By using the y^x key on a calculator, we can write the fifth root of 4 to the seventh power as an approximate decimal number.

In general, if b is a positive real number and x is any rational number, then b^x is a well-defined positive real number.

Can we assign any meaning to b^x where b is a positive real number and x is an *irrational* number? For example, can we evaluate $4^{\sqrt{2}}$? We can use the $\boxed{Y^x}$ key on a calculator to find values of 4^x for rational numbers x that approach $\sqrt{2}$:

Values of 4^x for rational values of x that approach $\sqrt{2}$.

x	1	1.4	1.41	1.414	1.4142	1.41421
4^x	4	6.964405...	7.061624...	7.100891...	7.102860...	7.102958

From these values, we see that as the exponent x approaches $\sqrt{2}$, the expression 4^x appears to approach a unique positive real number whose decimal expansion begins with 7.10. Hence, we define the value of $4^{\sqrt{2}}$ as the real number approached by 4^x as x takes on rational values that get closer and closer to $\sqrt{2}$. In general, if b is a positive real number and x is irrational, then we define b^x as the number approached by b^r as r takes on rational values that get closer and closer to x. We conclude that *for each real number x, b^x is a unique positive real number.*

The laws of exponents given in Section 1.2 apply to all real exponents. We restate them here and refer to them as the **properties of real exponents**.

Properties of Real Exponents

If the bases a and b are positive real numbers and the exponents x and y represent any real numbers, then

1. $b^0 = 1$

2. $b^{-x} = \dfrac{1}{b^x}$

3. $b^x b^y = b^{x+y}$

4. $(b^x)^y = b^{xy}$

5. $\dfrac{b^x}{b^y} = b^{x-y}$

6. $(ab)^x = a^x b^x$

7. $\left(\dfrac{a}{b}\right)^x = \dfrac{a^x}{b^x}$

8. $\dfrac{a^{-x}}{b^{-y}} = \dfrac{b^y}{a^x}$

Since $1^x = 1$ for all real x, we may also express property 2 as

$$b^{-x} = \left(\frac{1}{b}\right)^x$$

EXAMPLE 1   Use the properties of real exponents to write each expression as a constant or as an expression in the form b^x with $b > 0$.

(a) $\dfrac{5^\pi}{5^{x+\pi}}$ (b) $4^{-3x} \cdot 8^{2x+1}$

SOLUTION

(a) $\dfrac{5^\pi}{5^{x+\pi}} = 5^{\pi-(x+\pi)} = 5^{-x} = \left(\dfrac{1}{5}\right)^x$

(b) $4^{-3x} \cdot 8^{2x+1} = (2^2)^{-3x} \cdot (2^3)^{2x+1}$ **Write each factor with the same base**

$\qquad\qquad\quad = 2^{-6x} \cdot 2^{6x+3}$ **Multiply the exponents**

$\qquad\qquad\quad = 2^3$ or 8   **Add the exponents and simplify**   ◆

PROBLEM 1   Repeat Example 1 for $\dfrac{125^{x-2}}{25^{2x-3}}$.   ◆

◆ **Exponential Functions and Their Graphs**

We now use the properties of real exponents to define the *exponential function* with base b.

Exponential Function

If b is a real number such that $b > 0$ and $b \neq 1$, then the function f defined by

$$f(x) = b^x$$

is called an **exponential function** with base b. The domain of f is $(-\infty, \infty)$ and the range is $(0, \infty)$.

Note: We exclude 1 as a base for exponential functions, since $f(x) = 1^x = 1$ is a constant function. We exclude zero as a base, since 0^x is undefined when $x \le 0$. We exclude negative numbers as bases, since b^x for $b < 0$ is an imaginary number for infinitely many values of x, such as $x = \frac{1}{2}, \frac{1}{4}, \frac{1}{6}$, and so on.

EXAMPLE 2   Sketch the graph of each exponential function.

(a) $f(x) = 4^x$ (b) $g(x) = (\frac{1}{4})^x$

SOLUTION

(a) Selecting convenient inputs for x, we find the corresponding outputs $f(x)$ as shown in the following table:

x	-2	$-\frac{3}{2}$	-1	$-\frac{1}{2}$	0	$\frac{1}{2}$	1	$\frac{3}{2}$	2
$f(x)$	$\frac{1}{16}$	$\frac{1}{8}$	$\frac{1}{4}$	$\frac{1}{2}$	1	2	4	8	16

Plotting the points $(x, f(x))$ and connecting them in a smooth curve, we graph the exponential function $f(x) = 4^x$ as shown in Figure 6.1. Notice that the graph of this function is always increasing and its y-intercept is 1. Also, observe that $f(x)$ approaches zero $[f(x) \to 0]$ as x decreases without bound $(x \to -\infty)$. Hence, the x-axis is a horizontal asymptote.

(b) By the properties of real exponents,

$$g(x) = (\tfrac{1}{4})^x = 4^{-x}.$$

Now, by the y-axis reflection rule (Section 3.4), the graph of g must be the same as the graph of $f(x) = 4^x$ reflected about the y-axis. The graph of g is shown in Figure 6.2. Notice that the graph of this function is always

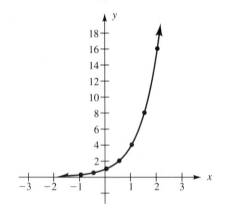

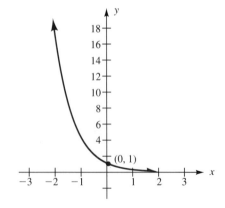

FIGURE 6.1

The graph of $f(x) = 4^x$ is an increasing function. Its y-intercept is 1 and the x-axis is a horizontal asymptote

FIGURE 6.2

The graph of $g(x) = (\frac{1}{4})^x = 4^{-x}$ is a decreasing function. Its y-intercept is 1 and the x-axis is a horizontal asymptote.

decreasing and its y-intercept is 1. Also, observe that $f(x)$ approaches zero $[f(x) \to 0]$ as x increases without bound $(x \to \infty)$. Hence, the x-axis is a horizontal asymptote.   ◆

PROBLEM 2   On the same coordinate plane, sketch the graphs of the exponential functions

$$f(x) = 5^x \quad \text{and} \quad g(x) = (\tfrac{1}{5})^x = 5^{-x}. \qquad ◆$$

Some important observations about the graph of the exponential function $f(x) = b^x$ will be useful in our following discussion:

 Characteristics of the Graph of $f(x) = b^x$

1. The y-intercept is 1, and the graph has no x-intercept.

2. The x-axis is a horizontal asymptote.

3. If $b > 1$, the graph of $f(x) = b^x$ is always increasing.

4. If $0 < b < 1$, the graph of $f(x) = b^x$ is always decreasing.

Knowing the basic shape of the exponential function $f(x) = b^x$ enables us to graph several other related functions by applying the shift rules and axis reflection rules, which we discussed in Section 3.4.

EXAMPLE 3   Sketch the graph of each function. Label the horizontal asymptote and the y-intercept.

(a) $F(x) = 4^{x+2}$ (b) $G(x) = 3 - 4^{-x}$

SOLUTION

(a)   By the horizontal shift rule (Section 3.4), the graph of $F(x) = 4^{x+2}$ is the same as the graph of $f(x) = 4^x$ shifted horizontally to the left 2 units. The y-intercept is

$$F(0) = 4^{0+2} = 4^2 = 16,$$

and the horizontal asymptote remains the x-axis. The graph of $F(x) = 4^{x+2}$ is shown in Figure 6.3.

(b)   By the x-axis reflection rule and the vertical shift rule (Section 3.4), the graph of $G(x) = 3 - 4^{-x}$ is the same as the graph of $g(x) = 4^{-x} = (\tfrac{1}{4})^x$ reflected about the x-axis and then shifted vertically upward 3 units. The y-intercept is

$$G(0) = 3 - 4^0 = 3 - 1 = 2.$$

Note that $G(x) \to 3$ as $x \to \infty$. Thus, $y = 3$ is the horizontal asymptote. The graph of $G(x) = 3 - 4^{-x}$ is shown in Figure 6.4.

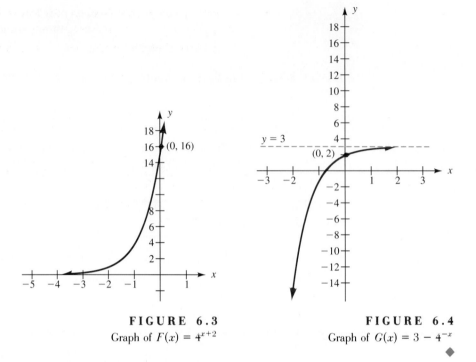

FIGURE 6.3
Graph of $F(x) = 4^{x+2}$

FIGURE 6.4
Graph of $G(x) = 3 - 4^{-x}$

Note: The graph of *G* shown in Figure 6.4 has an *x*-intercept between 0 and −1. To find the *x*-intercept, it is necessary to solve the equation

$$3 - 4^{-x} = 0 \qquad \text{or} \qquad 4^{-x} = 3$$

for *x*. However, none of our previous methods can be used to solve an equation in which the unknown appears as an exponent. In Section 6.2, we introduce the *logarithmic function*, which we can use to help solve this type of equation.

PROBLEM 3   State the domain and range of the function *G* defined in Example 3(b).   ◆

◆   **Base *e***

In applications of exponential functions, one particular irrational number occurs frequently as the base. This irrational number is denoted by the letter *e* and its value is

$$e = 2.71828 \ldots$$

The function *f* defined by

$$f(x) = e^x$$

is called the **exponential function with base e**. With a calculator, we can evaluate $f(x) = e^x$ for $x = 0, \pm 1, \pm 2,$ and ± 3 by using the $\boxed{e^x}$ key, the $\boxed{\text{Y}^x}$ key with $y = 2.718$, or the exponential table in Appendix B.

Table of values for $f(x) = e^x$

x	-3	-2	-1	0	1	2	3
$f(x)$	$e^{-3} \approx 0.05$	$e^{-2} \approx 0.14$	$e^{-1} \approx 0.37$	1	$e \approx 2.72$	$e^2 \approx 7.39$	$e^3 \approx 20.1$

Plotting these points, we sketch the graph of $f(x) = e^x$, as shown in Figure 6.5. The graph of $g(x) = e^{-x}$, which is the same as the graph of $f(x) = e^x$ reflected about the y-axis, is shown in Figure 6.6.

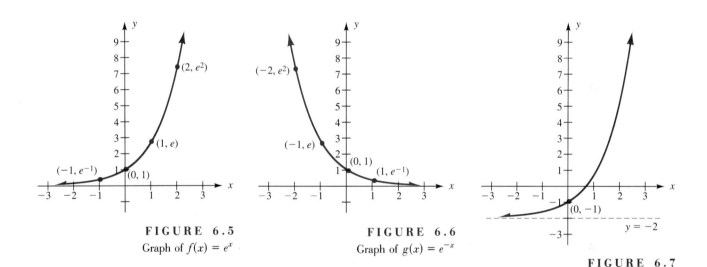

FIGURE 6.5
Graph of $f(x) = e^x$

FIGURE 6.6
Graph of $g(x) = e^{-x}$

FIGURE 6.7
Graph of $F(x) = e^x - 2$

EXAMPLE 4   Sketch the graph of the function F defined by $F(x) = e^x - 2$. Label the horizontal asymptote and the y-intercept.

SOLUTION   By the vertical shift rule (Section 3.4), the graph of $F(x) = e^x - 2$ is the same as the graph of $f(x) = e^x$ shifted vertically downward 2 units, as shown in Figure 6.7. The y-intercept is

$$F(0) = e^0 - 2 = 1 - 2 = -1.$$

Note that $F(x) \to -2$ as $x \to -\infty$. Thus, $y = -2$ is the horizontal asymptote.

◆

PROBLEM 4   Try to find the x-intercept for the graph of the function $F(x) = e^x - 2$ shown in Figure 6.7. What situation occurs?

◆

◆ **Application: Compound Interest**

In Section 2.2, we worked with word problems involving simple interest. Recall that the amount of simple interest i earned on a principal P invested at a certain rate of interest r per year over a time of t years is given by the formula

$$i = Prt.$$

Thus, the amount A in the account after t years is

$$A = P + Prt.$$

Compound interest is interest paid both on the principal and on any interest earned previously. If a certain principal P is deposited in a savings account at an interest rate r per year and interest is *compounded n times per year*, then interest is paid on both principal and any interest earned previously every $(1/n)$th of a year. The interest i accrued after the first compounding period is

$$i = Prt = Pr\frac{1}{n}.$$

Hence the amount A_1 in the account after the first compounding period is

$$A_1 = P + Prt = P + Pr\frac{1}{n} = P\left(1 + \frac{r}{n}\right).$$

The interest i accrued after the second compounding period is

$$i = A_1 r\frac{1}{n},$$

and the amount A_2 in the account after the second compounding period is

$$A_2 = A_1 + A_1 r\frac{1}{n}.$$

Now, replacing A_1 with $P\left(1 + \frac{r}{n}\right)$, we have

$$A_2 = P\left(1 + \frac{r}{n}\right) + P\left(1 + \frac{r}{n}\right)r\frac{1}{n}$$

$$= P\left(1 + \frac{r}{n}\right)\left[1 + \frac{r}{n}\right] \qquad \text{Factor out } P\left(1 + \frac{r}{n}\right) \text{ from each term}$$

$$= P\left(1 + \frac{r}{n}\right)^2 \qquad \text{Simplify}$$

Continuing in this manner, we can show that after 1 year (after n compounding periods) the amount A in the account is

$$A = P\left(1 + \frac{r}{n}\right)^n$$

and after t years the amount in the account is

$$A = P\left(1 + \frac{r}{n}\right)^{nt}$$

◆ Compound Interest
Formula for n
Compoundings Per Year

> If a certain principal P is deposited in a savings account at an interest rate r per year and interest is compounded n times per year, then the amount A in the account after t years is given by the formula
>
> $$A = P\left(1 + \frac{r}{n}\right)^{nt}.$$

If the number of compoundings per year increases *without bound*, then we have what is called **continuous compounding**. In the formula

$$A = P\left(1 + \frac{r}{n}\right)^{nt}$$

suppose we let $n/r = m$. Then $n = mr$, and by direct substitution and the properties of real exponents, we have

$$A = P\left(1 + \frac{r}{n}\right)^{nt} = P\left(1 + \frac{1}{m}\right)^{mrt} = P\left[\left(1 + \frac{1}{m}\right)^m\right]^{rt}.$$

Using a calculator, we can show that as m gets larger and larger, the expression $\left(1 + \dfrac{1}{m}\right)^m$ approaches a number whose decimal expansion begins with 2.718.

Values of $\left(1 + \dfrac{1}{m}\right)^m$ as m gets larger and larger

m	1	10	100	1000	10,000	100,000
$\left(1 + \dfrac{1}{m}\right)^m$	2	2.59374...	2.70481...	2.71692...	2.71814...	2.71826...

We say that as m increases without bound ($m \to \infty$) the limit of $(1 + 1/m)^m$ is the number e and we write

$$\lim_{m \to \infty}\left(1 + \frac{1}{m}\right)^m = e.$$

Hence, for continuous compounding, we have

$$A = P\left[\left(1 + \frac{1}{m}\right)^m\right]^{rt} = Pe^{rt}$$

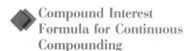 Compound Interest
Formula for Continuous
Compounding

> If a certain principal P is deposited in a savings account at an interest rate r per year and interest is compounded continuously, then the amount A in the account after t years is given by the formula
>
> $$A = Pe^{rt}.$$

EXAMPLE 5   Suppose that \$1000 is deposited in a bank account at an interest rate of 8% per year. Assuming the depositor makes no subsequent deposit or withdrawal, find the amount after 5 years if the interest is compounded **(a)** quarterly or **(b)** continuously.

SOLUTION

(a) If the interest is compounded quarterly, then it is compounded 4 times per year. Using the formula $A = P\left(1 + \frac{r}{n}\right)^{nt}$ with $P = 1000$, $r = 0.08$, $n = 4$, and $t = 5$, we have

$$A = 1000\left(1 + \frac{0.08}{4}\right)^{4 \cdot 5} = 1000(1.02)^{20} \approx 1485.95$$

Thus, if the interest is compounded quarterly, the amount in the account after 5 years is \$1485.95.

(b) To find the amount in the account when the interest is compounded continuously, we use the formula $A = Pe^{rt}$ with $P = 1000$, $r = 0.08$, and $t = 5$ as follows:

$$A = 1000e^{(0.08)5} = 1000e^{0.4} \approx 1491.82$$

Thus, if the interest is compounded continuously, the amount in the account after 5 years is \$1491.82.   ◆

PROBLEM 5   Repeat Example 5 if the interest is compounded monthly.   ◆

Exercises 6.1

 Basic Skills

In Exercises 1–14, use the properties of real exponents to write each expression as a constant or as an expression in the form b^x, where $b > 0$.

1. 1^{5x}

2. $1^{x - \pi}$

3. $(6^x)^{-2}$

4. $\left(\frac{1}{3}\right)^{-2x}$

5. $(4^{x-1})^0$

6. $(2^{2x})^{-2}$

7. $e(e^{x-1})$

8. $e^x \cdot e^{-x}$

9. $(3)(3^{2x-1})$

10. $\dfrac{36}{6^{2-x}}$

11. $(9^{3x-2})(27^{1-2x})$

12. $\dfrac{8^{2-x}}{4^{x+3}}$

13. $\left(\dfrac{5^{2x}}{25^{x-1}}\right)^{1/2}$

14. $(2^{-x} \cdot 4^x)^2$

In Exercises 15 and 16, sketch the graphs of the given equations on the same coordinate plane. What is the point of intersection of the graphs?

15. $y = 2^x$, $y = 3^x$, and $y = e^x$

16. $y = 2^{-x}$, $y = 3^{-x}$, and $y = e^{-x}$

Use the graphs of the equations in Exercises 15 and 16 in conjunction with the shift rules and axis reflection rules (Section 3.4) to sketch the graphs of the functions in Exercises 17–30. In each case, label the y-intercept and horizontal asymptote.

17. $f(x) = -2^x$

18. $f(x) = -3^{-x}$

19. $h(x) = 2^{x-2}$

20. $h(x) = e^{x+1}$

21. $F(x) = -(3^x + 2)$

22. $F(x) = 2^x + 1$

23. $G(x) = 2(\tfrac{1}{2})^x$

24. $G(x) = 9 \cdot 3^{-x}$

25. $H(x) = 1 - e^{-x}$

26. $H(x) = 2 - (\tfrac{1}{3})^{-x}$

27. $A(x) = 2 + 3^{2-x}$

28. $A(x) = 2^{x+1} - 3$

29. $h(x) = \dfrac{e}{e^x}$

30. $h(x) = 2 - e^{1-x}$

In Exercises 31–38, find the amount in the given account.

31. $5000 invested at 9% per year compounded annually for 4 years

32. $3000 invested at 7% per year compounded semiannually for 8 years

33. $10,000 invested at $8\tfrac{1}{2}$% per year compounded quarterly for 12 years

34. $800 invested at $9\tfrac{3}{4}$% per year compounded daily for 10 years

35. $2000 invested at 6% compounded continuously for 9 years

36. $15,000 invested at 10% compounded continuously for 6 years

37. $120,000 invested at 8% compounded continuously for 6 months

38. $250,000 invested at $12\tfrac{3}{4}$% compounded continuously for 8 months

39. Perform the indicated operations and simplify.

  (a) $(2^x + 2^{-x})(2^x - 2^{-x})$
  (b) $(3^x + 3^{-x})^2 - (3^x - 3^{-x})^2$

40. Given the function f defined by $f(x) = e^x$, determine if f is a one-to-one function and if there exists an inverse function f^{-1}. Sketch the graph of f^{-1} if it exists.

In Exercises 41-46, use factoring and the zero product property (Section 2.3) to help find the zeros of each function.

41. $F(x) = 2e^{-2x} - xe^{-2x}$

42. $G(x) = 2x^2e^{5x} - xe^{5x} - e^{5x}$

43. $g(x) = \dfrac{(x^2 - 4x + 4)e^x}{x + 1}$

44. $h(x) = \dfrac{x^2e^x + 2x\,e^x}{x - 3}$

45. $H(x) = x^2 3^{2x} - 9^x$

46. $f(x) = x^2(\tfrac{1}{2})^{3x} - x8^{-x}$

47. The graph of the exponential function $y = a^x$ passes through the point $(-2, 16)$. Determine the value of the base a.

48. Given that f is an exponential function defined by $f(x) = e^x$, show that

  (a) $f(a)f(b) = f(a + b)$ (b) $\dfrac{f(a)}{f(b)} = f(a - b)$
  (c) $[f(a)]^n = f(na)$

49. Complete the table, given that the function f is defined by $f(x) = (-2)^x$:

x	0	1	2	3	4	−1	−2	−3	−4
$f(x)$									

EXERCISE 49 *(continued)*

Why is it not possible to sketch the graph of f by plotting the points in this table and connecting them with a smooth curve?

50. The *hyperbolic sine function* f and the *hyperbolic cosine function* g are defined by

$$f(x) = \frac{e^x - e^{-x}}{2} \quad \text{and} \quad g(x) = \frac{e^x + e^{-x}}{2}$$

(a) Determine if f and g are even or odd functions.
(b) Sketch the graphs of f and g.

51. Suppose P is the air pressure (in pounds per square inch, or psi) at a certain height h in feet above sea level, and $P = ke^{-0.00004h}$, where k is a constant.

(a) Determine k if the air pressure at sea level is 15 psi.
(b) What is the pressure outside the cabin of an airplane whose altitude is 20,000 feet?

52. Suppose A is the number of moose living in a protected forest in Maine after a certain time t (in years since 1968), and $A = ke^{0.15t}$, where k is a constant.

(a) Determine k if 201 moose were present in 1988.
(b) What is the expected moose population in 1998?

53. The number A of bacteria present in a certain culture is a function of time t (in minutes) and is described by $A(t) = 18e^{0.2t}$.

(a) Find the number of bacteria initially present in the culture.
(b) Find the number of bacteria present in the culture after 10 minutes.
(c) Sketch the graph of the function A for $t \geq 0$.

54. The voltage V (in volts) in an electrical circuit is a function of time t (in milliseconds) and is given by $V(t) = 400e^{-t/10}$

(a) Find the initial voltage in the circuit.
(b) Find the voltage in the circuit after 30 milliseconds.
(c) Sketch the graph of the function V for $t \geq 0$.
(d) Describe the behavior of $V(t)$ as t increases without bound ($t \to \infty$).

55. The value V in dollars of a certain piece of machinery is a function of its age t in years and is given by $V(t) = 2000 + 10{,}000e^{-0.35t}$.

(a) What is the value of the machine when it is purchased?
(b) What is the value of the machine 4 years after it is purchased?
(c) Sketch the graph of the function V for $t \geq 0$.
(d) What is the significance of the horizontal asymptote?

56. For a certain secretarial student, the number N of words typed (correctly) per minute on a new word processor is a function of the time t (in hours) that the student has practiced on the machine and is given by $N(t) = 60(1 - e^{-0.05t})$.

(a) What is the student's rate of typing N after 10 hours of practice?
(b) Sketch the graph of the function N for $t \geq 0$.
(c) What is the best rate of typing this student can hope to achieve on this word processor?

Calculator Activities

57. Use a calculator to evaluate each expression. Round each answer to three significant digits.

(a) $3^{-\sqrt{2}}$ (b) $(\frac{4}{17})^{\sqrt{5}}$ (c) e^{π} (d) π^e

58. In calculus, it is shown that if $|x| \leq 1$, then

$$e^x \approx 1 + x + \frac{x^2}{2} + \frac{x^3}{6} + \frac{x^4}{24} + \frac{x^5}{120}.$$

Use this formula and a calculator to evaluate each expression.

(a) $e^{0.5}$ (b) $e^{-0.4}$ (c) e (d) e^{-1}

59. Given that $f(x) = e^x$,

(a) show that the *difference quotient*

$$\frac{f(x + \Delta x) - f(x)}{\Delta x} = e^x \left(\frac{e^{\Delta x} - 1}{\Delta x} \right).$$

(b) use a calculator to evaluate $\dfrac{e^{\Delta x} - 1}{\Delta x}$ for the values of Δx given in the table:

Δx	1	0.1	0.01	0.001	0.0001
$\dfrac{e^{\Delta x} - 1}{\Delta x}$					

(c) use your results from the table in part (b) to determine the number that $\dfrac{e^{\Delta x} - 1}{\Delta x}$ seems to approach as $\Delta x \to 0^+$.

📷 60. In statistics, the *normal probability distribution function* is given by

$$f(x) = \frac{1}{\sigma \sqrt{\pi}}\, e^{-(x-\mu)^2/2\sigma^2}$$

where μ is the mean and σ the standard deviation of the distribution. The graph of this function is usually referred to as a "bell-shaped" curve. By plotting points, sketch the graph of this bell-shaped curve for $\mu = 0$ and $\sigma = 1$. If you have access to a graphing calculator, check your results.

6.2   Logarithmic Functions

◆ Introductory Comments

As we discussed in Section 6.1, the exponential function $f(x) = b^x$ is always increasing when $b > 1$ and is always decreasing when $0 < b < 1$. Recall from Section 3.6 that every function f that is either an increasing function or a decreasing function is also a one-to-one function and has an inverse function f^{-1}. Using the method suggested in Section 3.6, we can attempt to find the inverse function of $f(x) = b^x$ by replacing x with $f^{-1}(x)$ and solving for $f^{-1}(x)$. If we proceed in this manner, we obtain

$$f(x) = b^x$$
$$f(f^{-1}(x)) = b^{f^{-1}(x)}$$
$$x = b^{f^{-1}(x)}$$

However, we have no algebraic procedure that we can use to solve this last equation for $f^{-1}(x)$. By convention, we solve this equation for $f^{-1}(x)$ by writing

$$f^{-1}(x) = \log_b x \longleftarrow \boxed{\text{Read "log base } b \text{ of } x\text{".}}$$

The function f^{-1} is called the *logarithmic function* and $\log_b x$ represents *the power to which b must be raised in order to obtain x.*

Recall from Section 3.6 that the graph of a function and its inverse are symmetric with respect to the line $y = x$. Hence, the graph of $f^{-1}(x) = \log_b x$ may be obtained from the graph of $f(x) = b^x$ by reflecting the graph of f in the line $y = x$, as shown in Figure 6.8.

Since the exponential function and logarithmic function are inverses of each other, the domain of the logarithmic function must be the range of the exponential function, namely, $(0, \infty)$. In other words, $\log_b x$ is defined only when x is positive.

Logarithmic Function

For $b > 0$ and $b \neq 1$, the **logarithmic function** with base b is defined as

$$y = \log_b x \qquad \text{if and only if} \qquad b^y = x, \qquad x > 0.$$

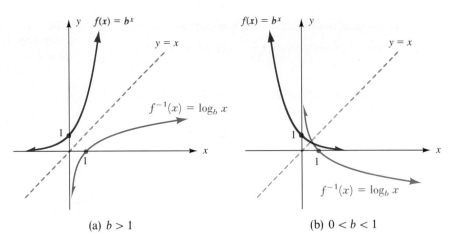

FIGURE 6.8
Like the graphs of any pair of inverse functions, the graphs of $f^{-1}(x) = \log_b x$ and $f(x) = b^x$ are symmetric with respect to the line $y = x$.

(a) $b > 1$ (b) $0 < b < 1$

◆ **Evaluating Logarithms**

Two logarithmic bases occur in most applied problems—base 10 and base e. We refer to a base 10 logarithm ($\log_{10}$) as a **common logarithm** and a base e logarithm ($\log_e$) as a **natural logarithm** or a **Napierian logarithm** (named after the Scottish mathematician John Napier, 1550–1617). The common logarithm of a positive number x is usually written **log x** (read "log of x"), and the natural logarithm of a positive number x is usually written **ln x** (read "el en of x"):

$$\log x \quad \text{means} \quad \log_{10} x$$

and

$$\ln x \quad \text{means} \quad \log_e x$$

Remember that *logarithms are exponents.*

EXAMPLE 1   Find the value of each logarithm, if it is defined.

(a) $\log_5 25$ **(b)** $\log \frac{1}{100}$ **(c)** $\log_9 27$ **(d)** $\ln(-1)$

SOLUTION

(a) Letting $y = \log_5 25$ and applying the definition of a logarithmic function, we have

$$5^y = 25.$$

Since $5^2 = 25$, we conclude that $y = 2$. Hence, $\log_5 25 = 2$.

(b) Letting $y = \log \frac{1}{100} = \log_{10} \frac{1}{100}$ and applying the definition of a logarithmic function, we have

$$10^y = \frac{1}{100}.$$

Since $10^{-2} = \frac{1}{100}$, we conclude that $y = -2$. Hence, $\log \frac{1}{100} = -2$.

(c) Letting $y = \log_9 27$ and applying the definition of a logarithmic function, we have

$$9^y = 27.$$

Since $9^{3/2} = 27$, we conclude that $y = \frac{3}{2}$. Hence, $\log_9 27 = \frac{3}{2}$.

(d) The domain of the logarithmic function $y = \log_b x$ is $(0, \infty)$. Thus, for $x \le 0$ the logarithmic function is undefined. Hence, we conclude

$$\ln(-1) = \log_e(-1) \quad \text{is undefined.} \qquad \blacklozenge$$

PROBLEM 1   Find the value of $\log_{1/2} 8$. $\qquad\qquad\qquad\qquad\qquad\qquad\blacklozenge$

Most calculators have $\boxed{\text{LOG}}$ and $\boxed{\text{LN}}$ keys. We can use these keys to evaluate common logarithms and natural logarithms, respectively. If you do not have access to a calculator, use the logarithmic tables in Appendix B. In Section 6.3, we illustrate a procedure that uses a calculator to evaluate logarithms to bases other than 10 or e.

EXAMPLE 2   Use a calculator to find the approximate value of each logarithm.

(a) $\log 25$ $\qquad$ (b) $\ln \frac{1}{2}$

SOLUTION

(a) The expression $\log 25$ represents the power x such that $10^x = 25$. We know that $10^1 = 10$ and $10^2 = 100$. Since 25 is between 10 and 100, we know that $\log 25$ must be some real number between 1 and 2. Using the $\boxed{\text{LOG}}$ key on a calculator, we approximate the value to four significant digits:

$$\log 25 \approx 1.398$$

(b) The expression $\ln \frac{1}{2}$ represents the power x such that $e^x = \frac{1}{2}$, or 0.5. Since $e^0 = 1$ and $e^{-1} = 1/e \approx 0.37$, we know that $\ln \frac{1}{2}$ must be some real number between 0 and -1. Using the $\boxed{\text{LN}}$ key on a calculator, we calculate that, to four significant digits, the approximate value of $\ln \frac{1}{2}$ is

$$\ln \tfrac{1}{2} = \ln 0.5 \approx -0.6931 \qquad\qquad\qquad\qquad \blacklozenge$$

PROBLEM 2   Use a calculator to verify the results we obtained in Example 1(b) and 1(d). $\blacklozenge$

◆ **Logarithmic Identities**

Next, we present four logarithmic identities. Each identity is a direct consequence of the definition of the logarithmic function, and in each case we assume $b > 0$ and $b \neq 1$.

1. If $y = \log_b 1$, then $b^y = 1$. Since $b^0 = 1$, we conclude that $\mathbf{\log_b 1 = 0}$.

2. If $y = \log_b b$, then $b^y = b$. Since $b^1 = b$, we conclude that $\mathbf{\log_b b = 1}$.

3. If $y = \log_b b^x$, then $b^y = b^x$. Since the exponential function is one-to-one, it follows that $y = x$. Hence, we conclude that $\mathbf{\log_b b^x = x}$.

4. If $y = \log_b x$, then $b^y = x$. Replacing y with $\log_b x$ in the equation $b^y = x$, we obtain $\mathbf{b^{\log_b x} = x}$.

In summary, we restate these identities.

  Logarithmic Identities

For all bases b with $b > 0$ and $b \neq 1$,

1. $\log_b 1 = 0$ 2. $\log_b b = 1$

3. $\log_b b^x = x$ for all real numbers x 4. $b^{\log_b x} = x$ provided $x > 0$

EXAMPLE 3   Find the value of each expression.

(a) $\log_{16} 1$ **(b)** $\ln e$ **(c)** $\log \sqrt{10}$ **(d)** $e^{\ln 6}$

SOLUTION

(a) Since $\log_b 1 = 0$ for any permissible base b, we have

$$\log_{16} 1 = 0.$$

(b) Since $\log_b b = 1$ for any permissible base b, we have

$$\ln e = \log_e e = 1.$$

(c) Since $\log_b b^x = x$ for any permissible base b and any real number x, we have

$$\log \sqrt{10} = \log_{10} 10^{1/2} = \tfrac{1}{2}.$$

(d) Since $b^{\log_b x} = x$ for any permissible base b and all positive numbers x, we have

$$e^{\ln 6} = e^{\log_e 6} = 6. \qquad\qquad ◆$$

PROBLEM 3   Find the value of $\log_3 3^{12}$.   ◆

In the next example, we apply the logarithmic identities to simplify expressions containing logarithms. This procedure is often employed in calculus.

EXAMPLE 4   Simplify each expression.

(a) $e^{\ln(t+1)}$ (b) $\log_2 16^{t-2}$

SOLUTION

(a) Since $b^{\log_b x} = x$, provided $x > 0$, we have

$$e^{\ln(t+1)} = e^{\log_e(t+1)} = t + 1 \quad \text{provided } t > -1.$$

(b) We begin by applying the properties of real exponents (Section 6.1) and matching the base of the exponential expression to the base of the logarithm:

$$\log_2 16^{t-2} = \log_2 (2^4)^{t-2} = \log_2 2^{4t-8}.$$

Now, since $\log_b b^x = x$ for any permissible base b and any real number x, we have

$$\log_2 16^{t-2} = \log_2 2^{4t-8} = 4t - 8. \qquad \blacklozenge$$

PROBLEM 4   Simplify $(\log_2 2)^{4t-8}$. $\qquad\qquad\qquad\qquad\qquad\qquad\qquad\qquad\qquad\qquad$ $\blacklozenge$

$\blacklozenge$ **Equating Logarithmic and Exponential Forms**

From our definition of the logarithmic function, we can state that the equation

$$\log_b u = v \qquad \text{is equivalent to} \qquad b^v = u.$$

The equation $\log_b u = v$ is said to be in **logarithmic form** and the equation $b^v = u$ is said to be in **exponential form**. If we change an equation from logarithmic form to exponential form, or vice versa, the base b remains the same in each case. That is, in logarithmic form the base is the subscript b, and in exponential form the base is the number b being raised to a power.

EXAMPLE 5   Change each equation from logarithmic form to exponential form, then find the value of the unknown.

(a) $\log_b 9 = \frac{2}{3}$ (b) $3 \log_2 (x^2 - 2x + 1) = 12$

SOLUTION

(a) The equation

$$\log_b 9 = \tfrac{2}{3} \qquad \text{is equivalent to} \qquad b^{2/3} = 9.$$

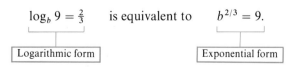

Logarithmic form   Exponential form

For all logarithmic bases b, we must have $b > 0$. Thus, we can solve for b as follows:

$$b^{2/3} = 9$$

$$(b^{2/3})^{3/2} = 9^{3/2} \qquad \text{Raise both sides to the 3/2 power}$$

$$b = 27 \qquad \text{Simplify}$$

(b) We begin by dividing both sides of the equation by 3 to obtain

$$\log_2(x^2 - 2x + 1) = 4.$$

The equation

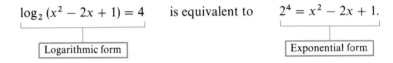

Solving for x, we find

$$x^2 - 2x + 1 = 2^4$$

$$x^2 - 2x - 15 = 0 \qquad \text{Subtract 16 from both sides}$$

$$(x - 5)(x + 3) = 0 \qquad \text{Factor the quadratic and solve}$$

$$x = 5 \quad \text{or} \quad x = -3 \qquad \blacklozenge$$

As with any equation we solve, it is good practice to check the solution of a logarithmic equation. To check Example 5(a), we proceed as follows:

$$\textit{Check:} \quad \log_{27} 9 = \frac{2}{3} \quad ? \qquad \text{Replace } b \text{ with 27}$$

$$27^{2/3} = 9 \quad ? \qquad \text{Change to exponential form}$$

$$9 = 9 \quad \checkmark \qquad \text{Simplify by using the definition of a rational exponent}$$

PROBLEM 5   Referring to Example 5(b), show that both $x = 5$ and $x = -3$ satisfy the equation

$$3 \log_2 (x^2 - 2x + 1) = 12. \qquad \blacklozenge$$

EXAMPLE 6   Change each equation from exponential form to logarithmic form, then find the value of the unknown.

(a) $10^{4x} = 30$ **(b)** $7e^{1 - 2t} = 28$

SOLUTION

(a)   The equation

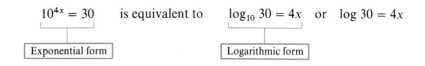

$$10^{4x} = 30 \quad \text{is equivalent to} \quad \log_{10} 30 = 4x \quad \text{or} \quad \log 30 = 4x$$

Solving for x, we find

$$4x = \log 30$$

$$x = \frac{\log 30}{4} \approx 0.3693$$

Solving for t, we find

(b)   We begin by dividing both sides of the equation by 7 to obtain

$$e^{1-2t} = 4.$$

The equation

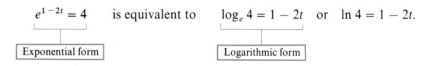

$$e^{1-2t} = 4 \quad \text{is equivalent to} \quad \log_e 4 = 1 - 2t \quad \text{or} \quad \ln 4 = 1 - 2t.$$

Solving for t, we find

$$\ln 4 = 1 - 2t$$

$$2t = 1 - \ln 4$$

$$t = \frac{1 - \ln 4}{2} \approx -0.1931$$

To check Example 6(a), we proceed as follows:

Check: $10^{4[(\log 30)/4]} = 30$?   **Replace x with $(\log 30)/4$**

$10^{\log 30} = 30$?   **Simplify**

$30 = 30$ √   **Apply logarithmic identity 4**

PROBLEM 6   Check Example 6(b).

◆ Applications: Exponential Growth and Decay

Functions that change by a fixed multiple over the same increment of time have the form

$$A(t) = A_0 e^{kt}$$

where $A(t)$ is the amount present at time t, A_0 is the original amount present at time 0, and k is a constant related to the rate of growth or decay. If $k > 0$, then $A(t) = A_0 e^{kt}$ is an **exponential growth function**, and if $k < 0$, then $A(t) = A_0 e^{kt}$ is called an **exponential decay function**.

The compound interest formula

$$A = Pe^{rt}$$

is an example of an exponential growth function. In Section 6.1, we used this formula to find the amount A in a savings account when a certain principal P is invested at an interest rate r per year and interest is compounded continuously. Now that we have defined logarithms, we can also use this formula to find the time t or the interest rate r for a given deposit to reach a particular amount.

E X A M P L E 7   A certain amount of money is deposited in a savings account paying 8% interest per year compounded continuously. Assuming the depositor makes no subsequent deposit or withdrawal, how long will it take for the money to double?

S O L U T I O N   If a certain principal P is deposited in a savings account at an interest rate of 8% per year and interest is compounded continuously, then the amount A in the account after t years is given by the formula

$$A = Pe^{0.08t}$$

We want to find the time t that it will take to *double P*. Thus, we replace A with $2P$ and solve for t as follows:

$$2P = Pe^{0.08t}$$

$$2 = e^{0.08t} \qquad \text{Divide both sides by } P, P \neq 0$$

$$0.08t = \ln 2 \qquad \text{Change to logarithmic form}$$

$$t = \frac{\ln 2}{0.08} \approx 8.66 \text{ years} \qquad \text{Divide both sides by 0.08}$$

Thus, it takes approximately 8 years 8 months to double the investment.   ◆

P R O B L E M 7   At what interest rate, compounded continuously, must money be invested if the amount is to double in 5 years 6 months?   ◆

Under controlled laboratory conditions a population P of living organisms increases exponentially as a function of time t, and the growth pattern is described by the **Malthusian model**

$$P(t) = P_0 e^{kt}$$

where P_0 is the initial population and k is a positive constant. The Malthusian model, named after Englishman Thomas Malthus (1766–1834), is another example of an exponential growth function.

EXAMPLE 8 Suppose 18 bacteria are present initially in a culture, and twenty minutes later 270 bacteria are present.

(a) Determine the Malthusian model that describes this growth pattern.

(b) Determine the time t (to the nearest minute) when 360 bacteria are present in the culture.

SOLUTION

(a) Since 18 bacteria are present initially, we know that $P_0 = 18$. Thus, starting with the Malthusian model $P(t) = P_0 e^{kt}$, we can write

$$P(t) = 18e^{kt}.$$

We must now determine the constant k. Using the fact that when $t = 20$, then $P(t) = 270$, we have

$$270 = 18e^{20k}$$

$$15 = e^{20k} \qquad \text{Divide both sides by 18}$$

$$20k = \ln 15 \qquad \text{Change to logarithmic form}$$

$$k = \frac{\ln 15}{20} \approx 0.1354 \qquad \text{Solve for } k$$

Replacing k with 0.1354 gives us the Malthusian model that describes this growth pattern. Thus, we have

$$P(t) = 18e^{0.1354t}.$$

(b) Using the exponential function developed in part (a), we replace $P(t)$ with 360 and solve for t as follows:

$$360 = 18e^{0.1354t}$$

$$20 = e^{0.1354t} \qquad \text{Divide both sides by 18}$$

$$0.1354t = \ln 20 \qquad \text{Change to logarithmic form}$$

$$t = \frac{\ln 20}{0.1354} \approx 22 \text{ minutes} \qquad \text{Solve for } t$$

◆

PROBLEM 8   Referring to Example 8, determine the time t (to the nearest minute) when 540 bacteria are present in the culture.   ◆

Radioactive materials decrease exponentially over time and their decay patterns can be described by exponential decay functions. When working with radioactive materials, the term *half-life* is often used. **Half-life** is defined as the time required for a given mass of a radioactive material to disintegrate to half its original mass.

EXAMPLE 9   Strontium-90, a waste product from nuclear reactors, has a half-life of 28 years. It is estimated that a certain quantity of this material will be safe to handle when its mass is $\frac{1}{1000}$ of its original amount. Determine the time required to store strontium-90 until it reaches this level of safety.

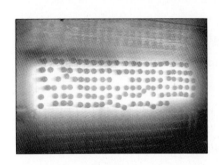

SOLUTION   If the half-life of strontium-90 is 28 years, then half of the original amount is present after 28 years. Thus, starting with the exponential decay function $A(t) = A_0 e^{kt}$, we replace $A(t)$ with $\frac{1}{2} A_0$ and t with 28, and then solve for k as follows:

$$\tfrac{1}{2}A_0 = A_0 e^{k(28)}$$

$$\tfrac{1}{2} = e^{k(28)} \qquad \text{Divide both sides by } A_0$$

$$28k = \ln \tfrac{1}{2} \qquad \text{Change to logarithmic form}$$

$$k = \frac{\ln \tfrac{1}{2}}{28} \approx -0.02476 \qquad \text{Solve for } k$$

Thus, we have

$$A(t) = A_0 e^{-0.02476t}$$

To determine the time t required to store strontium-90 until its mass reaches $\frac{1}{1000}$ of its original amount, we replace $A(t)$ with $0.001 A_0$, and solve for t as follows:

$$0.001 A_0 = A_0 e^{-0.02476t}$$

$$0.001 = e^{-0.02476t} \qquad \text{Divide both sides by } A_0, A_0 \neq 0$$

$$-0.02476t = \ln 0.001 \qquad \text{Change to logarithmic form}$$

$$t = \frac{\ln 0.001}{-0.02476} \approx 279 \text{ years} \qquad \text{Solve for } t \qquad ◆$$

PROBLEM 9   Referring to Example 9, how long does it take 30 grams of strontium-90 to decay to 27 grams?   ◆

Exercises 6.2

Basic Skills

In Exercises 1–30, find the value of each expression, if it is defined.

1. $\log_6 36$
2. $\log_2 8$
3. $\log_5 125$

4. $\log_3 81$
5. $\log_3 3$
6. $\log_7 (-7)$

7. $\log_2 \frac{1}{16}$
8. $\log_3 \frac{1}{27}$
9. $\log_{1/3} 9$

10. $\log_{1/2} 16$
11. $\log 100$
12. $\log \frac{1}{1000}$

13. $\ln 1$
14. $\log 1$
15. $\ln e^2$

16. $\ln \sqrt{e}$
17. $\log_{64} (-4)$
18. $\log_9 3$

19. $\log_4 8$
20. $\log_8 32$
21. $\log_{1/8} 16$

22. $\log_{27} \frac{1}{9}$
23. $4^{\log_4 3}$
24. $3^{\log_3 10}$

25. $e^{\ln 5}$
26. $10^{\log 2}$
27. $10^{\ln e}$

28. $e^{\log 10}$
29. $\log (\ln e)$
30. $\ln (\log 10)$

 In Exercises 31–46, simplify each expression. Assume that each variable is restricted to those real numbers that allow the logarithmic expression to be defined.

31. $\log_4 4^{x-3}$
32. $\log_3 3^{x+1}$

33. $6^{\log_6 (x+2)}$
34. $5^{\log_5 (x-4)}$

35. $e^{\ln (x^2+4)}$
36. $10^{\log (1/x)}$

37. $\ln e^{x^2+2x}$
38. $(\ln e)^{x^2+2x}$

39. $(\log 10)^{2-x^2}$
40. $\log 10^{2-x^2}$

41. $\log_2 8^{x+2}$
42. $\log_9 3^{2x-2}$

43. $e^{x+\ln x}$ [Hint: $a^{x+y} = a^x a^y$]

44. $e^{x-\ln x}$ [Hint: $a^{x-y} = a^x/a^y$]

45. $e^{2\ln x}$ [Hint: $a^{xy} = (a^x)^y$]

46. $e^{-\ln x}$ [Hint: $a^{xy} = (a^x)^y$]

In Exercises 47–60, change each equation to exponential form, then find the value of the unknown.

47. $\log_b 9 = 2$
48. $\log_b 125 = -3$

49. $\log_b 8 = -\frac{3}{4}$
50. $\log_b 64 = \frac{2}{3}$

51. $\log x = -2$
52. $\ln x = 3$

53. $2\log_4 x = 3$
54. $3\log_{1/8} x = -4$

55. $\ln (x - e) = 1$
56. $-3\log_8 (x + 9) = 1$

57. $\log_2 (2x^2 + 5x + 5) = 3$

58. $\ln (x^2 - 3x - 3) = 0$

59. $2 \log |x - 1| - 1 = 0$

60. $\log_3 \sqrt{2x^2 - 3x} = 1$

In Exercises 61–70, change each equation to logarithmic form, then find the value of the unknown.

61. $10^x = 5$
62. $e^x = 3$

63. $e^{3x} = \frac{2}{3}$
64. $10^{-2x} = 60$

65. $10^{3-2x} = 28$
66. $e^{2x-1} = 2$

67. $4e^{-7x} = 15$
68. $12(10^{1-x}) = 60$

69. $3 - 2(10^{-x}) = -117$
70. $1 + 2e^x = 9$

71. One thousand dollars is deposited in a savings account paying 7% interest per year compounded continuously. Assuming the depositor makes no subsequent deposit or withdrawal, find the time required (in years) for the investment to accumulate to the given amount:

    (a) $1500   (b) $6000   (c) $100,000

72. A certain amount of money is deposited into a savings account paying 10% interest per year compounded continuously. Assuming no subsequent deposit or withdrawal is made, find the time required for the amount to double.

73. When a child is born, his grandfather invests $10,000 for the child's college education. When the child is 18 years old, the balance in the account is $73,000. If interest is compounded continuously, find the interest rate (to the nearest percent) at which the investment is made. Assume no subsequent deposit or withdrawal is made and the interest rate does not change.

74. At what interest rate, compounded continuously, must money be invested if the amount is to quadruple in 18 years?

75. In 1982 foresters found 24 eagles living in a certain national park. In 1992 they counted 144 eagles living in this area.

    (a) Determine the Malthusian model that describes this growth pattern.
    (b) In what year is the eagle population expected to reach 300?

76. Approximately 10^4 bacteria are present in a culture. Five hours later, approximately 10^7 bacteria are present.

    (a) Determine the Malthusian model that describes this growth pattern.

EXERCISE 76 (continued)

(b) Approximately how many minutes does it take for this population to double in size?

77. When a plant or animal dies, the amount of carbon-14 in the plant or animal decreases exponentially according to the exponential decay function. Suppose in 1986 an archaeologist discovers a human skull in an ancient burial site and determines that 60% of the original carbon-14 is still present. If the half-life of carbon-14 is 5600 years, in approximately what year did this person die?

78. The radioactive substance iodine-131 decays from 30 grams to 25 grams in 50.5 hours.

(a) Determine the exponential decay function that describes this decay pattern.
(b) How many hours will it take for this radioactive substance to decay to 20 grams?
(c) Find the half-life (in days) of this radioactive substance.

 Critical Thinking

79. Find the base b of the logarithmic function $f(x) = \log_b x$ if the graph of this function passes through the given point.

(a) $(81, 4)$ (b) $(\frac{1}{8}, -\frac{3}{2})$

80. Between what two consecutive integers does the value of each logarithm lie?

(a) $\log_3 99$ (b) $\log_5 150$
(c) $\log_8 7$ (d) $\log_{1/2} 40$

81. Use interval notation to describe the domain of each function.

(a) $f(x) = \dfrac{1}{1 - \ln x}$

(b) $F(x) = \sqrt{10 - \log x}$

82. Given that $f(x) = e^x$ and $g(x) = \ln x$, find $f(g(x))$ and $g(f(x))$. What does this tell us about the functions f and g?

83. In a particular electrical circuit, the current i (in amperes) is a function of the time t (in seconds) after the switch is closed and is given by $i(t) = 20(1 - e^{-4.5t})$.

(a) Discuss the behavior of $i(t)$ as t increases without bound ($t \to \infty$).
(b) Find the time in milliseconds (ms) when $i(t) = 6$ amperes.

84. When limited resources such as food supply and space are taken into account, the population P of a living organism as a function of time t is more accurately described by the *logistic law* than by the Malthusian model. For the logistic law,

$$P(t) = \frac{cP_0}{P_0 + (c - P_0)e^{-kt}},$$

where P_0 is the initial population and c and k are positive constants. Suppose the population of wolves in a certain forest follows the logistic law with $c = 1500$, $k = 0.2$, and time t in years.

(a) Discuss the behavior of $P(t)$ as t increases without bound ($t \to \infty$).
(b) If $P_0 = 200$, find the time (in years) when $P(t) = 600$ wolves.

If a heated object of temperature T_0 is placed in a cooler medium that has a constant temperature of T_1, then the temperature T of the object at any time t is given by the formula

$$T = T_1 + (T_0 - T_1)e^{-kt}$$

where k is a positive constant. The formula is Newton's law of cooling, *named after Sir Isaac Newton (1642–1727). Use this formula to answer Exercises 85 and 86.*

85. A pizza baked at 400 °F is removed from the oven and placed in a room with a constant temperature of 70 °F. The pizza cools to 300 °F in 2 minutes.

(a) State the formula that describes this cooling process.
(b) Find the temperature of the pizza after it cools for 3 minutes.
(c) How soon after it is removed from the oven does the pizza cool to 150 °F?

86. A bottle of white zinfandel wine is stored in a room with a temperature of 70 °F. The manufacturer recommends that the wine is best served slightly chilled to 40 °F. At 4:00 P.M. the wine is placed in a refrigerator that has a constant temperature of 35 °F, and after 30 minutes the temperature of the wine has cooled to 50 °F. At what time should the wine be removed from the refrigerator if it is to be served at 40 °F?

Calculator Activities

Given the functions f and g defined by

$$f(x) = \ln(x^2 - 2x + 3) \quad \text{and} \quad g(x) = \log\frac{1}{x^2 + 2}$$

use a calculator to compute the functional values given in Exercises 87–92. Round each value to three significant digits.

87. $f(8.2)$

88. $g(11.5)$

89. $(f + g)(-5.22)$

90. $(f \cdot g)(-0.34)$

91. $(f \circ g)(4.4)$

92. $(g \circ f)(21.8)$

93. Use the $\boxed{\text{LN}}$ key on your calculator to help complete the following table.

x	2	18	24	π
y	3	4	15	10
$\ln(xy)$				
$\ln x + \ln y$				
$\ln(x/y)$				
$\ln x - \ln y$				
$\ln(x^y)$				
$y \ln x$				

(a) Compare the values in the table for $\ln(xy)$ and $\ln x + \ln y$. What do you conclude?

(b) Compare the values in the table for $\ln(x/y)$ and $\ln x - \ln y$. What do you conclude?

(c) Compare the values in the table for $\ln(x^y)$ and $y \ln x$. What do you conclude?

$\boxed{\frac{\Delta y}{\Delta x}}$ 94. In calculus, it is shown that if $|x| \leq 1$, then

$$\ln(x + 1) \approx x - \frac{x^2}{2} + \frac{x^3}{3} - \frac{x^4}{4} + \frac{x^5}{5} - \frac{x^6}{6}.$$

Use this formula and a calculator to evaluate each expression.

(a) $\ln 1.8$ (b) $\ln 0.2$ (c) $\ln 1.25$ (d) $\ln 0.75$

95. Chemists describe the acidity or alkalinity of a liquid by denoting its *pH*. An acid has pH < 7 and an alkaline has pH > 7. By definition,

$$pH = -\log[H^+],$$

where $[H^+]$ is the liquid's concentration of hydrogen ions, measured in moles per liter, mol/L. Determine the pH of the given liquid.

(a) milk: $[H^+] \approx 4.1 \times 10^{-7}$ mol/L

(b) apple juice: $[H^+] \approx 6.2 \times 10^{-4}$ mol/L

96. Refer to Exercise 95. Determine the hydrogen ion concentration in each liquid.

(a) vinegar: pH ≈ 2.5

(b) human blood: pH ≈ 7.4

6.3   Properties of Logarithms

In this section, we develop some important properties of logarithms. These properties enable us to rewrite certain functions that contain logarithms so that we may sketch the graph of these functions (Section 6.4). Also, these properties enable us to rewrite certain logarithmic equations so that we may solve these equations (Section 6.5).

◆ Rewriting Logarithmic Expressions

Recall from Section 6.1 three properties of real exponents for real numbers m and n:

1. $b^m b^n = b^{m+n}$ **2.** $\dfrac{b^m}{b^n} = b^{m-n}$ **3.** $(b^m)^n = b^{mn}$

By using logarithmic identity 4 from Section 6.2 ($b^{\log_b u} = u$, $u > 0$) and these three properties of real exponents, we can obtain three corresponding **properties of logarithms**:

$$b^{\log_b xy} = xy \qquad \text{Apply log identity 4 with } x > 0, \, y > 0$$
$$= b^{\log_b x} \cdot b^{\log_b y} \qquad \text{Rewrite using log identity 4}$$
$$= b^{\log_b x + \log_b y} \qquad \text{Add exponents}$$

Since the exponential function is one-to-one, we conclude that the exponents $\log_b xy$ and $\log_b x + \log_b y$ must be equal; that is,

$$\log_b xy = \log_b x + \log_b y$$

The logarithm of a product is the sum of the logarithms of the factors.

$$b^{\log_b (x/y)} = \frac{x}{y} \qquad \text{Apply log identity 4 with } x > 0, \, y > 0$$
$$= \frac{b^{\log_b x}}{b^{\log_b y}} \qquad \text{Rewrite using log identity 4}$$
$$= b^{\log_b x - \log_b y} \qquad \text{Subtract exponents}$$

Since the exponential function is one-to-one, we conclude that the exponents $\log_b (x/y)$ and $\log_b x - \log_b y$ must be equal; that is,

$$\log_b \frac{x}{y} = \log_b x - \log_b y$$

The logarithm of a quotient is the logarithm of the numerator minus the logarithm of the denominator.

$$b^{\log_b x^n} = x^n \qquad \text{Apply log identity 4 with } x > 0$$
$$= (b^{\log_b x})^n \qquad \text{Rewrite using log identity 4}$$
$$= b^{n \log_b x} \qquad \text{Multiply exponents}$$

Since the exponential function is one-to-one, we conclude that the exponents $\log_b x^n$ and $n \log_b x$ must be equal; that is,

$$\log_b x^n = n \log_b x$$

The logarithm of a quantity raised to a power is the product of the power and the logarithm of that quantity.

We now summarize these properties.

Properties of Logarithms

If $b > 0$, $b \neq 1$, then for positive real numbers x and y,

1. $\log_b xy = \log_b x + \log_b y$ **2.** $\log_b \dfrac{x}{y} = \log_b x - \log_b y$

3. $\log_b x^n = n \log_b x$

In the following example, we use the properties of logarithms to write a single logarithmic expression as the sum and difference of simpler logarithmic expressions. This procedure will be used in Section 6.4 to sketch the graphs of functions that contain logarithmic expressions.

E X A M P L E   1   Write each logarithmic expression as sums and differences of simpler logarithmic expressions without logarithms of products, quotients, and powers. Then simplify, if possible.

(a) $\log \sqrt[3]{10x^2}$ **(b)** $\ln \dfrac{\sqrt{x^2+4}}{xe^{3x}}$

S O L U T I O N

(a) Assuming that $x > 0$, we have

$$\log \sqrt[3]{10x^2} = \log (10x^2)^{1/3} \qquad \text{Change to a rational exponent}$$

$$= \tfrac{1}{3} \log (10x^2) \qquad \text{Apply log property 3}$$

$$= \tfrac{1}{3}(\log 10 + \log x^2) \qquad \text{Apply log property 1}$$

$$= \tfrac{1}{3}(\log 10 + 2 \log x) \qquad \text{Apply log property 3}$$

$$= \tfrac{1}{3}(1 + 2 \log x) \qquad \text{Evaluate log 10}$$

(b) Assuming that $x > 0$, we have

$$\ln \frac{\sqrt{x^2+4}}{xe^{3x}} = \ln \frac{(x^2+4)^{1/2}}{xe^{3x}} \qquad \text{Rewrite}$$

$$= \ln (x^2+4)^{1/2} - \ln xe^{3x} \qquad \text{Apply log property 2}$$

$$= \ln (x^2+4)^{1/2} - (\ln x + \ln e^{3x}) \qquad \text{Apply log property 1}$$

> Be sure to remember the parentheses.

$$= \tfrac{1}{2} \ln (x^2+4) - (\ln x + 3x \ln e) \qquad \text{Apply log property 3}$$

$$= \tfrac{1}{2} \ln (x^2+4) - \ln x - 3x \qquad \text{Evaluate ln } e \qquad \blacklozenge$$

Note: The logarithmic expression in Example 1(a) is defined for all real numbers except 0. We make the assumption that $x > 0$ so that the properties of logarithms may be applied. If we want to allow for the possibility that $x < 0$, then we write

$$\log \sqrt[3]{10x^2} = \tfrac{1}{3}(1 + 2 \log |x|).$$

Avoid using the three properties of logarithms in situations where they do not apply. Referring to Example 1(b), to apply log property 3 and write

$$\ln xe^{3x} \qquad \text{as} \qquad 3x \ln xe \quad \text{is WRONG}$$

$$\ln (xe)^{3x} = 3x \ln xe \quad \text{by log property 3.}$$

Several other common errors that occur when working with the properties of logarithms are mentioned in Exercises 1–4.

P R O B L E M 1   Show by using a specific example that $(\log_b x)^n$ and $n \log_b x$ are *not* equivalent logarithmic expressions.   ◆

The properties of logarithms can also be used in the reverse sense to write sums and differences of logarithmic expressions as a single logarithmic expression. For problems of this type, we begin by applying logarithmic property 3 and change any numerical coefficient to an exponent. Once we have removed numerical coefficients, we can apply the other two properties. The technique of writing sums and differences of logarithmic expressions as a single logarithm will be used in Section 6.5 to solve logarithmic equations.

E X A M P L E 2   Express each as a single logarithmic expression with a coefficient of 1. Then simplify, if possible.

(a) $\log 50 + 2 \log 4 - 3 \log 2$ **(b)** $\ln (x^2 - 1) - 2 \ln (x + 1)$

S O L U T I O N

(a) Beginning with logarithmic property 3, we have

$$\log 50 + 2 \log 4 - 3 \log 2$$

$$= \log 50 + \log 4^2 - \log 2^3 \qquad \text{Apply log property 3}$$

$$= \log 50 + \log 16 - \log 8 \qquad \text{Simplify}$$

$$= \log (50 \cdot 16) - \log 8 \qquad \text{Apply log property 1}$$

$$= \log \frac{50 \cdot 16}{8} \qquad \text{Apply log property 2}$$

$$= \log 100 = 2 \qquad \text{Reduce and evaluate}$$

(b)   Assuming that $x > 1$, we have

$$\ln (x^2 - 1) - 2 \ln (x + 1)$$

$$= \ln (x^2 - 1) - \ln (x + 1)^2 \qquad \text{Apply log property 3}$$

$$= \ln \frac{x^2 - 1}{(x + 1)^2} \qquad \text{Apply log property 2}$$

$$= \ln \frac{(x + 1)(x - 1)}{(x + 1)^2} \qquad \text{Factor the numerator}$$

$$= \ln \frac{x - 1}{x + 1} \qquad \text{Reduce} \qquad \blacklozenge$$

PROBLEM 2   Use the $\boxed{\text{LOG}}$ key on a calculator to evaluate $\log 50 + 2 \log 4 - 3 \log 2$. Your answer should agree with the result we obtained in Example 2(a). $\qquad \blacklozenge$

◆ **Change of Base Formula**

In some instances we may find it useful to change the base b logarithm, $\log_b x$, to an expression involving logarithms of another base a. If we let

$$\log_b x = y,$$

then we can change to exponential form and write

$$b^y = x.$$

Taking the base a logarithm of both sides of this equation, we obtain

$$\log_a b^y = \log_a x \qquad \text{Take the base } a \text{ logarithm of both sides}$$

$$y \log_a b = \log_a x \qquad \text{Apply log property 3}$$

$$y = \frac{\log_a x}{\log_a b} \qquad \text{Divide both sides by } \log_a b$$

Now, replacing y with $\log_b x$ gives us the **change of base formula**.

◆▷ Change of Base Formula

> If $\log_b x$ is defined, then
>
> $$\log_b x = \frac{\log_a x}{\log_a b}, \qquad a > 0, \quad a \neq 1.$$

We can use a calculator to evaluate logarithms to bases other than 10 or e by applying the change of base formula. This procedure is illustrated in the next example.

EXAMPLE 3   Use the change of base formula to find the approximate value of $\log_4 24$.

SOLUTION   The expression $\log_4 24$ represents the power to which 4 must be raised to get 24. We know that $4^2 = 16$ and $4^3 = 64$. Since 24 is between 16 and 64, $\log_4 24$ must be a real number between 2 and 3. Using the change of base formula with common logarithms, we write

$$\log_4 24 = \frac{\log 24}{\log 4}.$$

Now, using the $\boxed{\text{LOG}}$ key on a calculator, we obtain

$$\log_4 24 \approx 2.292$$

PROBLEM 3   Repeat Example 3 using the change of base formula with natural logarithms. You should obtain the same result.

If we replace x with a in the change of base formula, we obtain

$$\log_b a = \frac{1}{\log_a b}$$

or, equivalently,

$$(\log_b a)(\log_a b) = 1$$

EXAMPLE 4   Simplify each logarithmic expression.

(a) $\dfrac{1}{\log_2 12} + \dfrac{1}{\log_6 12}$ (b) $(\log_3 16)(\log_2 27)$

SOLUTION

(a) $\dfrac{1}{\log_2 12} + \dfrac{1}{\log_6 12} = \log_{12} 2 + \log_{12} 6$ Apply $\dfrac{1}{\log_a b} = \log_b a$

$= \log_{12}(2 \cdot 6)$ Apply log property 1

$= \log_{12} 12$ Simplify

$= 1$ Apply $\log_b b = 1$

(b) $(\log_3 16)(\log_2 27) = (\log_3 2^4)(\log_2 3^3)$ Rewrite

$= (4\log_3 2)(3\log_2 3)$ Apply log property 3

$= 4 \cdot 3[(\log_3 2)(\log_2 3)]$ Rearrange the factors

$= 12[1]$ Apply $(\log_b a)(\log_a b) = 1$

$= 12$ Simplify

PROBLEM 4   Repeat Example 4 for $(\log_2 5)(\log_5 8)$.

◆ Application: Logarithmic Scales

When physical quantities vary over a large range of values, it is convenient to work with **logarithmic scales** in order to obtain a more manageable set of numbers. Some examples of logarithmic scales are the **decibel scale** for measuring the magnitude of sound, the **brightness scale** for measuring the magnitude of a star, and the **Richter scale** (named after seismologist Charles F. Richter) for measuring the magnitude of an earthquake.

On the Richter scale, the magnitude R of an earthquake is given by

$$R = \log \frac{I}{I_0}$$

where I is the intensity of the earthquake and I_0 is the intensity of a zero-level earthquake having magnitude $R = 0$.

EXAMPLE 5   Seismologists estimate that the San Francisco earthquake of 1906 measured 8.3 on the Richter scale. How many times more intense was this earthquake than the Loma Prieta quake, which occurred during the 1989 World Series and measured 7.1 on the Richter scale?

SOLUTION Let

$$I_a = \text{intensity of the 1906 earthquake}$$

and

$$I_b = \text{intensity of the 1989 earthquake.}$$

Now, using the formula $R = \log \dfrac{I}{I_0}$ and logarithmic property 2, we have

$$8.3 = \log \frac{I_a}{I_0} \qquad \text{and} \qquad 7.1 = \log \frac{I_b}{I_0}$$

$$8.3 = \log I_a - \log I_0 \qquad\qquad 7.1 = \log I_b - \log I_0.$$

Solving these equations simultaneously by using the addition method (Section 4.5), we find

$$
\begin{array}{ll}
8.3 = \log I_a - \log I_0 & \\
\underline{7.1 = \log I_b - \log I_0} & \\
1.2 = \log I_a - \log I_b & \text{Subtract}
\end{array}
$$

$$1.2 = \log \frac{I_a}{I_b} \qquad\qquad \text{Apply log property 2}$$

$$\frac{I_a}{I_b} = 10^{1.2} \approx 16 \qquad\qquad \text{Change to exponential form}$$

Hence, the earthquake in 1906 was about 16 times as intense as the one in 1989.   ◆

PROBLEM 5   One of the strongest earthquakes ever recorded measured 8.9 on the Richter scale. It occurred in Japan in 1933. How many times more intense was this earthquake than the one that occurred in California during the 1989 World Series?   ◆

Exercises 6.3

 Basic Skills

In Exercises 1–4, show by using specific examples that the given logarithmic expressions are not *equivalent.*

1. $\log_b (x + y)$ and $\log_b x + \log_b y$

2. $\log_b (x - y)$ and $\log_b x - \log_b y$

3. $\dfrac{\log_b x}{\log_b y}$ and $\log_b \dfrac{x}{y}$

4. $(\log_b x)(\log_b y)$ and $\log_b xy$

In Exercises 5–18, write each expression as sums and differences of simpler logarithmic expressions without logarithms of products, quotients, and powers. Then simplify if possible. Assume that each variable is restricted to those real numbers that allow the properties of logarithms to be applied.

5. $\log_3 (27^2 \cdot 81)$

6. $\log 100 x^2$

7. $\log_2 [8x(x + 2)]$

8. $\log_5 \sqrt[4]{25x}$

9. $\ln \dfrac{xe^2}{10}$

10. $\log \dfrac{(x + 4)^2}{2x}$

11. $\log_3 \sqrt[3]{\dfrac{x^2}{27}}$

12. $\log_b \left(\dfrac{x^2 + 2}{x + 3}\right)^3$

13. $\log_b \dfrac{x^2}{\sqrt[3]{(x + 1)^2}}$

14. $\ln \dfrac{1}{2\sqrt{3x^2 + 2}}$

15. $\log_3 \dfrac{1}{9\sqrt{x^3 y^{2/3}}}$

16. $\log \dfrac{\sqrt[3]{1 + y^2}}{(x - 2)(x + 2)^{1/2}}$

17. $\ln \left(\dfrac{e^{x^2}}{e^x + 1}\right)^2$

18. $\ln \dfrac{xe^{-2x}}{\sqrt{2e^x - 1}}$

In Exercises 19–30, write each expression as a single logarithm with a coefficient of 1. Then simplify if possible. Assume that each variable is restricted to those real numbers that allow the properties of logarithms to be applied.

19. $\ln 2 + \ln 3 + \ln 4$

20. $\ln 6 - \ln 3 + \ln 5$

21. $\log 40 - (3 \log 2 - \log 20)$

22. $2 \log 6 - (2 \log 3 + \log 4)$

23. $-2 \log_3 6 + 4 \log_3 2 - \frac{1}{2} \log_3 16$

24. $5 \log_6 2 + \frac{2}{3} \log_6 27 + \frac{3}{2} \log_6 4$

25. $\log_5 (x - 1) - 2 \log_5 x + \log_5 (x + 3)$

26. $3 \ln x + \frac{2}{3} \ln (x - 1) + \ln 2$

27. $2 \ln (x - 2) - \ln (x^2 - 4)$

28. $2 \ln (x^3 + 1) - \ln (x^2 - x + 1)$

29. $\frac{1}{2}[\ln (x - 3) + \ln (x + 3)] - 2(\ln x - \ln 3)$

30. $3\left[\log \left(x + \sqrt{x^2 - 1}\right) + \log \left(x - \sqrt{x^2 - 1}\right)\right] - \log 2$

In Exercises 31–38, use the change of base formula to find the approximate value of each logarithm. Round the answer to four significant digits.

31. $\log_2 12$

32. $\log_4 9$

33. $\log_{12} 945$

34. $\log_{20} 1250$

35. $\log_3 \frac{2}{3}$

36. $\log_5 \frac{17}{5}$

37. $\log_{2/3} 34$

38. $\log_{1/2} \frac{3}{4}$

In Exercises 39–46, simplify each logarithmic expression.

39. $(\log_4 5)(\log_5 4)$

40. $(\log 7)(\log_7 10)$

41. $(\log_3 4)(\log_4 81)$

42. $(\log_5 16)(\log_2 125)$

43. $\log 20 + \dfrac{1}{\log_5 10}$

44. $\dfrac{1}{\log_4 10} + \dfrac{1}{\log_{25} 10}$

45. $\dfrac{1}{\log_a ab} + \dfrac{1}{\log_b ab}$

46. $\dfrac{\log_b x}{\log_{ab} x} - \dfrac{\log_b x}{\log_a x}$

47. The strongest earthquake ever recorded in the United States measured 8.6 on the Richter scale. It occurred in Alaska on Good Friday, March 27, 1964. How many times more intense was this earthquake than the one that occurred in California during the 1989 World Series (magnitude 7.1)?

48. One of the aftershocks from the earthquake during the 1989 World Series was 60 times less intense than the major earthquake (magnitude 7.1). What was the magnitude of the aftershock?

 Critical Thinking

49. Suppose the functions f and g are defined by

$$f(x) = \ln x^2 \quad \text{and} \quad g(x) = 2 \ln x.$$

Are f and g the same function? Explain.

50. Show that if $f(x) = \ln x$, then $f(1/x) = -f(x)$.

51. Given that $\log_b A = 2$ and $\log_b B = 3$, find the value of each logarithm.

    (a) $\log_A b$ (b) $\log_B b$
    (c) $\log_A b^2$ (d) $\log_B \sqrt{b}$
    (e) $\log_{AB} b$ (f) $\log_{AB}(1/b)$
    (g) $\log_{A/B} b$ (h) $\log_{B/A} b$

52. Find the fallacy in the following argument:

$$3 = \log_2 8$$
$$= \log_2 (4 + 4)$$
$$= \log_2 4 + \log_2 4$$
$$= 2 + 2$$
$$= 4$$

The magnitude m of a star is given by

$$m = -2.5 \log \frac{B}{B_0}$$

where B is the brightness of the star and B_0 is the brightness of a zero-level star having magnitude m = 0. Use this formula to answer Exercises 53 and 54.

53. The magnitude of our sun is -26.8 and the magnitude of Sirius, the brightest star in the heavens, is -1.5. How many times brighter is the sun than Sirius?

54. What is the difference in magnitude between two stars if the brightness of one is 50 times the brightness of the other?

The loudness L in decibels of a sound is given by

$$L = 10 \log \frac{I}{I_0}$$

where I is intensity of the sound and I_0 is the intensity of the faintest sound that can be heard. Use this fomula to answer Exercises 55 and 56.

55. The loudness of a whisper is 30 decibels and the loudness of a rock concert is 120 decibels. How many times more intense is the rock concert than the whisper?

56. If two sounds differ by 20 decibels, how many times more intense is the loudness of the more audible sound than the other sound?

 Calculator Activities

57. Use the $\boxed{\text{LOG}}$ key on your calculator to find the approximate values of

    $\log 6.2$, $\log 62$, $\log 620$,   and $\log 6200$.

    Compare the values you obtain. Do you observe a pattern? Show why this pattern develops by expressing 62, 620, and 6200 in scientific notation and then finding the common logarithms of these expressions by using the properties of logarithms.

58. Use the $\boxed{\text{LOG}}$ key on your calculator to find the approximate values of

    $\log 3.5$, $\log 0.35$, $\log 0.035$,   and $\log 0.0035$.

    Compare the values you obtain. Do you observe a pattern? Show why this pattern develops by expressing 0.35, 0.035, and 0.0035 in scientific notation and then finding the common logarithms of these expressions by using the properties of logarithms.

59. Evaluate the following expression by each of the given procedures:

$$\ln \pi + \ln\left(\frac{\sqrt{2}}{\pi}\right) + \frac{1}{2}\ln\left(\frac{3}{2}\right) - \ln\left(\frac{\sqrt{3}}{e}\right)$$

(a) Use the LN key on your calculator.
(b) Apply the properties of logarithms.

60. Use the change of base formula and a calculator to help complete the following table.

a	b	$\log_b(1/a)$	$\log_{1/b} a$
2	3		
18	4		
24	15		
π	10		

Compare the values of $\log_b(1/a)$ and $\log_{1/b} a$. Prove the relationship that you observe.

6.4   Graphs of Logarithmic Functions

◆ Introductory Comments

In Figure 6.8 of Section 6.2, we reflected the graph of the exponential function in the line $y = x$ to obtain the graph of its inverse, the logarithmic function. The graphs in Figure 6.9 show the basic shape of the logarithmic function $f(x) = \log_b x$ for $b > 1$ and for $0 < b < 1$.

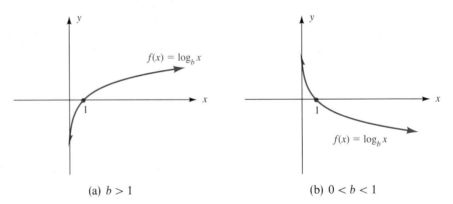

FIGURE 6.9
Graph of the logarithmic function
$f(x) = \log_b x$

(a) $b > 1$ (b) $0 < b < 1$

In this section, we study the characteristics of the graph of $f(x) = \log_b x$ and use these characteristics along with the properties of logarithms (Section 6.3) to help sketch the graphs of several related functions.

◆ Characteristics of the Graph of $f(x) = \log_b x$

In Figure 6.9(a) the graph of $f(x) = \log_b x$ for $b > 1$ is always increasing and its x-intercept is 1. Also, for $b > 1$, observe that $f(x)$ decreases without bound $[f(x) \to -\infty]$ as x approaches 0 from the right $(x \to 0^+)$. Hence, the y-axis is a vertical asymptote. In Figure 6.9(b) the graph of $f(x) = \log_b x$ for $0 < b < 1$ is always decreasing although its x-intercept remains 1. Also, for $0 < b < 1$, observe that $f(x)$ increases without bound $[f(x) \to \infty]$ as x approaches 0 from the right

$(x \rightarrow 0^+)$. Again, the y-axis is a vertical asymptote. We can summarize these features of the graph of the logarithmic function as follows.

 Characteristics of the Graph of $f(x) = \log_b x$

1. The x-intercept is 1, and the graph has no y-intercept.

2. The y-axis is a vertical asymptote.

3. If $b > 1$, the graph of $f(x) = \log_b x$ is always increasing.

4. If $0 < b < 1$, the graph of $f(x) = \log_b x$ is always decreasing.

To sketch the graph of a particular logarithmic function, we may simply use the basic characteristics of the general graph and plot a couple points.

EXAMPLE 1   Sketch the graph of each logarithmic function.

(a) $f(x) = \log_4 x$ (b) $g(x) = \log_{1/4} x$

SOLUTION

(a)   Since the base is 4, and $b = 4 > 1$, the graph of $f(x) = \log_4 x$ is always increasing. The x-intercept is 1, and since $f(x) \rightarrow -\infty$ as $x \rightarrow 0^+$, the y-axis is a vertical asymptote. Selecting a few convenient inputs for x, we find their corresponding outputs $f(x)$ as shown in the table:

x	
$f(x)$	

Plotting the points associated with these ordered pairs and connecting them to form a smooth curve gives us the graph of $f(x) = \log_4 x$, as shown in Figure 6.10.

(b)   Since the base b is $\frac{1}{4}$ and $0 < \frac{1}{4} < 1$, the graph of $g(x) = \log_{1/4} x$ is always decreasing. The x-intercept is 1, and, since $g(x) \rightarrow \infty$ as $x \rightarrow 0^+$, the y-axis is a vertical asymptote. Selecting a few convenient inputs for x, we find their corresponding outputs $g(x)$ as follows:

x	$\frac{1}{4}$	1	4
$g(x)$	1	0	-1

Plotting the points associated with these ordered pairs and connecting them to form a smooth curve gives us the graph of $g(x) = \log_{1/4} x$, as shown in Figure 6.11. ◆

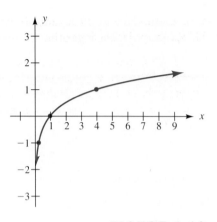

FIGURE 6.10

The function $f(x) = \log_4 x$ is an
increasing function.

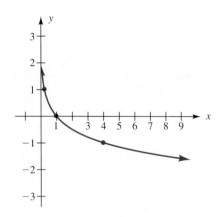

FIGURE 6.11

The function $g(x) = \log_{1/4} x$ is a
decreasing fuction.

◆

Note: Observe that the graph of $g(x) = \log_{1/4} x$ is the same as the graph of $f(x) = \log_4 x$ reflected about the x-axis. In general, if b is a real number such that $b > 0$ and $b \neq 1$, then the graph of $g(x) = \log_{1/b} x$ is the same as the graph of $f(x) = \log_b x$ reflected about the x-axis.

PROBLEM 1   Sketch the graph of the functions $f(x) = \log_5 x$ and $g(x) = \log_{1/5} x$ on the same coordinate plane.   ◆

◆ **Graphing Related Functions**

Knowing the basic shape of the logarithmic function $f(x) = \log_b x$ enables us to graph several other related functions by applying the shift rules and axis reflection rules that we discussed in Section 3.4.

EXAMPLE 2   Sketch the graph of each function. Label the vertical asymptote and any x- and y-intercepts.

(a) $F(x) = \log_4 (x + 2)$ (b) $G(x) = 1 + \log_4 (-x)$

SOLUTION

(a)   By the horizontal shift rule (Section 3.4), the graph of $F(x) = \log_4 (x + 2)$ is the same as the graph of $f(x) = \log_4 x$ shifted to the left 2 units, as shown in Figure 6.12. We find the y-intercept by evaluating $F(0)$, as follows:

$$F(0) = \log_4 (0 + 2) = \log_4 2 = \tfrac{1}{2}.$$

We find the x-intercept by solving the equation $F(x) = 0$, as follows:

$$\log_4 (x + 2) = 0$$

$$4^0 = x + 2 \qquad \text{Change to exponential form}$$

$$x = -1 \qquad \text{Solve for } x$$

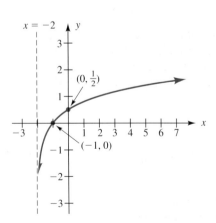

FIGURE 6.12

Graph of $F(x) = \log_4 (x + 2)$

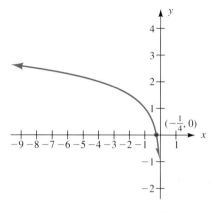

FIGURE 6.13

Graph of $G(x) = 1 + \log_4 (-x)$

Note that $F(x) \to -\infty$ as x approaches -2 from the right ($x \to -2^+$). Thus, the vertical asymptote is the line $x = -2$.

(b) By the y-axis reflection rule and the vertical shift rule (Section 3.4), the graph of $G(x) = 1 + \log_4 (-x)$ is the same as the graph of $f(x) = \log_4 x$ reflected about the y-axis and then shifted vertically upward 1 unit, as shown in Figure 6.13. We find the x-intercept by solving the equation $G(x) = 0$:

$$1 + \log_4 (-x) = 0$$

$$\log_4 (-x) = -1 \qquad \text{Subtract 1 from both sides}$$

$$4^{-1} = -x \qquad \text{Change to exponential form}$$

$$x = -\tfrac{1}{4} \qquad \text{Solve for } x$$

Note that $G(x) \to -\infty$ as x approaches 0 from the left ($x \to 0^-$). Hence, the y-axis is a vertical asymptote. ◆

PROBLEM 2   Repeat Example 2 for $H(x) = 1 - \log_4 x$. ◆

To sketch the graphs of other logarithmic functions, we apply the properties of logarithms (Section 6.3) in order to obtain an equivalent function whose graph is easily plotted.

EXAMPLE 3   Sketch the graph of $H(x) = \log_4 \dfrac{1}{x - 1}$. Label the vertical asymptote and any x- and y-intercepts.

SOLUTION   We begin by rewriting this function as follows:

$$H(x) = \log_4 \frac{1}{x - 1} = \log_4 1 - \log_4 (x - 1) \qquad \text{Apply log property 2}$$

$$= 0 - \log_4 (x - 1) \qquad \text{Evaluate } \log_4 1$$

$$= -\log_4 (x - 1) \qquad \text{Simplify}$$

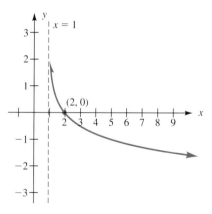

FIGURE 6.14

Graph of $H(x) = \log_4 \dfrac{1}{x - 1}$

Now by the horizontal shift rule and x-axis reflection rule (Section 3.4), the graph of $H(x) = -\log_4 (x - 1)$ is the same as the graph of $f(x) = \log_4 x$, shifted horizontally to the right 1 unit and then reflected about the x-axis, as shown in Figure 6.14. We find the x-intercept by solving the equation $H(x) = 0$:

$$\log_4 \frac{1}{x - 1} = 0$$

$$4^0 = \frac{1}{x - 1} \qquad \text{Change to exponential form}$$

$$x - 1 = 1 \qquad \text{Solve for } x$$

$$x = 2$$

Note that $H(x) \to \infty$ as $x \to 1^+$. Hence $x = 1$ is a vertical asymptote. ◆

PROBLEM 3   Repeat Example 3 for $F(x) = \log_4 16x$.   ◆

When applying the properties of logarithms to a logarithmic function, we must always preserve the domain of the original function. For example, consider the functions F and G defined by

$$F(x) = \log_4 x^2 \quad \text{and} \quad G(x) = 2 \log_4 x.$$

The domain of the function F is the set of all real numbers x with $x \neq 0$. However, the domain of the function G is the set of all positive real numbers x. Since the functions F and G have different domains, they are not the same function. To apply logarithmic property 3 to function F and preserve the same function, we must use the absolute value of x, and we write

$$F(x) = \log_4 x^2 = \log_4 |x|^2 = 2 \log_4 |x|.$$

EXAMPLE 4   Sketch the graph of $F(x) = \log_4 x^2$.

SOLUTION   As we discussed in the preceding paragraph,

$$F(x) = \log_4 x^2 = 2 \log_4 |x|.$$

If $x > 0$, then $|x| = x$. Hence, by the vertical stretch rule (Section 3.4), the graph of F for $x > 0$ is the same as the graph of $f(x) = \log_4 x$ stretched vertically by a factor of 2. If $x < 0$, then $|x| = -x$. Hence, by the vertical stretch rule and y-axis reflection rule (Section 3.4), the graph of F for $x < 0$ is the same as the graph of $f(x) = \log_4 x$ stretched vertically by a factor of 2 and then reflected about the y-axis. The graph of F is shown in Figure 6.15.

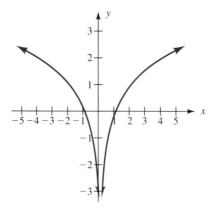

FIGURE 6.15
Graph of $F(x) = \log_4 x^2$

PROBLEM 4   Sketch the graph of $F(x) = \log_4 x^3$.   ◆

◆ **Application: The Retention Function**

In psychological experiments it has been shown that most of what we learn is rapidly forgotten, and the remainder of the learned material slowly recedes from memory. The function R that describes the percentage of the learned material we retain over a period of time t is called the **retention function**. The retention function is often defined in terms of logarithms, and its graph is called the **retention curve**.

EXAMPLE 5   In a psychological experiment, a student is asked to memorize a list of nonsense syllables by studying the list until one perfect repetition is performed from memory. At various times over the next month, the student is asked to recall the list of syllables. It is found that the retention function R associated with this experiment is given by

$$R(t) = 100 - 29 \log (27t + 1), \qquad 0 \leq t \leq 30,$$

where t is time in days.

(a)   What percentage of the learned material is retained initially (at $t = 0$)?

(b)   What percentage of the learned material is retained after 8 hours?

(c)   How much time has elapsed when 42% of the learned material is still retained?

(d)   Sketch the graph of the retention curve associated with this function.

SOLUTION

(a)   The percentage of the learned material that is retained initially is

$$R(0) = 100 - 29 \log [27(0) + 1]$$
$$= 100 - 29 \log 1 = 100\%$$

(b)   The percentage of the learned material retained after 8 hours ($t = \frac{1}{3}$ day) is

$$R(\tfrac{1}{3}) = 100 - 29 \log [27(\tfrac{1}{3}) + 1]$$
$$= 100 - 29 \log 10 = 71\%$$

(c)   To determine the time when 42% of the learned material is retained, we solve the equation $R(t) = 42$:

$$42 = 100 - 29 \log (27t + 1)$$
$$-58 = -29 \log (27t + 1) \qquad \text{Subtract 100 from both sides}$$
$$2 = \log (27t + 1) \qquad \text{Divide both sides by } -29$$
$$10^2 = 27t + 1 \qquad \text{Change to exponential form}$$
$$t = 3\tfrac{2}{3} \text{ days} \qquad \text{Solve for } t$$

(d)   The retention curve associated with this function is shown in Figure 6.16. Observe from the graph that forgetting is at first rapid and then slow.

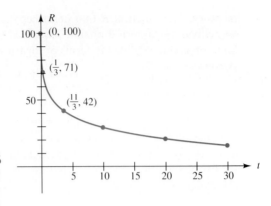

FIGURE 6.16
Retention curve of
$R(t) = 100 - 29 \log (27t + 1)$
for $0 \leq t \leq 30$

PROBLEM 5   Referring to Example 5, determine the percentage of the learned material that is retained after 10 days, 20 days, and 30 days. Check these values with the retention curve in Figure 6.16.

Exercises 6.4

    ### Basic Skills

In Exercises 1 and 2, sketch the graphs of the equations on the same coordinate plane.

1. $y = \log_2 x$, $y = \log_3 x$, $y = \log x$, and $y = \ln x$

2. $y = \log_{1/2} x$, $y = \log_{1/3} x$, $y = \log_{1/10} x$, and $y = \log_{1/e} x$

Use the graphs of the equations in Exercises 1 and 2 in conjunction with the shift rules and axis reflection rules (Section 3.4) to sketch the graphs of the functions in Exercises 3–16. Label any x- and y-intercepts and the vertical asymptote.

3. $F(x) = -\log_2 x$
4. $g(x) = -\log_{1/3} x$

5. $h(x) = \log_{1/10} (-x)$
6. $f(x) = -\ln (-x)$

7. $g(x) = \log_3 x + 2$
8. $G(x) = \log_2 x - 1$

9. $G(x) = \log_3 (x + 2)$
10. $H(x) = \log_2 (x - 1)$

11. $f(x) = -3 + \ln (x - 1)$

12. $f(x) = 2 + \log (x + 2)$

13. $H(x) = 2 - \log_2 (x + 3)$

14. $F(x) = 1 - \log (x - 4)$

15. $f(x) = \ln (3 - x) + 2$

16. $f(x) = 2 - \log_3 (2 - x)$

Determine which of the functions in Exercises 17–22 are the same function.

17. $f(x) = \ln x^2$, $g(x) = 2 \ln x$, and $h(x) = 2 \ln |x|$

18. $f(x) = \ln x^3$, $g(x) = 3 \ln x$, and $h(x) = 3 \ln |x|$

19. $f(x) = \log \sqrt{x + 1}$, $g(x) = \frac{1}{2} \log (x + 1)$, and $h(x) = \frac{1}{2} \log |x + 1|$

20. $f(x) = \log (x - 1)^4$, $g(x) = 4 \log (x - 1)$, and $h(x) = 4 \log |x - 1|$

21. $f(x) = \ln \dfrac{1}{x + 3}$, $g(x) = -\ln (x + 3)$, and $h(x) = -\ln |x + 3|$

22. $f(x) = \log \dfrac{10}{(x - 1)^2}$, $g(x) = 1 - 2 \log (x - 1)$, and $h(x) = 1 - 2 \log |x - 1|$

In Exercises 23–30, use the properties of logarithms to help sketch the graph of each function. Label any x- and y-intercepts and the vertical asymptote.

23. $f(x) = \frac{1}{2} \log x^4$
24. $f(x) = \ln x^3$

25. $f(x) = \ln \dfrac{1}{x + 3}$
26. $f(x) = \log_3 \dfrac{9}{x}$

27. $f(x) = \log_{1/2} [8(x - 1)]$

28. $f(x) = \log_2 (16x + 32)$

29. $f(x) = \log_3 \dfrac{1}{18 - 9x}$

30. $f(x) = \log \dfrac{10}{(x - 1)^2}$

31. In a psychological experiment, a student is asked to memorize a list of telephone numbers by studying the list until one perfect repetition is performed from memory. At various times over the next week, the student is asked to recall the list of numbers. It is found that the retention function R associated with this experiment is given by

$$R(t) = 100 - 31 \log (36t + 1), \qquad 0 \le t \le 7,$$

where t is time in days.

(a) What percentage of the learned material is retained initially (at $t = 0$)?

(b) What percentage of the learned material is retained after 6 hours?

(c) How much time has elapsed when 33% of the learned material is still retained?

(d) Sketch the graph of the retention curve associated with this function.

32. Students taking their last calculus course are given a final exam and the average score of the group is recorded. Each month thereafter, the students are given an equivalent exam and each time their average score is recorded. It is found that the average score S is a function of time t (in months) and is defined by

$$S(t) = 72 - 22 \log (t + 1), \qquad 0 \le t \le 12.$$

(a) What is the average score on the original exam?

(b) What is the average score six months later?

(c) How much time elapses before the average score decreases to 50?

(d) Sketch the graph of the function S.

Critical Thinking

In Exercises 33–44,

(a) find the inverse of each function.

(b) sketch the graphs of the function and its inverse on the same coordinate plane.

33. $f(x) = 6^x$

34. $g(x) = (\frac{1}{6})^x$

35. $h(x) = e^{x/2}$

36. $F(x) = 10^{2x/3}$

37. $G(x) = 1 - 10^{2x}$

38. $H(x) = 3e^{2x+1}$

39. $f(x) = \log_8 x$

40. $g(x) = \log_{1/8} x$

41. $h(x) = -\ln \dfrac{x}{2}$

42. $F(x) = 2 \log \dfrac{3x}{2}$

43. $G(x) = 1 + 2 \log 3x$

44. $H(x) = 3 \ln (2x - 1)$

45. The number N of computers sold by a certain company is a function of the amount x (in thousands of dollars) that is spent on advertising and is given by

$$N(x) = 1000[1 + \ln (x + 1)].$$

(a) If no money is spent on advertising, how many computers are sold?

(b) If \$100,000 is spent on advertising, approximately how many computers are sold?

(c) How much money in advertising must be spent to sell 4000 computers?

(d) Sketch the graph of the function N.

46. The time t in minutes it takes for a cup of hot tea to cool to a temperature of T degrees Fahrenheit when it is placed in a room whose temperature is maintained at 70 °F is given by

$$t(T) = -8 \ln \dfrac{T - 70}{100}, \qquad 70 < T \le 170.$$

(a) How long does it take for the tea to cool to 140 °F?

(b) How long does it take for the tea to cool to 100 °F?

(c) What is the temperature of the tea (to the nearest degree) when $t = 4$ minutes?

(d) Sketch the graph of the function t.

Calculator Activities

47. Given the function $f(x) = \dfrac{\ln x}{x}$,

(a) determine the domain of the function.

(b) determine any zeros of the function.

(c) explain the behavior of $f(x)$ as $x \to \infty$.

(d) explain the behavior of $f(x)$ as $x \to 0^+$.

(e) sketch the graph of the function.

(f) use a graphing calculator to generate the graph of this function, and trace to the relative maximum point. Zoom in on this maximum point to estimate the value of x where the maximum value of the function seems to occur.

48. Given the function $f(x) = \dfrac{1}{x \ln x}$,

(a) determine the domain of the function.
(b) determine any zeros of the function.
(c) explain the behavior of $f(x)$ as $x \to \infty$ and as $x \to 1^+$.
(d) explain the behavior of $f(x)$ as $x \to 0^+$ and as $x \to 1^-$.

(e) sketch the graph of the function.
(f) use a graphing calculator to generate the graph of this function, and trace to the relative maximum point in the interval (0, 1). Zoom in on this point to estimate the maximum value of this function in the interval (0, 1).

6.5   Logarithmic and Exponential Equations

◆ Introductory Comments

Equations that contain logarithmic expressions are referred to as **logarithmic equations**. In Section 6.2, we solved logarithmic equations containing a single logarithmic expression by changing the equation to exponential form. For example,

Logarithmic form		Exponential form

$$\log_3 (x - 12) = 2 \quad \text{is equivalent to} \quad 3^2 = x - 12$$

$$x = 3^2 + 12$$

$$x = 21$$

An equation in which the variable appears in an exponent is referred to as an **exponential equation**. In Section 6.2, we solved some exponential equations with base 10 and base e by changing the equation to logarithmic form. For example,

Exponential form		Logarithmic form

$$e^{x/2} = 9 \quad \text{is equivalent to} \quad \ln 9 = \frac{x}{2}$$

$$x = 2 \ln 9$$

$$x = \ln 81 \approx 4.394$$

In this section, we look at two types of equations:

1. A logarithmic equation that contains more than one logarithmic expression.
2. An exponential equation in which the base is different from 10 or e.

◆ Logarithmic Equations

To solve many logarithmic equations that contain more than one logarithmic expression, we use the following procedure.

Procedure for Solving Logarithmic Equations

1. Isolate the logarithmic expressions on one side of the equation.

2. Apply the properties of logarithms, and write the equation in logarithmic form.

3. Change to exponential form, and solve for the unknown.

4. Check the solutions. *This procedure may produce extraneous roots.*

E X A M P L E 1   Solve each logarithmic equation.

(a) $2 - \log x = \log 3$ (b) $\log_4 (x - 2) + 2 \log_4 x = 1 + \log_4 2x$

S O L U T I O N

(a) Using the given procedure, we have

$$2 - \log x = \log 3$$

$$2 = \log 3 + \log x \qquad \text{Isolate the logarithms on one side}$$

$$2 = \log 3x \qquad \text{Apply log property 1}$$

$$10^2 = 3x \qquad \text{Change to exponential form}$$

$$x = \frac{100}{3} \qquad \text{Solve for } x$$

Check: Replacing x with $\frac{100}{3}$ in the original equation, we have

$$2 - \log \tfrac{100}{3} = \log 3 \quad ?$$

$$2 - (\log 100 - \log 3) = \log 3 \quad ?$$

$$2 - (2 - \log 3) = \log 3 \quad ?$$

$$\log 3 = \log 3 \quad \checkmark$$

Thus, $x = \frac{100}{3}$ is a solution.

(b) Using the given procedure, we have

$$\log_4 (x - 2) + 2 \log_4 x = 1 + \log_4 2x$$

$$\log_4 (x - 2) + 2 \log_4 x - \log_4 2x = 1 \qquad \text{Isolate the logarithmic expressions}$$

$$\log_4 \frac{x^2(x - 2)}{2x} = 1 \qquad \text{Apply the log properties and write in logarithmic form}$$

$$\log_4 \frac{x(x - 2)}{2} = 1 \qquad \text{Reduce, } x \neq 0$$

EXAMPLE 1 *(continued)*

$$4^1 = \frac{x(x-2)}{2}$$ **Change to exponential form**

$$x^2 - 2x - 8 = 0$$ **Write in quadratic form and solve for x**

$$(x-4)(x+2) = 0$$

$$x = 4 \quad \text{or} \quad x = -2$$

Check I: $x = 4$

$$\log_4 2 + 2\log_4 4 = 1 + \log_4 8 \quad ?$$

$$\tfrac{1}{2} + 2 = 1 + \tfrac{3}{2} \quad ?$$

$$\tfrac{5}{2} = \tfrac{5}{2} \quad \checkmark$$

Check II: $x = -2$

$$\log_4(-4) + 2\log_4(-2) = 1 + \log_4(-4) \quad ?$$

Undefined

Thus, $x = 4$ is the only solution.   ◆

P R O B L E M 1   Solve the logarithmic equation $2\ln x - \ln 9 = 4$.   ◆

Many literal equations and formulas that contain logarithmic expressions may be solved by using this procedure.

E X A M P L E 2   Solve the literal equation $\ln(y+a) - 2\ln(x+b) = c$ for y. Assume a, b, and c are constants.

S O L U T I O N   We solve for y as follows:

$$\ln(y+a) - 2\ln(x+b) = c$$

$$\ln\frac{y+a}{(x+b)^2} = c$$ **Apply the log properties and write in logarithmic form**

$$e^c = \frac{y+a}{(x+b)^2}$$ **Change to exponential form**

$$y = e^c(x+b)^2 - a$$ **Solve for y**

Since c is a constant, e^c is also a constant. Relabeling e^c as the constant k gives us

$$y = k(x+b)^2 - a$$

where $k = e^c$.   ◆

P R O B L E M 2   Solve the literal equation $x + \ln y = \ln c$, where c is a constant, for y.   ◆

◆ Exponential Equations

We now look at exponential equations in which the bases are different from 10 or e. If both sides of an exponential equation can be written as powers of the same base, then the equation can be solved by equating the powers; that is,

$$\text{if}\quad b^x = b^y,\qquad \text{then}\quad x = y.$$

EXAMPLE 3   Solve the exponential equation $27^{x-2} = 9$.

SOLUTION   Both 27 and 9 can be written in terms of a base of 3. Hence,

$$27^{x-2} = 9$$

$$(3^3)^{x-2} = 3^2 \qquad \text{Write both sides with the same base}$$

$$3^{3(x-2)} = 3^2 \qquad \text{Multiply exponents}$$

$$3(x-2) = 2 \qquad \text{Equate exponents and solve for } x$$

$$3x = 8$$

$$x = \tfrac{8}{3}$$

Check: $27^{[(8/3)-2]} = 27^{2/3} = 9$ ✓   ◆

PROBLEM 3   Solve the exponential equation $4^{3x+2} = 8^{4x}$.   ◆

An alternate method for solving the exponential equation in Example 3 is to begin by taking the common (or natural) logarithm of both sides of the equation:

$$27^{x-2} = 9$$

$$\log 27^{x-2} = \log 9 \qquad \text{Take the common logarithm of both sides}$$

$$(x-2)\log 27 = \log 9 \qquad \text{Apply log property 3}$$

$$x - 2 = \frac{\log 9}{\log 27} \qquad \text{Divide both sides by } \log 27$$

$$x = 2 + \frac{\log 9}{\log 27} \qquad \text{Add 2 to both sides}$$

Now, since $9 = 3^2$ and $27 = 3^3$, we apply log property 3 and write

$$x = 2 + \frac{\log 9}{\log 27} = 2 + \frac{2\,\log 3}{3\,\log 3} = 2 + \frac{2}{3} = \frac{8}{3},$$

which agrees with the answer we obtained in Example 3.

This procedure is particularly useful when both sides of an exponential equation *cannot* be written as powers of the same base.

Procedure for Solving Exponential Equations

1. Take the common (or natural) logarithm of both sides of the equation.

2. Apply the properties of logarithms, and write the powers as coefficients of logarithms.

3. Solve for the unknown, and check the solution.

E X A M P L E 4   Solve each exponential equation.

(a) $6^x = 50$ (b) $3^{x-2} = 2^{-x}$

S O L U T I O N

(a) Since both sides of this equation cannot be written as powers of the same base, we apply the given procedure:

$$6^x = 50$$

$$\log 6^x = \log 50 \qquad \text{Take the common logarithm of both sides}$$

$$x \log 6 = \log 50 \qquad \text{Apply log property 3}$$

$$x = \frac{\log 50}{\log 6} \approx 2.183 \qquad \text{Divide both sides by log 6}$$

Use the LOG key on a calculator to find the approximate value.

You can check this solution with a calculator by using the Y^x key.

(b) Using the given procedure, we have

$$3^{x-2} = 2^{-x}$$

$$\log 3^{x-2} = \log 2^{-x} \qquad \text{Take the common logarithm of both sides}$$

$$(x - 2) \log 3 = -x \log 2 \qquad \text{Apply log property 3}$$

Be sure to remember the parentheses.

$$x \log 3 - 2 \log 3 = -x \log 2 \qquad \text{Multiply}$$

$$x \log 3 + x \log 2 = 2 \log 3 \qquad \text{Group the } x \text{ terms on one side}$$

$$x(\log 3 + \log 2) = 2 \log 3 \qquad \text{Factor out } x$$

$$x = \frac{2 \log 3}{\log 3 + \log 2} \qquad \text{Divide both sides by } (\log 3 + \log 2)$$

$$x = \frac{\log 9}{\log 6} \approx 1.226$$

Use the $\boxed{\text{LOG}}$ key on a calculator to find the approximate value.

◆

Note: To solve Example 4(a), we may also proceed as in Section 6.2 and change the exponential form $6^x = 50$ to the logarithmic form

$$x = \log_6 50.$$

Now to evaluate $\log_6 50$, we use the change of base formula (Section 6.3) with common logarithms to obtain

$$x = \log_6 50 = \frac{\log 50}{\log 6} \approx 2.183$$

P R O B L E M   4   Solve the equation $5^{x-1} = 325$.   ◆

◆ **Equations of Quadratic Type**

We can solve other logarithmic and exponential equations by recognizing them as equations of quadratic type.

E X A M P L E   5   Solve each equation.

(a) $(\log x)^2 = \log x^3$ **(b)** $e^x + 3e^{-x} = 4$

S O L U T I O N

(a)   We begin by rewriting the equation as follows:

$$(\log x)^2 = \log x^3$$

$$(\log x)^2 = 3 \log x \qquad \text{Apply log property 3}$$

$$(\log x)^2 - 3 \log x = 0 \qquad \text{Subtract } 3 \log x \text{ from both sides}$$

Now observe that this last equation is of quadratic type. Letting $u = \log x$, we obtain

$$u^2 - 3u = 0$$

$$u(u - 3) = 0 \qquad \text{Factor}$$

$$u = 0 \quad \text{or} \quad u = 3 \qquad \text{Apply the zero product property}$$

$$\log x = 0 \qquad \log x = 3 \qquad \text{Replace } u \text{ with } \log x$$

EXAMPLE 5 *(continued)*

$$x = 10^0 \qquad\qquad x = 10^3 \qquad \text{Change to exponential form}$$

$$x = 1 \qquad\qquad\quad x = 1000 \qquad \text{Simplify}$$

You can check both solutions.

(b) We begin by rewriting the equation as follows:

$$e^x + 3e^{-x} = 4$$

$$e^{2x} + 3 = 4e^x \qquad \text{Multiply both sides by } e^x$$

$$e^{2x} - 4e^x + 3 = 0 \qquad \text{Subtract } 4e^x \text{ from both sides}$$

Now observe that this last equation is of quadratic type. Letting $u = e^x$, we obtain

$$u^2 - 4u + 3 = 0$$

$$(u - 1)(u - 3) = 0 \qquad\qquad \text{Factor}$$

$$u = 1 \quad \text{or} \quad u = 3 \qquad\qquad \begin{array}{l}\text{Apply the zero product}\\ \text{property}\end{array}$$

$$e^x = 1 \qquad\qquad e^x = 3 \qquad\qquad \text{Replace } u \text{ with } e^x$$

$$x = \ln 1 \qquad\quad x = \ln 3 \qquad\qquad \text{Change to logarithmic form}$$

$$x = 0 \qquad\qquad x \approx 1.099 \qquad\qquad \text{Evaluate}$$

You can check both solutions.   ◆

PROBLEM 5   Solve the equation $\log \sqrt{x} = \sqrt{\log x}$ by squaring both sides.   ◆

◆ **Graphical Methods of Solutions**

If the techniques that we have discussed in this section are insufficient to solve a particular logarithmic or exponential equation, we can use a calculator or computer with graphing capabilities to find the approximate solutions.

EXAMPLE 6   Use a graphing calculator to find the approximate solutions of the equation $2^x + 3^x = 9$.

SOLUTION   The equation $2^x + 3^x = 9$ is equivalent to

$$2^x = 9 - 3^x.$$

To solve this equation graphically, we graph the equations

$$y = 2^x \qquad \text{and} \qquad y = 9 - 3^x$$

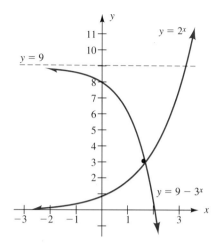

FIGURE 6.17

The solution of the equation $2^x + 3^x = 9$ is between $x = 1$ and $x = 2$.

on the same set of coordinate axes and determine the x-coordinate(s) of their intersection point(s). Using the techniques of graphing exponential functions (Section 6.1), we obtain the graphs shown in Figure 6.17. Note that the graphs intersect once, and the x-coordinate of the intersection point appears to be between $x = 1$ and $x = 2$.

To display these graphs on a calculator, we begin by choosing a viewing rectangle by pressing the RANGE key. Although we can select any Range values, Figure 6.17 suggests a reasonable viewing rectangle:

$$[-2, 4] \qquad \text{by} \qquad [-1, 10]$$

x-min $\quad$ x-max $\qquad$ y-min $\quad$ y-max

By entering the equations $y = 2^x$ and $y = 9 - 3^x$, we obtain the two graphs in the viewing rectangle. Next, we press the TRACE key and move the blinking cursor to the approximate point of intersection of the two curves. Reading the x-coordinate of this point, we find $x \approx 1.6$ to the nearest tenth. To obtain more precise values of x, we may zoom in on this intersection point by activating the Zoom feature and tracing to this point once again. By repeating this process of tracing and zooming, we find that the x-coordinate of the point of intersection is $x \approx 1.620$ to the nearest thousandth. In summary, we conclude that the solution to the equation $2^x + 3^x = 9$ is approximately 1.620. ◆

PROBLEM 6 Use the Y^x key on your calculator to check the approximate solution of the equation in Example 6. ◆

Exercises 6.5

 Basic Skills

In Exercises 1–12, solve each logarithmic equation.

1. $\log x + \log 5 = 1$

2. $\ln x = 1 + \ln 5$

3. $3 \ln (x + 1) - \ln 27 = 3$

4. $2 \log_6 (2 - x) = 2 - \log_6 3$

5. $\log_4 (x + 12) - \log_4 (x - 3) = 2$

6. $\log_3 x + \log_3 (x - 6) = 3$

7. $\log_5 x + \log_5 (x + 2) = \log_5 (x + 6)$

8. $\log 2 + \log (2x - 3) = \log 3x - \log 3$

9. $\log (x + 6) + 1 = 2 \log (3x - 2)$

10. $\log_2 (x - 1) + 2 \log_2 x = 2 + \log_2 3x$

11. $\frac{1}{2} \ln (3 - 2x) - \ln x = 0$

12. $\frac{1}{2} \log (x + 3) + \log 2 = 1$

In Exercises 13–20, solve each literal equation for y. Assume a, b, and c are constants.

13. $\ln y + 2x = \ln c$

14. $2(x + \ln x) = \ln y - \ln c$

15. $2 \log (x + a) = 1 - \log y$

16. $\log y = \log (x + y) + 2$

17. $\ln (x + y) - \ln (x - y) = c$

18. $\ln (x + 2y) + 2 \ln x = \ln y + c$

19. $\frac{1}{a} \ln y - \frac{1}{a} \ln (a - by) = c$

20. $-\frac{1}{4} \ln (y + 2) + \frac{1}{4} \ln (y - 2) = x + c$

In Exercises 21–36, solve each exponential equation.

21. $9^x = 27$ $\qquad\qquad$ 22. $16^x = \frac{1}{8}$

23. $2^{x-1} = 16^x$

24. $5^{x+3} = 25^{x-2}$

25. $7^x = 35$

26. $3^x = 36$

27. $3^{2x-1} = 40$

28. $6^{x+5} = 75$

29. $10^{2x-1} = 4^{-x}$

30. $5^{x-2} = 4^{2x+1}$

31. $e^{x/2} = 2^{x-1}$

32. $4^{-2x} = e^{x+1}$

33. $3 \cdot 2^{x-2} = 6^x$

34. $5 \cdot 3^{-x} = 4 \cdot 2^{x+2}$

35. $2^x - 6 \cdot 8^{3-x} = 0$

36. $\dfrac{10^{x-1}}{e^{2x}} - 4 = 0$

Rewrite each equation in Exercises 37–46 as an equation of quadratic type, then solve.

37. $(\ln x)^2 = \ln x^2$

38. $2[\log (x+1)]^2 - \log (x+1)^3 = 5$

39. $e^{4x} - 3e^{2x} = 4$

40. $2^{3x} + 4 \cdot 2^{-3x} = 5$

41. $3 \cdot 10^x - 10^{-x} = 2$

42. $\frac{1}{2}(e^x - 9e^{-x}) = 4$

43. $\dfrac{e^x + e^{-x}}{2} = 10$

44. $10^x + 10^{-x} = 6$

45. $2 \log_x 3 + \log_3 x = 3$

46. $2 \log_4 x - 3 \log_x 4 = 5$

 ## Critical Thinking

47. In Section 6.1 we developed the compound interest formula

$$A = P\left(1 + \frac{r}{n}\right)^{nt}$$

Solve for t and express the answer in terms of natural logarithms.

48. Solve for x in terms of b.
    (a) $\log_b (x - 3) = -1 + \log_b 5$
    (b) $2 + \log_b (4x + 1) = \log_b (2 - x)$

49. If y varies directly as the mth power of x, then

$$y = kx^m,$$

where k is the variation constant.

(a) Take the natural logarithm of both sides of this equation and show that

$$\ln y = m \ln x + \ln k.$$

(b) Construct the graph of $\ln y = m \ln x + \ln k$ in an XY-plane, where $X = \ln x$ and $Y = \ln y$. What type of graph do we obtain?

50. Find the fallacy in the following argument.

$$1 < 2$$

$$\frac{1}{4} < \frac{1}{2} \qquad \textbf{Divide both sides by 4}$$

$$\ln \frac{1}{4} < \ln \frac{1}{2} \qquad \textbf{Take the natural logarithm of both sides}$$

$$\ln \left(\frac{1}{2}\right)^2 < \ln \frac{1}{2} \qquad \textbf{Rewrite } \tfrac{1}{4}$$

$$2 \ln \frac{1}{2} < \ln \frac{1}{2} \qquad \textbf{Apply log property 3}$$

$$2 < 1 \qquad \textbf{Divide both sides by } \ln \tfrac{1}{2}$$

51. Because of acid rain, the population P of fish in a pond in New Hampshire is decreasing according to the equation

$$\log_2 P = -\tfrac{1}{3}t + \log_2 P_0,$$

where t is time in years after 1985 and P_0 is the original population in 1985.

(a) Solve the equation for P.
(b) In the year 1991, what percent of the original fish population remained?

52. In a series circuit containing a capacitor C, a resistance R, and a battery source E, the instantaneous current i at any time t is given by

$$\ln i = -\frac{t}{RC} + \ln E - \ln R.$$

(a) Solve the equation for i.
(b) What happens to the current i as t increases without bound ($t \to \infty$)?

Calculator Activities

 In Exercises 53–58, use a graphing calculator to find the approximate solutions of the equation. Round each answer to four significant digits.

53. $2x + \ln x = 5$ **54.** $e^{-x} - x^3 = 0$

55. $xe^x = 4$ **56.** $2^{x+1} + 5^x = 1$

57. $2^x = 3 - x^2$ **58.** $\ln (x + 1) = x^2 - 4x$

59. When a cable or rope is suspended between two points at the same height and allowed to hang under its own weight, it forms a curve called a **catenary** (after the Latin word for chain). The power line shown in the figure is an example of a catenary, and its equation is given by

$$y = 25(e^{x/50} + e^{-x/50}),$$

where x and y are measured in feet.

(a) Find the height of the power line at $x = 0$.
(b) Find x (to the nearest tenth of a foot) when $y = 75$ ft.

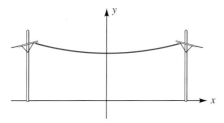

60. A student infected with the flu returns to a college campus of 1000 students. If no student leaves campus, then the time t (in days) it takes for N students to become infected with the flu is given by

$$t = \ln 999N - \ln (1000 - N).$$

(a) How many days does it take for 20 students to become infected?
(b) Determine the number of infected students after 6 days.

Chapter 6 Review

 Questions for Group Discussion

1. If $b > 0$, how is b^r defined when r is an irrational number?

2. Under what conditions is the *exponential function* $f(x) = b^x$ an increasing function? a decreasing function?

3. How is the graph of $f(x) = b^x$ related to the graph of $g(x) = b^{-x}$? Discuss the behavior of $f(x)$ and $g(x)$ as $x \to \infty$.

4. Are the functions $f(x) = e^{x + 1}$ and $g(x) = e \cdot e^x$ identical? Explain.

5. Give an example of an *exponential growth function* and an *exponential decay function*.

6. What is the meaning of the *base b logarithm*, $\log_b x$? When is $\log_b x$ undefined?

7. State the four *logarithmic identities*.

8. What is the base for *common logarithms*? for *natural logarithms*?

9. How are the expressions $\log_b a$ and $\log_a b$ related?

10. State the procedure for changing an equation from *logarithmic form* to *exponential form*, and vice versa.

11. Explain the general procedure for solving an *exponential equation*. Illustrate with an example.

12. List some characteristics of the graph of the *logarithmic function* $f(x) = \log_b x$.

13. Are the functions $f(x) = \ln (x + 1)^2$ and $g(x) = 2 \ln (x + 1)$ identical? Explain.

14. How is the graph of $f(x) = \log_b x$ related to the graph of $g(x) = \log_b (1/x)$?

15. What is the domain and range of the *natural logarithmic function* $f(x) = \ln x$?

16. What is the *inverse function* of the natural logarithmic function?

17. How can we rewrite the logarithm of a product? of a quotient? of a quantity raised to a power?

18. What is the value of $(1 + 1/n)^n$ as n increases without bound?

Review Exercises

In Exercises 1–6, use the properties of real exponents to write each expression as a constant or as an expression in the form b^x, where $b > 0$.

1. 1^{3x-2}

2. $(3^{x + \pi})^0$

3. $e^{2x} \cdot e^{1-2x}$

4. $(2^{-3x})^2$

5. $\dfrac{4^{x+3}}{8^{x+2}}$

6. $(3^{-x} \cdot 9^x)^2$

In Exercises 7–20, find the value of each expression.

7. $\log_7 49$

8. $\log_2 32$

9. $\log \sqrt[3]{100}$

10. $\ln e^3$

11. $\log_{1/2} 8$

12. $\log_{1/3} 9$

13. $\log_8 2$

14. $\log_{25} \frac{1}{5}$

15. $\log_8 \frac{1}{4}$

16. $\log_{16} 64$

17. $e^{\ln 4}$

18. $10^{\log 8}$

19. $3^{\log_7 7}$

20. $8^{\log_2 1}$

In Exercises 21–30, simplify each expression. Assume the variables are restricted to those real numbers that allow the logarithmic expressions to be defined.

21. $\log_7 7^{2x}$

22. $5^{\log_5 (2-x)}$

23. $e^{\ln (x^2 + 1)}$

24. $\ln e^{2-3x}$

25. $\log_3 27^{1-x}$

26. $\log_4 2^{4x-4}$

27. $e^{2x + \ln x}$

28. $e^{-\ln (x-1)}$

29. $(\log_3 x)(\log_x 3)$

30. $\dfrac{1}{\log_x \frac{x}{2}} - \dfrac{1}{\log_2 \frac{x}{2}}$

In Exercises 31–36, write each expression as sums and differences of simpler logarithmic expressions without logarithms of products, quotients, and powers. Simplify whenever possible. Assume that any variable is restricted to those real

numbers that allow for the properties of logarithms to be applied.

31. $\log_3 [9x^2(x - 1)]$

32. $\log_4 \left[4x \sqrt{x + 2} \right]$

33. $\log \dfrac{1}{x\sqrt{2x - 3}}$

34. $\log \dfrac{1000}{xy^{1/3}}$

35. $\ln \dfrac{xe^{-x^2}}{e^x - 1}$

36. $\ln \dfrac{4e^{2x}}{\sqrt{e^x + 2}}$

In Exercises 37–42, write each expression as a single logarithm with a coefficient of 1. Simplify whenever possible. Assume that any variable is restricted to those real numbers that allow for the properties of logarithms to be applied.

37. $\ln 12 - \ln 3 + \ln 2$

38. $-2 \log_2 6 + 3 \log_2 3 - \frac{1}{2} \log_2 9$

39. $3 \log_3 (x + 1) - \log_3 (x^2 - 1) + \log_3 (x - 1)$

40. $2 \ln x + \frac{1}{2} \ln (x + 1) + \ln 2$

41. $\log \dfrac{1}{\sqrt{x} - \sqrt{x - 1}} - \log \left(\sqrt{x} + \sqrt{x - 1} \right)$

42. $\log\left(\sqrt{x + 1} + 1 \right) + \log \left(\sqrt{x + 1} - 1 \right)$

In Exercises 43–64, solve each equation.

43. $\log_5 x = 1$

44. $\ln x = 10$

45. $\ln |x + 2| - 3 = 0$

46. $\log_3 (x^2 - 3x + 5) = 2$

47. $\log_3 (x - 7) - \log_3 (x + 1) = 2$

48. $\log 3 - \log (2 - x) = \log 2 - \log 2x$

49. $\frac{1}{2} \log (12 - x) - \log x = 0$

50. $\log_2 (3x - 2) + 2 = 2 \log_2 (x + 2)$

51. $\ln y = c + 2 \ln |x|$, where c is a constant, solve for y

52. $3(x - \ln x) = \ln (2y - 3) + \ln c$, where c is a constant, solve for y

53. $10^{2x} = 80$ **54.** $e^{2x-1} = 9$

55. $16^x = \frac{1}{8}$ **56.** $3^{x+3} = 27^{1-x}$

57. $3^{2x} = 4^{1-x}$ **58.** $e^{2x} = 2^{x+1}$

59. $2e^{-x} = 3^{2x-1}$ **60.** $5 \cdot 2^{-x} = 3 \cdot 5^{x+2}$

61. $x^2 e^{-x} - e^{-x} = 0$ **62.** $x \ln x - 3 \ln x = 0$

63. $\log_x 3 + 2\log_3 x = 3$ **64.** $e^x + 1 = 2e^{-x}$

In Exercises 65–80, sketch the graph of each function. Label any x- or y-intercept(s) and horizontal or vertical asymptote.

65. $f(x) = 5^x$ **66.** $g(x) = 6^{-x}$

67. $h(x) = e^{x-2}$ **68.** $F(x) = 2 + e^x$

69. $G(x) = -(2 + e^{x+1})$ **70.** $H(x) = 3^x + 3^{-x}$

71. $f(x) = 2^{|x+1|}$ **72.** $g(x) = x - e^x$

73. $h(x) = \log_{1/5} x$ **74.** $F(x) = \log_6(-x)$

75. $G(x) = -(2 + \log_4 x)$ **76.** $H(x) = \log_4(x+2)$

77. $f(x) = \ln \dfrac{1}{x+1}$ **78.** $g(x) = \ln\sqrt{1-x}$

79. $h(x) = \dfrac{\ln|x|}{x}$ **80.** $F(x) = x \ln x - x$

In Exercises 81–84, use the change of base formula and a calculator to evaluate each logarithm to the nearest thousandth.

81. $\log_3 45$ **82.** $\log_6 30$

83. $\log_{1/2} 100$ **84.** $\log_{2/3} \frac{1}{2}$

In Exercises 85–92, find the inverse of each function. Then sketch the graphs of the function and its inverse on the same coordinate plane.

85. $f(x) = 8^x$ **86.** $g(x) = e^{-2x}$

87. $h(x) = 2e^{3x-2}$ **88.** $F(x) = 1 - 8^{1-x}$

89. $G(x) = \log_{1/2}(2x)$ **90.** $H(x) = -\ln(x-1)$

91. $f(x) = 1 - \ln x^2, \quad x > 0$

92. $g(x) = \ln(x-1) - \ln(2x)$

93. Suppose that $5000 is invested at an interest rate of 9% per year. Assuming the investor makes no subsequent deposit or withdrawal, find the balance after 10 years if the interest is compounded (a) semiannually, (b) monthly, or (c) continuously.

94. Suppose that $100,000 is invested at an interest rate of $7\frac{1}{2}$% per year. Assuming the investor makes no subsequent deposit or withdrawal, find the balance after 9 months if the interest is compounded (a) quarterly, (b) daily, or (c) continuously.

95. Five hundred dollars is deposited into a savings account paying $6\frac{3}{4}$% interest per year compounded continuously. Assuming the depositor makes no subsequent deposit or withdrawal, find the time it takes for the deposit to accumulate to $800.

96. At what interest rate compounded continuously must money be invested if the amount is to triple in 12 years?

97. Initially, approximately 10^3 bacteria are present in a culture. Three hours later approximately 10^5 bacteria are present.

(a) Determine the Malthusian model that describes this growth pattern.
(b) Approximately how many minutes does it take for this population to double in size?

98. A radioactive substance decays from 60 grams to 50 grams in 4 years.

(a) Determine the exponential decay function that describes this decay pattern.
(b) How many years will it take for this radioactive substance to decay to 10 grams?
(c) Find the half-life of this radioactive substance.

99. One of the strongest earthquakes ever recorded measured 8.9 on the Richter scale. It occurred in Japan in 1933. How many times more intense was this earthquake than one that measures 6.0 on the Richter scale?

100. Students taking their last French course are given a final exam and the average score of the group is recorded. Each month thereafter, the students are given an equivalent exam, and each time their average score is recorded. It is found that the average score S is a function of time t (in months) and is given by $S(t) = 77 - 17 \ln(t+1)$.

(a) What is the average score on the original final exam?
(b) What is the average score three months later?
(c) How much time elapses before the average score decreases to 40?
(d) Sketch the graph of the function S.

101. When a hot metal object with a temperature of 400 °F is placed in a room whose temperature is 70 °F, the object cools at a rate such that its temperature T is a function of the time t (in minutes) that the object has been in the room and this rate of cooling is given by Newton's cooling law

$$T(t) = 70 + (400 - 70)e^{-kt},$$

where k is a constant.

EXERCISE 101 (continued)

(a) Determine the value of the constant k if the temperature of the object after 10 minutes is 290 °F.
(b) Find $T(60)$, the temperature of the object after one hour.
(c) How much time has elapsed when $T(t) = 150$ °F?
(d) Sketch the graph of the function T.

102. In a certain forest, the population P of foxes after t years is described by the logistic law

$$P(t) = \frac{25000}{100 + 150e^{-0.1t}}.$$

(a) Find $P(10)$, the number of foxes in the forest after ten years.
(b) Discuss the behavior of $P(t)$ as t increases without bound $(t \to \infty)$.
(c) Find the time in years when $P(t) = 210$ foxes.
(d) Sketch the graph of the function P.

Cumulative Review Exercises

Chapters 3, 4, 5, & 6

1. Find the zeros of the function f defined by $f(x) = a + \dfrac{b}{\ln x}$, where a and b are constants.

2. Find the equations of the asymptotes for each of the following hyperbolas.

    (a) $4x^2 - y^2 = 16$ (b) $y^2 - x^2 + 2x = 5$

3. Sketch the graphs of the ellipses $x^2 + 2y^2 = 18$ and $9x^2 + 4y^2 - 24y = 0$ on the same coordinate plane. Express the coordinates of their intersection points to three significant digits.

4. Find the x- and y-intercepts for the graph of the equation $y^2 = x(x - 2)^2$.

5. Explain the behavior of the function f defined by $f(x) = \dfrac{x}{x - 2}$ as x approaches the given value.

    (a) $x \to 2^-$ (b) $x \to 2^+$

6. Find the perimeter and area of the triangle formed by the intersecting lines $x + y = -3$, $3x - 4y = -9$, and $4x - 3y = 2$.

7. Factor completely $x^4 + 2x^3 + x^2 - 8x - 20$ over the real numbers.

8. Find the remainder when $2x^4 - 3x^2 + 4x - 3$ is divided by $x - a$.

9. Solve for x:

    (a) $\log_x 3 - \log_4 16 - \log_3 \frac{1}{9} - \ln e = 2 \log 1$
    (b) $x^{\ln x} - e^2 x = 0$

10. When the base is 4, find the logarithms of each number.

    (a) 64   (b) $\frac{1}{16}$ (c) 32   (d) $\frac{1}{8}$

11. Identify the graph of each equation.

    (a) $x + y = 9$ (b) $x + y^2 = 9$
    (c) $x^2 + y^2 = 9$ (d) $x^2 - y^2 = 9$

12. Given that $f(x) = \sqrt{x}$, express $\dfrac{f(x + h) - f(x)}{h}$ in simplest form with a rationalized numerator. Then evaluate this expression as $h \to 0$.

13. Sketch the graphs of the functions $f(x) = 3x$ and $g(x) = xe^x$ on the same coordinate plane. Then determine the intersection points of the two graphs.

14. Find the vertical and horizontal asymptotes for the graph of the function

$$f(x) = \frac{x^2 - 2}{x^2 - 7x + 6}.$$

Determine the value of x where the graph of f crosses its horizontal asymptote.

15. For the graph of the equation $y = f(x)$, what is the y-intercept?

16. Sketch the graph of the following quadratic functions. Label the vertex and x- and y-intercepts.

    (a) $f(x) = x^2 - 4x + 3$ (b) $g(x) = 1 - 4x - x^2$

17. Find the length of the line segment joining the intersection points of the curves $y = x + 1$ and $x^2 + y^2 = 25$.

18. Determine k so that the line $x - 3y + k = 0$ passes through the point $(3, -1)$.

19. Find the area of a triangle whose vertices are the points $(8, 8)$, $(-7, -2)$, and $(1, -6)$.

20. For $a > 0$, show that $a^x = e^{x \ln a}$. Use this fact to express each of the following expressions as an exponential with base e.

    (a) 10^x (b) 3^{-x} (c) 5^{3x} (d) 2^{x+1}

21. Determine whether the graph of each equation has symmetry with respect to the x-axis, y-axis, or origin.

    (a) $x^2 y = 9$ (b) $x + y = x^3$
    (c) $x^2 + y^2 - 6x = 16$ (d) $y = \pm\sqrt{9 - x^2}$

22. Determine the oblique asymptote for the graph of the equation $x^2 - 2xy - y = 0$. Then graph the equation.

23. Given the functions $f(x) = \ln(x + 1)$ and $g(x) = e^{-x} - 1$, find the indicated composite function. Then determine the domains of these functions.

    (a) $f \circ g$ (b) $g \circ f$

24. Given $f(x) = \ln x$, show that the difference quotient

$$\frac{f(x + \Delta x) - f(x)}{\Delta x} = \ln \left(1 + \frac{\Delta x}{x} \right)^{1/\Delta x}.$$

25. Find the equation of a line that passes through the point of intersection of the lines $3x - 2y = 7$ and $y = 7 - 2x$ and also satisfies the given condition.

    (a) is parallel to the x-axis
    (b) is perpendicular to $x + y = 8$
    (c) passes through $(5, -2)$
    (d) has a slope of -3

26. Given the function $f(x) = 2 - e^{-x}$, (a) show that f is one-to-one and (b) find the inverse function f^{-1}, and state its domain.

27. Solve each equation for y.

(a) $2 \ln x - \frac{1}{2} \ln y = 1$
(b) $x + 2 = e^{x - \ln y}$
(c) $2 \log_3 (y + 1) = 2 - \log_3 (y - 1)$

28. Discuss the nature of the roots of the equation $x^n - 1 = 0$, where n is a positive integer, and (a) n is even or (b) n is odd.

29. Sketch the graph of the equation $y = 1/x^2$. Then use the ideas of shifting and reflecting to help sketch the graphs of the following functions. Label all intercepts and asymptotes.

(a) $f(x) = 2 - \dfrac{1}{x^2}$ (b) $g(x) = \dfrac{1}{(x - 1)^2}$

30. A point is equidistant from $(2, 1)$ and $(-4, 3)$, and the slope of the straight line joining this point to the origin is 2. What are the coordinates of the point?

31. Determine the value of m so that the line $y = mx - 2$ passes through the point of intersection of the lines $y = 3x + 2$ and $y = x - 4$.

32. Determine the quadratic function with real coefficients that has $2 - 3i$ as one of its zeros.

33. Find the value of the account after 8 years. A sum of $10,000 is deposited in an account at a rate of 9% per year, compounded continuously.

34. The power P (in watts) dissipated in a certain electrical circuit is directly proportional to the square of the current i (in amperes) flowing through the circuit. Express P as a function of i if $P = 5$ watts when $i = 20$ milliamperes.

35. If a farmer picks his pumpkin crop today, he will have 30,000 pounds of pumpkins worth 20¢ per pound. If he waits, the crop will grow at 2000 pounds per week, but the price will drop by 1¢ per pound per week.

(a) Express the amount A of money (in dollars) that the farmer receives at the end of x weeks as a function of x.
(b) When is the most profitable time for the farmer to pick and sell the pumpkin crop?

36. A hot cup of coffee has an initial temperature of 200 °F when placed in a room with a temperature of 70 °F. After t minutes, the temperature T of the coffee has cooled to $T = 70 + 130e^{-0.38t}$. How many *seconds* does it take for the temperature of the coffee to reach 120 °F?

37. A rectangular field is twice as long as it is wide and is enclosed by x feet of fencing. Express the area A of the field as a function of x.

38. A tank contains 1000 gallons of water. The spigot is opened and water flows out of the tank at the uniform rate of $\frac{2}{3}$ gallons per minute.

(a) Express the amount A of water (in gallons) in the tank as a function of the time t (in minutes) that the spigot is opened.
(b) State the domain and range of the function defined in part (a).

39. If the acceleration of an object is constant, its velocity increases linearly with time. Suppose after 1 second a certain object with a constant acceleration has a velocity of 12 feet per second (ft/s), and after 5 seconds it has a velocity of 20 ft/s.

(a) Express the velocity v of the object as a function of time t.
(b) Sketch the graph of the function defined in part (a) and state the significance of the y-intercept.

40. The hypotenuse of a right triangle is 1 centimeter longer than one of its legs. The area of the triangle is 4 square centimeters. Estimate, to the nearest hundredth, the lengths of the sides of the triangle.

CHAPTER 7

Linear Systems and Matrices

The Centrum civic center in Worcester, Massachusetts, sold 12,000 tickets to a recent rock concert. The ticket prices were $12, $18, and $24, and the total income from ticket sales was $201,000. How many tickets of each type did the Centrum sell if the number of $12 tickets sold was twice the number of $24 tickets sold?

(For the solution, see Example 6 in Section 7.1.)

7.1 Systems of Linear Equations

◆ Introductory Comments

In Section 4.5, we graphed the linear equations

$$A_1 x + B_1 y = C_1 \quad \text{and} \quad A_2 x + B_2 y = C_2,$$

on the same coordinate plane. Figure 7.1 illustrates the three possibilities for the graphs of two linear equations.

FIGURE 7.1
The three possibilities for two lines when graphed on the same coordinate axes.

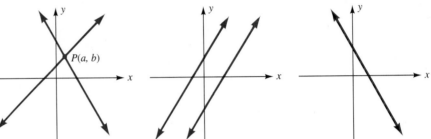

(a) The lines intersect at one point $P(a, b)$.

(b) The lines are parallel and do not intersect.

(c) The lines coincide and intersect at an infinite number of points.

We say that the pair of linear equations

$$A_1 x + B_1 y = C_1$$
$$A_2 x + B_2 y = C_2$$

represent a **system** of two linear equations in two unknowns x and y. The solution set of such a system is always one of three possibilities, as illustrated graphically in Figure 7.1:

1. The lines intersect at one point $P(a, b)$, as shown in Figure 7.1(a). In this case, the system has *exactly one solution*, namely, the ordered pair (a, b). Such a system is said to be **consistent**.

2. The lines are parallel and do not intersect, as shown in Figure 7.1(b). In this case, the system has *no solution*. Such a system is said to be **inconsistent**.

3. The lines coincide and intersect at an infinite number of points, as shown in Figure 7.1(c). In this case, the system has *infinitely many solutions*. Such a system is said to be **dependent**.

Although we can find the solution set of a system of two linear equations in two unknowns by using either the addition method or substitution (see Section 4.5), we are now interested in applying a systematic approach that can be extended to a system of linear equations in more than two unknowns. We refer to this new method as **Gaussian elimination** (named after the German mathematician Carl Friedrich Gauss). Gaussian elimination relies on generating **equivalent systems**, systems that have the same solution set. We can generate equivalent systems from a given system by performing any of the following rules:

Rules for Generating Equivalent Systems

1. Interchange the position of two equations.

2. Multiply both sides of one equation by a nonzero constant.

3. Add a multiple of one equation to another.

We begin by applying these rules to systems of two linear equations in two unknowns.

◆ **Systems of Two Linear Equations in Two Unknowns**

To solve the system

$$A_1 x + B_1 y = C_1 \quad \longleftarrow \boxed{\text{First equation}}$$
$$A_2 x + B_2 y = C_2 \quad \longleftarrow \boxed{\text{Second equation}}$$

by Gaussian elimination, we generate a chain of equivalent systems until we obtain a system in which *the coefficient of x in the second equation is 0*. The system is then said to be in **echelon form**. The solution set of a system in echelon form is easily found by using *back substitution*, a procedure illustrated in the next example.

EXAMPLE 1   Solve the system of linear equations by using Gaussian elimination.

(a) $2x - 5y = -6$ (b) $\frac{1}{2}x + y = -5$
     $-2x + 3y = 4$ $-3x + 5y = 8$

SOLUTION

(a) To generate an equivalent system in which the coefficient of x in the second equation is 0, we simply add the first equation to the second. We now have an equivalent system in echelon form:

$$\boxed{\text{Echelon form}} \left[\begin{array}{l} 2x - 5y = -6 \\ 0x - 2y = -2 \end{array} \right. \longleftarrow \begin{array}{l} \text{To obtain this equation,} \\ \text{add} \quad 2x - 5y = -6 \\ \text{to} \quad -2x + 3y = 4. \end{array}$$

Note that the second equation states that $-2y = -2$. Hence, $y = 1$. Now, using back substitution, we replace y with 1 in the first equation, and solve for x as follows:

$$2x - 5(1) = -6$$

$$x = -\tfrac{1}{2}$$

Since the solution of the system in echelon form is the ordered pair $(-\tfrac{1}{2}, 1)$, the solution of the original system is also $(-\tfrac{1}{2}, 1)$.

(b) Eliminating fractions, we multiply the first equation by 2 to obtain an equivalent system:

$$x + 2y = -10$$

$$-3x + 5y = 8$$

To generate an equivalent system in which the coefficient of x in the second equation is 0, we multiply the first equation by 3 and add the result to the second equation. We now have an equivalent system in echelon form.

$$\text{Echelon form} \quad \begin{cases} x + 2y = -10 \\ 0x + 11y = -22 \end{cases} \leftarrow \boxed{\begin{array}{l} \text{To obtain this equation,} \\ \text{add} \quad 3x + 6y = -30 \\ \text{to} \quad -3x + 5y = 8. \end{array}}$$

Note that the second equation states that $11y = -22$. Hence $y = -2$. Now using back substitution, we replace y with -2 in the first equation:

$$x + 2(-2) = -10$$

$$x = -6$$

Since the solution of the system in echelon form is the ordered pair $(-6, -2)$, the solution of the original system is also $(-6, -2)$. ◆

Note: The echelon form of the system of linear equations that we generate by Gaussian elimination is *not* unique. It depends on the sequence of operations that we employ. Referring to Example 1(a), we could interchange the two equations to obtain

$$-2x + 3y = 4$$

$$2x - 5y = -6,$$

and then replace the second equation by the sum of the first equation and the second to obtain

$$\text{Echelon form} \quad \begin{cases} -2x + 3y = 4 \\ 0x - 2y = -2. \end{cases}$$

Although this echelon form is different from the one we obtained in Example 1(a), its solution is still the ordered pair $(-\tfrac{1}{2}, 1)$.

If the ordered pair (a, b) is a solution of a linear system in two unknowns x and y, then *both* original equations in the system must be satisfied when x is replaced by a and y by b.

P R O B L E M   1   Check the system of equations in Examples 1(a) and 1(b).   ◆

When we generate equivalent systems in two unknowns x and y, the coefficients of x and y in one of the equations may both become 0. If

$$0x + 0y = 0,$$

then the system has infinitely many solutions and is dependent. If

$$0x + 0y = k, \qquad k \neq 0$$

then the system has no solution and is inconsistent.

E X A M P L E   2   Solve the system of linear equations.

(a) $\quad\begin{aligned} x - 3y &= 4 \\ -2x + 6y &= -3 \end{aligned}$ (b) $\quad\begin{aligned} 2x - 8y &= -4 \\ -5x + 20y &= 10 \end{aligned}$

S O L U T I O N

(a) To generate an equivalent system in which the coefficient of x in the second equation is 0, we multiply the first equation by 2 and add the result to the second equation. We now have an equivalent system in echelon form.

$$\boxed{\text{Echelon form}} \longrightarrow \left[\begin{aligned} x - 3y &= 4 \\ 0x + 0y &= 5 \longleftarrow \end{aligned} \right.$$

$$\boxed{\begin{aligned} &\text{To obtain this equation,} \\ &\text{add} \quad 2x - 6y = 8 \\ &\text{to} \quad\ -2x + 6y = -3. \end{aligned}}$$

Note that for all real numbers x and y, we have $0x + 0y = 0 \neq 5$. Thus, this system has no solution. Hence, we conclude that the original system is inconsistent.

(b) Multiplying the first equation by $\frac{1}{2}$, we obtain the equivalent system

$$x - 4y = -2$$
$$-5x + 20y = 10.$$

To generate an equivalent system in which the coefficient of x in the second equation is 0, we multiply the first equation by 5 and add the result to the second equation. We now have an equivalent system in echelon form.

Echelon form ⎡
$$x - 4y = -2$$
$$0x + 0y = 0$$

To obtain this equation,
add $5x - 20y = -10$
to $-5x + 20y = 10.$

Note that $0x + 0y = 0$ is true for all real numbers x and y. If we let $y = a$, where a is an arbitrary real number, then by back substitution we have

$$x - 4(a) = -2$$
$$x = 4a - 2.$$

Thus, this system has an infinite number of solutions of the form

$$(4a - 2, a), \qquad \text{where } a \text{ is any real number.}$$

Hence, we conclude that the original system is dependent.   ◆

PROBLEM 2   Referring to Example 2(b), find three of the infinitely many solutions by letting $a = -2$, $a = 0$, and $a = 5$.   ◆

◆ **Systems of n Linear Equations in n Unknowns**

A system of n linear equations in n unknowns $x_1, x_2, x_3, \ldots, x_n$ is called an $n \times n$ (read "n by n") linear system:

$$
\begin{aligned}
a_{11}x_1 + a_{12}x_2 + a_{13}x_3 + \cdots + a_{1n}x_n &= k_1 \quad \longleftarrow \boxed{\text{First equation}} \\
a_{21}x_1 + a_{22}x_2 + a_{23}x_3 + \cdots + a_{2n}x_n &= k_2 \quad \longleftarrow \boxed{\text{Second equation}} \\
a_{31}x_1 + a_{32}x_2 + a_{33}x_3 + \cdots + a_{3n}x_n &= k_3 \quad \longleftarrow \boxed{\text{Third equation}} \\
\vdots \qquad \vdots \qquad \vdots \qquad\quad \vdots \quad \vdots & \\
a_{n1}x_1 + a_{n2}x_2 + a_{n3}x_3 + \cdots + a_{nn}x_n &= k_n \quad \longleftarrow \boxed{\text{nth equation}}
\end{aligned}
$$

We use a *double subscript system* for the coefficients of each variable, where each of the two numbers in the subscript has a definite meaning. The first number corresponds to the equation number, and the second number corresponds to the unknown. Thus, a_{32} represents the coefficient of the second unknown (x_2) in the third equation, whereas a_{23} represents the coefficient of the third unknown (x_3) in the second equation.

To solve a system of n linear equations in n unknowns, we use Gaussian elimination to form an equivalent system in which the coefficients $a_{ij} = 0$ when $i > j$. That is, we generate an equivalent system in which the coefficients of x_1 in all equations after the first are 0; the coefficients of x_2 in all equations after the second are 0; the coefficients of x_3 in all equations after the third are 0, and so on. The system is then in *echelon form* and we can easily find its solution set by using back substitution.

As with a system of two linear equations in two unknowns, the solution set of a system of n linear equations in n unknowns is always one of three possibilities:

1.   exactly one solution, consistent system.

2.   no solution, inconsistent system.

3.   infinitely many solutions, dependent system.

When we use Gaussian elimination, it is usually advantageous to begin by generating an equivalent system in which the coefficient of the first unknown in the first equation is 1.

E X A M P L E   3   Solve the system of linear equations by using Gaussian elimination.

$$2x - y + 4z = -3$$
$$x + 2y - 3z = 1$$
$$3x + y - 3z = 6$$

S O L U T I O N   To obtain a coefficient of 1 for the first unknown in the first equation, we could multiply both sides of the first equation by $\frac{1}{2}$. However, this would require us to work with fractions in order to generate other equivalent systems. To avoid unnecessary fractional computations, we simply interchange the first and second equations to obtain the following equivalent system:

The leading coefficient of the first equation is now 1.	→	$x + 2y - 3z = 1$	Interchange the first and second equations.
		$2x - y + 4z = -3$	
		$3x + y - 3z = 6$	

Next, we generate another equivalent system in which the coefficients of x in the second and third equations are 0:

$$x + 2y - 3z = 1$$
$$0x - 5y + 10z = -5 \quad \leftarrow \text{To obtain this equation, add } -2 \text{ times the first equation to the second.}$$
$$0x - 5y + 6z = 3$$

To obtain this equation, add -3 times the first equation to the third.

Finally, from this system, we generate another equivalent system in which the coefficient of y in the third equation is 0. We now have an equivalent system in echelon form:

Echelon form	$x + 2y - 3z = 1$
	$0x - 5y + 10z = -5$
	$0x + 0y - 4z = 8$

To obtain this equation, add -1 times the second equation to the third.

The third equation now states that $-4z = 8$. Hence $z = -2$. Back-substituting -2 for z in the second equation yields $y = -3$. Finally, substituting -2 for z

and -3 for y in the first equation yields $x = 1$. Since the solution of the system in echelon form is $x = 1$, $y = -3$, and $z = -2$, the solution to the original system is also $x = 1$, $y = -3$, and $z = -2$. We record the solution as the *ordered triple* $(1, -3, -2)$. ◆

If $(1, -3, -2)$ is a solution of the system of linear equations in Example 3, then each of the original equations must be satisfied when $x = 1$, $y = -3$, and $z = -2$.

PROBLEM 3   Check the solution of the system in Example 3. ◆

EXAMPLE 4   Solve the system of linear equations by using Gaussian elimination.

$$\begin{aligned}
x + 2y \quad\;\; + t &= 3 \\
y + 2z \quad &= 3 \\
x \quad\;\; - 2z \quad &= 0 \\
3y - 4z + t &= 2
\end{aligned}$$

SOLUTION   Since the coefficients of x in the second and fourth equations are already 0, we begin by generating an equivalent system in which the coefficient of x in the third equation is 0:

$$\begin{aligned}
x + 2y \quad\;\; + t &= 3 \\
0x + \;\; y + 2z \quad &= 3 \\
0x - 2y - 2z - t &= -3 \\
0x + 3y - 4z + t &= 2
\end{aligned}$$

> To obtain this equation, add -1 times the first equation to the third.

Next, from this system, we generate another equivalent system in which the coefficients of y in the third and fourth equations are 0:

$$\begin{aligned}
x + 2y \quad\;\; + t &= 3 \\
0x + \;\; y + \;\; 2z \quad &= 3 \\
0x + 0y + \;\; 2z - t &= 3 \\
0x + 0y - 10z + t &= -7
\end{aligned}$$

> To obtain this equation, add 2 times the second equation to the third.

> To obtain this equation, add -3 times the second equation to the fourth.

Finally, we generate an equivalent system in which the coefficient of z in the fourth equation is 0. We now have a system in echelon form.

Echelon form
$$\begin{aligned}
x + 2y \quad\;\; + \;\; t &= 3 \\
0x + \;\; y + 2z \quad &= 3 \\
0x + 0y + 2z - \;\; t &= 3 \\
0x + 0y + 0z - 4t &= 8
\end{aligned}$$

> To obtain this equation, add 5 times the third equation to the fourth.

The fourth equation now states that $-4t = 8$. Hence $t = -2$. Using back substitution, we find that $z = \frac{1}{2}$, $y = 2$, and $x = 1$. Thus, the solution to the original system is the *ordered quadruple* $(1, 2, \frac{1}{2}, -2)$. You can check this solution. ◆

When we generate equivalent systems in n unknowns $x_1, x_2, x_3, \ldots, x_n$, the coefficients of $x_1, x_2, x_3, \ldots, x_n$ in one of the equations may all become 0. If

$$0x_1 + 0x_2 + 0x_3 + \cdots + 0x_n = 0,$$

then the system has infinitely many solutions and is dependent. However, if

$$0x_1 + 0x_2 + 0x_3 + \cdots + 0x_n = k, \qquad k \neq 0$$

then the system has no solution and is inconsistent.

PROBLEM 4 Solve the system of linear equations by using Gaussian elimination.

$$
\begin{aligned}
3x + 2y - z &= 4 \\
-3x - y \phantom{{}- z} &= -2 \\
2y - 2z &= 3
\end{aligned}
$$

◆

◆ **Nonsquare Systems**

Thus far we have worked with linear systems in which the number of equations is the same as the number of unknowns. Such a system is called a **square system**. If the number of equations is less than the number of unknowns, the system is a **nonsquare system**. A nonsquare system cannot have a unique solution.

EXAMPLE 5 Solve the system of linear equations using Gaussian elimination.

$$
\begin{aligned}
3x + 3y - 2z &= 3 \\
2x + y - z &= 2
\end{aligned}
$$

SOLUTION Remember, when using Gaussian elimination, it is advantageous to have a system in which the coefficient of the first unknown in the first equation is 1. Keeping this in mind, we add -1 times the second equation to the first equation to obtain the following equivalent system:

$$
\begin{aligned}
x + 2y - z &= 1 \\
2x + y - z &= 2
\end{aligned}
$$

> To obtain this equation, add $-2x - y + z = -2$ to $3x + 3y - 2z = 3$.

From this system, we generate another equivalent system in which the coefficient of x in the second equation is 0.

$$
\begin{aligned}
x + 2y - z &= 1 \\
0x - 3y + z &= 0
\end{aligned}
$$

> To obtain this equation, add -2 times the first equation to the second.

The second equation states that $-3y + z = 0$, or $z = 3y$. Replacing z with $3y$ in the first equation, we solve for x as follows:

$$x + 2y - (3y) = 1$$
$$x = y + 1$$

If we let $y = a$, where a is an arbitrary real number, then we have

$$x = a + 1, \qquad y = a, \qquad z = 3a.$$

Thus, this system has an infinite number of solutions of the form

$$(a + 1, a, 3a), \qquad \text{where } a \text{ is any real number.} \qquad \blacklozenge$$

We can always describe an infinite solution set in more than one way. Referring to Example 5, if we let $x = b$, where b is an arbitrary real number, then

$$x = b, \qquad y = b - 1, \qquad z = 3b - 3.$$

Hence, we can also state that every ordered triple of the form

$$(b, b - 1, 3b - 3), \qquad \text{where } b \text{ is a real number,}$$

is a solution of the system in Example 5.

PROBLEM 5 Describe the infinite solution set for the system in Example 5 by letting $z = c$, where c is an arbitrary real number. $\blacklozenge$

$\blacklozenge$ **Application: Tickets to a Rock Concert**

In Chapter 2 we solved many word problems containing two or more unknowns by assigning to one of the unknowns the variable x and recording the other unknowns in terms of x. We used this method to develop one equation with one unknown, and then solved for x. However, many word problems containing two or more unknowns are more naturally set up and solved by using a system of linear equations. The next example illustrates the procedure.

EXAMPLE 6 The Centrum civic center in Worcester, Massachusetts, sold 12,000 tickets to a recent rock concert. The ticket prices were \$12, \$18, and \$24, and the total income from ticket sales was \$201,000. How many tickets of each type did the Centrum sell if the number of \$12 tickets sold was twice the number of \$24 tickets sold?

SOLUTION Let

$$x = \text{number of \$12 tickets sold,}$$
$$y = \text{number of \$18 tickets sold,}$$

and

$$z = \text{number of } \$24 \text{ tickets sold.}$$

Since 12,000 tickets were sold, we have

$$x + y + z = 12{,}000.$$

Since the ticket prices were \$12, \$18, and \$24, and the total income from ticket sales was \$201,000, we have

$$12x + 18y + 24z = 201{,}000.$$

Lastly, since the Centrum sold twice as many \$12 tickets as \$24 tickets, we have

$$x = 2z, \qquad \text{or} \qquad x - 2z = 0.$$

Hence, the system of linear equations that describes this problem is

$$
\begin{aligned}
x + \;\; y + \;\; z &= 12{,}000 \\
12x + 18y + 24z &= 201{,}000 \\
x \qquad\quad - \;\; 2z &= 0
\end{aligned}
$$

The solution set of this linear system of equations may be found by using Gaussian elimination. In Problem 6, you are asked to verify that the solution is

$$x = 5000, \qquad y = 4500, \quad \text{and} \quad z = 2500.$$

In summary, the Centrum sold 5000 tickets at \$12, 4500 tickets at \$18, and 2500 tickets at \$24.   ◆

PROBLEM 6   Using Gaussian elimination, verify that the solution of the system of linear equations in Example 6 is $x = 5000$, $y = 4500$, and $z = 2500$.   ◆

Exercises 7.1

   Basic Skills

In Exercises 1–8, solve each linear system by using back substitution.

1. $3x - 4y = 5$
$2y = 1$

2. $5x + 2y = 1$
$-3y = 6$

3. $4x - \;\; y + 2z = 3$
$3y - \;\; z = 7$
$2z = 4$

4. $-x - 2y + 3z = 3$
$5y - 2z = -7$
$\tfrac{1}{2}z = 3$

5. $2x - \;\; y + \;\; z - \;\; t = -1$
$-3y \qquad\;\; - 2t = 0$
$2z - \;\; t = 1$
$-\tfrac{2}{3}t = -6$

6. $4x - 3y - 2z = -1$
$2y \quad\;\; + \;\; t = 5$
$z - \;\; t = 0$
$4t = 12$

7. $\begin{aligned} 2x_1 - 3x_2 \quad\quad\quad\quad + x_5 &= 0 \\ 2x_2 \quad - \ x_4 \quad\quad\ &= 0 \\ x_3 - 2x_4 - \ x_5 &= 0 \\ x_4 + 3x_5 &= 0 \\ 2x_5 &= 1 \end{aligned}$

8. $\begin{aligned} 3x_1 \quad\quad - 3x_3 \quad\quad\quad\quad &= -2 \\ 2x_2 + \ x_3 \quad\quad - x_5 &= 1 \\ x_3 + 2x_4 \quad\quad &= -2 \\ -2x_4 + x_5 &= 5 \\ x_5 &= 0 \end{aligned}$

In Exercises 9–36, solve each system of linear equations by using Gaussian elimination.

9. $\begin{aligned} x + 3y &= 7 \\ -x + 5y &= 1 \end{aligned}$

10. $\begin{aligned} x - 5y &= -2 \\ 3x + 4y &= 13 \end{aligned}$

11. $\begin{aligned} 2x - \ y &= 4 \\ x - 3y &= -3 \end{aligned}$

12. $\begin{aligned} 3x + 5y &= 4 \\ x - \ y &= -6 \end{aligned}$

13. $\begin{aligned} 6x - \ y &= 4 \\ 7x - 3y &= -10 \end{aligned}$

14. $\begin{aligned} 3x + 5y &= -5 \\ -2x - \ y &= 8 \end{aligned}$

15. $\begin{aligned} 3x - \ 9y &= 8 \\ -4x + 12y &= -5 \end{aligned}$

16. $\begin{aligned} \tfrac{1}{2}x - \ y &= -1 \\ -3x + 6y &= 6 \end{aligned}$

17. $\begin{aligned} \tfrac{1}{4}x + 5y &= -13 \\ -\tfrac{2}{3}x + 2y &= 0 \end{aligned}$

18. $\begin{aligned} \tfrac{3}{5}x - \tfrac{3}{10}y &= -9 \\ \tfrac{1}{3}x + \tfrac{1}{6}y &= 3 \end{aligned}$

19. $\begin{aligned} x \quad\quad + z &= 11 \\ y + z &= 5 \\ 4x + 2y \quad\quad &= 48 \end{aligned}$

20. $\begin{aligned} x + y + z &= 19 \\ 2x - y \quad\quad &= 3 \\ y - z &= 10 \end{aligned}$

21. $\begin{aligned} x + 3y - \ z &= 5 \\ 2x + 3y \quad\quad &= 4 \\ -x - \ y + 3z &= -1 \end{aligned}$

22. $\begin{aligned} x - 2y - 5z &= -1 \\ -3x + \ y - 2z &= 0 \\ 2x \quad\quad + 3z &= 1 \end{aligned}$

23. $\begin{aligned} 4x - 3y - \ z &= 0 \\ x - 3y + 2z &= 7 \\ 3x + 9y - \ z &= -2 \end{aligned}$

24. $\begin{aligned} 5x - 2y + \ z &= -2 \\ 3x - \ y - 3z &= -2 \\ x + 2y - \ z &= 4 \end{aligned}$

25. $\begin{aligned} 3x - \ 2y + \ z &= 3 \\ 4x + \ y \quad\quad &= 1 \\ 11y - 4z &= -9 \end{aligned}$

26. $\begin{aligned} 2x + 3y \quad\quad &= 2 \\ 8x \quad\quad - 4z &= 3 \\ 3y + \ z &= -1 \end{aligned}$

27. $\begin{aligned} 2x + 3y + \ z &= 7 \\ 5x - \ y - \ z &= 2 \\ 4x - \ y + 3z &= 9 \end{aligned}$

28. $\begin{aligned} 4x - \ y + 3z &= 3 \\ 3x - 4y + 2z &= 4 \\ 2x + 3y - \ z &= 1 \end{aligned}$

29. $\begin{aligned} 6x - 4y - 5z &= 4 \\ 3x - 2y + \ z &= 9 \\ \tfrac{3}{2}x - \tfrac{2}{3}y + 2z &= \tfrac{41}{6} \end{aligned}$

30. $\begin{aligned} 0.2x + 0.3y + 2z &= 3 \\ 3x - \ 2y + 4z &= 6 \\ 0.5x - \ y + 3z &= -0.5 \end{aligned}$

31. $\begin{aligned} x + y + z \quad\quad &= 1 \\ x \quad\quad + z + t &= 2 \\ x + y \quad\quad + t &= 6 \\ y + z + t &= 3 \end{aligned}$

32. $\begin{aligned} 3x + 2y + \ z \quad\quad &= 1 \\ x \quad\quad\quad + 3t &= -4 \\ y - 2z - \ t &= -2 \\ 2x + \ y - \ z \quad\quad &= -3 \end{aligned}$

33. $\begin{aligned} x_1 + \ x_2 + \ x_3 + \ x_4 &= 0 \\ 2x_1 - \ x_2 - \ x_3 - \ x_4 &= -3 \\ x_1 - 2x_2 - 3x_3 + \ x_4 &= -4 \\ 3x_1 - 2x_2 - \ x_3 + 2x_4 &= 3 \end{aligned}$

34. $\begin{aligned} 3x_1 + 2x_2 - 4x_3 + \ x_4 &= 2 \\ 2x_1 - \ x_2 - \ x_3 + 2x_4 &= 0 \\ -x_1 + 4x_2 - 2x_3 - 3x_4 &= -2 \\ 4x_1 - \ x_2 + 3x_3 - 2x_4 &= 4 \end{aligned}$

35. $\begin{aligned} x_1 + x_2 \quad\quad\quad\quad &= 2 \\ x_2 - x_3 + x_4 \quad\quad &= 5 \\ x_1 \quad\quad + x_3 \quad\quad + x_5 &= -1 \\ x_2 \quad\quad + x_4 \quad\quad &= 0 \\ x_4 + x_5 &= 2 \end{aligned}$

36. $\begin{aligned} x_1 + x_2 + x_3 + x_4 \quad\quad &= 12 \\ x_1 + x_2 + x_3 \quad\quad + x_5 &= 14 \\ x_1 + x_2 \quad\quad + x_4 + x_5 &= 16 \\ x_1 \quad\quad + x_3 + x_4 + x_5 &= 18 \\ x_2 + x_3 + x_4 + x_5 &= 20 \end{aligned}$

In Exercises 37–44, solve each nonsquare system of linear equations by using Gaussian elimination.

37. $\begin{aligned} 2x + 2y + z &= 4 \\ 4y - z &= 0 \end{aligned}$

38. $\begin{aligned} 3x - y + 4z &= 8 \\ y + 2z &= 1 \end{aligned}$

39. $\begin{aligned} x + 2y - z &= 1 \\ 3x + 2y - z &= 5 \end{aligned}$

40. $\begin{aligned} x - 3y + 2z &= -2 \\ 2x - 4y + 5z &= 1 \end{aligned}$

41. $\begin{aligned} 4x - 2y - 2z &= 2 \\ 3x + \ y - 2z &= -3 \end{aligned}$

42. $\begin{aligned} 2x + \ y \quad\quad &= 4 \\ 6x + 3y - z &= 3 \end{aligned}$

43. $\begin{aligned} 3x - 2y \quad\quad + 4t &= 4 \\ y - 2z + \ t &= 1 \\ x - 2y \quad\quad + 8t &= 0 \end{aligned}$

44. $\begin{aligned} 2x - 3y + \ z \quad\quad &= 1 \\ 3x - \ y \quad\quad - t &= 0 \\ y + 3z - 2t &= 3 \end{aligned}$

In Exercises 45–58, solve each problem by using a system of linear equations.

45. The perimeter of an isosceles triangle is 60 cm. The non-equal side is 15 cm less than one of the equal sides. What are the dimensions of the triangle?

46. It takes a ferryboat 4 hours to travel 20 miles upstream and 3 hours to make the return trip downstream. What is the speed of the boat and the speed of the river's current?

47. A car rental agency charges *a* dollars to rent a car plus *b* dollars per mile. Find *a* and *b* if the rental charge is $45 for 100 miles or $55 for 150 miles.

48. An investor buys two types of stocks; one costs $40 per share and the other costs $24 per share. Altogether she purchased 55 shares of stock for a total cost of $1720. How many shares of each stock did she purchase?

49. A total of $10,000 is invested, part of the money at 10% and the rest at 12%. If the total annual interest is $1090, how much money is invested at each rate?

50. A car radiator contains 15 quarts of a 20% antifreeze solution. How much 20% antifreeze solution must be drained and replaced with *pure* antifreeze to achieve a 60% antifreeze solution?

51. The sum of the three interior angles of any triangle is 180°. Suppose the largest angle is six times as large as the smallest angle and is equal to twice the sum of the two smaller angles. Find the three angles.

52. A contractor buys three parcels of land for $200,000. Parcel *A* costs three times as much as parcel *B* and $20,000 less than parcels *B* and *C* together. Find the cost of each parcel of land.

53. A woman invests $50,000 at the simple interest rates of 8%, 9%, and $9\frac{1}{2}$% per year. Altogether these investments earn $4550 per year. How much does she have invested at each rate if she has the same amount invested at 9% and $9\frac{1}{2}$%?

54. A merchant mixes three types of tea costing $2.40, $3.00, and $3.20 per pound in order to obtain 50 pounds of a blend that costs $2.92 per pound. How many pounds of each type does he use if he uses 10 pounds more of the $3.20 type than the $2.40 type?

55. The greens fee at a golf course is $15 for a child, $20 for an adult, and $18 for a senior citizen. On a certain day, 120 people played golf and the receipts totaled $2220. If twice as many seniors played as children, find the number of adults who played golf that day.

56. Three circular pulleys are tangent to each other, as shown in the sketch. If the center-to-center distances *AB*, *AC*, and *BC* are 18 cm, 22 cm, and 16 cm, respectively, find the radius of each pulley.

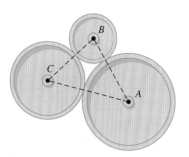

EXERCISE 56

57. A photography studio develops four sizes of prints (in inches); 2 × 3, 3 × 5, 5 × 8, and 8 × 10. After taking senior-class pictures, the studio offers the following four packages to the students:

Package	Number of prints			
	2 × 3	3 × 5	5 × 8	8 × 10
A	4	2	1	—
B	—	2	2	1
C	8	—	4	2
D	12	4	2	2

If package A sells for $12, package B for $24, package C for $46, and package D for $47, determine the cost per print for each size print.

58. The accompanying table shows the weight composition of four alloys. How many pounds of each alloy must be melted together to form 100 pounds of an alloy that is 20% copper, 46% gold, 21% tin and 13% zinc? [*Hint:* This system is dependent and has more than one set of solutions.]

Alloy	Copper	Gold	Tin	Zinc
A	20%	50%	20%	10%
B	—	60%	40%	—
C	40%	40%	—	20%
D	—	20%	50%	30%

Critical Thinking

59. Given the following system of linear equations,

$$x + 3y - 2z = B$$

$$-2x + y + Az = 8$$

$$3x - 5y - z = 12$$

find values of A and B for which the system is (a) inconsistent and (b) dependent.

60. Two complex numbers $z_1 = a_1 + b_1 i$ and $z_2 = a_2 + b_2 i$ are equal if and only if their real parts are equal $(a_1 = a_2)$ and their imaginary parts are equal $(b_1 = b_2)$. For each of the given equations, use this fact to help set up a system of equations and then solve for x and y.

(a) $(2 - 4i)x + (3 + i)y = -3 + 8i$

(b) $(-4 + i)(x + yi) + (6 - 3i)(2x - yi) = 4 - 40i$

61. Use a system of linear equations to find the values of A, B, and C such that the equation

$$A(x - 1)(x + 2) + B(x - 4)(x + 2)$$
$$+ C(x - 4)(x - 1) = x^2 + 8x$$

is an identity.

62. The graph of a circle, $x^2 + y^2 + Cx + Dy + E = 0$, passes through the points $(0, 3)$, $(-2, 0)$, and $(4, -4)$. Use a system of linear equations to determine the values for C, D, and E.

63. The graph of a quadratic function, $f(x) = ax^2 + bx + c$, is sketched in the figure. Use a system of linear equations to determine the values for a, b, and c.

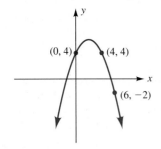

64. The graph of a cubic function, $f(x) = ax^3 + bx^2 + cx + d$, is shown in the sketch. Use a system of linear equations to determine the values for a, b, c, and d.

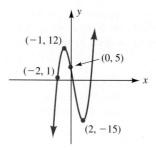

65. A group of 20 skiers spent \$260 to ski at Deer Mountain and \$180 to ski at Mt. Cabott. The cost of lift tickets at these ski areas are shown in the following table:

	Deer Mountain	Mt. Cabott
Adult	\$18	\$24
Student	\$15	\$15
Child	\$10	Free

How many adults, students, and children are in this group of 20 skiers?

66. The fueling facility at an airport uses three pumps to fill each jet with 1500 gallons of fuel before takeoff. When all three pumps are used, it takes 1 hour 15 minutes to load each plane. However, on a certain day, the first pump malfunctions and works at only half its normal rate. It now takes 1 hour 40 minutes to load each plane. When the malfunctioning pump is closed completely, it takes 2 hours 30 minutes to load each plane. What is the normal pumping rate (in gallons per hour) for the malfunctioning pump?

Calculator Activities

In Exercises 67–72, use a calculator to help solve each system of linear equations. Round each answer to three significant digits.

67. $63x - 92y = 1728$
$44x - 87y = 327$

68. $1.27x - 2.32y = 17.25$
$0.82x + 4.31y = -8.27$

69. $8.9x \qquad - 2.3z = 18.7$
$\qquad 6.2y + 4.4z = -32.6$
$5.7x - 3.2y \qquad = 14.4$

70. $0.62x - 1.42y \qquad = 12.2$
$2.85x - 0.92y + 1.84z = 18.6$
$\qquad 0.24y - 0.32z = 15.3$

71. $p + \quad q + \quad r = 5$
$72.1p - 23.2q + 43.9r = 189$
$\qquad 48.0q - 55.8r = 432$

72. $a \qquad - 14c \qquad = 220$
$\qquad 2b - 18c \qquad = -24$
$24a - 17b \qquad + 5d = 5540$
$\qquad 56c - \quad d = 125$

73. Two cables support an 800-pound weight, as shown in the figure. The tensions T_1 and T_2 (in pounds) in each cable may be found by solving the system of linear equations

$$0.4226T_1 + 0.7431T_2 = 800$$

$$0.9063T_1 - 0.6691T_2 = 0$$

Find T_1 and T_2, rounding each answer to three significant digits.

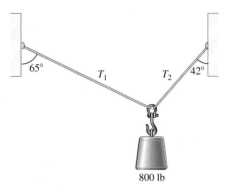

EXERCISE 73

74. For the electrical circuit shown in the figure, the branch currents i_1, i_2, and i_3 (in amperes) may be found by solving the system of linear equations

$$i_1 - \quad i_2 - \quad i_3 = 0$$
$$68i_1 \qquad + 48i_3 = 12.5$$
$$32i_2 - 48i_3 = -3.5$$

Determine the currents i_1, i_2, and i_3, rounding each answer to three significant digits.

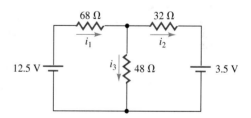

7.2 Using Matrices to Solve Systems of Linear Equations

In the previous section, we used Gaussian elimination to solve a system of linear equations. In this section, we modify this procedure slightly by introducing the idea of a *matrix*. The advantage of using matrices is that we do not need to keep writing the variables that appear in the linear system of equations. We begin by defining a matrix.

◆ **Defining a Matrix**

A rectangular array of numbers enclosed by a pair of brackets is called a **matrix** (plural, matrices). Each particular number in the matrix is referred to as an

element of the matrix. In this matrix

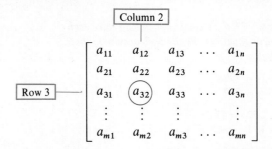

double subscripts are used for each element. Each of the two numbers in the subscript has a definite meaning: the first number corresponds to the **row** and the second number to the **column** in which the element stands. Thus, a_{32} represents the element in the *third row* and *second column*. We say that a matrix with m rows and n columns is of **dimension** $m \times n$ (read "m by n").

EXAMPLE 1   Determine the dimension of the given matrix. Then identify element a_{21}.

(a) $\begin{bmatrix} 2 & 0 & \pi \\ \sqrt{2} & -1 & 3 \end{bmatrix}$ (b) $\begin{bmatrix} 0 & \sqrt{3} \\ 8 & 4 \\ -2 & \pi \end{bmatrix}$

SOLUTION

(a)   The matrix

$$\begin{bmatrix} 2 & 0 & \pi \\ \sqrt{2} & -1 & 3 \end{bmatrix}$$

has two rows and three columns. Thus, it is of dimension 2×3. The element a_{21} corresponds to the number in the second row, first column. Hence, $a_{21} = \sqrt{2}$.

(b)   The matrix

$$\begin{bmatrix} 0 & \sqrt{3} \\ 8 & 4 \\ -2 & \pi \end{bmatrix}$$

has three rows and two columns. Thus, it is of dimension 3×2. The element a_{21} corresponds to the number in the second row, first column. Hence, $a_{21} = 8$.   ◆

PROBLEM 1   Referring to each of the matrices in Example 1(a) and 1(b), find the element that corresponds to a_{12}.   ◆

◆ **Augmented Matrix**

Now, consider the following system of m linear equations in n unknowns, x_1, $x_2, x_3, \ldots, x_n$.

$$
\begin{aligned}
a_{11}x_1 + a_{12}x_2 + a_{13}x_3 + \cdots + a_{1n}x_n &= k_1 \quad \longleftarrow \boxed{\text{First equation}} \\
a_{21}x_1 + a_{22}x_2 + a_{23}x_3 + \cdots + a_{2n}x_n &= k_2 \quad \longleftarrow \boxed{\text{Second equation}} \\
a_{31}x_1 + a_{32}x_2 + a_{33}x_3 + \cdots + a_{3n}x_n &= k_3 \quad \longleftarrow \boxed{\text{Third equation}} \\
\vdots \qquad \vdots \qquad \vdots \qquad\qquad \vdots \qquad \vdots & \\
a_{m1}x_1 + a_{m2}x_2 + a_{m3}x_3 + \cdots + a_{mn}x_n &= k_m \quad \longleftarrow \boxed{m\text{th equation}}
\end{aligned}
$$

For every such $m \times n$ linear system, there corresponds an enlarged or **augmented matrix** of dimension $m \times (n + 1)$ that is written as follows:

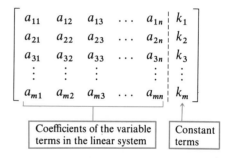

The vertical dashed line in the augmented matrix separates the coefficients of the variable terms in the linear system from the constant terms in the system.

An augmented matrix is in *echelon form* if the elements $a_{ij} = 0$ when $i > j$; that is, the element in the first column in every row after the first is 0; the element in the second column in every row after the second is 0; the element in the third column in every row after the third is 0; and so on.

EXAMPLE 2   Write the augmented matrix corresponding to the system of linear equations given in part (a). Write the system of linear equations corresponding to the augmented matrix given in part (b).

(a)
$$
\begin{aligned}
x + 3y - 2z &= 1 \\
-2x - y \phantom{{}+ 2z} &= 0 \\
3x \phantom{{}+ 3y} - z &= -2
\end{aligned}
$$

(b)
$$
\begin{bmatrix}
1 & 2 & 5 & \vdots & -5 \\
0 & 2 & -1 & \vdots & 3 \\
0 & 0 & 2 & \vdots & 4
\end{bmatrix}
$$

SOLUTION

(a)   Since missing unknowns must have coefficients of 0, we write the augmented matrix as follows:

$$
\begin{bmatrix}
1 & 3 & -2 & \vdots & 1 \\
-2 & -1 & 0 & \vdots & 0 \\
3 & 0 & -1 & \vdots & -2
\end{bmatrix}
$$

(b) This augmented matrix is in echelon form. Using the variables x, y, and z, we write the system of equations as

$$x + 2y + 5z = -5 \qquad\qquad x + 2y + 5z = -5$$
$$0x + 2y - z = 3 \qquad \text{or} \qquad 2y - z = 3$$
$$0x + 0y + 2z = 4 \qquad\qquad\qquad 2z = 4$$

Note that the system of linear equations is also in echelon form.   ◆

PROBLEM 2   Use back substitution to find the solution of the system of linear equations that correspond to the augmented matrix in Example 2(b).   ◆

◆ **Row Operations on a Matrix**

In the previous section we discussed three rules for generating equivalent systems of equations:

1. Interchange the position of two equations.
2. Multiply both sides of one equation by a nonzero constant.
3. Add a multiple of one equation to another.

For each of these rules, there corresponds an **elementary row operation** of a matrix.

Elementary Row Operations for a Matrix

1. Interchange the position of two rows.
2. Multiply all elements in a row by a nonzero constant.
3. Add a multiple of one row to another.

If one matrix is obtained from another by a sequence of elementary row operations, we say that the two matrices are **row equivalent**. If two matrices are row equivalent, we place the symbol $\sim$ between the two matrices. For example, when interchanging rows, we write

$$\begin{bmatrix} 2 & 5 \\ -3 & 4 \end{bmatrix} \sim \begin{bmatrix} -3 & 4 \\ 2 & 5 \end{bmatrix}$$

Read "is equivalent to"

In this text, we use the following notations when generating row-equivalent matrices:

$R_i \leftrightarrow R_j$ Interchange the ith row and jth row of a matrix.

$cR_i \to R_i$ Replace the ith row with a multiple of c times the ith row.

$cR_i + R_j \to R_j$ Replace the jth row with the sum of c times the ith row and the jth row.

We now state the **matrix method** which can be used to solve a system of linear equations.

Matrix Method for Solving a System of Linear Equations

Step 1   Write the augmented matrix for the system of linear equations.

Step 2   Use row operations to generate a row-equivalent matrix in echelon form.

Step 3   Write the system of linear equations that corresponds to the matrix in step 2.

Step 4   Solve the system of linear equations in step 3 by using back substitution.

EXAMPLE 3   Solve the system of linear equations by using the matrix method.

$$2x - 3y = 7$$
$$x + 4y = 9$$

SOLUTION   Beginning with the augmented matrix, we perform the following elementary row operations.

$$\begin{bmatrix} 2 & -3 & | & 7 \\ 1 & 4 & | & 9 \end{bmatrix} \sim \begin{bmatrix} 1 & 4 & | & 9 \\ 2 & -3 & | & 7 \end{bmatrix} \qquad R_1 \leftrightarrow R_2$$

$$\sim \begin{bmatrix} 1 & 4 & | & 9 \\ 0 & -11 & | & -11 \end{bmatrix} \qquad -2R_1 + R_2 \rightarrow R_2$$

This last matrix is in echelon form and corresponds to the following system of linear equations:

$$x + 4y = 9$$
$$0x - 11y = -11$$

The last equation in this system of equations states that $-11y = -11$. Hence $y = 1$. Using back substitution, we find that $x = 5$. Thus, the solution of the original system is $(5, 1)$. ◆

The echelon form of the matrix that we generate by row operations is not unique. It depends on the sequence of row operations we employ. Referring to Example 3, we could perform the following row operations on the augmented matrix:

$$\begin{bmatrix} 2 & -3 & | & 7 \\ 1 & 4 & | & 9 \end{bmatrix} \sim \begin{bmatrix} 1 & -\frac{3}{2} & | & \frac{7}{2} \\ 1 & 4 & | & 9 \end{bmatrix} \qquad \tfrac{1}{2}R_1 \rightarrow R_1$$

$$\sim \begin{bmatrix} 1 & -\frac{3}{2} & | & \frac{7}{2} \\ 0 & \frac{11}{2} & | & \frac{11}{2} \end{bmatrix} \qquad -R_1 + R_2 \rightarrow R_2$$

Although this echelon form is different from the one we obtained in Example 3, the solution to its corresponding linear system

$$x - \tfrac{3}{2}y = \tfrac{7}{2}$$
$$0x + \tfrac{11}{2}y = \tfrac{11}{2}$$

is still the ordered pair (5, 1).

PROBLEM 3   Solve the system of linear equations by using the matrix method.

$$3x + 2y = 4$$
$$2x - \; y = 5$$ ◆

EXAMPLE 4   Solve the system of linear equations by using the matrix method.

$$2x + 4y - 2z = -1$$
$$3y - 4z = 6$$
$$-2x - 2y + \; z = 0$$

SOLUTION   Beginning with the augmented matrix, we perform the following row operations:

$$\begin{bmatrix} 2 & 4 & -2 & | & -1 \\ 0 & 3 & -4 & | & 6 \\ -2 & -2 & 1 & | & 0 \end{bmatrix} \sim \begin{bmatrix} 2 & 4 & -2 & | & -1 \\ 0 & 3 & -4 & | & 6 \\ 0 & 2 & -1 & | & -1 \end{bmatrix} \quad R_1 + R_3 \rightarrow R_3$$

$$\sim \begin{bmatrix} 2 & 4 & -2 & | & -1 \\ 0 & 1 & -\tfrac{4}{3} & | & 2 \\ 0 & 2 & -1 & | & -1 \end{bmatrix} \quad \tfrac{1}{3}R_2 \rightarrow R_2$$

$$\sim \begin{bmatrix} 2 & 4 & -2 & | & -1 \\ 0 & 1 & -\tfrac{4}{3} & | & 2 \\ 0 & 0 & \tfrac{5}{3} & | & -5 \end{bmatrix} \quad -2R_2 + R_3 \rightarrow R_3$$

This last matrix is in echelon form and corresponds to the following system of linear equations:

$$2x + 4y - 2z = -1$$
$$0x + \; y - \tfrac{4}{3}z = 2$$
$$0x + 0y + \tfrac{5}{3}z = -5$$

The last equation in this system of equations states that $\tfrac{5}{3}z = -5$. Hence $z = -3$. Using back substitution, we find that $y = -2$ and $x = \tfrac{1}{2}$. Thus, the solution of the original system is $(\tfrac{1}{2}, -2, -3)$.   ◆

When we perform row operations on an augmented matrix, all the elements in one of the rows may become 0. If this occurs, the corresponding equation is

$$0x_1 + 0x_2 + 0x_3 + \cdots + 0x_n = 0,$$

and we conclude that the system has infinitely many solutions and, hence, is dependent. However, if all the elements in one of the rows except the last entry become 0, then the corresponding equation is

$$0x_1 + 0x_2 + 0x_3 + \cdots + 0x_n = k, \qquad k \neq 0.$$

We now conclude that the system has no solution and is inconsistent.

PROBLEM 4   Solve the system of linear equations by using the matrix method.

$$3x - 9y + 6z = 1$$
$$2x - y - z = 3$$
$$x - 2y + z = 1 \qquad \blacklozenge$$

We can also solve nonsquare systems by using the matrix method. Remember that a nonsquare system does not have a unique solution.

EXAMPLE 5   Solve the system of linear equations by using the matrix method.

$$3x + y + 6z = 3$$
$$4x + 3y + 3z = -1$$

SOLUTION   Beginning with the augmented matrix, we perform the following row operations.

$$\begin{bmatrix} 3 & 1 & 6 & \vdots & 3 \\ 4 & 3 & 3 & \vdots & -1 \end{bmatrix} \sim \begin{bmatrix} -1 & -2 & 3 & \vdots & 4 \\ 4 & 3 & 3 & \vdots & -1 \end{bmatrix} \qquad -R_2 + R_1 \to R_1$$

$$\sim \begin{bmatrix} -1 & -2 & 3 & \vdots & 4 \\ 0 & -5 & 15 & \vdots & 15 \end{bmatrix} \qquad 4R_1 + R_2 \to R_2$$

$$\sim \begin{bmatrix} -1 & -2 & 3 & \vdots & 4 \\ 0 & 1 & -3 & \vdots & -3 \end{bmatrix} \qquad -\tfrac{1}{5}R_2 \to R_2$$

The corresponding system of equations in echelon form is

$$-x - 2y + 3z = 4$$
$$0x + y - 3z = -3$$

The second equation states that $y - 3z = -3$, or $y = 3z - 3$. Replacing y with

$3z - 3$ in the first equation, we solve for x as follows:

$$-x - 2(3z - 3) + 3z = 4$$
$$-x - 6z + 6 + 3z = 4$$
$$-x = 3z - 2$$
$$x = 2 - 3z$$

If we let $z = a$, where a is an arbitrary real number, then we have

$$x = 2 - 3a, \qquad y = 3a - 3, \qquad z = a.$$

Thus, this system has an infinite number of solutions of the form

$$(2 - 3a, 3a - 3, a), \qquad \text{where } a \text{ is any real number.} \qquad \blacklozenge$$

PROBLEM 5 Referring to Example 5, find three of the infinitely many solutions of the system by letting $a = 0$, $a = 1$, and $a = 2$. $\blacklozenge$

◆ Gauss-Jordan Elimination

If the augmented matrix for a system of n linear equations in n unknowns $(x_1, x_2, x_3, \ldots, x_n)$ can be reduced to the form

$$\begin{bmatrix} 1 & 0 & 0 & \ldots & 0 & c_1 \\ 0 & 1 & 0 & \ldots & 0 & c_2 \\ 0 & 0 & 1 & \ldots & 0 & c_3 \\ \vdots & \vdots & \vdots & & \vdots & \vdots \\ 0 & 0 & 0 & \ldots & 1 & c_n \end{bmatrix}$$

by using row operations, then the equivalent system of equations becomes

$$
\begin{aligned}
x_1 + 0x_2 + 0x_3 + \cdots + 0x_n &= c_1 \\
0x_1 + x_2 + 0x_3 + \cdots + 0x_n &= c_2 \\
0x_1 + 0x_2 + x_3 + \cdots + 0x_n &= c_3 \\
\vdots \qquad \vdots \qquad \vdots \qquad\quad \vdots \quad \vdots & \\
0x_1 + 0x_2 + 0x_3 + \cdots + x_n &= c_n
\end{aligned}
\qquad \text{or} \qquad
\begin{cases}
x_1 = c_1 \\
x_2 = c_2 \\
x_3 = c_3 \\
\vdots \\
x_n = c_n
\end{cases}
$$

Note that back substitution is not required on a system in this form. The solution is simply $(c_1, c_2, c_3, \ldots, c_n)$. We refer to the procedure of reducing an augmented matrix to this special form as **Gauss-Jordan elimination**. The procedure is illustrated in the next example.

EXAMPLE 6   Solve the system of linear equations by using Gauss-Jordan elimination.

$$2x + 4y - 2z = -1$$
$$3y - 4z = 6$$
$$-2x - 2y + z = 0$$

SOLUTION   In Example 4, we used row operations on the augmented matrix of this system to reduce the matrix to the following echelon form:

$$\begin{bmatrix} 2 & 4 & -2 & | & -1 \\ 0 & 1 & -\frac{4}{3} & | & 2 \\ 0 & 0 & \frac{5}{3} & | & -5 \end{bmatrix}$$

Now, instead of using back substitution, we continue with row operations by first developing a *leading 1* in each row:

$$\begin{bmatrix} 2 & 4 & -2 & | & -1 \\ 0 & 1 & -\frac{4}{3} & | & 2 \\ 0 & 0 & \frac{5}{3} & | & -5 \end{bmatrix} \sim \begin{bmatrix} 1 & 2 & -1 & | & -\frac{1}{2} \\ 0 & 1 & -\frac{4}{3} & | & 2 \\ 0 & 0 & 1 & | & -3 \end{bmatrix} \quad \begin{array}{l} \frac{1}{2}R_1 \to R_1 \\ \\ \frac{3}{5}R_3 \to R_3 \end{array}$$

Next, we use row operations to form a matrix in which every column with a leading 1 has zeros in every position above and below the leading 1.

$$\begin{bmatrix} 1 & 2 & -1 & | & -\frac{1}{2} \\ 0 & 1 & -\frac{4}{3} & | & 2 \\ 0 & 0 & 1 & | & -3 \end{bmatrix} \sim \begin{bmatrix} 1 & 2 & 0 & | & -\frac{7}{2} \\ 0 & 1 & 0 & | & -2 \\ 0 & 0 & 1 & | & -3 \end{bmatrix} \quad \begin{array}{l} R_3 + R_1 \to R_1 \\ \frac{4}{3}R_3 + R_2 \to R_2 \end{array}$$

$$\sim \begin{bmatrix} 1 & 0 & 0 & | & \frac{1}{2} \\ 0 & 1 & 0 & | & -2 \\ 0 & 0 & 1 & | & -3 \end{bmatrix} \quad -2R_2 + R_1 \to R_1$$

Now, the corresponding system of equations is

$$\begin{array}{ll} x + 0y + 0z = \frac{1}{2} & x = \frac{1}{2} \\ 0x + y + 0z = -2 \quad \text{or} & y = -2 \\ 0x + 0y + z = -3 & z = -3 \end{array}$$

Thus, the solution to the original system of linear equations is $(\frac{1}{2}, -2, -3)$, which agrees with Example 4.   ◆

PROBLEM 6   Solve the system of linear equations in Problem 3 by using Gauss-Jordan elimination.   ◆

Exercises 7.2

   Basic Skills

In Exercises 1–6, find the dimension of each matrix.

1.
$$\begin{bmatrix} 2 & 3 & -1 \\ 4 & 1 & 6 \\ 2 & 1 & 0 \end{bmatrix}$$

2.
$$\begin{bmatrix} 2 & 1 \\ 4 & -2 \\ 0 & 2 \\ 0 & 0 \end{bmatrix}$$

3.
$$\begin{bmatrix} 1 & 9 & 8 & 3 \\ 2 & 0 & -3 & 1 \end{bmatrix}$$

4. $[1 \quad 2 \quad 4]$

5.
$$\begin{bmatrix} 1 \\ 9 \\ 8 \end{bmatrix}$$

6.
$$\begin{bmatrix} 3 & 4 & 5 & 6 \\ 2 & 9 & 0 & -1 \\ 0 & 0 & 4 & 8 \\ 1 & 2 & -4 & 0 \end{bmatrix}$$

7. Referring to the matrix in Exercise 1, find the element corresponding to

   (a) a_{12} (b) a_{33} (c) a_{21} (d) a_{23}

8. Referring to the matrix in Exercise 6, find the element corresponding to

   (a) a_{22} (b) a_{13} (c) a_{41} (d) a_{34}

In Exercises 9–12, write the augmented matrix corresponding to the given system of linear equations.

9. $\begin{aligned} 2x - 3y &= 1 \\ 4x + 2y &= -3 \end{aligned}$

10. $\begin{aligned} x - 2y - 3z &= 2 \\ 4x - 5y &= 4 \\ -2y + z &= 9 \end{aligned}$

11. $\begin{aligned} x - 5y + 2z &= 9 \\ 3x \quad\quad - z &= 4 \end{aligned}$

12. $\begin{aligned} 2y - z + t &= 2 \\ 2x - 3y + 4z - t &= 1 \\ x \quad\quad + z &= 0 \\ x \quad\quad\quad - 3t &= 6 \end{aligned}$

In Exercises 13–16, write the system of linear equations corresponding to the given augmented matrix.

13.
$$\left[\begin{array}{cc:c} -1 & 4 & 0 \\ 2 & 3 & 1 \end{array}\right]$$

14.
$$\left[\begin{array}{ccc:c} 3 & 0 & 2 & -1 \\ 0 & 2 & 4 & 3 \end{array}\right]$$

15.
$$\left[\begin{array}{ccc:c} 2 & 9 & 0 & -2 \\ -1 & 0 & 2 & 1 \\ -2 & 3 & -4 & 1 \end{array}\right]$$

16.
$$\left[\begin{array}{cccc:c} 3 & 0 & 0 & -1 & 5 \\ -1 & 2 & 0 & 2 & -3 \\ 2 & -1 & 7 & 0 & 1 \\ 0 & 0 & 4 & 1 & -2 \end{array}\right]$$

In Exercises 17–36, solve each system of linear equations by using the matrix method.

17. $\begin{aligned} x - y &= 7 \\ 5x + 6y &= 2 \end{aligned}$

18. $\begin{aligned} 6x - 5y &= -8 \\ 2x - 7y &= 8 \end{aligned}$

19. $\begin{aligned} 5x - 15y &= 5 \\ -4x + 12y &= -4 \end{aligned}$

20. $\begin{aligned} -2x + 6y &= -2 \\ 3x - 9y &= -3 \end{aligned}$

21. $\begin{aligned} 0.2x - 0.3y &= -7 \\ -1.1x + 1.8y &= 46 \end{aligned}$

22. $\begin{aligned} \tfrac{1}{2}x - \tfrac{1}{3}y &= 0 \\ 3x - \tfrac{3}{2}y &= 6 \end{aligned}$

23. $\begin{aligned} x - 3y \quad\quad &= 7 \\ 2x \quad\quad + z &= 5 \\ 4y + 2z &= -2 \end{aligned}$

24. $\begin{aligned} 4x \quad\quad - 2z &= -1 \\ 2x + 2y \quad\quad &= 3 \\ x - 3y + z &= -7 \end{aligned}$

25. $\begin{aligned} 2x - y - 2z &= 0 \\ 3x + 3y + z &= 6 \\ y + 3z &= -5 \end{aligned}$

26. $\begin{aligned} 5x - y + z &= 3 \\ 2x \quad\quad + 3z &= -5 \\ 2x - 3y - 4z &= 4 \end{aligned}$

27. $\begin{aligned} 2x + 3y - 2z &= 3 \\ 3x + 5y + z &= 2 \\ x + 2y - 5z &= 1 \end{aligned}$

28. $\begin{aligned} 3x - 4y + 19z &= 6 \\ x - y + 4z &= 2 \\ 5x - 2y - z &= 10 \end{aligned}$

29. $\begin{aligned} 4x - 3y - z &= 5 \\ 3x + 2y - 3z &= 2 \\ -2x + y - z &= 0 \end{aligned}$

30. $\begin{aligned} 2x - y + z &= 7 \\ 4x - 3y - 3z &= 11 \\ 5x + y - z &= 14 \end{aligned}$

31. $\begin{aligned} \tfrac{1}{2}x - 3y + z &= -9 \\ 2x - \tfrac{1}{3}y + z &= 2 \\ \tfrac{3}{2}x - 4y - z &= -8 \end{aligned}$

32. $\begin{aligned} 0.2x - 3y - 2.5z &= 10 \\ 3x + 0.6y + 0.2z &= 15 \\ -0.6x + 0.4y - 1.2z &= 5 \end{aligned}$

33. $\begin{aligned} x - 2y \quad\quad + t &= 1 \\ 2y - z + 3t &= 12 \\ x \quad\quad - 2z \quad\quad &= 2 \\ 2y \quad\quad + t &= 5 \end{aligned}$

34. $\begin{aligned} x + y \quad\quad - 2t &= 7 \\ 2x \quad\quad - z + t &= -4 \\ 3x + y + z \quad\quad &= 5 \\ 2y - z + 3t &= -2 \end{aligned}$

35. $\begin{aligned} x_1 + 2x_2 \quad\quad\quad\quad + x_5 &= 5 \\ x_2 - x_3 + x_4 \quad\quad &= 1 \\ x_1 \quad\quad + 3x_3 \quad\quad - x_5 &= -2 \\ 3x_2 \quad\quad - x_4 + 2x_5 &= 8 \\ x_3 + 2x_4 - x_5 &= -5 \end{aligned}$

36.
$$2x_1 + x_2 - x_3 \qquad + x_5 = -1$$
$$x_1 - 2x_2 \qquad + x_4 \qquad = 5$$
$$3x_2 - x_3 \qquad + 2x_5 = -5$$
$$3x_1 \qquad + 2x_3 - x_4 - x_5 = 7$$
$$4x_2 + 3x_3 + x_4 - x_5 = -1$$

In Exercises 37–44, solve each nonsquare system of linear equations by using the matrix method.

37.
$$x + y - z = -2$$
$$3x - 2y + 2z = 4$$

38.
$$2x + 3y + z = 2$$
$$x - 2y + 4z = -6$$

39.
$$3x + y - 2z = -2$$
$$7x + 2y + z = 3$$

40.
$$-11x - 2y + 5z = 6$$
$$5x + 5y - 2z = 0$$

41.
$$2x - y + z - 3t = 6$$
$$x \qquad + 2z + t = 3$$
$$2y + z + 5t = 0$$

42.
$$2x - y + z - 3t = 1$$
$$3x + y - 3z + t = 2$$
$$4x + 3y - 7z + 5t = 0$$

43.
$$3x - 2y + z + t = 3$$
$$2x + y - 3z + t = 2$$

44.
$$-2x - y + 3z + t = -1$$
$$3x + y - 2z - t = 1$$

In Exercises 45–52, solve each system of linear equations by using Gauss-Jordan elimination.

45.
$$x + 3y = 9$$
$$2x - y = -10$$

46.
$$3x - 2y = 6$$
$$6x + 5y = -6$$

47.
$$2x + 3y \qquad = 2$$
$$x \qquad - 2z = 7$$
$$2y - z = 3$$

48.
$$3x + y - 2z = -2$$
$$y - 3z = -1$$
$$2x + 3y + 5z = 3$$

49.
$$x - 2y + z = 1$$
$$3x - y + z = 6$$
$$-2x + y - 4z = 1$$

50.
$$5x - 3y + 4z = -1$$
$$3x + 8y - z = -4$$
$$2x - 11y + 4z = 2$$

51.
$$x + 3y \qquad + 2t = 3$$
$$3x \qquad + 2z - t = 3$$
$$3y - z + 2t = 1$$
$$x + 2y - 3z \qquad = -4$$

52.
$$5x - y + 2z \qquad = -3$$
$$2x - 5y \qquad - t = -3$$
$$4y - 3z + 4t = -3$$
$$-x \qquad + 4z - 2t = -3$$

 Critical Thinking

A system of linear equations is said to be homogeneous *if all the constant terms are zero. A homogeneous system of equations always has the trivial solution* $(0, 0, 0, \ldots, 0)$. *However, some systems also have nontrivial solutions. In Exercises 53 and 54, find the solution to the homogeneous system.*

53.
$$x + 3y - 2z = 0$$
$$3x - 2y + 5z = 0$$
$$-2x - y - z = 0$$

54.
$$3x - 5y + 5z = 0$$
$$x - 4y - 3z = 0$$
$$-4x + 15y + 10z = 0$$

In Exercises 55–60, use the indicated substitutions to change the given system of equations into a linear system. Apply the matrix method to the linear system of equations and then state the solution of the original system.

55.
$$\frac{2}{x} + \frac{3}{y} = 15$$
$$\frac{4}{x} - \frac{5}{y} = -14$$
(Let $u = 1/x$ and $v = 1/y$)

56.
$$\frac{1}{2x} + \frac{2}{3y} = \frac{3}{2}$$
$$\frac{2}{x} - \frac{1}{y} = \frac{13}{4}$$
(Let $u = 1/x$ and $v = 1/y$)

57.
$$\frac{1}{x} + \frac{1}{y} \qquad = 8$$
$$\frac{1}{y} + \frac{1}{z} = 15$$
$$\frac{1}{x} \qquad + \frac{1}{z} = 13$$
(Let $u = 1/x$, $v = 1/y$, and $w = 1/z$)

58.
$$\frac{1}{x} + \frac{1}{y} + \frac{1}{2z} = -1$$
$$\frac{1}{x} + \frac{1}{4y} - \frac{1}{z} = -1$$
$$\frac{2}{3x} + \frac{1}{2y} + \frac{1}{z} = \frac{1}{3}$$
(Let $u = 1/x$, $v = 1/y$, and $w = 1/z$)

59.
$$e^x - e^y + 2e^z = 0$$
$$2e^x - e^y + 3e^z = 3$$
$$5e^x - 2e^y - 3e^z = -1$$
(Let $u = e^x$, $v = e^y$, and $w = e^z$)

60.
$$\log_2 x + \log_2 y + \log_2 z = 9$$
$$\log_2 x^2 - \log_2 y + \log_2 z^3 = 16$$
$$\log_2 \frac{1}{x} + \log_2 y^3 - \log_2 z = 3$$
(Let $u = \log_2 x$, $v = \log_2 y$, and $w = \log_2 z$)

In Exercises 61 and 62, develop a system of linear equations and solve by the matrix method.

61. A cross-country ski trail that connects condominium A to condominium B has a warming hut 4 miles from A. Starting from condominium A, the trail goes uphill for 3 miles, level for 4 miles, and then downhill for 6 miles. Suppose a man can ski from A to B in 2 hours; from B to A in 2 hours 45 minutes; and from A to the warming hut and back again to A in 1 hour 30 minutes. What are this skier's rates of skiing uphill, on level ground, and downhill (in miles per hour, mph)?

62. When two resistors R_a and R_b are connected in parallel as shown in the figure, the total resistance R_t between points A and B may be found by using the equation

$$\frac{1}{R_t} = \frac{1}{R_a} + \frac{1}{R_b}.$$

Find the values of three resistors R_1, R_2, and R_3 if the total resistance is 60 Ω when R_1 and R_2 are connected in parallel, 100 Ω when R_2 and R_3 are connected in parallel, and 75 Ω when R_1 and R_3 are connected in parallel.

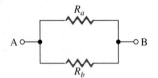

 Calculator Activities

In Exercises 63–68, use the matrix method in conjunction with a calculator to help solve each system of linear equations. Round each answer to three significant digits.

63. $56x - 35y = 3987$
    $33x - 79y = 398$

64. $2.22x - 5.38y = 19.22$
    $0.53x + 5.34y = -9.87$

65. $4.9x \qquad - 4.6z = 33.7$
    $\qquad 8.2y + 7.3z = -55.1$
    $9.8x - 6.9y \qquad = 37.4$

66. $0.73x - 3.81y \qquad = 22.3$
    $8.25x - 0.87y + 2.39z = 12.9$
    $\qquad 0.13y - 0.69z = 25.0$

67. $\quad a + \quad b + \quad c = 24$
    $52.9a - 19.2b + 48.1c = 228$
    $\qquad 58.0b - 28.1c = 329$

68. $\quad a \qquad - 45c + 21d = 540$
    $\qquad 25b - \quad 9c \qquad = -88$
    $-19a - 32b + 22c + \quad 9d = 237$
    $\qquad 56c - 88d = 567$

69. When an object is propelled vertically *upward* its distance s (in feet) above the ground after t seconds (s) is given by

$$s = -16t^2 + v_0 t + s_0,$$

where v_0 is the initial velocity of the object and s_0 is the initial distance above the ground. Use the matrix method to find v_0 and s_0 if $s = 98.3$ ft when $t = 1.2$ s and $s = 123.2$ ft when $t = 2.1$s. Round each answer to three significant digits.

70. The forces F_1, F_2, and F_3 (in pounds), which act on the beam shown in the sketch, may be found by solving the following system of linear equations:

$$0.7547F_1 \qquad - 0.4848F_3 = 0$$
$$0.6561F_1 - \quad F_2 + 0.8746F_3 = 200$$
$$-18.0F_2 + \quad 21.0F_3 = 2800$$

Using the matrix method, find F_1, F_2, and F_3, rounding each answer to three significant digits.

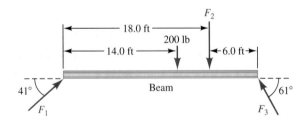

7.3 Matrix Algebra

◆ Introductory Comments

In the previous section we used matrices as an aid in solving systems of linear equations. Their use, however, far exceeds this single purpose. When English mathematician Arthur Cayley introduced matrices in 1858, he defined the idea of equality of matrices as well as the operations of addition and multiplication. In doing so, he created a new type of mathematics that has served as a model for various applications, from economics to quantum mechanics.

In this section, we discuss the algebraic operations of addition, subtraction, and multiplication with matrices. We use capital letters, such as A, B, C, and so on, to denote matrices, and write

$$A = [a_{ij}]$$

Read "the matrix with elements a_{ij}"

to denote the $m \times n$ matrix

$$\begin{bmatrix} a_{11} & a_{12} & a_{13} & \cdots & a_{1j} & \cdots & a_{1n} \\ a_{21} & a_{22} & a_{23} & \cdots & a_{2j} & \cdots & a_{2n} \\ a_{31} & a_{32} & a_{33} & \cdots & a_{3j} & \cdots & a_{3n} \\ \vdots & \vdots & \vdots & & \vdots & & \vdots \\ a_{i1} & a_{i2} & a_{i3} & \cdots & a_{ij} & \cdots & a_{in} \\ \vdots & \vdots & \vdots & & \vdots & & \vdots \\ a_{m1} & a_{m2} & a_{m3} & \cdots & a_{mj} & \cdots & a_{mn} \end{bmatrix}$$

where a_{ij} stands for the element in the ith row, jth column. The more compact notation $A = [a_{ij}]$ enables us to express the rules for matrix algebra in a form that is easy to read.

◆ Equality of Matrices

Two matrices are **equal** if and only if their corresponding elements are equal. More formally, we state the following definition of equality of matrices.

Equality of Matrices

> Two matrices $A = [a_{ij}]$ and $B = [b_{ij}]$ are **equal** if they both are of dimension $m \times n$ and
>
> $$a_{ij} = b_{ij}$$
>
> for all $i = 1, 2, 3, \ldots, m$ and $j = 1, 2, 3, \ldots, n$.

When working with matrices, do not confuse *equality* (written =) with *equivalency* (written ∼). For example, when interchanging rows, we write

$$\begin{bmatrix} 2 & 5 \\ -3 & 4 \end{bmatrix} \sim \begin{bmatrix} -3 & 4 \\ 2 & 5 \end{bmatrix}$$

However, since corresponding elements are *not* the same, we have

$$\begin{bmatrix} 2 & 5 \\ -3 & 4 \end{bmatrix} \neq \begin{bmatrix} -3 & 4 \\ 2 & 5 \end{bmatrix}$$

EXAMPLE 1 Determine the values of x, y, z, and w that make matrices A and B equal

$$A = \begin{bmatrix} x + y & 3z \\ y + z & x - w \end{bmatrix}, \qquad B = \begin{bmatrix} -2 & -6 \\ -5 & 7 \end{bmatrix}$$

SOLUTION For the matrices A and B to be equal, their corresponding elements must be equal; that is,

$$x + y = -2 \quad \longleftarrow \boxed{\text{First equation}}$$
$$3z = -6 \quad \longleftarrow \boxed{\text{Second equation}}$$
$$y + z = -5 \quad \longleftarrow \boxed{\text{Third equation}}$$
and
$$x - w = 7 \quad \longleftarrow \boxed{\text{Fourth equation}}$$

From the second equation, $3z = -6$, we find $z = -2$. Replacing z with -2 in the third equation, we find $y = -3$. Replacing y with -3 in the first equation, we find $x = 1$. Finally, replacing x with 1 in the fourth equation, we find $w = -6$. In summary, for the matrices A and B to be equal, we must have $x = 1$, $y = -3$, $z = -2$, and $w = -6$. ◆

PROBLEM 1 Determine the values of x and y that make matrices A and B equal.

$$A = \begin{bmatrix} x + y & 2 \\ -3 & 8 \end{bmatrix}, \qquad B = \begin{bmatrix} 4 & 2 \\ -3 & x - y \end{bmatrix}$$ ◆

◆ **Addition of Matrices**

To add two $m \times n$ matrices, we simply add the elements in corresponding positions. More formally, we state the following rule for matrix addition.

Matrix Addition

If $A = [a_{ij}]$ and $B = [b_{ij}]$ are matrices of dimension $m \times n$, then their **sum** $A + B$ is a matrix of dimension $m \times n$ defined by

$$A + B = [a_{ij} + b_{ij}].$$

EXAMPLE 2   Given the matrices

$$A = \begin{bmatrix} 6 & 5 & -2 \\ 4 & 0 & -1 \end{bmatrix}, \quad B = \begin{bmatrix} 3 & -5 & 2 \\ 0 & 1 & -3 \end{bmatrix}, \quad \text{and} \quad C = \begin{bmatrix} 2 & 3 \\ -1 & 1 \\ 0 & -2 \end{bmatrix}$$

determine each sum.

(a) $A + B$ **(b)** $C + C$

SOLUTION

(a) $A + B = \begin{bmatrix} 6+3 & 5+(-5) & -2+2 \\ 4+0 & 0+1 & -1+(-3) \end{bmatrix} = \begin{bmatrix} 9 & 0 & 0 \\ 4 & 1 & -4 \end{bmatrix}$

(b) $C + C = \begin{bmatrix} 2+2 & 3+3 \\ -1+(-1) & 1+1 \\ 0+0 & -2+(-2) \end{bmatrix} = \begin{bmatrix} 4 & 6 \\ -2 & 2 \\ 0 & -4 \end{bmatrix}$ ◆

We cannot add two matrices of different dimensions. For example, if

$$A = \begin{bmatrix} 6 & 5 & -2 \\ 4 & 0 & -1 \end{bmatrix} \quad \text{and} \quad C = \begin{bmatrix} 2 & 3 \\ -1 & 1 \\ 0 & -2 \end{bmatrix}$$

Dimension 2×3 Dimension 3×2

then we say that $A + C$ is undefined.

PROBLEM 2   Given the matrix B in Example 2, find $B + B$.   ◆

◆ Scalar Multiplication

Referring to Example 2(b), when we add matrix C to itself we obtain a matrix of the same dimension as C in which each element is twice the corresponding element of C. We refer to the matrix $2C$ as the one obtained by multiplying each element of C by 2. Hence, if C is a matrix, then

$$2C = C + C.$$

When working with matrices, we refer to the number 2 as a **scalar**. In this text, *scalars are real numbers*. To multiply a matrix by a scalar k, we simply multiply each element of the matrix by k.

◆ Scalar Multiplication

> If $A = [a_{ij}]$ is a matrix of dimension $m \times n$ and k is a scalar, then the **scalar multiple** kA is a matrix of dimension $m \times n$ defined by
>
> $$kA = [ka_{ij}].$$

EXAMPLE 3   Given the matrices

$$A = \begin{bmatrix} 6 & 5 & -2 \\ 4 & 0 & -1 \end{bmatrix} \quad \text{and} \quad B = \begin{bmatrix} 3 & -5 & 2 \\ 0 & 1 & -3 \end{bmatrix}$$

perform the indicated matrix operations.

(a) $\frac{1}{2}A$ (b) $-2A + 3B$

SOLUTION

(a) $\frac{1}{2}A = \begin{bmatrix} \frac{1}{2}(6) & \frac{1}{2}(5) & \frac{1}{2}(-2) \\ \frac{1}{2}(4) & \frac{1}{2}(0) & \frac{1}{2}(-1) \end{bmatrix} = \begin{bmatrix} 3 & \frac{5}{2} & -1 \\ 2 & 0 & -\frac{1}{2} \end{bmatrix}$

(b) $-2A + 3B = \begin{bmatrix} -12 & -10 & 4 \\ -8 & 0 & 2 \end{bmatrix} + \begin{bmatrix} 9 & -15 & 6 \\ 0 & 3 & -9 \end{bmatrix}$

$= \begin{bmatrix} -3 & -25 & 10 \\ -8 & 3 & -7 \end{bmatrix}$ ◆

PROBLEM 3   Given the matrices A and B in Example 3, determine $-2B + A$.   ◆

◆ **Subtraction of Matrices**

The matrix $-A$ (read "the negative of A") represents the scalar multiple $-1A$. Thus, $-A$ is a matrix of the same dimension as A in which each element is the *negative* of the corresponding element of A. For every matrix A, we have

$$A + (-A) = \mathbf{0},$$

where $\mathbf{0}$ represents the **zero matrix** of the same dimension as A. For example,

$$\text{if} \quad A = \begin{bmatrix} a_{11} & a_{12} \\ a_{21} & a_{22} \end{bmatrix}, \quad \text{then} \quad -A = \begin{bmatrix} -a_{11} & -a_{12} \\ -a_{21} & -a_{22} \end{bmatrix},$$

$$\text{and} \quad A + (-A) = \begin{bmatrix} 0 & 0 \\ 0 & 0 \end{bmatrix}$$

Zero matrix of
dimension 2×2

Recall from the definition of subtraction of real numbers (Section 1.1) that if a and b are real numbers, then $a - b = a + (-b)$. Similarly, for the subtraction of matrices, we have the following definition.

Subtraction of Matrices

If A and B are matrices of dimension $m \times n$, then

$$A - B = A + (-B).$$

EXAMPLE 4 Given the matrices

$$A = \begin{bmatrix} 2 & 5 \\ -3 & 0 \\ 1 & 4 \end{bmatrix} \quad \text{and} \quad B = \begin{bmatrix} 4 & 6 \\ -2 & -5 \\ 0 & 3 \end{bmatrix}$$

determine $2A - B$.

SOLUTION

$$2A - B = 2A + (-B) = \begin{bmatrix} 4 & 10 \\ -6 & 0 \\ 2 & 8 \end{bmatrix} + \begin{bmatrix} -4 & -6 \\ 2 & 5 \\ 0 & -3 \end{bmatrix} = \begin{bmatrix} 0 & 4 \\ -4 & 5 \\ 2 & 5 \end{bmatrix}$$ ◆

Many of the properties of matrices are similar to the properties of real numbers that we stated in Section 1.1. Next, we list eight properties of matrix addition and scalar multiplication.

Properties of Matrix Addition and Scalar Multiplication

If A, B, and C are $m \times n$ matrices, **0** is the zero $m \times n$ matrix, and c and d are scalars, then

1. $A + B = B + A$ Commutative property of matrix addition
2. $(A + B) + C = A + (B + C)$ Associative property of matrix addition
3. $A + \mathbf{0} = A$ Identity property of matrix addition
4. $A + (-A) = \mathbf{0}$ Inverse property of matrix addition
5. $(cd)A = c(dA)$ Associative property of scalar multiplication
6. $1A = A$ Identity property of scalar multiplication
7. $c(A + B) = cA + cB$ Distributive property of a scalar over matrix addition
8. $(c + d)A = cA + dA$ Distributive property of a matrix over scalar addition

PROBLEM 4 Using the matrices

$$A = \begin{bmatrix} 2 & 3 & 0 \\ -1 & 4 & 1 \end{bmatrix} \quad \text{and} \quad B = \begin{bmatrix} 1 & -3 & -1 \\ 5 & -2 & 2 \end{bmatrix},$$

show that $c(A + B) = cA + cB$ for any scalar c.   ◆

◆ Matrix Multiplication and Its Properties

Information displayed in a table can often be written as a matrix. For example, Table 7.1 shows the current inventory of a certain type of brass bed at three furniture store outlets in the greater Boston area.

Size	Brookline	Braintree	Cambridge
Twin	4	2	5
Double	1	6	0
Queen	6	0	2
King	0	2	3

Table 7.1 can be displayed as the matrix

$$B = \begin{matrix} \text{Brookline} & \text{Braintree} & \text{Cambridge} \\ \begin{bmatrix} 4 & 2 & 5 \\ 1 & 6 & 0 \\ 6 & 0 & 2 \\ 0 & 2 & 3 \end{bmatrix} & & \begin{matrix} \text{Twin} \\ \text{Double} \\ \text{Queen} \\ \text{King} \end{matrix} \end{matrix}$$

where the rows represent the sizes of the beds and the columns represent the locations of the outlets.

To understand matrix multiplication, let's suppose the cost of a twin bed is $300, a double bed is $350, a queen bed is $400, and a king bed is $600. We can describe the cost of each size of bed by a *row matrix*, a matrix with only one row:

$$A = [300 \quad 350 \quad 400 \quad 600].$$

Now the *product*

$$AB = [300 \quad 350 \quad 400 \quad 600] \begin{bmatrix} 4 & 2 & 5 \\ 1 & 6 & 0 \\ 6 & 0 & 2 \\ 0 & 2 & 3 \end{bmatrix}$$

represents the total cost of inventory for this type of bed at each outlet.

To find the total cost of the beds at the Brookline store, we multiply the cost of each size bed by the inventory at that store, then add the products. That is, we multiply the elements in the row matrix A by the corresponding elements in the first column of matrix B, then add the products as follows:

$$AB = [300 \quad 350 \quad 400 \quad 600] \begin{bmatrix} 4 & 2 & 5 \\ 1 & 6 & 0 \\ 6 & 0 & 2 \\ 0 & 2 & 3 \end{bmatrix} = [3950 \quad c_{12} \quad c_{13}]$$

$$300(4) + 350(1) + 400(6) + 600(0)$$

Similarly, this "row times column" procedure gives us the total cost of the bed inventory at the other two outlets.

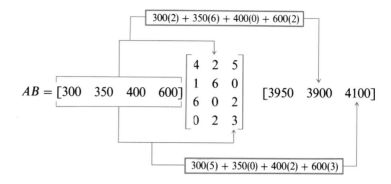

Thus, the 1×3 row matrix

$$
AB = \begin{array}{ccc} \text{Brookline} & \text{Braintree} & \text{Cambridge} \\ [\quad 3950 & 3900 & 4100 \quad] \end{array} \quad \text{Total cost}
$$

represents the total cost of the inventory of this type of bed at each outlet.

From this application we can make some observations about the product AB:

1. The product of two matrices is found by using a *row times column* procedure.

2. For the product AB to exist, the number of columns in matrix A must equal the number of rows in matrix B.

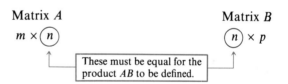

3. The product AB has exactly as many rows as A and exactly as many columns as B.

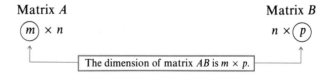

If matrix A has more than one row, then we repeat the row times column procedure using each row of matrix A. Thus, if

$$
A = \begin{bmatrix} a_{11} & a_{12} & a_{13} & \cdots & a_{1n} \\ a_{21} & a_{22} & a_{23} & \cdots & a_{2n} \\ \vdots & \vdots & \vdots & & \vdots \\ a_{i1} & a_{i2} & a_{i3} & \cdots & a_{in} \\ \vdots & \vdots & \vdots & & \vdots \\ a_{m1} & a_{m2} & a_{m3} & \cdots & a_{mn} \end{bmatrix} \quad \text{and} \quad B = \begin{bmatrix} b_{11} & b_{12} & \cdots & b_{1j} & \cdots & b_{1p} \\ b_{21} & b_{22} & \cdots & h_{2j} & \cdots & b_{2p} \\ b_{31} & b_{32} & \cdots & b_{3j} & \cdots & b_{3p} \\ \vdots & \vdots & & \vdots & & \vdots \\ b_{n1} & b_{n2} & \cdots & b_{nj} & \cdots & b_{np} \end{bmatrix}
$$

then the ith row, jth column element of the matrix AB is

$$a_{i1}b_{1j} + a_{i2}b_{2j} + a_{i3}b_{3j} + \cdots + a_{in}b_{nj}.$$

We now formalize the definition of matrix multiplication.

Matrix Multiplication

> If $A = [a_{ij}]$ is a matrix of dimension $m \times n$ and $B = [b_{ij}]$ is a matrix of dimension $n \times p$, then the **product** AB is a matrix of dimension $m \times p$ defined by
>
> $$AB = [c_{ij}], \qquad \text{where } c_{ij} = a_{i1}b_{1j} + a_{i2}b_{2j} + a_{i3}b_{3j} + \cdots + a_{in}b_{nj}.$$

EXAMPLE 5   Given the matrices

$$A = [2 \quad 3 \quad -1], \qquad B = \begin{bmatrix} 4 \\ 1 \\ -3 \end{bmatrix}, \qquad C = \begin{bmatrix} 2 & -1 \\ 3 & 4 \\ 0 & -2 \end{bmatrix}, \quad \text{and} \quad D = \begin{bmatrix} -3 & 2 \\ 1 & 4 \end{bmatrix}$$

determine each product.

(a) AB (b) BA (c) CD (d) DC

SOLUTION

(a) The product AB should have exactly as many rows as A and exactly as many columns as B. Thus, the product AB should be a 1×1 matrix. Using matrix multiplication, we find

$$AB = [2 \quad 3 \quad -1] \begin{bmatrix} 4 \\ 1 \\ -3 \end{bmatrix} = [(2)(4) + (3)(1) + (-1)(-3)] = [14]$$

(b) The product BA should have exactly as many rows as B and exactly as many columns as A. Thus, the product BA should be a 3×3 matrix. Using matrix multiplication, we find

$$BA = \begin{bmatrix} 4 \\ 1 \\ -3 \end{bmatrix} [2 \quad 3 \quad -1] = \begin{bmatrix} (4)(2) & (4)(3) & (4)(-1) \\ (1)(2) & (1)(3) & (1)(-1) \\ (-3)(2) & (-3)(3) & (-3)(-1) \end{bmatrix}$$

$$= \begin{bmatrix} 8 & 12 & -4 \\ 2 & 3 & -1 \\ -6 & -9 & 3 \end{bmatrix}$$

(c) Since the number of columns of matrix C is the same as the number of rows of matrix D, we know that the product CD is defined. Thus, we have

$$CD = \begin{bmatrix} 2 & -1 \\ 3 & 4 \\ 0 & -2 \end{bmatrix} \begin{bmatrix} -3 & 2 \\ 1 & 4 \end{bmatrix} = \begin{bmatrix} (2)(-3) + (-1)(1) & (2)(2) + (-1)(4) \\ (3)(-3) + (4)(1) & (3)(2) + (4)(4) \\ (0)(-3) + (-2)(1) & (0)(2) + (-2)(4) \end{bmatrix}$$

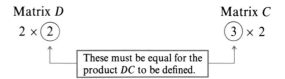

$$= \begin{bmatrix} -7 & 0 \\ -5 & 22 \\ -2 & -8 \end{bmatrix} \quad \longleftarrow \quad \boxed{\text{The product } CD \text{ has the same number of rows as } C \text{ and the same number of columns as } D.}$$

(d) Note that the number of columns of matrix D is *not* the same as the number of rows of matrix C.

$$\begin{array}{cc} \text{Matrix } D & \text{Matrix } C \\ 2 \times \textcircled{2} & \textcircled{3} \times 2 \end{array}$$

$$\boxed{\text{These must be equal for the product } DC \text{ to be defined.}}$$

Hence, the product DC is not defined. ◆

Caution Unlike multiplication of real numbers, matrix multiplication, in general, is not commutative. Observe from Example 5 that $AB \neq BA$ and $CD \neq DC$.

Although matrix multiplication is not commutative, many of its properties are similar to the properties of real numbers.

◆ Properties of Matrix Multiplication

If A, B, and C are matrices of the appropriate dimension such that each of the following is defined, then

1. $A(BC) = (AB)C$ Associative property of matrix multiplication
2. $A(B + C) = AB + AC$ Left distributive property of a matrix over matrix addition
3. $(B + C)A = BA + CA$ Right distributive property of a matrix over matrix addition
4. $c(AB) = (cA)B = A(cB)$, where c is a scalar.

PROBLEM 5   Given the matrices

$$A = \begin{bmatrix} 2 & 1 \\ 3 & 2 \end{bmatrix}, \quad B = \begin{bmatrix} -1 & 4 \\ 0 & 2 \end{bmatrix}, \quad \text{and} \quad C = \begin{bmatrix} 3 & -1 \\ -2 & 0 \end{bmatrix}$$

show that $A(B + C) = AB + AC$. ◆

◆ **Application: Basketball Statistics**

We conclude this section with an application of matrix algebra.

EXAMPLE 6   The given tables show the production (points scored, assists, and rebounds) for the starting five basketball players on the visiting team in the first two games of the championship series.

First Game

Player	Points	Assists	Rebounds
Canter	10	3	4
Filan	14	7	2
Kemen	18	5	19
Penta	22	5	9
Solomon	28	12	11

Second Game

Player	Points	Assists	Rebounds
Canter	14	7	2
Filan	12	9	6
Kemen	10	7	15
Penta	18	5	9
Solomon	42	10	7

(a) Display the production in the first game with matrix A and the production in the second game with matrix B.

(b) Find the matrix $\frac{1}{2}(A + B)$ and describe what it represents.

(c) Find the matrix $[1 \quad 1 \quad 1 \quad 1 \quad 1]A$, and describe what it represents.

SOLUTION

(a) Using the rows as the players and the columns as the production, we have

$$A = \begin{bmatrix} 10 & 3 & 4 \\ 14 & 7 & 2 \\ 18 & 5 & 19 \\ 22 & 5 & 9 \\ 28 & 12 & 11 \end{bmatrix}$$

$$B = \begin{bmatrix} 14 & 7 & 2 \\ 12 & 9 & 6 \\ 10 & 7 & 15 \\ 18 & 5 & 9 \\ 42 & 10 & 7 \end{bmatrix}$$

(b) $\frac{1}{2}(A + B) = \frac{1}{2}\left(\begin{bmatrix} 10 & 3 & 4 \\ 14 & 7 & 2 \\ 18 & 5 & 19 \\ 22 & 5 & 9 \\ 28 & 12 & 11 \end{bmatrix} + \begin{bmatrix} 14 & 7 & 2 \\ 12 & 9 & 6 \\ 10 & 7 & 15 \\ 18 & 5 & 9 \\ 42 & 10 & 7 \end{bmatrix}\right)$

$= \frac{1}{2}\begin{bmatrix} 24 & 10 & 6 \\ 26 & 16 & 8 \\ 28 & 12 & 34 \\ 40 & 10 & 18 \\ 70 & 22 & 18 \end{bmatrix} = \begin{bmatrix} 12 & 5 & 3 \\ 13 & 8 & 4 \\ 14 & 6 & 17 \\ 20 & 5 & 9 \\ 35 & 11 & 9 \end{bmatrix}$

The matrix $\frac{1}{2}(A + B)$ tells us the *average production* of each player for the first two games.

(c) $[1 \quad 1 \quad 1 \quad 1 \quad 1]A = [1 \quad 1 \quad 1 \quad 1 \quad 1]\begin{bmatrix} 10 & 3 & 4 \\ 14 & 7 & 2 \\ 18 & 5 & 19 \\ 22 & 5 & 9 \\ 28 & 12 & 11 \end{bmatrix}$

$= [10 + 14 + 18 + 22 + 28 \quad 3 + 7 + 5 + 5 + 12 \quad 4 + 2 + 19 + 9 + 11]$

$= [92 \quad 32 \quad 45]$

The matrix $[1 \quad 1 \quad 1 \quad 1 \quad 1]A$ tells us the *total production* (points scored, rebounds, and assists) in the first game for the starting five players. ◆

PROBLEM 6 Referring to Example 6, find the matrix $\frac{1}{2}[1 \quad 1 \quad 1 \quad 1 \quad 1](A + B)$, and describe what it represents. ◆

Exercises 7.3

Basic Skills

In Exercises 1–6, find the values of the unknowns that make matrices A and B equal.

1. $A = \begin{bmatrix} 2 & x \\ -3 & 4 \end{bmatrix}$, $B = \begin{bmatrix} 2 & -1 \\ y & 4 \end{bmatrix}$

2. $A = \begin{bmatrix} x + 3 & 2 \\ -4 & z \end{bmatrix}$, $B = \begin{bmatrix} 2x & y - 1 \\ -4 & 2 - z \end{bmatrix}$

3. $A = \begin{bmatrix} 2 & x - y & 3z \\ -1 & 3 & x + z \end{bmatrix}$,

   $B = \begin{bmatrix} 2 & 4 & -9 \\ -1 & 3 & 6 \end{bmatrix}$

4. $A = \begin{bmatrix} x + y & 2 \\ w & y - z \\ 4 & x - w \end{bmatrix}$, $B = \begin{bmatrix} 6 & 2 \\ -8 & 4 \\ 4 & 0 \end{bmatrix}$

5. $A = \begin{bmatrix} x + y & 2 \\ x - y & 4 \end{bmatrix}$, $B = \begin{bmatrix} 4 & w - z \\ 6 & 2w + z \end{bmatrix}$

6. $A = \begin{bmatrix} 2 & 0 & 1 \\ 3 & 2 & y - z \\ 5 & 0 & 2 \end{bmatrix}$,

   $B = \begin{bmatrix} 2 & x + y + z & 1 \\ 3 & 2 & -4 \\ x - y & 0 & 2 \end{bmatrix}$

In Exercises 7–46, perform the indicated matrix operations given that A, B, C, D, E, F, G, and H are defined as follows. If an operation is not defined, write undefined and state the reason.

$$A = \begin{bmatrix} 1 & 2 & 0 \\ -2 & 3 & -4 \end{bmatrix} \qquad B = \begin{bmatrix} 2 & -3 & -5 \\ -1 & 4 & 4 \end{bmatrix}$$

$$C = \begin{bmatrix} 2 & -1 \\ 3 & 2 \\ -1 & 4 \end{bmatrix} \qquad D = \begin{bmatrix} 6 & 0 \\ -3 & 9 \\ 0 & 3 \end{bmatrix}$$

$$E = \begin{bmatrix} 2 & -4 & -5 \end{bmatrix} \qquad F = \begin{bmatrix} 1 \\ -3 \\ 2 \end{bmatrix}$$

$$G = \begin{bmatrix} -1 & 1 \\ 1 & -1 \end{bmatrix} \qquad H = \begin{bmatrix} 5 & 5 \\ -2 & -2 \end{bmatrix}$$

7. $2A$

8. $-5G$

9. $-\frac{1}{3}D$

10. $\frac{3}{2}E$

11. $A + B$

12. $C + D$

13. $E + F$

14. $F + F$

15. $3G + H$

16. $2C + 5G$

17. $-4A + 3B$

18. $-\frac{1}{3}D + 2C$

19. $H - B$

20. $D - C$

21. $3G - 2H$

22. $\frac{1}{2}A - \frac{3}{2}B$

23. $-2(3E)$

24. $3A - 2A$

25. $\frac{1}{2}F - \frac{3}{2}F$

26. $2A + 2(B - A)$

27. AC

28. CB

29. GB

30. ED

31. DH

32. HD

33. EF

34. FE

35. HG

36. GH

37. $C(A + B)$

38. $CA + CB$

39. $(A - B)D$

40. $AD - BD$

41. $(EC)A$

42. $E(CA)$

43. $(BD)G$

44. $B(DG)$

45. $F(2E)$

46. $2(FE)$

Exercises 47–56 pertain to the car rental problem described as follows: A car rental agency rents subcompact, mid-size, and large cars. The rental rates (in dollars) charged by the agency are given in row matrix C.

	Subcompact	Mid-size	Large	
$C = [$	20	32	40 $]$	Cost ($)

Within a certain city, the agency has three terminals. The number of subcompact, mid-size, and large cars rented at each terminal on Monday, Tuesday, and Wednesday are given in matrices M, T, and W, respectively.

Monday

	Terminal 1	Terminal 2	Terminal 3	
	12	14	18	Subcompact
$M =$	8	14	10	Mid-size
	2	5	6	Large

Tuesday

	Terminal 1	Terminal 2	Terminal 3	
	10	8	9	Subcompact
$T =$	2	4	10	Mid-size
	0	1	6	Large

Wednesday

	Terminal 1	Terminal 2	Terminal 3	
	11	11	9	Subcompact
$W =$	5	3	4	Mid-size
	1	3	0	Large

47. Find the matrix $M + T + W$, and state what it represents.

48. Find the matrix $\frac{1}{3}(M + T + W)$, and state what it represents.

49. Find the matrix $M \begin{bmatrix} 1 \\ 1 \\ 1 \end{bmatrix}$, and state what it represents.

50. Find the matrix $[1 \quad 1 \quad 1]W$, and state what it represents.

51. Find the matrix $[1 \quad 1 \quad 1](M + T + W)$, and state what it represents.

52. Find the matrix $[1 \quad 1 \quad 1](M + T + W)\begin{bmatrix} 1 \\ 1 \\ 1 \end{bmatrix}$, and state what it represents.

53. Find the matrix CM, and state what it represents.

54. Find the matrix CT, and state what it represents.

55. Find the matrix $C(M + T)\begin{bmatrix} 1 \\ 1 \\ 1 \end{bmatrix}$, and state what it represents.

56. Find the matrix $\frac{1}{3}C(M + T + W)\begin{bmatrix} 1 \\ 1 \\ 1 \end{bmatrix}$, and state what it represents.

 Critical Thinking

57. The zero product property (Section 2.3) states that if a and b are real numbers and $ab = 0$, then either $a = 0$ or $b = 0$. This property, however, does not hold for matrices. Find a pair of 2×2 matrices A and B for which $AB = \mathbf{0}$, yet $A \neq \mathbf{0}$ and $B \neq \mathbf{0}$.

58. Using the idea of matrix multiplication and equality of matrices, determine the values of x and y such that

$$\begin{bmatrix} 2 & -3 \\ x & y \end{bmatrix}\begin{bmatrix} 4 & 3 \\ -6 & 5 \end{bmatrix} = \begin{bmatrix} 26 & -9 \\ 24 & -1 \end{bmatrix}$$

59. If A is an $m \times m$ matrix, then we define the *square* of matrix A by writing $A^2 = AA$. Given

$$A = \begin{bmatrix} 1 & -2 & 3 \\ 0 & 3 & -2 \\ 3 & 1 & 4 \end{bmatrix} \quad \text{and} \quad B = \begin{bmatrix} 2 & 0 & -1 \\ 1 & 3 & 4 \\ 0 & -1 & -1 \end{bmatrix}$$

perform each matrix operation:

(a) A^2 (b) B^2 (c) $A^2 - B^2$ (d) $(A + B)(A - B)$

Compare the answers from parts (c) and (d). Does $A^2 - B^2 = (A + B)(A - B)$?

60. Given that C and D are matrices, find the fallacy in the following argument.

$(C + D)^2 = (C + D)(C + D)$	Definition of $(C + D)^2$ (see Exercise 59)
$= (C + D)C + (C + D)D$	Left distributive property of a matrix over matrix addition
$= CC + DC + CD + DD$	Right distributive property of a matrix over matrix addition
$= C^2 + 2CD + D^2$	Rewrite

In Exercises 61 and 62, let

$$A = \begin{bmatrix} a_{11} & a_{12} \\ a_{21} & a_{22} \end{bmatrix}, \quad B = \begin{bmatrix} b_{11} & b_{12} \\ b_{21} & b_{22} \end{bmatrix},$$

$$A^T = \begin{bmatrix} a_{11} & a_{21} \\ a_{12} & a_{22} \end{bmatrix}, \quad \text{and} \quad B^T = \begin{bmatrix} b_{11} & b_{21} \\ b_{12} & b_{22} \end{bmatrix}.$$

Show that the given statement is true. (The matrices A^T and B^T are called the transposes *of matrices A and B, respectively. The transpose of a matrix is formed by switching its rows and columns.)*

61. $(AB)^T = B^T A^T$ **62.** $(A + B)^T = A^T + B^T$

 Calculator Activities

In Exercises 63–70, use a calculator to help perform the indicated matrix operations, given that A, B, C, and D are defined as follows:

$$A = \begin{bmatrix} 22.4 & 18.9 \\ -34.7 & 44.1 \end{bmatrix}$$

$$B = \begin{bmatrix} 28.5 & -36.6 \\ -11.1 & 42.9 \end{bmatrix}$$

$$C = \begin{bmatrix} 2.78 \\ -1.92 \end{bmatrix}$$

$$D = \begin{bmatrix} -0.743 & 0.667 & 1.234 \\ 2.003 & -3.775 & 0.651 \end{bmatrix}$$

63. $78.9C$ **64.** $-9.2D$

65. $A + 1.38B$ **66.** $18.7A - 92.6B$

67. AB

68. BA

69. $-1.9A(BC)$

70. $(A + B)D$

Exercises 71-74 pertain to the labor situation that follows:

An engineering firm employs three types of union workers: technicians, engineers, and senior engineers. The company and the union have agreed to a new contract that states that the hourly wages increase from

Old Hourly Wage ($)

$$A = \begin{bmatrix} 8.35 \\ 13.45 \\ 22.15 \end{bmatrix} \begin{array}{l} \text{Technician} \\ \text{Engineer} \\ \text{Senior engineer} \end{array}$$

to

New Hourly Wage ($)

$$B = \begin{bmatrix} 9.65 \\ 16.75 \\ 28.25 \end{bmatrix} \begin{array}{l} \text{Technician} \\ \text{Engineer} \\ \text{Senior engineer} \end{array}$$

and that each person works 37.5 hours per week. The company maintains offices in Boston, Atlanta, and Seattle, and the number of technicians, engineers, and senior engineers at each facility is

Technician   Engineer   'Senior Engineer

$$C = \begin{bmatrix} 81 & 67 & 19 \\ 57 & 42 & 16 \\ 122 & 88 & 31 \end{bmatrix} \begin{array}{l} \text{Boston} \\ \text{Atlanta} \\ \text{Seattle} \end{array}$$

71. Find the matrix 37.5 CB, and state what it represents.

72. Find the matrix $[1 \quad 1 \quad 1](37.5CB)$, and state what it represents.

73. Find the matrix $37.5(B - A)$, and state what it represents.

74. Find the matrix $37.5C(B - A)$, and state what it represents.

◆ 7.4   Inverse of a Matrix

◆   **Introductory Comments**

A matrix of dimension $n \times n$ is called a **square matrix**. In a square matrix, the elements $a_{11}, a_{22}, a_{33}, \ldots, a_{nn}$ are called its **main diagonal** elements. A square matrix with the digit 1 along its main diagonal and zeros elsewhere is called the **identity matrix** of dimension $n \times n$ and is denoted by

$$I_n = \begin{bmatrix} 1 & 0 & 0 & \ldots & 0 \\ 0 & 1 & 0 & \ldots & 0 \\ 0 & 0 & 1 & \ldots & 0 \\ . & . & . & \ldots & . \\ . & . & . & \ldots & . \\ . & . & . & \ldots & . \\ 0 & 0 & 0 & \ldots & 1 \end{bmatrix}$$

Anytime we multiply an $n \times n$ matrix by the identity matrix I_n, we obtain the original $n \times n$ matrix with which we started. We refer to this fact as the **identity property of matrix multiplication**.

◆Identity Property of Matrix Multiplication

> If A is a square matrix of dimension $n \times n$ and I_n is the $n \times n$ identity matrix, then
>
> $$AI_n = I_nA = A.$$

If the product of two square $n \times n$ matrices A and B is the identity matrix I_n, then we say that the matrices A and B are **inverses** of each other. Nonsquare matrices AB do not have inverses, since their products AB and BA (if they exist) are always of different dimensions and therefore are not equal.

In this section, we discuss a method for finding the inverse of a given matrix, and then show how this inverse may be used to solve a system of linear equations.

◆ Verifying Inverse Matrices

To verify that the square $n \times n$ matrices A and B are inverses of each other, we show that

$$AB = BA = I_n,$$

where I_n is the identity matrix of dimension $n \times n$.

EXAMPLE 1   Show that the matrices

$$A = \begin{bmatrix} 2 & -5 \\ -1 & 3 \end{bmatrix} \quad \text{and} \quad B = \begin{bmatrix} 3 & 5 \\ 1 & 2 \end{bmatrix}$$

are inverses of each other.

SOLUTION

$$AB = \begin{bmatrix} 2 & -5 \\ -1 & 3 \end{bmatrix}\begin{bmatrix} 3 & 5 \\ 1 & 2 \end{bmatrix} = \begin{bmatrix} 6 + (-5) & 10 + (-10) \\ -3 + 3 & -5 + 6 \end{bmatrix} = \begin{bmatrix} 1 & 0 \\ 0 & 1 \end{bmatrix}$$

and

Identity matrix of dimension 2×2

$$BA = \begin{bmatrix} 3 & 5 \\ 1 & 2 \end{bmatrix}\begin{bmatrix} 2 & -5 \\ -1 & 3 \end{bmatrix} = \begin{bmatrix} 6 + (-5) & -15 + 15 \\ 2 + (-2) & -5 + 6 \end{bmatrix} = \begin{bmatrix} 1 & 0 \\ 0 & 1 \end{bmatrix}$$

Since $AB = BA = I_2$, we conclude that the matrices A and B are inverses of each other.   ◆

PROBLEM 1   Are the matrices

$$A = \begin{bmatrix} 1 & 2 & 3 \\ 1 & 2 & 2 \\ -1 & -3 & -4 \end{bmatrix} \quad \text{and} \quad B = \begin{bmatrix} 2 & 1 & 2 \\ -2 & 1 & -1 \\ 1 & -1 & 0 \end{bmatrix}$$

inverses of each other?   ◆

◆ Finding the Inverse Matrix

We usually denote the inverse matrix of A by using the notation A^{-1} (read "A inverse"). We now state the **inverse property of matrix multiplication**.

◆ **Inverse Property of Matrix Multiplication**

> If A is a square $n \times n$ matrix and A^{-1} its inverse, then
>
> $$AA^{-1} = A^{-1}A = I_n,$$
>
> where I_n is the identity matrix of dimension $n \times n$.

If a is a real number, then $a^{-1} = 1/a$. However, if A is a matrix, the notation A^{-1} *does not mean* $1/A$. In fact, matrix division is not defined.

Given a square $n \times n$ matrix A we can find its inverse A^{-1}, provided that we can solve the system of linear equations that develops from the equation $AA^{-1} = I_n$. The procedure is illustrated in the next example.

EXAMPLE 2   Find the inverse of the matrix $A = \begin{bmatrix} 3 & 4 \\ 2 & 3 \end{bmatrix}$.

SOLUTION   We begin by letting the inverse matrix $A^{-1} = \begin{bmatrix} a_{11} & a_{12} \\ a_{21} & a_{22} \end{bmatrix}$.
Now, in order to find a_{11}, a_{12}, a_{21}, and a_{22}, we use the fact that $AA^{-1} = I_2$.

$$\begin{array}{ccc} A & A^{-1} & = \quad I_2 \end{array}$$

$$\begin{bmatrix} 3 & 4 \\ 2 & 3 \end{bmatrix}\begin{bmatrix} a_{11} & a_{12} \\ a_{21} & a_{22} \end{bmatrix} = \begin{bmatrix} 1 & 0 \\ 0 & 1 \end{bmatrix}$$

$$\begin{bmatrix} 3a_{11} + 4a_{21} & 3a_{12} + 4a_{22} \\ 2a_{11} + 3a_{21} & 2a_{12} + 3a_{22} \end{bmatrix} = \begin{bmatrix} 1 & 0 \\ 0 & 1 \end{bmatrix}$$

Since two matrices are equal if and only if their corresponding elements are equal, we can form the following two systems of equations.

System 1	System 2
$3a_{11} + 4a_{21} = 1$	$3a_{12} + 4a_{22} = 0$
$2a_{11} + 3a_{21} = 0$	$2a_{12} + 3a_{22} = 1$

To solve these systems, we use the Gauss-Jordan elimination method discussed in Section 7.2.

System 1 **System 2**

$$\begin{bmatrix} 3 & 4 & | & 1 \\ 2 & 3 & | & 0 \end{bmatrix} \qquad \begin{bmatrix} 3 & 4 & | & 0 \\ 2 & 3 & | & 1 \end{bmatrix} \qquad \begin{matrix} \text{Form the augmented} \\ \text{matrix for each system} \end{matrix}$$

$$\begin{bmatrix} 1 & 1 & | & 1 \\ 2 & 3 & | & 0 \end{bmatrix} \qquad \begin{bmatrix} 1 & 1 & | & -1 \\ 2 & 3 & | & 1 \end{bmatrix} \qquad -R_2 + R_1 \rightarrow R_1$$

$$\begin{bmatrix} 1 & 1 & | & 1 \\ 0 & 1 & | & -2 \end{bmatrix} \qquad \begin{bmatrix} 1 & 1 & | & -1 \\ 0 & 1 & | & 3 \end{bmatrix} \qquad -2R_1 + R_2 \rightarrow R_2$$

$$\begin{bmatrix} 1 & 0 & | & 3 \\ 0 & 1 & | & -2 \end{bmatrix} \qquad \begin{bmatrix} 1 & 0 & | & -4 \\ 0 & 1 & | & 3 \end{bmatrix} \qquad -R_2 + R_1 \rightarrow R_1$$

Hence, we conclude that $a_{11} = 3$, $a_{21} = -2$, $a_{12} = -4$, and $a_{22} = 3$ and that

$$A^{-1} = \begin{bmatrix} 3 & -4 \\ -2 & 3 \end{bmatrix}. \qquad \blacklozenge$$

PROBLEM 2   Referring to Example 2, verify that $AA^{-1} = A^{-1}A = I_2$. $\blacklozenge$

To solve the two systems of equations in Example 2 by Gauss-Jordan elimination, we used an identical sequence of elementary row operations on each augmented matrix. This suggests that we can shorten the procedure somewhat by solving the two systems at the same time. To do this, we begin by forming the 2×4 matrix

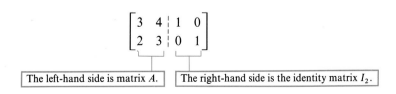

Now, performing the same sequence of elementary row operations as we did for each matrix in Example 2, we have

$$\begin{bmatrix} 3 & 4 & \vdots & 1 & 0 \\ 2 & 3 & \vdots & 0 & 1 \end{bmatrix} \sim \begin{bmatrix} 1 & 1 & \vdots & 1 & -1 \\ 2 & 3 & \vdots & 0 & 1 \end{bmatrix} \qquad -R_2 + R_1 \rightarrow R_1$$

$$\sim \begin{bmatrix} 1 & 1 & \vdots & 1 & -1 \\ 0 & 1 & \vdots & -2 & 3 \end{bmatrix} \qquad -2R_1 + R_2 \rightarrow R_2$$

$$\sim \begin{bmatrix} 1 & 0 & \vdots & 3 & -4 \\ 0 & 1 & \vdots & -2 & 3 \end{bmatrix} \qquad -R_2 + R_1 \rightarrow R_1$$

| The left-hand side is the identity matrix I_2. | The right-hand side is the inverse matrix A^{-1}. |

This procedure for finding an inverse works for any square matrix that has an inverse.

Procedure for Finding A^{-1}

To find the **inverse** of a square $n \times n$ matrix A, proceed as follows:

1. Write the $n \times 2n$ matrix $[A \vdots I_n]$ consisting of matrix A on the left and the identity matrix I_n on the right of the dashed line.

2. Use elementary row operations on the matrix $[A \vdots I_n]$ to obtain the matrix $[I_n \vdots B]$. The matrix B is the inverse matrix of A, that is, $B = A^{-1}$.

3. Check to see if $AA^{-1} = A^{-1}A = I_n$.

EXAMPLE 3   Find the inverse of $A = \begin{bmatrix} 1 & 2 & -3 \\ 2 & 3 & -4 \\ -1 & 0 & 3 \end{bmatrix}$.

SOLUTION   Beginning with the 3×6 matrix $[A \vdots I_3]$, we perform elementary row operations until we obtain the matrix $[I_3 \vdots B]$.

$$\begin{bmatrix} 1 & 2 & -3 & \vdots & 1 & 0 & 0 \\ 2 & 3 & -4 & \vdots & 0 & 1 & 0 \\ -1 & 0 & 3 & \vdots & 0 & 0 & 1 \end{bmatrix} \sim \begin{bmatrix} 1 & 2 & -3 & \vdots & 1 & 0 & 0 \\ 0 & -1 & 2 & \vdots & -2 & 1 & 0 \\ 0 & 2 & 0 & \vdots & 1 & 0 & 1 \end{bmatrix} \qquad \begin{matrix} -2R_1 + R_2 \rightarrow R_2 \\ R_1 + R_3 \rightarrow R_3 \end{matrix}$$

$$\sim \begin{bmatrix} 1 & 2 & -3 & \vdots & 1 & 0 & 0 \\ 0 & -1 & 2 & \vdots & -2 & 1 & 0 \\ 0 & 0 & 4 & \vdots & -3 & 2 & 1 \end{bmatrix} \qquad 2R_2 + R_3 \rightarrow R_3$$

$$\sim \begin{bmatrix} 1 & 2 & -3 & \vdots & 1 & 0 & 0 \\ 0 & 1 & -2 & \vdots & 2 & -1 & 0 \\ 0 & 0 & 1 & \vdots & -\frac{3}{4} & \frac{1}{2} & \frac{1}{4} \end{bmatrix} \quad \begin{array}{l} -R_2 \to R_2 \\ \frac{1}{4}R_3 \to R_3 \end{array}$$

$$\sim \begin{bmatrix} 1 & 2 & 0 & \vdots & -\frac{5}{4} & \frac{3}{2} & \frac{3}{4} \\ 0 & 1 & 0 & \vdots & \frac{1}{2} & 0 & \frac{1}{2} \\ 0 & 0 & 1 & \vdots & -\frac{3}{4} & \frac{1}{2} & \frac{1}{4} \end{bmatrix} \quad \begin{array}{l} 3R_3 + R_1 \to R_1 \\ 2R_3 + R_2 \to R_2 \end{array}$$

$$\sim \begin{bmatrix} 1 & 0 & 0 & \vdots & -\frac{9}{4} & \frac{3}{2} & -\frac{1}{4} \\ 0 & 1 & 0 & \vdots & \frac{1}{2} & 0 & \frac{1}{2} \\ 0 & 0 & 1 & \vdots & -\frac{3}{4} & \frac{1}{2} & \frac{1}{4} \end{bmatrix} \quad -2R_2 + R_1 \to R_1$$

We can now conclude that the inverse matrix of A is

$$A^{-1} = \begin{bmatrix} -\frac{9}{4} & \frac{3}{2} & -\frac{1}{4} \\ \frac{1}{2} & 0 & \frac{1}{2} \\ -\frac{3}{4} & \frac{1}{2} & \frac{1}{4} \end{bmatrix} \quad \text{or} \quad \frac{1}{4}\begin{bmatrix} -9 & 6 & -1 \\ 2 & 0 & 2 \\ -3 & 2 & 1 \end{bmatrix}.$$

To eliminate fractional elements, factor the scalar $\frac{1}{4}$ from each element of the matrix.

Check to see if $AA^{-1} = A^{-1}A = I_3$.   ◆

Since matrix A in Example 3 has an inverse, we say that A is **invertible**, or **nonsingular**. Not every square $n \times n$ matrix A has an inverse. If elementary row operations on the matrix $[A \mid I_n]$ yield a row of zeros on the A portion of this matrix, then we cannot write $[A \mid I_n]$ in the form $[I_n \mid B]$. If a matrix does not have an inverse, we say that it is **singular**.

PROBLEM 3   Show that matrix A is singular if $A = \begin{bmatrix} 1 & 2 & -3 \\ 1 & 0 & 2 \\ 2 & 2 & -1 \end{bmatrix}$.   ◆

◆ **Matrix Equations**

Consider the equation

$$AX = B,$$

where A and B are known matrices and X is the unknown matrix for which we wish to solve. We refer to such an equation as a **matrix equation**. Because many of the properties of matrices are similar to properties of real numbers, the process of solving a matrix equation is similar to the process for solving an algebraic equation. Two properties of real numbers do not have equivalent matrix properties and cannot be used when solving a matrix equation:

1. Matrix division is not defined.
2. Matrix multiplication is not commutative.

Since matrix division is undefined, we *cannot* solve the equation $AX = B$ for X by dividing both sides by A. Instead, we multiply both sides of the equation by the matrix A^{-1}. Since matrix multiplication is not commutative, we must be consistent in the order that we multiply. We can either write

$$A^{-1}(AX) = A^{-1}B \qquad \text{or} \qquad (AX)A^{-1} = BA^{-1}$$

| Multiply on the left by A^{-1} | Multiply on the right by A^{-1} |

If we multiply both sides of the equation on the left by A^{-1}, we can apply the associative property of matrix multiplication and solve for X as follows:

$$AX = B$$

$$A^{-1}(AX) = A^{-1}B \qquad \text{Multiply both sides of the equation on the left by } A^{-1}$$

$$(A^{-1}A)X = A^{-1}B \qquad \text{Apply associative property of matrix multiplication}$$

$$I_nX = A^{-1}B \qquad \text{Apply inverse property of matrix multiplication}$$

$$X = A^{-1}B \qquad \text{Apply identity property of matrix multiplication}$$

Thus, if A and B are known matrices and A is invertible, then we can find X by simply evaluating the product $A^{-1}B$. The next example illustrates how the properties of matrix addition, scalar multiplication, and matrix multiplication are used in solving a matrix equation.

EXAMPLE 4   Given

$$A = \begin{bmatrix} 3 & 4 \\ 2 & 3 \end{bmatrix}, \qquad B = \begin{bmatrix} -6 & 3 \\ 1 & 2 \end{bmatrix}, \quad \text{and} \quad C = \begin{bmatrix} 2 & 3 \\ -1 & 4 \end{bmatrix},$$

solve the equation $2AX + B = C$ for the matrix X.

SOLUTION   We begin by solving the equation for X as follows:

$$2AX + B = C$$

$$(2AX + B) + (-B) = C + (-B) \qquad \text{Add } -B \text{ to both sides}$$

$$2AX + \mathbf{0} = C - B \qquad \text{Apply associative and inverse properties of matrix addition}$$

$$2AX = C - B \qquad \text{Apply identity property of matrix addition}$$

$$AX = \tfrac{1}{2}(C - B) \qquad \text{Multiply both sides of the equation by the scalar } \tfrac{1}{2}$$

$$A^{-1}(AX) = \tfrac{1}{2}A^{-1}(C - B)$$

> Multiply both sides of the equation on the left by A^{-1}

$$I_nX = \tfrac{1}{2}A^{-1}(C - B)$$

> Apply associative and inverse properties of matrix multiplication

$$X = \tfrac{1}{2}A^{-1}(C - B)$$

> Apply identity property of matrix multiplication

From Example 2, we know that if

$$A = \begin{bmatrix} 3 & 4 \\ 2 & 3 \end{bmatrix}, \quad \text{then} \quad A^{-1} = \begin{bmatrix} 3 & -4 \\ -2 & 3 \end{bmatrix}.$$

Now substituting the matrices A^{-1}, B, and C, we have

$$X = \tfrac{1}{2}A^{-1}(C - B) = \frac{1}{2}\begin{bmatrix} 3 & -4 \\ -2 & 3 \end{bmatrix}\left(\begin{bmatrix} 2 & 3 \\ -1 & 4 \end{bmatrix} - \begin{bmatrix} -6 & 3 \\ 1 & 2 \end{bmatrix}\right)$$

$$= \frac{1}{2}\begin{bmatrix} 3 & -4 \\ -2 & 3 \end{bmatrix}\begin{bmatrix} 8 & 0 \\ -2 & 2 \end{bmatrix}$$

$$= \frac{1}{2}\begin{bmatrix} 32 & -8 \\ -22 & 6 \end{bmatrix}$$

$$= \begin{bmatrix} 16 & -4 \\ -11 & 3 \end{bmatrix}$$

Thus, we conclude $X = \begin{bmatrix} 16 & -4 \\ -11 & 3 \end{bmatrix}$. ◆

To check the solution of a matrix equation, we replace X in the original equation with the derived matrix to see if the equation becomes true. For Example 4, we need to determine if

$$\underset{2AX}{2\begin{bmatrix} 3 & 4 \\ 2 & 3 \end{bmatrix}\begin{bmatrix} 16 & -4 \\ -11 & 3 \end{bmatrix}} + \underset{B}{\begin{bmatrix} -6 & 3 \\ 1 & 2 \end{bmatrix}} = \underset{C}{\begin{bmatrix} 2 & 3 \\ -1 & 4 \end{bmatrix}}$$

PROBLEM 4 Check the matrix equation in Example 4. ◆

◆ **The Inverse Method**

The system of n linear equations in n unknowns

$$a_{11}x_1 + a_{12}x_2 + a_{13}x_3 + \cdots + a_{1n}x_n = k_1$$
$$a_{21}x_1 + a_{22}x_2 + a_{23}x_3 + \cdots + a_{2n}x_n = k_2$$
$$a_{31}x_1 + a_{32}x_2 + a_{33}x_3 + \cdots + a_{3n}x_n = k_3$$
$$\vdots \qquad \vdots \qquad \vdots \qquad \qquad \vdots \qquad \vdots$$
$$a_{n1}x_1 + a_{n2}x_2 + a_{n3}x_3 + \cdots + a_{nn}x_n = k_n$$

can be represented by the matrix equation

$$AX = B,$$

where

$$A = \begin{bmatrix} a_{11} & a_{12} & a_{13} & \cdots & a_{1n} \\ a_{21} & a_{22} & a_{23} & \cdots & a_{2n} \\ a_{31} & a_{32} & a_{33} & \cdots & a_{3n} \\ \vdots & \vdots & \vdots & & \vdots \\ a_{n1} & a_{n2} & a_{n3} & \cdots & a_{nn} \end{bmatrix}, \quad X = \begin{bmatrix} x_1 \\ x_2 \\ x_3 \\ \vdots \\ x_n \end{bmatrix}, \quad \text{and} \quad B = \begin{bmatrix} k_1 \\ k_2 \\ k_3 \\ \vdots \\ k_n \end{bmatrix}$$

We refer to matrix A as the **coefficient matrix**, matrix X as the **variable matrix**, and matrix B as the **constant matrix**. If the coefficient matrix A is invertible, then we can solve the equation $AX = B$ for X to obtain

$$X = A^{-1}B.$$

The product $A^{-1}B$ represents the solution to the system of linear equations. We refer to this technique of solving a system of linear equations as the **inverse method**.

◆ The Inverse Method

> If the matrix equation $AX = B$ represents a system of n linear equations in n unknowns and the coefficient matrix A is invertible, then the system has a unique solution given by
>
> $$X = A^{-1}B.$$
>
> If the coefficinet matrix A is not invertible, then the system is either dependent or inconsistent.

E X A M P L E 5   Use the inverse method to solve the given system of linear equations:

$$x + 2y - 3z = -1$$
$$2x + 3y - 4z = 2$$
$$-x \qquad + 3z = 5$$

S O L U T I O N   If we let

$$A = \begin{bmatrix} 1 & 2 & -3 \\ 2 & 3 & -4 \\ -1 & 0 & 3 \end{bmatrix}, \quad X = \begin{bmatrix} x \\ y \\ z \end{bmatrix}, \quad \text{and} \quad B = \begin{bmatrix} -1 \\ 2 \\ 5 \end{bmatrix},$$

Coefficient matrix   Variable matrix   Constant matrix then the matrix equation

$$
\begin{array}{ccc}
AX & = & B
\end{array}
$$

$$
\begin{bmatrix} 1 & 2 & -3 \\ 2 & 3 & -4 \\ -1 & 0 & 3 \end{bmatrix} \begin{bmatrix} x \\ y \\ z \end{bmatrix} = \begin{bmatrix} -1 \\ 2 \\ 5 \end{bmatrix}
$$

is equivalent to the given system of three equations in three unknowns. From Example 3, we know the coefficient matrix A is invertible and that

$$
A^{-1} = \begin{bmatrix} -\frac{9}{4} & \frac{3}{2} & -\frac{1}{4} \\ \frac{1}{2} & 0 & \frac{1}{2} \\ -\frac{3}{4} & \frac{1}{2} & \frac{1}{4} \end{bmatrix} \quad \text{or} \quad \frac{1}{4} \begin{bmatrix} -9 & 6 & -1 \\ 2 & 0 & 2 \\ -3 & 2 & 1 \end{bmatrix}.
$$

Thus, the system of equations has a unique solution given by

$$
X = A^{-1}B = \frac{1}{4} \begin{bmatrix} -9 & 6 & -1 \\ 2 & 0 & 2 \\ -3 & 2 & 1 \end{bmatrix} \begin{bmatrix} -1 \\ 2 \\ 5 \end{bmatrix} = \frac{1}{4} \begin{bmatrix} 16 \\ 8 \\ 12 \end{bmatrix} = \begin{bmatrix} 4 \\ 2 \\ 3 \end{bmatrix}.
$$

Since

$$
X = \begin{bmatrix} x \\ y \\ z \end{bmatrix} = \begin{bmatrix} 4 \\ 2 \\ 3 \end{bmatrix},
$$

we conclude that $x = 4$, $y = 2$, and $z = 3$.   ◆

PROBLEM 5   Determine if the given system of linear equations has a unique solution:

$$
\begin{array}{rcl}
x + 2y - 3z & = & -3 \\
x \quad\quad + 2z & = & 4 \\
2x + 2y - z & = & 1
\end{array}
$$ ◆

◆ **Application: Revenue from Sales**

In some applications, it is necessary to solve several systems of linear equations having the same coefficient matrix. The inverse method is especially useful for solving such systems.

EXAMPLE 6   A campus bookstore sells two types of calculators: the scientific type sells for $30 and the graphing type for $50. The following table shows the daily revenue earned from the sale of calculators for each of the first three days of the semester:

	Day 1	Day 2	Day 3
Number of calculators sold	22	31	35
Revenue earned	$740	$1150	$1250

Determine the number of each type calculator that is sold on each of these days.

SOLUTION   Let

$$x = \text{number of scientific calculators sold on a specific day,}$$

$$y = \text{number of graphing calculators sold on a specific day.}$$

If k_1 is the *total number* of calculators sold on a specific day and k_2 is the *total revenue* earned on that day, then we have

$$x + \quad y = k_1$$

$$30x + 50y = k_2$$

We can represent this system of linear equations by the matrix equation $AX = B$, where

$$A = \begin{bmatrix} 1 & 1 \\ 30 & 50 \end{bmatrix}, \quad X = \begin{bmatrix} x \\ y \end{bmatrix}, \quad \text{and} \quad B = \begin{bmatrix} k_1 \\ k_2 \end{bmatrix}.$$

In Problem 6 you are asked to verify that the inverse of the coefficient matrix A is

$$A^{-1} = \begin{bmatrix} \frac{5}{2} & -\frac{1}{20} \\ -\frac{3}{2} & \frac{1}{20} \end{bmatrix} \quad \text{or} \quad \frac{1}{20}\begin{bmatrix} 50 & -1 \\ -30 & 1 \end{bmatrix}$$

Thus, the solution to this system of linear equations is given by

$$X = A^{-1}B = \frac{1}{20}\begin{bmatrix} 50 & -1 \\ -30 & 1 \end{bmatrix}\begin{bmatrix} k_1 \\ k_2 \end{bmatrix},$$

where the values of k_1 and k_2 are in the given table.

Day 1:

$$X = \frac{1}{20}\begin{bmatrix} 50 & -1 \\ -30 & 1 \end{bmatrix}\begin{bmatrix} k_1 \\ k_2 \end{bmatrix} = \frac{1}{20}\begin{bmatrix} 50 & -1 \\ -30 & 1 \end{bmatrix}\begin{bmatrix} 22 \\ 740 \end{bmatrix} = \begin{bmatrix} 18 \\ 4 \end{bmatrix}$$

Thus, on the first day, 18 scientific and 4 graphing calculators are sold.

Day 2:

$$X = \frac{1}{20}\begin{bmatrix} 50 & -1 \\ -30 & 1 \end{bmatrix}\begin{bmatrix} k_1 \\ k_2 \end{bmatrix} = \frac{1}{20}\begin{bmatrix} 50 & -1 \\ -30 & 1 \end{bmatrix}\begin{bmatrix} 31 \\ 1150 \end{bmatrix} = \begin{bmatrix} 20 \\ 11 \end{bmatrix}$$

Thus, on the second day, 20 scientific and 11 graphing calculators are sold.

Day 3:

$$X = \frac{1}{20}\begin{bmatrix} 50 & -1 \\ -30 & 1 \end{bmatrix}\begin{bmatrix} k_1 \\ k_2 \end{bmatrix} = \frac{1}{20}\begin{bmatrix} 50 & -1 \\ -30 & 1 \end{bmatrix}\begin{bmatrix} 35 \\ 1250 \end{bmatrix} = \begin{bmatrix} 25 \\ 10 \end{bmatrix}$$

Thus, on the third day, 25 scientific and 10 graphing calculators are sold.   ◆

P R O B L E M   6   Verify that the inverse of the coefficient matrix A in Example 6 is

$$A^{-1} = \begin{bmatrix} \frac{5}{2} & -\frac{1}{20} \\ -\frac{3}{2} & \frac{1}{20} \end{bmatrix} \quad \text{or} \quad \frac{1}{20}\begin{bmatrix} 50 & -1 \\ -30 & 1 \end{bmatrix}$$ ◆

Exercises 7.4

Basic Skills

In Exercises 1–6, determine whether the matrices A and B are inverses of each other.

1. $A = \begin{bmatrix} 3 & -4 \\ -2 & 3 \end{bmatrix}$, $B = \begin{bmatrix} 3 & 4 \\ 2 & 3 \end{bmatrix}$

2. $A = \begin{bmatrix} 5 & -4 \\ -1 & 1 \end{bmatrix}$, $B = \begin{bmatrix} 1 & 4 \\ 1 & 5 \end{bmatrix}$

3. $A = \begin{bmatrix} 1 & 3 \\ 2 & 4 \end{bmatrix}$, $B = \begin{bmatrix} -2 & \frac{3}{2} \\ 1 & -\frac{1}{2} \end{bmatrix}$

4. $A = \begin{bmatrix} 2 & -15 \\ -1 & 6 \end{bmatrix}$, $B = \begin{bmatrix} -2 & -5 \\ -\frac{1}{3} & -\frac{2}{3} \end{bmatrix}$

5. $A = \begin{bmatrix} 1 & 2 & 0 \\ -1 & 4 & 2 \\ 0 & 3 & 3 \end{bmatrix}$, $B = \frac{1}{12}\begin{bmatrix} 6 & -6 & 4 \\ 3 & 3 & -2 \\ -3 & -3 & 6 \end{bmatrix}$

6. $A = \begin{bmatrix} 1 & 4 & -3 \\ 0 & 1 & 2 \\ -2 & -2 & 0 \end{bmatrix}$, $B = \frac{1}{18}\begin{bmatrix} -4 & -6 & -11 \\ 4 & 6 & 2 \\ -2 & 6 & -1 \end{bmatrix}$

In Exercises 7–28, find the inverse of matrix A. If A does not have an inverse, state that it is singular.

7. $A = \begin{bmatrix} 3 & 4 \\ -2 & -3 \end{bmatrix}$

8. $A = \begin{bmatrix} 6 & 7 \\ 5 & 6 \end{bmatrix}$

9. $A = \begin{bmatrix} 1 & 4 \\ 2 & 3 \end{bmatrix}$

10. $A = \begin{bmatrix} 4 & 2 \\ -1 & 2 \end{bmatrix}$

11. $A = \begin{bmatrix} 3 & 1 \\ 5 & -2 \end{bmatrix}$

12. $A = \begin{bmatrix} 4 & -3 \\ 2 & 0 \end{bmatrix}$

13. $A = \begin{bmatrix} 2 & 3 \\ -6 & -9 \end{bmatrix}$

14. $A = \begin{bmatrix} 5 & -2 \\ -10 & 4 \end{bmatrix}$

15. $A = \begin{bmatrix} 1 & 2 & -3 \\ 0 & 2 & -1 \end{bmatrix}$

16. $A = \begin{bmatrix} 3 & 5 \\ 0 & -2 \\ -2 & 3 \end{bmatrix}$

17. $A = \begin{bmatrix} 1 & -1 & 0 \\ 2 & -1 & 1 \\ 0 & 4 & 5 \end{bmatrix}$

18. $A = \begin{bmatrix} 1 & 0 & 2 \\ 0 & 1 & 4 \\ -1 & 0 & 2 \end{bmatrix}$

19. $A = \begin{bmatrix} 2 & 0 & 5 \\ 1 & -1 & 4 \\ -3 & 2 & -8 \end{bmatrix}$

20. $A = \begin{bmatrix} -3 & 9 & 0 \\ 4 & -20 & 1 \\ 1 & -5 & 0 \end{bmatrix}$

21. $A = \begin{bmatrix} -2 & -1 & -8 \\ 3 & 2 & 10 \\ 4 & 4 & 6 \end{bmatrix}$

22. $A = \begin{bmatrix} 3 & -8 & 12 \\ -2 & 6 & -8 \\ 5 & -8 & 12 \end{bmatrix}$

23. $A = \begin{bmatrix} 3 & 2 & 1 \\ -2 & 5 & 3 \\ 1 & 0 & -4 \end{bmatrix}$

24. $A = \begin{bmatrix} 5 & 1 & 2 \\ 2 & 4 & 1 \\ 3 & -3 & 1 \end{bmatrix}$

25. $A = \begin{bmatrix} 1 & 0 & 2 & 0 \\ 0 & 2 & 1 & 1 \\ 0 & 1 & 0 & 2 \\ 0 & 2 & 0 & 0 \end{bmatrix}$

26. $A = \begin{bmatrix} 1 & 1 & -2 & 3 \\ 0 & 1 & 0 & -3 \\ 1 & 2 & -4 & 0 \\ 0 & -1 & 2 & 1 \end{bmatrix}$

27. $A = \begin{bmatrix} 1 & 0 & 0 & 2 & 1 \\ 0 & 1 & 0 & -1 & 0 \\ 0 & 0 & 2 & -3 & 1 \\ 0 & 0 & -1 & 2 & 0 \\ 0 & 0 & 0 & -4 & -2 \end{bmatrix}$

28. $A = \begin{bmatrix} 1 & 0 & 2 & -1 & 0 \\ 2 & 1 & 0 & -2 & 4 \\ 0 & -1 & 5 & 0 & -2 \\ 0 & -1 & 3 & 2 & -3 \\ -1 & 1 & -6 & 0 & 2 \end{bmatrix}$

Given that

$$A = \begin{bmatrix} 2 & -3 \\ -3 & 5 \end{bmatrix}, \qquad B = \begin{bmatrix} 1 & 3 \\ 2 & 4 \end{bmatrix},$$

$$C = \begin{bmatrix} -4 & -3 \\ 5 & 3 \end{bmatrix},$$

solve the equations in Exercises 29–34 for the matrix X.

29. $A + X = B$ **30.** $2C - 3X = B$

31. $3AX = B$ **32.** $BX - 3C = B$

33. $AX + BX = 2C$ **34.** $2AX - 3B = CX + A$

Given that

$$A = \begin{bmatrix} 1 & 2 & 3 \\ -1 & -3 & 0 \\ 1 & 3 & 1 \end{bmatrix}, \qquad B = \begin{bmatrix} 2 & -2 & -3 \\ 1 & 0 & -2 \\ -2 & 2 & 4 \end{bmatrix},$$

$$C = \begin{bmatrix} 1 & 3 & -2 \\ -1 & -2 & 1 \\ 0 & -4 & 6 \end{bmatrix}$$

solve the equations in Exercises 35–40 for the matrix X.

35. $A - 2X = B$ **36.** $2A + X = C$

37. $2AX + C = 0$ **38.** $2A - CX = B$

39. $2AX - BX = 3C$ **40.** $BX - AX = CX - 2A$

In Exercises 41–52, use the inverse method to solve each system of linear equations.

41. $\begin{aligned} x - y &= 7 \\ 2x + 3y &= 11 \end{aligned}$ **42.** $\begin{aligned} 3x - 2y &= 5 \\ x + 2y &= -1 \end{aligned}$

43. $\begin{aligned} 5x - y &= 2 \\ 4x - y &= 0 \end{aligned}$ **44.** $\begin{aligned} 2x - 3y &= 0 \\ -3x + 5y &= 1 \end{aligned}$

45. $\begin{aligned} x - y + z &= -3 \\ 2x + y &= 1 \\ y - 3z &= 7 \end{aligned}$

46. $\begin{aligned} x + 3y &= 4 \\ y - 2z &= 3 \\ x - y + 3z &= -3 \end{aligned}$

47. $\begin{aligned} x - 3y + z &= -3 \\ 2x - 3z &= 7 \\ -x + y + 2z &= -4 \end{aligned}$

48. $\begin{aligned} 3x - y - 2z &= 0 \\ 2x + 8y &= -1 \\ x - y + 10z &= 27 \end{aligned}$

49. $\begin{aligned} 2x - y + z &= 4 \\ 3x + 2y - 3z &= 2 \\ 4x + 5y - 6z &= -3 \end{aligned}$

50. $\begin{aligned} -2x - y + 4z &= -7 \\ 3x + 2y - z &= 4 \\ 4x - 3y + 2z &= -1 \end{aligned}$

51. $\begin{aligned} x + 2y + t &= 3 \\ y + 2z &= 3 \\ x - 2z &= 0 \\ 3y - 4z + t &= 2 \end{aligned}$

52.
$$
\begin{aligned}
x - 2y \quad\quad\; + \; t &= 0 \\
3x - \; y + 4z - \; t &= -10 \\
3y - \; z + 2t &= 14 \\
x \quad\quad - 3z + 4t &= 19
\end{aligned}
$$

In Exercises 53–56, use systems of linear equations and the inverse method to solve each problem.

53. The nonequal side of an isosceles triangle is k_1 centimeters less than one of the equal sides, and the perimeter of the triangle is k_2 centimeters. Determine the dimensions of the triangle if k_1, k_2 are as given:

(a) $k_1 = 15$ cm and $k_2 = 60$ cm
(b) $k_1 = 36$ cm and $k_2 = 75$ cm
(c) $k_1 = 24$ cm and $k_2 = 112$ cm
(d) $k_1 = 32$ cm and $k_2 = 220$ cm

54. An investor buys two types of stocks: one costs $36 per share and the other costs $32 per share. Altogether she purchased k_1 shares of stock for a total cost of k_2 dollars. Find how many shares of each stock she purchased if k_1 and k_2 are as given:

(a) $k_1 = 30$ shares and $k_2 = \$1024$
(b) $k_1 = 72$ shares and $k_2 = \$2392$
(c) $k_1 = 67$ shares and $k_2 = \$2364$
(d) $k_1 = 224$ shares and $k_2 = \$8064$

55. Three circular pulleys are tangent to each other, as shown in the sketch. If the center-to-center distances AB,

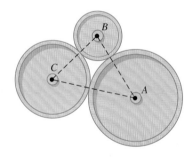

AC, and BC are k_1, k_2, and k_3, respectively, find the radius of each pulley given the following information:

(a) $k_1 = 18$ cm, $k_2 = 22$ cm, and $k_3 = 16$ cm
(b) $k_1 = 48$ cm, $k_2 = 34$ cm, and $k_3 = 26$ cm
(c) $k_1 = 73$ cm, $k_2 = 122$ cm, and $k_3 = 117$ cm
(d) $k_1 = 15$ cm, $k_2 = 11$ cm, and $k_3 = 17$ cm

56. A photography studio develops four sizes of prints (in inches): 2×3, 3×5, 5×8, and 8×10. After taking senior-class pictures, the studio offers the following four packages to the students:

	Number of prints			
Package	**2×3**	**3×5**	**5×8**	**8×10**
A	4	2	1	—
B	—	2	2	1
C	8	—	4	2
D	12	4	2	2

If package A sells for k_1 dollars, package B for k_2 dollars, package C for k_3 dollars, and package D for k_4 dollars, determine the cost per print for each size print, given the following information.

(a) $k_1 = \$12$, $k_2 = \$24$, $k_3 = \$46$, and $k_4 = \$47$
(b) $k_1 = \$7$, $k_2 = \$18$, $k_3 = \$36$, and $k_4 = \$36$
(c) $k_1 = \$14$, $k_2 = \$31$, $k_3 = \$62$, and $k_4 = \$55$
(d) $k_1 = \$15$, $k_2 = \$30$, $k_3 = \$57$, and $k_4 = \$54$

Critical Thinking

57. Find the value of x for which the given matrix is singular.

(a) $\begin{bmatrix} 3 & 2 \\ -4 & x \end{bmatrix}$ (b) $\begin{bmatrix} 1 & 4 & 3 \\ 2 & 6 & 4 \\ -3 & x & 1 \end{bmatrix}$

58. Describe the inverse of an $n \times n$ matrix having nonzero elements $a_{11}, a_{22}, a_{33}, \ldots, a_{nn}$ along its main diagonal and zeros elsewhere.

59. Suppose A, B, and C are $n \times n$ matrices and $AB = AC$. Is it possible to "cancel" matrix A from both sides of this equation and conclude that $B = C$? Explain.

60. Given matrices A and B, find each value:

$$
A = \begin{bmatrix} 3 & 5 \\ 4 & 7 \end{bmatrix} \quad \text{and} \quad B = \begin{bmatrix} 2 & -5 \\ -3 & 8 \end{bmatrix}
$$

EXERCISE 60 (*continued*)

(a) $(AB)^{-1}$ (b) $(BA)^{-1}$
(c) $B^{-1}A^{-1}$ (d) $A^{-1}B^{-1}$
(e) Compare your answers in parts (a)–(d). What do you observe? Prove that your observation is true for all $n \times n$ nonsingular matrices A and B.

61. Given the 2×2 matrix $A = \begin{bmatrix} a & b \\ c & d \end{bmatrix}$, where $ad - bc \neq 0$, find the product

$$\frac{1}{ad - bc} \begin{bmatrix} d & -b \\ -c & a \end{bmatrix} A.$$

What does this result tell us about the inverse of matrix A?

62. Using the result of Exercise 61, find the inverse of each matrix:

(a) $\begin{bmatrix} 2 & 1 \\ 3 & 7 \end{bmatrix}$ (b) $\begin{bmatrix} 4 & -5 \\ -9 & 4 \end{bmatrix}$

(c) $\begin{bmatrix} 3 & 2 \\ -4 & -3 \end{bmatrix}$ (d) $\begin{bmatrix} 15 & 12 \\ -9 & -8 \end{bmatrix}$

Calculator Activities

In Exercises 63–66, solve the system of linear equations for the given constant values, using the inverse method in conjunction with a calculator. Round each answer to three significant digits.

63. $2.4x - 3.9y = k_1$
$\quad 8.4x + 4.2y = k_2$

(a) $k_1 = 8.25, k_2 = 2.55$
(b) $k_1 = -12.3, k_2 = 19.8$
(c) $k_1 = -44.3, k_2 = -31.2$
(d) $k_1 = -0.882, k_2 = -0.982$

64. $18.4x - 12.6y = k_1$
$\quad 44.5x - 31.2y = k_2$

(a) $k_1 = 1.15, k_2 = 3.67$
(b) $k_1 = -18.9, k_2 = 22.8$
(c) $k_1 = -99.8, k_2 = -102.3$
(d) $k_1 = -0.0759, k_2 = -0.112$

65. $2.4x - 3.9y \qquad = k_1$
$\qquad 8.4y + 4.2z = k_2$
$\quad 3.2x \qquad - 1.9z = k_3$

(a) $k_1 = 1.32, k_2 = 1.98, k_3 = 3.91$
(b) $k_1 = -17.1, k_2 = 22.0, k_3 = 44.5$
(c) $k_1 = -32.0, k_2 = -22.8, k_3 = 63.9$
(d) $k_1 = -0.234, k_2 = -0.123, k_3 = -0.812$

66. $281x - 752y + 314z = k_1$
$\qquad 213y + 120z = k_2$
$\quad 191x - 348y \qquad = k_3$

(a) $k_1 = 5.35, k_2 = 1.90, k_3 = 6.13$
(b) $k_1 = -22.4, k_2 = 69.1, k_3 = 90.0$
(c) $k_1 = -124, k_2 = -458, k_3 = 200$
(d) $k_1 = -1260, k_2 = -7620, k_3 = -295$

67. A hardware store has two types of chain saws on sale. The 14-inch saw is on sale for $179.95 and the 16-inch

saw is on sale for $229.95. The following table shows the daily revenue earned from the chain saws for each of the first four days of the sale:

	Day 1	Day 2	Day 3	Day 4
Number of chain saws sold	12	9	16	11
Revenue earned	$2459.40	$1919.55	$3179.20	$2129.45

Determine the number of each type of chain saw that is sold on each of these days.

68. For the electrical circuit shown in the figure, the branch currents i_1, i_2, and i_3 (in amperes) may be found by solving the following system of linear equations:

$$i_1 - i_2 - i_3 = 0$$
$$8.8i_1 \qquad + 7.9i_3 = V_1$$
$$2.3i_2 - 7.9i_3 = -V_2$$

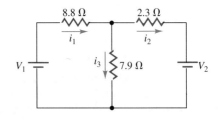

Determine the currents i_1, i_2, and i_3 for the given voltages, rounding each answer to three significant digits.

(a) $V_1 = 2.80$ volts, $V_2 = 3.50$ volts
(b) $V_1 = 22.3$ volts, $V_2 = 13.5$ volts
(c) $V_1 = 9.85$ volts, $V_2 = 2.25$ volts
(d) $V_1 = 34.5$ volts, $V_2 = 45.6$ volts

(*Hint:* Negative values for either i_1, i_2, or i_3 indicate that that current is flowing in the opposite direction of what is shown in the sketch.)

7.5 Determinants and Cramer's Rule

Associated with every square matrix A is a real number called the **determinant of A**, which we denote by writing $|A|$. In this section, we discuss methods for evaluating the determinant of a matrix. We then show how determinants may be used to help solve a system of linear equations.

◆ Determinant of a 2 × 2 Matrix

The 2×2 matrix

$$A = \begin{bmatrix} a_{11} & a_{12} \\ a_{21} & a_{22} \end{bmatrix}$$

has the **determinant**

$$|A| = \begin{vmatrix} a_{11} & a_{12} \\ a_{21} & a_{22} \end{vmatrix},$$

whose value we find by multiplying along the diagonals and subtracting the products as follows:

$$\begin{vmatrix} a_{11} & a_{12} \\ a_{21} & a_{22} \end{vmatrix} = a_{11}a_{22} - a_{21}a_{12}$$

Determinant of a 2 × 2 Matrix

The **determinant** of the 2×2 matrix

$$A = \begin{bmatrix} a_{11} & a_{12} \\ a_{21} & a_{22} \end{bmatrix}$$

is

$$|A| = \begin{vmatrix} a_{11} & a_{12} \\ a_{21} & a_{22} \end{vmatrix} = a_{11}a_{22} - a_{21}a_{12}.$$

EXAMPLE 1   Find the determinant of each matrix.

(a) $A = \begin{bmatrix} 9 & 4 \\ 3 & -2 \end{bmatrix}$ (b) $B = \begin{bmatrix} 0 & -2 \\ 3 & 1 \end{bmatrix}$

SOLUTION

(a) The determinant of matrix A is

$$|A| = \begin{vmatrix} 9 & 4 \\ 3 & -2 \end{vmatrix} = (9)(-2) - (3)(4) = -18 - 12 = -30.$$

(b) The determinant of matrix B is

$$|B| = \begin{vmatrix} 0 & -2 \\ 3 & 1 \end{vmatrix} = (0)(1) - (3)(-2) = 0 + 6 = 6 \qquad \blacklozenge$$

PROBLEM 1   Repeat Example 1 for the matrix $C = \begin{bmatrix} -3 & -2 \\ 6 & 4 \end{bmatrix}$. ◆

◆ **Minors and Cofactors**

In order to define the determinant for an $n \times n$ matrix ($n > 2$), we need to introduce the concepts of *minors* and *cofactors*.

Minors and Cofactors

> If A is an $n \times n$ matrix, then the **minor** of an element a_{ij}, denoted M_{ij}, is the determinant of the matrix that remains after deleting the row and column in which the element a_{ij} appears. The **cofactor** of an element a_{ij}, denoted C_{ij}, differs from M_{ij} at most in sign and is given by
>
> $$C_{ij} = (-1)^{i+j}M_{ij}.$$

Note: For the element a_{ij}, if $i + j$ is *even*, then the value of $(-1)^{i+j}$ is 1. If $i + j$ is *odd*, then the value of $(-1)^{i+j}$ is -1. For an $n \times n$ matrix, the value of $(-1)^{i+j}$ associated with the cofactor C_{ij} follows a pattern of alternating positive and negative signs:

$$\begin{bmatrix} + & - & + & - & + & \cdots \\ - & + & - & + & - & \cdots \\ + & - & + & - & + & \cdots \\ - & + & - & + & - & \cdots \\ + & - & + & - & + & \cdots \\ \vdots & \vdots & \vdots & \vdots & \vdots & \end{bmatrix}$$

EXAMPLE 2   Given the matrix

$$A = \begin{bmatrix} 1 & 3 & -2 \\ 3 & 0 & 1 \\ -1 & 2 & -4 \end{bmatrix},$$

determine the minor and cofactor of the elements (**a**) a_{31} and (**b**) a_{23}.

SOLUTION

(a)   To find M_{31}, the minor of a_{31}, we delete the third row and first column of matrix A as follows:

$$A = \begin{bmatrix} 1 & 3 & -2 \\ 3 & 0 & 1 \\ -1 & 2 & -4 \end{bmatrix}$$

Now, M_{31} is the determinant of the matrix that remains. Thus,

$$M_{31} = \begin{vmatrix} 3 & -2 \\ 0 & 1 \end{vmatrix} = 3 - 0 = 3.$$

The cofactor C_{31} of a_{31} is given by

$$C_{31} = (-1)^{3+1} M_{31} = M_{31} = 3.$$

(b)   To find M_{23}, the minor of a_{23}, we delete the second row and third column of matrix A as follows:

$$A = \begin{bmatrix} 1 & 3 & -2 \\ 3 & 0 & 1 \\ -1 & 2 & -4 \end{bmatrix}$$

Now, M_{23} is the determinant of the matrix that remains:

$$M_{23} = \begin{vmatrix} 1 & 3 \\ -1 & 2 \end{vmatrix} = 2 - (-3) = 5.$$

The cofactor C_{23} of a_{23} is given by

$$C_{23} = (-1)^{2+3} M_{23} = -M_{23} = -5. \qquad \blacklozenge$$

PROBLEM 2   Repeat Example 2 for the element a_{12}.   ◆

◆   **Determinant of an $n \times n$ Matrix**

We now define a procedure for finding the determinant of an $n \times n$ matrix. The procedure is called **expansion by cofactors**.

Expansion by Cofactors

The determinant of an $n \times n$ matrix can be found by multiplying each element in any row or column by its corresponding cofactor, and then adding the products.

For instance, to find the determinant of the 3 × 3 matrix

$$A = \begin{bmatrix} a_{11} & a_{12} & a_{13} \\ a_{21} & a_{22} & a_{23} \\ a_{31} & a_{32} & a_{33} \end{bmatrix},$$

we select any row or column for expansion by cofactors. If we choose the first row, then we have

$$|A| = a_{11}C_{11} + a_{12}C_{12} + a_{13}C_{13}$$

Now,

$$C_{11} = (-1)^{1+1} \begin{bmatrix} a_{11} & a_{12} & a_{13} \\ a_{21} & a_{22} & a_{23} \\ a_{31} & a_{32} & a_{33} \end{bmatrix} = \begin{vmatrix} a_{22} & a_{23} \\ a_{32} & a_{33} \end{vmatrix} = a_{22}a_{33} - a_{32}a_{23},$$

$$C_{12} = (-1)^{1+2} \begin{bmatrix} a_{11} & a_{12} & a_{13} \\ a_{21} & a_{22} & a_{23} \\ a_{31} & a_{32} & a_{33} \end{bmatrix} = -\begin{vmatrix} a_{21} & a_{23} \\ a_{31} & a_{33} \end{vmatrix} = -(a_{21}a_{33} - a_{31}a_{23}),$$

and

$$C_{13} = (-1)^{1+3} \begin{bmatrix} a_{11} & a_{12} & a_{13} \\ a_{21} & a_{22} & a_{23} \\ a_{31} & a_{32} & a_{33} \end{bmatrix} = \begin{vmatrix} a_{21} & a_{22} \\ a_{31} & a_{32} \end{vmatrix} = a_{21}a_{32} - a_{31}a_{22}$$

Thus,

$$\begin{aligned} |A| &= a_{11}C_{11} + a_{12}C_{12} + a_{13}C_{13} \\ &= a_{11}(a_{22}a_{33} - a_{32}a_{23}) - a_{12}(a_{21}a_{33} - a_{31}a_{23}) + a_{13}(a_{21}a_{32} - a_{31}a_{22}) \\ &= (a_{11}a_{22}a_{33} + a_{12}a_{23}a_{31} + a_{13}a_{21}a_{32}) \\ &\quad - (a_{12}a_{21}a_{33} + a_{11}a_{23}a_{32} + a_{13}a_{22}a_{31}) \end{aligned}$$

It is not necessary to memorize this formula. Instead, to find the determinant of a 3 × 3 matrix, we simply apply expansion by cofactors to any row or column. Since a zero element times its cofactor is zero, we can simplify the arithmetic by expanding about a row or column that contains the most zero elements.

EXAMPLE 3   Find the determinant of the matrix $A = \begin{bmatrix} 1 & 3 & -2 \\ 3 & 0 & 1 \\ -1 & 2 & -4 \end{bmatrix}$.

SOLUTION   Since $a_{22} = 0$, it is best to expand about the second row or second column. Choosing the second row for expansion by cofactors, we have

$$\begin{aligned} |A| &= a_{21}C_{21} + a_{22}C_{22} + a_{23}C_{23} \\ &= 3C_{21} + 0C_{22} + 1C_{23} \\ &= 3C_{21} + C_{23} \end{aligned}$$

Now,

$$C_{21} = (-1)^{2+1} \begin{bmatrix} 1 & 3 & -2 \\ 3 & 0 & 1 \\ -1 & 2 & -4 \end{bmatrix} = -\begin{vmatrix} 3 & -2 \\ 2 & -4 \end{vmatrix} = -[-12 - (-4)] = 8$$

and

$$C_{23} = (-1)^{2+3} \begin{bmatrix} 1 & 3 & -2 \\ 3 & 0 & 1 \\ -1 & 2 & -4 \end{bmatrix} = -\begin{vmatrix} 1 & 3 \\ -1 & 2 \end{vmatrix} = -[2 - (-3)] = -5$$

Thus,

$$|A| = 3C_{21} + C_{23} = (3)(8) + (-5) = 19. \qquad \blacklozenge$$

PROBLEM 3   Repeat Example 3 by choosing the second column for expansion by cofactors. You should obtain the same result. ◆

A **diagonal method** can be used to find the determinant of the 3×3 matrix

$$A = \begin{bmatrix} a_{11} & a_{12} & a_{13} \\ a_{21} & a_{22} & a_{23} \\ a_{31} & a_{32} & a_{33} \end{bmatrix}.$$

We first rewrite the first and second columns to the right of matrix A as follows:

$$\begin{bmatrix} a_{11} & a_{12} & a_{13} \\ a_{21} & a_{22} & a_{23} \\ a_{31} & a_{32} & a_{33} \end{bmatrix} \begin{matrix} a_{11} & a_{12} \\ a_{21} & a_{22} \\ a_{31} & a_{32} \end{matrix}$$

Now, we obtain the determinant of matrix A by adding the products of three diagonals and subtracting the products of three diagonals, as follows:

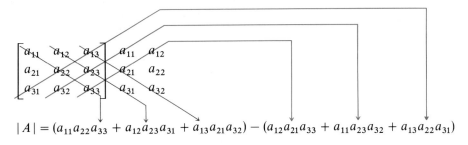

$$|A| = (a_{11}a_{22}a_{33} + a_{12}a_{23}a_{31} + a_{13}a_{21}a_{32}) - (a_{12}a_{21}a_{33} + a_{11}a_{23}a_{32} + a_{13}a_{22}a_{31})$$

Caution The diagonal method for finding the determinant can be applied only to a 3×3 matrix. In general, the diagonal method does not work for an $n \times n$ matrix.

EXAMPLE 4   Use the diagonal method to obtain the determinant of matrix A in Example 3.

SOLUTION   Rewriting the first and second columns to the right of matrix A, we find

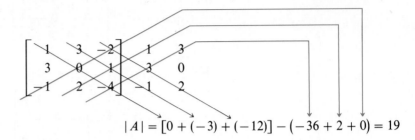

$$|A| = [0 + (-3) + (-12)] - (-36 + 2 + 0) = 19$$

This value agrees with the answer we obtained by expansion of cofactors.   ◆

PROBLEM 4   Find the determinant of the matrix

$$A = \begin{bmatrix} 2 & 3 & 1 \\ 3 & 1 & -2 \\ -3 & -1 & 0 \end{bmatrix}$$

by using (**a**) expansion by cofactors and (**b**) the diagonal method.   ◆

If we apply expansion by cofactors to a 4×4 matrix, we develop 3×3 determinants. To evaluate the 3×3 determinants, we can either continue with expansion by cofactors or apply the diagonal method.

EXAMPLE 5   Find the determinant of the matrix

$$A = \begin{bmatrix} 1 & -2 & 3 & -1 \\ 0 & 2 & 0 & 3 \\ 1 & 4 & 0 & -2 \\ 2 & 0 & 3 & 4 \end{bmatrix}$$

SOLUTION   Since both the second row and third column contain two zeros, we choose to expand about the second row or third column. Using the third column for expansion by cofactors, we have

$$|A| = a_{13}C_{13} + a_{23}C_{23} + a_{33}C_{33} + a_{43}C_{43}$$
$$= 3C_{13} + 0C_{23} + 0C_{33} + 3C_{43}$$
$$= 3C_{13} + 3C_{43}$$

Now, $C_{13} = (-1)^{1+3} \begin{bmatrix} 1 & -2 & 3 & -1 \\ 0 & 2 & 0 & 3 \\ 1 & 4 & 0 & -2 \\ 2 & 0 & 3 & 4 \end{bmatrix} = \begin{vmatrix} 0 & 2 & 3 \\ 1 & 4 & -2 \\ 2 & 0 & 4 \end{vmatrix},$

and $\qquad C_{43} = (-1)^{4+3} \begin{bmatrix} 1 & -2 & 3 & -1 \\ 0 & 2 & 0 & 3 \\ 1 & 4 & 0 & -2 \\ 2 & 0 & 3 & 4 \end{bmatrix} = - \begin{vmatrix} 1 & -2 & -1 \\ 0 & 2 & 3 \\ 1 & 4 & -2 \end{vmatrix},$

To evaluate these 3×3 cofactors, we can either continue with expansion by cofactors or apply the diagonal method. In Problem 5, you are asked to verify that

$$C_{13} = -40 \qquad \text{and} \qquad C_{43} = 20.$$

Thus, $\qquad |A| = 3C_{13} + 3C_{43} = 3(-40) + 3(20) = -60.$ ◆

PROBLEM 5 Referring to Example 5, verify that $C_{13} = -40$ and $C_{43} = 20$. ◆

◆ **Cramer's Rule**

Consider the system of two equations in two unknowns

$$a_{11}x + a_{12}y = k_1$$
$$a_{21}x + a_{22}y = k_2$$

Using any of the methods from the preceding sections, we solve this system and find that

$$x = \frac{a_{22}k_1 - a_{12}k_2}{a_{11}a_{22} - a_{21}a_{12}} \qquad \text{and} \qquad y = \frac{a_{11}k_2 - a_{21}k_1}{a_{11}a_{22} - a_{21}a_{12}}.$$

We can also use determinants to solve this system of linear equations. First, note that the coefficient matrix for this system is

$$A = \begin{bmatrix} a_{11} & a_{12} \\ a_{21} & a_{22} \end{bmatrix}.$$

Now, observe three facts about the values of x and y:

1. Both x and y have the same denominator, and this denominator is the determinant of the coefficient matrix A:

$$|A| = \begin{vmatrix} a_{11} & a_{12} \\ a_{21} & a_{22} \end{vmatrix} = a_{11}a_{22} - a_{21}a_{12}$$

2. The numerator of x is the determinant of the matrix A_x that is formed from the coefficient matrix A by replacing the coefficients of x (a_{11} and a_{21}) with the constant terms (k_1 and k_2):

$$|A_x| = \begin{vmatrix} k_1 & a_{12} \\ k_2 & a_{22} \end{vmatrix} = k_1a_{22} - k_2a_{12}$$

3. The numerator of y is the determinant of the matrix A_y that is formed from the coefficient matrix A by replacing the coefficients of y (a_{12} and a_{22}) with the constant terms (k_1 and k_2):

$$|A_y| = \begin{vmatrix} a_{11} & k_1 \\ a_{21} & k_2 \end{vmatrix} = a_{11}k_2 - a_{21}k_1$$

Putting these facts together, we have a rule that allows us to write the solution of a system of two linear equations in two unknowns in determinant form. We refer to this rule as **Cramer's rule**.

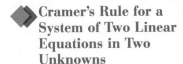

Cramer's Rule for a System of Two Linear Equations in Two Unknowns

Associated with the system

$$a_{11}x + a_{12}y = k_1$$
$$a_{21}x + a_{22}y = k_2$$

are the three determinants

$$|A| = \begin{vmatrix} a_{11} & a_{12} \\ a_{21} & a_{22} \end{vmatrix}, \quad |A_x| = \begin{vmatrix} k_1 & a_{12} \\ k_2 & a_{22} \end{vmatrix}, \quad \text{and} \quad |A_y| = \begin{vmatrix} a_{11} & k_1 \\ a_{21} & k_2 \end{vmatrix}.$$

The system has the unique solution

$$x = \frac{|A_x|}{|A|} \quad \text{and} \quad y = \frac{|A_y|}{|A|}, \quad \text{provided} \quad |A| \neq 0.$$

EXAMPLE 6   Solve the system of linear equations by using Cramer's rule.

$$x + 5y = 6$$
$$5x + 9y = -2$$

SOLUTION   Associated with this system are the three determinants

$$|A| = \begin{vmatrix} 1 & 5 \\ 5 & 9 \end{vmatrix} = 9 - 25 = -16, \qquad |A_x| = \begin{vmatrix} 6 & 5 \\ -2 & 9 \end{vmatrix} = 54 - (-10) = 64,$$

and

$$|A_y| = \begin{vmatrix} 1 & 6 \\ 5 & -2 \end{vmatrix} = -2 - 30 = -32.$$

Now, by Cramer's rule,

$$x = \frac{|A_x|}{|A|} = \frac{64}{-16} = -4 \quad \text{and} \quad y = \frac{|A_y|}{|A|} = \frac{-32}{-16} = 2.$$

Thus, the solution of the linear system is the ordered pair $(-4, 2)$.   ◆

If the determinant of the coefficient matrix A is zero, then Cramer's rule does *not* apply. If $|A| = 0$, and also $|A_x| = |A_y| = 0$, then the system is dependent (has an infinite number of solutions). If $|A| = 0$, but both $|A_x|$ and $|A_y|$ are not zero, then the system is inconsistent (has no solution).

PROBLEM 6   Use determinants to show that the given system of linear equations does not have a unique solution. Is the system dependent or inconsistent?

$$6x - 2y = -4$$
$$-9x + 3y = 6 \qquad \blacklozenge$$

For the $n \times n$ linear system

$$a_{11}x_1 + a_{12}x_2 + a_{13}x_3 + \cdots + a_{1n}x_n = k_1$$
$$a_{21}x_1 + a_{22}x_2 + a_{23}x_3 + \cdots + a_{2n}x_n = k_2$$
$$a_{31}x_1 + a_{32}x_2 + a_{33}x_3 + \cdots + a_{3n}x_n = k_3$$
$$\vdots \qquad \vdots \qquad \vdots \qquad \qquad \vdots \qquad \vdots$$
$$a_{n1}x_1 + a_{n2}x_2 + a_{n3}x_3 + \cdots + a_{nn}x_n = k_n$$

we can extend Cramer's rule to find $x_1, x_2, x_3, \ldots, x_n$, provided the coefficient matrix A of the system does not equal zero; that is, $|A| \neq 0$. For instance, to find x_j $(1 \leq j \leq n)$, we find the quotient of two determinants. The denominator is the determinant of the coefficient matrix A and the numerator is the determinant of the matrix A_j, which we obtain from A by replacing the jth column with the coefficients $k_1, k_2, k_3, \ldots, k_n$. Because of the enormous amount of arithmetic required, Cramer's rule is not an efficient method to use when solving a system of linear equations in which $n \geq 4$.

In the next example, we illustrate the procedure for solving a system of three equations in three unknowns by using Cramer's rule. The same system is solved in Example 4 of Section 7.2 by using matrices and elementary row operations. You should compare the two methods.

EXAMPLE 7   Solve the system of linear equations by using Cramer's rule.

$$2x + 4y - 2z = -1$$
$$3y - 4z = 6$$
$$-2x - 2y + z = 0$$

SOLUTION   Associated with this linear system are the four determinants

$$|A| = \begin{vmatrix} 2 & 4 & -2 \\ 0 & 3 & -4 \\ -2 & -2 & 1 \end{vmatrix}, \qquad |A_x| = \begin{vmatrix} -1 & 4 & -2 \\ 6 & 3 & -4 \\ 0 & -2 & 1 \end{vmatrix},$$

$$|A_y| = \begin{vmatrix} 2 & -1 & -2 \\ 0 & 6 & -4 \\ -2 & 0 & 1 \end{vmatrix}, \quad \text{and} \quad |A_z| = \begin{vmatrix} 2 & 4 & -1 \\ 0 & 3 & 6 \\ -2 & -2 & 0 \end{vmatrix}.$$

Using expansion by cofactors or the diagonal method, we find

$$|A| = 10, \qquad |A_x| = 5, \qquad |A_y| = -20, \quad \text{and} \quad |A_z| = -30.$$

Now, by Cramer's rule,

$$x = \frac{|A_x|}{|A|} = \frac{5}{10} = \frac{1}{2}, \qquad y = \frac{|A_y|}{|A|} = \frac{-20}{10} = -2,$$

and

$$z = \frac{|A_z|}{|A|} = \frac{-30}{10} = -3.$$

Thus, the solution of the linear system is the ordered triple $(\frac{1}{2}, -2, -3)$. ◆

PROBLEM 7 Repeat Example 7 for the linear system

$$
\begin{aligned}
2x \quad\quad - z &= 5 \\
-x + 3y \quad\quad &= 0 \\
6y - 3z &= 7
\end{aligned}
$$
◆

Exercises 7.5

Basic Skills

In Exercises 1–10, find the determinant of each matrix.

1. $\begin{bmatrix} 1 & 4 \\ 6 & 2 \end{bmatrix}$

2. $\begin{bmatrix} 9 & 0 \\ 1 & 2 \end{bmatrix}$

3. $\begin{bmatrix} 8 & -4 \\ -1 & 0 \end{bmatrix}$

4. $\begin{bmatrix} 5 & -2 \\ 3 & -4 \end{bmatrix}$

5. $\begin{bmatrix} -10 & -12 \\ 4 & -17 \end{bmatrix}$

6. $\begin{bmatrix} 12 & 6 \\ -15 & -18 \end{bmatrix}$

7. $\begin{bmatrix} \frac{1}{3} & -\frac{6}{5} \\ \frac{1}{2} & \frac{3}{5} \end{bmatrix}$

8. $\begin{bmatrix} 8 & -12 \\ -\frac{2}{3} & \frac{3}{4} \end{bmatrix}$

9. $\begin{bmatrix} a & u \\ 0 & b \end{bmatrix}$

10. $\begin{bmatrix} a & -a \\ -a & a \end{bmatrix}$

In Exercises 11–14, find the minor and cofactor of each element of the given matrix.

11. $\begin{bmatrix} 6 & 7 & 2 \\ -5 & 2 & -1 \\ 0 & 4 & 2 \end{bmatrix}$

12. $\begin{bmatrix} 1 & 3 & -4 \\ 9 & 2 & -3 \\ -1 & 0 & 7 \end{bmatrix}$

13. $\begin{bmatrix} 2 & 0 & -7 \\ -1 & 4 & 10 \\ -2 & 6 & 14 \end{bmatrix}$

14. $\begin{bmatrix} -8 & 4 & 12 \\ 0 & 9 & -7 \\ -16 & 3 & 8 \end{bmatrix}$

In Exercises 15–30, find the determinant of each matrix.

15. $\begin{bmatrix} 2 & 6 & 0 \\ 2 & -1 & 0 \\ -1 & 4 & 5 \end{bmatrix}$

16. $\begin{bmatrix} 0 & 0 & -5 \\ 9 & 3 & -2 \\ -3 & -4 & 1 \end{bmatrix}$

17. $\begin{bmatrix} 2 & 8 & -5 \\ 0 & 2 & -6 \\ 1 & -2 & 4 \end{bmatrix}$

18. $\begin{bmatrix} 4 & -5 & 8 \\ 9 & 2 & -2 \\ -1 & 0 & 3 \end{bmatrix}$

19. $\begin{bmatrix} 2 & -1 & 3 \\ 1 & 2 & 3 \\ 3 & -1 & 5 \end{bmatrix}$

20. $\begin{bmatrix} 3 & -2 & 1 \\ 2 & -2 & 1 \\ -2 & 1 & 1 \end{bmatrix}$

21. $\begin{bmatrix} i & j & k \\ 3 & 9 & 2 \\ 15 & 3 & -5 \end{bmatrix}$

22. $\begin{bmatrix} i & j & k \\ 8 & -13 & 9 \\ -1 & 9 & -4 \end{bmatrix}$

23. $\begin{bmatrix} a & u & v \\ 0 & b & x \\ 0 & 0 & c \end{bmatrix}$

24. $\begin{bmatrix} a & u & v & w \\ 0 & b & x & y \\ 0 & 0 & c & z \\ 0 & 0 & 0 & d \end{bmatrix}$

25. $\begin{bmatrix} 1 & 2 & 3 & 0 \\ -3 & 0 & 4 & 0 \\ 6 & -2 & 0 & 0 \\ 1 & 3 & 4 & -5 \end{bmatrix}$

26. $\begin{bmatrix} -1 & 2 & 0 & 4 \\ 0 & 0 & 0 & 2 \\ 2 & -3 & 4 & 0 \\ 1 & 0 & -3 & -2 \end{bmatrix}$

27. $\begin{bmatrix} 1 & -1 & 2 & 8 \\ 2 & 0 & 3 & -1 \\ -1 & 3 & 7 & 0 \\ 4 & 0 & 0 & 5 \end{bmatrix}$

28. $\begin{bmatrix} 0 & 3 & -4 & 1 \\ 9 & 1 & -1 & 0 \\ 2 & -2 & 0 & 6 \\ -1 & 6 & 3 & 0 \end{bmatrix}$

29. $\begin{bmatrix} 2 & -1 & 4 & 0 & 6 \\ 0 & 0 & -1 & 0 & 3 \\ -1 & 0 & 1 & 0 & 0 \\ 0 & 3 & 0 & -1 & 2 \\ 4 & 0 & 2 & 0 & 2 \end{bmatrix}$

30. $\begin{bmatrix} -3 & 0 & 0 & 1 & -2 \\ 1 & -3 & 4 & 0 & 2 \\ 0 & 0 & 2 & 0 & 0 \\ 5 & 0 & -3 & -4 & 1 \\ 0 & -1 & 2 & 0 & 0 \end{bmatrix}$

In Exercises 31-48, solve the system of linear equations by using Cramer's rule

31. $\begin{aligned} 2x - 3y &= 1 \\ x + 4y &= 6 \end{aligned}$

32. $\begin{aligned} 4x - 3y &= 0 \\ 2x - y &= -10 \end{aligned}$

33. $\begin{aligned} 9x - 4y &= -21 \\ 5x - 3y &= 18 \end{aligned}$

34. $\begin{aligned} 11x - 12y &= 13 \\ 8x - 9y &= 10 \end{aligned}$

35. $\begin{aligned} 5x - 20y &= -10 \\ -2x + 8y &= 4 \end{aligned}$

36. $\begin{aligned} 3x - 12y &= 6 \\ 4x - 16y &= -8 \end{aligned}$

37. $\begin{aligned} \tfrac{2}{3}x - \tfrac{1}{4}y &= 4 \\ \tfrac{5}{6}x + \tfrac{3}{8}y &= 16 \end{aligned}$

38. $\begin{aligned} \tfrac{3}{2}x + 3y &= -3 \\ -\tfrac{2}{5}x - \tfrac{1}{3}y &= -2 \end{aligned}$

39. $\begin{aligned} x + y + z &= -2 \\ 6y - 3z &= 3 \\ 3x \quad + 2z &= 0 \end{aligned}$

40. $\begin{aligned} 2x + y - z &= 100 \\ x + 3y + 3z &= -10 \\ 5x \quad + z &= 30 \end{aligned}$

41. $\begin{aligned} 2x - y + 4z &= 0 \\ 3x - 2y - 3z &= 50 \\ x + y + 5z &= -30 \end{aligned}$

42. $\begin{aligned} 7x - 2y + z &= -1 \\ -3x + y - 4z &= 3 \\ 4x + y - 8z &= 8 \end{aligned}$

43. $\begin{aligned} x - 2y + 2z &= 3 \\ 5x - 7y + 5z &= 8 \\ 3x - 9y + z &= 2 \end{aligned}$

44. $\begin{aligned} 2x - y + 5z &= 2 \\ 3x + 2y - z &= 4 \\ 4x + 5y - 7z &= -6 \end{aligned}$

45. $\begin{aligned} x - \tfrac{2}{5}y + \tfrac{2}{3}z &= 8 \\ \tfrac{1}{2}x + 2y \quad &= -11 \\ x \quad - \tfrac{5}{6}z &= -12 \end{aligned}$

46. $\begin{aligned} \tfrac{1}{2}x - \tfrac{1}{3}y + z &= 22 \\ \tfrac{2}{3}x + y \quad &= -8 \\ 3x + \tfrac{3}{2}y - \tfrac{2}{3}z &= -10 \end{aligned}$

47. $\begin{aligned} 2x - y \quad + t &= 9 \\ 3y + 2z \quad &= -6 \\ y - 2z + 4t &= -14 \\ 3x \quad - z + t &= -4 \end{aligned}$

48. $\begin{aligned} 3x - y + 2z \quad &= 12 \\ x \quad + 3z - 5t &= 28 \\ 2y + z - 4t &= 14 \\ 3x - y + 5z \quad &= 24 \end{aligned}$

Critical Thinking

In Exercises 49–52, find the value(s) of x for which the given determinant is equal to zero.

49. $\begin{vmatrix} x & -2 \\ -3 & 18 \end{vmatrix}$

50. $\begin{vmatrix} 2 & 4 & -1 \\ -6 & x & 3 \\ -4 & 5 & 3 \end{vmatrix}$

51. $\begin{vmatrix} x & 0 & 3 \\ 4 & -2 & x \\ 6 & 1 & -1 \end{vmatrix}$

52. $\begin{vmatrix} 1 & x & 0 & 3 \\ 0 & 4 & -1 & -2 \\ x & 0 & 0 & 3 \\ -2 & 1 & 0 & -6 \end{vmatrix}$

53. The triangle shown in the figure has vertices $(0, 0)$, (x_1, y_1), and (x_2, y_2).

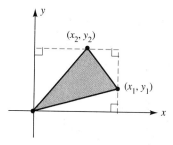

(a) Find the area of this triangle by subtracting the area of the three right triangles from the area of the dashed rectangle.

EXERCISE 53 *(continued)*

(b) Evaluate $\frac{1}{2}\begin{vmatrix} x_1 & y_1 \\ x_2 & y_2 \end{vmatrix}$, and compare your answer with part (a). What does this determinant represent?

54. Use the determinant in Exercise 53(b) to find the area of the triangle with the given vertices.

(a) $A(0, 0)$, $B(6, 2)$, $C(4, 4)$

(b) $A(0,0)$, $B\left(3\sqrt{2}, 3\right)$, $C\left(-\sqrt{2}, 3\right)$

55. Recall from Section 4.2 that the equation of a non-vertical line passing through two distinct points (x_1, y_1) and (x_2, y_2) is given by

$$y - y_1 = m(x - x_1),$$

where m is the slope of the line and

$$m = \frac{y_2 - y_1}{x_2 - x_1}.$$

(a) Replace m in the equation $y - y_1 = m(x - x_1)$ with $\frac{y_2 - y_1}{x_2 - x_1}$ and solve for y, expressing the answer as a single fraction with parentheses removed from the numerator.

(b) Solve the determinant equation

$$\begin{vmatrix} x & y & 1 \\ x_1 & y_1 & 1 \\ x_2 & y_2 & 1 \end{vmatrix} = 0$$

for y, and express the answer as a single fraction. Compare your answer to part (a). What does this determinant equation represent?

56. Use the determinant equation in Exercises 55(b) to find the equation of the line passing through the given points A and B. Write your answer in slope-intercept form, $y = mx + b$.

(a) $A(-2, 5)$, $B(2, -3)$

(b) $A\left(\sqrt{2}, -\sqrt{3}\right)$, $B\left(-\sqrt{3}, \sqrt{2}\right)$

Two or more lines are concurrent *if a single point lies on all of them. The common point where the lines intersect is called the* point of concurrency. *The three lines*

$$A_1x + B_1y + C_1 = 0, \qquad A_2x + B_2y + C_2 = 0,$$

and

$$A_3x + B_3y + C_3 = 0$$

have a point in common if

$$\begin{vmatrix} A_1 & B_1 & C_1 \\ A_2 & B_2 & C_2 \\ A_3 & B_3 & C_3 \end{vmatrix} = 0.$$

In Exercises 57 and 58, find

(a) *the value of C such that the three lines are concurrent.*

(b) *the point of concurrency for the three lines.*

57. $5x + 4y + C = 0$, $4x + 5y + 7 = 0$, $2x - y - 7 = 0$

58. $7x - 3y + C = 0$, $2x + 3y + 23 = 0$, $5x - 2y + 10 = 0$

Calculator Activities

In Exercises 59–62, use a calculator to help evaluate each determinant. Round the answer to three significant digits.

59. $\begin{vmatrix} 1.46 & -2.33 \\ -3.46 & 1.98 \end{vmatrix}$

60. $\begin{vmatrix} 22.3 & 0.00 & -10.4 \\ -61.4 & 18.3 & 13.8 \\ -41.7 & 51.4 & 32.1 \end{vmatrix}$

61. $\begin{vmatrix} 29.5 & 19.1 & 31.9 \\ 14.9 & -21.1 & -18.5 \\ 26.3 & 11.5 & 10.9 \end{vmatrix}$

62. $\begin{vmatrix} 0.67 & 0.92 & 0.18 & 0.00 \\ 0.00 & 4.34 & -1.90 & -0.62 \\ 1.25 & 0.00 & 0.00 & 3.95 \\ -2.38 & 1.98 & 0.32 & -6.23 \end{vmatrix}$

In Exercises 63–66, use Cramer's rule in conjunction with a calculator to solve each linear system of equations. Round each answer to three significant digits.

63. $0.25x - 1.41y = 7.42$
    $-1.81x + 2.40y = -9.60$

64. $-5.21x + 3.20y = -77.2$
    $0.54x + 2.22y = -32.7$

65. $1.35x - 5.32y + 2.21z = -25.0$
     $0.44x + 4.58y \qquad = 40.0$
     $-2.40x \qquad - 3.53z = -17.0$

66. $0.51x - 0.34y + \quad z = 22.4$
     $0.68x + \quad y \qquad = -8.09$
     $3.34x + 1.48y - 0.63z = -10.1$

67. Two cables support a 750-lb weight, as shown in the figure. The tensions T_1 and T_2 (in pounds) in each cable may be found by solving the following system of linear equations.

$$0.5150T_1 + 0.7880T_2 = 750$$

$$0.8572T_1 - 0.6157T_2 = 0$$

Use Cramer's rule to help find T_1 and T_2, rounding each answer to three significant digits.

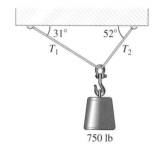

750 lb

68. The *Wheatstone bridge circuit*, shown in the sketch, is used in electrical measurement applications. The currents i_1, i_2, and i_m (in amperes) may be found by solving the following system of linear equations.

$$(10.2 + 31.5)i_1 \qquad\qquad - 31.5i_m = 125$$

$$(20.6 + 41.2)i_2 + \qquad 41.2i_m = 125$$

$$10.2i_1 + \qquad 41.2i_2 + (50.0 + 41.2)i_m = 125$$

Use Cramer's rule to help find the meter current i_m, rounding the answer to three significant digits.

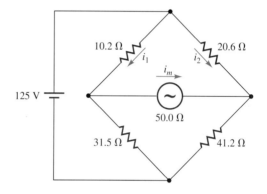

7.6 Properties of Determinants

In this section, we develop some properties of determinants by looking at the effects of elementary row and column operations on the value of a determinant. We then use these properties in conjunction with expansion by cofactors (from the preceding section) to help evaluate $n \times n$ determinants in which $n \geq 4$. We begin by discussing the determinant of a matrix in echelon form.

♦ Determinant of a Matrix in Echelon Form

The matrix

$$A = \begin{bmatrix} a_{11} & a_{12} \\ 0 & a_{22} \end{bmatrix}$$

is an example of a 2 × 2 matrix in echelon form and its determinant is given by

$$|A| = a_{11}a_{22} + (0)a_{12} = a_{11}a_{22}.$$

Note that the determinant of this matrix is simply the *product of the elements*

along its main diagonal. The matrix

$$A = \begin{bmatrix} a_{11} & a_{12} & a_{13} \\ 0 & a_{22} & a_{23} \\ 0 & 0 & a_{33} \end{bmatrix}$$

is an example of a 3×3 matrix in echelon form. To find the determinant of this matrix we can use either expansion by cofactors about the third row or the diagonal method (from the preceding section). Using either approach, we find

$$|A| = a_{11}a_{22}a_{33}.$$

The determinant of this matrix is also the *product of the elements along its main diagonal.* This observation is true for any $n \times n$ matrix in echelon form.

Determinant of a Matrix in Echelon Form

> Let A be an $n \times n$ matrix in echelon form. Then $|A|$ is the product of the elements along its main diagonal.

E X A M P L E   1   Find the determinant of the matrix

$$A = \begin{bmatrix} 3 & -1 & 0 & 4 & 6 \\ 0 & 2 & -2 & 5 & 0 \\ 0 & 0 & -1 & 7 & -2 \\ 0 & 0 & 0 & -2 & 7 \\ 0 & 0 & 0 & 0 & 1 \end{bmatrix}$$

S O L U T I O N   Since the matrix is in echelon form, the determinant of A is the product of the elements along its main diagonal. Thus,

$$|A| = (3)(2)(-1)(-2)(1) = 12.$$ ◆

P R O B L E M   1   Find the determinant for the matrix

$$A = \begin{bmatrix} 2 & -3 & 1 & 0 \\ 0 & 2 & 4 & -7 \\ 0 & 0 & -3 & 1 \\ 0 & 0 & 0 & -5 \end{bmatrix}.$$ ◆

◆ **Effects of Row Operations on a Determinant**

Consider the 2×2 matrix

$$A = \begin{bmatrix} a_{11} & a_{12} \\ a_{21} & a_{22} \end{bmatrix},$$

and the effects of elementary row operations on its determinant $|A|$:

1. $R_1 \leftrightarrow R_2$:

$$B = \begin{bmatrix} a_{21} & a_{22} \\ a_{11} & a_{12} \end{bmatrix}$$

$$\begin{aligned} |B| &= a_{21}a_{12} - a_{11}a_{22} \\ &= -(a_{11}a_{22} - a_{21}a_{12}) = -|A| \end{aligned}$$

If the rows of matrix A are interchanged to form the row-equivalent matrix B, then $|B| = -|A|$.

2. $cR_1 \rightarrow R_1$:

$$B = \begin{bmatrix} ca_{11} & ca_{12} \\ a_{21} & a_{22} \end{bmatrix}$$

$$\begin{aligned} |B| &= ca_{11}a_{22} - ca_{21}a_{12} \\ &= c(a_{11}a_{22} - a_{21}a_{12}) = c|A| \end{aligned}$$

If a row of matrix A is replaced by c times that row to form the row-equivalent matrix B, then $|B| = c|A|$.

3. $cR_1 + R_2 \rightarrow R_2$:

$$B = \begin{bmatrix} a_{11} & a_{12} \\ ca_{11} + a_{21} & ca_{12} + a_{22} \end{bmatrix}$$

$$\begin{aligned} |B| &= a_{11}(ca_{12} + a_{22}) - a_{12}(ca_{11} + a_{21}) \\ &= a_{11}a_{22} - a_{21}a_{12} = |A| \end{aligned}$$

If a row of matrix A is replaced by the sum of that row and c times another row to form the row-equivalent matrix B, then $|B| = |A|$.

We refer to these observations as the **row operation properties of determinants**. It can be shown that these three properties are true for any $n \times n$ matrix A. In Exercises 49–51, you are asked to verify these properties for a 3×3 matrix.

Row Operation Properties of Determinants

Suppose A is an $n \times n$ matrix.

1. If two rows of A are interchanged to form the row-equivalent matrix B, then $|B| = -|A|$.

2. If a row of A is replaced by c times that row to form the row-equivalent matrix B, then $|B| = c|A|$.

3. If a row of A is replaced by the sum of that row and c times another row to form the row-equivalent matrix B, then $|B| = |A|$.

Property 2 has the effect of factoring out a common factor from a row. For example,

$$\begin{vmatrix} 1 & 2 & 4 \\ -2 & 3 & 6 \\ 12 & 24 & 36 \end{vmatrix} = 12 \begin{vmatrix} 1 & 2 & 4 \\ -2 & 3 & 6 \\ 1 & 2 & 3 \end{vmatrix}$$

Factor out 12 from row 3

$\frac{1}{12}R_3 \rightarrow R_3$ Determinant is multiplied by 12

The determinant of a matrix can be found by using elementary row operations. *We write the matrix in echelon form, and record the effects of the row operations on the determinant. Once the matrix is in echelon form, we can easily find its determinant by multiplying the elements along its main diagonal.* The procedure is illustrated in the next example.

EXAMPLE 2   Use elementary row operations to find the determinant of the matrix

$$A = \begin{bmatrix} 2 & -2 & -3 \\ 0 & 6 & 3 \\ 1 & -2 & -5 \end{bmatrix}.$$

SOLUTION   We write the matrix in echelon form and record the effects of the row operations on the determinant as follows:

$$|A| = \begin{vmatrix} 2 & -2 & -3 \\ 0 & 6 & 3 \\ 1 & -2 & -5 \end{vmatrix} = - \begin{vmatrix} 1 & -2 & -5 \\ 0 & 6 & 3 \\ 2 & -2 & -3 \end{vmatrix}$$

$R_1 \leftrightarrow R_3$ Determinant changes sign

$$= - \begin{vmatrix} 1 & -2 & -5 \\ 0 & 6 & 3 \\ 0 & 2 & 7 \end{vmatrix}$$

$-2R_1 + R_3 \rightarrow R_3$

$$= -3 \begin{vmatrix} 1 & -2 & -5 \\ 0 & 2 & 1 \\ 0 & 2 & 7 \end{vmatrix}$$

$\frac{1}{3}R_2 \rightarrow R_2$ Determinant is tripled

Factor out 3 from row 2

$$= -3 \begin{vmatrix} 1 & -2 & -5 \\ 0 & 2 & 1 \\ 0 & 0 & 6 \end{vmatrix}$$

$-R_2 + R_3 \rightarrow R_3$

Now, multiplying the elements along the main diagonal, we have

$$|A| = -3[(1)(2)(6)] = -36. \qquad \blacklozenge$$

PROBLEM 2   Referring to Example 2, use expansion by cofactors about the first column to find $|A|$. Your answer should agree with that obtained in Example 2. $\blacklozenge$

♦ **Effects of Column Operations on a Determinant**

The elementary row operations for a matrix can also be performed on the columns of a matrix.

Elementary Column Operations for a Matrix

1. Interchange the position of two columns.

2. Multiply all elements in a column by a nonzero constant.

3. Add a multiple of one column to another.

If we obtain one matrix from another by a sequence of column operations, we say that the two matrices are **column equivalent**. In this text, we use the following notations when generating column-equivalent matrices:

$C_i \leftrightarrow C_j$:   Interchange the ith column and jth column of a matrix.

$cC_i \rightarrow C_i$:   Replace the ith column with a multiple of c times the ith column.

$cC_i + C_j \rightarrow C_j$:   Replace the jth column with the sum of c times the ith column and the jth column.

Replacing the word *row* with *column* in the row operation properties of determinants gives us the **column operation properties of determinants**.

Column Operation Properties of Determinants

Suppose A is an $n \times n$ matrix.

1. If two columns of A are interchanged to form the column-equivalent matrix B, then $|B| = -|A|$.

2. If a column of A is replaced by c times that column to form the column-equivalent matrix B, then $|B| = c|A|$.

3. If a column of A is replaced by the sum of that column and c times another column to form the column-equivalent matrix B, then $|B| = |A|$.

E X A M P L E 3   Use elementary column operations to find the determinant of the matrix

$$A = \begin{bmatrix} 3 & -2 & 1 & 1 \\ -2 & 3 & 4 & 14 \\ 1 & -4 & 4 & 0 \\ 1 & 2 & -2 & 0 \end{bmatrix}$$

SOLUTION   We write the matrix in echelon form and record the effects of the column operations on the determinant:

$$|A| = \begin{vmatrix} 3 & -2 & 1 & 1 \\ -2 & 3 & 4 & 14 \\ 1 & -4 & 4 & 0 \\ 1 & 2 & -2 & 0 \end{vmatrix} = - \begin{vmatrix} 1 & -2 & 1 & 3 \\ 14 & 3 & 4 & -2 \\ 0 & -4 & 4 & 1 \\ 0 & 2 & -2 & 1 \end{vmatrix} \quad \begin{array}{l} C_1 \leftrightarrow C_4 \\ \textbf{Determinant} \\ \textbf{changes sign} \end{array}$$

$$= - \begin{vmatrix} 1 & -1 & 1 & 3 \\ 14 & 7 & 4 & -2 \\ 0 & 0 & 4 & 1 \\ 0 & 0 & -2 & 1 \end{vmatrix} \quad C_3 + C_2 \rightarrow C_2$$

$$= - \begin{vmatrix} 3 & -1 & 1 & 3 \\ 0 & 7 & 4 & -2 \\ 0 & 0 & 4 & 1 \\ 0 & 0 & -2 & 1 \end{vmatrix} \quad -2C_2 + C_1 \rightarrow C_1$$

$$= - \begin{vmatrix} 3 & -1 & 7 & 3 \\ 0 & 7 & 0 & -2 \\ 0 & 0 & 6 & 1 \\ 0 & 0 & 0 & 1 \end{vmatrix} \quad 2C_4 + C_3 \rightarrow C_3$$

Now, multiplying the elements along the main diagonal, we have

$$|A| = -[(3)(7)(6)(1)] = -126. \qquad \blacklozenge$$

PROBLEM 3   Referring to Example 3, use expansion by cofactors about the fourth column to determine $|A|$. Your answer should agree with that obtained in Example 3.   ◆

◆   **Combining Row and Column Operations with Expansion by Cofactors**

In finding the determinant of an $n \times n$ matrix, where $n \geq 4$, we usually need to make several elementary row or column operations to put the matrix in echelon form. For these larger matrices, we can save time by simply generating a single row or column in which all but one element is zero. Expansion by cofactors (Section 7.5) can then be applied to that particular row or column in order to reduce the matrix to dimension $(n - 1) \times (n - 1)$. Now, working with the $(n - 1) \times (n - 1)$ matrix, we again use elementary row or column operations to generate a single row or column in which all but one element is zero. Expansion by cofactors can again be applied in order to reduce the matrix to dimension $(n - 2) \times (n - 2)$. This process is continued until we obtain a matrix whose determinant we can easily find. The procedure is illustrated in the next example.

EXAMPLE 4   Use elementary row and column operations in conjunction with expansion by cofactors to help find the determinant of the matrix

$$A = \begin{bmatrix} 2 & -1 & 3 & 4 & 2 \\ 1 & 0 & -1 & 0 & 1 \\ 3 & 2 & 2 & 1 & 0 \\ -1 & 0 & 3 & 1 & 2 \\ 1 & 1 & -2 & -1 & 3 \end{bmatrix}.$$

SOLUTION Since both the second row and second column already contain two zeros, it is advantageous to work with either the second row or second column in order to obtain a row or column in which all but one element is zero. Choosing the second column, we have

$$|A| = \begin{vmatrix} 2 & -1 & 3 & 4 & 2 \\ 1 & 0 & -1 & 0 & 1 \\ 3 & 2 & 2 & 1 & 0 \\ -1 & 0 & 3 & 1 & 2 \\ 1 & 1 & -2 & -1 & 3 \end{vmatrix} = \begin{vmatrix} 2 & -1 & 3 & 4 & 2 \\ 1 & 0 & -1 & 0 & 1 \\ 7 & 0 & 8 & 9 & 4 \\ -1 & 0 & 3 & 1 & 2 \\ 3 & 0 & 1 & 3 & 5 \end{vmatrix} \quad \begin{array}{l} 2R_1 + R_3 \to R_3 \\ \\ R_1 + R_5 \to R_5 \end{array}$$

Now, using expansion by cofactors about the second column, we have

$$|A| = (-1)(-1)^{1+2} \begin{vmatrix} 1 & -1 & 0 & 1 \\ 7 & 8 & 9 & 4 \\ -1 & 3 & 1 & 2 \\ 3 & 1 & 3 & 5 \end{vmatrix} = \begin{vmatrix} 1 & -1 & 0 & 1 \\ 7 & 8 & 9 & 4 \\ -1 & 3 & 1 & 2 \\ 3 & 1 & 3 & 5 \end{vmatrix}.$$

Since both the first row and third column already contain a zero, we should work with either the first row or third column in order to obtain a row or column in which all but one element is zero. Choosing the first row, we have

$$|A| = \begin{vmatrix} 1 & -1 & 0 & 1 \\ 7 & 8 & 9 & 4 \\ -1 & 3 & 1 & 2 \\ 3 & 1 & 3 & 5 \end{vmatrix} = \begin{vmatrix} 1 & 0 & 0 & 0 \\ 7 & 15 & 9 & -3 \\ -1 & 2 & 1 & 3 \\ 3 & 4 & 3 & 2 \end{vmatrix} \quad \begin{array}{l} C_1 + C_2 \to C_2 \\ -C_1 + C_4 \to C_4 \end{array}$$

Now, using expansion by cofactors about the first row, we have

$$|A| = 1(-1)^{1+1} \begin{vmatrix} 15 & 9 & -3 \\ 2 & 1 & 3 \\ 4 & 3 & 2 \end{vmatrix} = \begin{vmatrix} 15 & 9 & -3 \\ 2 & 1 & 3 \\ 4 & 3 & 2 \end{vmatrix}.$$

At this point, we can either evaluate this 3×3 determinant by the methods discussed in the preceding section, or continue with our procedure of generating a row or column in which all but one element is zero. Continuing with the latter, we have

$$|A| = \begin{vmatrix} 15 & 9 & -3 \\ 2 & 1 & 3 \\ 4 & 3 & 2 \end{vmatrix} = \begin{vmatrix} 0 & 0 & -3 \\ 17 & 10 & 3 \\ 14 & 9 & 2 \end{vmatrix}. \quad \begin{array}{l} 5C_3 + C_1 \to C_1 \\ 3C_3 + C_2 \to C_2 \end{array}$$

Thus,

$$|A| = -3(-1)^{1+3}\begin{vmatrix} 17 & 10 \\ 14 & 9 \end{vmatrix} = -3(153 - 140) = -39. \qquad \blacklozenge$$

PROBLEM 4   Repeat Example 4 by beginning with the second row instead of the second column. $\blacklozenge$

$\blacklozenge$ **Zero Determinants**

Suppose we have an $n \times n$ matrix A in which every element of either the ith row or jth column is zero. If the zeros are in the ith row and we expand about the ith row, we find

$$|A| = 0C_{i1} + 0C_{i2} + 0C_{i3} + \cdots + 0C_{ij} + \cdots + 0C_{in} = 0.$$

If the zeros are in the jth column and we expand about the jth column, we find

$$|A| = 0C_{1j} + 0C_{2j} + 0C_{3j} + \cdots + 0C_{ij} + \cdots + 0C_{nj} = 0.$$

In summary, *if every element in a row or column of an $n \times n$ matrix A is zero, then $|A| = 0$.*

When performing elementary row or column operations, we sometimes encounter a row or column in which every element is zero. If so, we simply conclude that the determinant is zero.

EXAMPLE 5   Use elementary row and column operations to help find the determinant of the matrix

$$A = \begin{bmatrix} -1 & 2 & 0 & 5 \\ 2 & -2 & -4 & 1 \\ 2 & -1 & -6 & -2 \\ 1 & -3 & 2 & 2 \end{bmatrix}$$

SOLUTION   Working with the first row, we have

$$|A| = \begin{vmatrix} -1 & 2 & 0 & 5 \\ 2 & -2 & -4 & 1 \\ 2 & -1 & -6 & -2 \\ 1 & -3 & 2 & 2 \end{vmatrix} = \begin{vmatrix} -1 & 0 & 0 & 0 \\ 2 & 2 & -4 & 11 \\ 2 & 3 & -6 & 8 \\ 1 & -1 & 2 & 7 \end{vmatrix} \quad \begin{array}{l} 2C_1 + C_2 \to C_2 \\ 5C_1 + C_4 \to C_4 \end{array}$$

Now, using expansion by cofactors about the first row, we find

$$|A| = -1(-1)^{1+1}\begin{vmatrix} 2 & -4 & 11 \\ 3 & -6 & 8 \\ -1 & 2 & 7 \end{vmatrix} = -\begin{vmatrix} 2 & -4 & 11 \\ 3 & -6 & 8 \\ -1 & 2 & 7 \end{vmatrix}.$$

Notice in the 3×3 determinant that the corresponding elements in the first and second columns are proportional. Replacing the second column with the sum of twice the first column and the second column, we obtain a column of zeros and conclude that $|A| = 0$:

$$|A| = - \begin{vmatrix} 2 & 0 & 11 \\ 3 & 0 & 8 \\ -1 & 0 & 7 \end{vmatrix} = 0 \qquad 2\mathrm{C}_1 + \mathrm{C}_2 \to \mathrm{C}_2$$

Since every element in the second column is zero, we conclude that $|A| = 0$.

◆

Note: As illustrated in Example 5, when finding the determinant of an $n \times n$ matrix A, it is important to watch for a pair of rows or columns in which corresponding elements are proportional. If such a pair occurs, we can easily show that $|A| = 0$. In summary, we list some conditions for which the determinant of an $n \times n$ matrix A is zero:

1. If all elements in a row or a column of matrix A are zero, then $|A| = 0$.

2. If corresponding elements in two rows or two columns of matrix A are proportional, then $|A| = 0$.

3. If corresponding elements in two rows or two columns of matrix A are identical, then $|A| = 0$.

PROBLEM 5   Find the determinant of the matrix

$$A = \begin{bmatrix} 1 & -3 & 4 & 6 \\ 2 & 4 & 2 & -1 \\ 0 & 2 & 5 & -6 \\ 6 & 12 & 6 & -3 \end{bmatrix}$$ ◆

Exercises 7.6

Basic Skills

In Exercises 1–6, find the determinant of the given matrix.

1. $\begin{bmatrix} 2 & 3 & -1 \\ 0 & 3 & 9 \\ 0 & 0 & -2 \end{bmatrix}$

2. $\begin{bmatrix} -3 & 9 & 0 \\ 0 & -4 & 5 \\ 0 & 0 & 4 \end{bmatrix}$

3. $\begin{bmatrix} 1 & -2 & 3 & 0 \\ 0 & 2 & 8 & 0 \\ 0 & 0 & -8 & 12 \\ 0 & 0 & 0 & -13 \end{bmatrix}$

4. $\begin{bmatrix} -4 & 0 & -8 & 4 \\ 0 & -4 & 5 & 10 \\ 0 & 0 & 13 & 0 \\ 0 & 0 & 0 & -5 \end{bmatrix}$

5. $\begin{bmatrix} 3 & 0 & 8 & 19 & 2 \\ 0 & -1 & 8 & 12 & 3 \\ 0 & 0 & -4 & 5 & 6 \\ 0 & 0 & 0 & -4 & -9 \\ 0 & 0 & 0 & 0 & 2 \end{bmatrix}$

6.
$$\begin{bmatrix} 3 & -2 & 4 & 0 & 9 \\ 0 & -1 & 9 & 3 & 9 \\ 0 & 0 & -5 & 3 & 2 \\ 0 & 0 & 0 & -12 & 3 \\ 0 & 0 & 0 & 0 & -14 \end{bmatrix}$$

16.
$$\begin{vmatrix} -3 & 4 & 5 & -8 & -15 \\ 12 & 0 & -1 & 0 & 3 \\ 9 & 3 & -2 & 1 & 6 \\ 0 & 5 & -7 & 18 & 21 \\ 1 & -2 & 0 & 4 & 0 \end{vmatrix} = 0$$

In Exercises 7–16, explain why each statement is true.

7.
$$\begin{vmatrix} 0 & 3 & 1 \\ 3 & 4 & -1 \\ 1 & 2 & -5 \end{vmatrix} = -\begin{vmatrix} 1 & 2 & -5 \\ 3 & 4 & -1 \\ 0 & 3 & 1 \end{vmatrix}$$

8.
$$\begin{vmatrix} 0 & 3 & 1 \\ 3 & 4 & -1 \\ 1 & 2 & -5 \end{vmatrix} = -\begin{vmatrix} 1 & 3 & 0 \\ -1 & 4 & 3 \\ -5 & 2 & 1 \end{vmatrix}$$

9.
$$\begin{vmatrix} 1 & 5 & -2 \\ 3 & -9 & 6 \\ 0 & 3 & -6 \end{vmatrix} = 2\begin{vmatrix} 1 & 5 & -1 \\ 3 & -9 & 3 \\ 0 & 3 & -3 \end{vmatrix}$$

10.
$$\begin{vmatrix} 1 & 5 & -2 \\ \frac{1}{3} & -1 & \frac{2}{3} \\ 0 & 3 & -6 \end{vmatrix} = \frac{1}{3}\begin{vmatrix} 1 & 5 & -2 \\ 1 & -3 & 2 \\ 0 & 3 & -6 \end{vmatrix}$$

11.
$$\begin{vmatrix} 1 & -1 & 6 & 7 \\ 2 & 3 & -7 & 0 \\ -3 & 2 & -3 & -1 \\ 4 & -3 & 2 & 1 \end{vmatrix} = \begin{vmatrix} 1 & -1 & 6 & 7 \\ 2 & 3 & -7 & 0 \\ 0 & -1 & 15 & 20 \\ 4 & -3 & 2 & 1 \end{vmatrix}$$

12.
$$\begin{vmatrix} 1 & -1 & 6 & 7 \\ 2 & 3 & -7 & 0 \\ -3 & 2 & -3 & -1 \\ 4 & -3 & 2 & 1 \end{vmatrix} = \begin{vmatrix} -11 & -1 & 6 & 7 \\ 16 & 3 & -7 & 0 \\ 3 & 2 & -3 & -1 \\ 0 & -3 & 2 & 1 \end{vmatrix}$$

13.
$$\begin{vmatrix} 2 & -1 & 0 & 2 \\ -1 & 2 & 0 & -4 \\ 8 & 0 & 0 & 9 \\ -5 & 9 & 0 & -1 \end{vmatrix} = 0$$

14.
$$\begin{vmatrix} 4 & -2 & 10 & 8 \\ 0 & 0 & 0 & 0 \\ -2 & 4 & 12 & 9 \\ 2 & 9 & 4 & -7 \end{vmatrix} = 0$$

15.
$$\begin{vmatrix} 4 & -2 & 6 & 0 & 14 \\ 1 & 2 & 3 & -5 & 1 \\ 2 & -1 & 3 & 0 & 7 \\ 1 & 3 & -4 & 9 & 10 \\ 0 & 0 & 3 & -5 & 6 \end{vmatrix} = 0$$

In Exercises 17–24, find the determinant of each matrix by writing the matrix in echelon form using elementary row operations only.

17.
$$\begin{bmatrix} 1 & 0 & 2 \\ 2 & -3 & 4 \\ -3 & 1 & 9 \end{bmatrix}$$

18.
$$\begin{bmatrix} 1 & 2 & -3 \\ 3 & 0 & 3 \\ -4 & 1 & -2 \end{bmatrix}$$

19.
$$\begin{bmatrix} 4 & 0 & -3 \\ 3 & -7 & 2 \\ 1 & -5 & 0 \end{bmatrix}$$

20.
$$\begin{bmatrix} 2 & -3 & 5 \\ -1 & 7 & 3 \\ 6 & -9 & 15 \end{bmatrix}$$

21.
$$\begin{bmatrix} 1 & -2 & 0 & 4 \\ 3 & 0 & 8 & -2 \\ 1 & 0 & -4 & 5 \\ 2 & -3 & -4 & 1 \end{bmatrix}$$

22.
$$\begin{bmatrix} 2 & -4 & 5 & 9 \\ 1 & -3 & 4 & 7 \\ 2 & -3 & 0 & 1 \\ 0 & -1 & 2 & 6 \end{bmatrix}$$

23.
$$\begin{bmatrix} 0 & 1 & -2 & 0 & 3 \\ 1 & 1 & -2 & 3 & 0 \\ -2 & 0 & 0 & -4 & 1 \\ 0 & 2 & -1 & 0 & 3 \\ -3 & 1 & 0 & -5 & 0 \end{bmatrix}$$

24.
$$\begin{bmatrix} 2 & -2 & 0 & -6 & 5 \\ 3 & -1 & 0 & -2 & 0 \\ -1 & 0 & 2 & 3 & -2 \\ 4 & 0 & -2 & 0 & 2 \\ 0 & 1 & -2 & -3 & 5 \end{bmatrix}$$

In Exercises 25–32, find the determinant of each matrix by writing the matrix in echelon form using elementary column operations only.

25.
$$\begin{bmatrix} 3 & 2 & 4 \\ 0 & 1 & 3 \\ 0 & -2 & 2 \end{bmatrix}$$

26.
$$\begin{bmatrix} 3 & -3 & 2 \\ 2 & 0 & 1 \\ 2 & 0 & -4 \end{bmatrix}$$

27.
$$\begin{bmatrix} 2 & 6 & 3 \\ 0 & 4 & 4 \\ 2 & -1 & 1 \end{bmatrix}$$

28.
$$\begin{bmatrix} 8 & 2 & -6 \\ 1 & 4 & 8 \\ -6 & 3 & 6 \end{bmatrix}$$

29.
$$\begin{bmatrix} 2 & 3 & 0 & 4 \\ 0 & 1 & 5 & 2 \\ 2 & -1 & 3 & 4 \\ -4 & 2 & -2 & 0 \end{bmatrix}$$

30. $\begin{bmatrix} 3 & 2 & -1 & 4 \\ 2 & -2 & 4 & 0 \\ -2 & 3 & 6 & 1 \\ 1 & 4 & -3 & 0 \end{bmatrix}$

31. $\begin{bmatrix} 2 & 1 & 0 & 2 & 0 \\ -2 & 0 & 2 & -2 & 2 \\ 0 & -6 & 3 & 1 & -1 \\ 0 & 0 & -3 & 3 & 1 \\ 0 & 0 & 1 & -1 & 0 \end{bmatrix}$

32. $\begin{bmatrix} 0 & 3 & 0 & -1 & 2 \\ 4 & 0 & -2 & 0 & -8 \\ 7 & 1 & 4 & -1 & 0 \\ 0 & -1 & 1 & 1 & 0 \\ -3 & 3 & 0 & 1 & 0 \end{bmatrix}$

In Exercises 33–40, use elementary row and column operations, along with expansion by cofactors, to find the determinant of each matrix.

33. $\begin{bmatrix} 1 & 2 & -3 \\ 0 & 3 & -4 \\ 2 & -2 & 1 \end{bmatrix}$
34. $\begin{bmatrix} 5 & -3 & 1 \\ -3 & 4 & 6 \\ 0 & 2 & -1 \end{bmatrix}$

35. $\begin{bmatrix} 7 & 4 & -3 & 2 \\ -3 & 5 & 2 & -1 \\ 4 & 0 & -2 & 0 \\ 3 & 4 & -1 & 3 \end{bmatrix}$

36. $\begin{bmatrix} 6 & -20 & 0 & -8 \\ -3 & -8 & 9 & 1 \\ 0 & 6 & -3 & 1 \\ -4 & 2 & 0 & -5 \end{bmatrix}$

37. $\begin{bmatrix} 2 & -5 & 6 & 12 \\ 1 & 4 & -3 & 5 \\ -3 & 3 & 9 & 3 \\ 0 & -9 & 11 & 4 \end{bmatrix}$

38. $\begin{bmatrix} 5 & 6 & -3 & 15 \\ 2 & -3 & 8 & 11 \\ -7 & 3 & 9 & -1 \\ 4 & -12 & 3 & 9 \end{bmatrix}$

39. $\begin{bmatrix} 2 & -1 & 8 & 0 & 0 \\ 0 & 4 & 0 & 0 & -2 \\ 2 & 2 & 7 & -1 & -3 \\ 0 & 0 & 4 & 2 & 1 \\ 0 & 2 & 0 & 4 & 9 \end{bmatrix}$

40. $\begin{bmatrix} 3 & 0 & 3 & 1 & -1 \\ 4 & 2 & 0 & 1 & 0 \\ -3 & 2 & 0 & 3 & -1 \\ 6 & 8 & 0 & 2 & -3 \\ 1 & -1 & 6 & -1 & 4 \end{bmatrix}$

Critical Thinking

Given that

$$\begin{vmatrix} a & b & c \\ 2 & 4 & 6 \\ \ln x & \ln y & \ln z \end{vmatrix} = 8,$$

state the value of each determinant in Exercises 41–46.

41. $\begin{vmatrix} a & b & c \\ \ln x & \ln y & \ln z \\ 1 & 2 & 3 \end{vmatrix}$

42. $\begin{vmatrix} c & b & a \\ 6 & 4 & 2 \\ \ln z & \ln y & \ln x \end{vmatrix}$

43. $\begin{vmatrix} a & b & c \\ 2 & 4 & 6 \\ \ln x^2 & \ln y^2 & \ln z^2 \end{vmatrix}$

44. $\begin{vmatrix} \ln x^3 & \ln y^3 & \ln z^3 \\ 2 & 4 & 6 \\ a-2 & b-4 & c-6 \end{vmatrix}$

45. $\begin{vmatrix} a & b & c-a \\ \ln x & \ln y & \ln (z/x) \\ 1 & 2 & 2 \end{vmatrix}$

46. $\begin{vmatrix} 1 & 3 & 5 \\ a & a+b & 2a+c \\ \ln x & \ln xy & \ln x^2 z \end{vmatrix}$

47. Evaluate the given determinant:

$$\begin{vmatrix} e^{-x} & 92 & 117 \\ 0 & e^x & -32 \\ 0 & -e^x & 5 \end{vmatrix}$$

48. Express the given determinant as the product of a pair of 2×2 determinants:

$$\begin{vmatrix} a & b & 0 & 0 \\ c & d & 0 & 0 \\ 0 & 0 & e & f \\ 0 & 0 & g & h \end{vmatrix}$$

In Exercises 49–52, use the 3×3 matrix

$$A = \begin{bmatrix} a_{11} & a_{12} & a_{13} \\ a_{21} & a_{22} & a_{23} \\ a_{31} & a_{32} & a_{33} \end{bmatrix}$$

to illustrate the given property of determinants.

49. If two rows of matrix A are interchanged to form the row-equivalent matrix B, then $|B| = -|A|$.

50. If a row of matrix A is replaced by c times that row to form the row-equivalent matrix B, then $|B| = c|A|$.

51. If a row of matrix A is replaced by the sum of that row and c times another row to form the row-equivalent matrix B, then $|B| = |A|$.

52. If the rows and columns of matrix A are interchanged to form the matrix A^T (the *transpose* of A), then $|A^T| = |A|$.

Calculator Activities

In Exercises 53–56, use elementary row or column operations to write each matrix in echelon form. Then find the determinant of the matrix, rounding the answer to three significant digits.

53. $\begin{bmatrix} 2.8 & 6.1 & 3.3 \\ 0.0 & 4.4 & 4.9 \\ 0.0 & -1.8 & 1.5 \end{bmatrix}$

54. $\begin{bmatrix} 1.26 & 2.32 & -6.95 \\ 8.05 & 4.98 & 4.89 \\ 0.00 & -3.86 & 6.96 \end{bmatrix}$

55. $\begin{bmatrix} 24.3 & 31.6 & 0.0 & 4.9 \\ 0.0 & 11.6 & 52.1 & 21.2 \\ 0.0 & -13.9 & 32.5 & 41.6 \\ 0.0 & 0.0 & -28.9 & -18.6 \end{bmatrix}$

56. $\begin{bmatrix} 327 & 228 & -1129 & 429 \\ 0 & -227 & 412 & 0 \\ -229 & 322 & 612 & 111 \\ 0 & 0 & -327 & 245 \end{bmatrix}$

7.7 Systems of Linear Inequalities and Linear Programming

◆ **Introductory Comments**

The following statements are **linear inequalities** in two variables x and y:

$$Ax + By > C \qquad Ax + By < C$$
$$Ax + By \geq C \qquad Ax + By \leq C$$

If a linear inequality in two variables is true when x is replaced by x_1 and y with y_1, then the ordered pair (x_1, y_1) is a **solution** of the linear inequality. For example, the ordered pair $(2, -1)$ is a solution of the linear inequality $3x - 2y > 7$, since

$$3(2) - 2(-1) > 7$$

$$8 > 7 \quad \text{is } true.$$

However, the ordered pair $(0, 1)$ is not a solution of this linear inequality, since

$$3(0) - 2(1) > 7$$

$$-2 > 7 \quad \text{is } \textit{false}.$$

Linear inequalities occur when we solve certain types of decision-making problems, called *linear programming problems*. Since these types of problems involve working with the graphs of linear inequalities, we begin with a discussion of graphing linear inequalities.

♦   **Graphing Linear Inequalities**

One way to illustrate *all* the ordered pairs that are solutions of a linear inequality is to draw its graph. We begin by graphing the equation $Ax + By = C$, using a dashed line if the inequality uses $<$ or $>$, or a solid line if it uses $\leq$ or $\geq$. This dashed or solid line divides the xy-plane into two half-planes. We then select any ordered pair from either half-plane as a *test point* in order to determine if it is a solution of the inequality. If the ordered pair we select is a solution, then every ordered pair in that half-plane is also a solution. Shading this half-plane gives us the *graph of the linear inequality*. The procedure is illustrated in the next example.

EXAMPLE 1   Graph the solution set of each linear inequality.

(a) $3x + y > 0$ (b) $3x - 2y \leq 6$

SOLUTION

(a) Since the inequality uses $>$, the graph of the corresponding equation $3x + y = 0$ is shown as the dashed line in Figure 7.2. Selecting $(1, 0)$ as an arbitrary test point for this inequality, we find that

$$3(1) + (0) > 0$$

$$3 > 0 \quad \text{is } \textit{true}.$$

Shading the half-plane that contains the point $(1, 0)$ gives us the graph of $3x + y > 0$ (see Figure 7.2). Every ordered pair in the shaded region, but not on the dashed line, is a solution of the inequality $3x + y > 0$.

(b) Since the inequality uses $\leq$, the graph of the corresponding equation $3x - 2y = 6$ is shown as the solid line in Figure 7.3. Selecting $(0, 0)$ as an arbitrary test point for this inequality, we find that

$$3(0) - 2(0) \leq 6$$

$$0 \leq 6 \quad \text{is } \textit{true}.$$

Shading the half-plane that contains the point $(0, 0)$ gives us the graph of $3x - 2y \leq 6$ (see Figure 7.3). Every ordered pair in the shaded region or on the solid line is a solution of the inequality $3x - 2y \leq 6$.

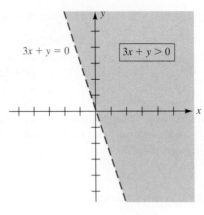

FIGURE 7.2

Graph of the solution set for the
inequality $3x + y > 0$

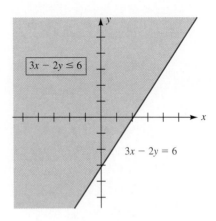

FIGURE 7.3

Graph of the solution set for the
inequality $3x - 2y \le 6$

◆

The following statements are linear inequalities in two variables x and y; written in slope-intercept form:

$$y < ax + b \qquad y > ax + b$$
$$y \le ax + b \qquad y \ge ax + b$$

The graph of the solution set of a linear inequality in slope-intercept form is the half-plane *above* the line $y = ax + b$ if the inequality uses $>$ or $\ge$, and the half-plane *below* the line $y = ax + b$ if the inequality uses $<$ or $\le$. Thus, an alternate method of graphing the solution set of the inequalities in Example 1 is to write the inequalities in slope-intercept form as follows:

(a) $3x + y > 0$ (b) $3x - 2y \le 6$

  $\ y > -3x$ $-2y \le -3x + 6$

  $\ y \ge \tfrac{3}{2}x - 3$

Be sure to reverse the inequality
when multiplying or dividing both
sides by a negative number.

The graph of the solution set for $y > -3x$ is the half-plane above the line $y = -3x$ as shown in Figure 7.2. The graph of the solution set for $y \ge \tfrac{3}{2}x - 3$ is the half-plane above the line $y = \tfrac{3}{2}x - 3$ as shown in Figure 7.3.

PROBLEM 1   Graph the solution set for each linear inequality.

(a) $y < -2$ (b) $y \le -2x + 5$ ◆

♦ **Graphing a System of Linear Inequalities**

When given several linear inequalities in two unknowns x and y to solve simultaneously, we say that we have a **system of linear inequalities**. If an ordered pair (x_1, y_1) satisfies *every* inequality in the system, then (x_1, y_1) is a solution of the system. *All* ordered pairs that satisfy the system can be illustrated graphically by shading the region in the xy-plane that is *common* to the graphs of all the inequalities in the system.

EXAMPLE 2   Graph the solution set of each system of linear inequalities.

(a) $3x + y > 0$ (b) $y \geq x - 3$
$3x - 2y \leq 6$ $x + 2y \leq 12$
                           $x \geq 0$
                           $y \geq 0$

SOLUTION

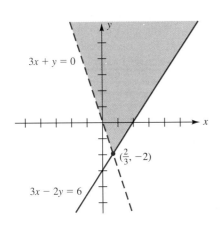

$3x + y = 0$

$\left(\frac{2}{3}, -2\right)$

$3x - 2y = 6$

FIGURE 7.4

Graph of the solution set for the system
$3x + y > 0$
$3x - 2y \leq 6$

(a) The shaded regions in Figures 7.2 and 7.3 represent the solution sets of the individual inequalities $3x + y > 0$ and $3x - 2y \leq 6$, respectively. Now, the solution set of this system of linear inequalities consists of all points common to both these shaded regions, as illustrated in Figure 7.4. We find the intersection point of the two lines, $\left(\frac{2}{3}, -2\right)$, by solving the system of equations $3x + y = 0$ and $3x - 2y = 6$ *simultaneously*. Every ordered pair within the shaded region of Figure 7.4 or along the portion of the solid line $3x - 2y = 6$ that touches this region, *except* the point $\left(\frac{2}{3}, -2\right)$, is a solution of this system of inequalities. The point $\left(\frac{2}{3}, -2\right)$ is not part of the solution set since $x = \frac{2}{3}$, $y = -2$ does not satisfy the inequality $3x + y > 0$.

(b) We begin by graphing each of the four individual inequalities on the same set of coordinate axes. The solution set of this system of linear inequalities is the region that is common to all these inequalities. The shaded region in Figure 7.5 shows the solution set. Each of the points where the sides of the shaded region intersect is called a *vertex*, and we find these points (the *vertices* of the graph) as follows:

1. Solving $y = x - 3$ and $y = 0$ simultaneously yields the vertex $(3, 0)$.
2. Solving $y = x - 3$ and $x + 2y = 12$ simultaneously yields the vertex $(6, 3)$.
3. Solving $x + 2y = 12$ and $x = 0$ simultaneously yields the vertex $(0, 6)$.
4. Solving $x = 0$ and $y = 0$ simultaneously yields the vertex $(0, 0)$.

Every ordered pair in the shaded region of Figure 7.5 or along the portions of the solid lines that touch this region is a solution of this system. ♦

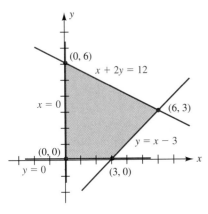

$(0, 6)$

$x + 2y = 12$

$x = 0$

$(6, 3)$

$y = x - 3$

$(0, 0)$

$y = 0$ $(3, 0)$

FIGURE 7.5

Graph of the solution set for the system
$y \geq x - 3$
$x + 2y \leq 12$
$x \geq 0$
$y \geq 0$

PROBLEM 2   Graph the solution set for the given system and label the vertices of the shaded region.

$$x - 2y \leq 4$$
$$x + 2y \leq 8$$
$$x \geq -3 \qquad \blacklozenge$$

◆ **An Introduction to Linear Programming**

Note that every line segment joining any two points within the shaded region of Figure 7.5 is contained in that region. A region with this characteristic is said to be **convex**. In **linear programming**, a branch of mathematics, we maximize or minimize a linear function f in two variables x and y over a convex region. The function f, defined by

$$f(x, y) = Ax + By,$$

is called the **objective function** with slope $-A/B$. The convex region over which we maximize or minimize the objective function is formed from a system of linear inequalities called the **constraints**.

Note: The linear programming problems that we discuss in this section represent an introduction to this branch of mathematics. More advanced linear programming problems use more than two unknowns and require sophisticated matrices and computer programming to find maximum and minimum values.

Suppose we wish to find the maximum and minimum values of an objective function f, defined by

$$f(x, y) = 2x + 3y,$$

over the convex region in Figure 7.5. Note that any line of this form has a slope of $-\frac{2}{3}$. Now, referring to Figure 7.6, observe the following:

1. The value of $f(x, y)$ remains constant along a line with slope $-\frac{2}{3}$, and this value may be determined by knowing one point on that line. Note that $f(0, 0) = 0$, $f(3, 0) = 6$, $f(0, 6) = 18$, and $f(6, 3) = 21$.
2. The value of $f(x, y)$ increases as lines with slope $-\frac{2}{3}$ move from left to right through the convex region.
3. The maximum value of $f(x, y)$ over the convex region is 21, and this occurs at a vertex of the region, namely, (6,3).
4. The minimum value of $f(x, y)$ over the convex region is 0, and this also occurs at a vertex of the region, namely, (0, 0).

In summary, we state the following **fundamental principle of linear programming**.

FIGURE 7.6
The maximum value of
$f(x, y) = 2x + 3y$ over the given convex
region is 21, and this occurs at the vertex
$(6, 3)$. The minimum value of
$f(x, y) = 2x + 3y$ over the given convex
region is 0, and this occurs at the vertex
$(0, 0)$.

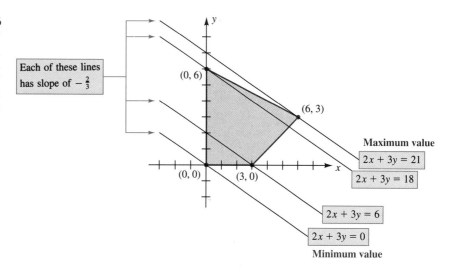

Each of these lines has slope of $-\frac{2}{3}$

$(0, 6)$

$(6, 3)$

Maximum value

$2x + 3y = 21$

$2x + 3y = 18$

$(0, 0)$ $(3, 0)$

$2x + 3y = 6$

$2x + 3y = 0$

Minimum value

◆ **Fundamental Principle of Linear Programming**

The maximum and minimum values (if they exist) of an objective function f defined by

$$f(x, y) = Ax + By$$

always occur at vertices of the convex region that is formed from a system of linear inequalities, called constraints.

EXAMPLE 3   Find the maximum and minimum values (if they exist) of the objective function f defined by $f(x, y) = 2x + 5y$, subject to the following set of constraints:

$$y \geq 7 - 2x$$
$$x + 3y \geq 7$$
$$x + \ y \geq 5$$
$$x \geq 0$$
$$y \geq 0$$

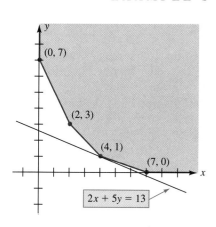

$(0, 7)$

$(2, 3)$

$(4, 1)$

$(7, 0)$

$2x + 5y = 13$

FIGURE 7.7
The minimum value of
$f(x, y) = 2x + 5y$ over this unbounded
convex region is 13, and it occurs when
$x = 4$, and $y = 1$. The function f has
no maximum value.

SOLUTION   We begin by graphing the system of linear inequalities given as constraints (see Figure 7.7). Note that the constraints form an *unbounded* convex region. We find the vertices of this convex region as follows:

1.   Solving $y = 7 - 2x$ and $x = 0$ simultaneously yields the vertex $(0, 7)$.

2.   Solving $y = 7 - 2x$ and $x + y = 5$ simultaneously yields the vertex $(2, 3)$.

3.   Solving $x + y = 5$ and $x + 3y = 7$ simultaneously yields the vertex $(4, 1)$.

4.   Solving $x + 3y = 7$ and $y = 0$ simultaneously yields the vertex $(7, 0)$.

According to the fundamental principle of linear programming, the maximum and minimum values of f (if they exist) occur at vertices of this convex region. Thus, we evaluate $f(x, y)$ at each vertex, as shown in the following table:

Vertex	Value of $f(x, y) = 2x + 5y$	
$(0, 7)$	$f(0, 7) = 2(0) + 5(7) = 35$	
$(2, 3)$	$f(2, 3) = 2(2) + 5(3) = 19$	
$(4, 1)$	$f(4, 1) = 2(4) + 5(1) = 13$	**Minimum value**
$(7, 0)$	$f(7, 0) = 2(7) + 5(0) = 14$	

Using this table and Figure 7.7, we observe that the minimum value of f is 13, and this occurs when $x = 4$ and $y = 1$. We cannot conclude from this table that the maximum value of f is 35. Note that values of $f(x, y)$ increase without bound as lines with slope $-\frac{2}{5}$ move from left to right through this unbounded convex region. Hence, we conclude a maximum value of f does not exist.   ◆

PROBLEM 3   Find the maximum and minimum values (if they exist) of the objective function f defined by $f(x, y) = 2x - 5y$ subject to the set of constraints given in Example 3.   ◆

◆ **Application: Maximizing a Profit**

Linear programming has many applications in business and economics. We conclude this section with an example showing how we can maximize a profit.

EXAMPLE 4   A ski company manufactures a slalom ski and a racing ski. The profit on each pair of slalom skis is $30 and on each pair of racing skis is $50. The company can produce at most 60 pairs of slalom skis per day and at most 40 pairs of racing skis per day. Production of a pair of slalom skis requires 2 hours of labor and production of a pair of racing skis requires 3 hours of labor. If the maximum number of hours available for the production of skis is 150 hours per day, determine the number of each type of ski that should be manufactured to maximize the profit.

SOLUTION   We begin by letting

$$x = \text{number of slalom skis produced}$$

and $$y = \text{number of racing skis produced.}$$

If P represents the profit, then

$$P(x, y) = 30x + 50y$$

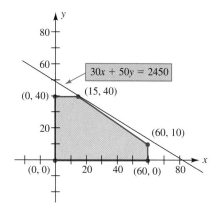

FIGURE 7.8
The maximum value of
$P(x, y) = 30x + 50y$ over this convex
region is 2450, and this occurs when
$x = 15$ and $y = 40$.

is the function we wish to maximize, subject to the following constraints:

$$x \geq 0$$

$$y \geq 0$$

$$x \leq 60$$

$$y \leq 40$$

$$2x + 3y \leq 150$$

The graph of the constraints is shown in Figure 7.8. According to the fundamental principle of linear programming, the maximum value of P must occur at a vertex of this convex region. Thus, we evaluate $P(x, y)$ at each vertex, as shown in the following table.

Vertex	Value of $P(x, y) = 30x + 50y$	
(0, 0)	$P(0, 0) = 30(0) + 50(0)$ $= 0$	
(0, 40)	$P(0, 40) = 30(0) + 50(40)$ $= 2000$	
(15, 40)	$P(15, 40) = 30(15) + 50(40) = 2450$	**Maximum value**
(60, 10)	$P(60, 10) = 30(60) + 50(10) = 2300$	
(60, 0)	$P(60, 0) = 30(60) + 50(0)$ $= 1800$	

From the table we observe that the maximum value of P is \$2450, and this occurs when $x = 15$ and $y = 40$ (see Figure 7.8). Thus, to maximize profit, the company should manufacture 15 pairs of slalom skis and 40 pairs of racing skis per day.

PROBLEM 4   Referring to Example 4, show that if the company's profit is \$30 on each pair of slalom skis, but is \$45 on each pair of racing skis, then we can find more than one way to maximize the profit.   ◆

Exercises 7.7

    Basic Skills

In Exercises 1–12, graph the solution set of each linear inequality.

1. $3x + 2y \geq 12$
2. $2x - 5y < 10$
3. $2x - 7y < 10$
4. $6x + 4y > 18$
5. $x > 3$
6. $y \leq -2$
7. $2y - 5 > 0$
8. $3x + 4 > 0$
9. $y \leq 2x - 3$
10. $y \geq 4 - 3x$
11. $y < \dfrac{2 - 3x}{4}$
12. $y \geq \dfrac{2x + 8}{3}$

In Exercises 13–30, graph the solution set of each system of linear inequalities. Label the vertices of the shaded region.

13. $x \geq 2$
    $y \leq 1$
14. $x + y < 4$
    $x > 3$
15. $x + 2y < 7$
    $x - y < 1$
16. $x - y \leq 2$
    $y < x + 3$
17. $x + 2y < 6$
    $x \geq 0$
    $y \geq 0$
18. $y < x$
    $x < 4$
    $y > -2$

19. $5x - 2y > -6$
$x - 2y \le 2$
$x + 2y \le 6$

20. $3x + 2y \ge -6$
$11x - 2y < 34$
$x - 4y \ge -16$

21. $\quad\quad y > x$
$x + 2y \ge 6$
$2x + y \le 15$

22. $\quad\quad 3y \ge x$
$x - 3y \ge -12$
$x \le 6$

23. $x + 3y \le 15$
$2y \ge x$
$y > 1$
$x \ge 0$

24. $2y \le x + 9$
$y \le -3x + 15$
$x > -3$
$y \ge 0$

25 $3x + 2y \ge 6$
$5x + 4y \le 20$
$x \ge 0$
$y \ge 0$

26. $x + y \ge 10$
$x + y \le 15$
$y \le 20$
$x \le 25$

27. $\quad\quad y \ge 2x - 8$
$x + y \le 7$
$x \le 4$
$x \ge 0$
$y \ge 0$

28. $x + 2y \ge 3$
$x - 2y \ge -3$
$3x + 2y \le 23$
$\quad\quad y \ge x - 6$
$x \ge 1$

29. $2x + y \ge 4$
$x + 2y \ge 5$
$x + 4y \ge 7$
$x \ge 0$
$y \ge 0$

30. $x + 4y \ge 14$
$x + 2y \ge 10$
$x + y \ge 7$
$x \ge 2$
$y \ge 0$

In Exercises 31–44, find the maximum and minimum values (if they exist) of the objective function subject to the given constraints.

31. Objective function: $f(x, y) = 3x + y$
Constraints: $x + y \le 6$
$x \le 3$
$x \ge 0$
$y \ge 0$

32. Objective function: $f(x, y) = 2x - 3y$
Constraints: $x + y \le 6$
$x \le 3$
$x \ge 0$
$y \ge 0$

33. Objective function: $f(x, y) = 4x - 5y$
Constraints: $y \le x + 1$
$y \ge x - 2$
$y \le 7$
$y \ge 2$

34. Objective function: $f(x, y) = 3x + 2y$
Constraints: $y \le x + 1$
$y \ge x - 2$
$y \le 7$
$y \ge 2$

35. Objective function: $f(x, y) = 2x - y$
Constraints: $2x + 3y \le 16$
$x - 4y \le -3$
$y \le 3x - 2$

36. Objective function: $f(x, y) = x + 5y$
Constraints: $2x + 3y \le 16$
$x - 4y \le -3$
$y \le 3x - 2$

37. Objective function: $f(x, y) = x + y$
Constraints: $x + 2y \le 10$
$y \le 6 - x$
$x \le 4$
$x \ge 0$
$y \ge 0$

38. Objective function: $f(x, y) = 5x + 2y$
Constraints: $x + 2y \le 10$
$y \le 6 - x$
$x \le 4$
$x \ge 0$
$y \ge 0$

39. Objective function: $f(x, y) = 5x + 2y$
Constraints: $\quad\quad y \le 27 - 3x$
$3x - 2y \le 18$
$3x + 4y \ge 18$
$3x - 5y \ge -9$

40. Objective function: $f(x, y) = 3x + y$
Constraints: $\quad\quad y \le 27 - 3x$
$3x - 2y \le 18$
$3x + 4y \ge 18$
$3x - 5y \ge -9$

41. Objective function: $f(x, y) = 3x + 4y$
Constraints: See Exercise 29

42. Objective function: $f(x, y) = 2x - y$
Constraints: See Exercise 29

43. Objective function: $f(x, y) = 7x - 2y$
Constraints: See Exercise 30

44. Objective function: $f(x, y) = 2x + 7y$
Constraints: See Exercise 30

In Exercises 45–52, use linear programming to solve each problem.

45. A doctor wants to buy bookcases for his office. One type of bookcase costs $100, holds 30 cubic feet of books, and takes up 6 square feet of floor space. Another type costs $200, holds 48 cubic feet of books, and takes up 9 square feet of floor space. The doctor wants to spend no more than $1200 for the bookcases and use no more than 66 square feet of office floor space. How many of each type bookcase should he buy to *maximize storage capacity*?

46. A sand and gravel company wishes to purchase 6-wheel and 10-wheel dump trucks to make deliveries to its customers. Each 6-wheel dump truck sells for $20,000, has an average monthly fuel cost of $100, and has a carrying capacity of 12 cubic yards. Each 10-wheel dump truck sells for $30,000, has an average monthly fuel cost of $300, and has a carrying capacity of 17 cubic yards. Suppose the company wishes to spend no more than $240,000 for the trucks and no more than $1500 per month for fuel. How many trucks of each type should the company purchase to *maximize the carrying capacity*?

47. A company manufactures oak tables and chairs. Each table requires 2 hours 15 minutes to assemble, 1 hour to finish, and 30 minutes to pack for shipping. Each chair requires 2 hours to assemble, 2 hours to finish, and 10 minutes to pack for shipping. The maximum number of labor hours available each day for assembling is 90 hours, for finishing is 80 hours, and for packing is 15 hours. If the manufacturer makes a profit of $80 on each table and $30 on each chair, determine the number of each that should be produced each day to *maximize the profit*. (Assume that all tables and chairs produced can be sold.)

48. A kennel raises Doberman pinschers and German shepherds. The kennel can raise no more than 40 dogs and wishes to have no more than 24 Doberman pinschers. The cost of raising a Doberman pinscher is $50, the cost of raising a German shepherd is $30, and the kennel can invest no more than $1500 for this purpose. If the kennel makes a profit of $150 per pinscher and $100 per shepherd, how many dogs of each type should be raised in order to *maximize the profit*? (Assume that all dogs can be sold.)

49. A grain company stocks two types of wild-bird seed: sunflower and cracked corn. The company buys the sunflower seed for $10 per bag and the cracked corn for $5 per bag. They sell the sunflower seed for $16 per bag and the cracked corn for $10 per bag. From past experience, the company knows that they will sell at least twice as many bags of sunflower seeds as bags of cracked corn. If the company does not wish to order more than $600 worth of seed, how many bags of each type should be ordered to *maximize the profit*? (Assume that all bags of seed can be sold.)

50. A woman inherits $48,000 from her grandfather's estate, subject to the following conditions: Part or all of the money must be invested either in treasury bills that pay 8¢ on each dollar or in bonds that pay 10¢ on each dollar. At least $10,000 must be invested in treasury bills and at most $30,000 in bonds. The amount invested in treasury bills must be no more than twice the amount invested in bonds. How much should she invest at each to *maximize her return*?

Critical Thinking

51. Given the convex region shown in the figure, find the slope *m* of an objective function $f(x, y) = Ax + By$ that has a maximum value at the given point(s):

(a) at both points *P* and *Q*
(b) at both points *Q* and *R*
(c) at point *P* only
(d) at point *Q* only
(e) at point *R* only

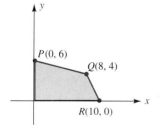

52. Given the convex region shown in the figure, find the slope *m* of an objective function $f(x, y) = Ax + By$ that has a minimum value at the given point(s):

(a) at both points *P* and *Q*
(b) at both points *Q* and *R*
(c) at point *P* only
(d) at point *Q* only
(e) at point *R* only

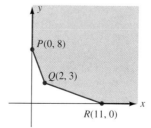

53. A landscaping company has two brands of lawn fertilizer with nutrient contents as shown in the following table:

	Brand A (kg per bag)	Brand B (kg per bag)
Nitrogen (N)	30	20
Phosphoric acid (P_2O_5)	2	4
Potash (K_2O)	1	4

It has been determined that a certain lawn needs at least 120 kg nitrogen, at least 16 kg phosphoric acid, and at least 12 kg potash. If brand A costs $22 per bag and brand B costs $18 per bag, how many bags of each brand should be used to *minimize the cost*? What is the minimum cost?

54. A computer company has two distribution centers, one in Boston and one in Providence. The company receives an order to ship 40 computers to college A, and 60 computers to college B. The cost of shipping each computer is shown in the following table:

	To college A	To college B
From Boston	$9	$8
From Providence	$8	$10

Suppose that the distribution center in Boston has 80 computers available for shipping and the distribution center in Providence has 30 computers available. How should the order be filled in order to *minimize the total shipping cost*? What is the minimum cost?

 Calculator Activities

In Exercises 55 and 56, graph the solution set of each system of linear inequalities. Label the coordinates of each vertex, rounding the answer to three significant digits.

55. $2.4x - 3.2y \leq 7.9$
$2.1x + 1.8y \leq 13.2$
$y \leq 3.4$
$x \geq 0$
$y \geq 0$

56. $y \geq 16.2 - 6.8x$
$y \geq 12.3 - 3.4x$
$y \geq 4.4 - 0.3x$
$x \geq 0$
$y \geq 0$

57. Find the maximum value of the objective function $f(x, y) = 3.2x + 5.4y$, subject to the constraints given in Exercise 55. Round the answer to three significant digits.

58. Find the minimum value of the objective function $f(x, y) = 14.5x + 3.6y$, subject to the constraints given in Exercise 56. Round the answer to three significant digits.

59. A company makes two brands of cereal P and Q, both of which are enriched with vitamins A and B. Cereal P costs 10¢ per ounce to manufacture and cereal Q costs 15¢ per ounce. The number of milligrams of each vitamin that is contained in each ounce of cereal is given in the following table:

Vitamin	Cereal P (mg/oz)	Cereal Q (mg/oz)
A	1.2	2.3
B	3.3	1.4

Suppose a woman needs at least 8.0 mg of vitamin A and at least 9.0 mg of vitamin B to satisfy her daily requirements. How many ounces of each type cereal should she consume if she wishes to *minimize the cost*? Round each answer to the nearest tenth of an ounce.

60. A farmer has 200 acres of land available and wishes to plant at least 40 acres of corn and beans. The cost of the corn seed is $115 per acre and the cost of the bean seed is $185 per acre. Past experience has shown that the cost of maintaining and harvesting the corn crop is $260 per acre, and the cost of maintaining and harvesting the bean crop is $145 pe acre. The farmer wants to spend no more than $12,500 for seed and no more than $16,500 on maintenance and harvest. If the expected profit from the corn crop and bean crop are $550 per acre and $520 per acre, respectively, how many acres of each crop should be planted to make the *largest profit*? Round each answer to the nearest tenth of an acre.

Chapter 7 Review

Questions for Group Discussion

1. State the three possibilities for the solution set of a system of n linear equations in n unknowns.

2. Explain various ways of generating *equivalent systems* of linear equations.

3. Can a *nonsquare system* of linear equations have a unique solution? Explain.

4. What is meant by the *dimension* of a matrix? Give an example of a matrix with dimension 4×3.

5. What is an *augmented matrix*? How is it formed?

6. Is the *echelon form* of an augmented matrix unique? Explain.

7. List three *elementary row operations* that can be performed on a matrix to generate a row-equivalent matrix. Illustrate each operation with an example.

8. Explain the procedure for solving a system of linear equations by using *Gauss-Jordan elimination*.

9. Give an example of two matrices that are equivalent, but not equal.

10. Is it possible to find the *sum of two matrices* having different dimensions? Explain.

11. Write the *zero matrix* of dimension 3×3.

12. List some properties of *matrix addition* and *scalar multiplication*. Illustrate each property with an example.

13. Explain the procedure for finding the *product of two matrices*. Under what conditions does the product exist?

14. If matrix A has dimension 3×2 and matrix B has dimension 2×5, what is the dimension of AB?

15. Is matrix multiplication commutative? associative? Illustrate with examples.

16. Suppose A and B are $n \times n$ matrices and $AB = I_n$, where I_n is the *identity matrix* of dimension $n \times n$. What is the relationship between A and B?

17. Explain the procedure for finding the *inverse* of a square matrix. Does every square matrix have an inverse?

18. What is the value of $|A|$ if the matrix A is *singular*? Explain.

19. When solving the matrix equation $AX + B = C$ for X, what restrictions must be placed on A, B, and C for the solution to exist?

20. What is the procedure for finding the *determinant* of a 2×2 matrix? of a 3×3 matrix? of a 4×4 matrix?

21. When is the *cofactor* of an element a_{ij} equal to the *minor* of that element?

22. Explain the procedure of using *Cramer's rule* to solve a system of linear equations. When does the rule not apply?

23. What is a *coefficient matrix*? How is it formed?

24. What are the effects of elementary row operations on the determinant of an $n \times n$ matrix?

25. List some conditions for which the determinant of an $n \times n$ matrix is zero.

26. How does the graph of $y > ax + b$ differ from the graph of $y < ax + b$?

27. Explain the procedure for graphing a *system of linear inequalities*. Illustrate with an example.

28. What is *linear programming*? Where do the maximum and minimum values of the objective function always occur?

Review Exercises

In Exercises 1–18 solve each system of linear equations.

1. $2x + 3y = -5$
$x - 5y = 17$

2. $3x - 4y = -1$
$-2x + 5y = 10$

3. $3x - 6y = 7$
$5x + 10y = 3$

4. $\frac{1}{2}x - y = 1$
$-\frac{3}{2}x + 3y = -3$

5. $x - y = 5$
$2x + 3y - z = -4$
$y + 3z = -6$

6. $2x + y = 1$
$x + z = -4$
$3y - 2z = 11$

7. $3x - 2y + z = -3$
$2x + y - 3z = 2$
$4y - 2z = 4$

8. $5x - 2y + 3z = 11$
$2x - y + 4z = 5$
$4x - 3y + 2z = 9$

9. $2x + 6y - 5z = 1$
$3x + 4y - 5z = -1$
$x + 8y - 5z = 3$

10. $-4x - 3y + z = 2$
$3x - y + 4z = -3$
$4x + 3y - z = 0$

11. $\frac{1}{2}x - 3y + 5z = 18$
$2x - \frac{3}{8}y - z = 15$
$3x + 5y + 6z = 4$

12. $0.2x - 1.6y - 3.2z = -12$
$1.4x + 2.3y - 0.7z = 104$
$9.8x - 2.4y + 1.1z = 113$

13. $x - 3y + z = 6$
$4x - 2y - z = -1$

14. $2x + y - 3z = 4$
$3x + 2y - z = 7$

15. $x + y + t = -1$
$2y - t = 2$
$x - z + 3t = 11$
$4y + 3z = 4$

16. $2x - y + 3z - t = 1$
$x + 3y + t = -5$
$2y + 4z + 5t = 1$
$4x - z + 2t = 2$

17. $3x - y + z = 0$
$4y - 4z + 3t = 12$
$2x - 2y - t = 0$
$5x - 3y - 2z + 5t = 12$

18. $x_1 + 2x_2 - x_5 = -7$
$x_2 - x_3 + x_4 = -6$
$x_1 + 2x_3 + x_5 = 11$
$3x_2 - 2x_4 = -4$
$4x_3 - x_4 - 2x_5 = 5$

In Exercises 19–34, perform the indicated matrix operations given that A, B, C, D, E, F, G, and H are defined as follows:

$$A = \begin{bmatrix} 4 & 5 \\ 3 & 4 \end{bmatrix} \qquad B = \begin{bmatrix} -2 & 3 \\ 1 & 0 \end{bmatrix}$$

$$C = \begin{bmatrix} 1 & 0 & -3 \\ -2 & 4 & 5 \end{bmatrix} \qquad D = \begin{bmatrix} 4 & 2 & -1 \\ 0 & -3 & -6 \end{bmatrix}$$

$$E = \begin{bmatrix} 2 & -3 \\ -1 & 5 \\ 3 & 2 \end{bmatrix} \qquad F = \begin{bmatrix} 3 & 5 \\ -2 & 7 \\ 1 & 0 \end{bmatrix}$$

$$G = \begin{bmatrix} 1 & 4 & -2 \end{bmatrix} \qquad H = \begin{bmatrix} -6 \\ 3 \\ -9 \end{bmatrix}$$

19. $-3G$ **20.** $\frac{1}{3}H$ **21.** $A + B$

22. $F - E$ **23.** $\frac{1}{2}A - \frac{3}{2}A$ **24.** $\frac{1}{3}(-3D)$

25. $2C - 3D$ **26.** $3(A + 2B)$ **27.** AB

28. CE **29.** GH **30.** HG

31. $(CF)A$ **32.** $-2(DH)$ **33.** $G(2E + F)$

34. $\frac{1}{2}(A^2 + B)$

In Exercises 35–40, find the inverse of each matrix (if it exists).

35. $\begin{bmatrix} 4 & 3 \\ 2 & 2 \end{bmatrix}$ **36.** $\begin{bmatrix} -6 & 3 \\ 5 & -2 \end{bmatrix}$

37. $\begin{bmatrix} 1 & 4 & 0 \\ 2 & 6 & -1 \\ 3 & 2 & 5 \end{bmatrix}$ **38.** $\begin{bmatrix} 2 & -1 & 3 \\ 3 & -5 & 8 \\ 4 & 6 & -2 \end{bmatrix}$

39. $\begin{bmatrix} 2 & 8 & -3 & 7 \\ 4 & 1 & -6 & 0 \\ 0 & 3 & 0 & -3 \\ -6 & 9 & 9 & -7 \end{bmatrix}$

40. $\begin{bmatrix} 4 & -9 & -12 & -3 \\ 0 & 4 & 9 & -4 \\ -3 & 7 & 10 & 2 \\ 2 & -2 & 0 & 7 \end{bmatrix}$

In Exercises 41–44, write each system of linear equations as a matrix equation $AX = B$. Then solve the system by using the inverse of the coefficient matrix A.

41. $3x + 7y = 1$
$\quad x + 2y = 1$

42. $-2x + 4y = 28$
$\quad 3x + 2y = 6$

43. $x - 2y + 3z = -9$
$2x + y - z = 13$
$3x + 4y - z = 23$

44. $2x + 3y + z = -3$
$5x - 2y + 3z = 13$
$3x - y - 4z = -10$

In Exercises 45–48, solve each matrix equation for X given that

$$A = \begin{bmatrix} 2 & 5 \\ 3 & 8 \end{bmatrix}, \qquad B = \begin{bmatrix} 2 & -2 \\ 3 & 0 \end{bmatrix},$$

and $\qquad C = \begin{bmatrix} -4 & 2 \\ 4 & -1 \end{bmatrix}.$

45. $\frac{1}{2}AX = B$

46. $B - AX = C$

47. $3AX - BX = 3C$

48. $2AX - B = A - CX$

In Exercises 49–58, find the determinant of each matrix.

49. $\begin{bmatrix} 2 & -3 \\ 5 & -2 \end{bmatrix}$

50. $\begin{bmatrix} 12 & 15 \\ -7 & 9 \end{bmatrix}$

51. $\begin{bmatrix} 1 & -2 & 0 \\ 2 & -3 & 1 \\ 6 & 0 & -2 \end{bmatrix}$

52. $\begin{bmatrix} 3 & -2 & 4 \\ 8 & 2 & -3 \\ -1 & 7 & 3 \end{bmatrix}$

53. $\begin{bmatrix} 3 & -4 & 7 \\ 0 & -2 & 3 \\ 0 & 0 & 5 \end{bmatrix}$

54. $\begin{bmatrix} 1 & 9 & 6 & -3 \\ 0 & -4 & 5 & 1 \\ 0 & 0 & -4 & 2 \\ 0 & 0 & 0 & -2 \end{bmatrix}$

55. $\begin{bmatrix} 2 & -3 & 0 & 1 \\ 1 & -4 & 2 & -5 \\ 0 & -1 & 0 & 2 \\ 2 & 6 & -4 & 0 \end{bmatrix}$

56. $\begin{bmatrix} 1 & 3 & -2 & 4 \\ 0 & 6 & 2 & -4 \\ 2 & 3 & -5 & 10 \\ 3 & 1 & 0 & 6 \end{bmatrix}$

57. $\begin{bmatrix} 1 & 0 & -2 & 0 & -3 \\ -2 & 0 & 4 & 1 & 0 \\ 0 & 3 & 0 & -2 & 1 \\ 3 & 0 & -6 & 0 & 3 \\ 0 & 1 & 0 & 3 & 4 \end{bmatrix}$

58. $\begin{bmatrix} 6 & 0 & 0 & 3 & 0 \\ 2 & -3 & 5 & -1 & 1 \\ 2 & -1 & 3 & 0 & 2 \\ 0 & -3 & 0 & -2 & 1 \\ 0 & -1 & 4 & 0 & 0 \end{bmatrix}$

In Exercises 59–62, solve each system of linear equations by using Cramer's rule.

59. $21x - 18y = -3$
$\quad 4x + 15y = 110$

60. $19x + 13y = -10$
$\quad 8x - 22y = 37$

61. $13x - 12y - 5z = 6$
$\quad 4x - 11y + 15z = 28$
$\quad 6x - 10y \quad\quad = 9$

62. $3x + 10y + 4z = -8$
$\quad 8x + 5y - 9z = 16$
$\quad 5x - 7y - 11z = 16$

In Exercises 63-66, graph the solution set of each system of linear inequality.

63. $\quad\quad y < x + 2$
$\quad 3x + 2y < 6$
$\quad\quad\quad x \le 6$

64. $3x + 4y \ge 12$
$\quad 6x + 5y \le 30$
$\quad\quad\quad x \ge 0$
$\quad\quad\quad y \ge 0$

65. $\quad\quad x \le 18 - 3y$
$\quad 3x + 2y \le 19$
$\quad 2x + y \le 12$
$\quad\quad\quad x \ge 0$
$\quad\quad\quad y \ge 0$

66. $4x + y \ge 8$
$\quad\quad y \ge 5 - x$
$\quad x + 5y \ge 13$
$\quad\quad\quad x \ge 0$
$\quad\quad\quad y \ge 0$

In Exercises 67 and 68, find the maximum and minimum values (if they exist) of the objective function subject to the given constraints.

67. Objective function: $f(x, y) = 2x + 3y$
Constraints: See Exercise 65

68. Objective function: $f(x, y) = 2x + y$
Constraints: See Exercise 66

69. Find k such that the given determinant is true:

(a) $\begin{vmatrix} k & 3 \\ 5 & 7 \end{vmatrix} = 6$ (b) $\begin{vmatrix} k & 2 \\ 5 & k \end{vmatrix} = 15$

(c) $\begin{vmatrix} 0 & k & 0 \\ -1 & 2 & 4 \\ 2 & 5 & k \end{vmatrix} = -12$ (d) $\begin{vmatrix} 2 & k & 3 \\ 0 & 3 & k \\ 5 & 1 & 1 \end{vmatrix} = 0$

70. Given that $A = \begin{bmatrix} a & b \\ c & d \end{bmatrix}$, where a, b, c, and d are real numbers, find a formula for A^{-1}. List any restrictions on the formula, then use it to determine the inverse of the matrices in Exercises 35 and 36.

In Exercises 71–76, use a system of linear equations to solve each problem.

71. A refinery has a "regular" gasoline that sells for 83¢ per gallon and a "super" gasoline that sells for 99¢ per

gallon. How many gallons of each type gasoline should be mixed together to form 12,000 gallons of a gasoline that sells for 86¢ per gallon?

72. A man has invested $10,000 in two bank accounts. One account gives an annual yield of 8% and the other an annual yield of 9%. The interest he earns from these accounts is taxable at the rate of 30%. After taxes, he earns $602 from these accounts for the year. How much is invested in each account?

73. Three types of tickets are available for baseball games: box seats for $16, grandstand seats for $10, and bleacher seats for $5. The seating capacity at the ball park is 32,500, and the park has 2500 more grandstand seats than bleacher seats. If a game is sold out and the revenue from ticket sales is $343,500, how many seats of each type are in the ball park?

74. At the baseball game, Kristen buys 2 hot dogs, 1 bag of peanuts, and 2 sodas for $8.50. Caryn buys 1 hot dog, 2 bags of peanuts, 1 ice-cream bar, and 1 soda for $7.00. David buys 4 hot dogs and 4 sodas for $16.00. Regina buys 1 hot dog, 2 ice-cream bars, and 1 bag of peanuts for $7.00. What is the price of a hot dog, a bag of peanuts, a soda, and an ice-cream bar?

75. A hiking trail, connecting campsite A to campsite B, has a pond $2\frac{1}{2}$ miles from campsite A. Starting from campsite A, the trail goes downhill for 1 mile, is level for 6 miles, and goes uphill for 2 miles. Suppose a woman can hike from campsite A to campsite B in 3 hours 15 minutes; from campsite B to campsite A in 3 hours; and from campsite A to the pond and back again to campsite A in 1 hour 45 minutes. Find her rates of walking (in miles per hour): downhill, on level ground, and uphill.

76. Find the equation of the parabola $x = ay^2 + by + c$ that passes through the points $(k_1, 1)$, $(k_2, -2)$, and $(k_3, 0)$, given the following information:

(a) $k_1 = 4$, $k_2 = 1$, and $k_3 = -1$
(b) $k_1 = -2$, $k_2 = 1$, and $k_3 = -3$
(c) $k_1 = 3$, $k_2 = 8$, and $k_3 = 4$
(d) $k_1 = 2$, $k_2 = 5$, and $k_3 = 0$

Exercises 77–80 pertain to the appliance sale problem described as follows:

An appliance store has a two-day sale on its three styles of microwave ovens. The sale prices (in dollars) are given in matrix C:

$$
\begin{array}{ccc}
\text{400-watt} & \text{500-watt} & \text{600-watt} \\
\text{model} & \text{model} & \text{model}
\end{array}
$$

$$C = [\quad 120 \qquad 160 \qquad 220 \quad] \quad \text{Cost (\$)}$$

Within the Boston area, the store has three outlets. The number of 400-watt, 500-watt, and 600-watt models of microwave ovens sold at each outlet during the two-day sale are given in matrices A and B:

First Day

$$
A = \begin{array}{c}
\\ \\ \\ \\
\end{array}
\begin{array}{ccc}
\text{Outlet 1} & \text{Outlet 2} & \text{Outlet 3} \\
\end{array}
$$

$$
A = \begin{bmatrix} 8 & 10 & 16 \\ 12 & 8 & 9 \\ 6 & 5 & 6 \end{bmatrix} \begin{array}{l} \text{400-watt} \\ \text{500-watt} \\ \text{600-watt} \end{array}
$$

Second Day

$$
\begin{array}{ccc}
\text{Outlet 1} & \text{Outlet 2} & \text{Outlet 3} \\
\end{array}
$$

$$
B = \begin{bmatrix} 5 & 12 & 8 \\ 7 & 3 & 6 \\ 8 & 3 & 6 \end{bmatrix} \begin{array}{l} \text{400-watt} \\ \text{500-watt} \\ \text{600-watt} \end{array}
$$

77. Find the matrix $A \begin{bmatrix} 1 \\ 1 \\ 1 \end{bmatrix}$, and state what it represents.

78. Find the matrix $[1 \quad 1 \quad 1](A + B)$, and state what it represents.

79. Find the matrix CB, and state what it represents.

80. Find the matrix $C(A + B) \begin{bmatrix} 1 \\ 1 \\ 1 \end{bmatrix}$, and state what it represents.

In Exercises 81 and 82, use linear programming to solve each problem.

81. A college football stadium has 27,500 seats, and of these, 7500 are midfield seats. Tickets for a midfield seat cost $20 each and tickets for all other seats cost $12 each. To make a profit, the college must make at least $300,000 from ticket sales. What is the minimum number of tickets that the college must sell to make a profit?

82. A manufacturer produces two types of log-home kits: a one-bedroom model and a two-bedroom model. The profit on each one-bedroom kit is $4500, and on each two-bedroom kit, $6000. The company can produce at most 4 one-bedroom kits per week and at most 3 two-bedroom kits per week. To produce a one-bedroom kit requires 200 hours of labor and to produce a two-bedroom kit requires 300 hours of labor. If the maximum number of labor hours available for the production of log-home kits is 1100 hours per week, determine the number of each type of kit that should be manufactured to maximize the profit. (Assume all log home kits can be sold.)

CHAPTER

8

Sequences, Series, and Probability

A father decides to deposit $150 each month in a savings account for his daughter's college education. The savings account earns interest at the rate of 9% per year, compounded monthly. How much money will be in the account at the end of 10 years if there is nothing in the account today?

(For the solution, see Example 7 in Section 8.4.)

8.1   An Introduction to Sequences and Series

When the phrase "a sequence of events" is used, it means the following of one event after another in a given order of succession. In mathematics we refer to a **sequence** as an ordered list of numbers, and the numbers in the list are called the *elements* of the sequence. If the sequence has a final (last) element, then it is a *finite sequence*. For instance, the numbers

$$2, 4, 6, 8, 10, 12$$

Last element

form a finite sequence. If the sequence does not have a last element, then it is an *infinite sequence*. For instance, the numbers

$$2, 4, 6, 8, 10, 12, \ldots$$

Three dots at the end indicate that there is no last element.

form an infinite sequence.

In this section, we introduce the idea of a sequence and notation that indicates the sum of the elements in the sequence. We begin with a formal definition of a sequence.

◆ Defining a Sequence

In mathematics, we define a *finite sequence function* as a function whose domain is the set of positive integers from 1 to *n*, and an *infinite sequence function* as a function whose domain is the set of positive integers. In either case, the elements in the range of the function form the sequence.

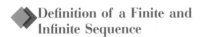
Definition of a Finite and Infinite Sequence

A function f, whose domain is the set of positive integers from 1 to n, is a finite sequence function. The elements in the range of f, taken in the order

$$f(1), f(2), f(3), \ldots, f(n),$$

form a **finite sequence**.

A function f, whose domain is the set of all positive integers, is an infinite sequence function. The elements in the range of f, taken in the order

$$f(1), f(2), f(3), \ldots,$$

form an **infinite sequence**.

Note: If the domain of a sequence function is not stated, we assume it to be the set of all positive integers, $\{1, 2, 3, \ldots\}$.

EXAMPLE 1

Find the sequence represented by each sequence function.

(a) $f(n) = 2n - 1$, domain: $\{1, 2, 3, 4\}$

(b) $g(n) = (-1)^{n+1}2^n$

SOLUTION

(a) The function f is a finite sequence function. The elements in the range of f are

$$f(1) = 2(1) - 1 = 1, \qquad f(2) = 2(2) - 1 = 3,$$
$$f(3) = 2(3) - 1 = 5, \qquad f(4) = 2(4) - 1 = 7.$$

The elements in the range of f, written in the order

$$1, 3, 5, 7,$$

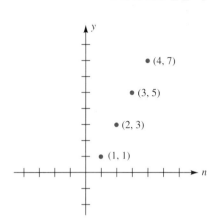

FIGURE 8.1

Graph of the sequence function
$f(n) = 2n - 1$ with domain $\{1, 2, 3, 4\}$

form a finite sequence. This sequence is the first four odd positive integers. The graph of this sequence function, which is shown in Figure 8.1, is an increasing function.

(b) The function g is an infinite sequence function, since the domain is understood to be the set of all positive integers. The first four elements in the range of g are

$$g(1) = (-1)^{1+1}2^1 = 2, \qquad g(2) = (-1)^{1+2}2^2 = -4,$$
$$g(3) = (-1)^{1+3}2^3 = 8, \qquad g(4) = (-1)^{1+4}2^4 = -16.$$

The elements in the range of g, written in the order

$$2, -4, 8, -16, \ldots,$$

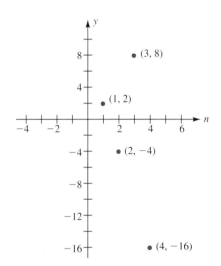

FIGURE 8.2

Graph of the first four elements of the
sequence function $g(n) = (-1)^{n+1}2^n$

with three dots (ellipses) at the end, form the infinite sequence. The graph of the first four elements of this sequence function is shown in Figure 8.2. This is neither an increasing nor a decreasing function. ◆

In the sequence in Example 1(b), the signs of the elements alternate between positive and negative value. We refer to a sequence with this characteristic as an **alternating sequence**.

PROBLEM 1

Repeat Example 1 for the sequence function f defined by

$$f(n) = (-1)^n 2^{n+1}.$$

Does this function define an alternating sequence? ◆

◆ **General Element of a Sequence**

It is customary to denote the first element of a sequence as a_1, the second element as a_2, the third element as a_3, and so on. The nth element of a sequence, denoted as a_n, is called the **general element** of the sequence. Thus, for a sequence function f, we have

$$f(1) = a_1 \qquad \text{First element}$$
$$f(2) = a_2 \qquad \text{Second element}$$
$$f(3) = a_3 \qquad \text{Third element}$$
$$\vdots \qquad \vdots \qquad\qquad \vdots$$
$$f(n) = a_n \qquad \text{General element}$$
$$\vdots \qquad \vdots$$

The general element of a sequence may be given **explicitly** in terms of n, or it may be expressed in terms of preceding elements. When it is expressed in terms of preceding elements, we say that the sequence is defined **recursively**.

EXAMPLE 2   Find the first six elements of a sequence whose general element is given by

(a) $a_n = \dfrac{1 + n}{n}$ **(b)** $a_n = \begin{cases} 2n & \text{if } n \text{ is odd} \\ 2a_{n-1} & \text{if } n \text{ is even} \end{cases}$

SOLUTION

(a)   This general element is defined explicitly in terms of n. For $n = 1, 2, 3, 4, 5, 6$, we have

$$a_1 = \frac{1+1}{1} = 2 \qquad a_2 = \frac{1+2}{2} = \frac{3}{2}$$
$$a_3 = \frac{1+3}{3} = \frac{4}{3} \qquad a_4 = \frac{1+4}{4} = \frac{5}{4}$$
$$a_5 = \frac{1+5}{5} = \frac{6}{5} \qquad a_6 = \frac{1+6}{6} = \frac{7}{6}$$

Thus, the first six elements of the sequence are $2, \frac{3}{2}, \frac{4}{3}, \frac{5}{4}, \frac{6}{5}, \frac{7}{6}$.

(b)   Note that when n is even, a_n is defined in terms of the preceding element a_{n-1}. Hence, this is an example of a recursively defined sequence. For $n = 1, 2, 3, 4, 5, 6$, we have

$$a_1 = 2(1) = 2 \qquad a_2 = 2a_1 = 2 \cdot 2 = 4$$
$$a_3 = 2(3) = 6 \qquad a_4 = 2a_3 = 2 \cdot 6 = 12$$
$$a_5 = 2(5) = 10 \qquad a_6 = 2a_5 = 2 \cdot 10 = 20$$

Thus, the first six elements of the sequence are 2, 4, 6, 12, 10, 20.   ◆

PROBLEM 2   Find the first six elements of the *Fibonacci sequence* [named after the Italian mathematician Leonardo Fibonacci (1180?–1250?)]. The general element is given by

$$a_n = \begin{cases} 1 & \text{if } n = 1 \\ 1 & \text{if } n = 2 \\ a_{n-1} + a_{n-2} & \text{if } n \geq 3 \end{cases}$$

◆

Given the first few elements of a sequence, we cannot *uniquely* determine the general element. For example, consider the infinite sequence

$$2, 4, 6, \ldots$$

These are the first three elements of the sequence for the general element given in Example 2(b). Thus, one possibility for the general element is

$$a_n = \begin{cases} 2n & \text{if } n \text{ is odd} \\ 2a_{n-1} & \text{if } n \text{ is even} \end{cases}$$

However, note that 2, 4, and 6 are even numbers, or multiples of 2. Thus, a second possibility for the general element is simply

$$a_n = 2n.$$

In the next example, we use **pattern recognition** to determine the general element of a given sequence. Keep in mind that the answers we give are not unique. They do, however, represent the most obvious choices.

EXAMPLE 3   Use pattern recognition to find the general element of each sequence.

(a) $1, 3, 5, 7, \ldots, 21$ (b) $1, 3, 9, 27, 81, 243$
(c) $\frac{2}{3}, \frac{3}{4}, \frac{4}{5}, \frac{5}{6}, \ldots$ (d) $-x, x^5, -x^9, x^{13}, \ldots$

SOLUTION

(a) We recognize the elements of the sequence $1, 3, 5, 7, \ldots, 21$ as odd, positive integers. Since these odd numbers are one less than the even numbers $2, 4, 6, 8, \ldots, 22$, which have a general element of $2n$, we may write the general element for these odd positive integers as

$$a_n = 2n - 1.$$

(b) The elements in the sequence $1, 3, 9, 27, 81, 243$ are the powers of three. Since we must obtain 1 when $n = 1$, we can write the general element as

$$a_n = 3^{n-1}.$$

(c) For the sequence $\frac{2}{3}, \frac{3}{4}, \frac{4}{5}, \frac{5}{6}, \ldots$, each denominator is one more than its corresponding numerator. Since we must obtain $\frac{2}{3}$ when $n = 1$, we can write the general element as

$$a_n = \frac{n+1}{n+2}.$$

(d) In the sequence $-x, x^5, -x^9, x^{13}, \ldots$, each exponent after the first is four more than the preceding exponent. This suggests an exponent containing $4n$. However, since we must obtain an exponent of 1 when $n = 1$, we write these exponents as $4n - 3$ for $n = 1, 2, 3, \ldots$. Also, since the *odd-numbered* elements are negative, we must include a factor of $(-1)^n$. Thus, we may write the general element for this sequence as

$$a_n = (-1)^n x^{4n-3} \qquad \blacklozenge$$

P R O B L E M 3   Referring to Example 3, verify that each general element gives the first four elements of its respective sequence. ◆

Note: For some sequences, it is not possible to find an equation that defines the general element. For example, for the sequence of prime numbers

$$2, 3, 5, 7, 11, 13, 17, 19, 23, 29, \ldots, a_n, \ldots,$$

where a_n is the nth prime number, we cannot write an equation that defines a_n.

◆ **Series**

The indicated sum of the elements in a sequence is called a **series**. Associated with the finite sequence

$$a_1, a_2, a_3, \ldots, a_n,$$

is the finite series

$$a_1 + a_2 + a_3 + \cdots + a_n$$

and associated with the infinite sequence

$$a_1, a_2, a_3, \ldots, a_n, \ldots$$

is the infinite series

$$a_1 + a_2 + a_3 + \cdots + a_n + \cdots$$

Note that each **term** in a series is the same as the corresponding element in its associated sequence. In a sequence, we refer to a_n as the general element, but in a series we refer to a_n as the **general term** of the series.

The finite series

$$a_1 + a_2 + a_3 + \cdots + a_n$$

and the infinite series

$$a_1 + a_2 + a_3 + \cdots + a_n + \cdots$$

are written here in **expanded form**. These series may also be written more compactly by using **sigma form**. The Greek letter Σ (read "sigma") is used as the summation symbol in sigma form.

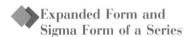Expanded Form and
Sigma Form of a Series

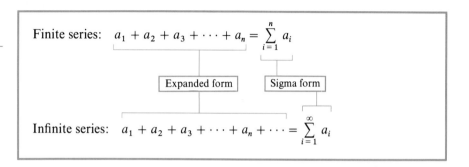

The letter i in the sigma form is called the **index of summation**, or **summation variable**. We often refer to the summation variable as a *dummy variable* because any other letter can be used without changing the series. The letters i, j, and k are commonly used as summation variables.

To change a series from sigma form to expanded form, we replace the summation variable (index) with successive integers that start with the integer written below Σ and end with the integer written above Σ. If ∞ is written above Σ, it indicates that the series is infinite and has no ending value.

EXAMPLE 4   Write each series in expanded form.

(a) $\displaystyle\sum_{i=1}^{4} (5i - 2)$ (b) $\displaystyle\sum_{j=1}^{\infty} (-1)^j j^2$

SOLUTION

(a) To express $\displaystyle\sum_{i=1}^{4} (5i - 2)$ [read "the sum of $(5i - 2)$ from $i = 1$ to $i = 4$"] in expanded form, we let the index i take on successive integers from $i = 1$ to $i = 4$. Thus,

$$\sum_{i=1}^{4} (5i - 2) = [5(1) - 2] + [5(2) - 2] + [5(3) - 2] + [5(4) - 2]$$

$$= \underbrace{3 + 8 + 13 + 18}_{\text{Expanded form}}$$

(b) To express $\displaystyle\sum_{j=1}^{\infty} (-1)^j j^2$ in expanded form, we let the index j take on *all* successive integers starting with $j = 1$. To do this, we write the first few terms of the series, followed by three dots (ellipses), then the nth term, followed by ellipses, as follows:

$$\sum_{j=1}^{\infty} (-1)^j j^2 = (-1)^1 1^2 + (-1)^2 2^2 + (-1)^3 3^2 + \cdots + (-1)^n n^2 + \cdots$$

$$= \underbrace{-1 + 4 - 9 + \cdots + (-1)^n n^2 + \cdots}_{\text{Expanded form}}$$

If a finite series contains many terms, and we wish to write the series in expanded form, then we usually write the first few terms followed by ellipses, the nth term followed by ellipses, and the last term. For example,

$$\sum_{i=1}^{40} (5i - 2) = 3 + 8 + 13 + \cdots + (5n - 2) + \cdots + \underset{\boxed{\text{We find the last term by evaluating } [5(40) - 2].}}{198}$$

PROBLEM 4   Repeat Example 4 for $\displaystyle\sum_{k=1}^{27} \frac{k}{k+1}$.

If the general term of a series in expanded form is known, then we can write the series in sigma form. The letter we use for the summation variable is immaterial—only the general term and the beginning and ending values of the summation variable are important.

EXAMPLE 5   Use pattern recognition to find the general term of each series. Then write each series in sigma form.

(a) $1 + 3 + 5 + 7 + 9$ (b) $2 + 5 + 8 + 11 + \cdots + 83$

(c) $\dfrac{1}{2} - \dfrac{1}{4} + \dfrac{1}{8} - \dfrac{1}{16} + \cdots$ (d) $2x^2 + 3x^3 + 4x^4 + 5x^5 + \cdots$

SOLUTION

(a)   We recognize the terms in the series $1 + 3 + 5 + 7 + 9$ as the first five odd, positive integers. Thus, from Example 3(a) we know that the general term of this finite series is

$$a_n = 2n - 1.$$

In our definition of sigma form for a finite series, the summation variable i begins at a value of 1 and ends at a value of n, where n is the number of terms in the finite series. Since there are five terms in this series, we write

$$1 + 3 + 5 + 7 + 9 = \sum_{i=1}^{5} (2i - 1)$$

Sigma form

(b) In the series $2 + 5 + 8 + 11 + \cdots + 83$, each term after the first is three
more than the preceding term. This suggests a general term containing $3n$.
However, since we must obtain 2 when $n = 1$, we write the general term of
this finite series as

$$a_n = 3n - 1.$$

Now we must find the upper value of the summation variable. To do this,
we set the general term equal to the last term in the series and solve for
the number of terms n, as follows:

$$3n - 1 = 83$$

$$3n = 84$$

$$n = 28$$

Thus, using i as the summation variable, we write

$$2 + 5 + 8 + 11 + \cdots + 83 = \sum_{i=1}^{28} (3i - 1)$$

Sigma form

(c) The terms of the series $\frac{1}{2} - \frac{1}{4} + \frac{1}{8} - \frac{1}{16} + \cdots$ are powers of one-half
with alternating signs. Powers of one-half can be written as $(\frac{1}{2})^n$ for
$n = 1, 2, 3, \ldots$. Since the even-numbered terms are negative, we must also
include the factor $(-1)^{n+1}$. Thus, the general term for this infinite series is

$$a_n = (-1)^{n+1} \left(\frac{1}{2}\right)^n \qquad \text{or} \qquad \frac{(-1)^{n+1}}{2^n}.$$

The upper value of the summation variable for an infinite series is
always ∞. Thus, using i as the summation variable, we have

$$\frac{1}{2} - \frac{1}{4} + \frac{1}{8} - \frac{1}{16} + \cdots = \sum_{i=1}^{\infty} \frac{(-1)^{i+1}}{2^i}.$$

Sigma form

(d) The coefficients of x in the series $2x^2 + 3x^3 + 4x^4 + 5x^5 + \cdots$ are consecutive integers. Since we must obtain 2 when $n = 1$, we can write these

coefficients as $n + 1$ for $n = 1, 2, 3, \ldots$. In each term, the exponent of x is the same as its coefficient. Thus, the general term for this infinite series is

$$a_n = (n + 1)x^{n+1}.$$

The upper value of the summation variable for an infinite series is always ∞. Thus, using i as the summation variable, we have

$$2x^2 + 3x^3 + 4x^4 + 5x^5 + \cdots = \underbrace{\sum_{i=1}^{\infty} (i + 1)x^{i+1}}_{\boxed{\text{Sigma form}}}$$

◆

Our definition of sigma form uses $i = 1$ as the lower value of the summation variable. We can use other lower values to describe the series in Example 5. Consider Example 5(a):

$$1 + 3 + 5 + 7 + 9 = \sum_{i=1}^{5} (2i - 1).$$

Writing the upper value as $i = 5$ instead of just 5 and performing a *shift of index* by letting $i = k + 1$, we obtain

$$\underbrace{\sum_{i=1}^{i=5} (2i - 1) = \sum_{k+1=1}^{k+1=5} [2(k + 1) - 1]}_{\boxed{\text{Replace each } i \text{ with } k + 1 \text{ and simplify.}}} = \sum_{k=0}^{k=4} (2k + 1).$$

Observe that

$$\underbrace{\sum_{k=0}^{4} (2k + 1)}_{\boxed{\begin{array}{l}\text{Sigma form with} \\ \text{a shift of index}\end{array}}} = 1 + 3 + 5 + 7 + 9 = \underbrace{\sum_{i=1}^{5} (2i - 1)}_{\boxed{\text{Sigma form}}}.$$

For consistency, when changing from expanded form to sigma form, we shall always use $i = 1$ as the lower value of the summation variable.

P R O B L E M 5   Referring to Example 5(d), perform a shift of index on the sigma form by letting $i = k - 1$.

◆

Exercises 8.1

Basic Skills

In Exercises 1–6, find the sequence represented by each sequence function. Graph the function and determine whether it is increasing, decreasing, or neither.

1. $f(n) = 6 - 3n$,   domain: $\{1, 2, 3, 4, 5\}$

2. $f(n) = n^2 - 1$,   domain: $\{1, 2, 3, 4, 5, 6\}$

3. $f(n) = (-1)^{n+1}n$,   domain: $\{1, 2, 3, 4\}$

4. $f(n) = \dfrac{n}{n+1}$,   domain: $\{1, 2, 3, 4, 5, 6, 7\}$

5. $f(n) = \dfrac{2^{2n-1}}{n^2}$

6. $f(n) = \dfrac{(-1)^n}{2^n - 1}$

In Exercises 7–20, find the first six elements of a sequence with the given general element.

7. $a_n = 5n - 4$

8. $a_n = 4 - 3n$

9. $a_n = 3(-2)^n$

10. $a_n = 9(\frac{2}{3})^{n-1}$

11. $a_n = 2 + \dfrac{1}{n}$

12. $a_n = \dfrac{n}{2^n}$

13. $a_n = 1 + (-1)^{n+1}$

14. $a_n = \dfrac{(-1)^n}{3n - 4}$

15. $a_n = \begin{cases} n & \text{if } n \text{ is odd} \\ 2n - 1 & \text{if } n \text{ is even} \end{cases}$

16. $a_n = \begin{cases} 0 & \text{if } n \text{ is odd} \\ n^2 - 1 & \text{if } n \text{ is even} \end{cases}$

17. $a_n = \begin{cases} 3 & \text{if } n = 1 \\ 2a_{n-1} & \text{if } n \geq 2 \end{cases}$

18. $a_n = \begin{cases} -1 & \text{if } n = 1 \\ (a_{n-1})^{n-1} & \text{if } n \geq 2 \end{cases}$

19. $a_n = \begin{cases} 1 & \text{if } n = 1 \\ 2 & \text{if } n = 2 \\ a_{n-1} - a_{n-2} & \text{if } n \geq 3 \end{cases}$

20. $a_n = \begin{cases} -5 & \text{if } n = 1 \\ 2 & \text{if } n = 2 \\ a_{n-1}a_{n-2} & \text{if } n \geq 3 \end{cases}$

In Exercises 21–32, use pattern recognition to find the general element of each sequence. (Answers are not unique.)

21. $1, 4, 9, 16, 25, \ldots$

22. $1, 8, 27, 81, 243, \ldots$

23. $2, -4, 6, -8, 10, -12$

24. $-1, 3, -5, 7, -9, \ldots, 27$

25. $3, 4, 5, 6, 7, \ldots, 29$

26. $1, 4, 7, 10, 13, \ldots$

27. $\frac{1}{2}, \frac{3}{4}, \frac{5}{6}, \frac{7}{8}, \frac{9}{10}, \ldots$

28. $\frac{1}{2}, \frac{3}{5}, \frac{5}{8}, \frac{7}{11}, \frac{9}{14}, \frac{11}{17}, \frac{13}{20}$

29. $2, 0.2, 0.02, 0.002, 0.0002, \ldots$

30. $0.31, 0.0031, 0.000031, 0.00000031, \ldots$

31. $-x, \dfrac{x^2}{3}, -\dfrac{x^3}{5}, \dfrac{x^4}{7}, -\dfrac{x^5}{9}, \ldots$

32. $x, -\dfrac{x^3}{5}, -\dfrac{x^5}{9}, -\dfrac{x^7}{13}, \dfrac{x^9}{17}, \ldots$

In Exercises 33–42, write each series in expanded form.

33. $\displaystyle\sum_{i=1}^{5} 2i$

34. $\displaystyle\sum_{i=1}^{4} (7i - 3)$

35. $\displaystyle\sum_{n=1}^{10} (-1)^n n^3$

36. $\displaystyle\sum_{j=1}^{7} (-1)^{j+1}(j^2 - 1)$

37. $\displaystyle\sum_{k=1}^{\infty} k^k$

38. $\displaystyle\sum_{k=1}^{\infty} \dfrac{1}{k}$

39. $\displaystyle\sum_{i=1}^{50} \dfrac{2i - 3}{2}$

40. $\displaystyle\sum_{i=1}^{25} \dfrac{i}{i^2 + 1}$

41. $\displaystyle\sum_{i=1}^{\infty} \dfrac{(-1)^i(2x^{i-1})}{i}$

42. $\displaystyle\sum_{i=1}^{\infty} \dfrac{(-1)^{i+1}x^i}{3i + 1}$

In Exercises 43–54, write the series in sigma form using $i = 1$ as the lower value of the summation variable.

43. $2 + 4 + 6 + 8 + 10 + 12 + 14$

44. $1 - 3 + 5 - 7 + 9 - 11 + 13 - 15$

45. $2 + 7 + 12 + 17 + \cdots + 147$

46. $3 + 8 + 13 + 18 + \cdots + 98$

47. $1 - 1 + 1 - 1 + 1 - 1 + \cdots$

48. $-1 + 4 - 9 + 16 - 25 + \cdots$

49. $-1 + \dfrac{1}{2} - \dfrac{1}{4} + \dfrac{1}{8} - \dfrac{1}{16} + \cdots + \dfrac{1}{512}$

50. $2 + \dfrac{3}{2} + \dfrac{4}{3} + \dfrac{5}{4} + \cdots + \dfrac{54}{53}$

51. $2 + \dfrac{4}{3} + \dfrac{8}{9} + \dfrac{16}{27} + \dfrac{32}{81} + \cdots$

52. $1 + \dfrac{4}{3} + \dfrac{16}{5} + \dfrac{64}{7} + \dfrac{256}{9} + \cdots$

53. $3x + 4x^3 + 5x^5 + 6x^7 + 7x^9$

54. $\dfrac{1}{x^4} - \dfrac{2}{x^{10}} + \dfrac{4}{x^{16}} - \dfrac{8}{x^{22}} + \cdots$

In Exercises 55–60, perform the indicated shift of index.

55. Rewrite $\displaystyle\sum_{i=1}^{4} (2i - 5)$ by letting $i = k + 1$.

56. Rewrite $\displaystyle\sum_{i=0}^{50} (-1)^{i+1}(i + 1)$ by letting $i = k - 1$.

57. Rewrite $\displaystyle\sum_{i=1}^{\infty} \dfrac{(-1)^i(2x^{i-1})}{i}$ using $k = 0$ as the lower value of the summation variable.

58. Rewrite $\displaystyle\sum_{i=1}^{\infty} \dfrac{(-1)^{i+1}x^i}{3i + 1}$ using $k = 2$ as the lower value of the summation variable.

59. Rewrite $\displaystyle\sum_{i=2}^{25} \dfrac{i}{i^2 + 1}$ using $k = 0$ as the lower value of the summation variable.

60. Rewrite $\displaystyle\sum_{i=0}^{10} \dfrac{(-1)^{i+1}3^{i+2}}{2i + 4}$ using $k = 2$ as the lower value of the summation variable.

Critical Thinking

61. One possibility for the general element of a sequence that begins with 1, 3, 5, is $a_n = 2n - 1$. Find another possibility for the general element of a sequence that begins with 1, 3, 5.

62. How does the graph of the sequence function $f(n) = 2n - 3$ differ from the graph of the linear function $f(x) = 2x - 3$?

63. Find the finite sequence function f indicated by each graph and state its domain.

64. Give a recursive definition for the general element of each infinite sequence.

    (a) 6, 10, 14, 18, 22, ...   (b) 8, 4, 2, 1, $\frac{1}{2}$, ...
    (c) 1, 4, 10, 22, 46, 94, ... (d) 1, 2, 5, 26, 677, ...

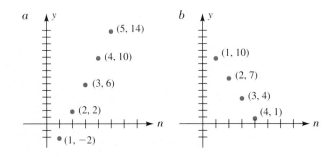

EXERCISE 63

Calculator Activities

65. Use a calculator to help express each series in expanded form.

    (a) $\displaystyle\sum_{i=1}^{4} (7.26i^2 - 3.95)$ (b) $\displaystyle\sum_{i=1}^{5} \dfrac{8.94 + i}{2i}$

66. Each element in the sequence with general element

$$a_n = \begin{cases} \dfrac{N}{2} & \text{if } n = 1 \\[2mm] \dfrac{1}{2}\left(a_{n-1} + \dfrac{N}{a_{n-1}}\right) & \text{if } n \geq 2 \end{cases}$$

gives a better approximation of $\sqrt{N}$ than the element that precedes it. For each of the given square roots, find the fifth element, a_5, rounding the answer to five significant digits. Compare a_5 to the value you obtain by using the $\boxed{\sqrt{}}$ key on your calculator.

    (a) $\sqrt{2}$ (b) $\sqrt{3}$ (c) $\sqrt{22}$ (d) $\sqrt{85}$

8.2   The Sum of a Series and Mathematical Induction

For every finite series, a number can be assigned as its *sum*. For example, we find the sum of the finite series

$$\sum_{i=1}^{4} (2i - 1)$$

by adding the terms in the series, as follows:

$$\sum_{i=1}^{4} (2i - 1) = 1 + 3 + 5 + 7 = 16 \longleftarrow \boxed{\text{Sum of the first four odd integers}}$$

Some, but not all, infinite series can be assigned a sum. For example, in Section 8.5, we show that

$$\sum_{i=1}^{\infty} (-1)^{i+1} \left(\frac{1}{2}\right)^{i} = \frac{1}{2} - \frac{1}{4} + \frac{1}{8} - \frac{1}{16} + \cdots = \frac{1}{3}$$

$$\boxed{\text{Sum of the infinite series}}$$

However, other infinite series such as

$$\sum_{i=1}^{\infty} i = 1 + 2 + 3 + 4 + \cdots$$

do not appear to add up to any finite number. The sum of an infinite series is usually studied in calculus. In this section, we limit our discussion to methods of finding the sum of finite series.

◆ Finding the Sum of a Series by Direct Computation

If the number of terms in the series is small, we can find the sum of the series by using direct computation and adding the terms in the series.

EXAMPLE 1   Find the sum of each series.

(a) $\displaystyle\sum_{i=1}^{8} 7$ **(b)** $\displaystyle\sum_{i=1}^{6} \frac{i}{2^{i}}$

SOLUTION

(a) $\displaystyle\sum_{i=1}^{8} 7 = 7 + 7 + 7 + 7 + 7 + 7 + 7 + 7 = 8(7) = 56$

$\boxed{\text{8 terms}}$

(b) $\displaystyle\sum_{i=1}^{6} \frac{i}{2^i} = \frac{1}{2} + \frac{2}{4} + \frac{3}{8} + \frac{4}{16} + \frac{5}{32} + \frac{6}{64}$

$\displaystyle = \frac{1(32) + 2(16) + 3(8) + 4(4) + 5(2) + 6}{64}$

$\displaystyle = \frac{120}{64} = \frac{15}{8}$ ◆

Referring to Example 1(a), for any series of the form

$$\sum_{i=1}^{n} c$$

where c is a constant, we have

$$\sum_{i=1}^{n} c = \underbrace{c + c + c + \cdots + c}_{n \text{ terms}} = nc$$

Formulas like this one enable us to quickly find the sum of a series containing a large number of terms.

P R O B L E M 1   Find the sum of the series $\displaystyle\sum_{i=1}^{1000} 4$.   ◆

◆ Proof by Mathematical Induction

The direct computation procedure for finding the sum of a series is quite tedious when the series contains a large number of terms. When we want to find the sum of a series with a large number of terms, it is advantageous to develop and apply a formula. In mathematics, we often use pattern recognition to help develop formulas. For example, suppose we wish to find a formula for the sum of the first n odd, positive integers. We can represent this sum by the series

$$\sum_{i=1}^{n} (2i - 1) = 1 + 3 + 5 + \cdots + (2n - 1)$$

We begin by calculating the sum of the odd, positive integers for the first few values of n, and observe any pattern that develops. Table 8.1 shows the sum of the first n odd, positive integers from $n = 1$ to $n = 6$.

The sums in Table 8.1 form the sequence

$$1, 4, 9, 16, 25, 36$$

Do you detect the pattern? From this sequence, we are led to believe that the sum of the first n odd, positive integers is n^2 or, in series notation

$$\sum_{i=1}^{n} (2i - 1) = 1 + 3 + 5 + \cdots + (2n - 1) = n^2.$$

Table 8.1

Sum of the first n odd, positive integers for $n = 1$ to $n = 6$

n	Sum of the first n odd, positive integers
1	$1 = 1$
2	$1 + 3 = 4$
3	$1 + 3 + 5 = 9$
4	$1 + 3 + 5 + 7 = 16$
5	$1 + 3 + 5 + 7 + 9 = 25$
6	$1 + 3 + 5 + 7 + 9 + 11 = 36$

However, choosing a formula based on a few observations does not guarantee the validity of the formula for all integers n. What we need is a "chain reaction" procedure, which will guarantee that once the formula is valid for a particular integer *n*, then the formula is valid for the next integer, and the next, and the next, and so on, indefinitely.

Consider the string of dominoes placed on end as shown in Figure 8.3. If we know that

1. the first domino is knocked over, and
2. if any domino in the string is knocked over, then the next one in line is also knocked over,

FIGURE 8.3
A string of dominoes placed on end

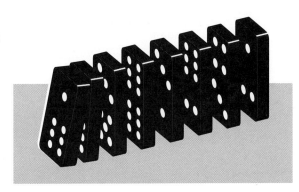

then we can conclude that all dominoes in the string will be knocked over. By analogy with the domino problem, we can state that a *mathematical statement*, such as

$$\sum_{i=1}^{n} (2i - 1) = 1 + 3 + 5 + \cdots + (2n - 1) = n^2,$$

is true for all positive integers *n*, provided that

1. the statement is true when $n = 1$, and
2. if the statement is true for $n = k$, where *k* is an arbitrary integer, then the statement is true for the next integer, namely, $n = k + 1$.

We refer to this chain reaction procedure as **proof by mathematical induction**.

Proof by Mathematical Induction

The following two-step procedure, called **proof by mathematical induction**, may be used to show that a mathematical statement is true for all positive integers *n*:

Step 1 Show that the statement is true when $n = 1$.

Step 2 Assume that the statement is true for an arbitrary positive integer *k* and show, by using this assumption, that the statement is also true when $n = k + 1$.

EXAMPLE 2 Use mathematical induction to prove that

$$\sum_{i=1}^{n}(2i-1)=1+3+5+\cdots+(2n-1)=n^2$$

for all positive integers n.

SOLUTION

Step 1 We must first show that the statement is true for $n=1$

$$1=1^2 \quad ?$$
$$1=1 \quad \text{is } true.$$

Step 2 Assuming that the statement is true for $n=k$, we must show that, on the basis of that assumption, the statement is true for $n=k+1$.

Assume: $\qquad 1+3+5+\cdots+(2k-1)=k^2$

Prove: $\quad 1+3+5+\cdots+(2k-1)+[2(k+1)-1]=(k+1)^2$

Proof:

$$1+3+5+\cdots+(2k-1)+[2(k+1)-1]=\{1+3+5+\cdots+(2k-1)\}+[2(k+1)-1]$$

Replace these terms with the *assumed* statement.

$$=\qquad k^2 \qquad +[2(k+1)-1]$$
$$=k^2+2k+1 \qquad\qquad\qquad \text{Combine like terms}$$
$$=(k+1)^2 \qquad\qquad\qquad\qquad \text{Factor}$$

Thus, the truth of the statement when $n=k$ implies the truth of the statement when $n=k+1$. Hence, by mathematical induction, we can state that

$$\sum_{i=1}^{n}(2i-1)=1+3+5+7+\cdots+(2n-1)=n^2$$

for all positive integers n. ◆

PROBLEM 2 Using the results of Example 2, find the sum of the series $\sum_{i=1}^{50}(2i-1)$. ◆

EXAMPLE 3 Use mathematical induction to prove that the sum of the first n positive integers is $\dfrac{n(n+1)}{2}$ by showing that

$$\sum_{i=1}^{n}i=1+2+3+\cdots+n=\frac{n(n+1)}{2}$$

for all positive integers n.

SOLUTION

Step 1   We must first show that the statement is true for $n = 1$:

$$1 = \frac{1(1 + 1)}{2} \quad ?$$

$$1 = 1 \quad \text{is } true.$$

Step 2   Assuming that the statement is true for $n = k$, we must show that, on the basis of that assumption, the statement is true for $n = k + 1$.

Assume: $$1 + 2 + 3 + 4 + \cdots + k = \frac{k(k + 1)}{2}$$

Prove: $$1 + 2 + 3 + 4 + \cdots + k + (k + 1) = \frac{(k + 1)[(k + 1) + 1]}{2}$$

Proof:

$$1 + 2 + 3 + 4 + \cdots + k + (k + 1) = [1 + 2 + 3 + 4 + \cdots + k] + (k + 1)$$

Replace these terms with the *assumed* statement.

$$= \frac{k(k + 1)}{2} + (k + 1)$$

$$= \frac{k(k + 1) + 2(k + 1)}{2} \qquad \text{Add fraction}$$

$$= \frac{(k + 1)(k + 2)}{2} \qquad \text{Factor}$$

$$= \frac{(k + 1)[(k + 1) + 1]}{2} \qquad \text{Rewrite}$$

Thus, the truth of the statement when $n = k$ implies the truth of the statement when $n = k + 1$. Hence, by mathematical induction, we can state that

$$\sum_{i=1}^{n} i = 1 + 2 + 3 + \cdots + n = \frac{n(n + 1)}{2}$$

for all positive integers n.   ◆

The formula for the sum of the positive integers along with formulas for the **sums of powers of the positive integers** are used in calculus to help find the area under a curve. Here we list the formulas for powers up to and including 3. Mathematical induction may be used to prove formulas 2 and 3 as well (see Exercises 13 and 14).

Formulas for the Sums of Powers of Positive Integers

1. $\displaystyle\sum_{i=1}^{n} i = 1 + 2 + 3 + \cdots + n = \frac{n(n+1)}{2}$

2. $\displaystyle\sum_{i=1}^{n} i^2 = 1^2 + 2^2 + 3^2 + \cdots + n^2 = \frac{n(n+1)(2n+1)}{6}$

3. $\displaystyle\sum_{i=1}^{n} i^3 = 1^3 + 2^3 + 3^3 + \cdots + n^3 = \left[\frac{n(n+1)}{2}\right]^2$

PROBLEM 3   Find the sum of each series. Check each answer by using direct computation.

(a) $\displaystyle\sum_{i=1}^{19} i$ (b) $\displaystyle\sum_{i=1}^{19} i^2$ (c) $\displaystyle\sum_{i=1}^{19} i^3$ ◆

◆ Summation Properties

We now state three fundamental **summation properties**. Each property can be proved by mathematical induction or by writing the property in expanded form and then applying the properties of real numbers (see Exercises 58–60).

Summation Properties

1. $\displaystyle\sum_{i=1}^{n} ca_i = c \sum_{i=1}^{n} a_i$ 2. $\displaystyle\sum_{i=1}^{n} (a_i + b_i) = \sum_{i=1}^{n} a_i + \sum_{i=1}^{n} b_i$

3. $\displaystyle\sum_{i=1}^{n} (a_i - b_i) = \sum_{i=1}^{n} a_i - \sum_{i=1}^{n} b_i$

We use these summation properties in conjunction with the formulas for the sums of powers of integers to find the sums of several other series. For example, suppose we wish to find a formula for the sum of the first n even, positive integers. We can represent this sum by the series

$$\sum_{i=1}^{n} 2i = 2 + 4 + 6 + \cdots + 2n$$

Now, applying summation property 1, we have

$$\sum_{i=1}^{n} 2i = 2 \sum_{i=1}^{n} i$$

However, from Example 3, we know that

$$\sum_{i=1}^{n} i = \frac{n(n+1)}{2}.$$

Hence,

$$\sum_{i=1}^{n} 2i = 2 \sum_{i=1}^{n} i = 2 \frac{n(n+1)}{2} = n(n+1).$$

Thus, a formula for the sum of the first n even, positive integers is

$$\sum_{i=1}^{n} 2i = 2 + 4 + 6 + \ldots + 2n = n(n+1).$$

E X A M P L E 4 Use the summation properties and the formulas for the sums of powers of integers to find the sum of the series

$$\sum_{i=1}^{9} (i^2 + 3i - 4)$$

S O L U T I O N Using the summation properties, we write

$$\sum_{i=1}^{9} (i^2 + 3i - 4) = \sum_{i=1}^{9} i^2 + 3 \sum_{i=1}^{9} i - \sum_{i=1}^{9} 4$$

Now, using the following facts,

$$\sum_{i=1}^{n} i^2 = \frac{n(n+1)(2n+1)}{6}, \qquad \sum_{i=1}^{n} i = \frac{n(n+1)}{2}, \quad \text{and} \quad \sum_{i=1}^{n} c = nc,$$

we have

$$\sum_{i=1}^{9} (i^2 + 3i - 4) = \sum_{i=1}^{9} i^2 + 3 \sum_{i=1}^{9} i - \sum_{i=1}^{9} 4$$
$$= \frac{9(9+1)[2(9)+1]}{6} + 3 \left[\frac{9(9+1)}{2} \right] - 9(4)$$
$$= 285 + 135 - 36$$
$$= 384 \qquad \blacklozenge$$

P R O B L E M 4 Repeat Example 4 for $\sum_{i=1}^{20} (5i - 2)$. $\qquad \blacklozenge$

◆ **Other Applications of Mathematical Induction**

Some mathematical statements are *not* true for the first $(j - 1)$ positive integers, but *are* true for integers greater than or equal to j. To prove such statements for $n \geq j$, we can *extend* mathematical induction by showing that

1. the statement is true when $n = j$, and
2. the truth of the statement $n = k$, for $k \geq j$, implies the truth of the statement $n = k + 1$.

EXAMPLE 5   Use mathematical induction to prove that the sum of the interior angles of an n-sided convex polygon is $(n - 2)180°$ for all positve integers $n \geq 3$.

SOLUTION

Step 1   We must first show that the statement is true for $n = 3$:

$$(3 - 2)180° = 180°$$ ← Sum of the interior angles of a triangle (a 3-sided polygon)

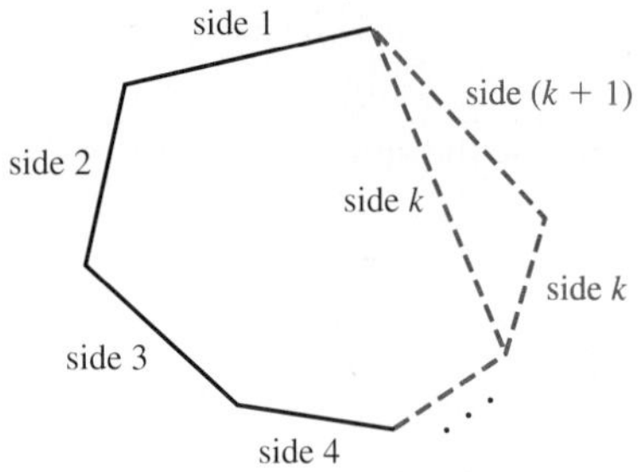

Hence, the statement is true for $n = 3$.

Step 2   Assuming that the statement is true for $n = k$, where $k \geq 3$, we must show that, on the basis of that assumption, the statement is true for $n = k + 1$.

Assume: $(k - 2)180°$ is the sum of the interior angles of a k-sided convex polygon.

Prove: $[(k + 1) - 2]180°$ is the sum of the interior angles of a $(k + 1)$-sided convex polygon.

Proof: By attaching a triangle to a k-sided convex polygon, as shown in Figure 8.4, we form a $(k + 1)$-sided convex polygon. Hence, the sum of the interior angles of the $(k + 1)$-sided polygon is

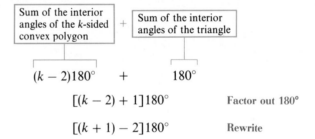

| Sum of the interior angles of the k-sided convex polygon | + | Sum of the interior angles of the triangle |

$$(k - 2)180°   +   180°$$

$$[(k - 2) + 1]180°$$ **Factor out 180°**

$$[(k + 1) - 2]180°$$ **Rewrite**

Thus, the truth of the statement when $n = k$, for $k \geq 3$, implies the truth of the statement when $n = k + 1$. Therefore, by mathematical induction, we can state that the sum of the interior angles of an n-sided convex polygon is $(n - 2)180°$ for all positive integers $n \geq 3$.   ◆

PROBLEM 5   Determine the sum of the interior angles of a 12-sided convex polygon.   ◆

FIGURE 8.4

Attaching a triangle to a k-sided convex polygon forms a $(k + 1)$-sided convex polygon.

(labels in figure: side 1, side 2, side 3, side 4, side k, side k, side $(k + 1)$)

Exercises 8.2

Basic Skills

In Exercises 1–10, use direct computation to find the sum of each series.

1. $\sum\limits_{i=1}^{10} 15$

2. $\sum\limits_{i=1}^{24} \dfrac{5}{6}$

3. $\sum\limits_{i=1}^{100} \pi$

4. $\sum\limits_{i=1}^{250} \sqrt{2}$

5. $\sum\limits_{i=1}^{8} (2 - i)$

6. $\sum\limits_{i=1}^{5} 3(-2)^i$

7. $\displaystyle\sum_{i=1}^{7} (2^i - 1)$

8. $\displaystyle\sum_{i=1}^{6} (i^2 + 1)$

9. $\displaystyle\sum_{i=1}^{5} \frac{(-1)^{i+1}2^i}{i}$

10. $\displaystyle\sum_{i=1}^{4} \frac{(-1)^i i^2}{i+1}$

27. $\dfrac{1}{2} + \dfrac{1}{4} + \dfrac{1}{8} + \dfrac{1}{16} + \cdots + \dfrac{1}{2^n} = 1 - \dfrac{1}{2^n}$

28. $\dfrac{3}{2} + \dfrac{5}{2^2} + \dfrac{7}{2^3} + \dfrac{9}{2^4} + \cdots + \dfrac{2n+1}{2^n} = 5 - \dfrac{2n+5}{2^n}$

In Exercises 11–28, prove each statement by using mathematical induction.

11. $2 + 6 + 10 + 14 + \cdots + (4n - 2) = 2n^2$

12. $1 + 4 + 7 + 10 + \cdots + (3n - 2) = \dfrac{n(3n-1)}{2}$

13. $1^2 + 2^2 + 3^2 + \cdots + n^2 = \dfrac{n(n+1)(2n+1)}{6}$

14. $1^3 + 2^3 + 3^3 + \cdots + n^3 = \left[\dfrac{n(n+1)}{2}\right]^2$

15. $1^2 + 3^2 + 5^2 + 7^2 + \cdots + (2n-1)^2$
$= \dfrac{n(2n-1)(2n+1)}{3}$

16. $2^2 + 4^2 + 6^2 + 8^2 + \cdots + (2n)^2$
$= \dfrac{2n(n+1)(2n+1)}{3}$

17. $2 \cdot 4 + 4 \cdot 6 + 6 \cdot 8 + 8 \cdot 10 + \cdots + 2n(2n+2)$
$= \dfrac{4n(n+1)(n+2)}{3}$

18. $1 \cdot 3 + 2 \cdot 5 + 3 \cdot 7 + 4 \cdot 9 + \cdots + n(2n+1)$
$= \dfrac{n(n+1)(4n+5)}{6}$

19. $\dfrac{1}{1 \cdot 2} + \dfrac{1}{2 \cdot 3} + \dfrac{1}{3 \cdot 4} + \dfrac{1}{4 \cdot 5} + \cdots + \dfrac{1}{n(n+1)} = \dfrac{n}{n+1}$

20. $\dfrac{1}{1 \cdot 3} + \dfrac{1}{3 \cdot 5} + \dfrac{1}{5 \cdot 7} + \dfrac{1}{7 \cdot 9} + \cdots + \dfrac{1}{(2n-1)(2n+1)}$
$= \dfrac{n}{2n+1}$

21. $2 + 2^2 + 2^3 + 2^4 + \cdots + 2^n = 2(2^n - 1)$

22. $3 + 3^2 + 3^3 + 3^4 + \cdots + 3^n = \frac{3}{2}(3^n - 1)$

23. $4 + 4^2 + 4^3 + 4^4 + \cdots + 4^n = \frac{4}{3}(4^n - 1)$

24. $5 + 5^2 + 5^3 + 5^4 + \cdots + 5^n = \frac{5}{4}(5^n - 1)$

25. $1 + 2 \cdot 2 + 3 \cdot 2^2 + 4 \cdot 2^3 + \cdots + n \cdot 2^{n-1}$
$= 1 + (n-1)2^n$

26. $1 + 2 \cdot 3 + 3 \cdot 3^2 + 4 \cdot 3^3 + \cdots + n \cdot 3^{n-1}$
$= \dfrac{1 + (2n-1)3^n}{4}$

In Exercises 29–46, use the formulas of Exercises 11–28 to find the sum of the given series.

29. $\displaystyle\sum_{i=1}^{20} (4i - 2)$

30. $\displaystyle\sum_{i=1}^{26} (3i - 2)$

31. $\displaystyle\sum_{i=1}^{36} i^2$

32. $\displaystyle\sum_{i=1}^{24} i^3$

33. $\displaystyle\sum_{i=1}^{25} (2i - 1)^2$

34. $\displaystyle\sum_{i=1}^{18} (2i)^2$

35. $\displaystyle\sum_{i=1}^{15} 2i(2i + 2)$

36. $\displaystyle\sum_{i=1}^{18} i(2i + 1)$

37. $\displaystyle\sum_{i=1}^{499} \frac{1}{i(i+1)}$

38. $\displaystyle\sum_{i=1}^{100} \frac{1}{(2i-1)(2i+1)}$

39. $\displaystyle\sum_{i=1}^{10} 2^i$

40. $\displaystyle\sum_{i=1}^{8} 3^i$

41. $\displaystyle\sum_{i=1}^{6} 4^i$

42. $\displaystyle\sum_{i=1}^{5} 5^i$

43. $\displaystyle\sum_{i=1}^{12} i \cdot 2^{i-1}$

44. $\displaystyle\sum_{i=1}^{9} i \cdot 3^{i-1}$

45. $\displaystyle\sum_{i=1}^{8} \frac{1}{2^i}$

46. $\displaystyle\sum_{i=1}^{10} \frac{2i+1}{2^i}$

In Exercises 47–54, use the summation properties and the formulas for the sums of powers of integers to find the sum of the given series.

47. $\displaystyle\sum_{i=1}^{50} 3i$

48. $\displaystyle\sum_{i=1}^{36} 5i^2$

49. $\displaystyle\sum_{i=1}^{24} (2i^2 - 5)$

50. $\displaystyle\sum_{i=1}^{10} (4 - i^3)$

51. $\displaystyle\sum_{i=1}^{20} i^2(i - 3)$

52. $\displaystyle\sum_{i=1}^{30} i(i + 4)$

53. $\displaystyle\sum_{i=1}^{18} (2i^2 - i + 1)$

54. $\displaystyle\sum_{i=1}^{12} (3i^3 - 2i - 6)$

In Exercises 55 and 56, use mathematical induction to prove the given statement.

55. A regular convex polygon with n sides $(n \geq 4)$ has $\frac{1}{2}n(n - 3)$ diagonals.

56. If n people are in a room $(n \geq 2)$ and every person shakes hand with every other person, then there are $\frac{1}{2}n(n - 1)$ handshakes.

Critical Thinking

57. Find the sum of each series, expressing the answer as a single logarithm.

(a) $\sum_{i=1}^{5} \log i$ (b) $\sum_{i=1}^{5} \log \dfrac{i+1}{i}$

In Exercises 58–60, prove each summation property by writing it in expanded form and then applying the properties of real numbers.

58. $\sum_{i=1}^{n} ca_i = c \sum_{i=1}^{n} a_i$

59. $\sum_{i=1}^{n} (a_i + b_i) = \sum_{i=1}^{n} a_i + \sum_{i=1}^{n} b_i$

60. $\sum_{i=1}^{n} (a_i - b_i) = \sum_{i=1}^{n} a_i - \sum_{i=1}^{n} b_i$

In Exercises 61–63,

(a) *use pattern recognition to state the general term of each series.*

(b) *find the sum of the first 20 terms of the series by applying the summation properties and the formulas for the sums of powers of integers.*

61. $1 \cdot 2 + 3 \cdot 4 + 5 \cdot 6 + 7 \cdot 8 + \cdots$

62. $1 \cdot 3 + 3 \cdot 5 + 5 \cdot 7 + 7 \cdot 9 + \cdots$

63. $1 \cdot 2 \cdot 3 + 4 \cdot 5 \cdot 6 + 7 \cdot 8 \cdot 9 + 10 \cdot 11 \cdot 12 + \cdots$

64. On the basis of the results from Exercises 21–24, make a conjecture about the sum of the series $\sum_{i=1}^{n} r^i$, where $r \neq 1$. Then prove it by using mathematical induction.

In Exercises 65–70, use mathematical induction to prove the given statement.

65. $x - 1$ is a factor of $x^n - 1$ for all positive integers n. [*Hint:* For step 2, add and subtract x.]

66. $a^n - b^n$ is divisible by $a - b$ for all positive integers n. [*Hint:* For step 2, add and subtract ab^k.)

67. The product of any two consecutive, positive integers n and $n + 1$ is an even number.

68. The sum of cubes of any three consecutive, positive integers n, $n + 1$, and $n + 2$ is a multiple of 9.

69. $2^n > 2n$ for all positive integers $n \geq 3$.

70. $2^n > n^2$ for all positive integers $n \geq 5$.

Calculator Activities

In Exercises 71–74, use a calculator to help find the sum of each series.

71. $\sum_{i=1}^{84} (9.75 - 1.02i)$

72. $\sum_{i=1}^{75} 0.04(5 + 0.08i)$

73. $\sum_{i=1}^{42} (3.6i^2 - 5.6i + 8.9)$

74. $\sum_{i=1}^{25} (2.78i^3 - 4.32i)$

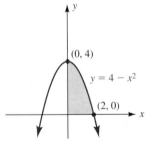

EXERCISE 75

$\dfrac{\Delta y}{\Delta x}$ *The series in Exercises 75 and 76 are used in calculus to approximate the area of the shaded region that is shown in the figure. Use the summation properties and the formulas for the sums of powers of integers to find the shaded area to the nearest tenth of a square unit.*

75. $\sum_{i=1}^{100} 0.02[4 - (0.02i)^2]$ **76.** $\sum_{i=1}^{100} 0.02[8 - (0.02i)^3]$

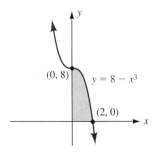

EXERCISE 76

◆ **Introductory Comments**

Consider the sequence

$$3, 7, 11, 15, 19, 23, \ldots$$

It appears that each element after the first is obtained by adding 4 to the preceding element. A sequence in which every element after the first is obtained by *adding* a fixed number d to the preceding element is called an *arithmetic sequence*.

◆ Arithmetic Sequence

> The sequence $a_1, a_2, a_3, \ldots, a_n, \ldots$ is called an **arithmetic sequence** if there exists a fixed number d such that, for all positive integers k,
>
> $$a_{k+1} = a_k + d.$$

It follows from the definition of an arithmetic sequence that the difference between any two successive elements must be the constant d, that is,

$$a_{k+1} - a_k = d.$$

We refer to the constant d as the **common difference** of the sequence. In this section we study this special type of sequence and its related series.

◆ **Generating an Arithmetic Sequence**

If the first element and the common difference of an arithmetic sequence are known, then we can generate the sequence.

EXAMPLE 1   Write the first five elements of an arithmetic sequence whose first element is $a_1 = -5$ and whose common difference is $d = 3$.

SOLUTION   If the first element is $a_1 = -5$ and the common difference is $d = 3$, then

$$a_2 = a_1 + d = -5 + 3 = -2, \qquad a_3 = a_2 + d = -2 + 3 = 1,$$
$$a_4 = a_3 + d = 1 + 3 = 4, \qquad\quad a_5 = a_4 + d = 4 + 3 = 7.$$

Thus, the first five elements of the arithmetic sequence are

$$-5, -2, 1, 4, 7$$

PROBLEM 1   Repeat Example 1 if the first element is $a_1 = 8$ and the common difference is $d = -5$.   ◆

◆ **Finding the General Element of an Arithmetic Sequence**

Our definition of an arithmetic sequence allows us to define the general element a_n in terms of the element that precedes it, namely, a_{n-1}. Thus, we can define the general element a_n by the recursive formula

$$a_n = a_{n-1} + d,$$

where d is a fixed number (the common difference). We now proceed to find a formula that defines the general element a_n *explicitly*. If the first element in an arithmetic sequence is a_1 and the common difference is d, then the first six elements of the sequence are as follows:

First element: a_1

Second element: $a_2 = a_1 + d$

Third element: $a_3 = a_2 + d = (a_1 + d) + d = a_1 + 2d$

Fourth element: $a_4 = a_3 + d = (a_1 + 2d) + d = a_1 + 3d$

Fifth element: $a_5 = a_4 + d = (a_1 + 3d) + d = a_1 + 4d$

Sixth element: $a_6 = a_5 + d = (a_1 + 4d) + d = a_1 + 5d$

Do you see the pattern? Note that each element is the sum of a_1 and a multiple of d, where the coefficient of d is one less than the number of the element. Thus, it appears that the general element of an arithmetic sequence can be written as

$$a_n = a_1 + (n-1)d.$$

We now verify this formula by using mathematical induction (see Section 8.2).

Step 1   We must first show that the statement is true for $n = 1$.

$$a_1 = a_1 + (1-1)d \quad ?$$

$$a_1 = a_1 \quad \text{is } true.$$

Step 2   Assuming that the statement is true for $n = k$, we must show that, on the basis of that assumption, the statement is true for $n = k + 1$.

Assume: $a_k = a_1 + (k-1)d$

Prove: $a_{k+1} = a_1 + [(k+1)-1]d$

Proof: Beginning with the definition of an arithmetic sequence, we have

$$a_{k+1} = \underset{\underbrace{}}{a_k} + d$$

Rewrite using the
assumed statement

$$= [a_1 + (k-1)d] + d$$

$$= a_1 + [(k-1)d + d] \qquad \text{Associative property}$$

$$= a_1 + [(k-1) + 1]d \qquad \text{Factor}$$

$$= a_1 + [(k+1) - 1]d \qquad \text{Commutative and associative properties}$$

Thus, the truth of the statement when $n = k$ implies the truth of the statement when $n = k + 1$. Hence, by mathematical induction, we have the following.

General Element of an
Arithmetic Sequence

> The **general element** of an arithmetic sequence with first term a_1 and common difference d is given by
>
> $$a_n = a_1 + (n-1)d.$$

E X A M P L E   2   Find the general element of the arithmetic sequence

$$23, 17, 11, 5, \dots$$

S O L U T I O N   The first element of this sequence is $a_1 = 23$. We can find the common difference d by selecting any two adjacent elements and computing the difference $a_{k+1} - a_k$. Choosing the third and fourth elements, we find

$$d = a_4 - a_3 = 5 - 11 = -6.$$

Thus, the general element of this arithmetic sequence is

$$a_n = a_1 + (n-1)d$$
$$= 23 + (n-1)(-6)$$
$$= 29 - 6n \qquad \blacklozenge$$

P R O B L E M   2   Find the general element of the arithmetic sequence $3, 7, 11, 15, \dots$.   ◆

Once the general element of an arithmetic sequence is known, we can find any element in the sequence. For example, to find the 25th element (a_{25}) of the arithmetic sequence in Example 2, we simply replace n with 25 in the general form $a_n = 29 - 6n$, as follows:

$$a_{25} = 29 - 6(25) = -121.$$

EXAMPLE 3   Find the 30th element a_{30} of an arithmetic sequence whose first element a_1 is 3 and whose common difference d is 4.

SOLUTION   For this arithmetic sequence, $a_1 = 3$ and $d = 4$. Thus, the general element of the arithmetic sequence is

$$a_n = 3 + (n-1)4,$$

and the 30th element is

$$a_{30} = 3 + (30-1)4 = 119.   \blacklozenge$$

PROBLEM 3   Find the first element a_1 of an arithmetic sequence whose 60th element a_{60} is -98 and whose common difference d is -4. $\blacklozenge$

EXAMPLE 4   Find the common difference d and the first element a_1 of an arithmetic sequence whose 5th element a_5 is 20 and 11th element a_{11} is 62.

SOLUTION   Using the formula $a_n = a_1 + (n-1)d$, we have

$$a_5 = a_1 + (5-1)d   \quad\text{and}\quad   a_{11} = a_1 + (11-1)d$$
$$20 = a_1 + 4d   \qquad\qquad   62 = a_1 + 10d$$

To find d and a_1, we solve these equations simultaneously, as follows:

$$
\begin{array}{lll}
a_1 + 4d = 20 & \xrightarrow{\text{Multiply by }-1} & -a_1 - 4d = -20 \\
\underline{a_1 + 10d = 62} & \xrightarrow{\text{Leave as is}} & \underline{a_1 + 10d = 62} \\
& & 6d = 42 \qquad \text{Add and solve for } d. \\
& & d = 7
\end{array}
$$

Substituting $d = 7$ into equation $a_1 + 4d = 20$, we find

$$a_1 + 4(7) = 20$$
$$a_1 = -8$$

Thus, the common difference is 7 and the first element is -8. $\blacklozenge$

PROBLEM 4   Write the general element of the arithmetic sequence described in Example 4. Then use the general element to verify that $a_5 = 20$ and $a_{11} = 62$. $\blacklozenge$

◆ Arithmetic Series

The indicated sum of the elements of an arithmetic sequence is called an **arithmetic series**. The arithmetic series associated with the finite arithmetic sequence

$$a_1, a_1 + d, a_1 + 2d, a_1 + 3d, \ldots, a_1 + (n-1)d$$

can be written in expanded form as

$$a_1 + (a_1 + d) + (a_1 + 2d) + (a_1 + 3d) + \cdots + [a_1 + (n-1)d],$$

or in sigma form as

$$\sum_{i=1}^{n} [a_1 + (i-1)d]$$

We now proceed to develop a formula that we can use to find the sum of a finite arithmetic series. If we represent the sum of these n terms by S_n, then

$$S_n = a_1 + (a_1 + d) + (a_1 + 2d) + (a_1 + 3d) + \cdots + [a_1 + (n-1)d].$$

Alternatively, we can find the sum by starting with the nth term a_n and subtracting the common difference d each time to get to the next term as follows:

$$S_n = a_n + (a_n - d) + (a_n - 2d) + (a_n - 3d) + \cdots + [a_n - (n-1)d].$$

If we add these two versions of S_n, term by term, the multiples of d drop out, and we obtain the sum of the series:

$$S_n = \quad a_1 \quad + (a_1 + d) + (a_1 + 2d) + (a_1 + 3d) + \cdots + [a_1 + (n-1)d]$$
$$+ \ S_n = \quad a_n \quad + (a_n - d) + (a_n - 2d) + (a_n - 3d) + \cdots + [a_n - (n-1)d]$$
$$\overline{2S_n = (a_1 + a_n) + (a_1 + a_n) + (a_1 + a_n) + (a_1 + a_n) + \cdots + \quad (a_1 + a_n)}$$

n terms

$$2S_n = n(a_1 + a_n)$$

$$S_n = n\left(\frac{a_1 + a_n}{2}\right)$$

Number of terms

Average value of the first and last terms

Sum of a Finite Arithmetic Series

The **sum S_n of a finite arithmetic series** with n terms is given by

$$S_n = n\left(\frac{a_1 + a_n}{2}\right),$$

where a_1 is the first term and a_n is the last term in the series.

EXAMPLE 5 Find the sum of each arithmetic series.

(a) $\displaystyle\sum_{i=1}^{40} (4i - 3)$ (b) $3 + 9 + 15 + \cdots + 273$

SOLUTION

(a) Since the summation variable i starts at 1 and ends at 40, the number of terms in this series is $n = 40$. Writing this series in expanded form, we have

$$\sum_{i=1}^{40} (4i - 3) = 1 + 5 + 9 + \cdots + 157$$

This is an arithmetic series with common difference $d = 4$, first term $a_1 = 1$, and last term $a_{40} = 157$. The average value of the first and last terms is

$$\frac{a_1 + a_{40}}{2} = \frac{1 + 157}{2} = \frac{158}{2} = 79.$$

Thus, the sum of this arithmetic series is

$$S_{40} = 40\left(\frac{a_1 + a_{40}}{2}\right) = 40(79) = 3160.$$

(b) The first term of the arithmetic series $3 + 9 + 15 + \ldots + 273$ is $a_1 = 3$, the last term is $a_n = 273$. The average value of the first and last terms is

$$\frac{3 + 273}{2} = \frac{276}{2} = 138.$$

To find the number of terms n in this series, we use the formula $a_n = a_1 + (n - 1)d$ with $a_n = 273$, $a_1 = 3$, and $d = 6$, then solve for n, as follows:

$$273 = 3 + (n - 1)6$$
$$273 = 6n - 3$$
$$276 = 6n$$
$$n = 46$$

Thus, the sum of the arithmetic series $3 + 9 + 15 + \cdots + 273$ is

$$S_{46} = 46(138) = 6348.$$

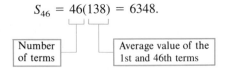

Number of terms   Average value of the 1st and 46th terms

If the summation variable i starts at a number other than 1, we must be certain to determine the correct number of terms in the series. For instance, the arithmetic series $\sum_{i=0}^{40}(4i - 3)$ has 41 terms, not 40.

PROBLEM 5   Find the sum of the arithmetic series $\sum_{i=0}^{40} (4i - 3)$.   ◆

If we replace a_n with $a_1 + (n-1)d$ in the formula $S_n = n\left(\dfrac{a_1 + a_n}{2}\right)$, we obtain

$$S_n = n\left(\frac{a_1 + [a_1 + (n-1)d]}{2}\right) = \frac{n}{2}[2a_1 + (n-1)d].$$

◆ **Sum of a Finite Arithmetic Series**

(Alternative Formula)

The **sum S_n of a finite arithmetic series** with n terms is given by

$$S_n = \frac{n}{2}[2a_1 + (n-1)d]$$

where d is the common difference and a_1 is the first term in the series.

EXAMPLE 6   Find the sum of the first 50 terms of an arithmetic series whose first term is 24 and second term is 20.

SOLUTION   Since we want the sum of the first 50 terms, we use $n = 50$. The first term is $a_1 = 24$, and the second term $a_2 = 20$. Thus, the common difference is

$$d = a_2 - a_1 = 20 - 24 = -4.$$

Now, using the formula $S_n = \dfrac{n}{2}[2a_1 + (n-1)d]$, we have

$$\begin{aligned}
S_{50} &= \tfrac{50}{2}[2(24) + (50-1)(-4)] \\
&= 25[48 + (49)(-4)] \\
&= -3700.
\end{aligned}$$

Hence, the sum of the first 50 terms of an arithmetic series with first term 24 and common difference -4 is -3700.   ◆

PROBLEM 6   The first term a_1 of an arithmetic series is 8 and the sum of the first 10 terms is 215. What is the common difference d?   ◆

◆ **Application: Installment Plan**

Arithmetic sequences and series are used in many types of applied problems. We conclude this section with an application concerning the payment of a debt by using an *installment plan*.

EXAMPLE 7   A stereo system costing $1200 is purchased using an installment plan. The plan consists of making 24 monthly payments in the amount of $50 plus $1\frac{1}{2}\%$ interest on the unpaid balance for that month.

(a) Find the amount of the first payment, the second payment, and the third payment.

(b) Find the amount of the 24th payment.

(c) Find the total amount paid for the stereo system.

SOLUTION

(a) The first monthly payment a_1 is $50 plus $1\frac{1}{2}\%$ of $1200, that is,

$$a_1 = \$50 + 0.015(\$1200) = \$68.00$$

The unpaid balance after the first payment is $1200 − $50 = $1150. Thus, the second payment a_2 is $50 + $1\frac{1}{2}\%$ of $1150, or

$$a_2 = \$50 + 0.015(\$1150) = \$67.25$$

The unpaid balance after the second payment is $1150 − $50 = $1100. Thus, the third payment a_3 is $50 + $1\frac{1}{2}\%$ of $1100, or

$$a_3 = \$50 + 0.015(\$1100) = \$66.50$$

Note that the first three payments form the arithmetic sequence

$$\$68.00, \$67.25, \$66.50$$

The common difference d of this sequence is

$$d = 1\frac{1}{2}\% \text{ of } (-\$50) = -\$0.75$$

(b) Realizing that each payment is an element of an arithmetic sequence, we can find the 24th payment a_{24} by using the formula $a_n = a_1 + (n-1)d$ as follows:

$$a_{24} = \$68 + (24 - 1)(\$-0.75) = \$50.75$$

(c) The total amount paid for the stereo system is represented by the arithmetic series

$$\$68.00 + \$67.25 + \$66.50 + \cdots + \$50.75$$

The sum of this series is

$$S_{24} = 24\left(\frac{a_1 + a_{24}}{2}\right) = 24\left(\frac{\$68.00 + \$50.75}{2}\right) = \$1425.$$

Thus, the total amount paid for the stereo under this installment plan is $1425. ◆

PROBLEM 7   Referring to Example 7, find the amount of the 15th payment. ◆

Exercises 8.3

    Basic Skills

In Exercises 1–6, write the first five elements of an arithmetic sequence whose first element a_1 and common difference d are given.

1. $a_1 = 3, d = 4$
2. $a_1 = -28, d = 12$
3. $a_1 = 2, d = -\frac{2}{3}$
4. $a_1 = -\frac{1}{2}, d = -\frac{3}{4}$
5. $a_1 = -1, d = \pi$
6. $a_1 = \sqrt{2}, d = -\sqrt{2}$

In Exercises 7–16, find the general element of the given arithmetic sequence.

7. $3, 9, 15, 21, \ldots$
8. $1, 8, 15, 22, \ldots$
9. $11, 4, -3, -10, \ldots$
10. $-2, -7, -12, -17, \ldots$
11. $-\frac{2}{3}, 1, \frac{8}{3}, \frac{13}{3}, \ldots$
12. $\frac{3}{4}, -2, -\frac{19}{4}, -\frac{15}{2}, \ldots$
13. $\frac{\pi}{6}, -\frac{\pi}{6}, -\frac{\pi}{2}, -\frac{5\pi}{6}, \ldots$
14. $\frac{\sqrt{3}}{2}, 2\sqrt{3}, \frac{7\sqrt{3}}{2}, 5\sqrt{3}, \ldots$
15. $x - 4, x - 2, x, x + 2, \ldots$
16. $x, y, 2y - x, 3y - 2x, \ldots$

In Exercises 17–26, information is given about an arithmetic sequence $a_1, a_2, a_3, \ldots, a_n$ with common difference d. Find the indicated unknown.

17. $a_1 = 4, d = 7;\quad a_{28} = ?$
18. $a_1 = 12, d = -9;\quad a_{32} = ?$
19. $a_{82} = 17, d = 4;\quad a_1 = ?$
20. $a_{31} = -5, d = -\frac{2}{5};\ a_1 = ?$
21. $a_2 = 9, a_{14} = 3;\quad d = ?$
22. $a_{13} = 46, a_4 = 100;\quad a_1 = ?$
23. $a_6 = 2.25, a_{16} = 6.25;\quad a_{80} = ?$
24. $a_5 = 72, a_{21} = -8;\quad a_{62} = ?$
25. $a_{19} - a_5 = -36;\quad d = ?$
26. $a_5 - a_{24} = 9;\quad d = ?$

In Exercises 27–36, find the sum of the arithmetic series.

27. $\sum_{i=1}^{35} 4i$
28. $\sum_{i=1}^{18} (2i - 5)$
29. $\sum_{i=1}^{58} (7 - 5i)$
30. $\sum_{i=1}^{100} \frac{3i + 2}{2}$
31. $\sum_{i=0}^{25} \left(\frac{i}{2} - 1\right)$
32. $\sum_{i=3}^{30} (0.6i + 0.8)$
33. $8 + 11 + 14 + 17 + \cdots + 170$
34. $-5 + 1 + 7 + 13 + \cdots + 283$
35. $2.4 + 1.8 + 1.2 + 0.6 + \cdots + (-18)$
36. $\frac{3}{2} + \frac{9}{4} + 3 + \frac{15}{4} + \cdots + 78$

In Exercises 37–46, information is given about an arithmetic series $a_1 + a_2 + a_3 + \cdots + a_n$ with sum S_n and common difference d. Find the indicated unknown.

37. $a_1 = 12, d = 5;\quad S_{31} = ?$
38. $a_1 = \frac{5}{4}, d = -\frac{1}{2};\quad S_{10} = ?$
39. $S_{32} = 448, a_1 = 2;\quad a_{32} = ?$
40. $S_{28} = -98, a_{28} = -181;\quad a_1 = ?$
41. $S_{54} = 11{,}583, a_1 = 2.5;\quad d = ?$
42. $S_{18} = -45, d = -9;\quad a_1 = ?$
43. $a_1 = -9, a_{40} = 147;\quad S_{62} = ?$
44. $a_5 = 15, a_{19} = 43;\quad S_{101} = ?$
45. $S_{24} = 432, a_1 = -28;\quad a_{33} = ?$
46. $S_{65} = -1560, d = -3;\quad a_{25} = ?$
47. Find the sum of the first 50 even, positive integers.
48. Find the sum of the first 80 odd, positive integers.
49. Find the number of multiples of three between 100 and 599.
50. Find the sum of the integers between 1 and 1000 that end in 9.
51. When writing his will, a man decides to leave $300,000

EXERCISE 51 (*continued*)

to his six children in such a way that each child receives $10,000 less than the next oldest child. How much is willed to the oldest child?

52. To drill an artesian well, a company charges $5 per foot until it reaches bedrock and, for each foot thereafter, 10¢ more than the preceding foot. How much does it cost to drill a well 350 feet deep if bedrock is reached at 150 feet?

53. An auditorium has 32 seats in the first row, and each row thereafter has 4 more seats than the preceding row. If the auditorium has 25 rows, how many seats are

(a) in the last row?   (b) in the auditorium?

54. An object falling freely from a position of rest travels 16 feet during the first second, 48 feet during the second second, 80 feet during the third second, and so on.

(a) Find the distance fallen during the tenth second.
(b) Find the total distance fallen after 10 seconds.

 Critical Thinking

55. Is the sequence $\ln a$, $\ln a^2$, $\ln a^3$, $\ln a^4$, ... an arithmetic sequence? If it is an arithmetic sequence, state the common difference.

56. Write the first five elements of the sequence defined recursively by

$$a_n = \begin{cases} 4 & \text{if } n = 1 \\ \dfrac{2a_{n-1} - 1}{2} & \text{if } n \geq 2 \end{cases}$$

Is this an arithmetic sequence? Express a_n explicitly in terms of n.

57. Find x if 10^{-2}, x, 10^{-4} are the first three elements of an arithmetic sequence.

58. Find x if $5x + 3$, $x + 4$, and $2x - 5$ are the first three elements of an arithmetic sequence.

59. If the numbers $a_1, a_2, a_3, \ldots, a_{n-1}, a_n$ form an arithmetic sequence, then $a_2, a_3, \ldots, a_{n-1}$ are called the $(n - 2)$ *arithmetic means* between a_1 and a_n. Insert four arithmetic means between 19 and 32.

60. The sum of the first four terms of an arithmetic series is 80. The fourth term is ten more than twice the second term. Find the first term and the common difference for this series.

61. A person borrows $20,000 from a loan shark and agrees to repay $34,800 by making payments of $300 the first month and, for each month thereafter, $100 more than the preceding month. (a) How many payments are required to repay this loan? (b) How much is the last payment?

62. A contractor has agreed to build an in-ground swimming pool in 20 days. If he does not complete the project on time, he must forfeit $150 for the first day over the given time allotment and, for each additional day thereafter, he must forfeit $20 more than the preceding day. How many days are required to put in the swimming pool if the contractor forfeits $1760?

 Calculator Activities

In Exercises 63 and 64, find the 79th element of the given arithmetic sequence.

63. $-8.32, -6.36, -4.40, -2.44, \ldots$

64. $837.6, 736.4, 635.2, 534.0, \ldots$

In Exercises 65–68, find the sum of each arithmetic series.

65. $\displaystyle\sum_{i=1}^{35} (2.78i - 19.4)$ 66. $\displaystyle\sum_{i=2}^{54} (437.3 - 17.6i)$

67. $0.0824 + 0.0044 + (-0.0736) + \cdots + (-1.5556)$

68. $1176.2 + 1989.1 + 2802.0 + \cdots + 22{,}311.6$

69. A sailboat costing $36,000 is purchased under an agreement to pay $600 per month for 60 months plus 1.25% interest on the unpaid balance for that month.

(a) Find the amount of the first payment, the second payment, and the third payment.
(b) Find the amount of the 60th payment.
(c) Find the total amount paid for the sailboat.

70. A woman buys a new automobile and agrees to pay $10,929.60 in 24 monthly installments that form an arithmetic sequence. After making the 12th payment, $\frac{1}{3}$ of the debt remains unpaid.

(a) Find the amount of the first payment.
(b) Find the amount of the 24th payment.

8.4   Geometric Sequences and Series

◆ Introductory Comments

Consider the sequence

$$2, 6, 18, 54, 162, 486, \ldots$$

It appears that each element after the first is obtained by multiplying the preceding element by 3. A sequence in which every element after the first is obtained by *multiplying* the preceding element by a nonzero fixed number r is called a *geometric sequence*.

◆ Geometric Sequence

> The sequence $a_1, a_2, a_3, \ldots, a_n, \ldots$ is called a **geometric sequence** if there exists a nonzero fixed number r such that for all positive integers k
>
> $$a_{k+1} = a_k r.$$

It follows from the definition of a geometric sequence that the ratio of any two successive elements must be the constant r, that is,

$$\frac{a_{k+1}}{a_k} = r$$

We refer to the constant r as the **common ratio** of the sequence. In this section we study this special type of sequence and its related series.

◆ Generating a Geometric Sequence

If the first element and the common ratio of a geometric sequence are known, then we can generate the sequence.

EXAMPLE 1   Write the first five elements of a geometric sequence whose first element is $a_1 = 8$ and whose common ratio is $r = -\frac{1}{2}$.

SOLUTION   If the first element is $a_1 = 8$ and the common ratio is $-\frac{1}{2}$, then

$$a_2 = a_1 r = 8(-\tfrac{1}{2}) = -4, \qquad a_3 = a_2 r = -4(-\tfrac{1}{2}) = 2,$$
$$a_4 = a_3 r = 2(-\tfrac{1}{2}) = -1, \qquad a_5 = a_4 r = -1(-\tfrac{1}{2}) = \tfrac{1}{2}.$$

Thus, the first five elements of the geometric sequence are

$$8, -4, 2, -1, \tfrac{1}{2} \qquad\qquad ◆$$

P R O B L E M 1 Repeat Example 1 if the first element is $a_1 = -2$ and the common ratio is $r = 6$. ◆

◆ Finding the General Element of a Geometric Sequence

Our definition of a geometric sequence allows us to define the general element a_n in terms of the element that precedes it, namely, a_{n-1}. Thus, we can define the general element a_n by the recursive formula

$$a_n = a_{n-1}r,$$

where r is a fixed number (the common ratio). We now proceed to find a formula that defines the general element a_n *explicitly*. If the first element in a geometric sequence is a_1 and the common ratio is r, then the first six elements of the sequence are as follows:

First element: a_1

Second element: $a_2 = a_1r$

Third element: $a_3 = a_2r = (a_1r)r = a_1r^2$

Fourth element: $a_4 = a_3r = (a_1r^2)r = a_1r^3$

Fifth element: $a_5 = a_4r = (a_1r^3)r = a_1r^4$

Sixth element: $a_6 = a_5r = (a_1r^4)r = a_1r^5$

Do you see the pattern? Note that each element is the product of a_1 and a power of r, where the exponent of r is one less than the number of the element. Thus, it appears that the general element of a geometric sequence can be written as

$$a_n = a_1r^{n-1}.$$

We now verify this formula by using mathematical induction (see Section 8.2).

Step 1   We must first show that the statement is true for $n = 1$.

$$a_1 = a_1r^{1-1} \quad ?$$
$$a_1 = a_1r^0 \quad ?$$
$$a_1 = a_1 \quad \text{is } true.$$

Step 2   Assuming that the statement is true for $n = k$, we must show that, on the basis of that assumption, the statement is true for $n = k + 1$.

Assume: $a_k = a_1r^{k-1}$

Prove: $a_{k+1} = a_1r^{(k+1)-1}$

Proof: Beginning with the definition of a geometric sequence, we have

$$a_{k+1} = \quad a_k r$$

Rewrite using the
assumed statement

$$= [a_1 r^{k-1}] r$$
$$= a_1 r^{(k-1)+1} \qquad \textbf{Adding exponents}$$
$$= a_1 r^{(k+1)-1} \qquad \textbf{Commutative and associative properties}$$

Thus, the truth of the statement when $n = k$ implies the truth of the statement when $n = k + 1$. Hence, by mathematical induction, we have the following.

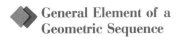

General Element of a Geometric Sequence

The **general element** of a geometric sequence with first element a_1 and common ratio r is given by

$$a_n = a_1 r^{n-1}.$$

EXAMPLE 2   Find the general element of the geometric sequence

$$2, -3, \tfrac{9}{2}, -\tfrac{27}{4}, \ldots$$

SOLUTION   The first element of this sequence is $a_1 = 2$. We can find the common ratio r by selecting any two adjacent elements and computing the ratio

$$\frac{a_{k+1}}{a_k}.$$

Choosing the first and second elements, we find

$$r = \frac{a_2}{a_1} = \frac{-3}{2} = -\frac{3}{2}.$$

Thus, the general element of this geometric sequence is

$$a_n = a_1 r^{n-1}$$
$$= 2(-\tfrac{3}{2})^{n-1}. \qquad \blacklozenge$$

PROBLEM 2   Find the general element of the geometric sequence

$$5, 15, 45, 135, \ldots \qquad \blacklozenge$$

Once the general element of a geometric sequence is known, we can find any element in the sequence. For example, to find the tenth element (a_{10}) of the geometric sequence in Example 2, we simply replace n with 10 in the general form

$a_n = 2(-\frac{3}{2})^{n-1}$, as follows:

$$a_{10} = 2\left(-\frac{3}{2}\right)^{10-1} = \frac{(-3)^9}{2^8} = -\frac{19{,}683}{256}.$$

EXAMPLE 3   Find the 13th element a_{13} of a geometric sequence whose first element a_1 is 3 and whose common ratio r is 4.

SOLUTION   For this geometric sequence, $a_1 = 3$ and $r = 4$. Thus, the general element of the geometric sequence is

$$a_n = 3(4)^{n-1},$$

and the 13th element is

$$a_{13} = 3(4)^{13-1} = 50{,}331{,}648. \qquad \blacklozenge$$

PROBLEM 3   Find the first element a_1 of a geometric sequence whose ninth element a_9 is $\frac{1}{81}$ and whose common ratio r is $\frac{1}{3}$. $\qquad \blacklozenge$

EXAMPLE 4   Find the common ratio r and the first element a_1 of a geometric sequence if the third element a_3 is 24 and the sixth element a_6 is 3.

SOLUTION   Using the formula $a_n = a_1 r^{n-1}$, we have

$$a_3 = a_1 r^{3-1} \qquad \text{and} \qquad a_6 = a_1 r^{6-1}$$
$$24 = a_1 r^2 \qquad\qquad\qquad 3 = a_1 r^5$$

To find a_1 and r, we solve the equations simultaneously by using substitution. Solving for a_1 in the equation $24 = a_1 r^2$, we obtain $a_1 = 24/r^2$. Now, substituting $24/r^2$ for a_1 in the other equation, $3 = a_1 r^5$, we obtain

$$3 = \left(\frac{24}{r^2}\right)r^5$$
$$3 = 24r^3$$
$$\tfrac{1}{8} = r^3$$
$$r = \tfrac{1}{2}$$

Substituting $r = \frac{1}{2}$ into the equation $3 = a_1 r^5$, we find

$$3 = a_1(\tfrac{1}{2})^5$$
$$3 = \frac{a_1}{32}$$
$$a_1 = 96$$

Thus, the common ratio is $\frac{1}{2}$ and the first element is 96. $\qquad \blacklozenge$

PROBLEM 4   Write the general element of the geometric sequence described in Example 4. Then use the general element to verify that $a_3 = 24$ and $a_6 = 3$.   ◆

◆ **Geometric Series**

The indicated sum of the elements of a geometric sequence is called a **geometric series**. The geometric series associated with the finite geometric sequence

$$a_1, a_1r, a_1r^2, a_1r^3, \ldots, a_1r^{n-1}$$

can be written in expanded form as

$$a_1 + a_1r + a_1r^2 + a_1r^3 + \cdots + a_1r^{n-1},$$

or in sigma form as

$$\sum_{i=1}^{n} a_1 r^{i-1}.$$

We now proceed to develop a formula that can be used to find the sum of a finite geometric series. If we represent the sum of these n terms by S_n, then

$$S_n = a_1 + a_1r + a_1r^2 + a_1r^3 + \cdots + a_1r^{n-2} + a_1r^{n-1}.$$

Multiplying both sides of this equation by $-r$, we obtain

$$-rS_n = -a_1r - a_1r^2 - a_1r^3 - a_1r^4 - \cdots - a_1r^{n-1} - a_1r^n.$$

If we add these two equations, all terms on the right-hand side drop out except for a_1 and $-a_1r^n$, and we obtain

$$S_n = a_1 + a_1r + a_1r^2 + a_1r^3 + \cdots + a_1r^{n-2} + a_1r^{n-1}$$
$$\underline{+ \quad -rS_n = \qquad -a_1r - a_1r^2 - a_1r^3 - a_1r^4 - \cdots - a_1r^{n-1} - a_1r^n}$$
$$S_n - rS_n = a_1 \qquad\qquad\qquad\qquad\qquad\qquad\qquad\qquad - a_1r^n$$

$$S_n(1 - r) = a_1(1 - r^n) \qquad\qquad \textbf{Factor both sides}$$

$$S_n = \frac{a_1(1 - r^n)}{1 - r}, \quad r \neq 1 \qquad \textbf{Divide both sides by } (1 - r)$$

 **Sum of a Finite Geometric Series**

The **sum S_n of a finite geometric series** with a common ratio r ($r \neq 1$) is given by

$$S_n = \frac{a_1(1 - r^n)}{1 - r},$$

where a_1 is the first term and n is the number of terms in the series.

Note: If the common ratio of a finite geometric series is $r = 1$, then the sum of the series $\sum_{i=1}^{n} a_1 r^{i-1}$ is

$$\underbrace{a_1 + a_1 + a_1 + \cdots + a_1}_{n \text{ terms}} = na_1.$$

E X A M P L E 5   Find the sum of each finite geometric series.

(a) $\displaystyle\sum_{i=1}^{10} 5(-2)^{i-1}$ (b) $\displaystyle\sum_{i=0}^{8} \left(\frac{1}{2}\right)^i$

S O L U T I O N

(a) Since the summation variable i starts at 1 and ends at 10, the number of terms in this series is $n = 10$. Writing this series in expanded form, we have

$$\sum_{i=1}^{10} 5(-2)^{i-1} = 5 - 10 + 20 - 40 + \cdots - 2560$$

This is a geometric series with common ratio $r = -2$ and first term $a_1 = 5$. Thus, the sum of this geometric series is

$$S_{10} = \frac{a_1(1 - r^{10})}{1 - r}$$

$$= \frac{5[1 - (-2)^{10}]}{1 - (-2)} = \frac{5(-1023)}{3} = -1705$$

(b) Note that the summation variable i ranges from 0 to 8. Thus the number of terms in this series is $n = 9$. Writing this series in expanded form, we have

$$\sum_{i=0}^{8} \left(\frac{1}{2}\right)^i = 1 + \frac{1}{2} + \frac{1}{4} + \cdots + \frac{1}{256}$$

This is a geometric series with common ratio $r = \frac{1}{2}$, and first term $a_1 = 1$. Thus, the sum of this geometric series is

$$S_9 = \frac{a_1(1 - r^9)}{1 - r}$$

$$= \frac{1[1 - (\frac{1}{2})^9]}{1 - \frac{1}{2}} = \frac{\frac{511}{512}}{\frac{1}{2}} = \frac{511}{256} \qquad \blacklozenge$$

P R O B L E M 5   Find the first term a_1 of a geometric series if the sum of the first 12 terms is 8190 and the common ratio r is 2. $\qquad \blacklozenge$

The formula for the sum of a finite geometric series can be rewritten as follows:

$$S_n = \frac{a_1(1 - r^n)}{1 - r} = \frac{a_1 - a_1 r^n}{1 - r} = \frac{a_1 - r(a_1 r^{n-1})}{1 - r}.$$

Since $a_n = a_1 r^{n-1}$, we can replace $a_1 r^{n-1}$ with a_n to obtain

$$S_n = \frac{a_1 - r a_n}{1 - r}.$$

Sum of a Finite Geometric Series

(Alternative Formula)

The **sum S_n of a finite geometric series** with a common ratio r ($r \neq 1$) is given by

$$S_n = \frac{a_1 - r a_n}{1 - r}$$

where a_1 is the first term and a_n is the last term in the series.

E X A M P L E   6   Find the sum of the geometric series

$$\tfrac{2}{3} + 2 + 6 + \cdots + 1458$$

S O L U T I O N   The common ratio for this geometric series is $r = 3$. The first term is $a_1 = \tfrac{2}{3}$, and the last term is $a_n = 1458$. Using the alternative formula for the sum of a finite geometric series, we have

$$S_n = \frac{a_1 - r a_n}{1 - r}$$

$$= \frac{\tfrac{2}{3} - 3(1458)}{1 - 3}$$

$$= \frac{\tfrac{2}{3}}{-2} - \frac{3(1458)}{-2}$$

$$= -\tfrac{1}{3} + 2187$$

$$= 2186\tfrac{2}{3} \qquad \blacklozenge$$

Referring to Example 6, we can find the number of terms n in this series by using the formula $a_n = a_1 r^{n-1}$ with $a_1 = \tfrac{2}{3}$, $a_n = 1458$, and $r = 3$. The equation becomes

$$1458 = \tfrac{2}{3}(3)^{n-1} \qquad \text{or} \qquad 3^{n-1} = 2187.$$

To solve this exponential equation (Section 6.5) for n, we begin by taking the log of both sides as follows:

$$\log 3^{n-1} = \log 2187$$

$$(n-1)\log 3 = \log 2187 \qquad \textbf{Property of logarithms}$$

$$n - 1 = \frac{\log 2187}{\log 3} \qquad \textbf{Divide both sides by log 3}$$

$$n = 1 + \frac{\log 2187}{\log 3} = 8 \qquad \textbf{Add 1 to both sides}$$

> Use the log key on your calculator to evaluate this expression.

Hence, the geometric series $\frac{2}{3} + 2 + 6 + \cdots + 1458$ has eight terms.   ◆

PROBLEM 6   Find the number of terms in the geometric series

$$3 + 12 + 48 + \cdots + 201{,}326{,}592 \qquad\qquad ◆$$

◆ **Application: Simple Annuities**

Recall (from Section 6.1) that if a certain principal P is deposited in a savings account at an interest rate r per year and interest is compounded n times per year, then the amount A in the savings account after t years is given by the formula

$$A = P\left(1 + \frac{r}{n}\right)^{nt}.$$

In this formula, r/n is the *interest rate per time interval*, which we denote by i, and nt is the *total number of compounding intervals*, which we denote by N. Thus, the compound interest formula is often written as

$$A = P(1 + i)^N.$$

A **simple annuity** is a sequence of equal periodic payments of R dollars that are made over N equal time intervals at an interest rate per time interval of i. By applying the formula $A = P(1 + i)^N$, we can find the amount accumulated on each periodic payment. Beginning with the Nth payment, we have the following information.

Payment	Amount Accumulated on Each Payment	Comment
Nth payment	R	Earns no interest
$(N-1)$st payment	$R(1+i)$	Earns interest on one time interval
$(N-2)$d payment	$R(1+i)^2$	Earns interest on two time intervals
$(N-3)$d payment	$R(1+i)^3$	Earns interest on three time intervals
$\vdots$	$\vdots$	$\vdots$
First payment	$R(1+i)^{N-1}$	Earns interest on $(N-1)$ time intervals

Now, to find the total amount S of an annuity immediately after the Nth payment, we simply sum the amounts accumulated on each payment as follows:

$$S = R + R(1+i) + R(1+i)^2 + R(1+i)^3 + \cdots + R(1+i)^{N-1}.$$

Note that this sum is a geometric series with first term $a_1 = R$ and common ratio $r = 1 + i$. Thus, using the formula for the sum of a finite geometric series, we have

$$S = \frac{a_1(1-r^n)}{1-r} = \frac{R[1-(1+i)^N]}{1-(1+i)} = \frac{R[(1+i)^N - 1]}{i}.$$

Amount of an Annuity

If equal periodic payments of R dollars are made over N equal time intervals at an interest rate per time interval of i, then the total **amount S of an annuity** immediately after the Nth payment is

$$S = \frac{R[(1+i)^N - 1]}{i}.$$

Saving plans, rent payments, and life-insurance payments are examples of an annuity. We conclude this section with an example concerning a savings plan.

EXAMPLE 7   A father decides to deposit $150 each month in a savings account for his daughter's college education. The savings account earns interest at the rate of 9% per year, compounded monthly. How much money will be in the account at the end of 10 years if there is nothing in the account today?

SOLUTION   For this annuity problem, we have equal monthly payments of $150 made over 10(12), or 120 months at an interest rate *per month* of $0.09 \div 12$, or 0.0075. Using the formula for the amount of an annuity with

$R = 150$, $N = 120$, and $i = 0.0075$, we obtain

$$S = \frac{R[(1 + i)^N - 1]}{i} = \frac{150[(1 + 0.0075)^{120} - 1]}{0.0075}$$

$$= 20{,}000[(1.0075)^{120} - 1]$$

> Use the y^x key on a calculator to approximate this power.

$$\approx 20{,}000[2.45136 - 1]$$

$$\approx 29{,}027.$$

Thus, after 10 years, approximately \$29,027 will be in this account.   ◆

PROBLEM 7   Repeat Example 7 if, at present, \$5000 is already in the account.   ◆

Exercises 8.4

 Basic Skills

In Exercises 1–6, write the first five elements of a geometric sequence whose first element a_1 and common ratio r are given.

1. $a_1 = 3, r = 4$

2. $a_1 = -2, r = 5$

3. $a_1 = \frac{1}{4}, r = -\frac{2}{3}$

4. $a_1 = -16, r = -\frac{3}{4}$

5. $a_1 = -1, r = \pi$

6. $a_1 = 1 + \sqrt{2}, r = 1 - \sqrt{2}$

In Exercises 7–16, find the general element of the given geometric sequence.

7. $1, 3, 9, 27, \ldots$

8. $1, -8, 64, -512, \ldots$

9. $-64, 16, -4, 1, \ldots$

10. $24, 36, 54, 81, \ldots$

11. $\dfrac{1}{4}, \dfrac{1}{6}, \dfrac{1}{9}, \dfrac{2}{27}, \ldots$

12. $-\dfrac{3}{16}, \dfrac{1}{4}, -\dfrac{1}{3}, \dfrac{4}{9}, \ldots$

13. $-\sqrt{2}, \sqrt{6}, -3\sqrt{2}, 3\sqrt{6}, \ldots$

14. $-3\sqrt{3}, -3, -\sqrt{3}, -1, \ldots$

15. $3, 6x^2, 12x^4, 24x^6, \ldots$

16. $2x^3, 10x^2y^2, 50xy^4, 250y^6, \ldots$

In Exercises 17–26, information is given about a geometric sequence $a_1, a_2, a_3, \ldots, a_n$ with common ratio r. Find the indicated unknown.

17. $a_1 = 4, r = 3; \quad a_9 = ?$

18. $a_1 = 12, r = -2; \quad a_{12} = ?$

19. $a_6 = 36, r = -6; \quad a_1 = ?$

20. $a_7 = -5, r = -\frac{5}{2}; \quad a_1 = ?$

21. $a_2 = -4, a_6 = -\frac{81}{4}; \quad r = ?$

22. $a_9 = \frac{3}{32}, a_4 = 3; \quad a_1 = ?$

23. $a_5 = \frac{8}{27}, a_8 = -1; \quad a_{13} = ?$

24. $a_5 = 0.5, a_9 = 312.5; \quad a_2 = ?$

25. $\dfrac{a_4}{a_8} = 16; \quad r = ?$

26. $\dfrac{a_5}{a_8} = -\dfrac{1}{27}; \quad r = ?$

In Exercises 27–36, find the sum of the geometric series.

27. $\displaystyle\sum_{i=1}^{6} 2(-3)^{i-1}$

28. $\displaystyle\sum_{i=1}^{9} -3(2)^{i-1}$

29. $\displaystyle\sum_{i=1}^{5} \frac{5}{3^{i+1}}$

30. $\displaystyle\sum_{i=0}^{10} (-2)^i$

31. $\displaystyle\sum_{i=0}^{8} 42(0.1)^i$

32. $\displaystyle\sum_{i=2}^{7} \frac{4^i}{3^{i-1}}$

33. $1 + 4 + 16 + 64 + \cdots + 1{,}048{,}576$

34. $-243 + 162 - 108 + 72 - \cdots + \frac{128}{9}$

35. $192 - 96 + 48 - 24 + \cdots - \frac{3}{8}$

36. $2 + \sqrt{2} + 1 + \dfrac{\sqrt{2}}{2} + \cdots + \dfrac{1}{8}$

In Exercises 37–46, information is given about a geometric series $a_1 + a_2 + a_3 + \cdots + a_n$ with sum S_n and common ratio r. Find the indicated unknown.

37. $a_1 = 16, r = -\frac{1}{2}$; $S_{11} = ?$

38. $a_1 = -\frac{5}{4}, r = 2$; $S_{10} = ?$

39. $S_6 = 728, r = 3$; $a_1 = ?$

40. $S_3 = 93, a_1 = 3$; $r = ?$

41. $a_3 = \frac{3}{4}, r = \frac{2}{3}$; $S_6 = ?$

42. $a_6 = 2048, r = -4$; $S_{10} = ?$

43. $a_1 = -64, a_4 = 27$; $S_7 = ?$

44. $a_3 = 24, a_6 = 3$; $S_{11} = ?$

45. $S_3 = 57, a_1 = 12$; $a_6 = ?$

46. $S_7 = 57\frac{7}{8}, r = -\frac{3}{2}$; $a_4 = ?$

47. The first three terms of a geometric series are 6, 12, and 24 and the sum of the series is 3066. How many terms are in the series?

48. A car radiator currently contains 3 gallons of the original antifreeze. One gallon of the original antifreeze is removed and replaced with one gallon of new anti-freeze. One year later one gallon of this mixture is removed and again replaced with one gallon of new antifreeze. If this process of removing a gallon and re-placing a gallon is repeated each year for three more years, how many gallons of the original antifreeze will remain in the radiator?

49. An automobile that costs $16,000 when new depreciates 50% in value each year. What is it worth at the end of five years?

50. A union contract states that the workers shall receive a 6% pay increase per year for the next *n* years. Suppose at the beginning of the first year of the contract, the workers are paid $28,540 per year. Express their annual pay *P* for each succeeding year as a function of *n*.

51. Suppose you have a choice between two scholarships. One pays $10,000 per year for four years. The other pays $10 the first month, $20 the second month, $40 the third month, and so on, doubling the amount each month, for one year only. Which offer would you accept?

52. How many ancestors (parents, grandparents, great-grandparents, and so on) do you have if you trace back ten generations?

Critical Thinking

53. Suppose $2x, 5x, 8x, 11x, \ldots$ is an arithmetic sequence.

   (a) What type of sequence is $e^{2x}, e^{5x}, e^{8x}, e^{11x}, \ldots$?
   (b) State the general element of the sequence in part (a).

54. Suppose $2x, 4x, 8x, 16x, \ldots$ is a geometric sequence.

   (a) What type of sequence is $\ln 2x, \ln 4x, \ln 8x, \ln 16x, \ldots$?
   (b) State the general element of the sequence in part (a).

55. Find *x* if $10^{-2}, x,$ and 10^{-4} are the first three elements of a geometric sequence.

56. Find *x* if $x - 4, 5 - 2x,$ and $4x - 1$ are the first three elements of a geometric sequence.

57. If the numbers $a_1, a_2, a_3, \ldots, a_{n-1}, a_n$ form a geometric sequence, then $a_2, a_3, \ldots, a_{n-1}$ are called the $(n - 2)$ *geometric means* between a_1 and a_n. Insert three geometric means between 16 and 81.

58. The sum of the first three terms of a geometric series is 63, and the third term is 45 more than the first. What is the first term of the series?

59. The product of the first three elements of a geometric sequence is -125 and their sum is 21. What is the common ratio of the sequence?

60. A ball is dropped from a height of 12 meters and re-bounds after hitting the ground each time $\frac{1}{3}$ the height from which it last fell. Find the *total distance* the ball has traveled when it hits the ground for the sixth time.

Calculator Activities

In Exercises 61 and 62, find the 12th element of the given geometric sequence. Round the answer to three significant digits.

61. $-8.36, -2.926, -1.0241, -0.358435, \ldots$

62. $6.4, -11.52, 20.736, -37.3248, \ldots$

In Exercises 63 and 64, find the sum of each geometric series. Round the answer to three significant digits.

63. $\displaystyle\sum_{i=1}^{18} 2.28(0.32)^{i-1}$ **64.** $\displaystyle\sum_{i=1}^{21} 13.3(-1.25)^{i-1}$

65. Find the number of elements in the geometric sequence $0.375, 1.5, 6, \ldots, 6,291,456$.

66. The population of a city that now has 125,400 inhabitants is increasing by $6\frac{1}{2}\%$ each year. What will be the population six years from now?

In Exercises 67 and 68, find the amount accumulated under each annuity. Round the answer to the nearest cent.

67. $100 per month for 120 months, at the rate of 8% per year, compounded monthly

68. $2500 per year for 20 years, at the rate of 9% per year, compounded yearly

69. Suppose your parents deposited $10,000 in an account on your first birthday and then deposited $1000 on each birthday thereafter. Furthermore, suppose that the account earned interest at the rate of 8.75%, compounded yearly. How much money would be in the account on your 18th birthday?

70. A woman decides to deposit $500 each month in an account that earns interest at the rate of 8.25% per year, compounded monthly. How much will she have in the account at the end of 5 years, if the account presently contains $22,420?

8.5 Infinite Geometric Series

The geometric series associated with the infinite geometric sequence

$$a_1, a_1r, a_1r^2, a_1r^3, \ldots, a_1r^{n-1}, \ldots$$

can be written in expanded form as

$$a_1 + a_1r + a_1r^2 + a_1r^3 + \cdots + a_1r^{n-1} + \cdots$$

or in sigma form as

$$\sum_{i=1}^{\infty} a_1 r^{i-1}$$

In some instances, we can find the sum of an infinite geometric series, and in other instances, we cannot. As we will show, it is the value of the common ratio r that determines whether an infinite geometric series can be assigned a sum.

◆ **Sequence of Partial Sums**

To understand what is meant by the sum of the infinite geometric series

$$a_1 + a_1r + a_1r^2 + a_1r^3 + \cdots + a_1r^{n-1} + \cdots,$$

we must first consider the finite sums

$$S_1 = a_1$$
$$S_2 = a_1 + a_1r$$
$$S_3 = a_1 + a_1r + a_1r^2$$
$$S_4 = a_1 + a_1r + a_1r^2 + a_1r^3$$
$$\vdots$$
$$S_n = a_1 + a_1r + a_1r^2 + a_1r^3 + \cdots + a_1r^{n-1} = \frac{a_1(1-r^n)}{1-r}, \quad r \neq 1.$$

We refer to the real number S_n as the **nth partial sum** of the infinite series

$$\sum_{i=1}^{\infty} a_1 r^{i-1},$$

and the infinite sequence

$$S_1, S_2, S_3, S_4, \ldots, S_n, \ldots$$

as the **sequence of partial sums**.

EXAMPLE 1   Find the sequence of partial sums for the infinite geometric series

$$\sum_{i=1}^{\infty} \left(\frac{1}{2}\right)^{i-1}$$

SOLUTION   Writing this series in expanded form, we have

$$\sum_{i=1}^{\infty} \left(\frac{1}{2}\right)^{i-1} = 1 + \frac{1}{2} + \frac{1}{4} + \frac{1}{8} + \cdots$$

This is an infinite geometric series with first term $a_1 = 1$ and common ratio $r = \frac{1}{2}$. The first four partial sums are

$$S_1 = 1,$$
$$S_2 = 1 + \tfrac{1}{2} = \tfrac{3}{2},$$
$$S_3 = 1 + \tfrac{1}{2} + \tfrac{1}{4} = \tfrac{7}{4},$$
$$S_4 = 1 + \tfrac{1}{2} + \tfrac{1}{4} + \tfrac{1}{8} = \tfrac{15}{8},$$

and the nth partial sum is

$$S_n = \frac{a_1(1 - r^n)}{1 - r} = \frac{1[1 - (\frac{1}{2})^n]}{1 - \frac{1}{2}} = 2[1 - (\tfrac{1}{2})^n]$$

Hence, the sequence of partial sums is

$$1, \tfrac{3}{2}, \tfrac{7}{4}, \tfrac{15}{8}, \ldots, 2[1 - (\tfrac{1}{2})^n], \ldots \qquad \blacklozenge$$

PROBLEM 1   Repeat Example 1 for the infinite geometric series $\displaystyle\sum_{n=0}^{\infty} \left(\frac{3}{2}\right)^n$. $\blacklozenge$

◆ **Sum of an Infinite Geometric Series**

Now, if the nth partial sum of a geometric series approaches a finite limit S as n increases without bound ($n \to \infty$), then the infinite series is said to **converge** to the sum S, and we write

$$\sum_{i=1}^{\infty} a_1 r^{i-1} = \lim_{n \to \infty} S_n = S.$$

Read "the limit of S_n
as n approaches ∞."

For instance, the sum of the series in Example 1 is

$$\sum_{i=1}^{\infty} \left(\frac{1}{2}\right)^{i-1} = \lim_{n \to \infty} S_n = \lim_{n \to \infty} 2[1 - (\tfrac{1}{2})^n]$$

Now, observe in Table 8.2 that $(\tfrac{1}{2})^n$ approaches 0 as n becomes larger and larger.

Table 8.2
Values of $(\tfrac{1}{2})^n$ as n becomes larger and larger

n	1	2	3	4	5	20
$(\tfrac{1}{2})^n$	$\frac{1}{2}$	$\frac{1}{4}$	$\frac{1}{8}$	$\frac{1}{16}$	$\frac{1}{32}$	$\frac{1}{1,048,576}$

Hence,

$$\lim_{n \to \infty} 2[1 - (\tfrac{1}{2})^n] = 2[1 - 0] = 2.$$

Since the nth partial sum approaches 2 as $n \to \infty$, the infinite geometric series $\sum_{i=1}^{\infty} (\tfrac{1}{2})^{i-1}$ is said to **converge** to the sum 2, and we write

$$\sum_{i=1}^{\infty} \left(\frac{1}{2}\right)^{i-1} = 2.$$

However, if the nth partial sum approaches positive infinity, negative infinity, or no specific limit as n increases without bound ($n \to \infty$), then the infinite geometric series cannot be assigned a sum and is said to **diverge**.

Let's examine what happens to the nth partial sum of an infinite geometric series as $n \to \infty$. We must consider four cases.

Case 1 If an infinite geometric series has a common ratio r such that $|r| < 1$, then $|r^n| \to 0$ as $n \to \infty$. Hence,

$$\sum_{i=1}^{\infty} a_1 r^{i-1} = \lim_{n \to \infty} S_n = \lim_{n \to \infty} \frac{a_1(1 - r^n)}{1 - r}$$
$$= \frac{a_1(1 - 0)}{1 - r}$$
$$= \frac{a_1}{1 - r} \quad \text{if } |r| < 1$$

Case 2 If an infinite geometric series has a common ratio r such that $|r| > 1$, then $|r^n| \to \infty$ as $n \to \infty$. Hence, $\lim_{n \to \infty} S_n$ does not exist and the series diverges.

Case 3 If an infinite geometric series has common ratio $r = 1$, then the sequence of partial sums is

$$a_1, 2a_1, 3a_1, \ldots, na_1, \ldots$$

Hence, $\lim_{n \to \infty} S_n$ does not exist and the series diverges.

Case 4   If an infinite geometric series has common ratio $r = -1$, then the sequence of partial sums is

$$a_1, 0, a_1, 0, a_1, 0, \ldots$$

Hence, $\lim_{n \to \infty} S_n$ does not exist and the series diverges.

In summary, it is the value of the common ratio r that determines whether an infinite geometric series converges or diverges.

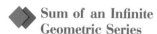

Sum of an Infinite Geometric Series

The infinite geometric series

$$\sum_{i=1}^{\infty} a_1 r^{i-1} = a_1 + a_1 r + a_1 r^2 + a_1 r^3 + \cdots + a_1 r^{n-1} + \cdots$$

converges if $|r| < 1$ and has the **sum**

$$S = \frac{a_1}{1 - r}$$

or *diverges* if $|r| \geq 1$ and has no finite sum.

EXAMPLE 2   Determine if the infinite geometric series converges or diverges. If it converges, find the sum.

(a) $\displaystyle\sum_{n=0}^{\infty} \left(\frac{3}{2}\right)^n$ **(b)** $\displaystyle\sum_{i=1}^{\infty} (-1)^{i+1} \left(\frac{1}{2}\right)^i$

SOLUTION

(a)   Writing this series in expanded form, we have

$$\sum_{n=0}^{\infty} \left(\frac{3}{2}\right)^n = 1 + \frac{3}{2} + \frac{9}{4} + \frac{27}{8} + \cdots$$

This is an infinite geometric series with first term $a_1 = 1$ and common ratio $r = \frac{3}{2}$. Since

$$|r| = \left|\frac{3}{2}\right| = \frac{3}{2} \geq 1,$$

the series diverges and has no finite sum.

(b)   Writing this series in expanded form, we have

$$\sum_{i=1}^{\infty} (-1)^{i+1} \left(\frac{1}{2}\right)^i = \frac{1}{2} - \frac{1}{4} + \frac{1}{8} - \frac{1}{16} + \cdots$$

This is an infinite geometric series with first term $a_1 = \frac{1}{2}$ and common ratio $r = -\frac{1}{2}$. Since

$$|r| = |-\tfrac{1}{2}| = \tfrac{1}{2} < 1,$$

the infinite geometric series converges and the sum S of the series is

$$S = \frac{a_1}{1-r} = \frac{\frac{1}{2}}{1-(-\frac{1}{2})} = \frac{\frac{1}{2}}{\frac{3}{2}} = \frac{1}{3}. \qquad \blacklozenge$$

PROBLEM 2   Repeat Example 2 for the infinite geometric series

$$1 + 0.1 + 0.01 + 0.001 + \cdots \qquad \blacklozenge$$

♦ **Rational Numbers Represented by Repeating Decimals**

In Section 1.1, we stated that all decimal numbers that either terminate or repeat the same block of digits are rational numbers. For example, the rational number represented by the terminating decimal 0.25 is $\frac{25}{100}$, or $\frac{1}{4}$. To find the rational number represented by a repeating decimal such as

$$0.\overline{12} = 0.121212\ldots$$

The block of digits under the bar is repeated indefinitely.

we can use the formula for the sum of an infinite geometric series. The procedure is outlined in the next example.

EXAMPLE 3   Find the rational number represented by each repeating decimal number.

(a) $0.\overline{12} = 0.121212\ldots$ (b) $6.3\overline{450} = 6.3450450450\ldots$

SOLUTION

(a)   The decimal number $0.121212\ldots$ can be written as the infinite series

$$0.12 + 0.0012 + 0.000012 + \cdots$$

This is an infinite geometric series with first term $a_1 = 0.12$ and common ratio $r = 0.01$. Since $r < 1$, the series converges to the sum

$$S = \frac{a_1}{1-r} = \frac{0.12}{1-0.01} = \frac{0.12}{0.99} = \frac{4}{33}$$

Hence, the rational number represented by the repeating decimal $0.121212\ldots$ is $\frac{4}{33}$.

(b)   The decimal number $6.3450450450\ldots$ can be written as

$$6.3 + [0.0450 + 0.0000450 + 0.0000000450 + \cdots]$$

The expression within the brackets is an infinite geometric series with first term $a_1 = 0.0450$ and common ratio $r = 0.001$. Since $r < 1$, the series converges to the sum

$$S = \frac{a_1}{1 - r} = \frac{0.0450}{1 - 0.001} = \frac{45}{999} = \frac{5}{111}.$$

Thus, the rational number represented by the repeating decimal $6.3450450450\ldots$ is

$$6.3 + \frac{5}{111} = \frac{63}{10} + \frac{5}{111} = \frac{7043}{1110}. \qquad \blacklozenge$$

You can check the results of Example 3(a) by using the division key on your calculator and dividing 4 by 33. The display should show the repeating decimal $0.12121212\ldots.$

PROBLEM 3 Use a calculator to check the results of Example 3(b). $\blacklozenge$

◆ Interval of Convergence

In some infinite series, each term contains a variable. Such series occur often in calculus and its applications. For example, consider the infinite series

$$1 + (cx) + (cx)^2 + (cx)^3 + \cdots$$

where x is a variable and c a constant with $c \neq 0$. Obviously, if $x = 0$, this series converges to 1. If $x \neq 0$, we can think of this series as an infinite geometric series with first term $a_1 = 1$ and common ratio $r = cx$. Depending on the real number that we substitute for x, the infinite geometric series either converges (if $|cx| < 1$) or diverges (if $|cx| \geq 1$). For an infinite series with variable terms, the subset of real numbers for which the series converges is the **interval of convergence**.

EXAMPLE 4 Find the interval of convergence and the sum of the infinite series

$$1 + 2x + 4x^2 + 8x^3 + \cdots$$

SOLUTION Obviously, if $x = 0$, the series converges to 1. If $x \neq 0$, we can think of this series as an infinite geometric series with first term $a_1 = 1$ and common ratio $r = 2x$. The infinite geometric series converges whenever

$$|2x| < 1$$
$$-1 < 2x < 1$$
$$-\tfrac{1}{2} < x < \tfrac{1}{2}$$

Thus, the interval of convergence for this infinite series is $(-\tfrac{1}{2}, \tfrac{1}{2})$.

We can find the sum S of this infinite series by using the formula for the sum of an infinite geometric series. Thus,

$$S = \frac{a_1}{1-r} = \frac{1}{1-2x}, \quad \text{provided } -\tfrac{1}{2} < x < \tfrac{1}{2}. \qquad \blacklozenge$$

PROBLEM 4   Use polynomial long division (Section 5.1) to show that

$$\frac{1}{1-2x} = 1 + 2x + 4x^2 + 8x^3 + \cdots \qquad \blacklozenge$$

◆ **Application: Hammering a Nail**

Infinite geometric series occur in many types of applied problems. We conclude this section with an application.

EXAMPLE 5   A nail $2\tfrac{1}{2}$ inches long is being hammered into a board. The first impact drives the nail $\tfrac{3}{4}$ inch into the board, and each additional impact drives the nail two-thirds the distance of the preceding impact. If the nail is hammered indefinitely, will its head ever be flush with the board?

SOLUTION   The first impact drives the nail $\tfrac{3}{4}$ inch into the board. The second impact drives the nail two-thirds of $\tfrac{3}{4}$ inch, or $\tfrac{1}{2}$ inch into the board. The third impact drives the nail two-thirds of $\tfrac{1}{2}$ inch, or $\tfrac{1}{3}$ inch into the board, and so on. The total distance that the nail is driven into the board can be represented by the infinite geometric series

$$\tfrac{3}{4} + \tfrac{1}{2} + \tfrac{1}{3} + \cdots$$

where $a_1 = \tfrac{3}{4}$ and $r = \tfrac{2}{3}$. Since $|r| < 1$, the sum S of this infinite geometric series is

$$S = \frac{a_1}{1-r} = \frac{\tfrac{3}{4}}{1-\tfrac{2}{3}} = \frac{\tfrac{3}{4}}{\tfrac{1}{3}} = \frac{9}{4} \quad \text{or} \quad 2\tfrac{1}{4}.$$

Thus, after infinitely many hits, the nail is driven into the board $2\tfrac{1}{4}$ inches. Since the nail is $2\tfrac{1}{2}$ inches long, $\tfrac{1}{4}$ inch remains sticking out from the board. Hence, the head of the nail will never be flush with the board. ◆

PROBLEM 5   Referring to Example 5, will the head of the nail ever be flush with the board if the initial impact drives the nail $\tfrac{7}{8}$ inch into the board? ◆

Exercises 8.5

Basic Skills

In Exercises 1–8,

(a) *find the sequence of partial sums for each infinite geometric series.*

(b) *find* $\lim\limits_{n\to\infty} S_n$, *if it exists.*

1. $\displaystyle\sum_{i=1}^{\infty}\left(\frac{2}{3}\right)^{i-1}$

2. $\displaystyle\sum_{i=1}^{\infty}\left(-\frac{3}{4}\right)^{i}$

3. $\displaystyle\sum_{i=0}^{\infty} 2(-3)^{i}$

4. $\displaystyle\sum_{i=2}^{\infty} 3(0.1)^{i-2}$

5. $-243 + 162 - 108 + 72 - \cdots$

6. $1 + 4 + 16 + 64 + \cdots$

7. $\dfrac{9}{8} + \dfrac{3}{4} + \dfrac{1}{2} + \dfrac{1}{3} + \cdots$

8. $15 - 5 + \dfrac{5}{3} - \dfrac{5}{9} + \cdots$

In Exercises 9–20, determine if the infinite geometric series converges or diverges. If it converges, find the sum of the series.

9. $\displaystyle\sum_{i=1}^{\infty} 4(3)^{i-1}$

10. $\displaystyle\sum_{i=0}^{\infty} 2\left(-\frac{3}{5}\right)^{i}$

11. $\displaystyle\sum_{i=2}^{\infty} (-0.2)^{i}$

12. $\displaystyle\sum_{i=1}^{\infty} 8\left(\frac{3}{2}\right)^{i-1}$

13. $\displaystyle\sum_{i=1}^{\infty} (-1)^{i+1}\left(\frac{5}{6}\right)^{i}$

14. $\displaystyle\sum_{i=1}^{\infty} \frac{2}{3^{i+1}}$

15. $1 - 2 + 4 - 8 + \cdots$

16. $3.375 + 2.25 + 1.5 + 1 + \cdots$

17. $-6 + 3 - \dfrac{3}{2} + \dfrac{3}{4} - \cdots$ 18. $1 - \dfrac{1}{\pi} + \dfrac{1}{\pi^2} - \dfrac{1}{\pi^3} + \cdots$

19. $5\sqrt{10} + 5\sqrt{2} + \sqrt{10} + \sqrt{2} + \cdots$

20. $-\sqrt{3} - \sqrt{6} - 2\sqrt{3} - 2\sqrt{6} - \cdots$

In Exercises 21–28, information is given about an infinite geometric series $a_1 + a_2 + a_3 + \cdots$ with sum S and common ratio r. Find the indicated unknown.

21. $S = -36, a_1 = -12; \quad r = ?$

22. $S = 24, r = -\frac{1}{2}; \quad a_1 = ?$

23. $a_3 = \frac{3}{4}, r = \frac{2}{3}; \quad S = ?$

24. $a_6 = 1.28, r = 0.4; \quad S = ?$

25. $a_1 = -64, a_4 = 27; \quad S = ?$

26. $a_3 = 24, a_6 = 3; \quad S = ?$

27. $S = 18, a_2 = 4; \quad a_1 = ?$

28. $S = 32, a_2 = 6; \quad r = ?$

In Exercises 29–38, find the rational number represented by each repeating decimal.

29. $0.33333\ldots$

30. $0.77777\ldots$

31. $0.090909\ldots$

32. $0.272727\ldots$

33. $2.1\overline{08}$

34. $34.\overline{303}$

35. $15.020312031\ldots$

36. $2.00010101\ldots$

37. $0.8\overline{015}$

38. $5.25\overline{144}$

$\dfrac{\Delta y}{\Delta x}$ *In Exercises 39–46, find the interval of convergence and the sum S of each infinite series.*

39. $x - \dfrac{x^2}{5} + \dfrac{x^3}{25} - \dfrac{x^4}{125} + \cdots$

40. $3 + 9x + 27x^2 + 81x^3 + \cdots$

41. $1 + 4x^2 + 16x^4 + 64x^6 + \cdots$

42. $x - 8x^4 + 64x^7 - 512x^{10} + \cdots$

43. $1 - (x - 2) + (x - 2)^2 - \cdots$

44. $2(x + 1) + 4(x + 1)^2 + 8(x + 1)^3 - \cdots$

45. $2x + 1 + \dfrac{1}{2x} + \dfrac{1}{4x^2} + \cdots$

46. $1 - \dfrac{2}{x} + \dfrac{4}{x^2} - \dfrac{8}{x^3} + \cdots$

47. A golf ball is dropped from a height of 18 meters and rebounds each time $\frac{2}{3}$ of the height from which it last fell, as shown by the arrows in the figure.

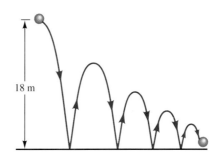

18 m

EXERCISE 47 *(continued)*

(a) Find the distance the ball travels *downward* before coming to rest.

(b) Find the distance the ball travels *upward* before coming to rest.

(c) Find the *total distance* the ball travels before coming to rest.

48. A lawn mower blade makes 450 revolutions during the first second after the power is shut off and $\frac{1}{16}$ as many revolutions during each succeeding second. How many revolutions does the blade make in coming to rest?

49. A man runs the 26-mile Boston Marathon at the rate of 10 miles the first hour, 6 miles the second hour, 3.6 miles the third hour, and so on. Does he ever finish the race?

50. A 24-foot steel pile is driven vertically into the soil to form part of a foundation. The first impact drives the pile 5 feet into the soil, and each additional impact drives the pile $\frac{4}{5}$ the distance of the preceding impact. Will the top of the pile ever be flush with the ground?

Critical Thinking

51. Determine the value of x if $\sum_{i=1}^{\infty} (3x)^{i-1} = \frac{2}{3}$.

52. Find the sum of each infinite series.

(a) $\sum_{i=1}^{\infty} 2\left[\left(\frac{2}{3}\right)^{i-1} + \left(\frac{3}{4}\right)^{i}\right]$

(b) $\sum_{i=0}^{\infty} \left(\frac{1}{2^i} - \frac{1}{3^i} + \frac{1}{4^i}\right)$

53. Find the sum of each infinite series, if it exists.

(a) $1 - 1 + 1 - 1 + 1 - 1 + \cdots$

(b) $(1 - 1) + (1 - 1) + (1 - 1) + \cdots$

(c) $1 + (-1 + 1) + (-1 + 1) + \cdots$

Does the associative property of real numbers seem to apply to infinite series?

54. Is it possible to have an infinite geometric series with first term 7 and sum 3? Explain.

55. Find the common ratio of an infinite geometric series in which each term is twice the sum of all the terms that follow it.

56. Find an infinite geometric series in which the first term is 4 and each term is 5 times the sum of all the terms that follow it.

57. A square has sides of length 24 cm. Inside this square, a second square is constructed by joining the midpoints of the sides of the first square. Inside the second square, a third square is constructed by joining the midpoints of the sides of the second square, and so on indefinitely, as shown in the sketch.

(a) What is the sum of the areas of all the squares?

(b) What is the sum of the perimeters of all the squares?

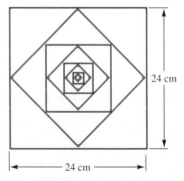

24 cm

24 cm

EXERCISE 57

58. In a right triangle ABC, angle A is 30°, angle B is 60°, angle C is 90°, and the length of side $\overline{AB}$ is 32 inches. Suppose $\overline{CD}$ is drawn perpendicular to $\overline{AB}$, $\overline{DE}$ is drawn perpendicular to $\overline{AC}$, $\overline{EF}$ is drawn perpendicular to $\overline{AB}$, $\overline{FG}$ is drawn perpendicular to $\overline{AC}$, and so on indefinitely, as shown in the sketch.

(a) What is the sum of all the perpendiculars to side $\overline{AC}$ if $\overline{BC}$ is considered the first of these perpendiculars?

(b) What is the sum of all the perpendiculars to side $\overline{AB}$?

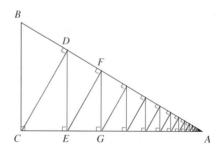

Calculator Activities

In Exercises 59–62, use a calculator to help find the sum of each infinite geometric series.

59. $\displaystyle\sum_{i=0}^{\infty} 138.6(-0.68)^i$ **60.** $\displaystyle\sum_{i=1}^{\infty} 875.52(0.24)^{i-1}$

61. $21.28 + 14.4704 + 9.839872 + \cdots$

62. $-26.46 + 2.1168 - 0.169344 + \cdots$

 The infinite geometric series in Exercises 63 and 64 are used in calculus to approximate the area of the shaded region that is shown in the figure. Find the shaded area to the nearest hundredth of a square unit.

63. $\displaystyle\sum_{i=1}^{\infty} 0.01e^{-0.01i}$

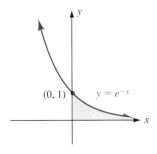

64. $\displaystyle\sum_{i=1}^{\infty} 0.01(3^{-0.01i})$

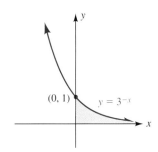

65. A pendulum swings through an arc 22.54 cm long. On each succeeding swing, the pendulum travels an arc that is 0.84 of the length of the preceding swing. How far does the pendulum travel before coming to rest?

66. A person with cancer is given 485.0 mg of an experimental drug the first week, and each succeeding week the patient is given 85% of the previous week's dosage. How many milligrams of this drug is the patient given? Round the answer to the nearest tenth of a milligram.

8.6 Factorials and the Binomial Theorem

◆ Introductory Comments

Consider the sequence defined recursively by

$$a_n = \begin{cases} 1 & \text{if } n = 1 \\ na_{n-1} & \text{if } n \geq 2 \end{cases}.$$

We have

$$a_1 = 1$$
$$a_2 = 2a_1 \quad = 2 \cdot 1$$
$$a_3 = 3a_2 \quad = 3 \cdot 2 \cdot 1$$
$$a_4 = 4a_3 \quad = 4 \cdot 3 \cdot 2 \cdot 1$$
$$a_5 = 5a_4 \quad = 5 \cdot 4 \cdot 3 \cdot 2 \cdot 1$$
$$\vdots \qquad \vdots$$
$$a_n = na_{n-1} = n \cdot (n-1) \cdot (n-2) \cdot (n-3) \cdots \cdot 2 \cdot 1$$

Note that a_n represents the product of all consecutive integers from 1, up to and including n. This special product is called **n factorial** and is denoted by **$n!$**.

 **n Factorial**

> If n is a positive integer, then **n factorial** ($n!$) is given by
>
> $$n! = n \cdot (n - 1) \cdot (n - 2) \cdot (n - 3) \cdots 2 \cdot 1.$$

We usually extend this definition to include **zero factorial** (0!). As we will observe in this section and the next, it is convenient to have

$$0! = 1$$

Factorials appear in diverse mathematical situations, such as special sequences that occur in calculus and counting techniques in probability theory. In this section, we discuss some important factorial expressions and develop the *binomial theorem*, which may be used to expand $(A + B)^n$, where n is a positive integer.

◆ **Simplifying Factorial Expressions**

To simplify many factorial expressions, we apply the fact that if n and $n - 1$ are consecutive positive integers, then

$$n \cdot (n - 1)! = n!$$

Consider these illustrations of simplified factorial expressions:

$$8 \cdot 7! = 8!$$

$$15 \cdot 14 \cdot 13 \cdot 12 \cdot 11! = 15!$$

$$\frac{10!}{7!} = \frac{10 \cdot 9 \cdot 8 \cdot \cancel{7!}}{\cancel{7!}} = 10 \cdot 9 \cdot 8 = 720$$

E X A M P L E 1   Given that k is a positive integer, simplify each factorial expression.

(a) $(k + 1)k!$ (b) $\dfrac{k!}{(k - 1)!}$

S O L U T I O N

(a) Since $k + 1$ and k are consecutive integers, we have

$$(k + 1)k! = (k + 1)!$$

SECTION 8.6 Factorials and the Binomial Theorem

(b) Rewriting the numerator, we have

$$\frac{k!}{(k-1)!} = \frac{k(k-1)!}{(k-1)!} = k.$$ ◆

PROBLEM 1 Simplify $\dfrac{(k+1)!}{(k-1)!}$.   ◆

◆ **Binomial Coefficients**

The factorial expression

$$\frac{n!}{k!\,(n-k)!}$$

occurs frequently in mathematics. We denote this factorial expression by using the symbol

$$\binom{n}{k}.$$

**Binomial Coefficient
Symbol**

If k is a nonnegative integer such that $k \le n$, then the **binomial coefficient symbol** $\binom{n}{k}$ is defined by

$$\binom{n}{k} = \frac{n!}{k!\,(n-k)!}.$$

EXAMPLE 2 Find the value of the binomial coefficients

$$\binom{5}{0}, \quad \binom{5}{1}, \quad \binom{5}{2}, \quad \binom{5}{3}, \quad \binom{5}{4}, \quad \text{and} \quad \binom{5}{5}.$$

SOLUTION Using $\binom{n}{k} = \dfrac{n!}{k!\,(n-k)!}$, we have

$$\binom{5}{0} = \frac{5!}{0!\,(5-0)!} = \frac{5!}{0! \cdot 5!} = \frac{1}{0!} = \frac{1}{1} = 1,$$

$$\binom{5}{1} = \frac{5!}{1!\,(5-1)!} = \frac{5!}{1! \cdot 4!} = \frac{5 \cdot 4!}{1! \cdot 4!} = \frac{5}{1} = 5,$$

$$\binom{5}{2} = \frac{5!}{2!\,(5-2)!} = \frac{5!}{2! \cdot 3!} = \frac{5 \cdot 4 \cdot 3!}{2! \cdot 3!} = \frac{20}{2} = 10,$$

EXAMPLE 2 (*continued*)

$$\binom{5}{3} = \frac{5!}{3!\,(5-3)!} = \frac{5!}{3! \cdot 2!} = \frac{5 \cdot 4 \cdot 3!}{3! \cdot 2!} = \frac{20}{2} = 10,$$

$$\binom{5}{4} = \frac{5!}{4!\,(5-4)!} = \frac{5!}{4! \cdot 1!} = \frac{5 \cdot 4!}{4! \cdot 1!} = \frac{5}{1} = 5,$$

and

$$\binom{5}{5} = \frac{5!}{5!\,(5-5)!} = \frac{5!}{5! \cdot 0!} = \frac{1}{0!} = \frac{1}{1} = 1.$$ ◆

From the results of Example 2, we can make some general observations about binomial coefficients:

1. We have

$$\binom{5}{0} = \binom{5}{5}, \quad \binom{5}{1} = \binom{5}{4}, \quad \text{and} \quad \binom{5}{2} = \binom{5}{3}.$$

In general, if k and n are nonnegative integers such that $k \le n$, then

$$\cdot \binom{n}{k} = \binom{n}{n-k}.$$

2. The binomial coefficients $\binom{5}{0}$ and $\binom{5}{5}$ both equal 1. In general, for any nonnegative integer n, we have

$$\binom{n}{0} = \binom{n}{n} = 1.$$

These are two **fundamental properties of binomial coefficients**.

Fundamental Properties of Binomial Coefficients

If k and n are nonnegative integers such that $k \le n$, then

1. $\binom{n}{k} = \binom{n}{n-k}$ 2. $\binom{n}{0} = \binom{n}{n} = 1$

PROBLEM 2 Find the value of each binomial coefficient.

(a) $\binom{9}{0}$ (b) $\binom{7}{7}$ (c) $\binom{12}{9}$ (d) $\binom{12}{3}$ ◆

◆ **The Binomial Theorem**

In Section 1.4, we developed formulas for the square and cube of a binomial. Recall that

$$(A + B)^2 = A^2 + 2AB + B^2$$

and

$$(A + B)^3 = A^3 + 3A^2B + 3AB^2 + B^3.$$

It is interesting to note that the coefficients of the terms in the expansion of $(A + B)^2$ are equivalent to

$$\binom{2}{0} = 1, \quad \binom{2}{1} = 2, \quad \text{and} \quad \binom{2}{2} = 1,$$

and the coefficients of the terms in the expansion of $(A + B)^3$ are equivalent to

$$\binom{3}{0} = 1, \quad \binom{3}{1} = 3, \quad \binom{3}{2} = 3, \quad \text{and} \quad \binom{3}{3} = 1.$$

Let's see if we can use pattern recognition to derive a formula for the expansion of $(A + B)^n$, where n is any positive integer.

Observation	Conjecture
1. The expansion of $(A + B)^2$ contains 3 terms. The first term is A^2 and the last term is B^2. The expansion of $(A + B)^3$ contains 4 terms. The first term is A^3 and the last term is B^3.	1. The expansion of $(A + B)^n$ should contain $(n + 1)$ terms. The first term should be A^n and the last term should be B^n.
2. In the expansion of $(A + B)^2$, the exponents on A decrease by 1 in each succeeding term after the first, and the sum of the exponents on A and B together in any term is 2. In the expansion of $(A + B)^3$, the exponents on A decrease by 1 in each succeeding term after the first, and the sum of the exponents on A and B together in any term is 3.	2. In the expansion of $(A + B)^n$, the exponents on A should decrease by 1 in each succeeding term after the first, and the sum of the exponents on A and B together in any term should be n.
3. The coefficients of the terms in the expansion of $(A + B)^2$ are equivalent to $$\binom{2}{0}, \binom{2}{1}, \binom{2}{2}.$$ The coefficients of the terms in the expansion $(A + B)^3$ are equivalent to $$\binom{3}{0}, \binom{3}{1}, \binom{3}{2}, \binom{3}{3}.$$	3. The coefficients of the terms in the expansion of $(A + B)^n$ should be equivalent to $$\binom{n}{0}, \binom{n}{1}, \binom{n}{2}, \binom{n}{3}, \cdots, \binom{n}{n}.$$

From our preceding conjectures, we write

$$(A + B)^n = \binom{n}{0}A^n + \binom{n}{1}A^{n-1}B + \binom{n}{2}A^{n-2}B^2 + \binom{n}{3}A^{n-3}B^3 + \binom{n}{4}A^{n-4}B^4 + \cdots + \binom{n}{n}B^n.$$

We refer to this formula as the **binomial theorem** and state it more compactly, using sigma form with lower value $i = 0$:

Binomial Theorem

If n is a positive integer, then

$$(A + B)^n = \sum_{i=0}^{n} \binom{n}{i} A^{n-i}B^i$$

EXAMPLE 3   Expand $(x - 2y)^5$ by using the binomial theorem.

SOLUTION   For $(x - 2y)^5$, we have $A = x$, $B = -2y$, and $n = 5$. Thus, by the binomial theorem, we obtain

$$(x - 2y)^5 = \sum_{i=0}^{5} \binom{5}{i}(x)^{5-i}(-2y)^i$$

$$= \binom{5}{0}x^5 + \binom{5}{1}x^4(-2y) + \binom{5}{2}x^3(-2y)^2$$

$$+ \binom{5}{3}x^2(-2y)^3 + \binom{5}{4}x(-2y)^4 + \binom{5}{5}(-2y)^5$$

Now, from Example 2 we know that

$$\binom{5}{0} = \binom{5}{5} = 1, \quad \binom{5}{1} = \binom{5}{4} = 5, \quad \text{and} \quad \binom{5}{2} = \binom{5}{3} = 10.$$

Hence,

$$(x - 2y)^5 = 1x^5 + 5x^4(-2y) + 10x^3(4y^2)$$
$$+ 10x^2(-8y^3) + 10x(16y^4) + 1(-32y^5)$$
$$= x^5 - 10x^4y + 40x^3y^2 - 80x^2y^3 + 160xy^4 - 32y^5$$

The **kth term** $(1 < k < n + 1)$ in the expansion of $(A + B)^n$ can be written as follows.

kth Term of $(A + B)^n$

$$\binom{n}{k-1} A^{n-(k-1)}B^{k-1}$$

PROBLEM 3   Find the fifth term in the expansion of $(x + y)^9$.

◆ Proof of the Binomial Theorem

We conclude this section by proving the binomial theorem using mathematical induction (Section 8.2.) The following three facts are needed in the proof:

1. $\dbinom{k}{0} = \dbinom{k+1}{0}$

2. $\dbinom{k}{k} = \dbinom{k+1}{k+1}$

3. $\dbinom{k}{r} + \dbinom{k}{r-1} = \dbinom{k+1}{r}$

The first two facts are direct consequences of binomial coefficient property 2. Since the values of $\dbinom{k}{0}$ and $\dbinom{k+1}{0}$ are both equal to 1, they are equal to each other. Similarly, since the values of $\dbinom{k}{k}$ and $\dbinom{k+1}{k+1}$ are both equal to 1, they are also equal to each other. To prove the third fact, we begin by adding the binomial coefficients:

$$\binom{k}{r} + \binom{k}{r-1} = \frac{k!}{r!\,(k-r)!} + \frac{k!}{(r-1)!\,[k-(r-1)]!}$$

$$= \frac{k!}{r!\,(k-r)!} + \frac{k!}{(r-1)!\,[(k-r)+1]!}$$

Since $r! = r(r-1)!$ and $[(k-r)+1]! = [(k-r)+1](k-r)!$, the LCD is $r!\,[(k-r)+1]!$. Continuing, we have

$$\binom{k}{r} + \binom{k}{r-1} = \frac{k!\,[(k-r)+1] + k!\,r}{r!\,[(k-r)+1]!} \qquad \text{Add fractions}$$

$$= \frac{k!\,[(k-r)+1+r]}{r!\,[(k-r)+1]!} \qquad \text{Factor}$$

$$= \frac{k!\,(k+1)}{r!\,[(k-r)+1]!} \qquad \text{Simplify}$$

$$= \frac{(k+1)!}{r!\,[(k+1)-r]!} \qquad \text{Rewrite}$$

$$= \binom{k+1}{r}$$

We now proceed with the proof of the binomial theorem using mathematical induction.

Step 1 We must first show that the binomial theorem is true for $n = 1$.

$$(A+B)^1 = \sum_{i=0}^{1} \binom{1}{i} A^{1-i} B^i \quad ?$$

$$= \binom{1}{0} A^1 B^0 + \binom{1}{1} A^0 B^1 \quad ?$$

$$= A + B \quad \text{is } true$$

Step 2 Assuming that the statement is true for $n = k$, we must show that, on the basis of that assumption, the statement is true for $n = k + 1$.

Assume:
$$(A + B)^k = \sum_{i=0}^{k} \binom{k}{i} A^{k-i} B^i$$

Prove:
$$(A + B)^{k+1} = \sum_{i=0}^{k+1} \binom{k+1}{i} A^{(k+1)-i} B^i$$

Proof:

$$(A + B)^{k+1} = (A + B)(A + B)^k$$

$$= (A + B) \sum_{i=0}^{k} \binom{k}{i} A^{k-i} B^i \qquad \text{Replace } (A + B)^k \text{ with the assumed statement}$$

$$= A \sum_{i=0}^{k} \binom{k}{i} A^{k-i} B^i + B \sum_{i=0}^{k} \binom{k}{i} A^{k-i} B^i \qquad \text{Distributive property}$$

$$= \sum_{i=0}^{k} \binom{k}{i} A^{(k+1)-i} B^i + \sum_{i=0}^{k} \binom{k}{i} A^{k-i} B^{i+1} \qquad \text{Summation property 1}$$

Now, writing in expanded form, we have

$$(A + B)^{k+1} = \left[\binom{k}{0} A^{k+1} + \binom{k}{1} A^k B + \binom{k}{2} A^{k-1} B^2 + \cdots + \binom{k}{k} A B^k \right]$$

$$+ \left[\binom{k}{0} A^k B + \binom{k}{1} A^{k-1} B^2 + \cdots + \binom{k}{k-1} A B^k + \binom{k}{k} B^{k+1} \right]$$

$$= \binom{k}{0} A^{k+1} + \left[\binom{k}{1} + \binom{k}{0} \right] A^k B + \left[\binom{k}{2} + \binom{k}{1} \right] A^{k-1} B^2$$

$$+ \cdots + \left[\binom{k}{k} + \binom{k}{k-1} \right] A B^k + \binom{k}{k} B^{k+1} \qquad \text{Collect like terms}$$

Finally, using the facts that

$$\binom{k}{0} = \binom{k+1}{0}, \qquad \binom{k}{k} = \binom{k+1}{k+1}, \quad \text{and} \quad \binom{k}{r} + \binom{k}{r-1} = \binom{k+1}{r}$$

we obtain

$$(A + B)^{k+1} = \binom{k+1}{0} A^{k+1} + \binom{k+1}{1} A^k B + \binom{k+1}{2} A^{k-1} B^2 + \cdots + \binom{k+1}{k} A B^k + \binom{k+1}{k+1} B^{k+1}$$

$$= \sum_{i=0}^{k+1} \binom{k+1}{i} A^{(k+1)-i} B^i$$

Thus, the truth of the statement when $n = k$ implies the truth of the statement when $n = k + 1$. Hence, by mathematical induction, we can state that

$$(A + B)^n = \sum_{i=0}^{n} \binom{n}{i} A^{n-i} B^i$$

for all positive integers n.

Exercises 8.6

Basic Skills

In Exercises 1–6, simplify each factorial expression.

1. (a) $10 \cdot 9 \cdot 8!$ 　　　(b) $k(k-1)(k-2)!$

2. (a) $\dfrac{15!}{14!}$ 　　　(b) $\dfrac{(k+1)!}{k!}$

3. (a) $\dfrac{21!}{21}$ 　　　(b) $\dfrac{k!}{k}$

4. (a) $\dfrac{12!}{11 \cdot 9!}$ 　　　(b) $\dfrac{(k+1)!}{k \cdot (k-2)!}$

5. (a) $\dfrac{1}{10!} + \dfrac{1}{9!}$ 　　　(b) $\dfrac{1}{k!} + \dfrac{1}{(k-1)!}$

6. (a) $\dfrac{1}{7! \cdot 11!} + \dfrac{1}{6! \cdot 12!}$ 　　　(b) $\dfrac{1}{k! \cdot (r-1)!} + \dfrac{1}{(k-1)! \cdot r!}$

In Exercises 7–20, evaluate each binomial coefficient expression.

7. $\dbinom{42}{1}$ 　　　　8. $\dbinom{75}{74}$

9. $\dbinom{19}{19}$ 　　　　10. $\dbinom{25}{25}$

11. $\dbinom{31}{0}$ 　　　　12. $\dbinom{15}{0}$

13. $\dbinom{6}{2}$ 　　　　14. $\dbinom{8}{3}$

15. $\dbinom{9}{6}$ 　　　　16. $\dbinom{10}{7}$

17. $\dbinom{9}{7}\dbinom{7}{3}$ 　　　18. $\dbinom{11}{8}\dbinom{10}{1}\dbinom{8}{8}$

19. $\dbinom{12}{5} + \dbinom{12}{4} - \dbinom{13}{5}$ 　　　20. $\dbinom{9}{4} - \dbinom{8}{4} - \dbinom{8}{3}$

In Exercises 21–30, use the binomial theorem to expand each expression.

21. $(x + y)^5$

22. $(m - n)^7$

23. $(2x - 1)^6$

24. $(1 + 3y)^4$

25. $(3a + 2b)^4$

26. $(5x - 2y)^5$

27. $(2x^2 - 1)^7$

28. $(1 + m^3)^8$

29. $(\sqrt{x} + 2)^8$

30. $(2x^{1/3} - 3y)^6$

In Exercises 31–38, find the indicated term of the expansion.

31. Fourth term of $(x + y)^{12}$

32. Third term of $(a - b)^{16}$

33. Third term of $(3n - 2)^7$

34. Fourth term of $(2p + 1)^9$

35. Fifth term of $(x + 3y^2)^8$

36. Fifth term of $(x - 2y^{-1/2})^8$

37. Seventh term of $(\sqrt{t} - 2)^{10}$

38. Next-to-last term of $(9x^2 - y^5)^{12}$

Critical Thinking

39. Show that each statement is true for all positive integers n.

(a) $\binom{n}{1} = n$ (b) $\binom{n}{n-1} = n$

40. Show that each statement is true for all positive integers n and k, where $k \le n$.

(a) $\binom{n}{k} - \binom{n-1}{k} = \binom{n-1}{k-1}$

(b) $\dfrac{n-k+1}{k} \cdot \binom{n}{k-1} = \binom{n}{k}$

41. Show that each statement is true for all positive integers n.

(a) $\binom{n}{0} + \binom{n}{1} + \binom{n}{2} + \cdots + \binom{n}{n} = 2^n$

(b) $\binom{n}{0} - \binom{n}{1} + \binom{n}{2} - \cdots + (-1)^n \binom{n}{n} = 0$

42. Use the binomial theorem to find the indicated power of the complex number.

(a) $(2 + 3i)^4$ (b) $(2 - i)^6$

43. Use the binomial theorem to expand each trinomial.

(a) $(x^2 + x + 1)^5$ (b) $(x^2 - x - 1)^5$

$\boxed{\frac{\Delta y}{\Delta x}}$ **44.** For each function f, defined as indicated, find the difference quotient

$$\frac{f(x + \Delta x) - f(x)}{\Delta x}.$$

(a) $f(x) = x^4$
(b) $f(x) = x^n$, n a positive integer.

45. In the seventeenth century, French mathematician Blaise Pascal discovered a triangular pattern for the coefficients of the terms in the expansion of $(A + B)^n$. The first seven rows of *Pascal's triangle* are given:

Row												Coefficients in the expansion of:
1						1						$(A + B)^0$
2					1		1					$(A + B)^1$
3				1		2		1				$(A + B)^2$
4			1		3		3		1			$(A + B)^3$
5		1		4		6		4		1		$(A + B)^4$
6	1		5		10		10		5		1	$(A + B)^5$
7	1	6		15		20		15		6	1	$(A + B)^6$

Each row in the triangle starts and ends with a coefficient of 1. The other coefficients in each row of the triangle can be obtained from the coefficients in the preceding row. Do you see the pattern? Using your observation, write the coefficients for each row:

(a) row 8   (b) row 9   (c) row 10

46. Find the sum of the numbers in each row of Pascal's triangle in Exercise 45. Do you observe a pattern to these sums? Guess a formula for the sum of the numbers in the nth row of Pascal's triangle, and verify your answer.

$\boxed{\frac{\Delta y}{\Delta x}}$ **47.** In calculus, it is shown that the expansion of $(1 + x)^n$, where n is a negative integer or a rational number and $|x| < 1$, is given by the *infinite binomial series*

$$1 + nx + \frac{n(n-1)}{2!}x^2 + \frac{n(n-1)(n-2)}{3!}x^3 + \cdots$$

Write the first four terms in the expansion of the given expression.

(a) $(1 + x)^{-1}$,   provided $|x| < 1$
(b) $(1 - y)^{-2}$,   provided $|y| < 1$
(c) $(1 - z)^{-1/2}$,   provided $|z| < 1$
(d) $(1 + x)^{1/3}$,   provided $|x| < 1$

48. Use mathematical induction to prove that $n! > 2^n$ for all integers $n \ge 4$.

Calculator Activities

In Exercises 49–52, use the $\boxed{n!}$ key on your calculator to help evaluate each binomial coefficient.

49. $\binom{27}{10}$ **50.** $\binom{52}{13}$

51. $\binom{32}{21}$ **52.** $\binom{100}{5}$

53. Evaluate $(1.02)^4$ (a) by expanding the binomial $(1 + 0.02)^4$ and (b) by using the $\boxed{y^x}$ key on your calculator.

54. Evaluate $(0.99)^5$ (a) by expanding the binomial $(1 - 0.01)^5$ and (b) by using the $\boxed{Y^x}$ key on your calculator.

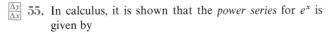 **55.** In calculus, it is shown that the *power series* for e^x is given by

$$e^x = 1 + x + \frac{x^2}{2!} + \frac{x^3}{3!} + \frac{x^4}{4!} + \cdots$$

Use the first ten terms of this series to estimate the value of e. Compare this value of e to the value you obtain by usng the $\boxed{e^x}$ key on your calculator with $x = 1$.

56. *Stirling's formula* states that a good approximation to $n!$, especially for large values of n, is given by

$$\sqrt{2\pi n}\left(\frac{n}{e}\right)^n.$$

(a) Use Stirling's formula and a calculator to estimate $5!$, $10!$, and $15!$.
(b) Use the $\boxed{n!}$ key on your calculator to find the values of $5!$, $10!$, and $15!$.
(c) Compute the error in the approximations of $5!$, $10!$, and $15!$ when using Stirling's formula.

8.7 Counting Techniques

In this section, we develop some counting techniques that enable us to determine the possible number of ways a certain procedure can be done, such as

the possible number of different license plate registrations consisting of three digits followed by three letters,

the possible number of different ways in which eight contestants in the 100-yard dash can finish in the first three positions, or

the possible number of different poker hands that can be dealt from a deck of 52 cards,

As you will see, factorials (from the preceding section) often develop when determining the possible number of ways in which something can be done. Since we apply the counting techniques discussed in this section to some elementary probability theory in the following section, we begin by discussing what is meant by the sample space and events of a random experiment.

◆ Sample Space and Events of a Random Experiment

A **random experiment** is one in which we can

(a) repeat the experiment under essentially unchanged conditions,

(b) describe the set of all possible outcomes of the experiment, and

(c) observe a definite pattern as the experiment is repeated a large number of times.

We refer to the set of all possible outcomes of a random experiment as its **sample space**, and we call each outcome in the sample space an **element**. In this text, we work with random experiments in which the sample space contains a *finite* number of elements. For example, the sample space S_1 for a random experiment of rolling a die (singular of dice) and observing the number that shows on the top face is

$$S_1 = \{1, 2, 3, 4, 5, 6\}.$$

Note the sample space S_1 contains six elements.

In some experiments, it is useful to construct a *tree diagram* to help list the elements in the sample space. The procedure is illustrated in the next example.

E X A M P L E 1   Determine the sample space associated with the random experiment of first tossing a coin and observing whether it lands showing heads (H) or tails (T), and then rolling a die and observing the number that appears on the top face.

S O L U T I O N   This experiment consists of two procedures:

> *First procedure*:   Tossing a coin
> *Second procedure:*   Rolling a die

The first procedure yields either H or T, and either of these outcomes will be followed by the second procedure, which yields a number from 1 to 6. The elements of the sample space can be found by following the branches of the tree diagram shown in Figure 8.5.

F I G U R E 8 . 5
Tree diagram for tossing a coin and then rolling a die.

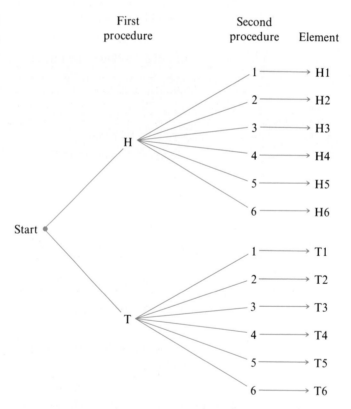

Hence, the sample space *S* associated with this experiment is

$$S = \{H1, H2, H3, H4, H5, H6, T1, T2, T3, T4, T5, T6\}.$$ ◆

Any subset of the sample space is called an **event** of the random experiment. For the sample space of Example 1, the event *A* of tossing a head and rolling an even number is the subset

$$A = \{\text{H2, H4, H6}\}.$$

PROBLEM 1   For the sample space of Example 1, determine the event B of tossing a tail and rolling a number divisible by 3.   ◆

◆ **Multiplication Principle**

Referring to Example 1, note that the sample space of this experiment contains 12 elements. We can find the total number of elements in this sample space, without actually listing each element, by multiplying the number of ways that the first procedure can occur, n_1, by the number of ways that the second procedure can occur, n_2. For Example 1, we have

$$n_1 n_2 = (2)(6) = 12 \text{ possible ways.}$$

This fundamental principle of counting is referred to as the **multiplication principle**.

◆◆ Multiplication Principle

> If a random experiment consists of k procedures in which the ith procedure may be performed in n_i ways ($i = 1, 2, 3, \ldots, k$), then the first procedure, followed by the second procedure, followed by the third procedure, . . . , followed by the kth procedure may be performed in
>
> $$n_1 \cdot n_2 \cdot n_3 \cdots n_k \text{ ways.}$$

EXAMPLE 2   How many elements are in the sample space of an experiment that consists of tossing a coin five times and observing the sequence of heads and tails?

SOLUTION   We can think of this experiment as consisting of five procedures, each of which is tossing a coin. Let's represent the procedures by the five boxes that follow:

1st   2d   3d   4th   5th

□   □   □   □   □

Each procedure has two outcomes (either heads or tails). Thus, by the multiplication principle, the sample space consists of

$$\boxed{2} \cdot \boxed{2} \cdot \boxed{2} \cdot \boxed{2} \cdot \boxed{2} = 2^5 \quad \text{or} \quad 32 \text{ elements.}$$ ◆

PROBLEM 2   How many ways can a true-false test of ten questions be answered?   ◆

EXAMPLE 3   The registration number on a Massachusetts license plate consists of three digits followed by three letters. How many different registrations are possible if digits and letters can be repeated?

SOLUTION   Let the general license plate registration be represented by the six boxes

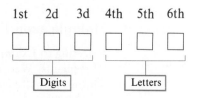

where the first three boxes require the selection of a digit from 0 to 9, and the last three boxes require the selection of a letter from A to Z. Since we have 10 digits from which to choose, each of the first three boxes can be filled in 10 ways. Since we have 26 letters from which to choose, each of the last three boxes can be filled in 26 ways. Thus, by the multiplication principle, the number of different license plate registrations that are possible is

$$\boxed{10} \cdot \boxed{10} \cdot \boxed{10} \cdot \boxed{26} \cdot \boxed{26} \cdot \boxed{26} = 17{,}576{,}000 \qquad \blacklozenge$$

PROBLEM 3   Repeat Example 3 if no license plate number has a first digit of zero.   ◆

◆ **Permutations**

An arrangement of a set of n distinct objects in a given order is called a **permutation**. Any two objects a and b have two permuations:

$$ab \quad \text{and} \quad ba$$

The three objects a, b, and c have six permutations:

$$abc, \quad acb, \quad bac, \quad bca, \quad cab, \quad cba$$

Now, suppose we wish to determine the number of permutations for n different objects. Consider the following n boxes:

1st   2d   3d $(n-1)$th nth

□ □ □ ⋯ □ □

We can fill the first box by choosing any of the n objects. Once we have made a choice for the first box, then we have $n-1$ choices for the second box. After choosing an object for the second box, we have $n-2$ choices for the third box, and so on, until we have but one choice for the nth box. Thus, by the multiplication principle, the number of permutations of n distinct objects is

$$n \cdot (n-1) \cdot (n-2) \cdots 2 \cdot 1 = n!$$

Permutations of *n* Objects

The number of permutations of *n* distinct objects is *n*!.

Now, suppose we wish to determine the number of permutations of *n* distinct objects taken *r* at a time, where $r \le n$. Consider the following *r* boxes:

1st   2d   3d   *r*th

$$\square \quad \square \quad \square \quad \cdots \quad \square$$

The first box can be filled in *n* ways, the second box in $(n-1)$ ways; the third box in $(n-2)$ ways; and the *r*th box in $[n-(r-1)]$ ways. Thus, by the multiplication principle, the number of permutations of *n* distinct objects taken *r* at a time, where $r \le n$, is

$$n \cdot (n-1) \cdot (n-2) \cdot \cdots \cdot [n-(r-1)].$$

Rewriting $[n-(r-1)]$ as $[(n-r)+1]$ and multiplying by $(n-r)!/(n-r)!$, we obtain

$$n \cdot (n-1) \cdot (n-2) \cdot \cdots \cdot [(n-r)+1] \cdot \frac{(n-r)!}{(n-r)!} = \frac{n!}{(n-r)!}$$

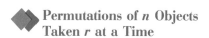 **Permutations of *n* Objects Taken *r* at a Time**

The number of permutations of *n* distinct objects taken *r* at a time, where $r \le n$, is denoted by $_nP_r$ and is given by

$$_nP_r = \frac{n!}{(n-r)!}$$

Note: If $r = n$, we obtain

$$_nP_n = \frac{n!}{(n-n)!} = \frac{n!}{0!} = \frac{n!}{1} = n!$$

Recall that *n*! is the number of permutations of *n* distinct objects.

E X A M P L E   4   At a track meet, eight contestants are entered in the 100-yard dash. Find the number of possible orders of finish for

(a)   all eight contestants   (b)   the first three positions.

S O L U T I O N

(a)   The number of permutations of eight distinct objects is 8!. Thus, the number of possible orders of finish for the race is

$$8! \quad \text{or} \quad 40{,}320 \text{ ways.}$$

(b) The number of permutations of 8 distinct objects taken three at a time is $_8P_3$. Thus, the number of possible orders of finish for the first three positions is

$$_8P_3 = \frac{8!}{(8-3)!} = \frac{8!}{5!} \quad \text{or} \quad 336 \text{ ways.} \qquad \blacklozenge$$

P R O B L E M   4   Repeat Example 4 for the first four positions. $\blacklozenge$

Suppose we wish to find the number of permutations that can be made from the letters of the word PAPA. If we distinguish between the P's by putting one in boldface type (**P**), and between the A's by putting one in boldface type (**A**) then we have 4!, or 24 permutations:

PAPA, PAPA, PAPA, PAPA,
PPAA, PPAA, PPAA, PPAA,
PAAP, PAAP, PAAP, PAAP,
APPA, APPA, APPA, APPA,
AAPP, AAPP, AAPP, AAPP,
APAP, APAP, APAP, APAP

> For each of these six groups, we have 2! ways to arrange the P's and 2! ways to arrange the A's.

However, if we write all the letters in the same style of type, only six of these 24 permutations are *distinguishable*:

PAPA,   PPAA,   PAAP,   APPA,   AAPP,   and   APAP.

Thus, for four letters, two of which are P's and two of which are A's, we have

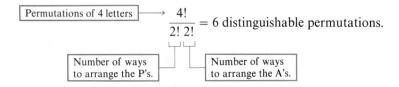

In general, we have the following rule for distinguishable permutations.

**Distinguishable
Permutations**
_____

> If n objects are made up of n_1 of one kind, n_2 of a second kind, ..., n_k of a kth kind, such that $n_1 + n_2 + \cdots + n_k = n$, then the number of **distinguishable permutations** of these n objects is given by
>
> $$\frac{n!}{n_1! \, n_2! \cdots n_k!}$$

EXAMPLE 5   In how many ways is it possible for a college football team to end the season with six wins and four losses?

SOLUTION   The team plays 10 games, six of which are wins (W), and four of which are losses (L). The number of distinguishable permutations for WWWWWWLLLL is

$$\frac{10!}{6!\,4!} = 210$$

Thus, a football team can end the season with 6 wins and 4 losses in 210 ways.   ◆

PROBLEM 5   Repeat Example 5 for a team that ends the season with three wins, five losses, and two ties.   ◆

◆   Combinations

The selection of r objects from a set of n distinct objects *without regard to order* is called a **combination** of n objects taken r at a time. We denote the number of such combinations by $_nC_r$. For example, consider choosing two objects from a set of four distinct objects a, b, c, and d. We have $_4P_2$, or 12 permutations:

$$
\left.
\begin{array}{ll}
ab, & ba \\
ac, & ca \\
ad, & da \\
bc, & cb \\
bd, & db \\
cd, & dc
\end{array}
\right\}
$$

Once 2 objects are chosen, we have 2! ways of arranging them.

Now, if we disregard order, then each of these groups is counted only once. Hence, the combinations of 4 objects taken 2 at a time are

$$ab, \quad ac, \quad ad, \quad bc, \quad bd, \quad \text{and} \quad cd.$$

The number of such combinations is

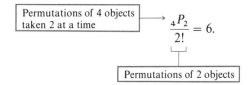

Permutations of 4 objects taken 2 at a time → $\dfrac{_4P_2}{2!} = 6.$

Permutations of 2 objects

In general, we have the following rule.

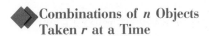

Combinations of n Objects Taken r at a Time

> The number of **combinations** of n distinct objects taken r at a time, with $r \leq n$, is denoted by $_nC_r$, and is given by
>
> $$_nC_r = \frac{_nP_r}{r!} = \frac{n!}{r!\,(n-r)!}.$$

Note: The formula for $_nC_r$ is the same as the binomial coefficient symbol $\binom{n}{r}$ that we defined in the preceding section. Thus, we often write

$$_nC_r \quad \text{as} \quad \binom{n}{r}.$$

Read "from n, choose r."

Caution

Do not confuse permutations and combinations. Remember, in a permutation, order is considered. In a combination, order does not matter. For example, a signal made up of three flags that may be obtained from eight differently colored flags is a permutation, because a rearrangement of the three flags makes a different signal (order matters). However, a committee of three members that may be obtained from eight people is a combination, because a rearrangement of the three members is the same committee (the order of the members does not change the committee).

E X A M P L E   6   In a certain poker game, five cards are dealt to a player from an ordinary deck of 52 cards.

(a) How many different possible poker hands can be dealt?

(b) How many poker hands consist of all hearts?

(c) How many poker hands consist of three aces?

S O L U T I O N   The order in which the five cards are dealt does not matter. Hence, we are working with a combination problem. A standard deck of 52 cards has four suits (hearts, diamonds, clubs, spades), and each suit contains 13 cards (2, 3, 4, . . . , 10, jack, queen, king, ace).

(a) The number of possible poker hands that can be dealt is the number of combinations of 52 objects taken 5 at a time:

$$_{52}C_5 \quad \text{or} \quad \binom{52}{5}$$

From 52 cards, choose 5.

Now, $_{52}C_5 = \dfrac{52!}{5!\,(52-5)!} = \dfrac{52!}{5!\,47!} = 2{,}598{,}960.$

Thus, 2,598,960 poker hands can be dealt.

(b)   The number of poker hands consisting of all hearts is the number of combinations of 13 objects taken 5 at a time:

$$_{13}C_5 \quad \text{or} \quad \begin{pmatrix} 13 \\ 5 \end{pmatrix}.$$

From 13 hearts, choose 5.

Now, $_{13}C_5 = \dfrac{13!}{5!\,(13-5)!} = \dfrac{13!}{5!\,8!} = 1287.$

Thus, a hand of all hearts can be dealt in 1287 ways.

(c)   The number of ways of being dealt three aces from the four aces in the deck is

$$_4C_3 = \begin{pmatrix} 4 \\ 3 \end{pmatrix} = \dfrac{4!}{3!\,1!} = 4 \text{ ways.}$$

From 4 aces, choose 3.

The number of ways of being dealt the remaining two cards from the 48 non-ace cards is

$$_{48}C_2 = \begin{pmatrix} 48 \\ 2 \end{pmatrix} = \dfrac{48!}{2!\,46!} = 1128 \text{ ways.}$$

From 48 non-aces, choose 2.

Thus, by the multiplication principle, a hand consisting of three aces can be dealt in

$$_4C_3 \cdot {_{48}C_2} = 4 \cdot 1128 = 4512 \text{ ways.} \qquad \blacklozenge$$

PROBLEM 6   How many poker hands consist of two queens and three kings? $\blacklozenge$

Exercises 8.7

Basic Skills

In Exercises 1–10, determine the sample space S associated with each random experiment.

  1. Selecting one letter from the first ten letters of the English alphabet

  2. Selecting one state from the six New England states

  3. Asking two men if they shave with a particular brand of shaving cream, and recording their response as yes (Y) or no (N)

4. Selecting three students from a math class, and classifying them as male (M) or female (F)

5. Selecting two cars, and noting whether the vehicles were manufactured in America (A), Europe (E), or Japan (J)

6. Tossing a coin, and then tossing it a second time if a tail (T) occurs on the first toss. If a head (H) appears on the first toss, then rolling a die once and observing the number on the top face

7. Selecting light bulbs from a box containing four non-defective light bulbs (N) and one defective light bulb (D) until the defective light bulb is selected

8. Tossing a coin until a head (H) or three tails (T) appear

9. Rolling two dice, one red and one green, and recording as an ordered pair (x, y) the numbers that appear on the top face of each die

10. Selecting three patients of a particular dentist, and recording as an ordered triple (x, y, z) which of two hygienists (1 and 2) clean their teeth

In Exercises 11–20, determine the indicated event of the given sample space.

11. For the sample space of Exercise 1, the event A of selecting a vowel

12. For the sample space of Exercise 2, the event B of selecting a state that ends with a vowel

13. For the sample space of Exercise 3, the event C of having both responses be the same

14. For the sample space of Exercise 4, the event A of selecting at least two female students

15. For the sample space of Exercise 5, the event B of selecting at least one vehicle manufactured in Japan

16. For the sample space of Exercise 6, the event C of tossing a head and rolling an odd number

17. For the sample space of Exercise 7, the event A of selecting at most one nondefective light bulb

18. For the same sample space of Exercise 8, the event B of tossing at most two tails

19. For the sample space of Exercise 9, the event C of having an ordered pair whose sum is 7

20. For the sample space of Exercise 10, the event A that all three patients have the same hygienist

In Exercises 21–38, use the multiplication principle and the formulas for permutations and combinations to help answer each question.

21. How many elements are in the sample space of an experiment that consists of rolling a die and observing the number that appears on top, then selecting a letter at random from the English alphabet?

22. How many elements are in the sample space of an experiment that consists of tossing a coin and observing whether it lands showing heads or tails, then selecting a card at random from a deck of 52 cards?

23. How many elements are in the sample space of an experiment that consists of tossing a die three times and observing the sequence of numbers that appear on the top face.

24. A bag contains one green ball, one red ball, one blue ball, and one white ball. An experiment consists of drawing three balls from this bag and observing the sequence of colors selected.

    (a) How many elements are in the sample space if each ball is selected, one after the other, without replacement?
    (b) How many elements are in the sample space if the ball is replaced in the bag after each draw?

25. There are six possible ski trails from the top of a mountain to a warming hut halfway down, and eight possible trails from the warming hut to the lodge at the bottom of the mountain. How many routes are possible from the top of the mountain to the lodge?

26. For lunch, the school cafeteria offers three choices of drinks, four choices of sandwiches, and two choices of desserts. How many different menus can be selected?

27. A foreign car manufacturer offers three choices of models, five choices of color, two choices of transmission style, and two different engines. How many cars must a dealer have on the lot if all possible choices are displayed?

28. How many ways can three girls and four boys sit in a row if the girls and boys must alternate?

29. A family of five has bought seats to a baseball game.

    (a) In how many different ways can they be seated in the five seats?
    (b) In how many ways can they be seated if the father must sit in the middle seat?

30. Signals are sent from a battleship by placing four flags on a vertical pole. How many different signals can be formed by selecting the four flags from (a) four different colored flags? (b) two identical red flags and two identical white flags?

31. How many distinct permutations can be made from the word ALGEBRA?

32. How many distinct permutations starting with *M* can be made from the word MISSISSIPPI?

33. How many ways can five apple trees, three pear trees, and two peach trees be planted along a driveway if it is not possible to distinguish between trees of the same type?

34. In how many ways is it possible for a person to toss a coin seven times and end up four heads and three tails?

35. From nine equally qualified students, how many ways are there to select three of them for scholarships?

36. A hockey lineup consists of two forwards, two defensemen, one center, and one goalie. How many lineups can a coach form from a team roster of six forwards, six defensemen, three centers, and three goalies?

37. From a collection of records consisting of three hard-rock records, three soft-rock records, and two jazz records, a disk jockey selects five records at random to play.

(a) How many different selections are possible?
(b) How many selections are possible with two soft-rock records?
(c) How many selections are possible with two hard-rock records and two soft-rock records if one certain jazz record must be played.

38. A basketball conference has eight teams. How many games are scheduled if each team plays every other team twice?

Critical Thinking

In Exercises 39–44, solve for n.

39. $_nP_2 = 182$

40. $_nP_3 = 1320$

41. $_nC_2 = 36$

42. $_nC_3 = 120$

43. $_nC_3 = {_nP_2}$

44. $_nC_3 - {_{n-1}P_2} = 0$

45. Show that $_nC_r = {_nC_{n-r}}$.

46. Show that $_nC_r + {_nC_{r+1}} = {_{n+1}C_{r+1}}$.

47. In how many different ways can *k* distinguishable trees be planted in a circle if $k = 2$? 3? 4? 5? ... *n*? [*Hint:* Two circular permutations are different only if corresponding objects are preceded or followed by different objects. Hence, simply rotating the objects clockwise or counterclockwise one or more positions does not make a different circular permutation.]

48. How many events can be formed, containing at least one member, from a sample space consisting of 2 elements? 3 elements? 4 elements? 5 elements? *n* elements?

Calculator Activities

In Exercises 49–56, use a calculator to help solve each problem.

49. A multiple-choice test has eight questions with four choices for each question.

(a) How many different ways can the test be answered?
(b) How many different ways can the test be answered with every question answered incorrectly?

50. License plate registration numbers in a certain state contain two letters followed by four digits. Find the number of possible license plate registrations that end with an odd digit if letters and digits can be repeated.

51. A college baseball team has 15 players. How many different ways can the 9 starting positions be filled if each player can play any position?

52. In how many ways can a kindergarten teacher assign seats to her 12 students if the classroom has 15 seats.

53. How many ways can 14 professors be assigned to teach 9 sections of college algebra if no professor is assigned more than one section?

54. The Greek alphabet contains 24 letters. How many fraternity names can be formed using (a) any three Greek letters? (b) three different Greek letters?

55. In the game of bridge, 13 cards are dealt to each of 4 players from a deck of 52 cards.

(a) How many different bridge hands are possible?
(b) How many bridge hands consist of 4 aces?
(c) How many bridge hands consist of 8 spades?

56. Suppose 54 Democrats and 46 Republicans sit in the United States Senate. How many ways can a committee of 4 Democrats and 3 Republicans be chosen?

8.8   An Introduction to Probability

◆ Introductory Comments

In the preceding section we discussed sample spaces and events of a random experiment. We also developed several counting techniques that allow us to find the number of elements in a sample space or event. In this section we assign a real number to the probability, or likelihood, of the occurrence of an event resulting from a random experiment. If all elements in the sample space are *equally likely* to occur, then we define this *probability of an event* as follows.

 Probability of an Event

> If S is a finite sample space of a random experiment in which all elements of S are equally likely to occur, and E is an event associated with the experiment, then the **probability of event E**, denoted by $P(E)$, is defined by
>
> $$P(E) = \frac{n(E)}{n(S)},$$
>
> where $n(E)$ is the number of elements in E and $n(S)$ is the number of elements in S.

Although entire textbooks have been written on the subject of probability, the intent of this section is simply to introduce some basic ideas. We begin by looking at some properties of the probability of an event.

◆ Properties of the Probability of an Event

Three important properties of the probability of an event are direct consequences of the definition:

1.   Since E is always a subset of the sample space S,

$$0 \leq n(E) \leq n(S)$$

$$0 \leq \frac{n(E)}{n(S)} \leq 1 \qquad \text{Divide each member by } n(S)$$

$$0 \leq P(E) \leq 1 \qquad \text{Definition of } P(E)$$

Hence, *the probability of an event is a real number between 0 and 1, inclusive.*

2.   If $E = S$, then $n(E) = n(S)$, and

$$P(E) = P(S) = \frac{n(S)}{n(S)} = 1.$$

Hence, *the probability of an event that is certain to occur is equal to 1.*

3. If E has no elements, then $n(E) = 0$, and

$$P(E) = P(\varnothing) = \frac{0}{n(S)} = 0.$$

The set that contains no element is called the *empty set* and is designated by $\varnothing$.

Hence, *the probability of an event that is impossible to occur is equal to* 0.

The more likely an event E is to occur, the closer $P(E)$ is to 1. On the other hand, the less likely an event E is to occur, the closer $P(E)$ is to 0.

EXAMPLE 1   Determine the probability of each event.

(a)   Obtaining a head when tossing a coin.

(b)   Obtaining an even number when rolling a die.

(c)   Obtaining a head and an even number when tossing a coin and then rolling a die.

SOLUTION

(a)   The sample space for this random experiment is $S = \{H, T\}$, where H is heads and T is tails. Hence, $n(S) = 2$. If E is the event of obtaining a head, then $E = \{H\}$. Hence, $n(E) = 1$ and

$$P(E) = \frac{n(E)}{n(S)} = \frac{1}{2}.$$

Thus, the probability of obtaining a head when tossing a coin is $\frac{1}{2}$.

(b)   The sample space for this random experiment is $S = \{1, 2, 3, 4, 5, 6\}$. Hence, $n(S) = 6$. If E is the event of obtaining an even number, then $E = \{2, 4, 6\}$. Hence, $n(E) = 3$ and

$$P(E) = \frac{n(E)}{n(S)} = \frac{3}{6} = \frac{1}{2}.$$

Thus, the probability of obtaining an even number when rolling a die is $\frac{1}{2}$.

(c)   From Example 1 of Section 8.7, the sample space S for this random experiment is

$$S = \{H1, H2, H3, H4, H5, H6, T1, T2, T3, T4, T5, T6\}.$$

Therefore, $n(S) = 12$. If E is the event of tossing a head and rolling an even number, then

$$E = \{H2, H4, H6\}.$$

Hence, $n(E) = 3$ and

$$P(E) = \frac{n(E)}{n(S)} = \frac{3}{12} = \frac{1}{4}.$$

Thus, the probability of tossing a head *and* rolling an even number is $\frac{1}{4}$. ◆

 Always try to select a sample space S for which every outcome is *equally likely to occur*. For instance, suppose a coin is tossed two times and A is the event that exactly one head is obtained.

Wrong analysis: All outcomes are *not* equally likely to occur.

$$S = \{0, 1, 2\} \rightarrow P(A) = \tfrac{1}{3}.$$

Each outcome represents the number of heads that can occur.   Incorrect answer

Correct analysis: All outcomes *are* equally likely to occur.

$$S = \{HH, HT, TH, TT\} \rightarrow P(A) = \tfrac{2}{4} \quad \text{or} \quad \tfrac{1}{2}$$

Each outcome represents the sequence of heads and tails that can occur.   Correct answer

PROBLEM 1   Find the probability of obtaining exactly one head when tossing three coins. ◆

The techniques of counting that we developed in Section 8.7 play an important part in many probability problems, as we illustrate in Example 2.

EXAMPLE 2   The registration number of a Massachusetts license plate consists of three digits followed by three letters. What is the probability that a license plate selected at random displays different digits and different letters?

SOLUTION   From Example 3 in Section 8.7, we have

$$n(S) = 10 \cdot 10 \cdot 10 \cdot 26 \cdot 26 \cdot 26 = 17{,}576{,}000.$$

digits   letters

If E is the event that a license plate registration contains different digits and different letters, then, by the multiplication principle, we have

$$n(E) = 10 \cdot 9 \cdot 8 \cdot 26 \cdot 25 \cdot 24 = 11{,}232{,}000.$$

digits   letters

Hence,

$$P(E) = \frac{n(E)}{n(S)} = \frac{11{,}232{,}000}{17{,}576{,}000} \approx 0.639.$$

Thus, the probability of randomly selecting a license plate containing different digits and different letters is approximately 0.639. ◆

PROBLEM 2   Referring to Example 2, what is the probability that a license plate selected at random displays a number that begins with an odd digit and ends with a vowel? ◆

EXAMPLE 3   In a certain poker game, five cards are dealt from a deck of 52 cards. What is the probability of being dealt a hand that includes three aces?

SOLUTION   From Example 6 of Section 8.7, we have

$$n(S) = {}_{52}C_5 = \binom{52}{5} = 2{,}598{,}960.$$

From 52 cards, choose 5.

If E is the event of being dealt a hand that includes three aces, then

$$n(E) = {}_4C_3 \cdot {}_{48}C_2 = \binom{4}{3} \cdot \binom{48}{2} = 4512.$$

From 4 aces, choose 3.   From 48 non-aces, choose 2.

Hence,

$$P(E) = \frac{n(E)}{n(S)} = \frac{4512}{2{,}598{,}960} \approx 0.0017.$$

Thus, the probability of being dealt three aces is approximately 0.0017. ◆

PROBLEM 3   Referring to Example 3, what is the probability of being dealt a five-card hand that consists of two queens and three kings? ◆

◆ **Addition Rule for Probabilities**

If A and B are events in a sample space S, then $A \cup B$ (read "A **union** B") is the event that occurs if and only if A *or* B (or both) occur, whereas $A \cap B$ (read "A **intersect** B") is the event that occurs if and only if A *and* B occur. **Venn diagrams** of the events $A \cup B$ and $A \cap B$ are shown in Figure 8.6.

As shown in Figure 8.6(a), if we add the number of elements in event A and the number of elements in event B, we find that we have added the number of

FIGURE 8.6
Venn diagrams of the union and
intersection of event A and event B.

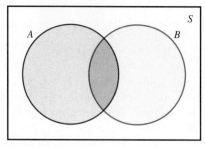

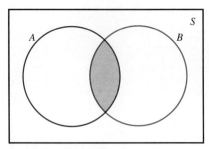

(a) The event $A \cup B$ (b) The event $A \cap B$

elements in $A \cap B$ twice. Thus, to find the number of elements in $A \cup B$, we can add the number of elements in event A and the number of elements in event B, and then subtract the number of elements in $A \cap B$ once:

$$P(A \cup B) = \frac{n(A \cup B)}{n(S)} = \frac{n(A) + n(B) - n(A \cap B)}{n(S)}$$

$$= \frac{n(A)}{n(S)} + \frac{n(B)}{n(S)} - \frac{n(A \cap B)}{n(S)}$$

$$= P(A) + P(B) - P(A \cap B).$$

We refer to this formula as the **addition rule for probabilities**. It is of particular use when finding the probability of event A or event B.

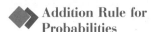

Addition Rule for Probabilities

If A and B are events in a sample space S, then

$$P(A \cup B) = P(A) + P(B) - P(A \cap B).$$

A or B A and B

EXAMPLE 4   A random experiment consists of tossing a coin and then rolling a die. What is the probability of tossing a head *or* rolling an even number?

SOLUTION   From Example 1, we know that the probability of tossing a head is $\frac{1}{2}$, the probability of rolling an even number is $\frac{1}{2}$, and the probability of tossing a head *and* rolling an even number is $\frac{1}{4}$. Letting A be the event of tossing a head and B the event of rolling an even number, we have

$$P(A) = \tfrac{1}{2}, \quad P(B) = \tfrac{1}{2}, \quad \text{and} \quad P(A \cap B) = \tfrac{1}{4}.$$

A and B

Now, using the addition rule for probabilities, we find

$$P(A \cup B) = P(A) + P(B) - P(A \cap B)$$
$$= \tfrac{1}{2} + \tfrac{1}{2} - \tfrac{1}{4}$$
$$= \tfrac{3}{4}$$

$\boxed{A \text{ or } B}$ $\boxed{A \text{ and } B}$

Thus, the probability of tossing a head *or* rolling an even number is $\tfrac{3}{4}$.   ◆

PROBLEM 4   What is the probability of obtaining an ace *or* a spade when drawing a single card from a deck of 52 cards?   ◆

Two events that have no element in common are said to be **mutually exclusive**. In general, if A and B are mutually exclusive events, then $A \cap B = \varnothing$, and $P(A \cap B) = P(\varnothing) = 0$. Hence, for mutually exclusive events, the addition rule for probabilities is

$$P(A \cup B) = P(A) + P(B).$$

EXAMPLE 5   A random experiment consists of rolling a pair of dice and observing the sum of the numbers that appear on the top faces. What is the probability of obtaining a 4 *or* a 5 when rolling a pair of dice?

SOLUTION   We can think of the sample space for this experiment as the ordered pairs (a, b) where a is the number that appears on the first die and b is the number that appears on the second die. Since a and b are integers from 1 to 6, by the multiplication principle, the sample space has $6 \cdot 6$, or 36 elements. Hence, $n(S) = 36$. If A is the event of rolling a sum of 4, then

$$A = \{(1, 3), (2, 2), (3, 1)\}.$$

Hence, $n(A) = 3$ and $P(A) = \tfrac{3}{36}$. If B is the event of rolling a sum of 5, then

$$B = \{(1, 4), (2, 3), (3, 2), (4, 1)\}.$$

Hence, $n(B) = 4$ and $P(B) = \tfrac{4}{36}$. Since it is impossible for the sum to be both 4 *and* 5 when a pair of dice are rolled, we conclude that A and B are mutually exclusive events. Hence,

$$P(A \cap B) = P(\varnothing) = 0.$$

Now, since A and B are mutually exclusive, we have

$$P(A \cup B) = P(A) + P(B) = \tfrac{3}{36} + \tfrac{4}{36} = \tfrac{7}{36}.$$

Thus, the probability of obtaining a 4 *or* a 5 when rolling a pair of dice is $\tfrac{7}{36}$.   ◆

In general, if $A_1, A_2, \ldots, A_k$ are mutually exclusive events, then

$$P(A_1 \cup A_2 \cup \cdots \cup A_k) = P(A_1) + P(A_2) + \cdots + P(A_k).$$

PROBLEM 5   Referring to Example 5, what is the probability of obtaining *at most* a sum of 5? [*Hint:* At most a sum of 5 means a sum of 2 *or* 3 *or* 4 *or* 5.]   ◆

Exercises 8.8

 Basic Skills

In Exercises 1–26, determine the probability of obtaining each event from the given experiment.

1. A vowel; selecting one letter from the first ten letters of the English alphabet

2. A state that begins with M; selecting one state from the six New England states

3. A multiple of three; rolling a single die

4. A red ball; drawing one ball from a bag that contains three red and two black balls

5. A face card; selecting one card from a deck of ordinary (52-card) playing cards

6. A red card that is not a face card; selecting one card from a deck of ordinary playing cards

7. Exactly three heads; tossing a coin three times

8. Exactly two tails; tossing a coin three times

9. A sum of 12; rolling a pair of dice

10. A sum of 8; rolling a pair of dice

11. One tail *or* one head; tossing a coin three times

12. A sum of 4 *or* 9; rolling a pair of dice

13. *At least* two tails; tossing a coin three times

14. *At most* a sum of 3; rolling a pair of dice

15. *At most* three heads; tossing a coin three times

16. A sum of *at least* 13; rolling a pair of dice

17. *At least* a 3, but *at most* a 6; drawing one card from a deck of ordinary playing cards

18. A sum of *at least* 7, but *at most* 10; rolling a pair of dice

19. An even number *and* a vowel; rolling a die and then selecting one letter from the English alphabet

20. A head *and* an ace; tossing a coin, then selecting one card from an ordinary deck of playing cards

21. An even number *or* a vowel; rolling a die, then selecting one letter from the English alphabet

22. A head *or* an ace; tossing a coin, then selecting one card from an ordinary deck of playing cards

23. A red card *or* a jack; drawing a single card from a deck of ordinary playing cards

24. A diamond *or* a face card; drawing a single card from a deck of ordinary playing cards

25. A sum that is an even number *or* a prime number; tossing a pair of dice

26. A sum that is an odd number *or* a multiple of three; tossing a pair of dice

In Exercises 27–32, use the counting techniques discussed in Section 8.7 to help answer each question.

27. A true-false test of five questions is given to a student that has no knowledge in the subject matter being tested. What is the probability that the student answers (a) every question correctly? (b) one question correctly?

28. A multiple-choice test of five questions with four choices for each question is given to a student that has no knowledge in the subject matter being tested. What is the probability that the student answers (a) every question correctly? (b) no question correctly?

29. Three girls and four boys are placed randomly in a row of seven seats. What is the probability that (a) the girls and boys sit in alternating seats? (b) the three girls sit together?

30. Suppose a permutation of the word *APRIL* is selected at random. What is the probability that (a) the word begins with a consonant? (b) the word has alternating consonants and vowels?

31. Nine equally qualified students, five boys and four girls, apply for three scholarships. What is the probability that (a) two girls receive scholarships? (b) no girl receives a scholarship?

32. From a collection of records consisting of three hard-rock records, three soft-rock records, and two jazz records, a disc jockey selects five records at random to play. What is the probability that he plays (a) exactly three soft-rock records? (b) at most two hard-rock records?

 Critical Thinking

33. Given that $P(A) = \frac{2}{5}$, $P(A \cup B) = \frac{3}{4}$, and $P(A \cap B) = \frac{1}{10}$, find $P(B)$.

34. Given that $P(A) = \frac{1}{3}$, $P(A \cup B) = 3P(B)$, and $P(A \cap B) = \frac{1}{8}$, find $P(B)$.

In Exercises 35 and 36, set up a Venn diagram to help solve each problem.

35. Of 100 high school seniors, 42 studied physics, 68 studied chemistry, and 30 studied both physics and chemistry. If one of these seniors is selected at random, what is the probability that the student (a) studied physics *or* chemistry? (b) studied neither subject?

36 At a certain hospital, it is known that $\frac{1}{4}$ of the doctors are under 35 years of age and that $\frac{4}{5}$ of the doctors are male. It is also known that $\frac{9}{10}$ of the doctors are under 35 years of age *or* male. What is the probability that a doctor selected at random is (a) under 35 years of age *and* male? (b) at least 35 years of age *and* female?

37. The *odds in favor* of an event occurring is the ratio of the probability that it does happen to the probability that it doesn't happen.

(a) What are the odds in favor of throwing a sum of 7 on a single throw of a pair of dice?

(b) If the odds in favor of a horse winning a race are 3 to 2, what is the probability that the horse wins the race?

38. The coefficients of the terms in the expansion of $(H + T)^n$ give the number of ways in which k heads H and $(n - k)$ tails T can occur when tossing a coin n times, $n \geq k$. For example, the second term in the expansion of $(H + T)^3$ tells us that there are

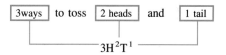

namely, HHT, HTH, THH. Use this idea to help find the probability of obtaining 7 heads and 3 tails when tossing a coin 10 times.

 Calculator Activities

In Exercises 39–44, use a calculator to help solve each problem. Round each answer to three significant digits.

39. Merchandise in a certain department store is coded with two letters followed by a two-digit number that does not begin with zero. What is the probability that an item selected at random has (a) different letters and an odd number? (b) repeated letters and a multiple of 5?

40. In a certain state lottery game, the player selects six different numbers from the integers 1 to 40. If all six numbers match, the player wins $1,000,000, and if any five of six numbers match, the player wins $1000. With one lottery ticket, what is the probability of winning (a) $1,000,000? (b) $1000?

41. A class reunion is held, and 40 married females, 40 married males, 20 single females, and 15 single males attend. Each person at the reunion is given a numbered ticket for a chance to win one of three door prizes. What is the probability that (a) three married females win door prizes? (b) two single females win door prizes?

42. In the game of bridge, 13 cards are dealt from an ordinary deck of playing cards. What is the probability of being dealt a hand that contains (a) 4 aces? (b) 3 queens and 3 kings?

43. In a certain poker game, five cards are dealt from a deck of ordinary playing cards. What is the probability of being dealt a hand that contains (a) two jacks *or* two queens? (b) at least one ace?

44. Of the 19 army jeeps being sold at an auction, two have worn-out transmissions. If a company makes a purchase of 4 of these jeeps, what is the probability that they purchase (a) no jeep with a worn-out transmission? (b) at least one jeep with a worn-out transmission?

Review Exercises

In Exercises 1 and 2, graph the first six elements of each sequence function, and determine whether the function is increasing, decreasing, or neither.

1. $f(n) = \dfrac{n+1}{n}$;   domain: $\{1, 2, 3, 4, 5, 6\}$

2. $f(n) = (-1)^{n+1}(2n)$

In Exercises 3–8, find the first six elements of a sequence with the given general element.

3. $a_n = 3n - 7$

4. $a_n = 6(-\tfrac{1}{2})^{n-1}$

5. $a_n = (-1)^{n+1}n!$

6. $a_n = \dfrac{n}{n+2}$

7. $a_n = \begin{cases} n-1 & \text{if } n \text{ is odd} \\ (a_{n-1})^2 & \text{if } n \text{ is even} \end{cases}$

8. $a_n = \begin{cases} 3 & \text{if } n = 1 \\ 5 & \text{if } n = 2 \\ a_{n-1} + a_{n-2} & \text{if } n \geq 3 \end{cases}$

In Exercises 9–22,

(a) determine whether each sequence appears to be arithmetic, geometric, or neither.

(b) use pattern recognition to find the general element of each sequence.

9. $-1, 3, -5, 7, -9, 11, -13, 15, \dots$

10. $2, -4, 6, -8, 10, -12, \dots$

11. $3, 9, 15, 21, 27, 33, 39, \dots$

12. $49, 38, 27, 16, 5, -6, -17, -28, \dots$

13. $768, 192, 48, 12, 3, \dots$

14. $5, 15, 45, 135, 405, 1215, \dots$

15. $\tfrac{1}{2}, -\tfrac{3}{4}, \tfrac{5}{6}, -\tfrac{7}{8}, \tfrac{9}{10}, \dots$ **16.** $3, \tfrac{4}{3}, 1, \tfrac{6}{7}, \tfrac{7}{9}, \dots$

17. $-4, 2, -1, \tfrac{1}{2}, -\tfrac{1}{4}, \tfrac{1}{8}, \dots$

18. $2500, 25, 0.25, 0.0025, 0.000025, \dots$

19. $\tfrac{13}{3}, 3, \tfrac{5}{3}, \tfrac{1}{3}, -1, \dots$ **20.** $-\tfrac{3}{2}, -\tfrac{1}{2}, \tfrac{1}{2}, \tfrac{3}{2}, \tfrac{5}{2}, \tfrac{7}{2}, \dots$

21. $0, 3, 8, 15, 24, 35, 48, 63, 80, \dots$

22. $1, 1, 2, 6, 24, 120, 720, \dots$

In Exercises 23–34,

(a) determine whether each series is arithmetic, geometric, or neither.

(b) find the sum of each series.

23. $\displaystyle\sum_{i=1}^{35} 2i$

24. $\displaystyle\sum_{i=1}^{20} (7 - 3i)$

25. $\displaystyle\sum_{i=2}^{10} 8\left(-\frac{3}{2}\right)^{i-1}$

26. $\displaystyle\sum_{i=0}^{12} (-3)^i$

27. $\displaystyle\sum_{i=1}^{18} 3i^2$

28. $\displaystyle\sum_{i=1}^{24} i(i - 3)$

29. $\displaystyle\sum_{i=1}^{\infty} 5\left(\frac{1}{2}\right)^{i-1}$

30. $\displaystyle\sum_{i=2}^{\infty} \frac{1}{(-2)^i}$

31. $\displaystyle\sum_{i=0}^{30} \left(\frac{i}{3} - 2\right)$

32. $\displaystyle\sum_{i=0}^{32} \frac{5i + 4}{6}$

33. $\displaystyle\sum_{i=1}^{16} (2i^3 - 3i + 5)$

34. $\displaystyle\sum_{i=1}^{6} \frac{(-1)^{i+1}2^i}{i}$

In Exercises 35–40, prove each statement using mathematical induction.

35. $2 + 8 + 14 + 20 + \cdots + (6n - 4) = 3n^2 - n$

36. $6 + 11 + 16 + 21 + \cdots + (5n + 1) = \dfrac{n(5n + 7)}{2}$

37. $1 \cdot 2 + 2 \cdot 3 + 3 \cdot 4 + 4 \cdot 5 + \cdots + n(n + 1)$
$= \dfrac{n(n + 1)(n + 2)}{3}$

38. $1 \cdot 2 \cdot 3 + 2 \cdot 3 \cdot 4 + 3 \cdot 4 \cdot 5 + 4 \cdot 5 \cdot 6 + \cdots$
$+ n(n + 1)(n + 2) = \dfrac{n(n + 1)(n + 2)(n + 3)}{4}$

39. $1^3 + 3^3 + 5^3 + 7^3 + \cdots + (2n - 1)^3 = n^2(2n^2 - 1)$

40. $\dfrac{1}{2} + \dfrac{2}{2^2} + \dfrac{3}{2^3} + \dfrac{4}{2^4} + \cdots + \dfrac{n}{2^n} = 2 - \dfrac{n + 2}{2^n}$

In Exercises 41–44, use the formulas of Exercises 35–40 to find the sum of the given series.

41. $\displaystyle\sum_{i=1}^{28} i(i + 1)$

42. $\displaystyle\sum_{i=1}^{36} i(i + 1)(i + 2)$

43. $\displaystyle\sum_{i=1}^{15} (2i - 1)^3$

44. $\displaystyle\sum_{i=1}^{12} \frac{i}{2^i}$

In Exercises 45 and 46, write each arithmetic series in sigma form using (a) $i = 1$ and (b) $i = 0$ as the lower value of the summation variable.

45. $5 + 9 + 13 + \cdots + 1001$

46. $88 + 76 + 64 + 52 + \cdots$

In Exercises 47 and 48, write each geometric series in sigma form using i = 1 as the lower value of the summation variable.

47. $243 - 162 + 108 - 72 + \cdots$

48. $2 + 6 + 18 + 54 + \cdots + 9,565,938$

In Exercises 49 and 50,

(a) *find the sequence of partial sums for each infinite geometric series.*

(b) *find* $\lim\limits_{n \to \infty} S_n$, *if it exists.*

49. $384 - 96 + 24 - 6 + \cdots$

50. $\sum\limits_{i=1}^{\infty} \left(\dfrac{5}{4}\right)^{i-1}$

51. What is the 26th element of an arithmetic sequence whose first element is 92 and whose common difference is -7?

52. What is the general element of an arithmetic sequence whose 6th element is 34 and 12th element is 52?

53. What is the sum of the first 60 terms of an arithmetic series whose first term is 9 and second term is 24?

54. The first term of an arithmetic series is -19 and the sum of the first 30 terms is 300. What is the 15th term of the series?

55. What is the first element of a geometric sequence whose 11th element is $\frac{3}{32}$ and whose common ratio is $\frac{1}{2}$?

56. What is the general element of a geometric sequence whose second element is 12 and eighth element is 8748?

57. Find x so that $x - 3$, $2x - 3$, and $6x + 1$ are the first three terms of a geometric series.

58. The sum of the first three terms of a geometric series is 26, and the third term is 16 more than the first term. What is the general term of the series?

59. What is the sum of an infinite geometric series whose first term is $\frac{27}{16}$ and fifth term is $\frac{1}{3}$?

60. What is the common ratio of an infinite geometric series in which each term is three times the sum of all the terms that follow it?

61. Find the rational number represented by each repeating decimal:
(a) $3.030303\ldots$ (b) $2.7\overline{306}$

62. Find the interval of convergence and the sum of each infinite series:

(a) $9 + 6x + 4x^2 + \cdots$
(b) $3 - 9(x - 1) + 27(x - 1)^2 - \cdots$

63. A Christmas tree farm has 20 trees in the first row, and each row thereafter has 6 more trees than the preceding row. If the farm has 116 rows of trees, how many trees does it have altogether?

64. A car costing $12,000 is purchased under an agreement to pay $250 per month plus $\frac{3}{4}\%$ interest on the unpaid balance for that month. Find the total amount paid for the car.

65. To save for a downpayment on a home, a married couple deposits $300 each month into an account that earns interest at the rate of 8% per year, compounded monthly. How much money will be in the account at the end of 3 years, if (a) the account is empty today? (b) the account has $6000 today?

66. The enrollment at a certain college that presently has 5411 students is decreasing by 4% each year. What will be the enrollment five years from now?

67. A boy pedals his bicycle up a 105-foot hill at the rate of 36 feet the first second, 24 feet the second second, 16 feet the third second, and so on. Does he ever reach the top of the hill?

68. A tennis ball is dropped from a height of 16 feet, and rebounds each time $\frac{1}{4}$ of the height from which it last fell. Find the total distance the ball travels before coming to rest.

In Exercises 69–74, evaluate each binomial coefficient expression.

69. $\dbinom{16}{1}$

70. $\dbinom{22}{0}$

71. $\dbinom{13}{11}$

72. $\dbinom{20}{4}$

73. $\dbinom{15}{5} + \dbinom{15}{4} - \dbinom{16}{5}$

74. $\dbinom{7}{2}\dbinom{3}{1} \bigg/ \dbinom{10}{3}$

75. Expand and simplify $(3x - 2)^6$.

76. Expand and simplify $(2 + a^{1/2})^8$.

77. Find the fourth term of $(a^{-1} + b)^{10}$.

78. Find the tenth term of $(4x^2 - 1)^{12}$.

79. At a certain college a freshman can choose from three math courses, four foreign language courses, three English courses, five science courses, and six social study courses. If a freshman's schedule must contain one course from each group, how many different schedules are possible?

80. How many ways can ten people be assigned to play the four infield positions of a baseball team, if each person can play any infield position?

81. How many ways can six sedans, three coupes, and four station wagons be lined up in a row at a dealership, if one does not distinguish between vehicles of the same type?

82. A college athletic director must hire three assistant football coaches from 12 equally qualified candidates. How many ways can he select the 3 coaches?

83. (a) Determine the sample space S associated with the following random experiment: Tossing a coin once and recording whether it is heads (H) or tails (T). If the coin is heads, then rolling a die once and observing the number that appears on the top face.
    (b) Is each element in S equally likely to occur? Explain.

84. What is the probability of obtaining exactly two heads when tossing a coin four times?

85. What is the probability of obtaining "doubles" (the same number appears on the top face of both dice) when rolling a pair of dice?

86. What is the probability of obtaining a tail *or* a diamond when tossing a coin, then drawing a single card from a deck of ordinary playing cards?

87. Of ten balls in a bag, three are red. A person selects three balls from the bag, one after the other, without replacement. What is the probability of obtaining (a) three red balls? (b) at most one red ball?

88. To pass a true-false quiz of ten questions, at least 7 questions must be answered correctly. What is the probability of passing the quiz if one guesses at each question?

Cumulative Review Exercises

Chapters 7 & 8

1. Find $f(1) + f(2) + f(3) + \cdots + f(20)$ for the given function f.

   (a) $f(x) = 2x$ (b) $f(x) = 2^x$

2. The sum of a finite arithmetic series with n terms is n^2. Find the sum of the first and last terms of the series.

3. (a) Use mathematical induction to show that

$$\sum_{i=1}^{n} i^4 = 1^4 + 2^4 + 3^4 + \cdots + n^4$$

$$= \frac{n(n+1)(2n+1)(3n^2 + 3n - 1)}{30}.$$

   (b) Find the sum of the series $1^4 + 2^4 + 3^4 + \cdots + 20^4$.

4. (a) Use mathematical induction to show that

$$\sum_{i=1}^{n} i^5 = 1^5 + 2^5 + 3^5 + \cdots + n^5$$

$$= \frac{n^2(n+1)^2(2n^2 + 2n - 1)}{12}.$$

   (b) Find the sum of the series $1^5 + 2^5 + 3^5 + \cdots + 20^5$.

5. The nth partial sum of an infinite series is $S_n = 4[1 - (\tfrac{2}{3})^n]$. What is the sum of the series?

6. How many consecutive multiples of 6, beginning with 6, must be taken so that their sum is 7650?

7. Solve each system of equations.

   (a) $\dfrac{2}{x} + \dfrac{3}{y} + \dfrac{1}{z} = 4$

   $\dfrac{3}{x} - \dfrac{5}{y} + \dfrac{2}{z} = -5$

   $\dfrac{4}{x} - \dfrac{6}{y} + \dfrac{3}{z} = -7$

   (b) $x - y \quad\; - 2w = 2$

   $\quad\; 2y + 3z + 2w = -8$

   $x \quad\;\; + 2z - 4w = 1$

   $x + 4y + z \quad\;\;\; = 6$

8. Given that the sequence a^2, b^2, c^2 is an arithmetic sequence, show that

$$\frac{1}{b+c}, \quad \frac{1}{a+c}, \quad \frac{1}{a+b}$$

   is also an arithmetic sequence.

9. Given the matrices

$$A = \begin{bmatrix} 7 & -2 & 6 \\ 2 & 3 & 5 \\ 5 & -1 & 4 \end{bmatrix}, \quad X = \begin{bmatrix} x \\ y \\ z \end{bmatrix}, \quad B = \begin{bmatrix} 5 \\ 2 \\ 5 \end{bmatrix},$$

   (a) find A^{-1} and (b) solve the matrix equation $AX = B$ for X.

10. Find x and y if $x + 3y$, $2x - y$, $x - 8y$, and $8x + 6$ are the first four elements in an arithmetic sequence.

11. Find x and y if 4, x, and y are the first three elements in an arithmetic sequence and x, y, and 18 are the first three elements in a geometric sequence.

12. Solve for x.

   (a) $\begin{vmatrix} x & 2 \\ 1 & 3 \end{vmatrix} = 7$ (b) $\begin{vmatrix} 2 & x & 3 \\ -2 & 4 & x \\ 5 & -3 & -2 \end{vmatrix} = -51$

13. Determine the nature of the sequence $\log a_1$, $\log a_2$, $\log a_3$, $\log a_4$, $\ldots$, if $a_1, a_2, a_3, a_4, \ldots$ is a geometric sequence with positive elements and common ratio r.

14. Determine the nature of the sequence 10^{a_1}, 10^{a_2}, 10^{a_3}, $10^{a_4}, \ldots$, if $a_1, a_2, a_3, a_4, \ldots$ is an arithmetic sequence with common difference d.

15. Perform the indicated operations, given that matrices A and B are defined as follows.

$$A = \begin{bmatrix} 2 & 6 & -1 & 0 \\ 5 & 5 & -4 & 2 \\ 0 & 8 & 2 & -3 \\ 6 & 8 & 2 & -3 \end{bmatrix}$$

$$B = \begin{bmatrix} 5 & -1 & 4 & -7 \\ 3 & -1 & 0 & 3 \\ 9 & 2 & 4 & 0 \\ -6 & -3 & 7 & 2 \end{bmatrix}$$

   (a) $A + B$ (b) $3B - 2A$ (c) AB (d) BA

16. Find the interval of convergence and the sum of the infinite geometric series

$$\frac{1}{x+1} + \frac{x}{(x+1)^2} + \frac{x^2}{(x+1)^3} + \cdots$$

17. The product of the first seven terms in a geometric series is 0.0002187. What is the fourth term of the series?

18. Find the sum of each infinite geometric series.

   (a) $2, 2 - \sqrt{2}, \dfrac{\sqrt{2} - 1}{\sqrt{2} + 1}, \ldots$ (b) $\displaystyle\sum_{i=1}^{\infty} \left[\left(\dfrac{1}{3}\right)^{i-1} + \left(\dfrac{1}{2}\right)^{i} \right]$

19. Evaluate each series.

   (a) $\displaystyle\sum_{k=0}^{6} \binom{6}{k}$ (b) $\displaystyle\sum_{k=0}^{6} \left[\binom{6}{k} + \binom{6}{6-k} \right]$

   (c) $\displaystyle\sum_{k=0}^{6} \left[\binom{6}{k} - \binom{6}{6-k} \right]$ (d) $\displaystyle\sum_{k=0}^{6} \left[\binom{6}{k} \binom{6}{6-k} \right]$

20. Find the middle term in the expansion of $(x^{-2} - 2\sqrt{x}\, y)^{20}$.

21. The sides of a right triangle form an arithmetic sequence with common difference 6. Find the lengths of the sides of the triangle.

22. How long does it take for an annuity to amount to a million dollars if $600 is invested each month into an account that earns interest at the rate of 9% per year, compounded monthly?

23. Suppose 12 points lie in a plane, but no 3 points lie on the same straight line.

   (a) How many different lines are determined by the 12 points?
   (b) How many different triangles are determined by the 12 points?

24. It is estimated that a new car costing $12,000 depreciates in value by 25% each year. What is its approximate value after seven years?

25. A mechanic connects wires from the distributor of a four-cylinder car to the spark plugs without knowing the firing order of the cylinders. What is the probability of connecting the wires in the correct order?

26. Two cards are drawn from a deck of 52 cards, one after the other, without replacement. What is the probability that both cards are greater than 3 but less than 7?

27. The probability that a town will build a playground in Gaffield Park is 0.8, the probability that it will build a playground in Miller Park is 0.3, and the probability is 0.9 that it will build a playground in either Gaffield Park or Miller Park or both. What is the probability that a playground will be built (a) in both parks? (b) in neither park?

28. An equilateral triangle has sides of length 12 cm. Inside this triangle, a second equilateral triangle is constructed by joining the midpoints of the sides of the first triangle. Inside the second triangle a third equilateral triangle is constructed by joining the midpoints of the sides of the second triangle, and so on, indefinitely.

   (a) What is the sum of the perimeters of all the equilateral triangles?
   (b) What is the sum of the areas of all the equilateral triangles?

29. The sum of the first four terms of a series is $\frac{25}{12}$. The first term is twice the second term, and the second term is twice the fourth term. Six times the third term is one less than twice the sum of the first two terms. Determine the first four terms of the series, and then use pattern recognition to describe the general term.

30. A company manufactures different types of prefinished oak flooring and packs the flooring in 30-square-foot cartons for shipping. The company receives an order for 65 cartons of a parquet flooring from customer A and an order for 90 cartons of the same parquet floor from customer B. The company has 100 cartons of this product at warehouse W_1 and 70 cartons at warehouse W_2. The shipping costs from the warehouses to the customers are given in the following table.

Warehouse	Customer	Shipping cost per carton
W_1	A	$1.25
W_2	A	$1.40
W_1	B	$1.30
W_2	B	$1.50

   (a) How should the orders be filled to minimize the shipping cost?
   (b) What is the minimum shipping cost?

A

Significant Digits

Consider measuring the width of the wooden block in Figure A.1 with a ruler marked in intervals of 0.1-inch. The width of the block appears to be 1.8 inches. However, does the end of the block fall exactly in the middle of the marking for 1.8 inches, or slightly to the left or right of this marking? If we use a powerful magnifying glass, we might attempt to answer this question, but we could never determine the *exact* width of the block. For this reason, we say that every number found by a measuring process is an *approximate number*.

FIGURE A.1
Measuring the width of a block with a ruler marked in 0.1-inch intervals.

The number 8 in the tenths position of the number 1.8 inches does have some *significance*, since the width seems to be closer to 1.8 inches than to either 1.7 inches or 1.9 inches. If a digit contributes to our knowledge of how good an approximation is, it is called a **significant digit**. Throughout this text, you are asked to round answers to a certain number of significant digits. The following may be used as a guide for this purpose.

Significant Digits

1. All nonzero digits are significant.

Examples: 475 has 3 significant digits, 12.827 has 5 significant digits.

2. Zeros between nonzeros digits are significant.

Examples: 6506 has 4 significant digits, 42.0072 has 6 significant digits.

3. Zeros appearing at the end of a decimal fraction are significant.

Examples: 7.00 has 3 significant digits, 76.40 has 4 significant digits.

4. Zeros at the beginning of a decimal fraction are *not* significant and serve only to locate the decimal point correctly.

Examples: 0.002 has 1 significant digit, 0.023 has 2 significant digits.

5. Zeros at the end of an integer are *not* significant unless a tilde (~) is placed above one of the zeros. The tilde is placed over the last zero that is significant.

Examples: 22,000 has 2 significant digits, 22,0̃00 has 4 significant digits.

If a number is written in scientific notation as

$$k \times 10^n, \quad \text{where} \quad 1 \le |k| < 10 \text{ and } n \text{ is an integer,}$$

then the number of significant digits in that number is the same as the number of significant digits in k.

Examples: 6.2×10^3 $= 6200$ has 2 significant digits.

6.20×10^3 $= 62\tilde{0}0$ has 3 significant digits.

6.2×10^{-3} $= 0.0062$ has 2 significant digits.

$6.20 \times 10^{-3} = 0.00620$ has 3 significant digits.

To avoid *rounding errors* when working with approximate numbers, we carry along a few extra significant digits through the calculating process and then round the final answer to the desired accuracy. For example, suppose we wish to find the area of the gable end of the house shown in Figure A.2 and round this answer to three significant digits. First, we find the area of the rectangular part of the gable, without rounding the answer, as follows:

$$A = lw = (24.26 \text{ ft})(9.78 \text{ ft}) \approx 237.2628 \text{ sq ft.}$$

Next, we find the area of the triangular part of the gable, without rounding the answer, as follows:

$$A = \tfrac{1}{2}bh = \tfrac{1}{2}(24.26 \text{ ft})(8.52 \text{ ft}) \approx 103.3476 \text{ sq ft.}$$

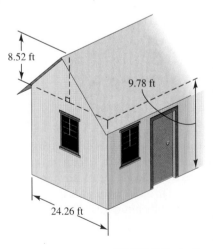

FIGURE A.2
Dimension of the gable end of a house

Finally, we add these areas and then round to the desired accuracy of three significant digits. Hence, the area of the gable end of the house is

$$237.2628 \text{ sq ft} + 103.3476 \text{ sq ft} = 340.6104 \text{ sq ft} \approx 341 \text{ sq ft.}$$

Round to 3 significant digits

If we round the areas of the rectangular and triangular parts to three significant digits and then add these areas, we accumulate a *rounding error*. Note that

$$237 \text{ sq ft} + 103 \text{ sq ft} = 34\tilde{0} \text{ sq ft,}$$

which is not the desired answer. When working with approximate numbers, we usually use a calculator to perform the basic operations. It is best to let the calculator store all the digits and then round the final display on the calculator to the desired accuracy. By using this procedure, you will be sure to avoid rounding errors.

Some exercises in this text that deal with approximate numbers may not specify a required number of significant digits in the final answer. For exercises of this nature, look at the data given in the problem and determine which approximate number has the *fewest* number of significant digits. It is common practice to round the final answer to the same number of significant digits as the approximate number with the fewest number of significant digits. For example, note in Figure A.2 that both 8.52 ft and 9.78 ft have the fewest number of significant digits, namely, three. Hence, it is acceptable to round the area of the gable to three significant digits as well.

B

Exponential and Logarithmic Tables

Table 1   Values of e^x and e^{-x}

x	e^x	e^{-x}	x	e^x	e^{-x}	x	e^x	e^{-x}	x	e^x	e^{-x}
0.00	1.0000	1.0000	0.25	1.2840	0.7788	2.0	7.3891	0.1353	4.5	90.017	0.0111
0.01	1.0101	0.9901	0.30	1.3499	0.7408	2.1	8.1662	0.1225	4.6	99.484	0.0101
0.02	1.0202	0.9802	0.35	1.4191	0.7047	2.2	9.0250	0.1108	4.7	109.95	0.0091
0.03	1.0305	0.9704	0.40	1.4918	0.6703	2.3	9.9742	0.1003	4.8	121.51	0.0082
0.04	1.0408	0.9608	0.45	1.5683	0.6376	2.4	11.023	0.0907	4.9	134.29	0.0074
0.05	1.0513	0.9512	0.50	1.6487	0.6065	2.5	12.182	0.0821	5.0	148.41	0.0067
0.06	1.0618	0.9418	0.55	1.7333	0.5769	2.6	13.464	0.0743	5.5	244.69	0.0041
0.07	1.0725	0.9324	0.60	1.8221	0.5488	2.7	14.880	0.0672	6.0	403.43	0.0025
0.08	1.0833	0.9231	0.65	1.9155	0.5220	2.8	16.445	0.0608	6.5	665.14	0.0015
0.09	1.0942	0.9139	0.70	2.0138	0.4966	2.9	18.174	0.0550	7.0	1096.6	0.0009
0.10	1.1052	0.9048	0.75	2.1170	0.4724	3.0	20.086	0.0498	7.5	1808.0	0.0006
0.11	1.1163	0.8958	0.80	2.2255	0.4493	3.1	22.198	0.0450	8.0	2981.0	0.0003
0.12	1.1275	0.8869	0.85	2.3396	0.4274	3.2	24.533	0.0408	8.5	4914.8	0.0002
0.13	1.1388	0.8781	0.90	2.4596	0.4066	3.3	27.113	0.0369	9.0	8103.1	0.0001
0.14	1.1503	0.8694	0.95	2.5857	0.3867	3.4	29.964	0.0334	10.0	22026	0.00005
0.15	1.1618	0.8607	1.0	2.7183	0.3679	3.5	33.115	0.0302			
0.16	1.1735	0.8521	1.1	3.0042	0.3329	3.6	36.598	0.0273			
0.17	1.1853	0.8437	1.2	3.3201	0.3012	3.7	40.447	0.0247			
0.18	1.1972	0.8353	1.3	3.6693	0.2725	3.8	44.701	0.0224			
0.19	1.2092	0.8270	1.4	4.0552	0.2466	3.9	49.402	0.0202			
0.20	1.2214	0.8187	1.5	4.4817	0.2231	4.0	54.598	0.0183			
0.21	1.2337	0.8106	1.6	4.9530	0.2019	4.1	60.340	0.0166			
0.22	1.2461	0.8025	1.7	5.4739	0.1827	4.2	66.686	0.0150			
0.23	1.2586	0.7945	1.8	6.0496	0.1653	4.3	73.700	0.0136			
0.24	1.2712	0.7866	1.9	6.6859	0.1496	4.4	81.451	0.0123			

Table 2   Values of $\ln x$

x	$\ln x$	x	$\ln x$	x	$\ln x$
		4.5	1.5041	9.0	2.1972
0.1	−2.3026	4.6	1.5261	9.1	2.2083
0.2	−1.6094	4.7	1.5476	9.2	2.2192
0.3	−1.2040	4.8	1.5686	9.3	2.2300
0.4	−0.9163	4.9	1.5892	9.4	2.2407
0.5	−0.6931	5.0	1.6094	9.5	2.2513
0.6	−0.5108	5.1	1.6292	9.6	2.2618
0.7	−0.3567	5.2	1.6487	9.7	2.2721
0.8	−0.2231	5.3	1.6677	9.8	2.2824
0.9	−0.1054	5.4	1.6864	9.9	2.2925
1.0	0.0000	5.5	1.7047	10	2.3026
1.1	0.0953	5.6	1.7228	11	2.3979
1.2	0.1823	5.7	1.7405	12	2.4849
1.3	0.2624	5.8	1.7579	13	2.5649
1.4	0.3365	5.9	1.7750	14	2.6391
1.5	0.4055	6.0	1.7918	15	2.7081
1.6	0.4700	6.1	1.8083	16	2.7726
1.7	0.5306	6.2	1.8245	17	2.8332
1.8	0.5878	6.3	1.8405	18	2.8904
1.9	0.6419	6.4	1.8563	19	2.9444
2.0	0.6931	6.5	1.8718	20	2.9957
2.1	0.7419	6.6	1.8871	25	3.2189
2.2	0.7885	6.7	1.9021	30	3.4012
2.3	0.8329	6.8	1.9169	35	3.5553
2.4	0.8755	6.9	1.9315	40	3.6889
2.5	0.9163	7.0	1.9459	45	3.8067
2.6	0.9555	7.1	1.9601	50	3.9120
2.7	0.9933	7.2	1.9741	55	4.0073
2.8	1.0296	7.3	1.9879	60	4.0943
2.9	1.0647	7.4	2.0015	65	4.1744
3.0	1.0986	7.5	2.0149	70	4.2485
3.1	1.1314	7.6	2.0281	75	4.3175
3.2	1.1632	7.7	2.0412	80	4.3820
3.3	1.1939	7.8	2.0541	85	4.4427
3.4	1.2238	7.9	2.0669	90	4.4998
3.5	1.2528	8.0	2.0794	95	4.5539
3.6	1.2809	8.1	2.0919	100	4.6052
3.7	1.3083	8.2	2.1041		
3.8	1.3350	8.3	2.1163		
3.9	1.3610	8.4	2.1282		
4.0	1.3863	8.5	2.1401		
4.1	1.4110	8.6	2.1518		
4.2	1.4351	8.7	2.1633		
4.3	1.4586	8.8	2.1748		
4.4	1.4816	8.9	2.1861		

Table 3   Values of log *x*

x	0	1	2	3	4	5	6	7	8	9
1.0	.0000	.0043	.0086	.0128	.0170	.0212	.0253	.0294	.0334	.0374
1.1	.0414	.0453	.0492	.0531	.0569	.0607	.0645	.0682	.0719	.0755
1.2	.0792	.0828	.0864	.0899	.0934	.0969	.1004	.1038	.1072	.1106
1.3	.1139	.1173	.1206	.1239	.1271	.1303	.1335	.1367	.1399	.1430
1.4	.1461	.1492	.1523	.1553	.1584	.1614	.1644	.1673	.1703	.1732
1.5	.1761	.1790	.1818	.1847	.1875	.1093	.1931	.1959	.1987	.2014
1.6	.2041	.2068	.2095	.2122	.2148	.2175	.2201	.2227	.2253	.2279
1.7	.2304	.2330	.2355	.2380	.2405	.2430	.2455	.2480	.2504	.2529
1.8	.2553	.2577	.2601	.2625	.2648	.2672	.2695	.2718	.2742	.2765
1.9	.2788	.2810	.2833	.2856	.2878	.2900	.2923	.2945	.2967	.2989
2.0	.3010	.3032	.3054	.3075	.3096	.3118	.3139	.3160	.3181	.3201
2.1	.3222	.3243	.3263	.3284	.3304	.3324	.3345	.3365	.3385	.3404
2.2	.3424	.3444	.3464	.3483	.3502	.3522	.3541	.3560	.3579	.3598
2.3	.3617	.3636	.3655	.3674	.3692	.3711	.3729	.3747	.3766	.3784
2.4	.3802	.3820	.3838	.3856	.3874	.3892	.3909	.3927	.3945	.3962
2.5	.3979	.3997	.4014	.4031	.4048	.4065	.4082	.4099	.4116	.4133
2.6	.4150	.4166	.4183	.4200	.4216	.4232	.4249	.4265	.4281	.4298
2.7	.4314	.4330	.4346	.4362	.4378	.4393	.4409	.4425	.4440	.4456
2.8	.4472	.4487	.4502	.4518	.4533	.4548	.4564	.4579	.4594	.4609
2.9	.4624	.4639	.4654	.4669	.4683	.4698	.4713	.4728	.4742	.4757
3.0	.4771	.4786	.4800	.4814	.4829	.4843	.4857	.4871	.4886	.4900
3.1	.4914	.4928	.4942	.4955	.4969	.4983	.4997	.5011	.5024	.5038
3.2	.5051	.5065	.5079	.5092	.5105	.5119	.5132	.5145	.5159	.5172
3.3	.5185	.5198	.5211	.5224	.5237	.5250	.5263	.5276	.5289	.5302
3.4	.5315	.5328	.5340	.5353	.5366	.5378	.5391	.5403	.5416	.5428
3.5	.5441	.5453	.5465	.5478	.5490	.5502	.5514	.5527	.5539	.5551
3.6	.5563	.5575	.5587	.5599	.5611	.5623	.5635	.5647	.5658	.5670
3.7	.5682	.5694	.5705	.5717	.5729	.5740	.5752	.5763	.5775	.5786
3.8	.5798	.5809	.5821	.5832	.5843	.5855	.5866	.5877	.5888	.5899
3.9	.5911	.5922	.5933	.5944	.5955	.5966	.5977	.5988	.5999	.6010
4.0	.6021	.6031	.6042	.6053	.6064	.6075	.6085	.6096	.6107	.6117
4.1	.6128	.6138	.6149	.6160	.6170	.6180	.6191	.6201	.6212	.6222
4.2	.6232	.6243	.6253	.6263	.6274	.6284	.6294	.6304	.6314	.6325
4.3	.6335	.6345	.6355	.6365	.6375	.6385	.6395	.6405	.6415	.6425
4.4	.6435	.6444	.6454	.6464	.6474	.6484	.6493	.6503	.6513	.6522
4.5	.6532	.6542	.6551	.6561	.6571	.6580	.6590	.6599	.6609	.6618
4.6	.6628	.6637	.6646	.6656	.6665	.6675	.6684	.6693	.6702	.6712
4.7	.6721	.6730	.6739	.6749	.6758	.6767	.6776	.6785	.6794	.6803
4.8	.6812	.6821	.6830	.6839	.6848	.6857	.6866	.6875	.6884	.6893
4.9	.6902	.6911	.6920	.6928	.6937	.6946	.6955	.6964	.6972	.6981
5.0	.6990	.6998	.7007	.7016	.7024	.7033	.7042	.7050	.7059	.7067
5.1	.7076	.7084	.7093	.7101	.7110	.7118	.7126	.7135	.7143	.7152
5.2	.7160	.7168	.7177	.7185	.7193	.7202	.7210	.7218	.7226	.7235
5.3	.7243	.7251	.7259	.7267	.7275	.7284	.7292	.7300	.7308	.7316
5.4	.7324	.7332	.7340	.7348	.7356	.7364	.7372	.7380	.7388	.7396

x	0	1	2	3	4	5	6	7	8	9

Table 3   Values of log x *(continued)*

x	0	1	2	3	4	5	6	7	8	9
5.5	.7404	.7412	.7419	.7427	.7435	.7443	.7451	.7459	.7466	.7474
5.6	.7482	.7490	.7497	.7505	.7513	.7520	.7528	.7536	.7543	.7551
5.7	.7559	.7566	.7574	.7582	.7589	.7597	.7604	.7612	.7619	.7627
5.8	.7634	.7642	.7649	.7657	.7664	.7672	.7679	.7686	.7694	.7701
5.9	.7709	.7716	.7723	.7731	.7738	.7745	.7752	.7760	.7767	.7774
6.0	.7782	.7789	.7796	.7803	.7810	.7818	.7825	.7832	.7839	.7846
6.1	.7853	.7860	.7868	.7875	.7882	.7889	.7896	.7903	.7910	.7917
6.2	.7924	.7931	.7938	.7945	.7952	.7959	.7966	.7973	.7980	.7987
6.3	.7993	.8000	.8007	.8014	.8021	.8028	.8035	.8041	.8048	.8055
6.4	.8062	.8069	.8075	.8082	.8089	.8096	.8102	.8109	.8116	.8122
6.5	.8129	.8136	.8142	.8149	.8156	.8162	.8169	.8176	.8182	.8189
6.6	.8195	.8202	.8209	.8215	.8222	.8228	.8235	.8241	.8248	.8254
6.7	.8261	.8267	.8274	.8280	.8287	.8293	.8299	.8306	.8312	.8319
6.8	.8325	.8331	.8338	.8344	.8351	.8357	.8363	.8370	.8376	.8382
6.9	.8388	.8395	.8401	.8407	.8414	.8420	.8426	.8432	.8439	.8445
7.0	.8451	.8457	.8463	.8470	.8476	.8482	.8488	.8494	.8500	.8506
7.1	.8513	.8519	.8525	.8531	.8537	.8543	.8549	.8555	.8561	.8567
7.2	.8573	.8579	.8585	.8591	.8597	.8603	.8609	.8615	.8621	.8627
7.3	.8633	.8639	.8645	.8651	.8657	.8663	.8669	.8675	.8681	.8686
7.4	.8692	.8698	.8704	.8710	.8716	.8722	.8727	.8733	.8739	.8745
7.5	.8751	.8756	.8762	.8768	.8774	.8779	.8785	.8791	.8797	.8802
7.6	.8808	.8814	.8820	.8825	.8831	.8837	.8842	.8848	.8854	.8859
7.7	.8865	.8871	.8876	.8882	.8887	.8893	.8899	.8904	.8910	.8915
7.8	.8921	.8927	.8932	.8938	.8943	.8949	.8954	.8960	.8965	.8971
7.9	.8976	.8982	.8987	.8993	.8998	.9004	.9009	.9015	.9020	.9025
8.0	.9031	.9036	.9042	.9047	.9053	.9058	.9063	.9069	.9074	.9079
8.1	.9085	.9090	.9096	.9101	.9106	.9112	.9117	.9122	.9128	.9133
8.2	.9138	.9143	.9149	.9154	.9159	.9165	.9170	.9175	.9180	.9186
8.3	.9191	.9196	.9201	.9206	.9212	.9217	.9222	.9227	.9232	.9238
8.4	.9243	.9248	.9253	.9258	.9263	.9269	.9274	.9279	.9284	.9289
8.5	.9294	.9299	.9304	.9309	.9315	.9320	.9325	.9330	.9335	.9340
8.6	.9345	.9350	.9355	.9360	.9365	.9370	.9375	.9380	.9385	.9390
8.7	.9395	.9400	.9405	.9410	.9415	.9420	.9425	.9430	.9435	.9440
8.8	.9445	.9450	.9455	.9460	.9465	.9469	.9474	.9479	.9484	.9489
8.9	.9494	.9499	.9504	.9509	.9513	.9518	.9523	.9528	.9533	.9538
9.0	.9542	.9547	.9552	.9557	.9562	.9566	.9571	.9576	.9581	.9586
9.1	.9590	.9595	.9600	.9605	.9609	.9614	.9619	.9624	.9628	.9633
9.2	.9638	.9643	.9647	.9652	.9657	.9661	.9666	.9671	.9675	.9680
9.3	.9685	.9689	.9694	.9699	.9703	.9708	.9713	.9717	.9722	.9727
9.4	.9731	.9736	.9741	.9745	.9750	.9754	.9759	.9763	.9768	.9773
9.5	.9777	.9782	.9786	.9791	.9795	.9800	.9805	.9809	.9814	.9818
9.6	.9823	.9827	.9832	.9836	.9841	.9845	.9850	.9854	.9859	.9863
9.7	.9868	.9872	.9877	.9881	.9886	.9890	.9894	.9899	.9903	.9908
9.8	.9912	.9917	.9921	.9926	.9930	.9934	.9939	.9943	.9948	.9952
9.9	.9956	.9961	.9965	.9969	.9974	.9978	.9983	.9987	.9991	.9996
x	0	1	2	3	4	5	6	7	8	9

Note: To evaluate log x for $0 < x < 1$ or $x > 10$, rewrite x using scientific notation and then apply the properties of logarithms. For instance,

1. $\log 0.25 = \log (2.5 \times 10^{-1}) = \log 2.5 + \log 10^{-1} = 0.3979 - 1 = -0.6021$

2. $\log 25 = \log (2.5 \times 10^{1}) = \log 2.5 + \log 10^{1} = 0.3979 + 1 = 1.3979$

<h1>Solutions to Problems and Answers to Odd-Numbered Exercises</h1>

CHAPTER 1

Section 1.1

 Problems

1. (a) By the inverse property, $-(x+2)+(x+2) = \boxed{0}$.
   (b) By the subtraction property of fractions,
   $\dfrac{x-4}{3} = \dfrac{x}{3} - \boxed{\dfrac{4}{3}}$.
   (c) By negatives in quotients and the fundamental property of fractions, $\dfrac{-(x+4)}{-2x(x+4)} = \dfrac{\boxed{x+4}}{2x\,(x+4)} = \boxed{\dfrac{1}{2x}}$.

2. (a) $a \le 8$ (b) $-4 \le b < 0$

3. (a) $BA = \left|\dfrac{5}{3} - \dfrac{-13}{4}\right| = \left|\dfrac{59}{12}\right| = \dfrac{59}{12}$
   (b) $BA = \left|\sqrt{3} - 2\right| = -\left(\sqrt{3} - 2\right) = 2 - \sqrt{3}$, since $\sqrt{3} - 2 < 0$

Exercises

1. $3x + 6 = 6 + \boxed{3x}$ 3. $\boxed{(x+y)} + [-(x+y)] = 0$

5. $(2x+3)\boxed{0} = 0$ 7. $\dfrac{-(t+5)}{2} = -\dfrac{\boxed{t+5}}{2}$

9. $9m + 6n = \boxed{3}(3m + 2n)$ 11. $5 - 3x = -(\boxed{3x} - 5)$

13. $\left(\pi - \sqrt{2}\right)x = \boxed{\pi x} - \sqrt{2}\,x$

15. $x[6(2+y)] = \boxed{6x}(y+2)$

17. $(x-3)(y+z) = (x-3)\boxed{y} + (x-3)\boxed{z}$

19. $\dfrac{x-2}{\pi} = \dfrac{x}{\pi} - \boxed{\dfrac{2}{\pi}}$

21. $\dfrac{x}{y} + \dfrac{m}{n} = \dfrac{\boxed{xn}}{yn} + \dfrac{\boxed{my}}{yn} = \dfrac{\boxed{xn+my}}{yn}$

23. $\dfrac{2x-2y}{3x-3y} = \dfrac{\boxed{2}\,(x-y)}{\boxed{3}\,(x-y)} = \boxed{\dfrac{2}{3}}$

25. $\dfrac{1}{y+2}[x(y+2)] = \left[\boxed{\dfrac{1}{y+2}}(y+2)\right]x = \boxed{1}x = \boxed{x}$

A 9

27. $\dfrac{2\sqrt{3}}{xt+t} = \dfrac{2}{x+1} \cdot \boxed{\dfrac{\sqrt{3}}{t}}$ **29.** $x < 0$ **31.** $a \le 7$
33. $2 < p \le 10$ **35.** $0 < c < 8$ **37.** $-2 \le t \le 0$
39. 15   **41.** $\frac{29}{24}$ **43.** $\pi - \sqrt{2}$ **45.** $3 - \pi$
47. $x - \pi$ **49.** 1   **51.** $-5 < x < 5$
53. $|a - 7| \ge 3$ **55.** $|1 - d| < |d|$ **57.** $\frac{157}{50}, \pi, \frac{22}{7}$, 3.145, $\sqrt{10}$, 3.2

Section 1.2

Problems

(a) $(-2x^5)^4 = (-2)^4(x^5)^4 = 16x^{20}$

(b) $\dfrac{3x+4}{(3x+4)^5} = \dfrac{1}{(3x+4)^{5-1}} = \dfrac{1}{(3x+4)^4}$

$\dfrac{4^{-2}x^{-3}y^4}{4x^{-4}y^{-3}} = \dfrac{x^4y^3y^4}{4\cdot 4^2 x^3} = \dfrac{xy^7}{4^3}$ or $\dfrac{xy^7}{64}$

$\dfrac{9.3 \times 10^7 \text{ mi}}{1.86 \times 10^5 \text{ mi/s}} = 5 \times 10^2 \text{ s} = 500 \text{ s} \cdot \dfrac{1 \text{ min}}{60 \text{ s}} \approx 8.33 \text{ min}$

$P = \dfrac{(6 \times 10^5)^2}{3 \times 10^1} = 12 \times 10^9 \text{ W} = 12{,}000 \text{ MW}$

Exercises

1. 36   **3.** -16 **5.** $\frac{1}{64}$ **7.** $-\frac{1}{64}$ **9.** $\frac{81}{16}$
11. $\frac{1}{81}$ **13.** $\frac{2}{9}$ **15.** $-\frac{1}{4}$ **17.** $\frac{7}{12}$ **19.** $\frac{6}{5}$
21. $-\dfrac{8}{xy^3}$ **23.** $\dfrac{10}{(1-2x)^6}$ **25.** $-50x^4y^{13}$ **27.** $\dfrac{16}{y^4}$
29. $2x^3y^4$ **31.** $\dfrac{1}{x+3}$ **33.** $(x-2)^8$
35. $(2y+3)^2$ **37.** $\dfrac{1}{18mn^2}$ **39.** $8p^3(q-r)^7$
41. 0.0000000069   **43.** $-175{,}000{,}000{,}000$
45. 0.000392   **47.** 5.43×10^4 **49.** 1.3×10^{-7}
51. 2.40×10^5 **53.** 2×10^3 **55.** 3.1×10^{-18}
57. 3.125×10^{48} **59.** 2.4×10^{-6} **61.** 12 years
63. 6.7 µA   **65.** 33,100 sq m   **67.** 1   **69.** $2x^{-2}$
71. $\frac{1}{3}x^{-5}$ **73.** Since $a/0$ is undefined, we cannot
have $1 = (a/0)^0$.   **75.** (a) 1.119   (b) 0.4522
(c) 42,3õ0   (d) 0.002709   **77.** (a) \$643.70
(b) approximately \$231,732

Section 1.3

Problems

1. $(-27)^{-2/3} = \dfrac{1}{(-27)^{2/3}} = \dfrac{1}{(\sqrt[3]{-27})^2} = \dfrac{1}{(-3)^2} = \dfrac{1}{9}$

2. (a) $\sqrt[5]{(x^2+2)^2} = (x^2+2)^{2/5}$
   (b) $(2x^2)^{3/7} = \sqrt[7]{(2x^2)^3} = \sqrt[7]{8x^6}$

3. (a) $[(-2)^4x^4]^{1/4} = [(-2)^4]^{1/4}(x^4)^{1/4} = |-2|x = 2x$
   (b) $[(-2)^4]^{1/4}(x^4)^{1/4} = |-2||x| = 2|x|$

4. (a) $\sqrt[3]{32(x+y)^3} = \sqrt[3]{8(x+y)^3 \cdot 4} = 2(x+y)\sqrt[3]{4}$
   (b) $\sqrt[4]{\sqrt[3]{x^2y^6}} = \sqrt[12]{x^2y^6} = \sqrt[6]{xy^3}$

5. $\dfrac{\sqrt[3]{5x^2}}{\sqrt[3]{2y}} = \dfrac{\sqrt[3]{5x^2}}{\sqrt[3]{2y}} \cdot \dfrac{\sqrt[3]{4y^2}}{\sqrt[3]{4y^2}} = \dfrac{\sqrt[3]{20x^2y^2}}{2y}$

Exercises

1. 2   **3.** -2 **5.** 343   **7.** 25   **9.** $\dfrac{1}{1000}$
11. $-\dfrac{1}{27}$ **13.** $\dfrac{1}{4}$ **15.** $\dfrac{2}{3}$ **17.** -1 **19.** 8
21. $(2a)^{1/2}$ **23.** $(y^2-3)^{4/7}$ **25.** $x(x^2+y^2)^{-1/2}$
27. $x^{3/2}$ **29.** $\sqrt{x}$ **31.** $\sqrt[4]{27m^6}$ or $m\sqrt[4]{27m^2}$
33. $\sqrt[5]{(x+y)^4}$ **35.** $\dfrac{2}{\sqrt[3]{x^2}}$ **37.** $4x$ **39.** $\dfrac{4}{x^2}$
41. $(x+y)^{17/6}$ **43.** $\dfrac{y^2}{16x^2}$ **45.** $\dfrac{x^{4/3}}{(x+4)^{1/4}}$ **47.** $\dfrac{y^2}{2}$
49. $\dfrac{x}{(x^2-1)^{1/2}}$ **51.** $2|x| + 2x$ **53.** $x+1$
55. $2xy\sqrt[3]{2x}$ **57.** $7(a+b)\sqrt{a^2+b^2}$ **59.** $\dfrac{3xy\sqrt[3]{2y}}{5z}$
61. $\sqrt[3]{5xy^2}$ **63.** $\dfrac{x\sqrt{2x}}{y}$ **65.** $2 - 2\sqrt[3]{2} + 2\sqrt[3]{5}$
67. $\dfrac{\sqrt[3]{x+y}}{4x}$ **69.** $|x+1|$ **71.** $\dfrac{42a}{5\sqrt{6a}}$
73. $\dfrac{x+2}{\sqrt[3]{3x(x+2)}}$ **75.** $\dfrac{2pq}{\sqrt[4]{24p^2q}}$ **77.** $\dfrac{\sqrt{x+3}}{x+3}$
79. $\dfrac{\sqrt[3]{12x^2y}}{3xy}$ **81.** $\dfrac{\sqrt[4]{24(t+9)}}{6(t+9)}$ **83.** $\sqrt[6]{675}$
85. $\sqrt[6]{x^3y^4}$ **87.** (a) b^2 (b) b^2 **89.** $\dfrac{\pi\sqrt{2L}}{4}$
91. (a) 2.290   (b) -0.8680 (c) 0.001123
(d) 0.0001136   **93.** approximately 8.00 ft

Section 1.4

Problems

1. (a) $\frac{1}{2}xy + \frac{3}{4}xy + xy = \left(\frac{2}{4} + \frac{3}{4} + \frac{4}{4}\right)xy = \frac{9}{4}xy$

   (b) $\sqrt[3]{8x} - \sqrt[3]{27x} = 2\sqrt[3]{x} - 3\sqrt[3]{x} = -\sqrt[3]{x}$

2. (a) polynomial of degree 1 (b) not a polynomial because of the variable in the radicand

3. $\left(5x^2 - 3x\sqrt{2x}\right) - \left(x^2 - x\sqrt{50x}\right)$
   $= 5x^2 - 3x\sqrt{2x} - x^2 + 5x\sqrt{2x} = 4x^2 + 2x\sqrt{2x}$

4. $$
\begin{array}{r}
m^3 - m^2 + 3m + 4 \\
m^2 + 2m - 3 \\
\hline
m^5 - m^4 + 3m^3 + 4m^2 \\
2m^4 - 2m^3 + 6m^2 + 8m \\
-3m^3 + 3m^2 - 9m - 12 \\
\hline
m^5 + m^4 - 2m^3 + 13m^2 - m - 12
\end{array}
$$

5. $(2x^2 - 3)(4x^4 + 6x^2 + 9)$
   $= 8x^6 + 12x^4 + 18x^2 - 12x^4 - 18x^2 - 27$
   $= 8x^6 - 27$

6. (a) $(3x - 5y)^2 = (3x)^2 + 2(3x)(-5y) + (-5y)^2$
   $= 9x^2 - 30xy + 25y^2$

   (b) $(x^{-4} + 4)^3 = (x^{-4})^3 + 3(x^{-4})^2(4) + 3(x^{-4})(4)^2 + (4)^3$
   $= x^{-12} + 12x^{-8} + 48x^{-4} + 64$

7. $x^2 - [y(x + 2y) - (x + y)(2y - x)]$
   $= x^2 - [xy + 2y^2 - (2y^2 + xy - x^2)] = 0$

Exercises

1. $x^2 - x^5$ 3. $x^3 + 3x^2 + xy^2 - 6x^2y + y^3$

5. $x - 3xy$ 7. $-4x^{2/3} - \dfrac{2}{3x^{2/3}}$ 9. $\sqrt{10x}$

11. polynomial of degree 6   13. not a polynomial because of the negative exponent on y 15. polynomial of degree 2   17. polynomial of degree 8   19. not a polynomial because of the variables in the radicand   21. $7m^2 - 5m - 9$

23. $\dfrac{2}{x^2} + 6y + x^2$ 25. $-2x\sqrt{3x}$

27. $-32m^5n + 8m^4 - 4m^3 + 4m^2$ 29. $\sqrt{x + x^2} - 1 - x$

31. $3x^3 + 7x^2 - 27xy - 63y$ 33. $2n^3 + n^2 - 12n + 4$

35. $16y^2 + \dfrac{8y^3}{x} - 2x - y$ 37. $x^3\sqrt{3} - 9$

39. $12x^2 - xy - y^2$ 41. $x - 2\sqrt{x} - 15$

43. $-4n^2 - 9n + 4$ 45. $x^{1/3} - x^{2/3} - 8$ 47. m^3

49. $-\dfrac{1}{a^3} + 2a + ab^4$ 51. $9x^2 - y^2$ 53. $x^3 + 27$

55. $8x^3 - 1$ 57. $x - 3y$ 59. $m - n$

61. $m^2 + 12m + 36$ 63. $4a^2 - 12ab^2 + 9b^4$

65. $\dfrac{9}{x^6} + \dfrac{30}{x^3} + 25$ 67. $4x - 28y\sqrt{x} + 49y^2$

69. $2x + 2\sqrt{x^2 - 1}$ 71. $x^3 + 9x^2 + 27x + 27$

73. $64x^3 - 144x^2y^2 + 108xy^4 - 27y^6$

75. $\dfrac{1}{x^9} + \dfrac{3}{2x^6y} + \dfrac{3}{4x^3y^2} + \dfrac{1}{8y^3}$

77. $1 + 3x\sqrt[3]{(1 - x^3)^2} + 3x^2\sqrt[3]{1 - x^3}$ 79. degree of $P + Q$ is m; degree of $P - Q$ is m; degree of PQ is $m + n$; degree of P^2 is $2m$ 81. $5x^2 + 12x - 10$

83. $10\pi r - 25\pi$ 85. $1.06x$ 87. (a) 19   (b) -5

89. (a) $(6.32 \times 10^{-3})x^2 - (3.46 \times 10^{-4})x^3 + (4.72 \times 10^{-6})x^4$
   (b) 0.53 inch

Section 1.5

Problems

1. (a) $3\sqrt{-80} - \frac{1}{3}\sqrt{-45} = 3 \cdot 4i\sqrt{5} - \frac{1}{3} \cdot 3i\sqrt{5} = 11i\sqrt{5}$

   (b) $\dfrac{9}{\sqrt{-36}} = \dfrac{9}{6i} = \dfrac{3}{2i} = -\dfrac{3}{2}i$

2. (a) $i^{22} = (i^4)^5 i^2 = (1)^5(-1) = -1$

   (b) $\dfrac{1}{i^{51}} = \dfrac{1}{(i^4)^{12}i^3} = \dfrac{1}{(1)^{12}(-i)} = \dfrac{1}{-i} = i$

3. $(2 + 3i) - (5 - 4i) + (-3 + i)$
   $= (2 - 5 - 3) + (3 + 4 + 1)i = -6 + 8i$

4. $(1 + 3i)(2 + 5i) = 2 + 11i + 15i^2 = -13 + 11i$

5. $\dfrac{1 + \sqrt{-9}}{1 - \sqrt{-9}} = \dfrac{1 + 3i}{1 - 3i} \cdot \dfrac{1 + 3i}{1 + 3i} = \dfrac{1 + 6i + 9i^2}{1 - 9i^2} = \dfrac{-8 + 6i}{10}$
   $= -\dfrac{4}{5} + \dfrac{3}{5}i$

Exercises

1. $9i$ 3. -18 5. $3i\sqrt{5}$ 7. 3   9. $-\dfrac{3}{2}i$

11. $-48i$ 13. -5 15. $2i$ 17. 0   19. $108i$

21. $15 + 3i$ 23. $-2 + 33i$ 25. $24 + 16i$

27. $30 + 19i$ 29. $-\dfrac{3}{5} - \dfrac{4}{5}i$ 31. $-\dfrac{\sqrt{6}}{5} + \dfrac{2}{5}i$

33. $2 - 4i$ 35. $-99 - 20i$ 37. $-9 - 46i$

39. $-\dfrac{21}{2} + \dfrac{51}{2}i$ 41. $\dfrac{7}{25} - \dfrac{1}{25}i$

43. Since $(1 + 0i)(a + bi) = a + bi$, $1 + 0i$ is the multiplicative identity of $a + bi$.   45. (a) 0   (b) 0

47. (a) 0   (b) 0   49. Since $\left(\dfrac{\sqrt{2} + \sqrt{2}\,i}{2}\right)^2 = i$,

$\dfrac{\sqrt{2} + \sqrt{2}\,i}{2}$ is a square root of i.   51. (a) $4.24i$

(b) $22.5i$ (c) $12.4i$ (d) $0.936i$ 53. 1.07 A

Section 1.6

◆ Problems

1. $2y(x-1)^3 + y(x-1)^4 = y(x-1)^3[2+(x-1)]$
$$= y(x-1)^3(x+1)$$

2. $(9-3y)+(2xy-6x) = 3(3-y)-2x(3-y)$
$$= (3-y)(3-2x)$$

3. (a) $x^2 - 2xy - 48y^2 = (x-8y)(x+6y)$
   (b) $y^4 - 7y^2 + 10 = (y^2-2)(y^2-5)$

4. $5x^2 + 13xy + 6y^2 = (5x+3y)(x+2y)$

5. (a) $(x+y)^2 - 25y^4 = [(x+y)+5y^2][(x+y)-5y^2]$
   (b) $64 - x^3 = (4-x)(16+4x+x^2)$

6. (a) $(x^2+4x+8)(x^2-4x+8)$
   $= x^4 - 4x^3 + 8x^2 + 4x^3 - 16x^2 + 32x + 8x^2 - 32x + 64$
   $= x^4 + 64$
   (b) $(x^2+x+1)(x^2-x+1)$
   $$= x^4 - x^3 + x^2 + x^3 - x^2 + x + x^2 - x + 1$$
   $$= x^4 + x^2 + 1$$

7. $x^6 + x^3 - 2$
$$= (x + \sqrt[3]{2})(x^2 - x\sqrt[3]{2} + \sqrt[3]{4})(x-1)(x^2+x+1)$$

◆ Exercises

1. $ab(3a-1)$ 3. $3xy(2x^2-3x+y)$
5. $(x-1)(5-x)$ 7. $m(m+3)^3(6m+19)$
9. $(a+b)(c+d)$ 11. $(x-3)(x^2+2)$
13. $(2-x)(3-y)$ 15. $(5-x^2)(3-2y)$
17. $(a+5)(a-2)$ 19. $(x-9y)(x-2y)$
21. $(12+x)(2+x)$ 23. $(x^2+6)(x^2-3)$
25. $(2x+3)(x+1)$ 27. $(3a-4b)(a+5b)$
29. $(4x+1)(x-3)$ 31. $(5-2t)(3-2t)$
33. $(4x-5y)(3x-2y)$ 35. $(2n^2+5)(n^2+1)$
37. $(x^3-2y^2)(6x^3-5y^2)$ 39. $(y^6+2)(8y^6-3)$
41. $(x+9)(x-9)$ 43. $(5x+y)(5x-y)$
45. $(2x^2-5x+10)(2x^2+5x+10)$
47. $(t+2)(t^2-2t+4)$
49. $(3x^{-1}+2y^2)(9x^{-2}-6x^{-1}y^2+4y^4)$
51. $(2-5n)(4+10n+25n^2)$ 53. $(t+4)^2$
55. $(2a^2-3b^2)^2$ 57. $(x+y-2)(x-y-2)$
59. $(2x+2y+3)(2x-2y+3)$
61. $(x^2+2x+2)(x^2-2x+2)$
63. $(a^2+3a+5)(a^2-3a+5)$
65. $(x^2+y^2+xy)(x^2+y^2-xy)$
67. $(5m^2+7n^2+3mn)(5m^2+7n^2-3mn)$
69. $4xy(x-3)(x-4)$ 71. $2x(3x+2)(x-2)$
73. $5a(a+3)(a-3)$ 75. $2xy(2x+1)(4x^2-2x+1)$
77. $2(n+2)(n-2)(n^2+4)$
79. $(2x+3)(2x-3)(x+1)(x-1)$
81. $3a(a^3+3)(a-1)(a^2+a+1)$

83. $2y(x+1)(2x+3)(2x-3)$
85. $(x+1)(x-1)(x^2+x+1)(x^2-x+1)$
87. $a(a^2+2)(a^2-2)(a^2+2a+2)(a^2-2a+2)$
89. $(x+\sqrt{10})(x-\sqrt{10})$
91. $(a+b\sqrt[3]{5})(a^2-ab\sqrt[3]{5}+b^2\sqrt[3]{25})$
93. $(x^2+3+x\sqrt{6})(x^2+3-x\sqrt{6})$
95. $(x+3+\sqrt{3})(x+3-\sqrt{3})$
97. $(t^2+2)(t+\sqrt{2})(t-\sqrt{2})(t^2+2t+2)(t^2-2t+2)$
99. $(x+5i)(x-5i)$
101. $(t+\sqrt{6})(t-\sqrt{6})(t+i\sqrt{6})(t-i\sqrt{6})$
103. $2(x+3)^2(4x+3)$ 105. $10(2x+1)(3x+1)(3x-1)^3$
107. $2\pi rt(h+r)$ 109. 891,000   111. 1200
113. 10,000   115. $(3.2+A)(3.2-A)(1+A^2)$

Section 1.7

◆ Problems

1. $\dfrac{x-2}{x+5} = \dfrac{x-2}{x+5} \cdot \dfrac{3x-2}{3x-2} = \dfrac{3x^2-8x+4}{3x^2+13x-10}$

2. $\dfrac{16-x^2}{2x^3-5x^2-12x} = \dfrac{(4+x)(4-x)}{x(x-4)(2x+3)} = \dfrac{-(4+x)}{x(2x+3)}$

3. $\dfrac{x^2}{2x^2-5x+2} \div \dfrac{4x}{4-x^2}$
$$= \dfrac{x^2}{(2x-1)(x-2)} \cdot \dfrac{(2+x)(2-x)}{4x} = \dfrac{x(2+x)}{4(1-2x)}$$

4. $\dfrac{2x^2-9}{2x^2+11x+5} - \dfrac{x^2+16}{2x^2+11x+5} = \dfrac{(2x^2-9)-(x^2+16)}{2x^2+11x+5}$
$$= \dfrac{x^2-25}{(2x+1)(x+5)}$$
$$= \dfrac{x-5}{2x+1}$$

5. $\dfrac{x+1}{x^2-2x+1} - \dfrac{1}{x-1} = \dfrac{x+1}{(x-1)^2} - \dfrac{x-1}{(x-1)^2}$
$$= \dfrac{(x+1)-(x-1)}{(x-1)^2}$$
$$= \dfrac{2}{(x-1)^2}$$

6. $\dfrac{\dfrac{3}{x+h}-\dfrac{3}{x}}{h} = \dfrac{\dfrac{3x-3(x+h)}{x(x+h)}}{h} = \dfrac{-3h}{x(x+h)} \cdot \dfrac{1}{h} = \dfrac{-3}{x(x+h)}$

7. $\dfrac{1}{\sqrt{x+h}+\sqrt{x}} = \dfrac{1}{\sqrt{x+h}+\sqrt{x}} \cdot \dfrac{\sqrt{x+h}-\sqrt{x}}{\sqrt{x+h}-\sqrt{x}}$
$$= \dfrac{\sqrt{x+h}-\sqrt{x}}{(x+h)-x} = \dfrac{\sqrt{x+h}-\sqrt{x}}{h}$$

8. $(2 - x)^{-1/2} - x(2 - x)^{-3/2} = (2 - x)^{-3/2}[(2 - x)^1 - x]$
$$= (2 - x)^{-3/2}(2 - 2x)$$
$$= \frac{2 - 2x}{(2 - x)^{3/2}}$$

◆ Exercises

1. $6m^2n - 2mn$ **3.** $2x - 1$ **5.** $x^2 - 9$

7. $12xy - 4y^2$ **9.** $t^2 - 4$ **11.** $\dfrac{1}{1 - 2x}$

13. $-\dfrac{1}{3}$ **15.** $\dfrac{2x + 1}{x - 1}$ **17.** $-\dfrac{n + 1}{n + 6}$ **19.** $\dfrac{1}{x + 3}$

21. $\dfrac{3 + x + y}{3 - x + y}$ **23.** $\dfrac{3(x + 1)}{x - 1}$ **25.** $\dfrac{2 + n}{1 - n}$

27. $\dfrac{2x}{x + y}$ **29.** $-\dfrac{3}{3b + a}$ **31.** $n + 2$ **33.** $\dfrac{x}{3}$

35. $\dfrac{m}{m^2 - 9}$ **37.** $\dfrac{x}{(x - 1)^3}$ **39.** $\dfrac{4 - 2x^2}{x(x + 2)^2}$

41. $\dfrac{-2}{(1 + a)(1 - a)}$ **43.** $\dfrac{1}{a - 1}$ **45.** $\dfrac{x^2 + y^2}{xy}$

47. $-\dfrac{a + b + c}{bc}$ **49.** $\dfrac{a}{1 - a}$ **51.** $\dfrac{x^3(x + 3)}{5 - x}$

53. $\dfrac{x^2 + 4}{2(x + 2)}$ **55.** $\dfrac{a - b}{a - 2\sqrt{ab} + b}$ **57.** $\dfrac{-1}{3 + \sqrt{t + 3}}$

59. $\dfrac{2x + h}{\sqrt{(x + h)^2 + 1} + \sqrt{x^2 + 1}}$ **61.** $\dfrac{3 + 2\sqrt{3x} + x}{3 - x}$

63. $x + \sqrt{x^2 - 1}$ **65.** $\dfrac{mn(\sqrt{m} - \sqrt{n})}{m - n}$ **67.** $\dfrac{16t - 6}{(2 - 3t)^4}$

69. $\dfrac{5 - 16n}{5(1 - 4n)^{6/5}}$ **71.** $\dfrac{3 - 5t}{(3 - 4t)^{3/4}}$ **73.** $\dfrac{ab}{b - a}$

75. $\dfrac{x(x + 6)}{x + 5}$ **77.** $\dfrac{n + 1}{n - 1}$ **79.** $\dfrac{5t^2 - 1}{t^{4/3}}$

81. $\dfrac{1}{4x^{3/4}(x + 1)^{5/4}}$ **83.** The expressions are *not* equal

when $x = 1$.   **85.** (a) $y + 7 - \dfrac{5}{y}$ (b) $\dfrac{14}{t + 7} - t - 7$

87. $\dfrac{2E}{2R + r}$ **89.** $\dfrac{120 - 60\sqrt{n} - 4n + n^2}{4 - n}$

91. approximately 105.9 cm   **93.** (a) $\dfrac{43}{30}$

(b) $1.433\overline{3}. . .$

Chapter 1 Review Exercises

1. $8m + 3 = 3 + \boxed{8m}$ **3.** $6 - (-3x) = \boxed{3x} + 6$
5. $-3[-(x + \sqrt{5})] = \boxed{3}(x + \sqrt{5})$
7. $\dfrac{x}{3} - 2 = \dfrac{x}{3} - \dfrac{\boxed{6}}{3} = \dfrac{\boxed{x - 6}}{3}$ **9.** $\dfrac{3x + 5}{2} = \boxed{\dfrac{1}{2}}(3x + 5)$

11. $\dfrac{3}{8 - x} = \dfrac{\boxed{-3}}{x - 8}$ **13.** $a \geq 7$ **15.** $-10 < c < 0$

17. 15 **19.** $-50x^6y^{13}$ **21.** $\dfrac{1}{(2x + 1)^4}$ **23.** $x - 1$

25. $x + 1$ **27.** $\dfrac{y^6}{8x^{3/2}}$ **29.** 3×10^{-24}

31. $(x + 2)^{2/3}$ **33.** $\dfrac{1}{\sqrt[6]{x + 4}}$ **35.** $4(x + y)\sqrt{2(x + y)}$

37. $2y\sqrt[4]{4x^3}$ **39.** $3|xy|$ **41.** $2a^2 + a$

43. $2\sqrt{10x}$ **45.** $2x^6 + 4x^5 - 3x^4 - 13x^3 - 2x^2 + 9x + 3$

47. $30x^4 - 13x^2y - 10y^2$ **49.** $4t^2 - 25$

51. $x^3 - 64$ **53.** $9x^2 - 42xy + 49y^2$

55. $8x^3 - 60x^2 + 150x - 125$

57. $x^6 - 27x^4 + 243x^2 - 729$ **59.** $-3y^2$ **61.** $x^2 + 3$

63. $3y(y - 2)^3(3y - 4)$ **65.** $(m - 3)(m^2 + 1)$

67. $(3x + 4y)(2x - 3y)$ **69.** $(2x^2 - 3)(x^2 - 5)$

71. $(x + 5)(x - 3)$ **73.** $(5x + 3y^2)(5x - 3y^2)$

75. $(2x - 1)(4x^2 + 2x + 1)$ **77.** $(t + 2)^3$

79. $2xy(4x - 3)(x + 1)$ **81.** $2xy(1 + 2x)(1 + 3x)(1 - 3x)$

83. $3x(x + 3)^2$ **85.** $\dfrac{n + 9}{3n - 2}$ **87.** $\dfrac{1}{\ }$ **89.** $-6x$

91. $\dfrac{21y^2 - 10x^3}{24xy}$ **93.** $\dfrac{-x(2x + 3)}{3y(x + 3)}$ **95.** $\dfrac{-3}{x + 3}$

97. $\dfrac{12}{x(x + 4)^2}$ **99.** $\dfrac{2x}{(x + 2y)(x - 2y)^2}$ **101.** a

103. $\dfrac{x - 5}{\sqrt[4]{3x(x - 5)}}$ **105.** $2x + \sqrt{4x^2 - 1}$ **107.** $22i\sqrt{5}$

109. -64 **111.** $12 - i$ **113.** $47 + i$

115. 40 meters   **117.** 103,000 square feet

CHAPTER 2

Section 2.1

◆ Problems

1. $(n - 1)^2 + 3(n - 3) = (n + 1)(n + 4)$
$$n^2 - 2n + 1 + 3n - 9 = n^2 + 5n + 4$$
$$-4n = 12$$
$$n = -3$$

2. Multiplying both sides by 24 (the LCD), we obtain
$$2(n - 2) - 3(n - 5) = 24(n - 1)$$
$$2n - 4 - 3n + 15 = 24n - 24$$
$$35 = 25n$$
$$n = \tfrac{7}{5}$$

3. Division by zero is undefined.

4. Grouping x terms on the *left-hand side,* we obtain

$$ax + bx - cx = -ab$$

$$x(a + b - c) = -ab$$

$$x = \frac{-ab}{a + b - c} \cdot \frac{-1}{-1} = \frac{ab}{c - a - b}$$

5. $A = \dfrac{h}{2}(b_1 + b_2)$

$$\frac{2A}{h} = b_1 + b_2$$

$$b_1 = \frac{2A}{h} - b_2$$

 Exercises

1. -7 3. -4 5. -4 7. $-\frac{3}{2}$ 9. 5

11. -2 13. -3 15. 2   17. $\frac{2}{5}$ 19. 3

21. $\frac{7}{6}$ 23. -11 25. 8   27. $-\frac{1}{5}$ 29. 1

31. no solution   33. $x = \dfrac{k}{m - n}$ 35. $x = \dfrac{ak - a}{b - 3k}$

37. $x = \dfrac{2mn}{n - m}$ 39. $x = \dfrac{ac - bc}{a - 2b + c}$ 41. $x = -\dfrac{k}{5}$

43. $x = -\dfrac{a}{2}$ 45. (a) $E = IR$ (b) $R = \dfrac{E}{I}$

47. (a) $L_0 = \dfrac{L}{1 + \mu\Delta t}$ (b) $\Delta t = \dfrac{L - L_0}{L_0\mu}$

49. (a) $r = \dfrac{A - P}{Pt}$ (b) $P = \dfrac{A}{1 + rt}$

51. (a) $t = \dfrac{mv - mv_0}{F}$ (b) $m = \dfrac{Ft}{v - v_0}$

53. (a) $t_1 = \dfrac{Smt_2 - H}{Sm}$ (b) $m = \dfrac{H}{St_2 - St_1}$

55. $A - B = 0$ is an equivalent equation; $A/B = 1$ and

$A^2 = B^2$ are not equivalent equations.   57. $\dfrac{dy}{dx} = \dfrac{7 - y}{x + 1}$

59. only (d)   61. $x \approx 0.576$ 63. $x \approx 40.8$

65. $x \approx 2.77$ 67. 48.6 °F

Section 2.2

 Problems

1. Let $x =$ time (in days) allowed in contract to complete
        job, $x \le 28$, and
    $28 - x =$ time (in days) beyond that in contract.

$$250x - 100(28 - x) = 4200$$

$$250x - 2800 + 100x = 4200$$

$$350x = 7000$$

$$x = 20 \text{ days}$$

2. Let $x =$ amount of water (in liters) to be evaporated from
        the 8% solution and
    $25 - x =$ total amount (in liters) in required 10% solution.

$$\frac{10}{100}(25 - x) = \frac{8}{100}(25)$$

$$10(25 - x) = 8(25)$$

$$-10x = -50$$

$$x = 5 \text{ liters}$$

3. Let $x =$ height of flagpole (in feet) and
    $x + 4 =$ length of rope (in feet)

$$(x + 4)^2 = x^2 + 20^2$$

$$x^2 + 8x + 16 = x^2 + 400$$

$$8x = 384$$

$$x = 48 \text{ feet}$$

4. Let $x =$ amount (in dollars) invested at 7% and
    $27{,}500 - x =$ amount (in dollars) invested at 9%.

$$\frac{7}{100}x + \frac{9}{100}(27{,}500 - x) = 2325$$

$$7x + 9(27{,}500 - x) = 232{,}500$$

$$-2x = -15{,}000$$

$$x = \$7500$$

5. Let $x =$ number of units that must be sold.

$$400x - (300x + 1500) = 0$$

$$100x = 1500$$

$$x = 15 \text{ units}$$

6. Let $x =$ crew's speed (in mph) in still water,
    $x + 2 =$ speed downstream, and
    $x - 2 =$ speed upstream

$$\frac{4}{x + 2} = \frac{3}{x - 2}$$

$$4(x - 2) = 3(x + 2)$$

$$x = 14 \text{ mph.}$$

Hence, the team will not qualify.

7. Let x = amount of time (in hours) for both spillways to fill the bog.

$$\frac{1}{48} + \frac{1}{36} = \frac{1}{x}$$

$$3x + 4x = 144$$

$$x = 20\tfrac{4}{7} \text{ hours}$$

◆ **Exercises**

1. $-50, -49, -48$ **3.** $16, 14$ **5.** $16\tfrac{1}{3}$ ft by $32\tfrac{2}{3}$ ft
7. $35°$ **9.** 12 yr and 36 yr **11.** 13 in. **13.** 15
bills of each denomination **15.** 9728 box seats
17. $16,250 at 8% and $3,750 at 12% **19.** $\tfrac{1}{4}$ liter
21. $750 **23.** 8 yr **25.** 80 lb of 0.010% type and 40
lb of 0.025% type **27.** $6000 at 6% and $2000 at 9%
29. in $10\tfrac{1}{2}$ hr; depth is $61\tfrac{1}{2}$ in. **31.** $3\tfrac{1}{2}$ mi **33.** $37\tfrac{1}{2}$ hr
35. $280,000 **37.** $6\tfrac{1}{4}$ yr **39.** 441 sq ft for square
room, 459 sq ft for rectangular room **41.** $200
43. $38\tfrac{1}{4}$ ft **45.** 380 hertz and 456 hertz **47.** 61.7 ft
49. 0.280 mi

Section 2.3

◆ **Problems**

1. $x^2 - x - 12 = 3(x + 2)(x - 2)$

$\quad x^2 - x - 12 = 3x^2 - 12$

$\quad\quad 2x^2 + x = 0$

$\quad\quad x(2x + 1) = 0$

$\quad\quad x = 0 \quad\text{or}\quad x = -\tfrac{1}{2}$

2. $(x + 2)^2 - 9 = 0$

$\quad\quad (x + 2)^2 = 9$

$\quad\quad x + 2 = \pm\sqrt{9}$

$\quad\quad\quad x = -2 \pm 3$

$\quad x = 1 \quad\text{or}\quad x = -5$

3. (a) $\quad 4y^2 = 3y + 27$

$\quad\quad y^2 - \tfrac{3}{4}y = \tfrac{27}{4}$

$\quad\quad y^2 - \tfrac{3}{4}y + \tfrac{9}{64} = \tfrac{27}{4} + \tfrac{9}{64}$

$\quad\quad (y - \tfrac{3}{8})^2 = \tfrac{441}{64}$

$\quad\quad y - \tfrac{3}{8} = \pm\tfrac{21}{8}$

$\quad\quad y = 3 \quad\text{or}\quad y = -\tfrac{9}{4}$

$\quad$ (b) $\quad 4y^2 = 3y + 27$

$\quad\quad 4y^2 - 3y - 27 = 0$

$\quad\quad (4y + 9)(y - 3) = 0$

$\quad\quad y = -\tfrac{9}{4} \quad\text{or}\quad y = 3$

4. (a) *Check I:* $\quad (1 + 2i)^2 + 5 = 2(1 + 2i) \quad ?$

$\quad\quad (1 + 4i + 4i^2) + 5 = 2 + 4i \quad ?$

$\quad\quad (-3 + 4i) + 5 = 2 + 4i \quad ?$

$\quad\quad 2 + 4i = 2 + 4i \quad \checkmark$

$\quad$ *Check II:* $\quad (1 - 2i)^2 + 5 = 2(1 - 2i) \quad ?$

$\quad\quad (1 - 4i + 4i^2) + 5 = 2 - 4i \quad ?$

$\quad\quad (-3 - 4i) + 5 = 2 - 4i \quad ?$

$\quad\quad 2 - 4i = 2 - 4i \quad \checkmark$

$\quad$ (b) $\quad 2\left(\dfrac{\sqrt{10}}{2}\right)^2 - 2\sqrt{10}\left(\dfrac{\sqrt{10}}{2}\right) + 5 = 2\left(\dfrac{10}{4}\right) - 10 + 5 = 0$

5. Let x = uniform width (in feet) of the additional parking area.

$$(220 + x)(100 + x) = 2(220)(100)$$

$$x^2 + 320x - 22,000 = 0$$

$$x = \frac{-320 \pm \sqrt{(320)^2 - 4(1)(-22,000)}}{2(1)} = \frac{-320 \pm \sqrt{190,400}}{2}$$

$$x \approx 58.2 \text{ ft}$$

◆ **Exercises**

1. $-4, 2$ **3.** $0, 2$ **5.** $-1, \tfrac{5}{2}$ **7.** $-6, -3$
9. $-4, 2$ **11.** $5, -4$ **13.** $2, -\tfrac{3}{7}$ **15.** $4, 2$

17. 2 **19.** 4 **21.** ± 6 **23.** $\pm\dfrac{\sqrt{15}}{2}$ **25.** $\pm\tfrac{7}{2}i$

27. $\dfrac{-3 \pm 4\sqrt{2}}{4}$ **29.** $5 \pm 3i\sqrt{2}$ **31.** $-6 \pm 4\sqrt{3}$

33. $2 \pm 2i$ **35.** $\dfrac{3 \pm 3\sqrt{5}}{2}$ **37.** $\tfrac{7}{5}, -\tfrac{3}{5}$ **39.** $\tfrac{3}{4}, -\tfrac{9}{4}$

41. $\dfrac{1 \pm 2\sqrt{7}}{3}$ **43.** $\dfrac{5 \pm i\sqrt{15}}{2}$ **45.** $\dfrac{11 \pm 5\sqrt{5}}{2}$

47. $\tfrac{3}{4}, -\tfrac{4}{5}$ **49.** $\dfrac{2 \pm i}{5}$ **51.** $\dfrac{-5 \pm 3\sqrt{5}}{2}$

53. $\dfrac{\sqrt{7} \pm 5i}{4}$ **55.** $\pm\dfrac{\sqrt{5}}{2}$ **57.** -3 **59.** $1 \pm \sqrt{3}$

61. 18, 14 **63.** base is 9 in., height is 12 in.
65. width is 5 ft, length is 7 ft **67.** $4
69. large pump, 40 min; small pump, 60 min

71. $0, -\dfrac{b}{a}$ **73.** $-a \pm b$ **75.** $a, -(a + 2b)$

77. $0, \dfrac{a + b}{b}$ **79.** $a, -\dfrac{1}{a}$ **81.** $k = 64$

83. $k = \pm 2\sqrt{6}$ **85.** $k = 8$ **87.** $700
89. 12 students **91.** $-0.991, 3.34$ **93.** $-0.927, 1.12$
95. approximately 3.57 ft **97.** approximately 10.4 knots,
12.6 knots

Section 2.4

 Problems

1. Since no real number to the fourth power can be -16, this equation has *no real solution.*

2. Raising both sides to the fourth power, we obtain

$$(x-3)^3 = 2^4$$
$$x - 3 = \sqrt[3]{2^4}$$
$$x = 3 + 2\sqrt[3]{2}$$

3. Rewrite as $\sqrt{3y} = y - 6$. Now, square both sides and solve:

$$3y = y^2 - 12y + 36$$
$$y^2 - 15y + 36 = 0$$
$$(y - 3)(y - 12) = 0$$
$$y = 3 \text{ (discard)} \quad \text{or} \quad y = 12$$

4. Rewrite as $\sqrt{x^2 + 16} = 9 - \sqrt{x^2 + 7}$. Now square both sides as follows:

$$(x^2 + 16) = 81 - 18\sqrt{x^2 + 7} + (x^2 + 7)$$
$$\sqrt{x^2 + 7} = 4$$
$$x^2 + 7 = 4^2$$
$$x = \pm 3$$

5. $3t^2(t-3)^{1/2} - 4t(t-3)^{3/2} = 0$

$$t(t-3)^{1/2}[3t - 4(t-3)^1] = 0$$
$$t(t-3)^{1/2}(12 - t) = 0$$
$$t = 0, \quad t = 3, \quad \text{or} \quad t = 12$$

6. Isolating the radical, we have $2\sqrt{y} = y - 1$. Now square both sides as follows:

$$4y = y^2 - 2y + 1$$
$$y^2 - 6y + 1 = 0$$
$$y = \frac{-(-6) \pm \sqrt{(-6)^2 - 4(1)(1)}}{2(1)}$$
$$= \frac{6 \pm \sqrt{32}}{2} = 3 \pm 2\sqrt{2}$$

Only $3 + 2\sqrt{2}$ is a solution.

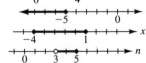 **Exercises**

1. ± 3 **3.** $3\sqrt[3]{2}$ **5.** $1 \pm 2\sqrt[4]{3}$ **7.** $\pm\frac{1}{5}$ **9.** -2
11. $-\frac{2}{3}$ **13.** 121 **15.** $\pm 2\sqrt{2}$ **17.** No real solution   **19.** $\frac{4}{9}$ **21.** No real solution   **23.** $5, -3, 1$
25. $\frac{7}{9}$ **27.** ± 2 **29.** 1 **31.** -3 **33.** $0. -6$
35. 8 **37.** $8, 4$ **39.** 3 **41.** $0, \pm 1$
43. $0, \pm 2\sqrt{2}$ **45.** $0, \dfrac{3 \pm \sqrt{17}}{2}$ **47.** $0, \dfrac{3 \pm \sqrt{17}}{4}$

49. $\frac{3}{2}$ **51.** $0, -2, -5$ **53.** $0, 2$ **55.** $0, 8$
57. $\frac{3}{2}$ **59.** $0, 4$ **61.** $2, -1$ **63.** $\pm\sqrt{-1 + \sqrt{6}}$
65. ± 1 **67.** $\frac{1}{8}, -64$ **69.** $\frac{1}{16}$ **71.** $4, -2$
73. $\frac{65}{32}$ **75.** $-1 \pm \sqrt{17}$ **77.** 2 **79.** $x = 2a \pm 1$
81. 17 cm   **83.** 100 units   **85.** 1.95 **87.** -695
89. ± 1.95 **91.** 170 ft

Section 2.5

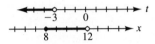 **Problems**

1. (a) $(-\infty, -3)$
   (b) $[8, 12)$

2. Multiplying both sides by 8, we obtain

$$4(y + 1) - 5(y - 1) > 8$$
$$4y + 4 - 5y + 5 > 8$$
$$-y > -1$$
$$y < 1$$

Hence, the solution set is $(-\infty, 1)$.

3.
$$5 < 3(3 - 2y) \le 6$$
$$\tfrac{5}{3} < \; 3 - 2y \; \le 2$$
$$-\tfrac{4}{3} < \; -2y \; \le -1$$
$$\tfrac{2}{3} > \; y \; \ge \tfrac{1}{2}$$

Hence, the solution set is $\left[\tfrac{1}{2}, \tfrac{2}{3}\right)$.

4.
$$2x - 3 < 7 \le 5 - x$$

$2x - 3 < 7$	*and*	$7 \le 5 - x$
$2x < 10$		$x + 7 \le 5$
$x < 5$		$x \le -2$

Hence, the solution set is $(-\infty, -2]$.

5. Let $x = $ time (in minutes) of a call, $x \ge 3$

$$31.10 \le 12.50 + 1.55(x - 3) \le 69.85$$
$$18.60 \le \quad 1.55(x - 3) \quad \le 57.35$$
$$12 \le \quad x - 3 \quad \le 37$$
$$15 \le \quad x \quad \le 40$$

Hence, the time of the calls ranged from 15 min to 40 min.

Exercises

1. $(4, \infty)$

3. $(-\infty, -5]$

5. $[-4, 1]$

7. $(3, 5]$

9. $(-\infty, 3)$ **11.** $(-\infty, -5)$ **13.** $(-\infty, 2]$
15. $(-7, \infty)$ **17.** $[2, \infty)$ **19.** $\left(-\frac{3}{2}, \infty\right)$
21. $(-\infty, 10]$ **23.** $\left[-\frac{24}{7}, \infty\right)$ **25.** $\left(\frac{1}{2}, 4\right)$
27. $[-4, 16)$ **29.** $\left[\frac{11}{24}, \frac{3}{4}\right]$ **31.** $(-3, 1)$
33. $\left[\frac{3}{2}, 3\right)$ **35.** $\left(-\infty, -\frac{1}{2}\right)$ **37.** more than 175
miles **39.** between 62% and 100%, inclusive
41. between 4 and 5 inches, exclusive **43.** 3, 5, 7 or
1, 3, 5 **45.** more than 30,000 books sold
47. (a) $x > \dfrac{c - b}{a}$ (b) $x < \dfrac{c - b}{a}$
49. (a) $x \geq ac - y$ (b) $x \leq ac - y$
51. (a) $\left(-\infty, -\frac{8}{3}\right)$ (b) $\left(\frac{2}{5}, \infty\right)$ **53.** Since $x > 5$, we
have $5 - x < 0$, Hence, when dividing both sides by $5 - x$, we
must reverse the inequality. **55.** $(-\infty, -0.147]$
57. $(-1.04, -0.165)$ **59.** $(-0.445, 0.479]$
61. from 0 ft to approximately 9.50 ft

Section 2.6

◆ Problems

1.
$$x^5 < 16x$$
$$x^5 - 16x < 0$$
$$x(x + 2)(x - 2)(x^2 + 4) < 0$$

Critical values: $-2, 0, 2$
Algebraic signs:

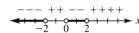

Solution set: $(-\infty, -2) \cup (0, 2)$

2.
$$\frac{x^2 + 2x - 4}{x + 2} \geq 1$$
$$\frac{x^2 + 2x - 4}{x + 2} - 1 \geq 0$$
$$\frac{(x + 3)(x - 2)}{x + 2} \geq 0$$

Critical values: $-3, -2, 2$
Algebraic signs:

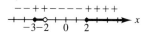

Solution set: $[-3, -2) \cup [2, \infty)$

3. When the baseball hits the ground, $d = 0$. Thus, we solve
$0 = 64t - 16t^2$ for t as follows:
$$0 = 16t(4 - t)$$
$$t = 0 \quad \text{or} \quad t = 4$$
Hence, the baseball hits the ground 4 seconds after it is hit.

◆ Exercises

1. $(-3, 3)$ **3.** $(-\infty, 0] \cup [12, \infty)$ **5.** $[-1, 9]$
7. $\left(-\infty, -\frac{7}{2}\right) \cup \left(\frac{1}{2}, \infty\right)$ **9.** $\left(\dfrac{11 - 5\sqrt{5}}{2}, \dfrac{11 + 5\sqrt{5}}{2}\right)$
11. $\left[-4, -\frac{2}{3}\right]$ **13.** $\left(-\infty, 2 - 2\sqrt{3}\right] \cup \left[0, 2 + 2\sqrt{3}\right]$
15. $(-\infty, -3) \cup \left(-\sqrt{2}, 0\right) \cup \left(\sqrt{2}, \infty\right)$ **17.** $\left(-\sqrt{5}, \sqrt{5}\right)$
19. $\left[-2, \frac{1}{2}\right] \cup \left[\frac{2}{3}, \infty\right)$ **21.** $\left(0, \frac{1}{2}\right)$ **23.** $(-3, 1)$
25. $\left[-\frac{7}{5}, -\frac{1}{2}\right)$ **27.** $(-\infty, -4] \cup [1, 2)$
29. $\left(-3 - \sqrt{7}, -5\right) \cup \left(-3 - \sqrt{7}, 5\right)$
31. $(-8, -4) \cup (0, 4)$ **33.** $(-3, 2)$ **35.** $(-3, 4]$
37. between 2 m and 11 m **39.** 0, 2, 4 or 2, 4, 6 or
4, 6, 8 **41.** when $p > \$3$ **43.** $5 \text{ s} < t < 15 \text{ s}$
45. $(-4, -2) \cup (3, 5)$ **47.** $(-\infty, -3] \cup [-1, \infty)$
49. $[3, 8]$ **51.** $(-2, 4)$ **53.** $\left(-\frac{3}{2}, 1\right]$
55. $k \leq -2\sqrt{6}$ or $k \geq 2\sqrt{6}$ **57.** $k \leq 0$ or $k \geq 4$
59. $(-\infty, -1.82) \cup (1.82, \infty)$ **61.** $(-1.31, 1.31)$
63. $(-0.883, -0.611) \cup (0, \infty)$ **65.** $0 \, \Omega < R_2 \leq 36.2 \, \Omega$

Section 2.7

◆ Problems

1. For $|1 - 2x| = 3x + 6$, we solve $1 - 2x = \pm(3x + 6)$
and discard values of x for which $3x + 6 < 0$.
$$1 - 2x = 3x + 6 \quad \text{or} \quad 1 - 2x = -(3x + 6)$$
$$-5x = 5 \qquad\qquad 1 - 2x = -3x - 6$$
$$x = -1 \qquad\qquad x = -7 \quad \text{(discard)}$$

2. The equation $|4x - 3| = |2 - x|$ is equivalent to
$4x - 3 = \pm(2 - x)$.
$$4x - 3 = 2 - x \quad \text{or} \quad 4x - 3 = -(2 - x)$$
$$5x = 5 \qquad\qquad 3x = 1$$
$$x = 1 \qquad\qquad x = \frac{1}{3}$$

3. The inequality $|-6x| - 4 > 5$ is equivalent to
$|-6x| > 9$, which in turn is equivalent to
$$-6x < -9 \quad \text{or} \quad -6x > 9$$
$$x > \tfrac{3}{2} \qquad\qquad x < -\tfrac{3}{2}$$
Hence, the solution set is $\left(-\infty, -\frac{3}{2}\right) \cup \left(\frac{3}{2}, \infty\right)$.

4. The inequality $\left|\dfrac{y-2}{y}\right| < 3$ is equivalent to

$$-3 < \dfrac{y-2}{y} < 3.$$

$$-3 < \dfrac{y-2}{y} \qquad \text{and} \qquad \dfrac{y-2}{y} < 3$$

$$0 < \dfrac{y-2}{y} + 3 \qquad\qquad \dfrac{y-2}{y} - 3 < 0$$

$$\dfrac{4y-2}{y} > 0 \qquad\qquad \dfrac{-2y-2}{y} < 0$$

Critical values: $\tfrac{1}{2}$, 0   Critical values: -1, 0

Algebraic signs: Algebraic signs:

Solution set: $(-\infty, -1) \cup \left(\tfrac{1}{2}, \infty\right)$

Exercises

1. ± 4 **3.** $\tfrac{1}{5}$, 1   **5.** 3   **7.** 0, 2, 4
9. No solution   **11.** $-\tfrac{25}{7}, -\tfrac{31}{7}$ **13.** $-4, -\tfrac{4}{7}$ **15.** $\tfrac{3}{8}$
17. 0, 7   **19.** $3, -1-\sqrt{10}$ **21.** $-1, 15$
23. $\pm 1, \pm 9$ **25.** $\pm 3, 4$ **27.** No solution
29. $(-3, 3)$ **31.** $\left(-\infty, -\tfrac{3}{2}\right) \cup \left(\tfrac{3}{2}, \infty\right)$
33. $(-\infty, 1) \cup (4, \infty)$ **35.** $\left[\tfrac{5}{2}, \tfrac{11}{2}\right]$
37. $\left(-\infty, -\tfrac{14}{3}\right] \cup \left[\tfrac{10}{3}, \infty\right)$
39. $(-\infty, -3] \cup \left[-\sqrt{3}, \sqrt{3}\right] \cup [3, \infty)$
41. $\left(2 - 2\sqrt{2}, 2\right) \cup \left(2, 2 + 2\sqrt{2}\right)$
43. $(-\infty, 0) \cup \left(0, \tfrac{1}{3}\right) \cup (1, \infty)$ **45.** $\left(-\infty, \tfrac{9}{10}\right] \cup \left[\tfrac{15}{2}, \infty\right)$
47. $x = \dfrac{-b \pm 1}{a}$ **49.** $\left(-\infty, -\tfrac{1}{3}\right] \cup [2, \infty)$ **51.** $\delta = \dfrac{\epsilon}{2}$
53. $\delta = 2\epsilon$ **55.** (a) $|x - 1000| \le 5$ (b) $[995, 1005]$
57. $5.90, -1.44$ **59.** $-0.738, -0.858$
61. $(-1.46, -0.131)$ **63.** $(-\infty, -11.1] \cup [15.3, \infty)$

Chapter 2 Review Exercises

1. 2   **3.** $-\tfrac{1}{2}$ **5.** $\dfrac{2ab}{a+b}$ **7.** $\tfrac{1}{6}$ **9.** $\tfrac{20}{3}, -4$
11. $8, -2$ **13.** 7   **15.** ± 4 **17.** $\pm 4i$
19. $2 \pm \sqrt{14}$ **21.** $1 \pm 2i$ **23.** $\sqrt{6}, -\dfrac{\sqrt{6}}{2}$
25. $3 \pm \sqrt{5}$ **27.** $\pm \dfrac{3\sqrt{3}}{8}$ **29.** $\tfrac{1}{9}$ **31.** 5
33. $-\dfrac{\sqrt{3}}{2} a$ **35.** $0, -3, 11$ **37.** $0, 4, -2$
39. $\pm 2, \pm 3$

41. $-\tfrac{1}{5}$ **43.** 256   **45.** $\pm \dfrac{2\sqrt{6}}{9}$ **47.** $0, \tfrac{2}{3}$
49. $\pm\sqrt{3 + 2\sqrt{3}}$ **51.** $L = \dfrac{Rd^2}{\mu + \mu t}$
53. $r = \dfrac{-\pi h + \sqrt{\pi^2 h^2 + 2\pi A}}{2\pi}$ **55.** $(-\infty, -3)$
57. $[2, \infty)$ **59.** $[-10, 8]$ **61.** $(-\infty, -1) \cup \left(\tfrac{13}{3}, \infty\right)$
63. $\left[-3, \tfrac{2}{3}\right]$ **65.** $\left[-1 - \sqrt{5}, 0\right] \cup \left[-1 + \sqrt{5}, \infty\right)$
67. $(-\infty, -2] \cup [4, \infty)$ **69.** $(-\infty, -4) \cup \left(-\tfrac{2}{3}, 0\right) \cup (2, \infty)$
71. $\left(\tfrac{6}{5}, 3\right) \cup (3, 6)$ **73.** $(-\infty, -7] \cup \left[-\tfrac{1}{3}, \infty\right)$
75. \$240,000   **77.** \$8400 at 8% and \$4100 at 9%
79. 8 inches   **81.** $1\tfrac{1}{2}$ inches   **83.** 24 mph
85. $\pm 10\sqrt{26}$ **87.** 10 hr for smaller pipeline and 6 hr for
larger pipeline   **89.** 64   **91.** When yearly sales are
greater than \$240,000   **93.** $(-\infty, -1) \cup (0, 1)$
95. after 200 months of service

CUMULATIVE REVIEW EXERCISES FOR CHAPTERS 1 & 2

1. (a) $(x + 1)(x - 4)(2x + 3)$ (b) $(x - 2a)(x + 2a - 4y)$
3. 5   **5.** (a) $0, \pm 2\sqrt{a - 1}$ (b) $2a - 1 \pm \sqrt{1 - a^2}$
7. 8   **9.** (a) $k = 0$ (b) $k = 9$ (c) $k < 9$
(d) $k > 9$ **11.** 30 rows   **13.** $-2 + 2i$ **15.**
(a) 1   (b) $\dfrac{4}{x^{1/2} + 1}$ **17.** 10 ft for real wheel and 6 ft
for front wheel   **19.** (a) $\dfrac{1}{b - c}$ (b) $\dfrac{x^2 - x + 1}{x^2}$
21. $9 + 6\sqrt{2}$ **23.** all squares from 9 in. by 9 in. to 11
in. by 11 in.

CHAPTER 3

Section 3.1

Problems

1.

2. $PQ = \sqrt{[1 -(-3)]^2 + (-2 - 1)^2} = \sqrt{25} = 5$

3. $\left(\dfrac{-4 + 5}{2}, \dfrac{-3 + (-1)}{2}\right) = \left(\tfrac{1}{2}, -2\right)$

4. $AB = \sqrt{[2 - (-2)]^2 + (5 - 3)^2} = \sqrt{20} = 2\sqrt{5}$

$AC = \sqrt{(4 - 2)^2 + (1 - 5)^2} = \sqrt{20} = 2\sqrt{5}$

$BC = \sqrt{[4 - (-2)]^2 + (1 - 3)^2} = \sqrt{40} = 2\sqrt{10}$

Since $AB = AC$, we conclude that triangle ABC is isosceles.
Also since $(AB)^2 + (AC)^2 = (BC)^2$, we conclude that triangle
ABC is a right triangle with hypotense $\overline{BC}$.
Area $= \tfrac{1}{2}\left(2\sqrt{5}\right)\left(2\sqrt{5}\right) = 10$ square units.

 Exercises

1. III   **3.** positive, negative   **5.** x-axis
7. $A(2, 1), B(-4, 4), C(-3, -1), D(4, -2), E(3, 0), F(0, -3)$
9. (a) 5   (b) $\left(\tfrac{5}{2}, 4\right)$ **11.** (a) $\sqrt{34}$ (b) $\left(-\tfrac{1}{2}, \tfrac{11}{2}\right)$
13. (a) $\sqrt{113}$ (b) $\left(-1, -\tfrac{3}{2}\right)$ **15.** (a) $\tfrac{5}{12}$
(b) $\left(-\tfrac{5}{8}, \tfrac{5}{6}\right)$ **17.** (a) $3\sqrt{2}$ (b) $\left(-2, \dfrac{\sqrt{2}}{2}\right)$
19. $d = \sqrt{x^2 + y^2}$ **21.** $3\sqrt{13} + 3\sqrt{5}$
23. $(AB)^2 = (AC)^2 + (BC)^2$, where $AB = 5\sqrt{2}, AC = 2\sqrt{10}$,
$BC = \sqrt{10}$; area = 10 square units
25. $AB = AC = BC = 2\sqrt{2}$; area $= 2\sqrt{3}$ square units
27. $\left(-1, \tfrac{3}{2}\right), (2, -3), \left(5, -\tfrac{15}{2}\right)$ **29.** $2\sqrt{17}, 5\sqrt{2}, \sqrt{74}$
31. $\left(0, \tfrac{7}{9}\right)$ **33.** $\left(\dfrac{a}{2}, \dfrac{b}{2}\right)$ **35.** $\dfrac{\sqrt{a^2 + b^2}}{2}$
37. $(a + b, c)$ **39.** $\left(\dfrac{a + b}{2}, \dfrac{c}{2}\right)$
41. $\left(\tfrac{14}{3}, 2\right)$ and $\left(\tfrac{10}{3}, 0\right)$ **43.** (a) Approximately 4.10
(b) (1.28, 1.60)   **45.** (a) Approximately 90.0
(b) $(-23.45, -38.4)$ **47.** radius ≈ 4.33 units,
area ≈ 59.0 square units

Section 3.2

 Problems

1.

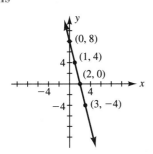

2. *Symmetric* with respect to the x-axis; x-intercept 4;
y-intercepts ± 2

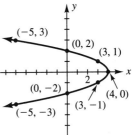

3.

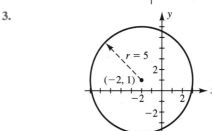

4.

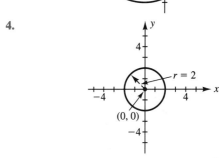

5. Completing the square, we have $(x + 2)^2 + (y + 3)^2 = 0$.
Since $r^2 = 0$, the graph of this equation is the single point
$(-2, -3)$.

 Exercises

1. x-intercept $\tfrac{5}{3}$; y-intercept 2   **3.** x-intercepts ± 6;
y-intercept 12   **5.** x-intercepts 0 and 2; y-intercept 0
7. x-intercepts 8 and -2; y-intercepts ± 4
9. x-intercepts $\pm\sqrt{2}$; no y-intercept   **11.** No x-intercept;
y-intercept 3   **13.** Symmetric with respect to the y-axis
15. Symmetric with respect to the origin   **17.** Symmetric
with respect to the x-axis   **19.** Symmetric with respect to
the origin   **21.** No symmetry   **23.** Symmetric with
respect to the x-axis, y-axis, and origin
25.

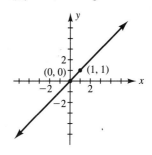

27.

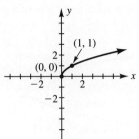

37.

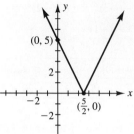

29.

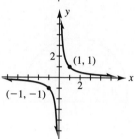

39.

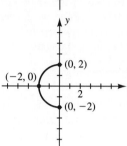

31.

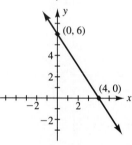

41.

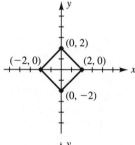

33.

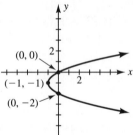

43.

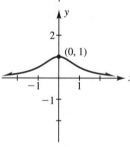

35.

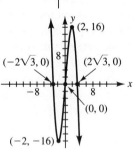

45.

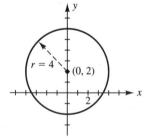

47.

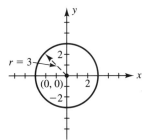

49.

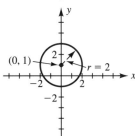

51.

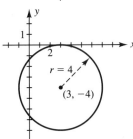

53.

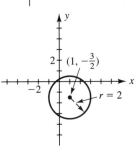

55. The equation does not define a circle and has no graph.
57. $x^2 + y^2 = 4$ **59.** $x^2 + y^2 - 4x + 6y + 11 = 0$
61. $x^2 + y^2 = 25$ **63.** $x^2 + y^2 - 8y + 11 = 0$
65. The graph of $y = |x| + c$, where $c > 0$, is the same as the graph of $y = |x|$, but is shifted vertically upward c units.
67. The graph of $y = \sqrt{x + c}$, where $c < 0$, is the same as the graph of $y = \sqrt{x}$, but is shifted horizontally to the right $|c|$ units. **69.** (a) $x^2 + y^2 - 10x + 6y + 25 = 0$ (b) 27π
71. $(x - 5)^2 + (y - 5)^2 = 25$, $(x + 5)^2 + (y - 5)^2 = 25$, $(x + 5)^2 + (y + 5)^2 = 25$, $(x - 5)^2 + (y + 5)^2 = 25$
73. x-intercept is 0.896; y-intercept is -2.66
75. No x-intercept; y-intercepts are 6.08 and 9.78
77. Center is (3.61, 2.42); radius is 2.56; x-intercepts are 2.78 and 4.44; no y-intercept **79.** (a) $(x - 8.2)^2 + y^2 = 14.3^2$
(b) 12.4 ft

Section 3.3

Problems

1. Solving the equation for y, we obtain $y = \pm\sqrt{4 - 2x}$. Since there are *two* outputs for each input value $x < 2$, we conclude the equation $2x + y^2 = 4$ does *not* define y as a function of x. Also, note that the graph of this equation fails the vertical line test.

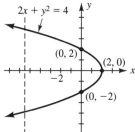

2. $G(x - h) = (x - h)^2 - 4(x - h) + 3$
$= x^2 - 2xh + h^2 - 4x + 4h + 3$

3. $h(1) = (1) - 1 = 0$

4. The radicand $2 - x$ must be positive. Thus,
$$2 - x > 0$$
$$-x > -2$$
$$x < 2$$
Hence, the domain is $(-\infty, 2)$

5. $h(-x) = (-x) - (-x)^2 = -x - x^2$. Since $h(-x) \neq h(x)$ and $h(-x) \neq -h(x)$, we conclude that the function h is neither even nor odd.

6. Domain $(-\infty, \infty)$; range $(-\infty, \infty)$

Exercises

1. Defines y as a function of x **3.** Does not define y as a function of x **5.** Defines y as a function of x **7.** Does not define y as a function of x **9.** Defines y as a function of x **11.** Defines y as a function of x **13.** Defines y as a function of x **15.** 13 **17.** 5 **19.** $9 - 2\sqrt{2}$
21. $|2ab + 3|$ **23.** $t - 4$ **25.** $n^2 - n + 1$
27. $\sqrt{x}$ **29.** $\sqrt{1 + x^2} - 4$ **31.** $4x - 2\sqrt{x - 2} - 7$
33. $4x - 34\sqrt{x} + 73$ **35.** (a) 9 (b) -11 (c) 4
(d) 0 **37.** (a) 7 (b) 3 (c) 5
(d) $x + 1$ if $x > -1$; 0 if $x = -1$; $-(x + 1)$ if $x < -1$
39. $(-\infty, \infty)$ **41.** $(-\infty, \infty)$ **43.** $(-\infty, 4]$
45. $[-4, 4]$ **47.** $(-\infty, -2) \cup (-2, \infty)$
49. $(-\infty, -2) \cup (-2, 2) \cup (2, \infty)$
51. $(-\infty, -5) \cup (-5, 2) \cup (2, \infty)$ **53.** $\left[\frac{1}{2}, 4\right) \cup (4, \infty)$
55. Odd **57.** Even **59.** Odd **61.** Even
63. Neither **65.** $\frac{5}{3}$ **67.** 8, -5 **69.** $4 \pm 2\sqrt{3}$
71. $\frac{3}{2}$ **73.** ± 2 **75.** (a) $(-\infty, \infty)$ (b) $[-2, \infty)$

(c) ± 2 (d) Even   **77.** (a) $(-\infty, \infty)$ (b) $[-1, \infty)$
(c) $0, -2$ (d) Neither   **79.** (a) $\left[-\sqrt{5}, \sqrt{5}\right]$
(b) $\left[0, \sqrt{5}\right]$ (c) $\pm\sqrt{5}$ (d) Even   **81.** (a) $(-\infty, \infty)$
(b) $[-2, 0)$ (c) None (d) Even   **83.** (a) $[-2, 2]$
(b) $[-22, 22]$ (c) $0, \pm\sqrt[4]{5}$ (d) Odd   **85.** 2

87. $2x + \Delta x$ **89.** $2x + 2 + \Delta x$ **91.** $\dfrac{-1}{x(x + \Delta x)}$

93. (a) $r(25) = 7$ m, which represents the radius of the oil spill after 25 min. (b) 13 min   **95.** The population is one-fourth as large   **97.** 90.6   **99.** -0.915
101. 2.40   **103.** ± 1.83 **105.** (a) $C(12.4) = \$19.48$, which represents the cost of a taxicab fare when the cab is driven 12.4 miles.   (b) 22.4 mi

Section 3.4

 Problems

1. The graph of $h(x) = x^3 - 1$ is obtained from the graph of $f(x) = x^3$ shifted vertically *downward* 1 unit:

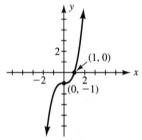

2. The graph of $h(x) = (x - 1)^3$ is obtained from the graph of $f(x) = x^3$ shifted horizontally to the *right* 1 unit:

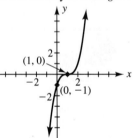

3. The graph of $F(x) = (x + 1)^2 - 4$ is obtained from the graph of $f(x) = x^2$ shifted vertically downward 4 units and horizontally to the left 1 unit:

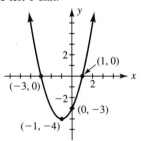

4. The graph of $F(x) = -|x + 2|$ is obtained by reflecting the graph of $f(x) = |x|$ about the x-axis and then shifting this graph to the left 2 units:

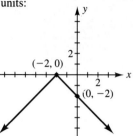

5. The graph of $F(x) = 4x - 1$ is obtained by stretching the graph of $f(x) = x$ by a factor of 4 and then shifting this graph downward 1 unit.

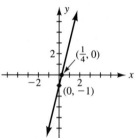

6. (a) Continuous; no break in the graph on this interval
(b) Continuous; no break in the graph on this interval
(c) Discontinuous; break at $x = 3$
(d) Discontinuous; break at $x = 3$

 Exercises

1.

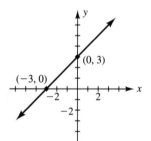

3.

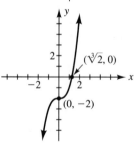

5.
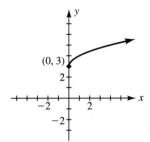

7.
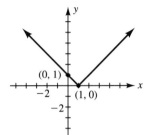

9.
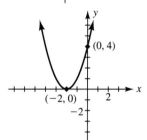

11.
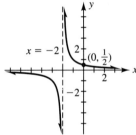

13.
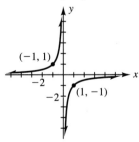

15.
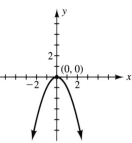

17.
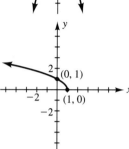

19.
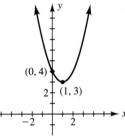

21.
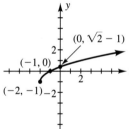

23.
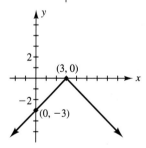

25.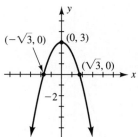

27.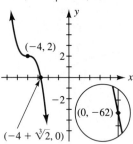

29.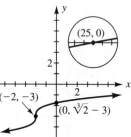

31.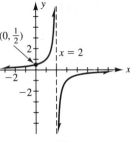

33.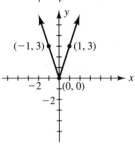

35.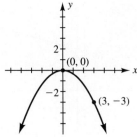

37.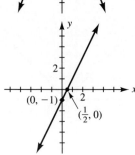

39.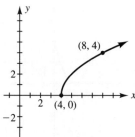

41.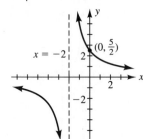

43. Increasing   **45.** Decreasing   **47.** Neither

49. Increasing

51.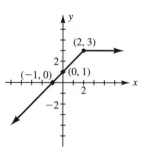

Increasing: $(-\infty, 3)$

Constant: $[3, \infty)$

53.

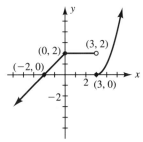

Increasing: $(-\infty, 0) \cup [3, \infty)$
Constant: $[0, 3)$

55.

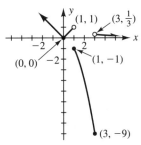

Increasing: $(0, 1)$
Decreasing: $(-\infty, 0) \cup [1, \infty)$

57. (a) Continuous (b) Discontinuous
(c) Discontinuous (d) Continuous

59.

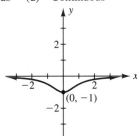

61.

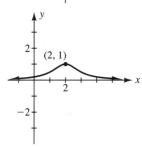

63. $y = 1 + \sqrt{4 - x^2}$ **65.** $y = -2 + \sqrt{4 - (x + 1)^2}$

67. (a)

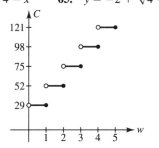

(b) 75 cents

69. (a)

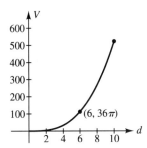

(b) 36π, Volume when diameter is 6

71.

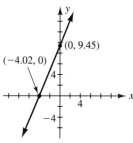

73.

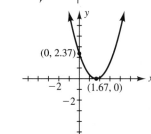

75.

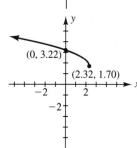

77. Horizontal shift rule

Section 3.5

◆ Problems

1. (a) $(g - f)(0) = g(0) - f(0) = -1 - 2 = -3$

(b) $\left(\dfrac{g}{f}\right)(1) = \dfrac{g(1)}{f(1)} = \dfrac{0}{\sqrt{3}} = 0$

2. (a) $(g - f)(2) = -2(2)^2 - \frac{1}{2} = -\frac{17}{2}$

(b) $\left(\dfrac{f}{g}\right)(-1) = \dfrac{3(-1)^3}{(-1)^3 - 1} = \dfrac{-3}{-2} = \dfrac{3}{2}$

3. $(f \circ g)(0) = f(g(0)) = f(0) = -2$

4. (a) $f(9) = 3$.

Hence, $(g \circ f)(9) = g(f(9)) = g(3) = 3^2 - 4 = 5$

     (b) $(g \circ f)(9) = 9 - 4 = 5$

5. One possibility is $f(x) = \sqrt{x}$ and $g(x) = \dfrac{1}{x+1}$.

◆ Exercises

1. $4 + \sqrt{2}$ **3.** 2 **5.** Undefined   **7.** $2\sqrt{2}$

9. 0 **11.** Undefined   **13.** 2 **15.** -1

17. (a) $(f + g)(x) = x^2 + 2x + 2$ (b) $(-\infty, \infty)$

19. (a) $(h + f)(x) = \sqrt{x} + 2x + 1$ (b) $[0, \infty)$

21. (a) $(H - G)(x) = \dfrac{x+1}{x(x-1)}$

(b) $(-\infty, 0) \cup (0, 1) \cup (1, \infty)$

23. (a) $(h \cdot F)(x) = \sqrt{x^3 - 4x}$ (b) $[2, \infty)$

25. (a) $(g \cdot G)(x) = \dfrac{x^2 + 1}{x}$ (b) $(-\infty, 0) \cup (0, \infty)$

27. (a) $\left(\dfrac{f}{H}\right)(x) = \dfrac{2x^2 - x - 1}{2}$ (b) $(-\infty, 1) \cup (1, \infty)$

29. (a) $(f \circ g)(x) = 2x^2 + 3$ (b) $(-\infty, \infty)$

31. (a) $(G \circ f)(x) = \dfrac{1}{2x + 1}$ (b) $\left(-\infty, -\tfrac{1}{2}\right) \cup \left(-\tfrac{1}{2}, \infty\right)$

33. (a) $(F \circ h)(x) = \sqrt{x - 4}$ (b) $[4, \infty)$

35. (a) $(G \circ H)(x) = \dfrac{x - 1}{2}$ (b) $(-\infty, 1) \cup (1, \infty)$

37. (a) $(H \circ f)(x) = \dfrac{1}{x}$ (b) $(-\infty, 0) \cup (0, \infty)$

39. (a) $(f \circ f)(x) = 4x + 3$ (b) $(-\infty, \infty)$

41. (a) $[f \cdot (g + h)](x) = 2x^3 + 2x + 2x\sqrt{x} + x^2 + 1 + \sqrt{x}$

(b) $[0, \infty)$ **43.** (a) $[f \circ (g \circ h)](x) = 2x + 3$ (b) $[0, \infty)$

45. $\tfrac{2}{3}$ **47.** $\tfrac{1}{2}$ **49.** $F(x) = (g \circ f)(x)$

51. $H(x) = (h \circ g)(x)$ **53.** $Q(x) = (g \circ g)(x)$

55. $h(x) = (f \circ g)(x)$ where $f(x) = 1/x$ and $g(x) = \sqrt{x}$.

57. $P(x) = (f \circ g)(x)$ where $f(x) = x^2$ and $g(x) = 2x^3 - 1$.

59. $h(x) = (f \circ g)(x)$ where $f(x) = |x|$ and $g(x) = 3x + 2$.

61. $H(x) = (f \circ g)(x)$ where $f(x) = 2x^2 + 3x$ and $g(x) = x - 1$.

63. 4 **65.** 3 **67.** 2 **69.** $f(x) = \tfrac{1}{3}x - 2$

71. (a) $Z(L) = (f \circ g)(L)$ where $f(L) = \sqrt{20 + L^2}$ and $g(L) = 12L$.   (b) $Z(L) = [(f \circ g) \circ h](L)$ where $f(L) = \sqrt{L}$, $g(L) = 20 + L^2$, and $h(L) = 12L$.   **73.** 11.2

75. $73{,}700$ **77.** -0.684

79. (a) $(C \circ n)(t) = 1012.5t - 16.2t^2$ (b) Production cost as a function of the number of hours t that the factory operates (c) $\$7213$ (d) Approximately 4 hr 30 min

Section 3.6

◆ Problems

1. $f(g(x)) = f\left(\sqrt[5]{x} + 2\right) = \left[\left(\sqrt[5]{x} + 2\right) - 2\right]^5 = \left(\sqrt[5]{x}\right)^5 = x$

$g(f(x)) = g[(x - 2)^5] = \sqrt[5]{(x - 2)^5} + 2 = (x - 2) + 2 = x$

2. $H(a) = H(b)$ implies $a^2 - 1 = b^2 - 1$

$$a^2 = b^2$$

$$a = b \quad \text{if } a \geq 0, b \geq 0$$

Also, note that the graph of $H(x) = x^2 - 1$ with $x \geq 0$ passes the horizontal line test.

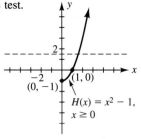

$H(x) = x^2 - 1$, $x \geq 0$

3. $f(f^{-1}(x)) = f\left(\sqrt[3]{x + 1}\right) = \left(\sqrt[3]{x + 1}\right)^3 - 1$

$= (x + 1) - 1 = x$

$f^{-1}(f(x)) = f^{-1}(x^3 - 1) = \sqrt[3]{(x^3 - 1) + 1} = \sqrt[3]{x^3} = x$

4. Interchanging x and y and solving for y, we obtain

$$x = \frac{1}{y + 1}$$

$$y + 1 = \frac{1}{x}$$

$$y = \frac{1}{x} - 1 \text{ is the inverse function.}$$

5. $f(f^{-1}(x)) = 4 - [f^{-1}(x)]^2$

$x = 4 - [f^{-1}(x)]^2$

$[f^{-1}(x)]^2 = 4 - x$

$f^{-1}(x) = \sqrt{4 - x}$

◆ Exercises

11. (a) One-to-one   **13.** (a) Not one-to-one

(b) $[0, \infty)$ is one restriction.   **15.** (a) One-to-one

17. (a) Not one-to-one   (b) $[-2, \infty)$ is one restriction.

19. (a) Not one-to-one   (b) $[2, \infty)$ is one restriction.

21. (a) $f^{-1}(x) = \dfrac{x}{2}$ (b)

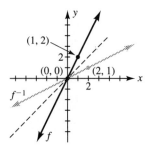

23. (a) $f^{-1}(x) = 3 - x$ (b)

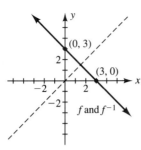

25. (a) $f^{-1}(x) = \sqrt[3]{2 - x}$ (b)

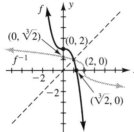

27. (a) $f^{-1}(x) = x^3 + 4$ (b)

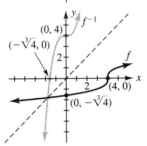

29. (a) $f^{-1}(x) = \dfrac{1}{x} - 1$ (b)

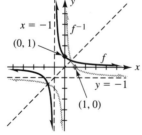

31. (a) $f^{-1}(x) = (x - 1)^2 + 4, \; x \geq 1$ (b)

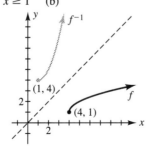

33. (a) $f^{-1}(x) = \sqrt{x} + 1$ (b)

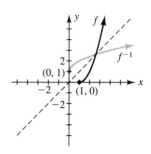

35. (a) $f^{-1}(x) = \sqrt{9 - x^2}, \; 0 \leq x \leq 3$ (b)

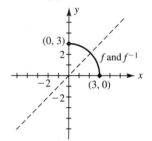

37. (a) $f^{-1}(x) = -\sqrt{x - 3}$ (b)

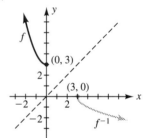

39. (a) $f^{-1}(x) = \begin{cases} \sqrt{x} & \text{if } x \geq 0 \\ x/2 & \text{if } x < 0 \end{cases}$ (b)

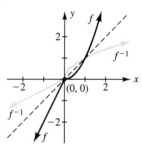

41.

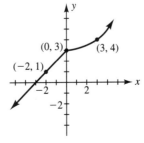

x	-2	0	3
$f^{-1}(x)$	1	3	4

43.

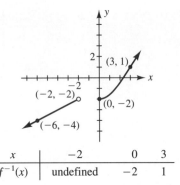

x	-2	0	3
$f^{-1}(x)$	undefined	-2	1

45. $f^{-1}(4) = 14$ **47.** $(f^{-1} \circ g)(-2) = -13$

49. $(f \circ g)^{-1}(x) = (g^{-1} \circ f^{-1})(x) = \dfrac{3x + 1}{3}$

51. $(-\infty, 0) \cup (0, \infty)$ **53.** $\left(-\infty, \frac{2}{3}\right) \cup \left(\frac{2}{3}, \infty\right)$

55. $g^{-1}(r) = \dfrac{\sqrt{\pi r}}{2\pi}, \quad r > 0$ **57.** $(f \circ g^{-1})(S) = \dfrac{S\sqrt{\pi S}}{6\pi},$

the volume of a sphere as a function of its surface area.

59. 1.92   **61.** 1.10   **63.** 0.173   **65.** (a) f and g are inverses, since $(f \circ g)(x) = x$.   (b) f and g are inverses, since $(f \circ g)(x) = x$.

Section 3.7

◆ Problems

1. The maximum area of 2500 square feet occurs when $l = 50$ ft and $w = 100 - l = 50$ ft.

2. The volume V of water (in cubic inches) is given by $V = 8t$, where t is the time (in minutes). The time t it takes to fill the funnel is $\dfrac{64\pi \text{ cu in.}}{8 \text{ cu in./min}} = 8\pi$ min. Hence, the domain of $V = 8t$ is $[0, 8\pi]$.

3. $T = 168{,}000/40 = \$4200$

4. $t = 240/50 = 4.8$ hours

5. Since the electrical resistance R of a wire varies directly as its length l and inversely as the square of its radius r, we write $R = \dfrac{kl}{r^2}$. Doubling both l and r, we obtain

$R = \dfrac{k(2l)}{(2r)^2} = \dfrac{1}{2} \cdot \dfrac{kl}{r^2}$. Thus, the resistance becomes one half as large.

◆ Exercises

1. $d = \dfrac{c}{\pi}$ **3.** (a) $C = 40 + \dfrac{n}{5}$

5. (a) $S = \dfrac{d}{\sqrt{2}}$ (b) $A = \dfrac{d^2}{2}$ (c) $P = 2\sqrt{2}\,d$

7. $W = 125{,}000 + 4000t$

9. (a) $V = \frac{4}{3}\pi h^3$ (b) $V = \frac{1}{6}\pi r^3$

11. (a) $d = \sqrt{2 - 2x}$ (b) $[-1, 1]$

13. (a) $A = \dfrac{x\sqrt{x}}{2}$ (b) $P = x + \sqrt{x} + \sqrt{x^2 + x}$

15. $S = 2x^2 + \dfrac{256}{x}$

17. (a) $V = 9h^2$ (b)

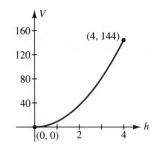

19. (a) $d = 30\sqrt{9 + t^2}$ (b) Domain $[0, 3]$; range $[90, 90\sqrt{2}]$

21. (a) $d = F/50$ (b) Domain $[0, 2400]$

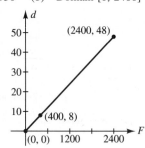

23. (a) $W = 125l$ (b) Domain $[0, \infty)$

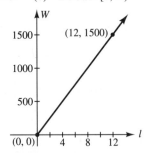

25. (a) $N = 10{,}000/d$ (b) Domain $(0, \infty)$

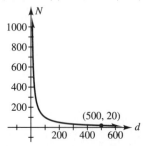

27. (a) $w = 300/f$ (b) Domain $(0, \infty)$

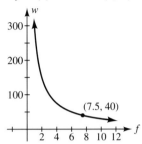

29. (a) $F = 160{,}000/d^2$ (b) Domain $(0, \infty)$

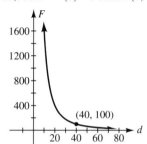

31. The force is tripled.   **33.** $A = 30r - 2r^2 - \dfrac{\pi r^2}{2}$

35. $d = \begin{cases} 30t & \text{if } t \le 3 \\ 30\sqrt{9 + (t-3)^2} & \text{if } 3 < t \le 6 \end{cases}$ **37.** 25 seconds

39. (a) $A = \dfrac{x\sqrt{262.44 - x^2}}{2}$ (b) Approximately 64.6

square units   **41.** (a) $A = 1000x - 2x^2$ (b) $(0, 500)$
(c) 250 ft by 500 ft   **43.** (a) $T \approx 2.007\sqrt{l}$
(b) Approximately 3.11 seconds

Chapter 3 Review Exercises

1. (a) 5   (b) $\left(2, \frac{9}{2}\right)$ (c) $(x-2)^2 + \left(y - \frac{9}{2}\right)^2 = \frac{25}{4}$
3. (a) $2\sqrt{5}$ (b) $(0, 2)$ (c) $x^2 + (y-2)^2 = 5$
5. (a) $\sqrt{97}$ (b) $\left(\frac{3}{2}, -3\right)$ (c) $\left(x - \frac{3}{2}\right)^2 + (y+3)^2 = \frac{97}{4}$
7. $AB = BC = CD = DA = 2\sqrt{2}$; perimeter $= 8\sqrt{2}$; area $= 8$
9. $a/2$, half as long as OA **11.** $\dfrac{\sqrt{(a-b)^2 + c^2}}{2}$, half as

long as AB
13. (a) x-intercept 0
      y-intercept 0
(b) Symmetry with
    respect to
    origin

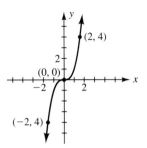

(d) Defines y as a
    function of x
15. (a) x-intercept 0
      y-intercept 0
(b) Symmetry with
    respect to the
    x-axis

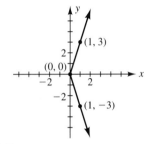

(d) Does *not* define y
    as a function of x
17. (a) No x-intercept
      No y-intercept
(b) Symmetry with
    respect to the
    y-axis

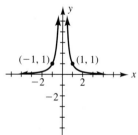

(d) Defines y as a
    function of x
19. (a) x-intercept 3
      y-intercept 2
(b) No symmetry

(c)
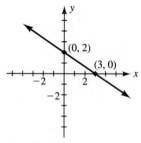

(c)
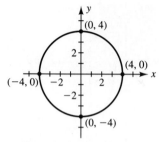

(d) Defines y as a
     function of x

21. (a) x-intercepts ± 3
           y-intercept -3

(b) Symmetry with
     respect to the
     y-axis

(d) Does *not* define y
     as a function of x

27. (a) x-intercepts $-1, 7$
           y-intercepts $\pm\sqrt{7}$

(b) Symmetry with
     respect to the
     x-axis

(c)
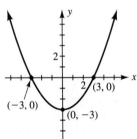

(c)
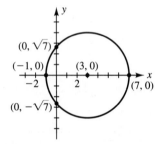

(d) Defines y as a
     function of x

23. (a) x-intercepts ± 2
           y-intercepts ± 1

(b) Symmetry with
     respect to the
     x-axis, y-axis,
     and origin

(d) Does *not* define y
     as a fucntion of x

29. (a) x-intercepts 2, 8
           y-intercept 4

(b) No symmetry

(c)
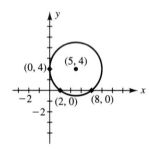

(c)
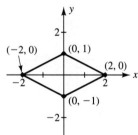

(d) Does *not* define y
     as a function of x

25. (a) x-intercepts ± 4
           y-intercepts ± 4

(b) Symmetry with
     respect to the
     x-axis, y-axis,
     and origin

(d) Does *not* define y
     as a function of x

31. 1   **33.** 5   **35.** $2x^2 - 13x + 21$

37. -5 **39.** 5   **41.** $18x^4 + 45x^2 + 28$

43. $\dfrac{3x + 1}{x - 1}$ **45.** $\dfrac{x - 2}{3}$ **47.** 2

49. $4x - 1 + 2\Delta x$

51. (a) $(-\infty, \infty)$ (b) -3 (c) Neither

(d)
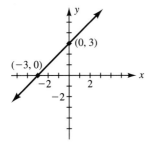

(e)   Increasing on $(-\infty, \infty)$ (f)   One-to-one
(g) $(-\infty, \infty)$
53.   (a) $(-\infty, \infty)$ (b)   0   (c)   Odd
(d)
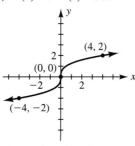

(e)   Increasing on $(-\infty, \infty)$ (f)   One-to-one
(g) $(-\infty, \infty)$ **55.**   (a) $(-\infty, \infty)$ (b) ± 3 (c)   Even
(d)
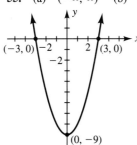

(e)   Decreasing: $(-\infty, 0)$; increasing: $(0, \infty)$
(f)   Not one-to-one   (g) $[-9, \infty)$
57.   (a) $(-\infty, \infty)$ (b) -3 (c)   Neither
(d)
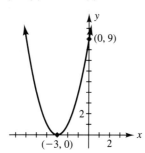

(e)   Decreasing: $(-\infty, -3)$; increasing: $(-3, \infty)$
(f)   *Not* one-to-one   (g) $[0, \infty)$ **59.**   (a) $(-\infty, \infty)$
(b) -2 (c)   Neither

(d)
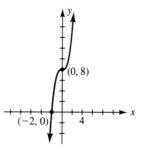

(e)   Increasing on $(-\infty, \infty)$ (f)   One-to-one   (g) $(-\infty, \infty)$
61.   (a) $(-\infty, \infty)$ (b)   2 and 4   (c)   Neither
(d)
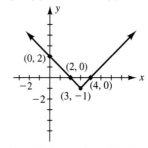

(e)   Decreasing: $(-\infty, 3)$; increasing: $(3, \infty)$
(f)   *Not* one-to-one   (g) $[-1, \infty)$ **63.**   (a) $(-\infty, \infty)$
(b)   0 and 6   (c)   Neither
(e)   Increasing: $(-\infty, 3)$; decreasing: $(3, \infty)$
(f)   *Not* one-to-one   (g) $(-\infty, 9]$ **65.**   (a) $[-10, 10]$
(b) ± 10 (c)   Even
(e)   Increasing: $[-10, 0)$; decreasing: $(0, 10]$
(f)   *Not* one-to-one   (g) $[0, 10]$ **67.**   (a) $(-\infty, \infty)$
(b)   0 and $\pm\sqrt{2}$ (c)   Even
(e)   Increasing: $(-\infty, -1) \cup (0, 1)$;
decreasing: $(-1, 0) \cup (1, \infty)$ (f)   *Not* one-to-one
(g) $(-\infty, 1]$ **69.**   (a) $(-\infty, 3]$ (b)   None
(c)   Neither   (d)
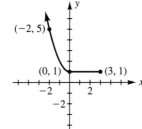

(e)   Decreasing: $(-\infty, 0]$; constant: $(0, 3]$ (f)   *not* one-to-one
(g) $[1, \infty)$ **71.**
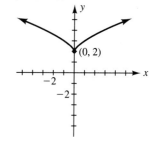

73.

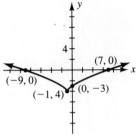

75.

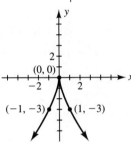

77. (a) $f^{-1}(x) = -\sqrt{x}$ (b)

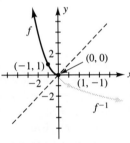

79. (a) $f^{-1}(x) = \sqrt[3]{3 - x}$ (b)

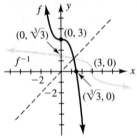

81. (a) $f^{-1}(x) = x^2 + 2,\ x \geq 0$ (b)

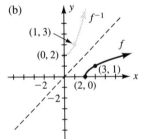

83. (a) $f^{-1}(x) = \begin{cases} \sqrt[3]{x} & \text{if } x < 1 \\ 2x - 1 & \text{if } x \geq 1 \end{cases}$

(b)

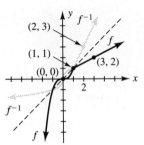

85. $(-\infty, 3]$ **87.** 3   **89.** 4   **91.** Yes;

(a) -2 (b) 0 (c) -6 **93.** $f^{-1}(x) = \dfrac{a - xb}{x - 1}$

95. (a) \$63.43 (b) 425.6 miles

97. $C = \begin{cases} 0.015n & \text{if } n \leq 1000 \\ 15 + 0.02(n - 1000) & \text{if } n > 1000 \end{cases}$

99. (a) $A = 3x - x^3$ (b) $\left[0, \sqrt{3}\right]$

101. (a) $T = V/25$ (b) Domain $[0, \infty)$

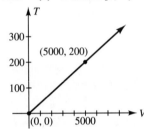

103. (a) $G = \dfrac{n(2000 - n)}{19{,}200}$ (b) Approximately 39 field

mice per acre

CHAPTER 4

Section 4.1

◆ Problems

1. (a) $m = \dfrac{2 - 4}{-1 - 3} = \dfrac{-2}{-4} = \dfrac{1}{2}$

(b) $m = \dfrac{-3 - 4}{2 - (-2)} = \dfrac{-7}{4} = -\dfrac{7}{4}$

(c) $m = \dfrac{3 - 3}{-2 - 3} = \dfrac{0}{-5} = 0$;

Relabeling the points does not change the value of the slope.

2.

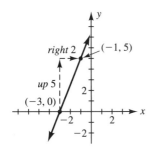

3. slope is 3, y-intercept is $-\frac{1}{2}$

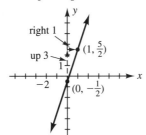

4. *x-intercept;* $3x + 8(0) = 12$
                              $3x = 12$
                              $x = 4$
    y-intercept; $3(0) + 8y = 12$
                              $8y = 12$
                              $y = \frac{3}{2}$

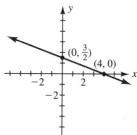

5. $0.33x - 4656.50 = 15000$
             $0.33x = 19656.50$
                 $x = \$59{,}565.15$

Exercises

1. 2   **3.** $-\frac{2}{3}$ **5.** undefined   **7.** -4

9. $\dfrac{3 - b}{4}$ **11.** -1

13.

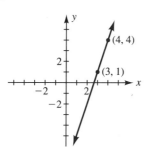

15.

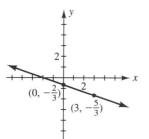

17.

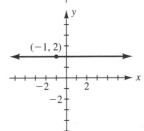

19. $m = 3$, $b = 0$

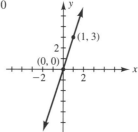

21. $m = -\frac{1}{3}$, $b = 0$

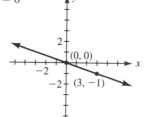

23. $m = 2$, $b = 3$

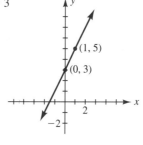

25. $m = -4$, $b = 1$

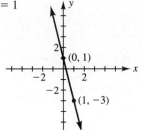

27. $m = \dfrac{1}{2}$, $b = -1$

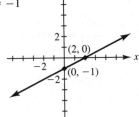

29. $m = -\dfrac{3}{4}$, $b = \dfrac{1}{4}$

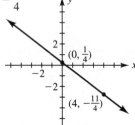

31.

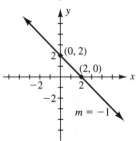

33.

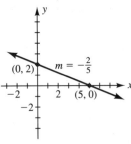

35.

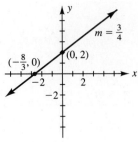

37.

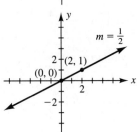

39.

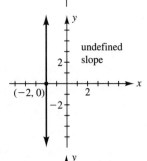

41.

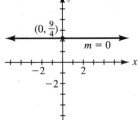

43.

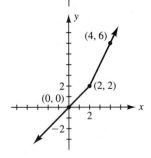

45.

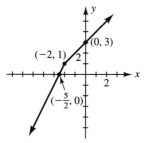

47.

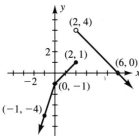

49. (a) $m = \frac{9}{5}, b = 32$ (b)

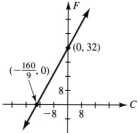

51. (a) $m = -3, b = 44$ (b)

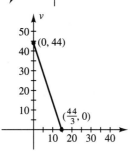

(c) $t = \frac{44}{3}$ is the time when the car comes to rest.

53. $(-5, 5)$ and $(3, -1)$ **55.** (a) 1 (b) 9 (c) -5

57. $f^{-1}(x) = \dfrac{x - b}{m}$ **59.** $m \approx 0.250$ **61.** $m \approx -1.05$

63. $m \approx 0.736$ **65.** $m \approx 0.156$

67. y-intercept 6; x-intercept $\frac{24}{7} \approx 3.4$

69. x-intercept $-\frac{57}{14} \approx -4.1$; y-intercept $\frac{19}{6} \approx 3.2$

71. (a) $C = \begin{cases} 3 & \text{if } x \le 30 \\ 0.06x + 1.2 & \text{if } 30 < x \le 100 \\ 0.04x + 3.2 & \text{if } 100 < x \le 300 \\ 0.01x + 12.2 & \text{if } x > 300 \end{cases}$

(b)

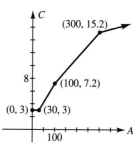

(c) $12.80   (d) 272 kilowatt hours

Section 4.2

Problems

1. $y - 0 = \frac{1}{2}[x - (-3)]$, or $y = \frac{1}{2}x + \frac{3}{2}$

2.
$$y - 1 = 1(x - 2)$$
$$y - 1 = x - 2$$
$$y = x - 1$$

3.
$$\frac{x}{1/2} + \frac{y}{3} = 1$$
$$2x + \frac{y}{3} = 1$$
$$6x + y - 3 = 0$$

4.
$$y - 0 = \frac{2}{3}[x - (-3)]$$
$$y = \frac{2}{3}x + 2$$

5. $m = \dfrac{0 - (-3)}{3 - 2} = 3$. Hence, the equation of the perpendicular line is
$$y - (-4) = -\frac{1}{3}(x - 3)$$
$$y = -\frac{1}{3}x - 3.$$

6. The V-intercept ($11,000) represents the value of the car when new.

Exercises

1. (a) $3x - 4y + 12 = 0$ (b) $y = \frac{3}{4}x + 3$

3. (a) $3x - y + 5 = 0$ (b) $y = 3x + 5$

5. (a) $7x + 3y - 28 = 0$ (b) $y = -\frac{7}{3}x + \frac{28}{3}$

7. (a) $3x + y = 0$ (b) $y = -3x$

9. (a) $3x + y - 11 = 0$ (b) $y = -3x + 11$

11. (a) $7x + 15y - 6 = 0$ (b) $y = -\frac{7}{15}x + \frac{2}{5}$

13. (a) $23x + 18y - 3 = 0$ (b) $y = -\frac{23}{18}x + \frac{1}{6}$

15. (a) $2x - 3y - 6 = 0$ (b) $y = \frac{2}{3}x - 2$

17. (a) $24x + y - 6 = 0$ (b) $y = -24x + 6$

19. (a) $2y - 3 = 0$ (b) $y = \frac{3}{2}$

21. (a) $x - 3 = 0$
(b) can't be written in slope-intercept form

23. (a) $x - 5y - 10 = 0$ (b) $y = \frac{1}{5}x - 2$

25. (a) $x + y - 5 = 0$ (b) $y = -x + 5$

27. (a) $x + 2y + 2 = 0$ (b) $y = -\frac{1}{2}x - 1$

29. (a) $3x - y + 8 = 0$ (b) $y = 3x + 8$

31. (a) $2x - 3y - 13 = 0$ (b) $y = \frac{2}{3}x - \frac{13}{3}$

33. (a) $x - 4y + 12 = 0$ (b) $y = \frac{1}{4}x + 3$

35. (a) $3x - 4y + 3 = 0$ (b) $y = \frac{3}{4}x + \frac{3}{4}$

37. (a) $2x - y = 0$ (b) $y = 2x$

39. (a) $r = \frac{5}{2}T - 100$ (b) 0 chirps per min

41. (a) $V = 60{,}000 - 9000x$ (b) \$24,000

43. (a) $V = 15{,}000x + 50{,}000$ (b) \$230,000

45. (a) $y = -\frac{1}{3}x - \frac{16}{3}$ or $x + 3y + 16 = 0$

(b) $y = \frac{4}{3}x - \frac{11}{6}$ or $8x - 6y - 11 = 0$ **47.** $f(x) = \frac{4}{3}x + \frac{13}{3}$

49. $g(x) = 3x - 7$ **51.** $\frac{12}{5}$

53. (a) -2 (b) 0 (c) -6 (d) 4   **55.** $y = -6x + 4$

57. $y = -\dfrac{A}{B}x + \left(\dfrac{A}{B}x_1 + y_1\right)$ **59.** $y = -2.10x + 7.48$

61. $y = 4.76x - 31.4$

63. Both lines have slope 1.72 and, thus, are parallel.

65. (a) $w = -0.040625T + 4.57185$ (b) Approximately 3.15 mm   (c) Approximately 112.5°F

Section 4.3

◆ Problems

1. *vertex:* $(3, -1)$

x-intercept: $\quad 0 = (x - 3)^2 - 1$

$$1 = (x - 3)^2$$
$$\pm 1 = x - 3$$
$$x = 4 \quad \text{or} \quad x = 2$$

y-intercept: $\quad f(0) = (0 - 3)^2 - 1 = 8$

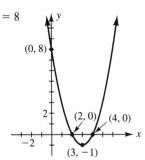

2. The basic form is $y = a(x - 4)^2 + 1$. Since the graph passes through $(2, 3)$, we have

$$3 = a(2 - 4)^2 + 1$$
$$3 = 4a + 1$$
$$a = \tfrac{1}{2}$$

Hence, the equation is $y = \frac{1}{2}(x - 4)^2 + 1$.

3. Using the quadratic formula, we find

$$x = \frac{-4 \pm \sqrt{(4)^2 - 4(-2)(-5)}}{2(-2)} = \frac{-4 \pm \sqrt{-24}}{-4}$$
$$= 1 \pm \frac{\sqrt{6}}{2}i,$$

which are not real numbers.

4. Since $a = -3 < 0$, the parabola opens downward with a maximum value at $x = -\dfrac{0}{2(-3)} = 0$. Hence, the maximum value is $y = 3 - 3(0)^2 = 3$ and it occurs when $x = 0$.

5. If fencing all four sides, then $100 = 2l + 2w$, or

$$l = 50 - w.$$

Hence

$$A = lw = (50 - w)(w) = 50w - w^2.$$

The graph of this equation is a parabola that opens downward with a maximum value at

$$w = -\frac{50}{2(-1)} = 25 \text{ ft.}$$

Hence,

$$l = 50 - 25 = 25 \text{ ft,}$$

and the largest area that can be enclosed is

$$(25 \text{ ft})(25 \text{ ft}) = 625 \text{ sq ft.}$$

◆ Exercises

1.

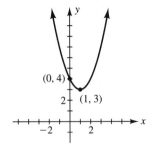

3.

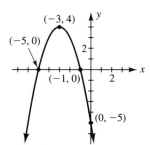

5.

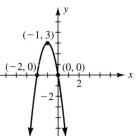

7.

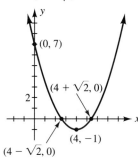

9.

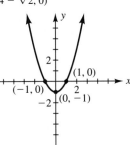

11.

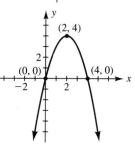

13.

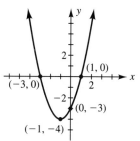

15.

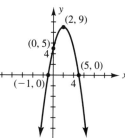

17.

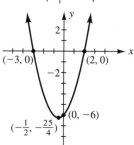

19.

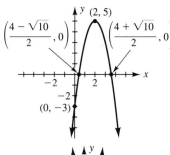

21.

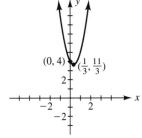

23.

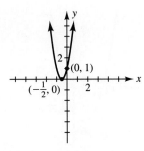

25. $y = \frac{1}{4}x^2$ **27.** $y = -\frac{1}{2}x^2 + 2$ **29.** $y = -\frac{2}{9}x^2 + \frac{4}{3}x$

31. $y = \frac{2}{3}x^2 + \frac{4}{3}x + 2$ **33.** maximum value of 16

35. minimum value of $\frac{3}{4}$ **37.** maximum value of 9

39. (a) 2 sec   (b) 144 ft   (c) 5 sec

41. (a) 50 pairs   (b) \$2300   **43.** 62 and 62

45. 11 in. by 11 in.   **47.** $33\frac{1}{3}$ ft by 100 ft   **49.** $\dfrac{a+b}{2}$

51. $y = \frac{1}{2}(x - 1)^2 - 2$ **53.** $c = 0$

55. (a) \$12   (b) \$3600   **57.** (0.744, 3.78)

59. (2.63, 1.51)   **61.** $y \approx 0.425(x - 2.67)^2 - 1.98$

63. No effect; the axis of symmetry of $y = x^2 + 4x + c$ is $x = -2$ for *all* values of c.   **65.** approximately 343 watts

Section 4.4

 Problems

1. Solving for y, we obtain $y = \frac{1}{2}x^2 - 2x + 5$, which is the equation of a parabola with a *vertical* axis of symmetry.
Vertex: (2, 3)
x-intercepts: none
y-intercept: 5

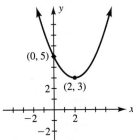

2. The y-intercepts are obtained by letting $x = 0$:

$$\frac{(0+1)^2}{16} + \frac{(y-3)^2}{4} = 1$$

$$\frac{(y-3)^2}{4} = \frac{15}{16}$$

$$(y-3)^2 = \frac{15}{4}$$

$$y = \frac{6 \pm \sqrt{15}}{2}$$

3. Completing the square, we have

$$3(x-1)^2 + 2(y+2)^2 = -4.$$

Since $-4 < 0$, this equation does *not* define an ellipse. No point (x, y) with real coordinates can satisfy this equation.

4. If the transverse axis is vertical, we have

$$2b = 6 \quad \text{or} \quad b = 3,$$

and

$$\pm\frac{3}{a} = \pm\frac{1}{2} \quad \text{or} \quad a = 6.$$

Hence, the equation is $\dfrac{(y-1)^2}{9} - \dfrac{x^2}{36} = 1$.

5. Completing the square, we obtain

$$(x-3)^2 - 4(y-1)^2 = 0.$$

Since the right-hand side of this equation is zero, we conclude this equation does not define a hyperbola. The graph of this equation is the pair of intersecting lines $y = \frac{1}{2}x - \frac{1}{2}$ and $y = -\frac{1}{2}x + \frac{5}{2}$.

 Exercises

1.

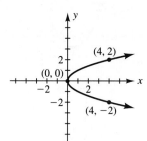

3.

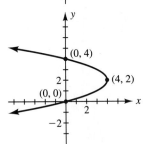

5.

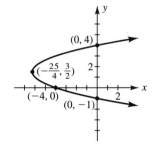

7.

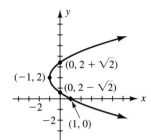

9.

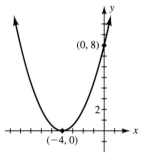

11.

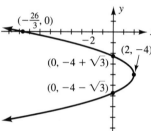

13.

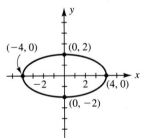

15.

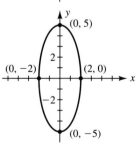

17.

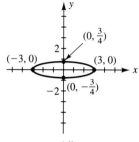

19.

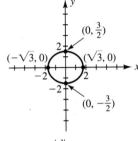

21.

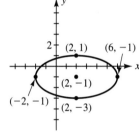

23.

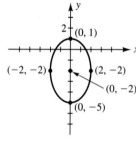

25.

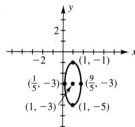

27.

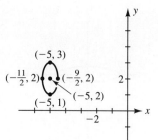

29.

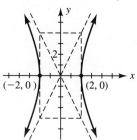

31.

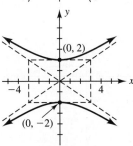

33.

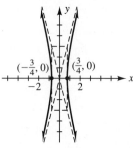

35.

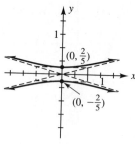

37.

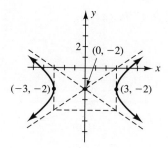

39.

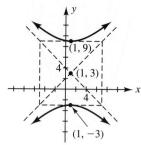

41.

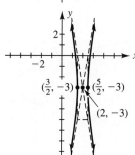

43.

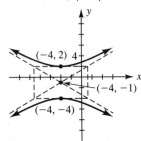

45. $y^2 - 8x = 0$ **47.** $4x^2 + 25y^2 - 25 = 0$

49. $64x^2 + 9y^2 - 256x + 36y + 148 = 0$

51. $16y^2 - x^2 - 256 = 0$ **53.** $x^2 - y^2 - 4 = 0$

55. Ellipse   **57.** Parabola   **59.** Hyperbola

61. Ellipse   **63.** Degenerate circle   **65.** Degenerate hyperbola

67. Domain $[-3, 3]$; range $[0, 2]$

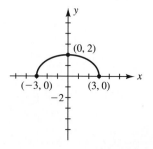

69. Domain $[-2, 2]$; range $[-6, 0]$

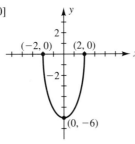

71. Domain $(-\infty, -2] \cup [2, \infty)$
Range $[0, \infty)$

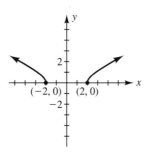

73. Domain $(-\infty, \infty)$
Range $(-\infty, -3]$

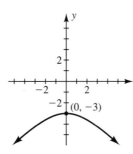

75. $8\sqrt{2} \approx 11.3$ ft

77. (a) $(0.979, -1.78)$ (b) $(-0.584, 2.87)$

79. (a) $(1.61, 0), (-1.61, 0)$ (b) $(0, 2.86), (0, 0.304)$

81. $\dfrac{(x - 3.54)^2}{1.44} + \dfrac{(y + 2.07)^2}{12.96} = 1$

83. Approximately 8.1 m

Section 4.5

 Problems

1. $5\left(\dfrac{14}{13}\right) + \left(-\dfrac{31}{13}\right) = \dfrac{70}{13} - \dfrac{31}{13} = \dfrac{39}{13} = 3$

and $3\left(\dfrac{14}{13}\right) - 2\left(-\dfrac{31}{13}\right) = \dfrac{42}{13} + \dfrac{62}{13} = \dfrac{104}{13} = 8.$

2. (a) Substituting $2x - 3$ for y, we have

$$2x - 3 = 4x + 5$$
$$-8 = 2x$$
$$x = -4$$

Hence, $y = 2(-4) - 3 = -11$. The lines intersect at the point $(-4, -11)$.

(b) By the addition method, we have

$$
\begin{array}{lll}
y = 2x - 3 & \longrightarrow & -y = -2x + 3 \\
y = 4x + 5 & \longrightarrow & \underline{y = 4x + 5} \\
& & 0 = 2x + 8 \qquad \text{Add} \\
& & x = -4
\end{array}
$$

Hence, $y = 2(-4) - 3 = -11$. The lines intersect at the point $(-4, -11)$.

3. Substituting $x^2 - 2x + 3$ for y, we obtain

$$4x - (x^2 - 2x + 3) = 6$$
$$x^2 - 6x + 9 = 0$$
$$(x - 3)^2 = 0$$
$$x = 3$$

Hence, $y = (3)^2 - 2(3) + 3 = 6$. The line is tangent to the parabola at the point $(3, 6)$.

4. Substituting x^2 for y, we obtain

$$x^2 = 4x - x^2$$
$$2x^2 - 4x = 0$$
$$2x(x - 2) = 0$$
$$x = 0 \quad \text{and} \quad x = 2$$

Hence, $y = 0^2 = 0$ and $y = 2^2 = 4$. The parabolas intersect at $(0, 0)$ and $(2, 4)$.

5.
$$x^6 + 4x^2 - 16 = 0$$
$$(x^6 - 8) + 4x^2 - 8 = 0$$
$$(x^2 - 2)(x^2 + 2x + 4) + 4(x^2 - 2) = 0$$
$$(x^2 - 2)(x^2 + 2x + 8) = 0$$
$$
\underset{\displaystyle x = \pm\sqrt{2}}{x^2 - 2 = 0} \quad \text{or} \quad \underset{\displaystyle \underbrace{}_{\text{No real solution}}}{x^2 + 2x + 8 = 0}
$$

Hence, the *exact*, coordinates of the intersection points are $P\left(-\sqrt{2}, -2\sqrt{2}\right)$ and $Q\left(\sqrt{2}, 2\sqrt{2}\right)$.

 Exercises

1. $\left(2, \frac{2}{5}\right)$ **3.** $\left(-\frac{1}{2}, -1\right)$ **5.** $\left(-\frac{1}{2}, \frac{11}{2}\right)$ **7.** $(-3, -23)$

9. No intersection point   **11.** $\left(\frac{7}{3}, \frac{5}{6}\right)$

13. $(-3, -6), (1, 2)$ **15.** $(2, 1), \left(\frac{9}{8}, -\frac{3}{4}\right)$

17. $(2, 5), \left(-\frac{3}{2}, \frac{3}{2}\right)$ **19.** $\left(\frac{1}{2}, -\frac{7}{4}\right), (5, 5)$

21. $(-2, -3), \left(\frac{1}{4}, \frac{21}{8}\right)$ **23.** $(7, 4)$

25. $\left(\dfrac{-11 + \sqrt{41}}{4}, \dfrac{-11 + \sqrt{41}}{8}\right), \left(\dfrac{-11 - \sqrt{41}}{4}, \dfrac{-11 - \sqrt{41}}{8}\right)$

27. No intersection point   **29.** (0, 0), (4, 4)

31. (0, 3), (3, 0)   **33.** $\left(4, 2\sqrt{5}\right), \left(4, -2\sqrt{5}\right)$

35. (3, 4), $\left(-\frac{24}{5}, \frac{7}{5}\right)$ **37.** $\left(-\frac{1}{2}, -2\right)$, (1, 1)

39. (4, 0), (8, 2)   **41.** No intersection point

43. $\left(1, 3\sqrt{3}\right), \left(1, -3\sqrt{3}\right), \left(-1, 3\sqrt{3}\right), \left(-1, -3\sqrt{3}\right)$

45. (0, 1), $\left(-\frac{8}{5}, -\frac{3}{5}\right)$ **47.** $\left(2, \sqrt{2}\right)$ **49.** (0, 2), (0, −2)

51. $\left(\frac{5}{4}, -\frac{3}{4}\right)$ **53.** (9, 3), (9, −3)   **55.** (1, 2), $\left(\frac{2}{5}, -\frac{8}{5}\right)$

57. (−2, −2), $\left(-\frac{32}{3}, \frac{20}{3}\right)$ **59.** (5, −1), (2, 2), (−1, −7)

61. (1, 5), (9, 1), (−1, 1)   **63.** $4\sqrt{5}$

65. $P\left(-1, \frac{9}{2}\right), Q(4, 2), R\left(4, -\frac{7}{4}\right), S\left(-1, -\frac{1}{2}\right)$

67. $P(2, 1), Q(4, 5), R(4, -5)$

69. $P(-2, 5), Q(1, 2), R(-2, -3), S\left(\frac{7}{2}, -3\right)$

71. (3.61, 33.38)   **73.** (1.22, 2.77), (−1.39, 9.86)

75. (3.47, 16.84), (0.00, −17.19)

77. (−1.879, −6.638), (0.347, 0.042)

79. (−1.195, 3.817), (2.930, 2.723)

81. (1.115, 0.897), (3.934, 0.254)

83. (−1.328, 6.220), (0.816, 0.886)

Chapter 4 Review Exercises

1.

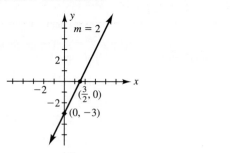

3.

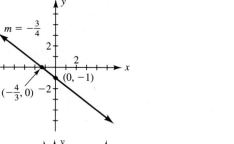

5.

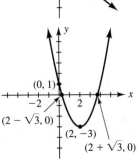

7.

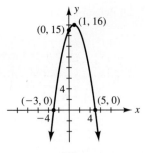

9.

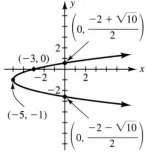

11.

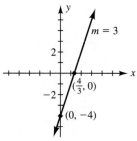

13.

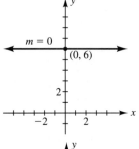

15.

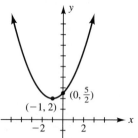

17.

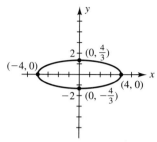

19.

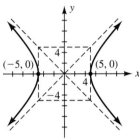

21.

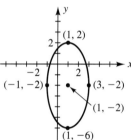

23.

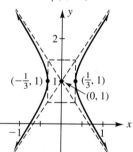

25. $3x - 4y - 8 = 0$ **27.** $3x - y + 10 = 0$

29. $3x + 2y + 4 = 0$ **31.** $x^2 - y - 4 = 0$

33. $2y^2 + 12y - 9x = 0$ **35.** $4x^2 + 9y^2 - 24x = 0$

37. $9y^2 - 4x^2 - 81 = 0$ **39.** $11x + 2y - 20 = 0$

41. $\left(\frac{1}{2}, -3\right)$ **43.** $(4, 6), (1, 3)$

45. $\left(\sqrt{11}, 5\right), \left(-\sqrt{11}, 5\right), (0, -6)$

47. $\left(1, \sqrt{2}\right), \left(1, -\sqrt{2}\right), \left(-1, \sqrt{2}\right), \left(-1, -\sqrt{2}\right)$

49. $\left(6, \frac{3}{2}\right), \left(-\frac{10}{3}, -\frac{5}{6}\right)$ **51.** $(1.104, 1.344)$

53. $x = \dfrac{C_1 B_2 - C_2 B_1}{A_1 B_2 - A_2 B_1}, y = \dfrac{A_1 C_2 - A_2 C_1}{A_1 B_2 - A_2 B_1}$

55. (a) $y = \dfrac{B}{A} x$ (b) $\left(\dfrac{-AC}{A^2 + B^2}, \dfrac{-BC}{A^2 + B^2}\right)$

57. $(-2, 2), \left(\frac{4}{3}, \frac{16}{3}\right), \left(-\frac{1}{3}, \frac{7}{6}\right), \left(3, \frac{9}{2}\right)$

59. minimum value of $\frac{11}{3}$ **61.** -2

63. (a) $V = -400x + 2400$ (b) 6 years

65. $C(x) = \begin{cases} 120 & \text{if } x \leq 5000 \\ 0.018x + 30 & \text{if } 5000 < x \leq 15{,}000 \\ 0.011x + 135 & \text{if } 15{,}000 < x \leq 30{,}000 \\ 0.003x + 375 & \text{if } x > 30{,}000 \end{cases}$

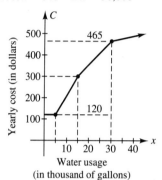

67. 200 ft along tar road by 125 ft along gravel road

69. 30 meters

CHAPTER 5

Section 5.1

Problems

1. $(x - 2)(2x^2 - x + 1) + (-2)$
$= (2x^3 - x^2 + x - 4x^2 + 2x - 2) + (-2)$
$= 2x^3 - 5x^2 + 3x - 4$

2.

$$3x^2 + 0x - 2 \overline{\big)6x^3 - 3x^2 + 0x + 0}$$
$$\underline{6x^3 + 0x^2 - 4x}$$
$$-3x^2 + 4x$$
$$\underline{-3x^2 + 0x + 2}$$
$$4x - 2$$

Hence, $\dfrac{6x^3 - 3x^2}{3x^2 - 2} = 2x - 1 + \dfrac{4x - 2}{3x^2 - 2}$.

3. $1\overline{\big)1 \quad -5 \quad \ 7 \quad -3}$
$$\underline{\quad 1 \quad -4 \quad \ 3}$$
$$1 \quad -4 \quad \ 3 \quad \ 0$$

Hence, $\dfrac{x^3 - 5x^2 + 7x - 3}{x - 1} = x^2 - 4x + 3$.

4. $(x + 3)(2x^3 - 6x^2 + 3x) + 1$
$= (2x^4 - 6x^3 + 3x^2 + 6x^3 - 18x^2 + 9x) + 1$
$= 2x^4 - 15x^2 + 9x + 1$

5.
$$3)\overline{\begin{array}{rrrrr} 1 & -3 & -4 & 13 & -8 \\ & 3 & 0 & -12 & 3 \end{array}}$$
$$\begin{array}{rrrrr} 1 & 0 & -4 & 1 & -5 \end{array}$$

Hence, the remainder is -5.

6. $P\left(-\frac{1}{2}\right) = 4\left(-\frac{1}{2}\right)^3 - 6\left(-\frac{1}{2}\right)^2 + 4\left(-\frac{1}{2}\right) - 3$

$$= -\frac{1}{2} - \frac{3}{2} - 2 - 3$$

$$= -7$$

7.
$$1 - i)\overline{\begin{array}{rrrr} 3 & -8 & 10 & -4 \\ & 3 - 3i & -8 + 2i & 4 \end{array}}$$
$$\begin{array}{rrrr} 3 & -5 - 3i & 2 + 2i & 0 \end{array}$$

Since the remainder is 0, we conclude $1 - i$ is a zero of P.

◆ **Exercises**

1. $3x^2 + x + 2 + \dfrac{-1}{x - 1}$ **3.** $3x^2 - 4x - 1 + \dfrac{-4x + 6}{x^2 - 2x + 3}$

5. $2x^2 + x + 2 + \dfrac{-1}{2x - 1}$ **7.** $x^2 + 1 + \dfrac{-3x^2 + 4}{2x^3 + 1}$

9. $x^2 - x + 1 + \dfrac{-1}{x + 1}$ **11.** $-3x^3 - x - 1$

13. $-3 + \dfrac{6}{2x + 1}$ **15.** $2x + 2 + \dfrac{6x - 2}{(2x - 1)^2}$

17. $2x - 1 + \dfrac{3}{x - 1}$ **19.** $x^2 + 3x - 8 + \dfrac{20}{x + 2}$

21. $2x^3 + x^2 + x + 4$ **23.** $4x^3 + x^2 - 6x + 4 + \dfrac{-24}{x + 6}$

25. $-x^4 - 2x^3 - 4x^2 - 8x - 16 + \dfrac{-29}{x - 2}$

27. $3x^2 - 6x + 15$ **29.** $2x^3 - 4x^2 + 10x - 44 + \dfrac{118}{x + \frac{5}{2}}$

31. $P(1) = 8$

33. $F(-2) = 7$ **35.** $P(4) = 25$ **37.** $Q\left(\frac{1}{2}\right) = 0$

39. $f\left(-\frac{1}{3}\right) = 0$ **41.** $h(1 - i) = 7 - 4i$

43. $p\left(\sqrt{2}\right) = 12$ **45.** No   **47.** Yes   **49.** Yes

51. Yes   **53.** No   **55.** Yes   **57.** Yes   **59.** Yes

61. $x^4 + 7x^3 + 7x^2 + 11x + 18$ **63.** $2x^2 - 3x + 2$

65. $8x^3 + x^2 + 2x - 2$ **67.** $-x^3 - 2x + 3$

69. $k = 16$ **71.** $k = 2$ **73.** $14.5x^2 + 2.5x - 45.5$

75. $275x^3 + 495x^2 + 765x + 750$ **77.** -5.0582

79. 455 meters

Section 5.2

◆ **Problems**

1.
$$2i)\overline{\begin{array}{rrrr} 2 & -3 & 8 & -12 \\ & 0 + 4i & -8 - 6i & 12 \end{array}}$$
$$\begin{array}{rrrr} 2 & -3 + 4i & -6i & 0 \end{array}$$

Since $P(2i) = 0$, we conclude that $x - 2i$ is a factor of $P(x) = 2x^3 - 3x^2 + 8x - 12$.

2. $P(x) = a(x - 1)^2(x + 1)^2$

$$= a[(x - 1)(x + 1)]^2$$

$$= a(x^2 - 1)^2$$

$$= a(x^4 - 2x^2 + 1).$$

3.
$$4)\overline{\begin{array}{rrrr} 1 & -1 & -10 & -8 \\ & 4 & 12 & 8 \end{array}}$$
$$\begin{array}{rrrr} 1 & 3 & 2 & 0 \end{array}$$

Hence, $P(x) = (x - 4)(x^2 + 3x + 2) = (x - 4)(x + 2)(x + 1)$. By the factor theorem, the zeros of P are 4, -2, and -1.

4. $(x - 3)^2[x - (1 + 2i)][x - (1 - 2i)]$

$$= (x^2 - 6x + 9)[x^2 - (1 + 2i)x - (1 - 2i)x + (1 + 2i)(1 - 2i)]$$

$$= (x^2 - 6x + 9)(x^2 - 2x + 5)$$

$$= x^4 - 8x^3 + 26x^2 - 48x + 45$$

5. $P(x) = [x - (2 + i)][x - (2 - i)][x - (1 + i)][x - (1 - i)]$

◆ **Exercises**

1. No   **3.** No   **5.** Yes   **7.** Yes   **9.** No

11. Yes

13. (a) fifth degree   (b) 0 of multiplicity two, 2 of multiplicity three

15. (a) eighth degree   (b) 0 of multiplicity three, -1 of multiplicity two, 4 of multiplicity three

17. (a) fourth degree   (b) 0 of multiplicity two, 2 of multiplicity two

19. (a) fourth degree   (b) -1 of multiplicity two, 1 of multiplicity two

21. (a) third degree   (b) -2 of multiplicity two, 2

23. (a) fourth degree   (b) $\pm 2i$, ± 2

25. $P(x) = a(x^3 - 2x^2 - x + 2)$ **27.** $P(x) = a(x^4 - 1)$

29. $P(x) = a(x^5 - 2x^3)$

31. (a) $P(x) = (x - 1)(x - 1)(x + 3)$

(b) 1 of multiplicity two, -3

33. (a) $P(x) = (2x + 1)(3x - 4)(x + 1)$ (b) $-\frac{1}{2}, \frac{4}{3}, -1$

35. (a) $P(x) = (x - 2)^2(x + i)(x - i)$

(b) 2 of multiplicity two, $\pm i$

37. (a) $P(x) = (2x - 3)(x - 3)(x - 2)^2$

(b) $\frac{3}{2}$, 3, 2 of multiplicity two

39. (a) $P(x) = (x + 2i)(x - 2i)(x - 4)(x + 1)$
(b) $\pm 2i, 4, -1$
41. (a) $P(x) = [x - (3 - i)][x - (3 + i)]\left[x - \left(-1 + \sqrt{6}\right)\right]$
$\left[x - \left(-1 - \sqrt{6}\right)\right]$ (b) $3 \pm i, -1 \pm \sqrt{6}$ **43.** $1, 6, -2$
45. -2 of multiplicity two, $3 \pm i$ **47.** $2 \pm i, \frac{4}{3}, -1$
49. $x^3 + 27 = (x + 3)(x^2 - 3x + 9)$; roots of $x^3 + 27 = 0$
are -3 and $\dfrac{3 \pm 3i\sqrt{3}}{2}$.
51. For $f(x) = x^n - a^n$, $f(-a) = 0$ when n is even, but
$f(-a) = -2a^n$ when n is odd. Hence, $x + a$ is a factor of
$x^n - a^n$ only when n is even.
53. $k = -7$ **55.** 1 or 3 times **57.** 0, 2, 4, or 6 times
59. 1, 3, 5, 7, or 9 times
61. $f(x) = -2x^3 + 4x^2 + 10x - 12$
63. $0.240, 1.30, -2.30$ **65.** $-3.25, 8.00, \pm 1.24$
67. $-0.240, 3.44, -1.04$
69. Approximately 5 hr 55 min after high tide

Section 5.3

Problems

1. The possible rational zeros of
$P(x) = 3x^3 + 22x^2 + 25x + 6$ are
$$\frac{\pm 1}{\pm 1}, \frac{\pm 2}{\pm 1}, \frac{\pm 3}{\pm 1}, \frac{\pm 6}{\pm 1} \quad \text{and} \quad \frac{\pm 1}{\pm 3}, \frac{\pm 2}{\pm 3}, \frac{\pm 3}{\pm 3}, \frac{\pm 6}{\pm 3}.$$
which simplify to $\pm 1, \pm 2, \pm 3, \pm 6, \pm\frac{1}{3}, \pm\frac{2}{3}$.

2. According to Descartes' rule of signs, the function P has
no positive real zero and either one or three negative real zeros.
Checking the possible negative rational zeros from Problem 1,
we find

$$\begin{array}{r|rrrr}
-1 & 3 & 22 & 25 & 6 \\
   &   & -3 & -19 & -6 \\
\hline
   & 3 & 19 & 6 & \textcircled{0}
\end{array}$$

Hence, $P(x) = (x + 1)(3x^2 + 19x + 6) = (x + 1)(x + 6)(3x + 1)$.
By the factor theorem, the zeros of P are -1, -6, and $-\frac{1}{3}$.

3.
$$\begin{array}{r|rrrrr}
-2 & 2 & -5 & -2 & 5 & -24 \\
   &   & -4 & 18 & -32 & 54 \\
\hline
   & 2 & -9 & 16 & -27 & 30
\end{array}$$

Since the numbers in the final row alternate signs, we conclude
that -2 is a lower bound for the zeros of P.

$$\begin{array}{r|rrrrr}
3 & 2 & -5 & -2 & 5 & -24 \\
  &   & 6 & 3 & 3 & 24 \\
\hline
  & 2 & 1 & 1 & 8 & 0
\end{array}$$

Since the numbers in the final row are nonnegative, we
conclude that 3 is an upper bound for the zeros of P.

4. Possible rational zeros: $\pm 1, \pm 2, \pm 3, \pm 5, \pm 6, \pm 10,$
$\pm 15, \pm 30.$

Types of zeros: Either zero or two positive real zeros and
exactly one negative real zero.

$$\begin{array}{r|rrrr}
2 & 1 & 0 & -19 & 30 \\
  &   & 2 & 4 & -30 \\
\hline
  & 1 & 2 & -15 & \textcircled{0}
\end{array}$$

Hence, $P(x) = (x - 2)(x^2 + 2x - 15) = (x - 2)(x - 3)(x + 5)$.
By the factor theorem, the zeros of P are 2, 3, and -5.

5. Possible rational zeros: $\pm 1, \pm 2, \pm 3, \pm 4 \pm 6 \pm 12, \pm\frac{1}{3},$
$\pm\frac{2}{3}, \pm\frac{4}{3}.$
Types of zeros: Either zero or two positive real zeros and zero
or two negative real zeros.

$$\begin{array}{r|rrrr}
1 & 3 & 1 & 10 & -26 & 12 \\
  &   & 3 & 4 & 14 & -12 \\
\hline
\frac{2}{3} & 3 & 4 & 14 & -12 & \textcircled{0} \\
  &   & 2 & 4 & 12 \\
\hline
  & 3 & 6 & 18 & \textcircled{0}
\end{array}$$

Hence, $P(x) = (x - 1)\left(x - \frac{2}{3}\right)(3x^2 + 6x + 18)$
$\qquad\qquad = (x - 1)(3x - 2)(x^2 + 2x + 6)$

Since $x^2 + 2x + 6$ is prime over the reals, the only rational
zeros of P are 1 and $\frac{2}{3}$.

6. When $x = \frac{3}{2}$ inches,
$$\text{length} = 15 - 2x = 15 - 2\left(\tfrac{3}{2}\right) = 12 \text{ inches}$$
and $\quad$ width $= 8 - 2x = 8 - 2\left(\tfrac{3}{2}\right) = 5$ inches.
When $x \approx 1.84$ inches,
$$\text{length} = 15 - 2x \approx 15 - 2(1.84) = 11.32 \text{ inches}$$
and $\quad$ width $= 8 - 2x \approx 8 - 2(1.84) = 4.32$ inches.

Exercises

1. (a) $\pm 1, \pm 2, \pm 3, \pm 6$
(b) No positive real zero; 1 or 3 negative real zeros
3. (a) $\pm 1, \pm 5, \pm\dfrac{1}{2}, \pm\dfrac{5}{2}$
(b) 0, 2, or 4 positive real zeros; no negative real zero
5. (a) $\pm 1, \pm 2, \pm 4, \pm 8, \pm\frac{1}{2}, \pm\frac{1}{4}$
(b) 0 or 2 positive real zeros; 1 negative real zero
7. (a) $\pm 1, \pm 2, \pm 3, \pm 4, \pm 6, \pm 12, \pm\frac{1}{3}, \pm\frac{2}{3}, \pm\frac{4}{3}$
(b) 1 positive real zero; 1 negative real zero
9. (a) $\pm 1, \pm 3, \pm\frac{1}{2}, \pm\frac{3}{2}, \pm\frac{1}{3}, \pm\frac{1}{4}, \pm\frac{3}{4}, \pm\frac{1}{6}, \pm\frac{1}{12}$
(b) 1 positive real zero; 0 or 2 negative real zeros
19. (a) 1, 2, 3 (b) $P(x) = (x - 1)(x - 2)(x - 3)$
21. (a) $4, -2, \frac{1}{2}$ (b) $f(x) = (x - 4)(x + 2)(2x - 1)$
23. (a) $-\frac{1}{4}$ (b) $H(x) = (4x + 1)(x^2 + 2x + 6)$
25. (a) $\frac{1}{2}, \frac{2}{3}, -\frac{3}{2}$ (b) $P(x) = (2x - 1)(3x - 2)(2x + 3)$

27. (a) $-2, \frac{3}{2}$ (b) $f(x) = (x + 2)(2x - 3)(x^2 + 3)$

29. (a) $\frac{8}{3}$ (b) $g(x) = (3x - 8)(x^3 - 2x^2 - 1)$

31. (a) -1 of multiplicity two, $-2, -\frac{1}{2}$

(b) $P(x) = (x - 1)^2(x + 2)(2x + 1)$

33. (a) ± 2 (b) $f(x) = (x - 2)(x + 2)(x^3 + x^2 - 2x + 4)$

35. (a) 1 of multiplicity two, -2 of multiplicity two, $-\frac{1}{2}$

(b) $F(x) = (x - 1)^2(x + 2)^2(2x + 1)$

37. More than 7 units   **39.** 4 ft by 3 ft by $1\frac{1}{2}$ ft

41. $\frac{1}{3}$ **43.** $\frac{2}{3}, -2$ **45.** 2   **47.** $k = -2, 4$

49. 5 inches

51. (a)

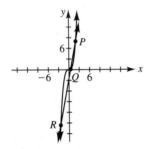

(b) $P(2, 8),\ Q\left(\frac{1}{2}, \frac{1}{8}\right),\ R\left(-\frac{5}{2}, -\frac{125}{8}\right)$

53. (a)

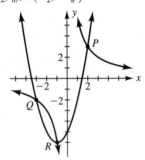

(b) $P(2, 3),\ Q(-3, -2),\ R(-1, -6)$

55. $1\frac{1}{2}$ cm or approximately 0.177 cm

Section 5.4

◆ Problems

1. $P(1) = -2$ and $P(2) = 7$.
 opposite signs

Hence, the function P has at least one real zero in the interval $[1, 2]$.

2. $P(1.3) = -0.203$ and $P(1.4) = 0.544$

$P(1.32) = -0.060032$ and $P(1.33) = 0.012637$

$P(1.328) = -0.00196045$ and $P(1.329) = 0.00533429$.

Hence, the irrational zero is 1.33 (nearest hundredth).

3. $P(-1) = -2$ and $P(0) = 1$.
opposite signs

Hence, the negative real zero of P is in the interval $[-1, 0]$. Using a calculator or computer with graphing capabilities, we find that the approximate value of this zero is -0.86 to the nearest hundredth.

4. *Step 1:* Possible rational zeros: $\pm 1, \pm 2, \pm 4, \pm 8$.

$$
\begin{array}{r|rrrrr}
-2 & 1 & 2 & 1 & -2 & -8 \\
   &   & -2 & 0 & -2 & 8 \\
\hline
   & 1 & 0 & 1 & -4 & ⓪
\end{array}
$$

Step 2: $P(x) = (x + 2)(x^3 + x - 4)$.
Thus, -2 is a rational zero.
The function Q defined by $Q(x) = x^3 + x - 4$ has exactly one positive real zero and no negative real zero.
Step 3: Since $Q(1) = -2$ and $Q(2) = 6$, we conclude a positive irrational zero lies in the interval $[1, 2]$.
Step 4: Using the method of successive approximations or a calculator with graphing capabilities, we find that this irrational zero is 1.38 (nearest hundredth). The remaining two zeros of P are imaginary zeros.

5. Since $P(179) = -29.041$ and $P(180) = 3000$, we conclude that the other positive real zero of P is in the interval $[179, 180]$. However, this zero is not considered a solution to the given problem since the domain of P is $(0, 60)$.

◆ Exercises

1. (b) 0.45   **3.** (b) 2.29   **5.** (b) -2.46

7. (b) -0.78 **9.** (b) 1.37   **11.** 0.13, 0.93

13. 0.10, 0.46, 0.83   **15.** 1.38   **17.** -1.55

19. $-1.88, 0.78, 4.11$ **21.** 2.09, $-1.15, -0.28$

23. 0.69, 1.78   **25.** $-1, 1.13$ **27.** $-\frac{3}{2}, -0.20$

29. $-1, 2, 1.22$ **31.** $\pm 2, -3.26, -0.28, 0.54$

33. 1, -1.25 **35.** Approximately 5.1 years

37. Approximately 4.44 in. by 4.44 in. by 2.24 in.

39. (a)

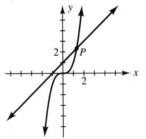

(b) $P(1.32, 2.32)$

41. (a)

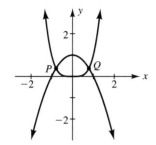

(b) $P(-0.79, 0.38)$, $Q(0.79, 0.38)$

43. (a)

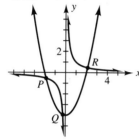

(b) $P(-1.86, -0.54)$, $Q(-0.25, -3.94)$, $R(2.11, 0.47)$
45. Approximately 3.85 cm
47. Approximately 1.3788, $-0.6894 \pm 1.5575i$
49. (a) $x^3 - 12x + 16 = 0$ (b) 3 of multiplicity two, -3

Section 5.5

◆ **Problems**

1. (a) $a_n = -2 < 0$ and $n = 6$ is even. Hence, the graph of P goes down to the left as $x \to -\infty$ and down to the right as $x \to \infty$.
    (b) $a_n = 4 > 0$ and $n = 3$ is odd. Hence, the graph of P goes down to the left as $x \to -\infty$ and up to right as $x \to \infty$.

2. It is impossible to make such a sketch.
3. *Step 1:* For the function P defined by $P(x) = x^3 + 2x^2 + 5x$, $a_n = 1$ and $n = 3$. Hence, the graph of P goes down to the left and up to the right with either $3 - 1 = 2$ relative extrema or no relative extremum.
    Step 2: y-intercept: $P(0) = 0^3 + 2(0)^2 + 5(0) = 0$
    Step 3: x-intercept: $0 = x^3 + 2x^2 + 5x$
                                $0 = x(x^2 + 2x + 5)$
                                $x = 0$ or $\underline{x^2 + 2x + 5 = 0}$
                                            no real solution
    Step 4:

x	-2	-1	$-\frac{1}{2}$	$\frac{1}{2}$	1	2
$P(x)$	-10	-4	$-\frac{17}{8}$	$\frac{25}{8}$	8	26

Step 5:

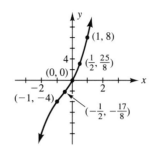

4. $P\left(\frac{2}{3}\right) = 2\left(\frac{2}{3}\right)^3 - 11\left(\frac{2}{3}\right)^2 + 12\left(\frac{2}{3}\right) + 9 = 12\frac{19}{27}$, which is larger than $P(1) = 12$.

5. To find the distances x where the bending moment is 375 lb-ft, we must solve the equation $x^3 - 100x + 375 = 0$. Possible rational zeros: $\pm 1, \pm 3, \pm 5, \pm 15, \pm 25, \pm 75, \pm 125, \pm 375$. Testing, we find that 5 is a rational zero:

$$\begin{array}{r|rrr} 5) & 1 & 0 & -100 & 375 \\ & & 5 & 25 & -375 \\ \hline & 1 & 5 & -75 & ⓪ \end{array}$$

Hence, $(x - 5)(x^2 + 5x - 75) = 0$

$x = 5$ or $x^2 + 5x - 75 = 0$

$$x = \frac{5 \pm \sqrt{5^2 - 4(1)(-75)}}{2(1)}$$

$$= \frac{-5 \pm 5\sqrt{13}}{2} \approx 6.51 \text{ or } -11.51$$

Thus, the bending moment is 375 lb-ft at exactly 5 ft or at approximately 6.51 ft.

◆ **Exercises**

1. (a) Down to the left as $x \to -\infty$, up to the right as $x \to \infty$
(b) Two
3. (a) Down to the left as $x \to -\infty$, down to the right as $x \to \infty$
(b) Three
5. (a) Up to the left as $x \to -\infty$, down to the right as $x \to \infty$
(b) Six
7. (a) Up to the left as $x \to -\infty$, up to the right as $x \to \infty$
(b) Five
9.

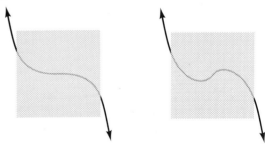

11.

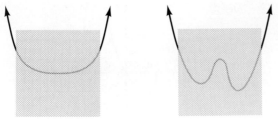

13.

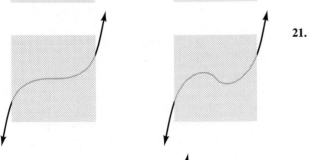

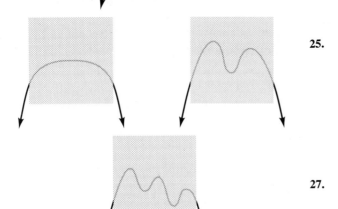

15.

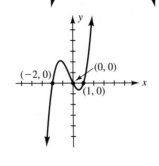

17.

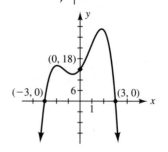

19.

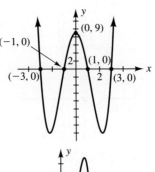

21.

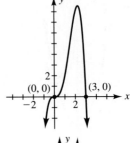

23.

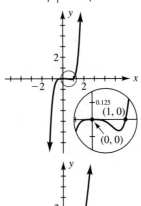

25.

27.

29.

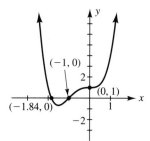

31.

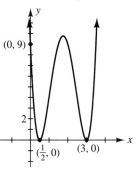

33.

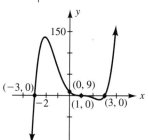

35.

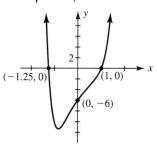

37.

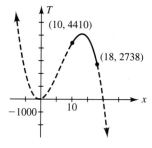

39. (a)

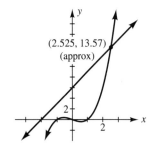

(b)

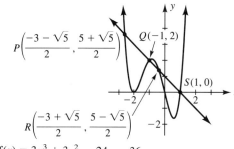

41. $f(x) = 3x^3 + 3x^2 - 24x - 36$

43. (a)

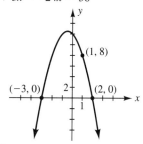

(b) $P(x) = -2x^2 - 2x + 12$

45. (a)

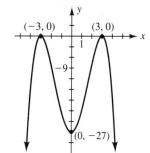

(b) $P(x) = -\frac{1}{3}x^4 + 6x^2 - 27$

47. (a) $V = 64x - 28x^2 + 3x^3$

(b)

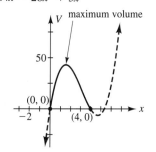

(c) Approximately 1.51 cm

Section 5.6

 Problems

1. The zeros of the denominator
$D(x) = x^2 - 2x - 8 = (x + 2)(x - 4)$ are -2 and 4.
Thus, the domain of f is $(-\infty, -2) \cup (-2, 4) \cup (4, \infty)$.

2. Vertical asymptote: $x^2 - 2x + 1 = 0$
$$(x - 1)^2 = 0$$
$$x = 1$$

Horizontal asymptote:

$$f(x) = \frac{2x - 4}{x^2 - 2x + 1} = \frac{\dfrac{2}{x} - \dfrac{4}{x^2}}{1 - \dfrac{2}{x} + \dfrac{1}{x^2}}$$

As $|x| \to \infty$, $f(x) \to 0$. Hence, $y = 0$, is the horizontal asymptote.

3. From Problem 2, we know that $x = 1$ is the vertical asymptote. As $x \to 1^-$, $f(x) \to -\infty$, and as $x \to 1^+$, $f(x) \to -\infty$. From Problem 2, $y = 0$ is the horizontal asymptote. As $x \to \infty$, $f(x) \to 0^+$, and as $x \to -\infty$, $f(x) \to 0^-$. Also, we have
x-intercept: $2x - 4 = 0$ y-intercept: $f(0) = -4$
$$x = 2$$

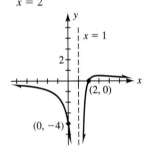

4. $f(x) = \dfrac{2x^3 + 3}{x^2} = 2x + \dfrac{3}{x^2}$

As $|x| \to \infty$, $\dfrac{3}{x^2} \to 0$. Hence, $f(x) \to 2x$. Thus, the line $y = 2x$ is an oblique asymptote.

5. Since $f(x) = \dfrac{x^2 + 2x}{x + 2} = \dfrac{x(x + 2)}{x + 2}$, we reduce to lowest terms and obtain the new function $F(x) = x$. The graph of f is the same as the graph of F, except it has a hole at $(-2, -2)$ as shown.

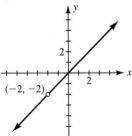

6. To determine the number of kits that are produced if the average cost per kit is \$16,640, we solve the equation

$$16{,}640 = \frac{141{,}120 + 12{,}000x + 20x^2}{x}$$
$$16{,}640x = 141{,}120 + 12{,}000x + 20x^2$$
$$20x^2 - 4640x + 141{,}120 = 0$$
$$x^2 - 232x + 7056 = 0$$
$$x = \frac{-(-232) \pm \sqrt{(-232)^2 - 4(1)(7056)}}{2(1)} = 36 \text{ or } 196 \text{ kits}$$

 Exercises

1. (a) $x = -2$ (b) as $x \to -2^-$, $f(x) \to \infty$, as $x \to -2^+$, $f(x) \to -\infty$ **3.** (a) none

5. (a) $x = \frac{1}{2}, x = -3$

(b) as $x \to \frac{1}{2}^-$, $f(x) \to \infty$; as $x \to \frac{1}{2}^+$, $f(x) \to -\infty$
as $x \to -3^-$, $f(x) \to -\infty$; as $x \to -3^+$, $f(x) \to \infty$

7. (a) $x = 2$ (b) as $x \to 2^-$, $f(x) \to -\infty$; as $x \to 2^+$, $f(x) \to \infty$

9. (a) $x = 2, x = 4, x = -1$

(b) as $x \to 2^-$, $f(x) \to -\infty$; as $x \to 2^+$, $f(x) \to \infty$
as $x \to 4^-$, $f(x) \to \infty$; as $x \to 4^+$, $f(x) \to -\infty$
as $x \to -1^-$, $f(x) \to -\infty$; as $x \to -1^+$, $f(x) \to \infty$

11. (a) $y = 1$ (b) as $x \to \infty$, $f(x) \to 1^+$; as $x \to -\infty$, $f(x) \to 1^-$ (c) does not cross

13. (a) $y = \frac{1}{2}$ (b) as $x \to \infty$, $f(x) \to \frac{1}{2}^-$;

as $x \to -\infty$, $f(x) \to \frac{1}{2}^+$ (c) crosses at $\left(\frac{13}{7}, \frac{1}{2}\right)$

15. None exist

17. (a) $y = -3$ (b) as $x \to \infty$, $f(x) \to -3^-$; as $x \to -\infty$,

$f(x) \to -3^+$ (c) crosses at $\left(-\frac{1}{10}, -3\right)$

19. $y = 0$ (b) as $x \to \infty$, $f(x) \to 0^+$; as $x \to -\infty$, $f(x) \to 0^+$

(c) crosses at $\left(\frac{1}{3}, 0\right)$

21.

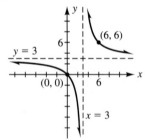

23.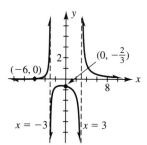

25.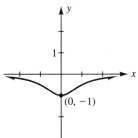

27.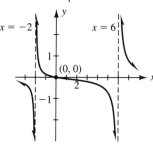

29.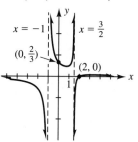

31.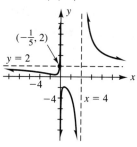

33.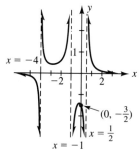

35.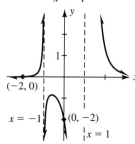

37. (a) $y = x$ (b) Does not cross
(c)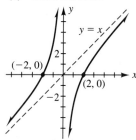

39. (a) $y = x + 3$ (b) Does not cross
(c)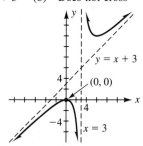

41. (a) $y = x - 6$ (b) Does not cross
(c)

43. (a) $y = x + 3$ (b) Crosses at $\left(-\frac{20}{13}, \frac{19}{13}\right)$
(c)

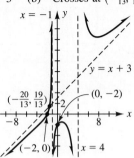

45.

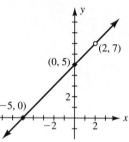

47.

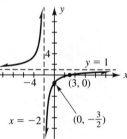

49.

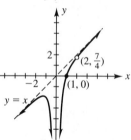

51. No. The procedure of dividing numerator and denominator by the highest power of x that appears in the rational function yields at most one horizontal asymptote.

53. One example is $f(x) = \dfrac{x^2 - 6x + 8}{x^3}$; its graph crosses the horizontal asymptote $y = 0$ at $(2, 0)$ and $(4, 0)$.

55. (a)

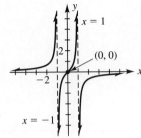

(b)

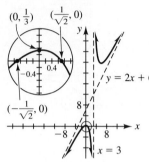

57. (a) $C(x) \to \infty$; a prohibitive cost
(b)

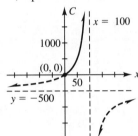

59. (b)

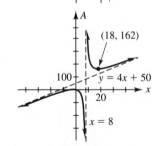

(c) $x = 18$ inches by $y = 9$ inches

61. (a) $C(x) = \dfrac{81{,}000 + (400x + 0.1x^2)}{x}$

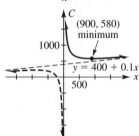

(b) Approximately 900 stoves

Section 5.7

 Problems

1. $\dfrac{2}{x} + \dfrac{1}{x-1} + \dfrac{-3}{x+2}$

$$= \dfrac{2(x-1)(x+2) + 1(x)(x+2) - 3(x)(x-1)}{x(x-1)(x+2)}$$

$$= \dfrac{2(x^2 + x - 2) + (x^2 + 2x) - 3(x^2 - x)}{x(x-1)(x+2)}$$

$$= \dfrac{7x - 4}{x(x-1)(x+2)} = f(x)$$

2. Equating the x terms and using the fact that $A = 1$ and $C = -3$, we have

$$0 = 6A + 3B + C$$
$$0 = 6(1) + 3B + (-3)$$
$$-3 = 3B$$
$$B = -1$$

3. $\dfrac{x-8}{x^3 + 4x} = \dfrac{x-8}{x(x^2+4)} = \dfrac{A}{x} + \dfrac{Bx+C}{x^2+4}$

Decomposition equation:

$$x - 8 = A(x^2 + 4) + (Bx + C)x$$
or $\qquad x - 8 = (A + B)x^2 + Cx + 4A.$

Letting $x = 0$, we find $-8 = 4A$, or $A = -2$. Equating the x^2 terms gives us $A + B = 0$, or $B = 2$. Finally, equating the x terms gives us $C = 1$. Thus,

$$f(x) = \dfrac{x-8}{x^3+4x} = \dfrac{-2}{x} + \dfrac{2x+1}{x^2+4}.$$

4. $\dfrac{6x^3}{x^4 + 2x^2 + 1} = \dfrac{6x^3}{(x^2+1)^2} = \dfrac{Ax+B}{x^2+1} + \dfrac{Cx+D}{(x^2+1)^2}$

Decomposition equation:

$$6x^3 = (Ax + B)(x^2 + 1) + (Cx + D)$$
or $\qquad 6x^3 = Ax^3 + Bx^2 + (A+C)x + (B+D).$

Equating the x^3 terms, we have $A = 6$. Equating the x^2 terms gives us $B = 0$. Equating the x terms, we obtain $A + C = 0$, or $C = -6$. Finally, equating the constant terms, we find $D = 0$. Thus,

$$f(x) = \dfrac{6x^3}{x^4 + 2x^2 + 1} = \dfrac{6x}{x^2+1} + \dfrac{-6x}{(x^2+1)^2}.$$

5. Using long division, we find

$$\begin{array}{r} 1 \\ x^2 - 2x + 1 \overline{) x^2 - 3x - 8} \\ \underline{x^2 - 2x + 1} \\ -x - 9 \end{array}$$

Hence, $\dfrac{x^2 - 3x - 8}{x^2 - 2x + 1} = 1 + \dfrac{-x - 9}{x^2 - 2x + 1}$

Now, we have $\dfrac{-x-9}{x^2 - 2x + 1} = \dfrac{-x-9}{(x-1)^2} = \dfrac{A}{x-1} + \dfrac{B}{(x-1)^2}.$

Decomposition equation:

$$-x - 9 = A(x - 1) + B$$
or $\qquad -x - 9 = Ax + (-A + B).$

Letting $x = 1$ gives us $B = -10$. Equating the x-terms, we find $A = -1$. Thus,

$$f(x) = \dfrac{x^2 - 3x - 8}{x^2 - 2x + 1} = 1 + \dfrac{-1}{x-1} + \dfrac{-10}{(x-1)^2}.$$

 Exercises

1. $f(x) = \dfrac{-1}{x+3} + \dfrac{1}{x-3}$ **3.** $f(x) = \dfrac{-2}{x+2} + \dfrac{3}{x+4}$

5. $f(x) = \dfrac{1}{x} + \dfrac{-1}{x-2} + \dfrac{1}{x+1}$

7. $f(x) = \dfrac{-1}{x} + \dfrac{-1}{x^2} + \dfrac{1}{x-1}$

9. $f(x) = \dfrac{-2}{x} + \dfrac{3}{x-2} + \dfrac{-4}{(x-2)^2}$

11. $f(x) = \dfrac{1}{x} + \dfrac{2x-1}{x^2+x+1}$

13. $f(x) = \dfrac{-1}{x+1} + \dfrac{x-1}{x^2-x+1}$

15. $f(x) = \dfrac{1}{2x+3} + \dfrac{2x-3}{x^2+9}$

17. $f(x) = \dfrac{-1}{x+1} + \dfrac{1}{x-1} + \dfrac{5}{x^2+4}$

19. $f(x) = \dfrac{3x-1}{x^2+5} + \dfrac{2x+1}{x^2+1}$

21. $f(x) = \dfrac{x}{x^2+1} + \dfrac{-5x}{(x^2+1)^2}$

23. $f(x) = \dfrac{1}{x} + \dfrac{-x+2}{x^2+4} + \dfrac{-7x-8}{(x^2+4)^2}$

25. $f(x) = 3x - 3 + \dfrac{8}{x-2} + \dfrac{1}{x-1}$

27. $f(x) = 1 + \dfrac{1}{x} + \dfrac{2}{x^2} + \dfrac{-1}{x-1}$

29. $f(x) = x + \dfrac{-2}{x} + \dfrac{4}{x+2} + \dfrac{4}{x-2}$

31. $f(x) = x^2 - 2 + \dfrac{1}{x^2} + \dfrac{-x+3}{x^2+2}$

33. $f(x) = \dfrac{1}{x-1} + \dfrac{-5}{x-2} + \dfrac{5}{x-3}$

35. $f(x) = \dfrac{1}{x-2} + \dfrac{-x+2}{x^2+x+3}$

37. $f(x) = \dfrac{-3}{x-1} + \dfrac{1}{(x-1)^2} + \dfrac{4}{(x-1)^3}$

39. $f(x) = \dfrac{2}{x+3} + \dfrac{-1}{x+1} + \dfrac{4}{(x-2)^2}$

41. $f(x) = \dfrac{165}{x} + \dfrac{-282}{x-244}$

43. $f(x) = \dfrac{18}{x-14} + \dfrac{21x-15}{x^2+21}$

45. $f(x) = 12x + 35 + \dfrac{13}{x+17} + \dfrac{28}{x-17}$

Chapter 5 Review Exercises

1. $3x + 2 + \dfrac{-6x+2}{x^2+1}$ **3.** $2x^3 - 2x^2 + 4x - 5 + \dfrac{9}{x+2}$

5. $6x^3 + 9x + 3 + \dfrac{-3}{x - \frac{1}{3}}$

7. $x^3 - x - 1 + \dfrac{x^2+2x+2}{x^3+x+1}$ **9.** 6   **11.** 0

13. Yes   **15.** No   **17.** Yes

19. (a) 1 positive real zero, 0 or 2 negative real zeros

21. (a) 0 or 2 positive real zeros, 0 or 2 negative real zeros

23. (a) 0 or 2 positive real zeros, 1 negative real zero

25. (a) $P(x) = (x-2)(x-4)(x+1)$ (b) 2, 4, −1

27. (a) $P(x) = (3x-4)\big[x - \big(1 + \sqrt{6}\big)\big]\big[x - \big(1 - \sqrt{6}\big)\big]$

(b) $\frac{4}{3}, 1 + \sqrt{6}, 1 - \sqrt{6}$

29. (a) $P(x) = (2x-1)(x+3)(x+2i)(x-2i)$

(b) $\frac{1}{2}, -3, \pm 2i$

31. (a) $P(x) = (x-1)(x+1)(x-2)(2x-1)(2x+3)$

(b) $\pm 1, 2, \frac{1}{2}, -\frac{3}{2}$

33. (a) $P(x) = (x+i)(x-i)[x-(2+2i)][x-(2-2i)]$
(b) $\pm i, 2 \pm 2i$

35. (a) $P(x) = [x-(2+i)][x-(2-i)](2x-1)(x+2i)$
$(x - 2i)$

(b) $2 \pm i, \frac{1}{2}, \pm 2i$ **37.** 1.67   **39.** $\frac{5}{2}, -1.15$

41. −5.98, −0.92   **43.** 2 of multiplicity two, 4.17

45.

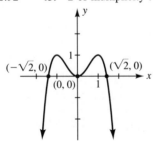

47.

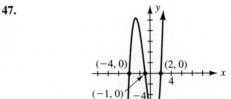

49.

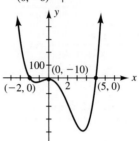

51.

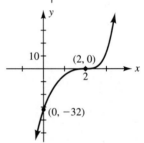

53.

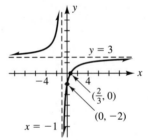

55.

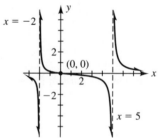

57.

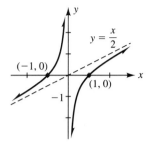

$y = \dfrac{x}{2}$

$(-1, 0)$ $(1, 0)$

-1

59.

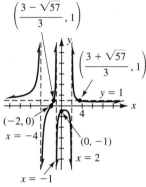

$\left(\dfrac{3 - \sqrt{57}}{3}, 1 \right)$

$\left(\dfrac{3 + \sqrt{57}}{3}, 1 \right)$

$y = 1$

$(-2, 0)$

$x = -4$

$(0, -1)$

$x = 2$

$x = -1$

61. $f(x) = \dfrac{5}{x+1} + \dfrac{5}{x-1}$ **63.** $f(x) = \dfrac{1}{x-4} + \dfrac{2x-5}{x^2+1}$

65. $f(x) = \dfrac{2}{x} + \dfrac{-2x}{x^2+3} + \dfrac{-6x}{(x^2+3)^2}$

67. (a) $P(x) = \frac{1}{2}(x^3 - 2x^2 - 5x + 6)$

(b) $P(x) = 2(x^3 - 2x^2 - 5x + 6)$

69.

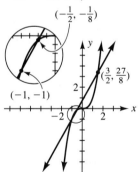

$\left(-\frac{1}{2}, -\frac{1}{8} \right)$

$\left(\frac{3}{2}, \frac{27}{8} \right)$

$(-1, -1)$

-2 2

73. (a) No positive real zero, one negative real zero

(b) One positive real zero, no negative real zero

75. $y = \frac{2}{3}$; yes; $\left(\frac{5}{3}, \frac{2}{3} \right)$ **77.** $2\frac{1}{2}$ feet   **79.** 4 centimeters

81. (a) $V = 4x^3 - 70x^2 + 300x$

(b) 5 cm or approximately 1.10 cm

(c)

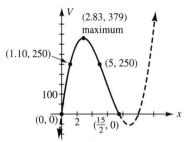

V $(2.83, 379)$ maximum

$(1.10, 250)$ $(5, 250)$

100

$(0, 0)$ 2 $(\frac{15}{2}, 0)$ x

(d) Approximately 2.83 centimeters

83. (b)

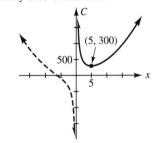

C

$(5, 300)$

500

5 x

(c) In meters, the dimensions are 5 by 5 by 4.

CHAPTER 6

Section 6.1

Problems

1. $\dfrac{125^{x-2}}{25^{2x-3}} = \dfrac{(5^3)^{x-2}}{(5^2)^{2x-3}} = \dfrac{5^{3x-6}}{5^{4x-6}} = 5^{-x} = \left(\dfrac{1}{5} \right)^x$

2.

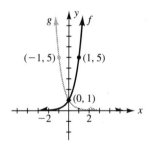

g y f

$(-1, 5)$ $(1, 5)$

$(0, 1)$

-2 2 x

3. Referring to the graph of $G(x) = 3 - 4^{-x}$ (see Figure 6.4), we determine that the domain is $(-\infty, \infty)$ and the range is $(-\infty, 3)$.

4. The x-intercept is found by solving the equation

$$0 = e^{-x} - 2 \quad \text{or} \quad e^{-x} = 2.$$

However we do not yet have a procedure for solving an equation in which the unknown appears as an exponent.

5. If interest is compounded monthly, then $n = 12$. Hence,

$$A = 1000 \left(1 + \frac{0.08}{12} \right)^{12 \cdot 5} \approx \$1489.85.$$

Exercises

1. 1   **3.** $\left(\frac{1}{36}\right)^x$ **5.** 1   **7.** e^x **9.** 9^x **11.** $\frac{1}{3}$
13. 5

15.

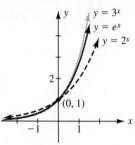

17.

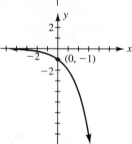

19.

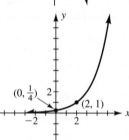

21.

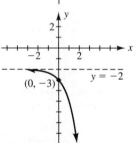

23.

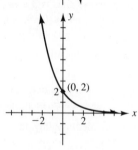

25.

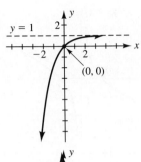

27.

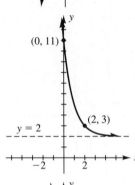

29.

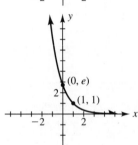

31. $7057.91   **33.** $27,437.17   **35.** $3432.01
37. $124,897.29   **39.** (a) $4^x - 4^{-x}$ (b) 4   **41.** 2
43. 2   **45.** ± 1 **47.** $\frac{1}{4}$
49.

x	0	1	2	3	4	-1	-2	-3	-4
$f(x)$	1	-2	4	-8	16	$-1/2$	$1/4$	$-1/8$	$1/16$

The function f is not defined when $x = 1/n$, where n is an
even integer. Thus, we cannot connect these points to form
the graph.
51. (a) $k = 15$ (b) 6.74 psi
53. (a) 18 bacteria   (b) 133 bacteria
(c)

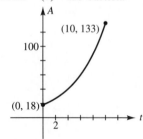

55. (a) $12,000   (b) $4466
(c)

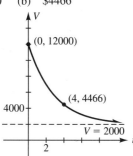

(d) The machine's scrap value is $2000.
57. (a) 0.211   (b) 0.0393   (c) 23.1   (d) 22.5
59. (b)

Δx	1	0.1	0.01	0.001	0.0001
$\dfrac{e^{\Delta x} - 1}{\Delta x}$	1.71828	1.05171	1.00502	1.00050	1.00005

(c) $\dfrac{e^{\Delta x} - 1}{\Delta x}$ approaches 1 as $\Delta x \to 0^+$

Section 6.2

◆ Problems

1. Since $\left(\frac{1}{2}\right)^{-3} = 8$, we conclude that $\log_{1/2} 8 = -3$.

2. Using the key on a calculator, we find that for log 0.01 the display reads -2. Using the $\boxed{\text{LN}}$ key on a calculator, we find that for ln (-1) the display shows *error*.

3. Since $\log_b b^x = x$ for all real x, we have $\log_3 3^{12} = 12$.

4. $(\log_2 2)^{4t-8} = (1)^{4t-8} = 1$

5. $3 \log_2 (5^2 - 2(5) + 1) = 3 \log_2 16 = 3 \cdot 4 = 12$ and
$3 \log_2 [(-3)^2 - 2(-3) + 1] = 3 \log_2 16 = 3 \cdot 4 = 12.$

6. $7e^{1-2[(1-\ln 4)/2]} = 7e^{\ln 4} = 7 \cdot 4 = 28$

7. $2 = e^{r(5.5)}$
$r(5.5) = \ln 2$
$r = \dfrac{\ln 2}{5.5} \approx 0.126$ or   12.6%

8. $540 = 18e^{0.1354t}$
$30 = e^{0.1354t}$
$0.1354t = \ln 30$
$t = \dfrac{\ln 30}{0.1354} \approx 25$ minutes

9. $27 = 30e^{-0.02476t}$
$0.9 = e^{-0.02476t}$
$-0.02476t = \ln 0.9$
$t = \dfrac{\ln 0.9}{-0.02476} \approx 4.26$ years

◆ Exercises

1. 2   **3.** 3   **5.** 1   **7.** -4 **9.** -2 **11.** 2
13. 0   **15.** 2   **17.** Undefined   **19.** $\frac{3}{2}$ **21.** $-\frac{4}{3}$
23. 3   **25.** 5   **27.** 10   **29.** 0   **31.** $x - 3$
33. $x + 2$ **35.** $x^2 + 4$ **37.** $x^2 + 2x$ **39.** 1
41. $3x + 6$ **43.** xe^x **45.** x^2 **47.** $b = 3$
49. $b = \frac{1}{16}$ **51.** $x = \frac{1}{100}$ **53.** $x = 8$ **55.** $x = 2e$
57. $x = \frac{1}{2}, -3$ **59.** $x = 1 \pm \sqrt{10}$
61. $x = \log 5 \approx 0.699$ **63.** $x = \dfrac{\ln \frac{2}{3}}{3} \approx -0.135$
65. $x = \dfrac{3 - \log 28}{2} \approx 0.7764$ **67.** $x = -\dfrac{\ln \frac{15}{4}}{7} \approx -0.1888$
69. $x = -\log 60 \approx -1.778$
71. (a) 5.8 yr   (b) 25.6 yr   (c) 65.8 yr   **73.** 11%
75. (a) $A(t) = 24e^{0.1792t}$ (b) 1996
77. Approximately 2140 B.C.   **79.** (a) 3   (b) 4
81. (a) $(0, e) \cup (e, \infty)$ (b) $(0, 10^{10}]$
83. (a) $i(t) \to 20$ amperes   (b) Approximately 79 ms
85. (a) $T = 70 + 330e^{-0.1805t}$ (b) 262°F
(c) About 7.85 min   **87.** 3.99   **89.** 2.24   **91.** 2.01
93. (a) It appears that ln $(xy) = \ln x + \ln y$
(b) It appears that ln $(x/y) = \ln x - \ln y$
(c) It appears that ln $(x^y) = y \ln x$
95. (a) pH ≈ 6.4 (b) pH ≈ 3.2

Section 6.3

◆ Problems

1. For example, $(\log_2 8)^2 \neq 2 \log_2 8$.
Note that $(\log_2 8)^2 = 3^2 = 9$, whereas $2 \log_2 8 = 2 \cdot 3 = 6$.

2. For most scientific calculators, the keying sequence is as follows:
$\boxed{50}$ $\boxed{\text{LOG}}$ $\boxed{+}$ $\boxed{2}$ $\boxed{\times}$ $\boxed{4}$ $\boxed{\text{LOG}}$ $\boxed{-}$ $\boxed{3}$ $\boxed{\times}$ $\boxed{2}$ $\boxed{\text{LOG}}$ $\boxed{=}$

3. $\log_4 24 = \dfrac{\ln 24}{\ln 4} \approx 2.292$

4. $(\log_2 5)(\log_5 8) = \left(\dfrac{1}{\log_5 2}\right)(\log_5 2^3)$
$= \left(\dfrac{1}{\log_5 2}\right)[3 \log_5 2] = 3$

5. Letting I_a = the intensity of the 1933 earthquake and
I_b = the intensity of the 1989 earthquake,
we have

$$8.9 = \log I_a - \log I_0$$
$$\underline{7.1 = \log I_b - \log I_0}$$
$$1.8 = \log I_a - \log I_b$$

Hence, $I_a/I_b = 10^{1.8} \approx 63$ times as intense.

 Exercises

1. $\log_2(8 + 8) = \log_2 16 = 4$, whereas

$\log_2 8 + \log_2 8 = 3 + 3 = 6$.   **3.** $\dfrac{\log_2 8}{\log_2 2} = \dfrac{3}{1} = 3$, whereas

$\log_2 \dfrac{8}{2} = \log_2 4 = 2$.   **5.** 10   **7.** $3 + \log_2 x + \log_2 (x + 2)$

9. $2 + \ln x - \ln 10$ **11.** $\frac{2}{3} \log_3 x - 1$

13. $2 \log_b x - \frac{2}{3} \log_b (x + 1)$

15. $-\left(2 + \frac{3}{2} \log_3 x + \frac{1}{3} \log_3 y\right)$ **17.** $2x^2 - 2 \ln (e^x + 1)$

19. $\ln 24$ **21.** 2   **23.** -2

25. $\log_5 \dfrac{x^2 + 2x - 3}{x^2}$ **27.** $\ln \dfrac{x - 2}{x + 2}$

29. $\ln \dfrac{9\sqrt{x^2 - 9}}{x^2}$ **31.** 3.585   **33.** 2.757

35. -0.3691 **37.** -8.697 **39.** 1   **41.** 4

43. 2   **45.** 1   **47.** About 32 times more intense

49. No. The domain of f is $(-\infty, 0) \cup (0, \infty)$ and the domain
of g is $(0, \infty)$.

51. (a) $\frac{1}{2}$ (b) $\frac{1}{3}$ (c) 1   (d) $\frac{1}{6}$ (e) $\frac{1}{5}$ (f) $-\frac{1}{5}$

(g) -1 (h) 1   **53.** About 1.32×10^{10} times brighter

55. 10^9 times more intense

57. $\log (6.2 \times 10^k) = \log 6.2 + \log 10^k = \log 6.2 + k$

59. (a) 1   (b) $\ln e = 1$

Section 6.4

 Problems

1.

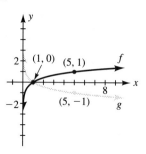

2. The graph of $H(x) = 1 - \log_4 x$ is the same as the graph
of $f(x) = \log_4 x$ reflected about the x-axis and then shifted
vertically upward 1 unit.

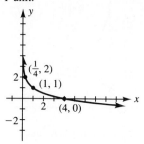

3. $\log_4 16x = \log_4 16 + \log_4 x = 2 + \log_4 x$. Thus, the
graph of $F(x) = \log_4 16x$ is the same as the graph of
$f(x) = \log_4 x$ shifted upward 2 units.

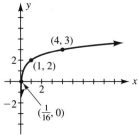

4. $F(x) = \log_4 x^3 = 3 \log_4 x$. Hence, the graph of F may be
obtained by stretching the graph of $f(x) = \log_4 x$ vertically by a
factor of 3.

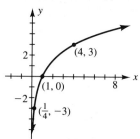

5. $R(10) = 100 - 29 \log [27(10) + 1]$
$= 100 - 29 \log 271 \approx 29.4\%$
$R(20) = 100 - 29 \log [27(20) + 1]$
$= 100 - 29 \log 541 \approx 20.7\%$
$R(30) = 100 - 29 \log [27(30) + 1]$
$= 100 - 29 \log 811 \approx 15.6\%$

Exercises

1.

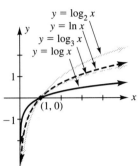

3.

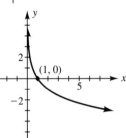

5.

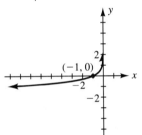

7.

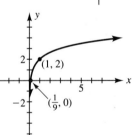

9.

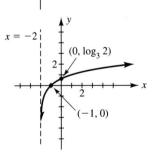

11.

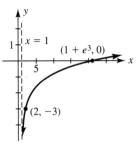

13.

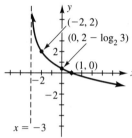

15.

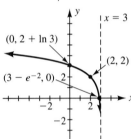

17. f and h **19.** f and g **21.** f and g

23.

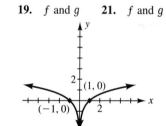

25.

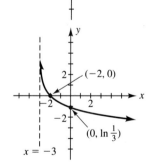

27.

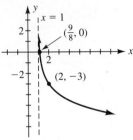

29.

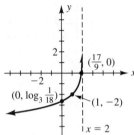

31. (a) 100% (b) 69% (c) About 4 days

(d)

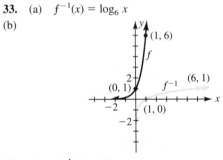

33. (a) $f^{-1}(x) = \log_6 x$

(b)

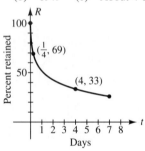

35. (a) $h^{-1}(x) = 2 \ln x$

(b)

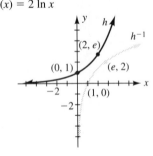

37. (a) $G^{-1}(x) = \frac{1}{2} \log(1 - x)$

(b)

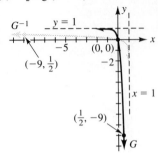

39. (a) $f^{-1}(x) = 8^x$

(b)

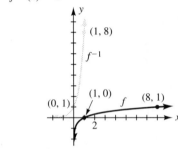

41. (a) $h^{-1}(x) = 2e^{-x}$

(b)

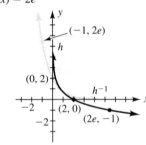

43. (a) $G^{-1}(x) = \frac{1}{3}[10^{(x-1)/2}]$

(b)

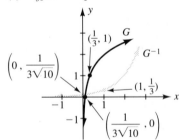

45. (a) 1000 (b) 5615 (c) About $19,000

(d)

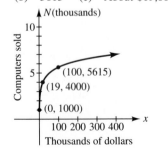

47. (a) $(0, \infty)$ (b) there are no zeros (c) $f(x) \to 0$
(d) $f(x) \to -\infty$
(e)

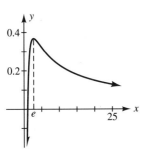

(f) $x = e$

Section 6.5

◆ Problems

1. $2 \ln x - \ln 9 = 4$

$$\ln \frac{x^2}{9} = 4$$

$$e^4 = \frac{x^2}{9}$$

$$x^2 = 9e^4$$

$$x = 3e^2$$

2. $x + \ln y = \ln c$

$$\ln y - \ln c = -x$$

$$\ln \frac{y}{c} = -x$$

$$e^{-x} = \frac{y}{c}$$

$$y = ce^{-x}$$

3. $4^{3x+2} = 8^{4x}$

$$(2^2)^{3x+2} = (2^3)^{4x}$$

$$2^{6x+4} = 2^{12x}$$

$$6x + 4 = 12x$$

$$x = \tfrac{2}{3}$$

4. $5^{x-1} = 325$

$$(x - 1) \log 5 = \log 325$$

$$x - 1 = \frac{\log 325}{\log 5}$$

$$x = 1 + \frac{\log 325}{\log 5} \approx 4.594$$

5. $\log \sqrt{x} = \sqrt{\log x}$

$$\tfrac{1}{2} \log x = \sqrt{\log x}$$

$$\tfrac{1}{4} (\log x)^2 = \log x$$

$$(\log x)^2 - 4 \log x = 0$$

$$\log x (\log x - 4) = 0$$

$$\log x = 0 \quad \text{or} \quad \log x - 4 = 0$$

$$x = 1 \qquad\qquad x = 10^4$$

6. On most scientific calculators, the keying sequence is
$2 \boxed{Y^x} 1.620 \boxed{+} 3 \boxed{Y^x} 1.620$. The display reads $9.002135812 \approx 9$.

◆ Exercises

1. 2   **3.** $3e - 1$ **5.** 4   **7.** 2   **9.** 4   **11.** 1

13. $y = ce^{-2x}$ **15.** $y = \dfrac{10}{(x + a)^2}$ **17.** $y = \dfrac{kx - x}{1 + k}$,

where $k = e^c$ **19.** $y = \dfrac{ka}{1 + kb}$, where $k = e^{ac}$ **21.** $\tfrac{3}{2}$

23. $-\tfrac{1}{3}$ **25.** $\dfrac{\ln 35}{\ln 7} \approx 1.827$ **27.** $\dfrac{\ln 120}{\ln 9} \approx 2.179$

29. $\dfrac{1}{2 + \log 4} \approx 0.384$ **31.** $\dfrac{\ln 4}{\ln 4 - 1} \approx 3.589$

33. $\dfrac{\ln \left(\tfrac{3}{4} \right)}{\ln 3} \approx -0.2619$ **35.** $\dfrac{\ln 3072}{\ln 16} \approx 2.896$ **37.** $1, e^2$

39. $\dfrac{\ln 4}{2}$ **41.** 0   **43.** $\ln \left(10 \pm 3\sqrt{11} \right) \approx \pm 2.993$

45. $3, 9$ **47.** $t = \dfrac{\ln (A/P)}{n \ln [1 + (r/n)]}$ **49.** (a) A straight
line   **51.** (a) $P = P_0 \, 2^{-(1/3)t}$ (b) 25% **53.** 2.123
55. 1.202   **57.** $-1.637, 1.000$

59. (a) 50 ft (b) $50 \ln \left(\dfrac{3 \pm \sqrt{5}}{2} \right) \approx \pm 48.1$ ft

Chapter 6 Review Exercises

1. 1   **3.** e **5.** $\left(\tfrac{1}{2} \right)^x$ **7.** 2   **9.** $\tfrac{2}{3}$

11. -3 **13.** $\tfrac{1}{3}$ **15.** $-\tfrac{2}{3}$ **17.** 4   **19.** 3

21. $2x$ **23.** $x^2 + 1$ **25.** $3 - 3x$ **27.** xe^{2x}
29. 1   **31.** $2 + 2 \log_3 x + \log_3 (x - 1)$

33. $-\log x - \tfrac{1}{2} \log (2x - 3)$

35. $-x^2 + \ln x - \ln (e^x - 1)$ **37.** $\ln 8$

39. $\log_3 (x + 1)^2$ **41.** 0   **43.** 5   **45.** $-2 \pm e^3$

47. No solution   **49.** 3   **51.** $y = kx^2$, where $k = e^c$

53. $\dfrac{\log 80}{2} \approx 0.9515$ **55.** $-\tfrac{3}{4}$ **57.** $\dfrac{\log 4}{\log 36} \approx 0.3869$

59. $\dfrac{\ln 6}{1 + \ln 9} \approx 0.5604$ **61.** ± 1 **63.** $3, \sqrt{3}$

65.

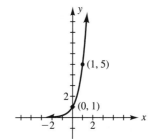

67.

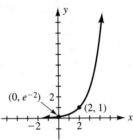

69.

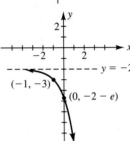

71.

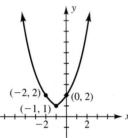

73.

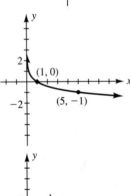

75.

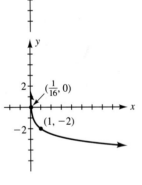

77.

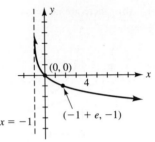

79.

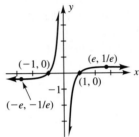

81. 3.465 **83.** -6.644 **85.** $f^{-1}(x) = \log_8 x$

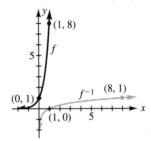

87. $h^{-1}(x) = \dfrac{2 + \ln (x/2)}{3}$

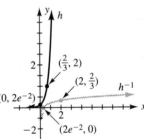

89. $G^{-1}(x) = 2^{-(x+1)}$

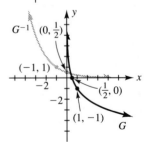

91. $f^{-1}(x) = \sqrt{e^{1-x}}$

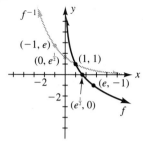

93. (a) $12,058.57 (b) $12,256.79 (c) $12,298.02

95. Approximately 7 years

97. (a) $A(t) = 10^3 e^{1.535t}$, t in hours

(b) Approximately 27 minutes

99. About 794 times more intense **101.** (a) $k \approx 0.0405$

(b) Approximately 99°F (c) About 35 minutes

(d)

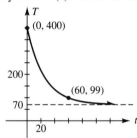

CUMULATIVE REVIEW EXERCISES FOR CHAPTERS 3, 4, 5, & 6

1. $e^{-b/a}$

3.

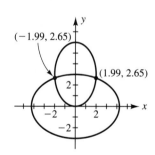

5. (a) $f(x) \to -\infty$ (b) $f(x) \to \infty$

7. $(x + 2)(x - 2)(x^2 + 2x + 5)$ **9.** (a) 3 (b) e^2, e^{-1}

11. (a) Line (b) Parabola (c) Circle

(d) Hyperbola

13.

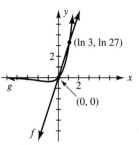

15. $f(0)$ **17.** $7\sqrt{2}$ **19.** 70 square units

21. (a) Symmetric with respect to the y-axis

(b) Symmetric with respect to the origin

(c) Symmetric with respect to the x-axis

(d) Symmetric with respect to the x-axis, y-axis, and origin

23. (a) $(f \circ g)(x) = -x$; domain, $(-\infty, \infty)$

(b) $(g \circ f)(x) = \dfrac{-x}{x + 1}$; domain, $(-1, \infty)$

25. (a) $y = 1$ (b) $y = x - 2$ (c) $3x + 2y = 11$

(d) $y = -3x + 10$ **27.** (a) $y = e^{-2}x^4$ (b) $y = \dfrac{e^x}{x + 2}$

(c) $y = 2$

29. (a)

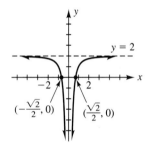

(b)

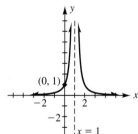

31. $m = \frac{5}{3}$ **33.** $20,544.33

35. (a) $A = 6000 + 100x - 20x^2$ (b) in $2\frac{1}{2}$ weeks

37. $A = \dfrac{x^2}{18}$

39. (a) $v = 2t + 10$

(b)

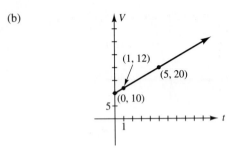

The y-intercept represents the initial velocity.

CHAPTER 7

Section 7.1

◆ Problems

1. (a) Replacing x with $-\frac{1}{2}$ and y with 1 in *both* original equations, we find

$$2\left(-\frac{1}{2}\right) - 5(1) = -1 - 5 = -6$$

$$-2\left(-\frac{1}{2}\right) + 3(1) = 1 + 3 = 4$$

(b) Replacing x with -6 and y with -2 in *both* original equations, we find

$$\frac{1}{2}(-6) + (-2) = -3 - 2 = -5$$

$$-3(-6) + 5(-2) = 18 - 10 = 8$$

2. For $a = -2$, we have $(4(-2) - 2, (-2)) = (-10, -2)$. For $a = 0$, we have $(4(0) - 2, (0)) = (-2, 0)$. For $a = 5$, we have $(4(5) - 2, (5)) = (18, 5)$. Thus, three of the infinitely many solutions are $(-10, -2)$, $(-2, 0)$, and $(18, 5)$.

3. Replacing x with 1, y with -3, and z with -2 in *all* three original equations, we find

$$2(1) - (-3) + 4(-2) = 2 + 3 - 8 = -3,$$

$$(1) + 2(-3) - 3(-2) = 1 - 6 + 6 = 1,$$

$$3(1) + (-3) - 3(-2) = 3 - 3 + 6 = 6.$$

4.
$$3x + 2y - z = 4$$
$$-3x - y + 0z = -2$$
$$0x + 2y - 2z = 3$$

add eq. 1
to eq. 2 ⟶
$$3x + 2y - z = 4$$
$$0x + y - z = 2$$
$$0x + 2y - 2z = 3$$

add -2(eq. 2)
to eq. 3 ⟶
$$3x + 2y - z = 4$$
$$0x + y - z = 2$$
$$0x + 0y + 0z = -1$$

In the third equation, the coefficients of x, y, and z are zero, but the coefficient on the right-hand side $-1 \neq 0$. Hence, the system is *inconsistent* and has *no solution*.

5. If we let $z = c$, where c is an arbitrary real number, then

$$z = 3y \quad \text{implies} \quad c = 3y \rightarrow y = \frac{c}{3},$$

$$\text{and } x = y + 1 \quad \text{implies} \quad x = \frac{c}{3} + 1 \rightarrow x = \frac{c+3}{3}.$$

Hence, every ordered triple of the form $\left(\dfrac{c+3}{3}, \dfrac{c}{3}, c\right)$ is a solution of the system in Example 5.

6.
$$x + y + z = 12,000$$
$$12x + 18y + 24z = 201,000$$
$$x + 0y - 2z = 0$$

add -12(eq 1) to eq 2 ⟶
add $-$(eq 1) to eq 3 ⟶
$$x + y + z = 12,000$$
$$0x + 6y + 12z = 57,000$$
$$0x - y - 3z = -12,000$$

Now, interchanging equations #2 and #3, we find

$$x + y + z = 12,000$$
$$0x - y - 3z = -12,000$$
$$0x + 6y + 12z = 57,000$$

add 6(eq 2)
to eq 3 ⟶
$$x + y + z = 12,000$$
$$0x - y - 3z = -12,000$$
$$0x + 0y - 6z = -15,000$$

Finally, using back substitution, gives us $z = 2500$, $y = 4500$, $x = 5000$.

◆ Exercises

1. $\left(\frac{7}{3}, \frac{1}{2}\right)$ **3.** $\left(\frac{1}{2}, 3, 2\right)$ **5.** $\left(-\frac{3}{2}, -6, 5, 9\right)$

7. $\left(-\frac{11}{8}, -\frac{3}{4}, -\frac{5}{2}, -\frac{3}{2}, \frac{1}{2}\right)$ **9.** $(4, 1)$ **11.** $(3, 2)$

13. $(2, 8)$ **15.** Inconsistent system, no solution

17. $(-6, -2)$ **19.** $(10, 4, 1)$ **21.** $(-1, 2, 0)$

23. $\left(\frac{2}{3}, -\frac{1}{9}, 3\right)$ **25.** Dependent system,

infinite number of solutions of the form $\left(\dfrac{5-a}{11}, \dfrac{4a-9}{11}, a\right)$

27. $(1, 1, 2)$ **29.** $(1, -2, 2)$ **31.** $(1, 2, -2, 3)$

33. $(-1, -4, 4, 1)$ **35.** Dependent system, infinite number of solutions of the form $(4 - a, a - 2, -5, 2 - a, a)$

37. $(2 - 3a, a, 4a)$, where a is a real number

39. $(2, a, 2a + 1)$, where a is a real number

41. $(3a + 5, a, 5a + 9)$, where a is a real number

43. $(2a + 2, 5a + 1, 3a, a)$, where a is a real number

45. Equal sides are 25 cm, nonequal side is 10 cm

47. $a = \$25$, $b = \$0.20$ **49.** $\$5500$ at 10%, $\$4500$ at

51. 120°, 40°, 20° **53.** $6000 at 8%, $22,000 at 9%,
$22,000 at $9\frac{1}{2}$% **55.** 20 children, 60 adults, 40 seniors
57. $0.75 for 2×3 print, $2.00 for 3×5 print, $5.00 for
5×8 print, $10.00 for 8×10 print
59. (a) $A = \frac{3}{2}$, B is any real number except -28

(b) $A = \frac{3}{2}$, $B = -28$ **61.** $A = \frac{8}{3}$, $B = -1$, $C = -\frac{2}{3}$
63. $a = -\frac{1}{2}$, $b = 2$, $c = 4$ **65.** No adult, 12 students, 8
children; or 5 adults, 4 students, 11 children
67. (83.9, 38.7) **69.** 1.11, -2.53, -3.85)
71. (6.68, 3.26, -4.94) **73.** $T_1 = 56\tilde{0}$ lb, $T_2 = 758$ lb

Section 7.2

Problems

1. For the matrix in Example 1(a), $a_{12} = 0$. For the matrix in
Example 1(b), $a_{12} = \sqrt{3}$.

2. $2z = 4$ implies $z = 2$.
$2y - z = 3$ implies $2y - (2) = 3$ or $y = \frac{5}{2}$.
$x + 2y + 5z = -5$ implies $x + 2(\frac{5}{2}) + 5(2) = -5$ or $x = -20$.
Hence, the solution is $(-20, \frac{5}{2}, 2)$.

3. $\begin{bmatrix} 3 & 2 & \vdots & 4 \\ 2 & -1 & \vdots & 5 \end{bmatrix}$

$\sim \begin{bmatrix} 1 & 3 & \vdots & -1 \\ 2 & -1 & \vdots & 5 \end{bmatrix}$ $-R_2 + R_1 \rightarrow R_1$

$\sim \begin{bmatrix} 1 & 3 & \vdots & -1 \\ 0 & -7 & \vdots & 7 \end{bmatrix}$ $-2R_1 + R_2 \rightarrow R_2$

Solving the corresponding system of linear equations by using
back substitution, we find $y = -1$ and $x = 2$. Thus, the solution
of the system of equations is $(2, -1)$.

4. $\begin{bmatrix} 3 & -9 & 6 & \vdots & 1 \\ 2 & -1 & -1 & \vdots & 3 \\ 1 & -2 & 1 & \vdots & 1 \end{bmatrix}$

$\sim \begin{bmatrix} 1 & -2 & 1 & \vdots & 1 \\ 2 & -1 & -1 & \vdots & 3 \\ 3 & -9 & 6 & \vdots & 1 \end{bmatrix}$ $R_1 \leftrightarrow R_3$

$\sim \begin{bmatrix} 1 & -2 & 1 & \vdots & 1 \\ 0 & 3 & -3 & \vdots & 1 \\ 0 & -3 & 3 & \vdots & -2 \end{bmatrix}$ $-2R_1 + R_2 \rightarrow R_2$ $-3R_1 + R_3 \rightarrow R_3$

$\sim \begin{bmatrix} 1 & -2 & 1 & \vdots & 1 \\ 0 & 3 & -3 & \vdots & 1 \\ 0 & 0 & 0 & \vdots & -1 \end{bmatrix}$ $R_2 + R_3 \rightarrow R_3$

Since all elements in the last row except the last entry are 0, we
conclude the system is inconsistent and has no solution.

5. For $a = 0$, we have $((2 - 3(0), 3(0) - 3, (0)) = (2, -3, 0)$.
For $a = 1$, we have $((2 - 3(1), 3(1) - 3, (1)) = (-1, 0, 1)$.
For $a = 2$, we have $((2 - 3(2), 3(2) - 3, (2)) = (-4, 3, 2)$.

Thus, three of the infinitely many solutions are $(2, -3, 0)$,
$(-1, 0, 1)$, and $(-4, 3, 2)$.

6. $\begin{bmatrix} 3 & 2 & \vdots & 4 \\ 2 & -1 & \vdots & 5 \end{bmatrix}$

$\sim \begin{bmatrix} 1 & 3 & \vdots & -1 \\ 2 & -1 & \vdots & 5 \end{bmatrix}$ $-R_2 + R_1 \rightarrow R_1$

$\sim \begin{bmatrix} 1 & 3 & \vdots & -1 \\ 0 & -7 & \vdots & 7 \end{bmatrix}$ $-2R_1 + R_2 \rightarrow R_2$

$\sim \begin{bmatrix} 1 & 3 & \vdots & -1 \\ 0 & 1 & \vdots & -1 \end{bmatrix}$ $-\frac{1}{7}R_2 \rightarrow R_2$

$\sim \begin{bmatrix} 1 & 0 & \vdots & 2 \\ 0 & 1 & \vdots & -1 \end{bmatrix}$ $-3R_2 + R_1 \rightarrow R_1$

Hence, the solution is $(2, -1)$.

Exercises

1. 3×3 **3.** 2×4 **5.** 3×1
7. (a) 3 (b) 0 (c) 4 (d) 6
9. $\begin{bmatrix} 2 & -3 & \vdots & 1 \\ 4 & 2 & \vdots & -3 \end{bmatrix}$ **11.** $\begin{bmatrix} 1 & -5 & 2 & \vdots & 9 \\ 3 & 0 & -1 & \vdots & 4 \end{bmatrix}$
13. $-x + 4y = 0$
$2x + 3y = 1$
15. $2x + 9y = -2$
$-x + 2z = 1$
$-2x + 3y - 4z = 1$

17. $(4, -3)$
19. Dependent system, infinite number of solutions of the form
$(3a + 1, a)$ **21.** $(40, 50)$ **23.** $(1, -2, 3)$
25. $(-1, 4, -3)$ **27.** $(\frac{23}{4}, -3, -\frac{1}{4})$ **29.** $(\frac{1}{10}, -\frac{11}{10}, -\frac{13}{10})$
31. $(2, 3, -1)$ **33.** $(0, 1, -1, 3)$
35. $(1, 2, -1, -2, 0)$
37. $(0, a, a + 2)$, where a is a real number
39. $(7 - 5a, 17a - 23, a)$, where a is a real number
41. $(a + 3, -2a, -a, a)$, where a is a real number
43. $(1 + 3a - 4b, a, b, 11b - 7a)$, where a and b are real
numbers **45.** $(-3, 4)$ **47.** $(1, 0, -3)$
49. $(\frac{5}{2}, 0, -\frac{3}{2})$ **51.** $(\frac{2}{3}, -\frac{1}{3}, \frac{4}{3}, \frac{5}{3})$
53. $(-a, a, a)$, where a is a real number **55.** $(\frac{2}{3}, \frac{1}{4})$
57. $(\frac{1}{3}, \frac{1}{5}, \frac{1}{10})$ **59.** (ln 2, ln 4, 0)
61. 3 mph uphill, 8 mph on level ground, 12 mph downhill
63. (92.1, 33.4) **65.** (2.21, -2.29, -4.98)
67. $(-5.31, 13.4, 15.9)$ **69.** $v_0 = 80.5$ ft/s, $s_0 = 24.8$ ft

Section 7.3

 Problems

1. For the matrices A and B to be equal, their corresponding elements must be equal. Hence,

$$x + y = 4$$
$$x - y = 8$$

Solving this system of equations, we find $x = 6, y = -2$.

2. $B + B =$

$$\begin{bmatrix} 3 & -5 & 2 \\ 0 & 1 & -3 \end{bmatrix} + \begin{bmatrix} 3 & -5 & 2 \\ 0 & 1 & -3 \end{bmatrix} = \begin{bmatrix} 6 & -10 & 4 \\ 0 & 2 & -6 \end{bmatrix}$$

3. $-2B + A = \begin{bmatrix} -6 & 10 & -4 \\ 0 & -2 & 6 \end{bmatrix} + \begin{bmatrix} 6 & 5 & -2 \\ 4 & 0 & -1 \end{bmatrix}$

$$= \begin{bmatrix} 0 & 15 & -6 \\ 4 & -2 & 5 \end{bmatrix}$$

4. $c(A + B) = c\left(\begin{bmatrix} 2 & 3 & 0 \\ -1 & 4 & 1 \end{bmatrix} + \begin{bmatrix} 1 & -3 & -1 \\ 5 & -2 & 2 \end{bmatrix} \right)$

$$= c \begin{bmatrix} 3 & 0 & -1 \\ 4 & 2 & 3 \end{bmatrix} = \begin{bmatrix} 3c & 0 & -c \\ 4c & 2c & 3c \end{bmatrix}$$

and

$$cA + cB = c \begin{bmatrix} 2 & 3 & 0 \\ -1 & 4 & 1 \end{bmatrix} + c \begin{bmatrix} 1 & -3 & -1 \\ 5 & -2 & 2 \end{bmatrix}$$

$$= \begin{bmatrix} 2c & 3c & 0 \\ -c & 4c & c \end{bmatrix} + \begin{bmatrix} c & -3c & -c \\ 5c & -2c & 2c \end{bmatrix}$$

$$= \begin{bmatrix} 3c & 0 & -c \\ 4c & 2c & 3c \end{bmatrix}$$

Hence, $c(A + B) = cA + cB$ for any scalar c.

5. $A(B + C) = \begin{bmatrix} 2 & 1 \\ 3 & 2 \end{bmatrix} \left(\begin{bmatrix} -1 & 4 \\ 0 & 2 \end{bmatrix} + \begin{bmatrix} 3 & -1 \\ -2 & 0 \end{bmatrix} \right)$

$$= \begin{bmatrix} 2 & 1 \\ 3 & 2 \end{bmatrix} \begin{bmatrix} 2 & 3 \\ -2 & 2 \end{bmatrix}$$

$$= \begin{bmatrix} 2 & 8 \\ 2 & 13 \end{bmatrix}$$

and

$$AB + AC = \begin{bmatrix} 2 & 1 \\ 3 & 2 \end{bmatrix} \begin{bmatrix} -1 & 4 \\ 0 & 2 \end{bmatrix} + \begin{bmatrix} 2 & 1 \\ 3 & 2 \end{bmatrix} \begin{bmatrix} 3 & -1 \\ -2 & 0 \end{bmatrix}$$

$$= \begin{bmatrix} -2 & 10 \\ -3 & 16 \end{bmatrix} + \begin{bmatrix} 4 & -2 \\ 5 & -3 \end{bmatrix}$$

$$= \begin{bmatrix} 2 & 8 \\ 2 & 13 \end{bmatrix}$$

Hence, $A(B + C) = AB + AC$.

6. Using property 4 of matrix multiplication and the result of Example 6(b), we have

$\frac{1}{2}[1 \quad 1 \quad 1 \quad 1 \quad 1] (A + B) = [1 \quad 1 \quad 1 \quad 1 \quad 1] \frac{1}{2}(A + B)$

$$= [1 \quad 1 \quad 1 \quad 1 \quad 1] \begin{bmatrix} 12 & 5 & 3 \\ 13 & 8 & 4 \\ 14 & 6 & 17 \\ 20 & 5 & 9 \\ 35 & 11 & 9 \end{bmatrix}$$

$$= [94 \quad 35 \quad 42]$$

The matrix $\frac{1}{2}[1 \quad 1 \quad 1 \quad 1 \quad 1] (A + B)$ represents the average production (points scored, assists, and rebounds) for the first two games.

 Exercises

1. $x = -1, y = -3$ **3.** $x = 9, y = 5, z = -3$
5. $x = 5, y = -1, w = 2, z = 0$

7. $\begin{bmatrix} 2 & 4 & 0 \\ -4 & 6 & -8 \end{bmatrix}$ **9.** $\begin{bmatrix} -2 & 0 \\ 1 & -3 \\ 0 & -1 \end{bmatrix}$

11. $\begin{bmatrix} 3 & -1 & -5 \\ -3 & 7 & 0 \end{bmatrix}$

13. Undefined, we cannot add matrices with different dimensions

15. $\begin{bmatrix} 2 & 8 \\ 1 & -5 \end{bmatrix}$ **17.** $\begin{bmatrix} 2 & -17 & -15 \\ 5 & 0 & 28 \end{bmatrix}$

19. Undefined, we cannot subtract matrices with different dimensions

21. $\begin{bmatrix} -13 & -7 \\ 7 & 1 \end{bmatrix}$ **23.** $[-12 \quad 24 \quad 30]$ **25.** $\begin{bmatrix} -1 \\ 3 \\ -2 \end{bmatrix}$

27. $\begin{bmatrix} 8 & 3 \\ 9 & -8 \end{bmatrix}$ **29.** $\begin{bmatrix} -3 & 7 & 9 \\ 3 & -7 & -9 \end{bmatrix}$

31. $\begin{bmatrix} 30 & 30 \\ -33 & -33 \\ -6 & -6 \end{bmatrix}$ **33.** $[4]$ **35.** $\begin{bmatrix} 0 & 0 \\ 0 & 0 \end{bmatrix}$

37. $\begin{bmatrix} 9 & -9 & -10 \\ 3 & 11 & -15 \\ -15 & 29 & 5 \end{bmatrix}$ **39.** $\begin{bmatrix} -21 & 60 \\ -3 & -33 \end{bmatrix}$

41. $[57 \quad -96 \quad 120]$ **43.** $\begin{bmatrix} -63 & 63 \\ 66 & -66 \end{bmatrix}$

45. $\begin{bmatrix} 4 & -8 & -10 \\ -12 & 24 & 30 \\ 8 & -16 & -20 \end{bmatrix}$ **47.** $\begin{bmatrix} 33 & 33 & 36 \\ 15 & 21 & 24 \\ 3 & 9 & 12 \end{bmatrix}$, total

number of subcompact, mid-size, and large cars that were rented at each terminal on Monday, Tuesday, and Wednesday

49. $\begin{bmatrix} 44 \\ 32 \\ 13 \end{bmatrix}$, total number of subcompact, mid-size, and large

cars that were rented on Monday

51. [51 63 72], total number of cars that were rented at each terminal on Monday, Tuesday, and Wednesday

53. [576 928 920], total revenue (in $) from cars that were rented at each terminal on Monday

55. [3756], total revenue (in $) from cars that were rented on Monday and Tuesday

57. One example is $A = \begin{bmatrix} a & a \\ b & b \end{bmatrix}$, $B = \begin{bmatrix} -1 & 1 \\ 1 & -1 \end{bmatrix}$, where a and b are real numbers.

59. (a) $\begin{bmatrix} 10 & -5 & 19 \\ -6 & 7 & -14 \\ 15 & 1 & 23 \end{bmatrix}$ (b) $\begin{bmatrix} 4 & 1 & -1 \\ 5 & 5 & 7 \\ -1 & -2 & -3 \end{bmatrix}$

(c) $\begin{bmatrix} 6 & -6 & 20 \\ -11 & 2 & -21 \\ 16 & 3 & 26 \end{bmatrix}$ (d) $\begin{bmatrix} 5 & -2 & 34 \\ -1 & 2 & -22 \\ 6 & 0 & 27 \end{bmatrix}$

From parts (c) and (d), we see that $A^2 - B^2 \neq (A + B)(A - B)$.

63. $\begin{bmatrix} 219.342 \\ -151.448 \end{bmatrix}$ **65.** $\begin{bmatrix} 61.73 & -31.608 \\ -50.018 & 103.302 \end{bmatrix}$

67. $\begin{bmatrix} 428.61 & -9.03 \\ -1478.46 & 3161.91 \end{bmatrix}$ **69.** $\begin{bmatrix} -2296.85946 \\ 19,343.8734 \end{bmatrix}$

71. $\begin{bmatrix} 91,524.375 \\ 63,958.125 \\ 132,264.375 \end{bmatrix}$, new weekly payroll (in $) at each facility

73. $\begin{bmatrix} 48.75 \\ 123.75 \\ 228.75 \end{bmatrix}$, weekly pay increase (in $) for each technician, engineer, and senior engineer

Section 7.4

 Problems

1. $AB =$

$\begin{bmatrix} 1 & 2 & 3 \\ 1 & 2 & 2 \\ -1 & -3 & -4 \end{bmatrix}\begin{bmatrix} 2 & 1 & 2 \\ -2 & 1 & -1 \\ 1 & -1 & 0 \end{bmatrix} = \begin{bmatrix} 1 & 0 & 0 \\ 0 & 1 & 0 \\ 0 & 0 & 1 \end{bmatrix}$

and

$BA = \begin{bmatrix} 2 & 1 & 2 \\ -2 & 1 & -1 \\ 1 & -1 & 0 \end{bmatrix}\begin{bmatrix} 1 & 2 & 3 \\ 1 & 2 & 2 \\ -1 & -3 & -4 \end{bmatrix} = \begin{bmatrix} 1 & 0 & 0 \\ 0 & 1 & 0 \\ 0 & 0 & 1 \end{bmatrix}$,

Since

$AB = BA = I_3$, we conclude that A and B are inverses of each other.

2. $AA^{-1} = \begin{bmatrix} 3 & 4 \\ 2 & 3 \end{bmatrix}\begin{bmatrix} 3 & -4 \\ -2 & 3 \end{bmatrix}$

$= \begin{bmatrix} 9 - 8 & -12 + 12 \\ 6 - 6 & -8 + 9 \end{bmatrix} = \begin{bmatrix} 1 & 0 \\ 0 & 1 \end{bmatrix}$,

$A^{-1}A = \begin{bmatrix} 3 & -4 \\ -2 & 3 \end{bmatrix}\begin{bmatrix} 3 & 4 \\ 2 & 3 \end{bmatrix}$

$= \begin{bmatrix} 9 - 8 & 12 - 12 \\ -6 + 6 & -8 + 9 \end{bmatrix} = \begin{bmatrix} 1 & 0 \\ 0 & 1 \end{bmatrix}$

3. $\begin{bmatrix} 1 & 2 & -3 & \vdots & 1 & 0 & 0 \\ 1 & 0 & 2 & \vdots & 0 & 1 & 0 \\ 2 & 2 & -1 & \vdots & 0 & 0 & 1 \end{bmatrix}$

$\sim \begin{bmatrix} 1 & 2 & -3 & \vdots & 1 & 0 & 0 \\ 0 & -2 & 5 & \vdots & -1 & 1 & 0 \\ 0 & -2 & 5 & \vdots & -2 & 0 & 1 \end{bmatrix}$

$\sim \begin{bmatrix} 1 & 2 & -3 & \vdots & 1 & 0 & 0 \\ 0 & -2 & 5 & \vdots & -1 & 1 & 0 \\ 0 & 0 & 0 & \vdots & -1 & -1 & 1 \end{bmatrix}$,

Since we obtain a row of zeros on the A portion of this matrix, we conclude that A is singular.

4. $2AX + B = 2\begin{bmatrix} 3 & 4 \\ 2 & 3 \end{bmatrix}\begin{bmatrix} 16 & -4 \\ -11 & 3 \end{bmatrix} + \begin{bmatrix} -6 & 3 \\ 1 & 2 \end{bmatrix}$

$= 2\begin{bmatrix} 4 & 0 \\ -1 & 1 \end{bmatrix} + \begin{bmatrix} -6 & 3 \\ 1 & 2 \end{bmatrix}$

$= \begin{bmatrix} 8 & 0 \\ -2 & 2 \end{bmatrix} + \begin{bmatrix} -6 & 3 \\ 1 & 2 \end{bmatrix} = \begin{bmatrix} 2 & 3 \\ -1 & 4 \end{bmatrix} = C$

5. The coefficient matrix for this system of equations is the same as matrix A defined in Problem 3. Since matrix A is singular (see Problem 3), A^{-1} does not exist and the matrix equation

$\begin{bmatrix} 1 & 2 & -3 \\ 1 & 0 & 2 \\ 2 & 2 & -1 \end{bmatrix}\begin{bmatrix} x \\ y \\ z \end{bmatrix} = \begin{bmatrix} -3 \\ 4 \\ 1 \end{bmatrix}$

does not have a unique solution for x, y, and z. The system of equations in Problem 5 is a dependent system with infinitely many solutions.

6. $\begin{bmatrix} 1 & 1 & \vdots & 1 & 0 \\ 30 & 50 & \vdots & 0 & 1 \end{bmatrix} \sim \begin{bmatrix} 1 & 1 & \vdots & 1 & 0 \\ 0 & 20 & \vdots & -30 & 1 \end{bmatrix}$

$\sim \begin{bmatrix} 1 & 1 & \vdots & 1 & 0 \\ 0 & 1 & \vdots & -\frac{3}{2} & \frac{1}{20} \end{bmatrix} \sim \begin{bmatrix} 1 & 0 & \vdots & \frac{5}{2} & -\frac{1}{20} \\ 0 & 1 & \vdots & -\frac{3}{2} & \frac{1}{20} \end{bmatrix}$.

Hence $A^{-1} = \begin{bmatrix} \frac{5}{2} & -\frac{1}{20} \\ -\frac{3}{2} & \frac{1}{20} \end{bmatrix}$ or $\frac{1}{20}\begin{bmatrix} 50 & -1 \\ -30 & 1 \end{bmatrix}$.

 Exercises

1. A and B are inverses **3.** A and B are inverses

5. A and B are inverses

7. $\begin{bmatrix} 3 & 4 \\ -2 & -3 \end{bmatrix}$ **9.** $\frac{1}{5}\begin{bmatrix} -3 & 4 \\ 2 & -1 \end{bmatrix}$ **11.** $\frac{1}{11}\begin{bmatrix} 2 & 1 \\ 5 & -3 \end{bmatrix}$

13. Singular   **15.** Singular

17. $\begin{bmatrix} -9 & 5 & -1 \\ -10 & 5 & -1 \\ 8 & -4 & 1 \end{bmatrix}$ **19.** $\frac{1}{5}\begin{bmatrix} 0 & -10 & -5 \\ 4 & 1 & 3 \\ 1 & 4 & 2 \end{bmatrix}$

21. $\begin{bmatrix} -14 & -13 & 3 \\ 11 & 10 & -2 \\ 2 & 2 & -\frac{1}{2} \end{bmatrix}$ **23.** $\frac{1}{75}\begin{bmatrix} 20 & -8 & -1 \\ 5 & 13 & 11 \\ 5 & -2 & -19 \end{bmatrix}$

25. $\frac{1}{4}\begin{bmatrix} 4 & -8 & 4 & 6 \\ 0 & 0 & 0 & 2 \\ 0 & 4 & -2 & -3 \\ 0 & 0 & 2 & -1 \end{bmatrix}$

27. $\begin{bmatrix} 1 & 0 & 0 & 0 & \frac{1}{2} \\ 0 & 1 & -1 & -2 & -\frac{1}{2} \\ 0 & 0 & -2 & -5 & -1 \\ 0 & 0 & -1 & -2 & -\frac{1}{2} \\ 0 & 0 & 2 & 4 & \frac{1}{2} \end{bmatrix}$ **29.** $X = \begin{bmatrix} -1 & 6 \\ 5 & -1 \end{bmatrix}$

31. $X = \frac{1}{3}\begin{bmatrix} 11 & 27 \\ 7 & 17 \end{bmatrix}$ **33.** $X = \frac{1}{27}\begin{bmatrix} -72 & -54 \\ 22 & 12 \end{bmatrix}$

35. $X = \frac{1}{2}\begin{bmatrix} -1 & 4 & 6 \\ -2 & -3 & 2 \\ 3 & 1 & -3 \end{bmatrix}$

37. $X = \frac{1}{2}\begin{bmatrix} -10 & -59 & 67 \\ 3 & 19 & -22 \\ 1 & 6 & -7 \end{bmatrix}$

39. $X = \frac{1}{20}\begin{bmatrix} -22 & -174 & 218 \\ 19 & 93 & -101 \\ -6 & -42 & 54 \end{bmatrix}$ **41.** $\left(\frac{32}{5}, -\frac{3}{5}\right)$

43. $(2, 8)$ **45.** $(0, 1, -2)$ **47.** $\left(\frac{1}{2}, \frac{1}{2}, -2\right)$

49. $(1, -5, -3)$ **51.** $\left(1, 2, \frac{1}{2}, -2\right)$

53. (a) Equal sides 25 cm, nonequal side 10 cm

(b) Equal sides 37 cm, nonequal side 1 cm

(c) Equal sides $45\frac{1}{3}$ cm, nonequal side $21\frac{1}{3}$ cm

(d) Equal sides 84 cm, nonequal side 52 cm

55. (a) 6 cm, 10 cm, 12 cm   (b) 20 cm, 28 cm, 6 cm

(c) 34 cm, 39 cm, 83 cm,   (d) $10\frac{1}{2}$ cm, $4\frac{1}{2}$ cm, $6\frac{1}{2}$ cm

57. (a) $x = -\frac{8}{3}$ (b) $x = -2$

59. Only if matrix A is invertible

61. The product is $\begin{bmatrix} 1 & 0 \\ 0 & 1 \end{bmatrix}$. The inverse of $A = \begin{bmatrix} a & b \\ c & d \end{bmatrix}$

is $A^{-1} = \dfrac{1}{ad - bc}\begin{bmatrix} d & -b \\ -c & a \end{bmatrix}$.

63. (a) $(1.04, -1.47)$ (b) $(0.597, 3.52)$

(c) $(-7.18, 6.94)$ (d) $(-0.176, 0.118)$

65. (a) $(1.10, 0.338, -0.205)$ (b) $(6.82, 8.58, -11.9)$

(c) $(4.05, 10.7, -26.8)$ (d) $(-0.198, -0.0617, 0.0942)$

67. Day 1: six 14-inch saws and six 16-inch saws; day 2: three 14-inch and six 16-inch; day 3: ten 14-inch and six 16-inch; day 4: eight 14-inch, three 16-inch

Section 7.5

Problems

1. $|C| = \begin{vmatrix} -3 & -2 \\ 6 & 4 \end{vmatrix} = (-3)(4) - (-2)(6) = 0$

2. $M_{12} = \begin{vmatrix} 3 & 1 \\ -1 & -4 \end{vmatrix} = (3)(-4) - (1)(-1) = -11$

$C_{12} = (-1)^{1+2}M_{12} = -M_{12} = 11$

3. $|A| = 3C_{12} + 2C_{32} = -3\begin{vmatrix} 3 & 1 \\ -1 & -4 \end{vmatrix} - 2\begin{vmatrix} 1 & -2 \\ 3 & 1 \end{vmatrix}$

$= -3(-11) - 2(7) = 19$

4. (a) $|A| = -3C_{31} - 1C_{32} = -3\begin{vmatrix} 3 & 1 \\ 1 & -2 \end{vmatrix} + 1\begin{vmatrix} 2 & 1 \\ 3 & -2 \end{vmatrix}$

$= -3(-7) + 1(-7) = 14$

(b) $|A| = (0 + 18 - 3) - (-3 + 4 + 0) = 15 - 1 = 14$

5. Using the diagonal method,

$C_{13} = (0 - 8 + 0) - (24 + 0 + 8) = -40$

and

$C_{43} = -[(-4 - 6 + 0) - (-2 + 12 + 0)] = 20.$

6. Since $|A| = \begin{vmatrix} 6 & -2 \\ -9 & 3 \end{vmatrix} = 0, |A_x| = \begin{vmatrix} -4 & -2 \\ 6 & 3 \end{vmatrix} = 0,$

and $|A_y| = \begin{vmatrix} 6 & -4 \\ -9 & 6 \end{vmatrix} = 0,$

we conclude that the system is dependent and has an infinite number of solutions.

7. $|A| = \begin{vmatrix} 2 & 0 & -1 \\ -1 & 3 & 0 \\ 0 & 6 & -3 \end{vmatrix} = -12,$

$|A_x| = \begin{vmatrix} 5 & 0 & -1 \\ 0 & 3 & 0 \\ 7 & 6 & -3 \end{vmatrix} = -24,$

$|A_y| = \begin{vmatrix} 2 & 5 & -1 \\ -1 & 0 & 0 \\ 0 & 7 & -3 \end{vmatrix} = -8,$ and

$|A_z| = \begin{vmatrix} 2 & 0 & 5 \\ -1 & 3 & 0 \\ 0 & 6 & 7 \end{vmatrix} = 12$

Therefore, $x = \dfrac{|A_x|}{|A|} = \dfrac{-24}{-12} = 2$ $y = \dfrac{|A_y|}{|A|} = \dfrac{-8}{-12} = \dfrac{2}{3}$

and

$$z = \dfrac{|A_z|}{|A|} = \dfrac{12}{-12} = -1.$$

Thus, the solution of the linear system is the ordered triple

$\left(2, \frac{2}{3}, -1\right)$

 Exercises

1. -22 **3.** -4 **5.** 218 **7.** $\frac{4}{5}$ **9.** ab

11. $M_{11} = 8, C_{11} = 8; M_{12} = -10, C_{12} = 10; M_{13} = -20,$
$C_{13} = -20; M_{21} = 6, C_{21} = -6; M_{22} = 12, C_{22} = 12; M_{23} = 24,$
$C_{23} = -24; M_{31} = -11, C_{31} = -11; M_{32} = 4, C_{32} = -4;$
$M_{33} = 47, C_{33} = 47$ **13.** $M_{11} = -4, C_{11} = -4; M_{12} = 6,$
$C_{12} = -6; M_{13} = 2, C_{13} = 2; M_{21} = 42, C_{21} = -42; M_{22} = 14,$
$C_{22} = 14; M_{23} = 12, C_{23} = -12; M_{31} = 28, C_{31} = 28; M_{32} = 13,$
$C_{32} = -13; M_{33} = 8, C_{33} = 8$ **15.** -70 **17.** -46
19. 1 **21.** $-51i + 45j - 126k$ **23.** abc **25.**
-370 **27.** 440 **29.** 20 **31.** $(2, 1)$

33. $\left(-\frac{135}{7}, -\frac{267}{7}\right)$ **35.** Dependent system, Cramer's rule
does not apply   **37.** $(12, 16)$ **39.** $(2, -1, -3)$
41. $(5, -10, -5)$

43. $\left(\frac{1}{5}, 0, \frac{7}{5}\right)$ **45.** $(-2, -5, 12)$ **47.** $\left(\frac{3}{7}, -\frac{46}{7}, \frac{48}{7}, \frac{11}{7}\right)$
49. $x = \frac{1}{3}$ **51.** $x = 8, -6$

53. (a) $\dfrac{x_1 y_2 - x_2 y_1}{2}$ (b) The area of a triangle with vertices
$(0, 0), (x_1, y_1),$ and $(x_2, y_2).$

55. (a) $y = \dfrac{x_2 y_1 - x_1 y_2 + x y_2 - x y_1}{x_2 - x_1}$
(b) The equation of a nonvertical line passing through two distinct points $(x_1, y_1),$ and $(x_2, y_2).$
57. (a) $C = 2$ (b) $(2, -3)$ **59.** -5.17
61. $10,300$
63. $(-2.19, -5.65)$ **65.** $(30.2, 5.83, -15.7)$
67. $T_1 = 465$ lb, $T_2 = 648$ lb

Section 7.6

 Problems

1. Since matrix A is in echelon form, we have
$|A| = (2)(2)(-3)(-5) = 60.$

2. $|A| = 2C_{11} + 1C_{31} = 2\begin{vmatrix} 6 & 3 \\ -2 & -5 \end{vmatrix} + 1\begin{vmatrix} -2 & -3 \\ 6 & 3 \end{vmatrix}$

$$= 2(-24) + 1(12) = -36.$$

3. $|A| = 1C_{14} + 14C_{24}$

$= -1\begin{vmatrix} -2 & 3 & 4 \\ 1 & -4 & 4 \\ 1 & 2 & -2 \end{vmatrix} + 14\begin{vmatrix} 3 & -2 & 1 \\ 1 & -4 & 4 \\ 1 & 2 & -2 \end{vmatrix}$

$= -1(42) + 14(-6) = -126.$

4. Beginning with the second row, we have

$|A| = \begin{vmatrix} 2 & -1 & 3 & 4 & 2 \\ 1 & 0 & -1 & 0 & 1 \\ 3 & 2 & 2 & 1 & 0 \\ -1 & 0 & 3 & 1 & 2 \\ 1 & 1 & -2 & -1 & 3 \end{vmatrix}$

$= \begin{vmatrix} 2 & -1 & 5 & 4 & 0 \\ 1 & 0 & 0 & 0 & 0 \\ 3 & 2 & 5 & 1 & -3 \\ -1 & 0 & 2 & 1 & 3 \\ 1 & 1 & -1 & -1 & 2 \end{vmatrix}$ $\begin{array}{l} C_1 + C_3 \to C_3, \\ -C_1 + C_5 \to C_5 \end{array}$

Using expansion by cofactors about the second row gives us

$|A| = 1C_{12} = -1\begin{vmatrix} -1 & 5 & 4 & 0 \\ 2 & 5 & 1 & -3 \\ 0 & 2 & 1 & 3 \\ 1 & -1 & -1 & 2 \end{vmatrix}$

$= -1\begin{vmatrix} -1 & 5 & 4 & 0 \\ 0 & 15 & 9 & -3 \\ 0 & 2 & 1 & 3 \\ 0 & 4 & 3 & 2 \end{vmatrix}$ $\begin{array}{l} 2R_1 + R_2 \to R_2 \\ R_1 + R_4 \to R_4 \end{array}$

Finally, using expansion by cofactors about the first column and then the diagonal method, we have

$|A| = -1(-1)\begin{vmatrix} 15 & 9 & -3 \\ 2 & 1 & 3 \\ 4 & 3 & 2 \end{vmatrix} = -1(-1)(-39) = -39.$

5. $|A| = \begin{vmatrix} 1 & -3 & 4 & 6 \\ 2 & 4 & 2 & -1 \\ 0 & 2 & 5 & -6 \\ 6 & 12 & 6 & -3 \end{vmatrix}$

$= \begin{vmatrix} 1 & -3 & 4 & 6 \\ 0 & 10 & -6 & -13 \\ 0 & 2 & 5 & -6 \\ 0 & 30 & -18 & -39 \end{vmatrix}$ $\begin{array}{l} -2R_1 + R_2 \to R_2 \\ -6R_1 + R_4 \to R_4 \end{array}$

Observe that corresponding elements in rows 2 and 4 are proportional. Hence, we conclude that $|A| = 0.$

 Exercises

1. -12 **3.** 208 **5.** -96 **7.** Interchanging rows 1 and 3 changes the sign of the determinant.   **9.** Factoring out 2 from column 3 produces a product that is equal to the value of the original determinant.   **11.** Adding three times row 1 to row 3 does not change the value of the determinant.

13. Since every element in column 3 is zero, the value of the determinant is 0.   **15.** Since corresponding elements in rows 1 and 3 are proportional, the value of the determinant is zero.
17. -45 **19.** 64   **21.** 368   **23.** 0   **25.** 24
27. 40   **29.** 304   **31.** 32   **33.** -3 **35.** -2
37. -240 **39.** 0   **41.** -4 **43.** 16   **45.** -4
47. -27 **53.** 43.2   **55.** 48,100

Section 7.7

◆ Problems

1. (a)

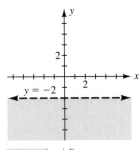

(b)

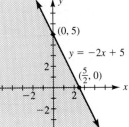

2.

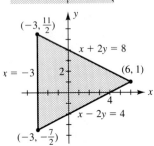

3. There are no maximum or minimum values when lines with slope 2/5 pass through the unbounded region shown in Figure 7.7.

4. The profit P is now given by $P(x,y) = 30x + 45y$.

Vertex	Value of $P(x,y) = 30x + 45y$
$(0, 0)$	0
$(0, 40)$	1800
$(15, 40)$	2250
$(60, 10)$	2250
$(60, 0)$	1800

Maximum value occurs when $x = 15$, $y = 40$, or when $x = 60$, $y = 10$.

Hence, to maximize profits, the company should manufacture either 15 pairs of slalom skis and 40 pairs of racing skis, or 60 pairs of slalom skis and 10 pair of racing skis.

◆ Exercises

1.

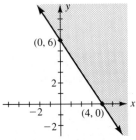

3.

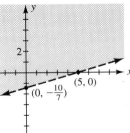

5.

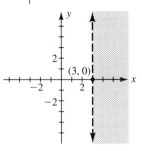

7.

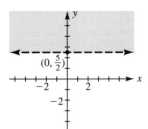

9.

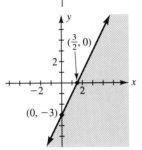

11.

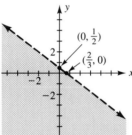

13.

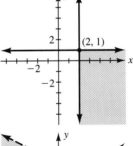

15.

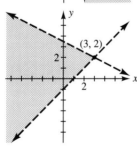

17.

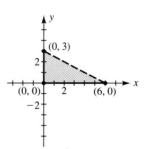

19.

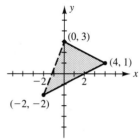

21.

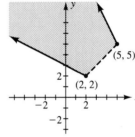

23.

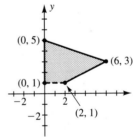

25.

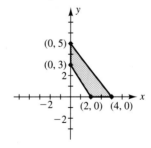

27.

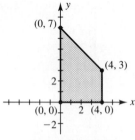

29.

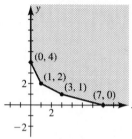

31. Maximum value, 12; minimum value, 0
33. Maximum value 6; minimum value, -11
35. Maximum value, 8; minimum value, 0   **37.** Maximum
value, 6; minimum value, 0   **39.** Maximum value, 47;
minimum value, 16   **41.** No maximum value; minimum
value 11   **43.** No maximum value; no minimum value
45. Eight $100-cases, two $200 cases   **47.** 24 tables, 18
chairs   **49.** 24 bags of cracked corn, 48 bags of sunflower
seeds
51. (a) $m = -\frac{1}{4}$ (b) $m = -2$ (c) $-\frac{1}{4} < m \le 0$
(d) $-2 < m < -\frac{1}{4}$ (e) $m < -2$ or $m > 0$ **53.** 2 bags
of brand A and 3 bags of brand B; minimum cost is $98.
55.

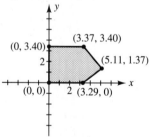

57. 29.1   **59.** 1.6 oz. of type P, 2.6 oz. of type Q

Chapter 7 Review Exercises

1. $(2, -3)$ **3.** $\left(\frac{22}{15}, -\frac{13}{30}\right)$ **5.** $(2, -3, -1)$
7. $\left(-\frac{1}{3}, \frac{2}{3}, -\frac{2}{3}\right)$ **9.** Dependent system, infinite number of
solutions of the form $(2a - 2, a, 2a - 1)$
11. $\left(7, -4, \frac{1}{2}\right)$ **13.** $(a + 1, a, 2a + 5)$, where a is a real
number   **15.** $(-11, 4, -4, 6)$ **17.** $\left(\frac{3}{4}, \frac{1}{4}, -2, 1\right)$

19. $[-3 \quad -12 \quad 6]$ **21.** $\begin{bmatrix} 2 & 8 \\ 4 & 4 \end{bmatrix}$ **23.** $\begin{bmatrix} -4 & -5 \\ -3 & -4 \end{bmatrix}$
25. $\begin{bmatrix} -10 & -6 & -3 \\ -4 & 17 & 28 \end{bmatrix}$ **27.** $\begin{bmatrix} -3 & 12 \\ -2 & 9 \end{bmatrix}$ **29.** $[24]$
31. $\begin{bmatrix} 15 & 20 \\ 18 & 27 \end{bmatrix}$ **33.** $[-23 \quad 59]$ **35.** $\begin{bmatrix} 1 & -\frac{3}{2} \\ -1 & 2 \end{bmatrix}$
37. $\frac{1}{20} \begin{bmatrix} -32 & 20 & 4 \\ 13 & -5 & -1 \\ 14 & -10 & 2 \end{bmatrix}$ **39.** Singular
41. $(5, -2)$ **43.** $(4, 2, -3)$ **45.** $X = \begin{bmatrix} 2 & -32 \\ 0 & 12 \end{bmatrix}$
47. $X = \begin{bmatrix} 82 & -\frac{65}{2} \\ -20 & 8 \end{bmatrix}$ **49.** 11   **51.** -14
53. -30 **55.** -16 **57.** 0   **59.** $(5, 6)$
61. $\left(\frac{1}{4}, -\frac{3}{4}, \frac{5}{4}\right)$
63.

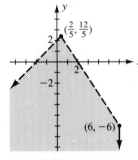

65.

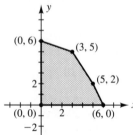

67. Minimum value, 0; maximum value, 21
69. (a) $k = 3$ (b) $k = \pm 5$ (c) $k = -6, -2$
(d) $k = 3, -\frac{13}{5}$ **71.** 9750 gal of regular gasoline and
2250 gal of super gasoline   **73.** 11,000 box seats, 12,000
grandstand seats, 9500 bleacher seats   **75.** 4 mph downhill,
3 mph on level ground, 2 mph uphill   **77.** $\begin{bmatrix} 34 \\ 29 \\ 17 \end{bmatrix}$, total
number of 400-watt, 500-watt, and 600-watt microwave ovens
sold during the first day of the sale

79. [3480 2580 3240], revenue (in $) from sales of microwave ovens at each outlet on second day of sale
81. 20,000 seats

CHAPTER 8

Section 8.1

 Problems

1. $f(1) = (-1)^1 2^2 = -4$, $f(2) = (-1)^2 2^3 = 8$,
$f(3) = (-1)^3 2^4 = -16$, $f(4) = (-1)^4 2^5 = 32$, and so on.
The elements
$$-4, 8, -16, 32, \ldots$$
represent an alternating sequence.

2. $a_1 = 1$, $a_2 = 1$, $a_3 = 1 + 1 = 2$, $a_4 = 2 + 1 = 3$,
$a_5 = 3 + 2 = 5$, $a_6 = 5 + 3 = 8$, and so on. Hence, the first six elements of this infinite sequence are
$$1, 1, 2, 3, 5, 8, \ldots$$

3. (a) For $a_n = 2n - 1$, $a_1 = 2(1) - 1 = 1$,
$a_2 = 2(2) - 1 = 3$, $a_3 = 2(3) - 1 = 5$, $a_4 = 2(4) - 1 = 7$.
  (b) For $a_n = 3^{n-1}$, $a_1 = 3^0 = 1$, $a_2 = 3^1 = 3$,
$a_3 = 3^2 = 9$, $a_4 = 3^3 = 27$.
  (c) For $a_n = \dfrac{n + 1}{n + 2}$, $a_1 = \dfrac{1 + 1}{1 + 2} = \dfrac{2}{3}$,
$a_2 = \dfrac{2 + 1}{2 + 2} = \dfrac{3}{4}$, $a_3 = \dfrac{3 + 1}{3 + 2} = \dfrac{4}{5}$, $a_4 = \dfrac{4 + 1}{4 + 2} = \dfrac{5}{6}$.
  (d) For $a_n = (-1)^n x^{4n-3}$, $a_1 = (-1)^1 x^{4-3} = -x$,
$a_2 = (-1)^2 x^{8-3} = x^5$, $a_3 = (-1)^3 x^{12-3} = -x^9$,
$a_4 = (-1)^4 x^{16-3} = x^{13}$.

4. $\displaystyle\sum_{k=1}^{27} \dfrac{k}{k + 1} = \dfrac{1}{2} + \dfrac{2}{3} + \dfrac{3}{4} + \ldots + \dfrac{n}{n + 1} + \ldots + \dfrac{27}{28}$

5. Replacing i with $k - 1$, we obtain

$$\sum_{I=1}^{\infty} (i + 1)x^{i+1} = \sum_{k-1=1}^{k=\infty} [(k - 1) + 1]x^{(k-1)+1} = \sum_{k=2}^{\infty} kx^k.$$

 Exercises

1. 3, 0, −3, −6, −9; decreasing function

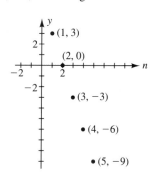

3. 1, −2, 3, −4; neither increasing nor decreasing

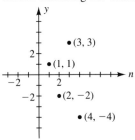

5. $2, 2, \frac{32}{9}, 8, \frac{512}{25}, \ldots$; neither, but increasing for $n \geq 2$

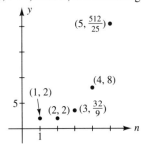

7. 1, 6, 11, 16, 21, 26   **9.** −6, 12, −24, 48, −96, 192

11. $3, \frac{5}{2}, \frac{7}{3}, \frac{9}{4}, \frac{11}{5}, \frac{13}{6}$ **13.** 2, 0, 2, 0, 2, 0   **15.** 1, 3, 3, 7, 5, 11   **17.** 3, 6, 12, 24, 48, 96   **19.** 1, 2, 1, −1, −2, −1   **21.** $a_n = n^2$ **23.** $a_n = (-1)^{n+1}(2n)$

25. $a_n = n + 2$ **27.** $a_n = \dfrac{2n - 1}{2n}$ **29.** $a_n = 2(0.1)^{n-1}$

31. $a_n = (-1)^n \dfrac{x^n}{2n - 1}$ **33.** $2 + 4 + 6 + 8 + 10$

35. $-1 + 8 - 27 + 64 - 125 + 216 - 343 + 512 - 729 + 1000$ **37.** $1 + 4 + 27 + 256 + 3125 + \ldots$

39. $-\dfrac{1}{2} + \dfrac{1}{2} + \dfrac{3}{2} + \dfrac{5}{2} + \ldots + \dfrac{2n - 3}{2} + \ldots + \dfrac{97}{2}$

41. $-2 + x - \dfrac{2x^2}{3} + \dfrac{x^3}{2} - \dfrac{2x^4}{5} + \ldots$

43. $\displaystyle\sum_{i=1}^{7} 2i$ **45.** $\displaystyle\sum_{i=1}^{30} (5i - 3)$ **47.** $\displaystyle\sum_{i=1}^{\infty} (-1)^{i+1}$

49. $\displaystyle\sum_{i=1}^{10} \dfrac{(-1)^i}{2^{i-1}}$ **51.** $\displaystyle\sum_{i=1}^{\infty} \dfrac{2^i}{3^{i-1}}$ **53.** $\displaystyle\sum_{i=1}^{5} (i + 2)x^{2i-1}$

55. $\displaystyle\sum_{k=0}^{3} (2k - 3)$ **57.** $\displaystyle\sum_{k=0}^{\infty} \dfrac{(-1)^{k+1}(2x^k)}{k + 1}$

59. $\displaystyle\sum_{k=0}^{23} \dfrac{k + 2}{(k + 2)^2 + 1}$ **61.** One possibility is

$a_n = (2n - 1) + (n - 1)(n - 2)(n - 3)$, which gives us the sequence 1, 3, 5, 13, 33, . . .

63. (a) $f(n) = 4n - 6$; domain: {1, 2, 3, 4, 5}
(b) $f(n) = -3n + 13$; domain: {1, 2, 3, 4}

65. (a) $3.31 + 25.09 + 61.39 + 112.21$
(b) $4.97 + 2.735 + 1.99 + 1.6175 + 1.394$

Section 8.2

 Problems

1. $\displaystyle\sum_{i=1}^{1000} 4 = 1000(4) = 4000$

2. $\displaystyle\sum_{i=1}^{50} (2i - 1) = (50)^2 = 2500$

3. (a) $\displaystyle\sum_{i=1}^{19} i = \frac{19(19 + 1)}{2} = 190$

   (b) $\displaystyle\sum_{i=1}^{19} i^2 = \frac{19(19 + 1)[2(19) + 1]}{6} = 2470$

   (c) $\displaystyle\sum_{i=1}^{19} i^3 = \left[\frac{19(19 + 1)}{2}\right]^2 = 36{,}100$

4. $\displaystyle\sum_{i=1}^{20} (5i - 2) = 5\sum_{i=1}^{20} i - \sum_{i=1}^{20} 2$

   $\displaystyle = 5\left[\frac{20(20 + 1)}{2}\right] - 20(2) = 1090$

5. The sum of the interior angles of a 12-sided convex polygon is $(12 - 2)180° = 1800°$.

 Exercises

1. 150   3. 100π 5. −20   7. 247   9. $\frac{76}{15}$
29. 800   31. 16,206   33. 20,825   35. 5440
37. $\frac{499}{500}$ 39. 2046   41. 5460   43. 45,057
45. $\frac{255}{256}$ 47. 3825   49. 9680   51. 35,490
53. 4065   57. (a) log 120   (b) log 6
61. (a) $a_n = (2n - 1)(2n) = 4n^2 - 2n$ (b) 11,060
63. (a) $a_n = (3n - 2)(3n - 1)(3n) = 27n^3 - 27n^2 + 6n$
(b) 1,114,470   71. −2822.4   73. 87,423
75. 5.3 square units

Section 8.3

 Problems

1. If $a_1 = 8$ and $d = -5$, then

   $a_2 = a_1 + d = 8 - 5 = 3$, $a_3 = a_2 + d = 3 - 5 = -2$,

   $a_4 = a_3 + d = -2 - 5 = -7$, $a_5 = a_4 + d = -7 - 5 = -12$

Thus, the first five elements of this arithmetic sequence are $8, 3, -2, -7, -12$.

2. For the sequence $3, 7, 11, 15, \ldots$, we have $a_1 = 3$ and $d = 4$. Hence, the general element is

   $a_n = a_1 + (n - 1)d$

   $= 3 + (n - 1)4$

   $= 4n - 1$

3. Using $a_n = a_1 + (n - 1)d$ with $n = 60$, we have

   $a_{60} = a_1 + 59d$

   Hence,

   $a_1 = a_{60} - 59d = -98 - (59)(-4) = 138$

4. If $a_1 = -8$ and $d = 7$, then

   $a_n = a_1 + (n - 1)d$

   $= -8 + (n - 1)7$

   $= 7n - 15$

   Hence, $a_5 = 7(5) - 15 = 20$ and $a_{11} = 7(11) - 15 = 62$.

5. $\displaystyle\sum_{i=0}^{40} (4i - 3) = 41\left(\frac{-3 + 157}{2}\right) = 3157$

6. Using $S_n = \frac{n}{2}[2a_1 + (n - 1)d]$ with $n = 10$, we have

   $S_{10} = \frac{10}{2}[2a_1 + (10 - 1)d]$

   $215 = 5[2(8) + 9d]$

   $43 = 16 + 9d$

   $27 = 9d$

   $d = 3$

7. Using $a_n = a_1 + (n - 1)d$, we find
$a_{15} = \$68 + (15 - 1)(-\$0.75) = \$57.50$

 Exercises

1. $3, 7, 11, 15, 19$ 3. $2, \frac{4}{3}, \frac{2}{3}, 0, -\frac{2}{3}$
5. $-1, -1 + \pi, -1 + 2\pi, -1 + 3\pi, -1 + 4\pi$
7. $a_n = 6n - 3$ 9. $a_n = 18 - 7n$ 11. $a_n = \dfrac{5n - 7}{3}$
13. $a_n = \dfrac{(3 - 2n)\pi}{6}$ 15. $a_n = x + (2n - 6)$
17. $a_{28} = 193$ 19. $a_1 = -307$ 21. $d = -\frac{1}{2}$
23. $a_{80} = 31.85$ 25. $d = -\frac{18}{7}$ 27. 2520
29. −8149   31. $\frac{273}{2}$ 33. 4895   35. −273
37. $S_{31} = 2697$ 39. $a_{32} = 26$ 41. $d = 8$
43. $S_{62} = 7006$ 45. $a_{33} = 100$ 47. 2550
49. 166   51. $75,000
53. (a) 128 seats   (b) 2000 seats
55. Yes; common difference-ln a
57. $x = 0.00505$ 59. $\frac{108}{5}, \frac{121}{5}, \frac{134}{5}, \frac{147}{5}$
61. (a) 24 payments   (b) $2600   63. 144.56
65. 1072.4   67. −16.2052
69. (a) $a_1 = \$1050, a_2 = \$1042.50, a_3 = \$1035$
(b) $607.50   (c) $49,725

Section 8.4

 Problems

1. If $a_1 = -2$ and $r = 6$, then

$$a_2 = a_1 r = -2 \cdot 6 = -12; \qquad a_3 = a_2 r = -12 \cdot 6 = -72,$$

$$a_4 = a_3 r = -72 \cdot 6 = -432, \qquad a_5 = a_4 r = -432 \cdot 6 = -2592.$$

Thus, the first five elements of this geometric sequence are
$$-2, -12, -72, -432, -2592$$

2. For the sequence 5, 15, 45, 135, . . . , we have $a_1 = 5$ and $r = 3$. Thus, the general element is
$$a_n = a_1 r^{n-1} = 5(3)^{n-1}$$

3. Using $a_n = a_1 r^{n-1}$ with $n = 9$, we have $a_9 = a_1 r^8$. Hence,

$$a_1 = \frac{a_9}{r^8} = \frac{\frac{1}{81}}{\left(\frac{1}{3}\right)^8} = \frac{\left(\frac{1}{3}\right)^4}{\left(\frac{1}{3}\right)^8} = 81.$$

4. If $a_1 = 96$ and $r = \frac{1}{2}$, then

$$a_n = a_1 r^{n-1} = 96\left(\frac{1}{2}\right)^{n-1}$$

Hence,

$$a_3 = 96\left(\frac{1}{2}\right)^2 = 24 \text{ and } a_6 = 96\left(\frac{1}{2}\right)^5 = 3$$

5. Using $S_n = \dfrac{a_1(1 - r^n)}{1 - r}$ with $n = 12$, we have

$$S_{12} = \frac{a_1(1 - r^{12})}{1 - r}$$

Hence,

$$a_1 = \frac{S_{12}(1 - r)}{1 - r^{12}} = \frac{8190\,(1 - 2)}{1 - 2^{12}} = 2$$

6. Using $a_n = a_1 r^{n-1}$ with $a_1 = 3$, $r = 4$, and $a_n = 201{,}326{,}592$, we find

$$201{,}326{,}592 = 3(4)^{n-1}$$

$$67{,}108{,}864 = 4^{n-1}$$

$$\log 67{,}108{,}864 = (n - 1) \log 4$$

$$n - 1 = \frac{\log 67{,}108{,}864}{\log 4}$$

$$n = \frac{\log 67{,}108{,}864}{\log 4} + 1 = 14$$

Hence, the series has 14 terms.

7. The amount of \$5000 earns interest at the rate of 9%, compounded monthly, for 10 years. Using the *compound interest formula* (Section 6.1), we find

$$A = P\left(1 + \frac{r}{n}\right)^{nt} = \$5000\left(1 + \frac{0.09}{12}\right)^{120}$$

$$= \$5000(1.0075)^{120}$$

$$= \$12{,}257$$

Adding this amount to the amount of the annuity (\$29,027) yields a total amount of \$41,284.

 Exercises

1. 3, 12, 48, 192, 768   **3.** $\frac{1}{4}, -\frac{1}{6}, \frac{1}{9}, -\frac{2}{27}, \frac{4}{81}$

5. $-1, -\pi, -\pi^2, -\pi^3, -\pi^4$ **7.** $a_n = 3^{n-1}$

9. $a_n = -64\left(-\frac{1}{4}\right)^{n-1}$ **11.** $a_n = \frac{1}{4}\left(\frac{2}{3}\right)^{n-1}$

13. $a_n = -\sqrt{2}\left(-\sqrt{3}\right)^{n-1}$ **15.** $a_n = 3(2x^2)^{n-1}$

17. $a_9 = 26{,}244$ **19.** $a_1 = -\frac{1}{216}$ **21.** $r = \pm\frac{3}{2}$

23. $a_{13} = \frac{243}{32}$ **25.** $r = \pm\frac{1}{2}$ **27.** -364 **29.** $\frac{605}{729}$

31. 46.66666662   **33.** 1,398,101   **35.** $\frac{1023}{8}$

37. $S_{11} = \frac{683}{64}$ **39.** $a_1 = 2$ **41.** $S_6 = \frac{665}{144}$

43. $S_7 = -\frac{2653}{64}$ **45.** $a_6 = \frac{729}{8}, -\frac{9375}{8}$ **47.** 9 terms

49. \$500   **51.** Choose the scholarship that doubles the amount each month for one year. It pays \$40,950, whereas the other scholarship pays only \$40,000.

53. (a) Geometric sequence with $r = e^{3x}$
(b) $a_n = e^{(3n-1)x}$ **55.** $x = 10^{-3}$ **57.** 24, 36, 54

59. $r = -\frac{1}{5}, -5$ **61.** -0.0000807 **63.** 3.35

65. 13   **67.** \$18,294.60   **69.** \$81,397.92

Section 8.5

 Problems

1. Since $\displaystyle\sum_{n=0}^{\infty} \left(\frac{3}{2}\right)^n = 1 + \frac{3}{2} + \frac{9}{4} + \frac{27}{8} + \dots$, we have

$$S_1 = 1, \qquad S_2 = 1 + \frac{3}{2} = \frac{5}{2},$$

$$S_3 = 1 + \frac{3}{2} + \frac{9}{4} = \frac{19}{4}, \qquad S_4 = 1 + \frac{3}{2} + \frac{9}{4} + \frac{27}{8} = \frac{65}{8}$$

$$S_n = \frac{a_1\,(1 - r^n)}{1 - r} = \frac{1\left[1 - \left(\frac{3}{2}\right)^n\right]}{1 - \frac{3}{2}} = -2\left[1 - \left(\frac{3}{2}\right)^n\right].$$

Thus, the sequence of partial sums is

$$1, \frac{5}{2}, \frac{19}{4}, \frac{65}{8}, \dots, -2\left[1 - \left(\frac{3}{2}\right)^n\right], \dots$$

2. For the series $1 + 0.1 + 0.01 + 0.001 + \dots$, we have $a_1 = 1$ and $r = 0.1$. Since $|r| = |0.1| = 0.1 < 1$, this series converges and the sum S is

$$S = \frac{a_1}{1 - r} = \frac{1}{1 - 0.1} = \frac{1}{0.9} = \frac{10}{9}$$

3. On most scientific calculators, the keying sequence is
$$7043 \;\boxed{\div}\; 1110 \;\boxed{=}$$
The display should show the repeating decimal
6.3450450450 . . .

4.

$$1 - 2x \overline{)1 + 2x + 4x^2 + 8x^3 + \cdots}$$

$$\underline{1 - 2x}$$
$$2x$$
$$\underline{2x - 4x^2}$$
$$4x^2$$
$$\underline{4x^2 - 8x^3}$$
$$8x^3$$
$$\underline{8x^3 - 16x^4}$$
$$16x^4$$

5. If the initial impact drives the nail $\frac{7}{8}$ inch, then the total distance the nail could be driven is given by

$$\frac{7}{8} + \frac{7}{12} + \frac{7}{18} + \frac{7}{27} + \cdots$$

The sum of this infinite geometric series is

$$S = \frac{a_1}{1 - r} = \frac{\frac{7}{8}}{1 - \frac{2}{3}} = \frac{21}{8} = 2\frac{5}{8} \text{ inches.}$$

Since the nail is $2\frac{1}{2}$ inches long and $2\frac{5}{8} > 2\frac{1}{2}$, we know that it is possible for the head of the nail to be flush with the board. In fact, this will occur during the eighth impact.

Exercises

1. (a) $1, \frac{5}{3}, \frac{19}{9}, \frac{65}{27}, \ldots, 3\left[1 - \left(\frac{2}{3}\right)^n\right], \ldots$ (b) 3

3. (a) $2, -4, 14, -40, \ldots, \frac{1}{2}[1 - (-3)^n], \ldots$

(b) Does not exist

5. (a) $-243, -81, -189, -117, \ldots,$

$-\frac{729}{5}\left[1 - \left(-\frac{2}{3}\right)^n\right], \ldots$ (b) $-\frac{729}{5}$

7. (a) $\frac{9}{8}, \frac{15}{8}, \frac{19}{8}, \frac{65}{24}, \ldots, \frac{27}{8}\left[1 - \left(\frac{2}{3}\right)^n\right], \ldots$ (b) $\frac{27}{8}$

9. Diverges and has no finite sum

11. Converges to the sum $\frac{1}{30}$ **13.** Converges to the sum $\frac{5}{11}$

15. Diverges and has no finite sum

17. Converges to the sum -4 **19.** Converges to the sum

$\frac{25}{4}(\sqrt{10} + \sqrt{2})$ **21.** $r = \frac{2}{3}$ **23.** $S = \frac{81}{16}$

25. $S = -\frac{256}{7}$ **27.** $a_1 = 6, 12$ **29.** $\cdot\frac{1}{3}$ **31.** $\frac{1}{11}$

33. $\frac{78}{37}$ **35.** $\dfrac{500,627}{33,330}$ **37.** $\dfrac{2669}{3330}$

39. Interval of convergence $(-5, 5)$; $S = \dfrac{5x}{x + 5}$

41. Interval of convergence $\left(-\frac{1}{2}, \frac{1}{2}\right)$; $S = \dfrac{1}{1 - 4x^2}$

43. Interval of convergence $(1, 3)$; $S = \dfrac{1}{x - 1}$ **45.** Interval

of convergence $\left(-\infty, -\frac{1}{2}\right) \cup \left(\frac{1}{2}, \infty\right)$; $S = \dfrac{4x^2}{2x - 1}$

47. (a) 54 m (b) 36 m (c) 90 m

49. No, he covers only 25 miles. **51.** $x = -\frac{1}{6}$

53. (a) Does not exist (b) 0 (c) 1; The associative property of real numbers does not apply to infinite series.

55. $r = \frac{1}{3}$

57. (a) 1152 sq cm (b) $(192 + 92\sqrt{2})$ cm

59. 82.5 **61.** 66.5

63. 1.00 sq. unit **65.** 140.875 cm

Section 8.6

Problems

1. $\dfrac{(k + 1)!}{(k - 1!} = \dfrac{(k + 1)(k)(k - 1)!}{(k - 1)!} = k^2 + k$

2. (a) $\begin{pmatrix} 9 \\ 0 \end{pmatrix} = \dfrac{9!}{0! \, 9!} = 1$

(b) $\begin{pmatrix} 7 \\ 7 \end{pmatrix} = \dfrac{7!}{7! \, 0!} = 1$

(c) $\begin{pmatrix} 12 \\ 9 \end{pmatrix} = \dfrac{12!}{9! \, 3!} = 220$

(d) $\begin{pmatrix} 12 \\ 3 \end{pmatrix} = \dfrac{12!}{3! \, 9!} = 220$

$\begin{pmatrix} 9 \\ 4 \end{pmatrix} x^{9-4} y^4 = 126 x^5 y^4$

Exercises

1. (a) 10! (b) $k!$ **3.** (a) 20! (b) $(k - 1)!$

5. (a) $\dfrac{11}{10!}$ (b) $\dfrac{1 + k}{k!}$ **7.** 42 **9.** 1 **11.** 1

13. 15 **15.** 84 **17.** 1260 **19.** 0

21. $x^5 + 5x^4y + 10x^3y^2 + 10x^2y^3 + 5xy^4 + y^5$

23. $64x^6 - 192x^5 + 240x^4 - 160x^3 + 60x^2 - 12x + 1$

25. $81a^4 + 216a^3b + 216a^2b^2 + 96ab^3 + 16b^4$

27. $128x^{14} - 448x^{12} + 672x^{10} - 560x^8 + 280x^6 - 84x^4 + 14x^2 - 1$ **29.** $x^4 + 16x^3\sqrt{x} + 112x^3 + 448x^2\sqrt{x} + 1120x^2 + 1792x\sqrt{x} + 1792x + 1024\sqrt{x} + 256$

31. $220x^9y^3$ **33.** $20,412n^5$ **35.** $5670x^4y^8$

37. $13,440t^2$ **43.** (a) $x^{10} + 5x^9 + 15x^8 + 30x^7 + 45x^6 + 51x^5 + 45x^4 + 30x^3 + 15x^2 + 5x + 1$

(b) $x^{10} - 5x^9 + 5x^8 + 10x^7 - 15x^6 - 11x^5 + 15x^4 + 10x^3 - 5x^2 - 5x - 1$

45. (a) 1, 7, 21, 35, 35, 21, 7, 1

(b) 1, 8, 28, 56, 70, 56, 28, 8, 1

(c) 1, 9, 36, 84, 126, 126, 84, 36, 9, 1

47. (a) $1 - x + x^2 - x^3 + \ldots$ (b) $1 + 2y + 3y^2 +$
$4y^3 + \ldots$ (c) $1 + \dfrac{z}{2} + \dfrac{3z^2}{8} + \dfrac{5z^3}{16} + \ldots$ (d) $1 +$
$\dfrac{x}{3} - \dfrac{x^2}{9} + \dfrac{5x^3}{81} - \ldots$ **49.** 8,436,285 **51.** 129,024,480
53. (a) 1.08243216 (b) 1.08243216 **55.** Power
series yields $e \approx 2.718281526$; $\boxed{e^x}$ key on a calculator yields
$e \approx 2.718281828$

Section 8.7

 Problems

1. The event B of tossing a tail and rolling a number
divisible by 3 is $B = \{T3, T6\}$.

2. Each of the ten questions can be answered in two ways.
Thus, the test can be answered in $2^{10} = 1024$ ways.

3. If no license plate registration has a first digit of zero, then
we have only nine possibilities for the first digit. Hence, the
number of different license plate registrations that are possible is
$9 \cdot 10 \cdot 10 \cdot 26 \cdot 26 \cdot 26 = 15,818,400$.

4. The number of possible orders of finish for the first four
positions is
$$_8P_4 = \frac{8!}{(8-4)!} = \frac{8!}{4!} = 1680.$$

5. If a team plays 10 games, then it can end the season with
3 wins, 5 loses, and 2 ties in
$$\frac{10!}{3!\,5!\,2!} = 2520 \text{ ways.}$$

6. By the multiplication principle, a hand containing two
queens and three kings can be dealt in
$$\binom{4}{2} \cdot \binom{4}{3} = \frac{4!}{2!\,2!} \cdot \frac{4!}{3!\,1!} = 24 \text{ ways.}$$

 Exercises

1. $S = \{a, b, c, d, e, f, g, h, i, j\}$
3. $S = \{YY, YN, NY, NN\}$ **5.** $S = \{AE, AJ, EA, JA, EJ,$
JE, AA, EE, JJ$\}$ **7.** $S = \{D, ND, NND, NNND, NNNND\}$
9. $S = \{(1, 1), (1, 2), (1, 3), (1, 4), (1, 5), (1, 6), (2, 1), (2, 2),$
$(2, 3), (2, 4), (2, 5), (2, 6), (3, 1), (3, 2), (3, 3), (3, 4), (3, 5),$
$(3, 6), (4, 1), (4, 2), (4, 3), (4, 4), (4, 5), (4, 6), (5, 1), (5, 2),$
$(5, 3), (5, 4), (5, 5), (5, 6), (6, 1), (6, 2), (6, 3), (6, 4), (6, 5),$
$(6, 6)\}$
11. $A = \{a, e, i\}$ **13.** $C = \{YY, NN\}$
15. $B = \{AJ, JA, EJ, JE, JJ\}$ **17.** $A = \{D, ND\}$
19. $C = \{(1, 6), (6, 1), (2, 5), (5, 2), (3, 4), (4, 3)\}$
21. 156 **23.** 216 **25.** 48 **27.** 60
29. (a) 120 (b) 24 **31.** 2520 **33.** 2520

35. 84 **37.** (a) 56 (b) 30 (c) 9 **39.** $n = 14$
41. $n = 9$ **43.** $n = 8$ **47.** $(n-1)!$
49. (a) 65,536 (b) 6561 **51.** 1,816,214,400
53. 726,485,760
55. (a) 635,013,559,600 (b) 1,677,106,640
(c) 740,999,259

Section 8.8

 Problems

1. Let
$S = \{HHH, HHT, HTH, THH, TTH, THT, HTT, TTT\}$.
If E is the event of obtaining exactly one head, then
$E = \{TTH, THT, HTT\}$. Hence, $P(E) = \dfrac{n(E)}{n(S)} = \dfrac{3}{8}$.

2. If E is the event of selecting a license plate whose number
begins with an odd digit and ends with a vowel, then

number of odd digits	number of vowels

$$P(E) = \frac{n(E)}{n(S)} = \frac{5 \cdot 10 \cdot 10 \cdot 26 \cdot 26 \cdot 5}{10 \cdot 10 \cdot 10 \cdot 26 \cdot 26 \cdot 26} = \frac{5}{52}.$$

3. If E is the event of being dealt a hand that consists of two
queens and three kings, then
$$P(E) = \frac{n(E)}{n(S)} = \frac{\binom{4}{2}\binom{4}{3}}{\binom{52}{5}} = \frac{1}{108,290}.$$

4. If A is the event of obtaining an ace and B is the event of
obtaining a spade, then
$$P(A \cup B) = P(A) + P(B) - P(A \cap B)$$
$$= \tfrac{4}{52} + \tfrac{13}{52} - \tfrac{1}{52}$$
$$= \tfrac{4}{13}$$

5. Let A be the event that the sum is 2, B be the event that
the sum is 3, C be the event that the sum is 4, and D be the
event that the sum is 5. Since these events are mutually
exclusive, we conclude that
$$P(A \cup B) = P(A) + P(B) + B(C) + P(D)$$
$$= \frac{1}{36} + \frac{2}{36} + \frac{3}{36} + \frac{4}{36} = \frac{5}{18}.$$

 Exercises

1. $\frac{3}{10}$ **3.** $\frac{1}{3}$ **5.** $\frac{3}{13}$ **7.** $\frac{1}{8}$ **9.** $\frac{1}{36}$ **11.** $\frac{3}{4}$
13. $\frac{1}{2}$ **15.** 1 **17.** $\frac{4}{13}$ **19.** $\frac{5}{52}$ **21.** $\frac{31}{52}$
23. $\frac{7}{13}$ **25.** $\frac{8}{9}$ **27.** (a) $\frac{1}{32}$ (b) $\frac{5}{32}$
29. (a) $\frac{1}{35}$ (b) $\frac{1}{7}$ **31.** (a) $\frac{5}{14}$ (b) $\frac{5}{42}$

33. $P(B) = \frac{9}{20}$ **35.** (a) $\frac{4}{5}$ (b) $\frac{1}{5}$
37. (a) $\frac{1}{5}$ (b) $\frac{3}{5}$
39. (a) 0.481   (b) 0.00769
41. (a) 0.0400   (b) 0.0731
43. (a) 0.0793   (b) 0.341

Chapter 8 Review Exercises

1. Decreasing function

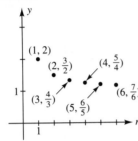

3. $-4, -1, 2, 5, 8, 11$ **5.** $1, -2, 6, -24, 120, -720$
7. $0, 0, 2, 4, 4, 16$
9. (a) Neither   (b) $a_n = (-1)^n (2n - 1)$
11. (a) Arithmetic   (b) $a_n = 6n - 3$
13. (a) Geometric   (b) $a_n = 768 \left(\frac{1}{4}\right)^{n-1}$
15. (a) Neither   (b) $a_n = \dfrac{(-1)^{n+1}(2n - 1)}{2n}$
17. (a) Geometric   (b) $a^n = -4\left(-\frac{1}{2}\right)^{n-1}$
19. (a) Arithmetic   (b) $a_n = \dfrac{17 - 4n}{3}$
21. (a) Neither   (b) $a_n = n^2 - 1$
23. (a) Arithmetic   (b) 1260
25. (a) Geometric   (b) $-\frac{12,117}{64}$
27. (a) Neither   (b) 6327
29. (a) Geometric   (b) 10
31. (a) Arithmetic   (b) 93
33. (a) Neither   (b) 36,664   **41.** 8120
43. 101,025
45. (a) $\displaystyle\sum_{i=1}^{250} (4i + 1)$ (b) $\displaystyle\sum_{i=0}^{249} (4i + 5)$
47. $\displaystyle\sum_{i=1}^{\infty} 243 \left(-\frac{2}{3}\right)^{i-1}$
49. (a) $384, 288, 312, 306, \ldots, \frac{1536}{5}\left[1 - \left(-\frac{1}{4}\right)^n\right], \ldots$
(b) $\frac{1536}{5}$ **51.** -83 **53.** 27,090

55. 96   **57.** $x = 4, -\frac{3}{2}$ **59.** $\frac{81}{16}$ or $\frac{81}{80}$
61. (a) $\frac{100}{33}$ (b) $\frac{3031}{1110}$ **63.** 42,340
65. (a) \$12,160.67   (b) \$19,782.09   **67.** Yes, during the 9th second   **69.** 16   **71.** 78   **73.** 0
75. $729x^6 - 2916x^5 + 4860x^4 - 4320x^3 + 2160x^2 - 576x + 64$ **77.** $120a^{-7}b^3$ **79.** 1080   **81.** 60,060
83. (a) $S = \{T, H1, H2, H3, H4, H5, H6\}$ (b) No. Throwing a tail (T) is six times more likely to occur than each of the other elements in S.
85. $\frac{1}{6}$ **87.** (a) $\frac{1}{120}$ (b) $\frac{49}{60}$

CUMULATIVE REVIEW EXERCISES FOR CHAPTERS 7 & 8

1. (a) 420   (b) 2,097,150   **3.** (b) 722,666   **5.** 4
7. (a) $\left(\frac{1}{2}, 1, -\frac{1}{3}\right)$ (b) $\left(1, 2, -3, -\frac{3}{2}\right)$
9. (a) $A^{-1} = \dfrac{1}{17}\begin{bmatrix} -17 & -2 & 28 \\ -17 & 2 & 23 \\ 17 & 3 & -25 \end{bmatrix}$ (b) $(3, 2, -2)$
11. $x = 8, y = 12; x = \frac{1}{2}, y = -3$ **13.** Arithmetic sequence with common difference $\log r$
15. (a) $\begin{bmatrix} 7 & 5 & 3 & -7 \\ 8 & 4 & -4 & 5 \\ 9 & 10 & 6 & -3 \\ 0 & 5 & 9 & -1 \end{bmatrix}$
(b) $\begin{bmatrix} 11 & -15 & 14 & -21 \\ -1 & -13 & 8 & 5 \\ 27 & -10 & 8 & 6 \\ -30 & -25 & 17 & 12 \end{bmatrix}$
(c) $\begin{bmatrix} 19 & -10 & 4 & 4 \\ -8 & -24 & 18 & -16 \\ 60 & 5 & -13 & 18 \\ 90 & -1 & 11 & -24 \end{bmatrix}$
(d) $\begin{bmatrix} -37 & 1 & -7 & 7 \\ 19 & 37 & 7 & -11 \\ 28 & 96 & -9 & -8 \\ -15 & 21 & 36 & -33 \end{bmatrix}$
17. 0.3   **19.** (a) 64   (b) 128   (c) 0   (d) 924
21. 18, 24, 30   **23.** (a) 66   (b) 220   **25.** $\frac{1}{24}$
27. (a) 0.2   (b) 0.1
29. $1 + \frac{1}{2} + \frac{1}{3} + \frac{1}{4} + \cdots + \dfrac{1}{n} + \cdots$

INDEX

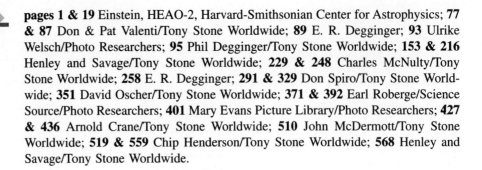